SOME DERIVED UNITS *(continued)*

Physical Quantity	Name of Unit	Symbol	SI Equivalent
Force	newton	N	$= kg \cdot m \cdot s^{-2}$
	dyne	dyn	$= 10^{-5}$ N
Frequency	hertz	Hz	$= s^{-1}$
Length	ångström	Å	$= 10^{-10}$ m
	nanometer	nm	$= 10^{-9}$ m
	micrometer	μm	$= 10^{-6}$ m
	millimeter	mm	$= 10^{-3}$ m
	centimeter	cm	$= 10^{-2}$ m
	kilometer	km	$= 10^{3}$ m
Mass	ton (metric)	t	$= 10^{3}$ kg
Pressure	pascal	Pa	$= N \cdot m^{-2} = kg \cdot m^{-1} \cdot s^{-2}$
	bar	bar	$= 10^{5}$ Pa
	atmosphere	atm	= 101325 Pa
	torr	torr	= 133.32 Pa
	millimeter mercury	mm Hg	$\simeq$ 133.32 Pa
	pounds per square inch	psi	$\simeq 6.89 \times 10^{3}$ Pa
Time	minute	min	= 60 s
	hour	h	= 3600 s
	day	d	= 86400 s
	year (365.25d)	yr	= 31557600 s
Viscosity	centipoise	cp	$= 10^{-3}\ kg \cdot m^{-1} \cdot s^{-1}$
Volume	liter	L	$= 10^{-3}\ m^{3}$
	milliliter	mL	$= 10^{-6}\ m^{3}$
	microliter	μL	$= 10^{-9}\ m^{3}$

ENVIRONMENTAL ORGANIC CHEMISTRY

ENVIRONMENTAL ORGANIC CHEMISTRY

RENÉ P. SCHWARZENBACH
Swiss Federal Institute of Technology (ETH)
Zürich, Switzerland
and
Swiss Federal Institute for Water Resources
and Water Pollution Control (EAWAG)
Dübendorf, Switzerland

PHILIP M. GSCHWEND
Department of Civil and Environmental Engineering
Massachusetts Institute of Technology
Cambridge, Massachusetts

DIETER M. IMBODEN
Swiss Federal Institute of Technology (ETH)
Zürich, Switzerland
and
Swiss Federal Institute for Water Resources
and Water Pollution Control (EAWAG)
Dübendorf, Switzerland

A Wiley-Interscience Publication
JOHN WILEY & SONS, INC.
New York / Chichester / Brisbane / Toronto / Singapore

This text is printed on acid-free paper.

Library of Congress Cataloging in Publication Data:

Schwarzenbach, René P., 1945–
Environmental organic chemistry / by René P. Schwarzenbach, Philip M.
Gschwend, Dieter M. Imboden.
p. cm.
"A Wiley-Interscience Publication."
Includes bibliographical references and index.
ISBN 0-471-83941-8 (cloth)
1. Organic compounds—Environmental aspects. 2. Water chemistry.
I. Gschwend, P. M. II. Imboden, Dieter M., 1943– . III. Title.
TD196.073S39 1992
628.1'68—dc20 92-10737

Printed in the United States of America

10

CONTENTS

PREFACE

Never drive home from a stimulating Gordon Research Conference with somebody who teaches a similar course as you do. Suddenly, your remote intention of turning your lecture notes into a modest little textbook becomes serious reality. This happened to us several years ago at the border between New Hampshire and Massachusetts. Writing this book has been an adventure ever since, not only because of our spatial separation (we had no fax machines at that time), but also because of the multidisciplinary approach that we decided to take. Coming from different scientific backgrounds, we had to learn a lot from each other to achieve what we hope is a homogeneous textbook. This experience has been a very rewarding one for us, and, unlike people who build a house together and then split for life, we have become even closer friends during this project.

What is this book all about? Does it contain everything you always wanted to know about the behavior of organic chemicals in the environment and more? Not quite, but we hope a great deal of it. Our major goal is to provide an understanding of how molecular interactions and macroscopic transport phenomena determine the distribution in space and time of organic compounds released into the natural environment. We hope to do this by teaching the reader to utilize the *structure* of the chemical to deduce that chemical's physical *properties* and *reactivities.* Emphasis is placed on *quantification* of processes at each level. By considering each of the processes that act on chemicals one at a time, we try to build understandings that can be combined in mathematical models to evaluate organic compound fates in the environment.

Who should read and use this book, or at least keep it on their bookshelf? Maybe with a little bit of wishful thinking, we are inclined to answer this question with "everybody who has to deal with organic pollutants in the environment." We have tried, wherever possible, to divide the various chapters or topics into a more elementary and a more advanced part, hoping to make this book useful for beginners as well as for people with more expertise. At many points, we have tried to explain concepts from the very beginning level (e.g., chemical potential) so that individuals who do not recall (or never had) their basic chemistry can develop insights and understand the origin and limits of modeling calculations and correlation equations. We have also incorporated references throughout to help people who want to follow particular topics further. Finally, we have attempted to help environmental practitioners to see how to arrive at quantitative results for particular cases of interest to them. Hence, this book should serve as a text for introductory courses in environmental organic chemistry, as well as a source of information for risk and hazard assessment of organic chemicals in the environment.

Those who have ever written a textbook know that the authors play only a minor role in the realization of the final product. Without the help of many of our colleagues, co-workers, and students, it would have taken at least another decade to finish this book. We thank all of of them, but particularly Béatrice Schwertfeger, Karen Hronek, Tom Carlin, Heidi Bolliger, Vreni Graf, John MacFarlane for typing the text and for preparing the figures; and to Patricia Colberg, Dieter Diem, and numerous students at the University of Minnesota, ETH-EAWAG, and MIT for reviewing the manuscript. We are also very grateful to François Morel and John Westall for comments on our sorption chapter, to Jürg Hoigné and Paul Tratnyek for their review of photochemistry, and to Josef Zeyer and Colleen Cavanaugh for their inputs on biochemical transformations. We are especially grateful to Dieter Diem, in particular, for his efforts on the index. Finally, we thank our families for coping with what must have seemed our endless preoccupation with *THE BOOK*!

We hope that with this textbook we can make a small contribution to the education of environmental scientists and engineers and, thus, to a better protection of our environment.

RENÉ P. SCHWARZENBACH
Dübendorf, Zürich, Switzerland

PHILIP M. GSCHWEND
Cambridge, Massachusetts

DIETER M. IMBODEN
Dübendorf, Zürich, Switzerland

ENVIRONMENTAL ORGANIC CHEMISTRY

CHAPTER 1

INTRODUCTION

Society's ever-expanding utilization of materials, energy, and space is accompanied by an increasing flux of anthropogenic organic chemicals to the environment. We term chemicals *anthropogenic* if they are introduced into the environment primarily or exclusively as a consequence of human activities. Hence, anthropogenic compounds encompass naturally occurring (e.g., petroleum components) as well as synthetic chemicals. Considering that the global consumption of mineral oil exceeds 3 billion tons per year, and that the production of synthetic organic chemicals has increased exponentially over the past several decades (Fig. 1.1), it is obvious that the contamination of water, soil, and air with such compounds is and will continue to be a major issue in environmental protection. As has been documented by numerous studies and is illustrated by Figure 1.2, many synthetic organic compounds, although applied or introduced to confined locations, become widely dispersed even to the "ends of the earth."

When addressing the issue of anthropogenic organic chemicals in the environment, one tends to emphasize the consequences of spectacular accidents or the problems connected with hazardous waste management (e.g., waste water treatment, waste incineration, and dump sites). These are significant problems, but of at least equivalent importance is the chronic contamination of the environment due to the use of chemicals. According to the Organization of Economic Cooperation and Development (OECD) there are presently about 70,000 (mostly organic) synthetic chemicals in daily use, and this number increases continuously. Although some of these "everyday" chemicals (e.g., pharmaceutical and cosmetic products, food additives) are not of direct environmental concern, numerous compounds are continuously introduced into the environment in large quantities (e.g., solvents, components of detergents, dyes and varnishes, additives in plastics and textiles, chemicals used for construction, antifouling

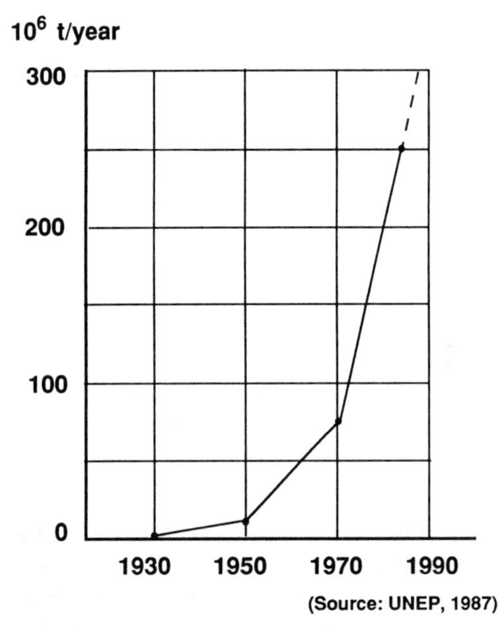

Figure 1.1 Development of the global annual production of synthetic organic chemicals and materials since 1930 (Weber, 1987). In the area previously called West Germany, for example, 25 millions tons of synthetic organic matter were produced in 1980, that is, 400 kg per capita or 100 g/m^2. For comparison, primary production by plants in industrialized countries is estimated to be about 300 g organic matter per square meter per year (Stumm et al., 1983).

agents, herbicides, insecticides, and fungicides). Hence, in addition to problems related to accidents and waste management, a major present and future task encompasses identification and possibly replacement of those widely used synthetic chemicals that may present unexpected hazards to us. Furthermore, new chemicals must be environmentally compatible—we must ensure that these compounds do not upset important processes and cycles of ecosystems. All of these tasks require knowledge of (1) the processes that govern the transport and transformations of anthropogenic chemicals in the environment and (2) the effects of such chemicals on organisms (including humans), organism communities, and whole ecosystems. The first topic is the theme of this book. Our focus is on anthropogenic *organic* chemicals, and we discuss these from the perspective of *aquatic* environments: groundwater, streams and rivers, ponds and lakes, and estuaries and oceans. We note, however, that the ubiquity of water on Earth and its interactions with soils, sediment beds, organisms, and the atmosphere implies that understanding chemical fates in aquatic realms closely corresponds to delineation of their fates in the environment as a whole.

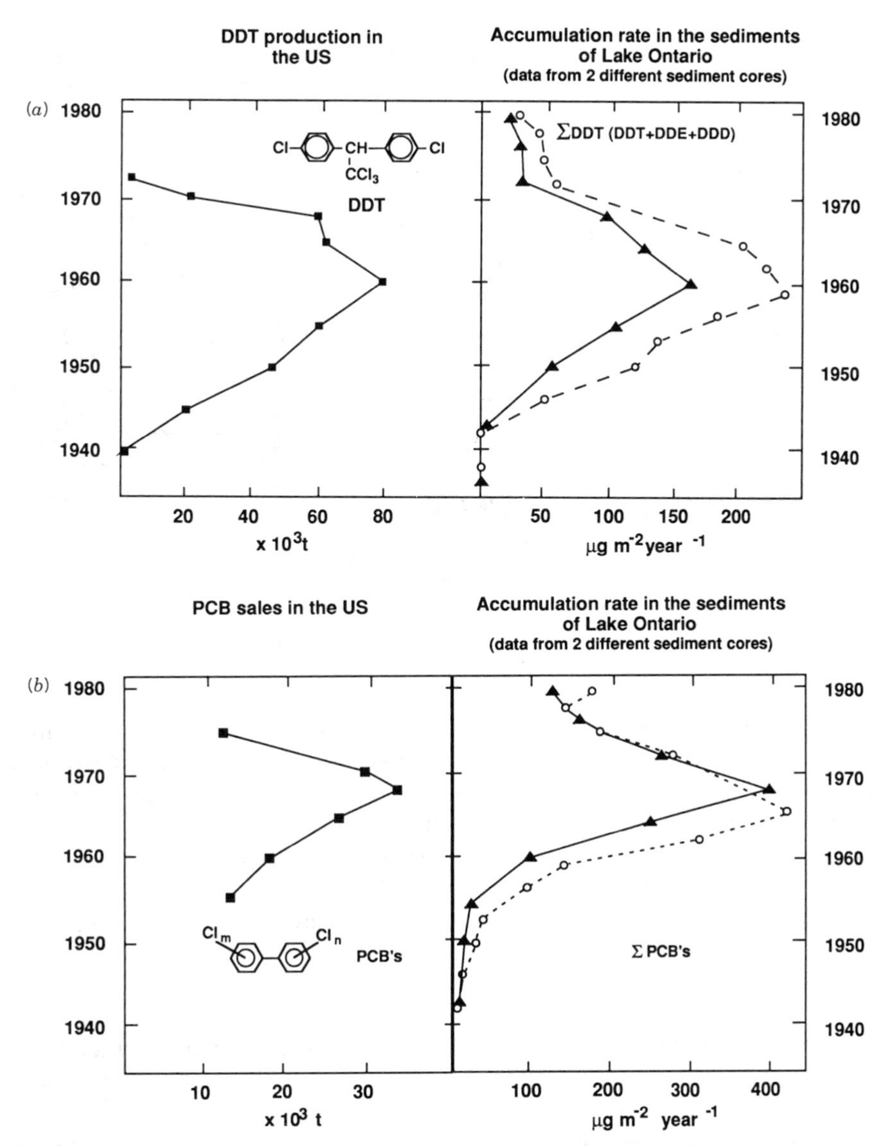

Figure 1.2 Historical records of the sales/production volumes of (*a*) DDT and (*b*) PCBs, and the similarity of these time-varying trends to the accumulation rates of these chemicals in the sediments of Lake Ontario (from Eisenreich et al., 1989).

As is exemplified in Figure 1.3 for a lake system, an organic chemical that is introduced into the environment is subjected to various physical, chemical, and biological processes. These processes can be divided into two major categories—those that leave the structure of a chemical (i.e., its "identity") unchanged and those that transform the chemical into one or several products of different environmental behavior and effect(s).

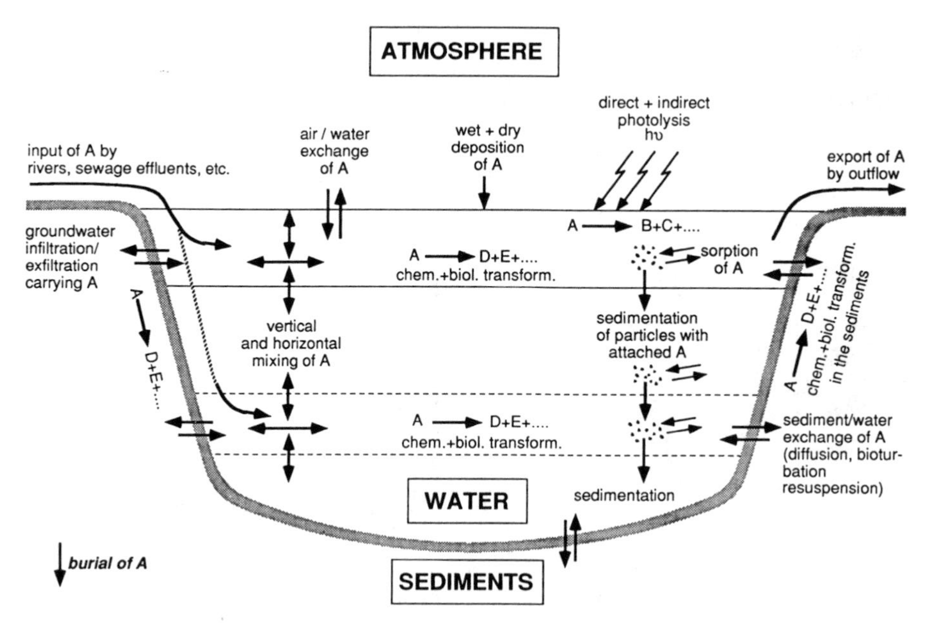

Figure 1.3 Processes that determine the distribution, residence time, and sinks of an anthropogenic organic compound in a lake.

The first category of processes includes transport and mixing phenomena within a given environmental compartment (e.g., in a water body) as well as transfer processes between different phases and/or compartments (e.g., water/air exchange, sorption and sedimentation, sediment/water exchange). Alterations of the structure of a compound may occur by chemical, photochemical, and/or biological (especially microbial) transformation reactions. It is important to recognize that, in a given environmental system, all of these processes may occur simultaneously and, therefore, that different processes may strongly influence each other.

The major goals of this book are:

1. To demonstrate how chemical *structures* cause the molecular interactions that govern the various transfer and reaction *processes* which organic compounds undergo in the environment, particularly in the hydrosphere.
2. To illustrate how principles of chemistry, physics, and biology can be used *to quantify* these processes (through rate and equilibrium coefficients) in a macroscopic environmental system.
3. To provide a *modeling* framework for evaluating the interrelationship and relative importance of the transfer mechanisms, reaction processes, and transport phenomena (i.e., water flow, mixing), and for assessing their impact on the distribution and residence time of a compound in a given natural water body.

This book is intended primarily for students in environmental sciences and engineering who are not specialists in the field. Thus, basic principles are emphasized and simplified pictures of molecular interactions and processes are given to help the inexperienced reader to enhance her or his intuitive perception of relationships between chemical structure and behavior of an organic compound in the environment. We have also chosen to write the text in a somewhat colloquial style to enhance the "palatability" of these fundamental discussions; we hope that the "professionals" among our readers will make an allowance for this effort to teach. We have, however, also tried to incorporate as many current concepts, information, and data as possible to make this book useful to those people who actually deal with the assessment of anthropogenic compound behavior in the environment. Therefore, numerous literature citations have been included, which is rather unusual for a textbook. We also intend for this to facilitate the efforts of graduate students pursuing particular topics in greater detail. Finally, although the major focus of this book is on synthetic organic chemicals, the basic principles discussed apply as well to naturally occurring organic compounds.

A SHORT GUIDE TO THIS BOOK

Figure 1.4 depicts schematically the various aspects to be considered when evaluating, assessing, or predicting the dynamic behavior of anthropogenic organic chemicals in the environment. The scheme also reflects the general buildup and contents of this textbook.

In Chapters 2 and 3 we start with a brief review of a few important concepts of organic and physical chemistry that are used throughout the book. These two chapters are aimed particularly at students who need to refresh their memory with respect to some common terminology and relationships used in organic chemistry (Chapter 2) and in thermodynamics (Chapter 3). Other basic concepts of chemistry, physics, and microbiology are introduced in later chapters.

Chapter 4 is the first of five chapters dealing with the molecular interactions that govern phase transfer processes of organic chemicals in the environment. We start by considering the molecular forces that determine the equilibrium partitioning of a compound between the gas phase and its pure liquid or solid phase. The entity that describes this equilibrium process, the vapor pressure of the compound, is one of the key parameters used for quantification of the partitioning behavior of a compound in the environment. Another important entity, the aqueous activity coefficient of a compound, is discussed in Chapter 5. Here we learn what structural features determine how much the compound likes (or dislikes) to be dissolved in water. We also see how the aqueous activity coefficient of a compound and its water solubility are interrelated, and how they are influenced by the presence of other water constituents (e.g., dissolved salts and organic cosolvents). Chapter 6 deals with air/water equilibrium partitioning; it builds upon the concepts derived in Chapters 4 and 5.

In Chapter 7 molecular factors are evaluated that determine the equilibrium distribution of an organic compound between an organic solvent and water, particularly between *n*-octanol and water. The discussion is also used to demonstrate the development of a linear free-energy relationship useful for predicting physical chemical partitioning behavior from chemical structure. Finally, the insights gained in this

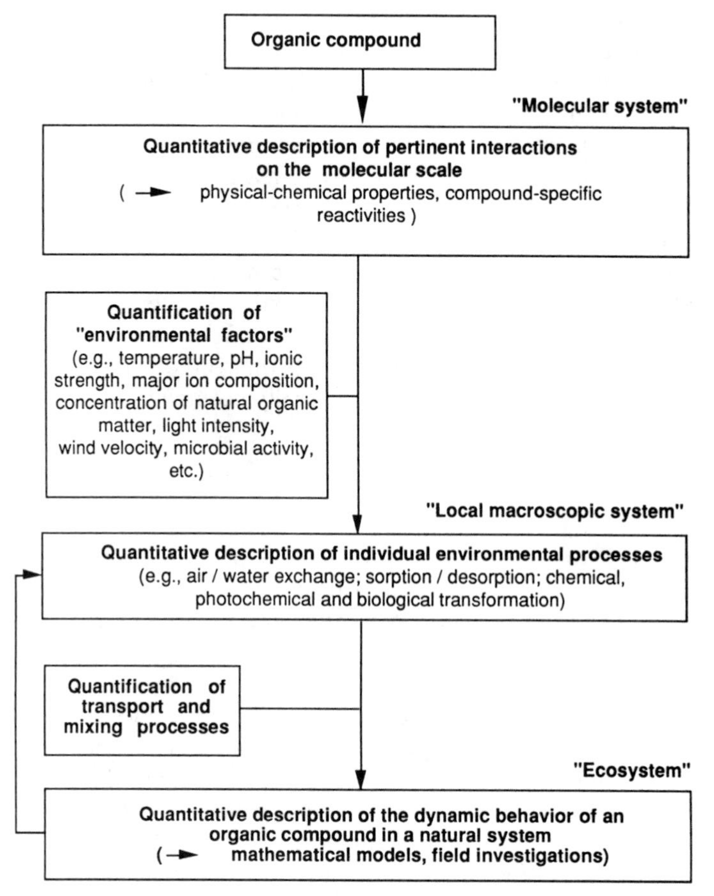

Figure 1.4 General scheme for evaluation of the environmental behavior of anthropogenic organic compounds.

chapter are very important for the subsequent treatment of partitioning processes of organic compounds between water and natural organic phases.

The discussion of basic aspects of partitioning processes is concluded with Chapter 8, which deals with compounds that may undergo proton transfer reactions in aqueous solution: organic acids and bases. Because ionic organic molecules show a different partitioning behavior than their neutral counterparts, the ability to deduce the fraction of a chemical present as an ionized species is necessary to predict phase equilibria of organic acids and bases. Furthermore, the contents of this chapter also enable us to illustrate a linear free-energy treatment of transformation reactions of related organic compounds.

In Chapter 9 we begin to worry about time scales of processes, especially the mathematical descriptions of mass transfers. In this chapter we are interested in the factors that determine how fast organic molecules are moving around in a given phase (e.g., water or air) owing to molecular diffusion. Knowledge about molecular diffusion is pertinent for many of the processes discussed in the subsequent chapters. Since the mathematical framework used to describe transport by molecular diffusion is the same as that for turbulent mixing in macroscopic systems (e.g., eddy diffusion and hydrodynamics dispersion), the latter processes are also addressed. These processes are important when considering the dynamic behavior of organic compounds in natural systems (Chapter 15).

With this background, we are in a position to tackle phase transfer processes in natural systems. In Chapter 10 we deal with the rate of transfer of organic compounds between the atmosphere and a water body. Transfer of compounds between solid phases and water (i.e., sorption processes) is the topic of Chapter 11. In both chapters the molecular aspects of the phase transfer as well as concepts to describe these processes in a macroscopic natural system are discussed.

Chapters 12–14 are devoted to abiotic and biological transformation processes. In Chapter 12, after a short review of chemical kinetics, two major categories of chemical reactions of anthropogenic organic compounds in aqueous solution are discussed: reactions with inorganic nucleophiles (particularly hydrolysis reactions) and redox reactions. Emphasis is placed on structure–reactivity relationships. Chapter 13 deals with direct and indirect (i.e., reaction with photooxidants) photolysis of organic compounds. Here the focus is on quantification of these processes rather than on a discussion of reaction pathways. In Chapter 14 some aspects of microbial transformations of anthropogenic compounds in the environment are addressed. The aim of this chapter is to provide some insights into the strategies applied by microorganisms to breakdown xenobiotic ("foreign to organisms") organic chemicals and to demonstrate concepts useful in quantifying such microbial transformation reactions.

The final chapter of this book, Chapter 15, is devoted to modeling concepts for organic compounds in aquatic systems. This chapter, designed as a brief introduction to the conceptual and mathematical framework of modeling, combines many of the items discussed in the preceding chapters. Using simple models, we demonstrate how the interplay of the various transport, mixing, phase transfer, and transformation processes in a given system may be evaluated and quantified.

In summary, this textbook has been designed to acquaint the reader with the basic principles of organic compound behavior in aquatic environments, and to provide conceptual tools and pertinent information necessary to evaluate and describe quantitatively the dynamics of anthropogenic organic chemicals in a given natural water body. Special emphasis is placed on the *interrelationship between chemical structure and environmental behavior* of organic compounds. The information contained in this book has been collected from many areas of basic and applied science and engineering, including chemistry, physics, biology, geology, limnology, oceanography, pharmacology, agricultural sciences, as well as chemical, civil, and environmental engineering. This reflects the multidisciplinary approach that must be taken when studying the dynamics of organic compounds in the environment. However, there are still numerous gaps in our knowledge, which will become apparent in the text. It is, therefore, our hope that this book will also motivate students to become active in this important field of research.

CHAPTER 2

AN INTRODUCTION TO ENVIRONMENTAL ORGANIC CHEMICALS

2.1 INTRODUCTION

When confronted with the plethora of natural and man-made organic chemicals released into the environment (Blumer, 1975; Stumm et al., 1983), many of us may feel overwhelmed. How can we ever hope to assess all of the things that happen to each of the substances in this menagerie, encompassing so many compound names, formulas, properties, and reactivities? It is the premise of our discussions in this book that each chemical's *structure*, which dictates that compound's "personality," provides a systematic basis with which to understand and predict chemical behavior in the environment. Thus in order to quantify the dynamics of organic compounds in the macroscopic world, we will need to learn to visualize organic molecules in the microscopic environments in which they exist. As a first step, we review some of the terminology and basic chemical concepts of organic chemistry used throughout this book. For readers with no background in organic chemistry, it may be useful to consult the introductory chapters of an organic chemistry textbook in addition to this chapter. On the other hand, professional chemists might skip this chapter.

2.2 THE MAKEUP OF ORGANIC COMPOUNDS

Elemental Composition—Molecular and Structural Formula

To understand the nature and reactivity of organic molecules, we first consider the pieces of which organic molecules are made. This involves both the various atoms and the chemical bonds linking them. First, we note that most of the millions of known natural and synthetic (man-made) organic compounds are combinations of only a few elements, namely carbon (C), hydrogen (H), oxygen (O), nitrogen (N), sulfur (S), phosphorus (P), and the halogens fluorine (F), chlorine (Cl), bromine (Br), and iodine (I). The chief reason for the almost unlimited number of stable organic molecules that can be built from these few elements is the ability of carbon to form stable carbon-to-carbon bonds. This permits all kinds of *carbon skeletons* to be made, even when the carbon atoms are also bound to other elements. Fortunately, despite the extremely large number of existing organic chemicals, knowledge of a few governing rules about the nature of the elements and chemical bonds present in organic molecules will enable us to understand important relationships between the structure of a given compound and its properties and reactivities. These properties and reactivities, in large part, determine the compound's behavior in the environment.

At this point we need to explain what the organic chemist means when using the term "structure" of an organic compound. First, when describing a compound, we have to specify which elements it contains. This information is given by the *elemental composition* of the compound. For example, a chlorinated hydrocarbon, as the name implies, consists of chlorine, hydrogen, and carbon. The next question we then have to address is how many atoms of each of these elements are present in one molecule. The answer to that question is given by the *molecular formula*, for example, four carbon atoms, nine hydrogen atoms, and one chlorine atom: C_4H_9Cl. The molecular formula allows us to calculate the *molecular mass* (or molecular weight) of the compound, which is the sum of the masses of all atoms present in the molecule. The atomic masses of the elements of interest to us are given in Table 2.1. Note that for many of the elements, there exist naturally occurring stable *isotopes*. Isotopes are atoms that have the same number of protons and electrons (which determines their chemical nature), but different numbers of neutrons in the nucleus, thus giving rise to different atomic masses. Examples of elements exhibiting isotopes that have a significant natural abundance are carbon ($^{13}C:^{12}C = 0.011:1$), sulfur ($^{34}S:^{32}S = 0.044:1$), chlorine ($^{37}Cl:^{35}Cl = 0.32:1$), and bromine ($^{81}Br:^{79}Br = 0.98:1$). Consequently, the atomic masses given in Table 2.1 represent averaged values of the naturally occurring isotopes of a given element (e.g., average carbon is 1.1% at 13 amu + 98.9% at 12 amu = 12.011 amu). Using these atomic mass values, we obtain a molecular mass (or molecular weight) of 92.57 amu for a compound with the molecular formula of C_4H_9Cl. If we express the molecular mass in grams, we obtain the quantity which is referred to as 1 mole (abbreviated as mol) of that compound, which corresponds to 6.02×10^{23} units, here molecules (Avogadro's number). For example, 92.57 μg of our model compound would be equal to 10^{-6} moles (1 μmol) or 6.02×10^{17} molecules.

Given the molecular formula, we now have to describe how the different atoms are connected to each other and how they are spatially (or sterically) arranged. This description is called the *structure* of the compound. Depending on the number and types of atoms, there may be many different ways to arrange a given set of atoms which yield different structures. Such related compounds are referred to as *isomers*. Before we can examine how many isomeric compounds exist with a given molecular formula (e.g., C_4H_9Cl), we have to recall some of the rules concerning the number and nature

TABLE 2.1 Atomic Mass, Electronic Configuration, and Typical Number of Covalent Bonds of the Most Important Elements Present in Organic Molecules

	Element[a]			Number of Electrons in Shell						Number of Covalent Bonds Commonly Occurring in Organic Molecules
Name	Symbol	Number	Mass[b] (amu)	K	L	M	N	O	Net Charge of Kernel	
Hydrogen	H	1	1.008	1					1+	1
Helium	He	2		2					0	
Carbon	C	6	12.011	2	4				4+	4
Nitrogen	N	7	14.007	2	5				5+	3, (4)[c]
Oxygen	O	8	15.999	2	6				6+	2, (1)[d]
Fluorine	F	9	18.998	2	7				7+	1
Neon	Ne	10		2	8				0	
Phosphorus	P	15	30.974	2	8	5			5+	3, 5
Sulfur	S	16	32.06	2	8	6			6+	2,4,6(1)[d]
Chlorine	Cl	17	35.453	2	8	7			7+	1
Argon	Ar	18		2	8	8			0	
Bromine	Br	35	79.904	2	8	18	7		7+	1
Krypton	Kr	36		2	8	18	8		0	
Iodine	I	53	126.905	2	8	18	18	7	7+	1
Xenon	Xe	54		2	8	18	18	8	0	

[a]The underlined elements are the noble gases.
[b]Based on the assigned atomic mass of $^{12}C = 12.000$ amu; abundance-averaged values of the naturally occurring isotopes.
[c]Positively charged atom.
[d]Negatively charged atom.

of bonds that each of the various elements present in organic molecules may form. To this end, let's first examine the electronic characteristics of the atoms involved.

Both theory and experiment indicate that the electronic structures of the noble gases [helium (He), neon (Ne), argon (Ar), krypton (Kr), xenon (Xe), and radon (Rn)] are especially stable (i.e., nonreactive); these atoms are said to contain "filled shells" (Table 2.1). Much of the chemistry of the elements present in organic molecules is understandable in terms of a simple model describing the tendencies of the atoms to attain such "filled-shell" conditions by gaining, losing, or, most importantly, sharing electrons. The first shell (K-shell) holds only two electrons (helium structure); the second (L-shell) holds eight; the third (M-shell) can ultimately hold 18, but a stable configuration is reached when the shell is filled with eight electrons (argon structure). Thus, among the elements present in organic molecules, hydrogen requires two electrons to fill its outer shell (one it supplies, the other it must get somewhere else), while the other important atoms of organic chemistry require eight, that is, an octet configuration (see Table 2.1). It is important to realize that the number of electrons supplied by a particular atom in its outer shell (the so-called valence electrons) chiefly determines the chemical nature of an element, although some significant differences between elements exhibiting the same number of outer-shell electrons do exist. This is due in large part to the different energetic status of the electrons in the various shells, reflecting the distance of the electrons from the positively charged nucleus. We address the differences between such elements (e.g., nitrogen and phosphorus, oxygen and sulfur) at various stages during our discussions.

Covalent Bonding

The means by which the atoms in organic molecules customarily complete their outer-shell or *valence-shell* octet is by sharing electrons with other atoms, thus forming so-called *covalent bonds.* A covalent bond is composed of a pair of electrons, one electron contributed by each of the two bonded atoms. The covalent bond may thus be characterized as a mutual deception in which each atom, though contributing only one electron to the bond, "feels" it has both electrons in its effort to fill its outer shell. Thus we visualize the bonds in an organic compound structure as electron pairs localized between two positive atomic nuclei; the electrostatic attraction of these nuclei to these electrons holds the atoms together. The simple physical law of the attraction of unlike charges and the repulsion of like charges is the most basic force in chemistry, and it will help us to explain many chemical phenomena.

Using the simple concept of electron sharing to complete the valence-shell octet, we can now easily deduce from Table 2.1 that H, F, Cl, Br, and I should form one bond (monovalent atoms), O and S two (bivalent), N and P three (trivalent), and C four (tetravalent) bonds in a neutral organic molecule. With a few exceptions, which we will address later, this concept is valid for the majority of cases that are of interest to us. For our compound with the molecular formula of C_4H_9Cl, we are now ready to draw all the possible structural isomers. Figure 2.1 shows that there are four different possibilities. With this example we also take the opportunity to get acquainted with some of the common conventions used to symbolize molecular constitutions and/or structures

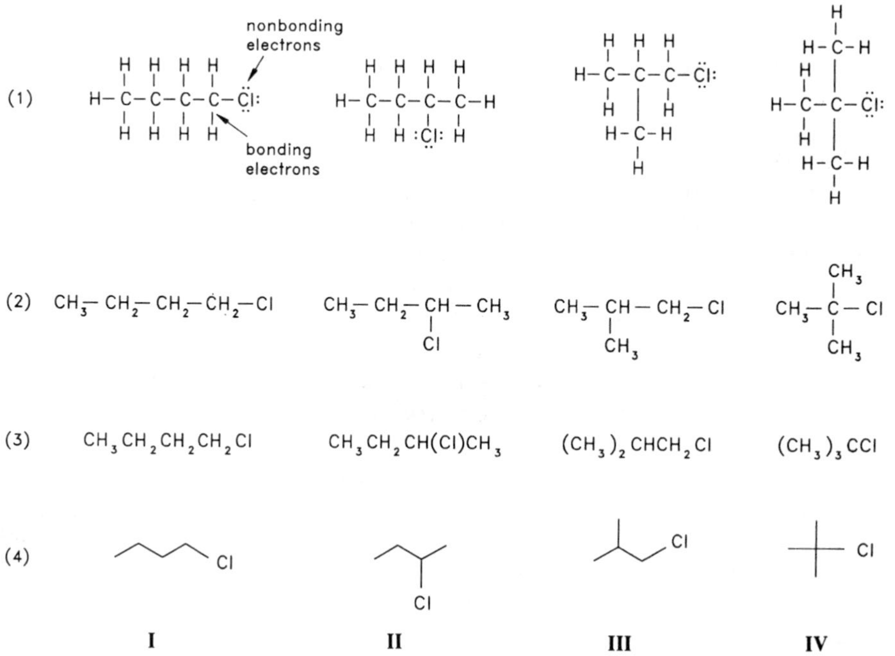

(1) nonbonding electrons; bonding electrons

(2) $CH_3-CH_2-CH_2-CH_2-Cl$ $CH_3-CH_2-CH(Cl)-CH_3$ $CH_3-CH(CH_3)-CH_2-Cl$ $CH_3-C(CH_3)_2-Cl$

(3) $CH_3CH_2CH_2CH_2Cl$ $CH_3CH_2CH(Cl)CH_3$ $(CH_3)_2CHCH_2Cl$ $(CH_3)_3CCl$

(4) Cl Cl Cl Cl

I II III IV

Figure 2.1 Conventions for symbolizing the molecular structures of the four butyl chloride isomers.

(if the steric arrangement is also indicated). Since it is clearer to separate shared and unshared electron pairs, the former (the actual covalent bond) are written as straight lines connecting two atomic symbols, while the unshared valence electrons are represented by pairs of dots (line 1 in Fig. 2.1). This representation clearly shows the nuclei and all of the electrons we must visualize. To simplify the drawing, all lines indicating bonds to hydrogen as well as the dots for unshared (nonbonding) electrons are frequently not shown (line 2). For further convenience we may, in many cases, eliminate all the bond lines without loss of clarity, as illustrated in line 3. Note that branching is indicated by using parentheses in this case. Finally, especially when dealing with compounds exhibiting a large number of carbon atoms, it is very convenient to just sketch the carbon skeleton as depicted in line 4. Each line is thus a skeletal bond and is assumed to have carbons at each end unless another element is shown. Furthermore, no carbon–hydrogen bonds are indicated, but are assumed present as required to make up full bonding (four bonds) at each carbon atom. To distinguish the various carbon–carbon bonds, bond lines are placed at about 120°, roughly resembling the true physical bond angle (see below).

At this point in our discussion about chemical bonds and structural formulas, we should stress that structural isomers may exhibit very different properties and reactivities. For example, the rates of hydrolysis (reaction with water, see Chapter 12) of the four butyl chlorides shown in Figure 2.1 are quite different. While the hydrolytic half-life (time required for the concentration to drop by a factor of 2) of compounds I and III is about 1 yr at 25°C, it is approximately 1 month for compound II, and only 30 sec for compound IV. When we compare the two possible isomers with the molecular formula C_2H_6O,

$$CH_3—CH_2—OH \quad (\mathbf{V}) \qquad\qquad CH_3—O—CH_3 \quad (\mathbf{VI})$$

we can again find distinct differences between these two compounds. The well-known compound **V** (ethanol) is a liquid at ambient conditions while compound **VI** (dimethyl ether) is a gas. These examples should remind us that differences in the arrangement of a single collection of atoms may mean very different environmental behavior; thus we must learn what it is about compound structure that dictates such differences.

So far we have dealt only with single bonding between two atoms. There are, however, many cases in which atoms with more than one "missing" electron in their outer shell form *double bonds* or, sometimes, even *triple bonds*; that is, two atoms share either two or even three pairs of electrons to complete their valence shells. A few examples of compounds exhibiting double or triple bonds are given in Figure 2.2. We note that a double bond is indicated by a double line and, logically, a triple bond by three parallel lines between the corresponding atoms. We also note from Figure 2.2 that there are compounds with ring structures (that may or may not exhibit double

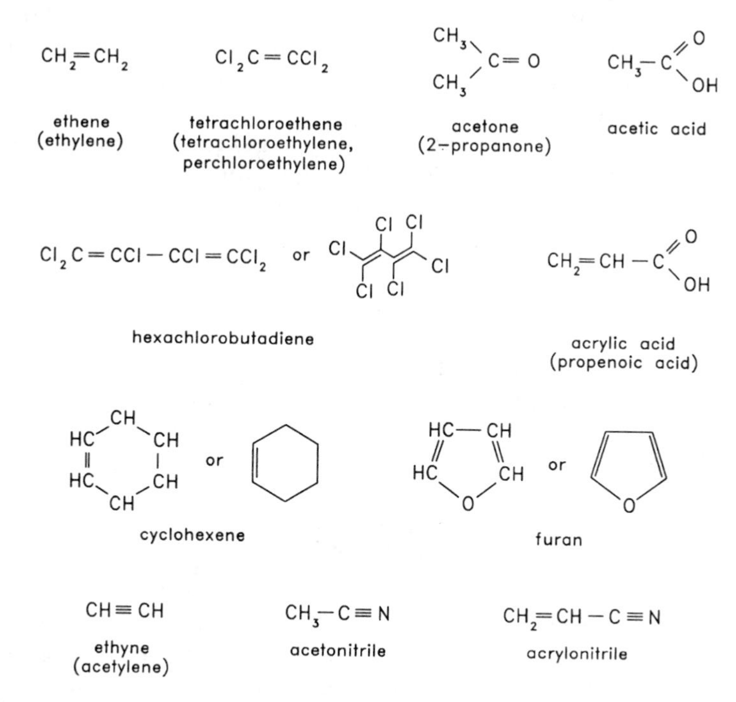

Figure 2.2 Some sample molecules exhibiting double and/or triple bonds.

bonds). Such rings are usually composed predominantly of carbon atoms, but they may also contain so-called heteroatoms (i.e., elements other than carbon or hydrogen such as O, N) in the ring.

Bond Energies (Enthalpies) and Bond Lengths—The Concept of Electronegativity

An important aspect of chemical bonding that we need to address is the strength of a chemical bond in organic molecules; that is, we should have a general idea of the energy involved in holding like and unlike atoms together in a covalent bond. The most convenient measure of bond energy is indicated by the bond dissociation enthalpy, ΔH_{AB}. For a diatomic molecule, this is defined as the *heat* change of the *gas phase* reaction:

$$\text{A—B} \longrightarrow \text{A}\cdot + \cdot\text{B} \qquad \Delta H_{AB} \tag{2-1}$$

at constant pressure and temperature (e.g., 1 atm and 25°C). Here ΔH_{AB} also contains the differences in translational, rotational (only AB), and vibrational (only AB) energies between *educt* (A—B) and *products* (A·, B·). Unfortunately, it is not possible to directly measure bond dissociation (or formation) enthalpies for each of the different bonds present in a molecule containing more than one bond; they have to be determined indirectly, commonly through thermochemical studies of evolved heat (calorimetric measurements) in reactions such as combustion. These studies yield ΔH values for overall reactions, and the individual bond dissociation (or formation) enthalpies have to be deduced from this data in various ways. The results are commonly shown in tables as average strengths for a particular type of bond, valid for gas phase reactions at 25°C and 1 atm. Table 2.2 summarizes average bond enthalpies (and bond lengths) of some important covalent bonds. From these data, some general conclusions about covalent bonds can be drawn, and a very useful concept can be derived to evaluate the uneven distribution of the electrons in a chemical bond, the concept of *electronegativity*.

When visualizing a chemical bond, it is appropriate to imagine that the "electron cloud" or averaged electron position located between the two nuclei is, in general, distorted toward the atom that has the higher attraction for the electrons, that is, the atom that is more *electronegative*. This results in the accumulation of negative charge at one end of the bond (denoted as $\delta -$) and correspondingly a deficiency at the other end (denoted as $\delta +$):

$$-\overset{|}{\underset{|}{\text{C}}}{}^{\delta+}\text{—X}^{\delta-}$$

Among the elements present in organic molecules, we intuitively (and correctly) predict that the smaller the atom (hence allowing a closer approach of the bonding electrons to the positively charged nucleus) and the higher the net charge of the *kernel* (nucleus plus the electrons of the inner, filled shells, see Table 2.1), the greater will be

TABLE 2.2 Average Bond Lengths (Å) and Average Bond Enthalpies ($kJ \cdot mol^{-1}$) of Some Important Covalent Bonds[a]

Bond	Length/Enthalpy	Bond	Length/Enthalpy	Bond	Length/Enthalpy
		Diatomic Molecules			
H—H	0.74/436	F—F	1.42/155	O=O	1.21/498
H—F	0.92/566	Cl—Cl	1.99/243	N≡N	1.10/946
H—Cl	1.27/432	Br—Br	2.28/193		
H—Br	1.41/367	I—I	2.67/152		
H—I	1.60/298				
		Covalent Bonds in Organic Molecules			
Single Bonds[b]					
H—C	1.11/415	C—C	1.54/348	C—F	1.38/486
H—N	1.00/390	C—N	1.47/306	C—Cl	1.78/339
H—O	0.96/465	C—O	1.41/360	C—Br	1.94/281
H—S	1.33/348	C—S	1.81/275	C—I	2.14/216
Double and triple bonds					
C=C	1.34/612	C=O[d]	1.20/737	C≡C	1.20/838
C=N	1.28/608	C=O[e]	1.20/750	C≡N	1.16/888
C=S[c]	1.56/536	C=O[f]	1.16/804		

[a] Bond length/bond enthalpy. Note that 1 Å equals 0.1 nm.
[b] Bond lengths are given for bonds in which none of the partner atoms is involved in a double or triple bond. In such cases bond lengths are somewhat shorter.
[c] In carbon disulfide.
[d] In aldehydes.
[e] In ketones.
[f] In carbon dioxide.

that atom's tendency to attract electrons. Hence, as indicated in Table 2.3, within a row in the Periodic Table (e.g., from C to F), electronegativity increases with increasing kernel charge, and within a column (e.g., from F to I), electronegativity decreases with increasing kernel size. The most commonly used quantitative scale to express electronegativity (Table 2.3) has been devised by Pauling (1960). On this scale, a value of 4.0 is arbitrarily assigned to the most electronegative atom, fluorine, and a value of 1.0 to lithium. The difference in electronegativity between two atoms A and B is calculated from the extra bond energy in A—B versus the mean bond energies of A—A and B—B in which the electrons should be equally shared. The reason for deriving relative electronegativities based on bond energies is that we interpret the extra bond strength in such a polarized bond to be due to the attraction of the partial positive and negative charges.

Let us follow up a little bit on the importance of charge separation in bonds involving atoms of different electronegativity, for example, C and N, O, or Cl. The extent of partial ionic character in such *polar* covalent bonds is a key factor in determining a compound's behavior and reactivity in the environment. The polariza-

TABLE 2.3 Electronegativities of Atoms According to the Scale Devised by Pauling (1960)

Charge of kernel:	+1	+4	+5	+6	+7	
	H					Increasing size of kernel
	2.2					↓
		C	N	O	F	
		2.5	3.0	3.5	4.0	
			P	S	Cl	
			2.2	2.5	3.0	
					Br	
					2.8	
					I	
					2.5	

tion in bonds is important in directing the course of chemical reactions in which either these bonds themselves or other bonds in the vicinity are broken. Furthermore, the partial charge separation makes each bond between dissimilar atoms a *dipole*. The (vector) sum of all bond dipoles in a structure yields the total dipole moment of the molecule, an entity that can be measured. However, it is the dipole moments of individual bonds that are most important with respect to the interactions of a given compound with its molecular surroundings.

From Table 2.3 it can be seen that according to Pauling's scale, carbon is slightly more electron-attracting than hydrogen. It should be noted, however, that the electron-attracting power of an atom in isolation differs from that attached to electron-attracting or electron-donating substituents in an organic molecule. For example, many experimental observations indicate that carbon in $—CH_3$ or other groups containing only carbon and hydrogen is significantly less electron-attracting than hydrogen, while the $—CF_3$ group is, as expected, found to be far more electron-attracting than hydrogen. In conclusion, we should be aware that the electronegativity values in Table 2.3 represent only a rough scale of the relative electron-attracting power of the elements. Hence in bonds between atoms of similar electronegativity, the direction and extent of polarization will also depend on the type of substitution at the two atoms.

One special result of the polarization of bonds to hydrogen which we should mention at this point is so-called *hydrogen bonding*. As indicated in Table 2.1, hydrogen does not possess any inner electrons isolating its nucleus (consisting of just one proton) from the bonding electrons. Thus, in bonds of hydrogen with highly electronegative atoms such as N or O, the bonding electrons are drawn strongly to the electronegative atom, leaving the proton exposed at the outer end of the covalent bond. This relatively bare proton can now attract another electron-rich center, especially heteroatoms with nonbonded electrons, and form a *hydrogen* bond as schematically indicated below by the dotted line:

$$\underset{\delta-}{—O}\underset{\delta+}{—H} \cdots\cdots\cdots\cdots\cdots \underset{\delta-}{:X}—$$

If the electron-rich center forms part of the same molecule, one speaks of an *intra*molecular hydrogen bond; if not, the association is referred to as an *inter*molecular hydrogen bond. Although such hydrogen bonds are, in general, relatively weak (15–20 $kJ \cdot mol^{-1}$), they are, as we will see throughout this book, of enormous importance with respect to the arrangements and interactions of molecules.

We now return to Table 2.2 to note a few simple generalities about bond lengths and bond strengths in organic molecules. As one can see, bond lengths of first-row elements (C, N, O, F) with hydrogen are all around 1 Å. Bonds involving larger atoms (S, P, Cl, Br, I) are longer and weaker. Finally, double and triple bonds are shorter and stronger than the corresponding single bonds; and we notice that the bond enthalpies of double and triple bonds are somewhat less than twice and three times, respectively, the values of the single bonds.

To get an appreciation of the magnitude of bond energies, it is illustrative to compare bond enthalpies to the energy of molecular motion (translational, vibrational, and rotational) which, at room temperature, is typically on the order of a few tens of kilojoules per mole. As can be seen from Table 2.2, most bond energies in organic molecules are much larger than this, and, therefore, organic compounds are, in general, stable to thermal disruption at ambient temperatures. At high temperatures, however, the energy of intramolecular motion increases and can then exceed certain bond energies. This leads to a thermally induced disruption of bonds, a process that is commonly referred to as pyrolysis (heat splitting).

It is important to realize that the stability of organic compounds in the environment is due to the relatively high energy (of activation) needed to break bonds and not because the atoms in a given molecule are present in their lowest possible energetic state (and, therefore, would not react with other chemical species). Hence, organic compounds are nonreactive for kinetic, not thermodynamic, reasons. We will discuss the energetics and kinetics of chemical reactions in detail later in this book (Chapters 3 and 12). Here a simple example helps to illustrate this point. From daily experience we know that heat can be gained when burning, for example, natural gas, gasoline, fuel oil, or wood. As we also know, all these fuels are stable under environmental conditions until we light a match, and then provide the necessary initial activation energy to break bonds. Once the reaction has started, enough heat is liberated to keep it going. The amount of heat liberated when burning, for example, methane gas in a stove, can be estimated from the bond enthalpies given in Table 2.2. The process that occurs is the reaction of the hydrocarbon, methane, with oxygen to yield CO_2 and H_2O:

$$CH_4(g) + 2O_2(g) \longrightarrow CO_2(g) + 2H_2O(g) \qquad (2\text{-}2)$$

In this gas phase reaction we break four C—H and two O=O "double" bonds and we make two C=O and four O—H bonds. Hence, we have to invest $(4 \times 415) + (2 \times 498) = +2656\ kJ \cdot mol^{-1}$, and we gain $(2 \times 804) + (4 \times 465) = -3468\ kJ \cdot mol^{-1}$. The estimated ΔH value (heat of reaction) at 25°C for reaction 2-2 is therefore, $-812\ kJ \cdot mol^{-1}$ (the experimental value is $-890\ kJ \cdot mol^{-1}$), a quite impressive amount of energy. We note that by convention we use a minus sign to indicate that the reaction is *exothermic*; that is, heat is given off to the outside. A positive sign is assigned to ΔH

if the reaction consumes heat by taking energy into the product structures; these reactions are then called *endothermic*.

Although we have only considered the enthalpy (and not the free energy) change of reaction 2-2, the example illustrates that energy can be gained when nonpolar bonds, as commonly encountered in organic molecules, are broken and polar bonds, such as those in carbon dioxide and water, are formed. Such reactions, which involve the transfer of electrons between different chemical species, are referred to as *redox reactions*. In general, redox reactions form the basis for the energy production of all organisms. From this point of view we can consider organic compounds as energy sources. In the context of this book, it will be of great interest to see how microorganisms utilize organic pollutants as "food," and thus help eliminate such compounds from the environment. We address this topic in Chapter 14.

Oxidation State of the Atoms in an Organic Molecule

When dealing with molecular transformations, it is important to know whether or not electrons have been transferred between the reactants. For evaluating the number of electrons transferred, it is necessary to examine the oxidation states of all atoms involved in the reaction. Of particular interest to us will be the (average) oxidation state of carbon, nitrogen, and sulfur in a given organic molecule, since these are the elements most frequently involved in redox reactions.

From inorganic chemistry, we recall that the terms oxidation and reduction refer, respectively, to the loss and gain of electrons at an atom or ion. An oxidation state of zero is assigned to the uncharged element; a loss of Z electrons is then an oxidation to an oxidation state of $+Z$. Similarly, a gain of electrons leads to an oxidation state lower by an amount equal to the number of gained electrons. A simple example is the oxidation of sodium by chlorine, resulting in the formation of sodium chloride:

$$\mathrm{Na}^{\circ}\cdot \xrightarrow{\text{oxidation}} \mathrm{Na}^{+1} + e^{-}$$

$$:\ddot{\underset{..}{\mathrm{Cl}}}^{\circ}\cdot + e^{-} \xrightarrow{\text{reduction}} :\ddot{\underset{..}{\mathrm{Cl}}}:^{-1}$$

To bridge the gap between full electron transfer in ionic redox reactions (as shown in the example above) and the situation with the shared electrons encountered in covalent bonds, one can formally assign the possession of the electron pair in a covalent bond to the more electronegative atom of the two bonded atoms. By doing so, one can then count the electrons on each atom as one would with simple inorganic ions. For any atom in an organic molecule, the oxidation state may be computed by adding 0 for each bond to an identical atom; -1 for each bond to a less electronegative atom or for each negative charge on the atom; and $+1$ for each bond to a more electronegative atom or for each positive charge. We note that in C—S, C—I, and even C—P bonds, the electrons are attributed to the heteroatom although the electronegativities of these heteroatoms are very similar to that of carbon. Table 2.4 gives some examples of the oxidation states of carbon and other elements in some simple organic molecules.

TABLE 2.4 Oxidation States of the Elements in Some Simple Organic Molecules

Compound Name	Structural Formula	Average[a] Oxidation State of the Elements
Methane	CH_4	C(−IV), H(+I)
Butane	$CH_3-CH_2-CH_2-CH_3$	C(−5/2), H(+I)
Ethene	$CH_2=CH_2$	C(−II), H(+I)
Ethanol	CH_3-CH_2-OH	C(−II), H(+I), O(−II)
Dimethyl sulfide	CH_3-S-CH_3	C(−II), H(+I), S(−II)
Dimethyl sulfoxide	$CH_3-S(=O)-CH_3$	C(−II), H(+I), O(−II), S(O)
Dimethyl amine	$CH_3-NH-CH_3$	C(−II), H(+I), N(−III)
Nitromethane	$CH_3-N^{\oplus}(-O^{\ominus})=O$	C(−II), H(+I), N(+III), O(−II)
Acetaldehyde	$CH_3-C(=O)H$	C(−I), H(+I), O(−II)
Benzosulfonic acid	$C_6H_5-S(=O)_2-OH$	C(−2/3), H(+I), O(−II), S(+IV)
Dichloromethane	CH_2Cl_2	C(O), Cl(−I)
Acetic acid	$CH_3-C(=O)OH$	C(O), H(+I), O(−II)
Tetrachloroethene	$Cl_2C=CCl_2$	C(+II), Cl(−I)
Trichloroacetic acid	$Cl_3C-C(=O)OH$	C(+III), H(+I), O(−II), Cl(−I)
Carbon dioxide	$O=C=O$	C(+IV), O(−II)

[a] Sum of the oxidation states of a given-element divided by the number of atoms of that element.

Note that Roman instead of Arabic numbers are frequently used to express the oxidation state of a covalently bound atom. From Table 2.4 we can see that in organic molecules (including CO_2) carbon may be present in all its possible oxidation states, that is, between −IV and +IV. Other elements that may exhibit a variety of oxidation states include nitrogen (−III to +III or even +V in nitrate), sulfur (−II to +IV or even +VI in sulfate), and phosphorus (−III to +V in phosphate). In contrast, only one oxidation state is generally assigned to hydrogen (+I) and to the halogens (−I). The most common oxidation state of oxygen is −II, although oxygen may also exhibit an oxidation state of −I (e.g., in peroxides: R—O—O—R′, where R and R′ are carbon-centered substituents).

The Spatial Arrangement of the Atoms in Organic Molecules

To describe the geometric or steric arrangement of the atoms in a molecule, in addition to bond lengths, we need to know something about the angles between the bonds, the sizes of the atoms, and their freedom to move within the molecule, in particular, rotations about bonds.

Bond Angles

A simple but very effective rule that we may apply when considering *bond angles* in molecules is that the electrons accept the closeness to one another because of pairing, but that each pair of electrons, shared (i.e., involved in a chemical bond) or unshared, wants to stay as far as possible from the other pairs of electrons. This means that in the case of a carbon atom with four single bonds, the bonds will generally tend to point toward the corners of a *tetrahedron*. In the symmetrical case, that is, when a carbon is bound to four identical *substituents* [i.e., atoms or groups of atoms as —H in CH_4, or

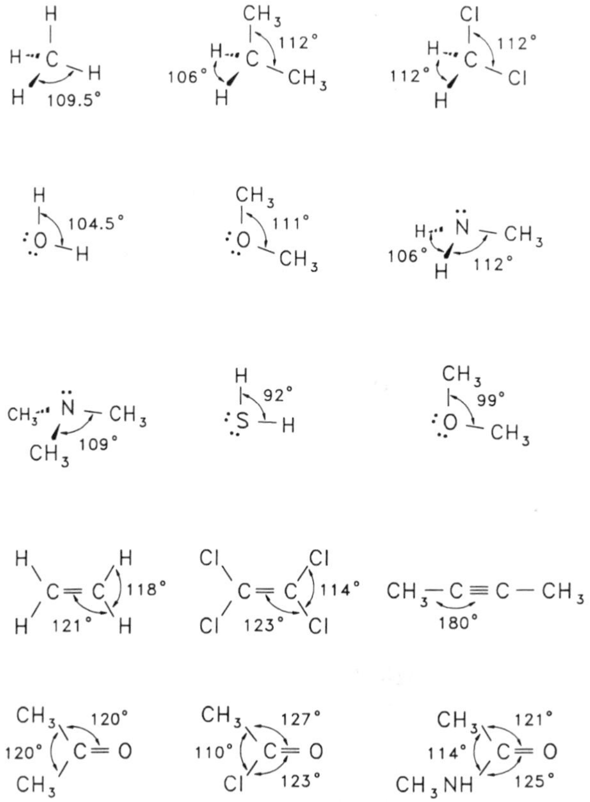

Figure 2.3 Examples of bond angles in some simple molecules (from Hendrickson et al., 1970, and March, 1985).

—Cl in CCl_4, or $—CH_3$ in $C(CH_3)_4$], the bond angles are, as expected, 109.5°. In most cases, however, each carbon atom is bound to different substituents, which leads to minor variations in the bond angles, as illustrated by some examples given in Figure 2.3. For *saturated* carbon atoms, that is, carbon atoms not involved in a double or triple bond, the C—C—C bond angles are typically about 112°, except for ring systems containing less than six ring atoms, where bond angles may be considerably smaller. With respect to the heteroatoms N, O, P, and S, we see from the examples given in Figure 2.3 that the nonbonded electron pairs behave as if they point to imaginary substituents, thus also giving rise to approximately tetrahedral geometry (provided that the heteroatoms are also only single-bonded to other atoms).

Stereoisomerism

We should note that the association of electrons in a single, or sigma (σ), bond does not prevent rotation about the axis of the linkage:

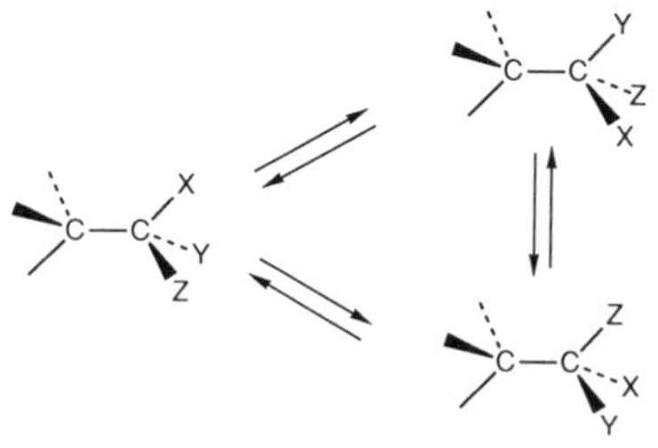

Such rotation does not disrupt the bonding electron pair (i.e., it does not break the bond), and therefore under ambient temperatures the substituents attached to two carbons bonded by a sigma bond are usually not "frozen" in position with respect to one another. Thus the spatial arrangement of groups of atoms connected by such a single bond may change from time to time owing to such rotation, but such geometric distributions of the atoms in the structure are usually not separable from one another since interconversions occur during separation. However, as discussed below, even if fast rotations about a single bond occur, *stereoisomerism* is possible. *Stereoisomers* are compounds made up of the same atoms bonded by the same sequence of bonds (i.e., compounds having the same constitution) but having different three-dimensional structures which are not interchangeable.

One case of stereoisomerism that is of utmost importance in living systems is due to *chirality*. A molecule (or any other object) is called *chiral* if the image and mirror image of the molecule (as of the object) are distinguishable; that is, if they are not superimposable. For example, if in a molecule a carbon atom is bound to four different substituents (an "asymmetric" carbon center), two structural isomers are possible that are alike in every respect but one: they are mirror images of each other.

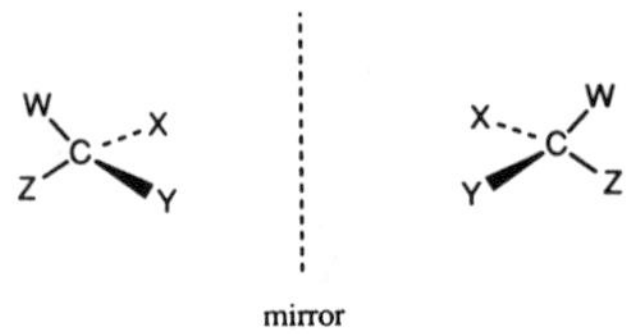

mirror

Nonidentical, mirror-image molecules are called *enantiomers* or *optical isomers* (because they rotate the plane of polarized light in opposite directions). In general, we may say that enantiomers have identical properties in a symmetrical molecular environment, but that their behavior may differ quite significantly in an asymmetrical environment. Most importantly, they may react at very different rates with other chiral species. This is the reason why many compounds are biologically active, while their enantiomers are not. For example, the "R-form" of Mecoprop (**VII**) is an active herbicide, whereas the "S-form" is biologically inactive [Bosshardt, 1988; for details on nomenclature (R, S-forms) see any organic chemistry textbook]:

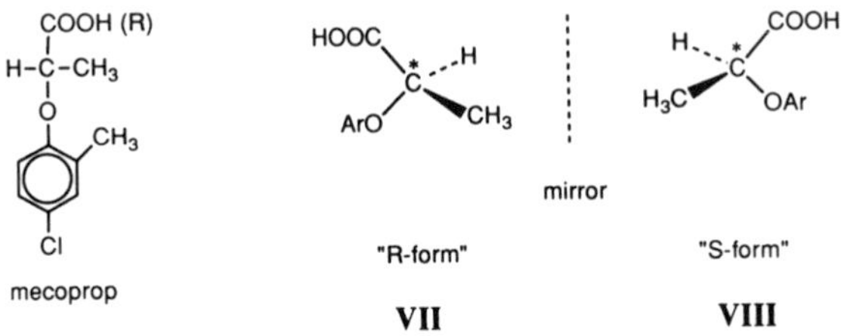

mecoprop

mirror

"R-form"

VII

"S-form"

VIII

Compounds that contain more than one asymmetric center can exist in more than two stereoisomeric forms. Such compounds that are not enantiomers are referred to as *diastereomers*. Diastereomers have different physical and chemical properties in any molecular environment. For example, many natural products contain 2–10 asymmetric centers per molecule, and molecules such as starch and proteins contain hundreds. Thus, organisms may build large molecules that exhibit highly stereoselective sites, which is important for many biochemical reactions including the transformations of organic pollutants (see Chapter 14).

Another important form of diastereoisomerism results from restricted rotation about bonds such as encountered with double bonds and/or ring structures. When considering the geometry of a *double bond*, we imagine a combination of two different types of bonds between two atoms. One of the bonds would be equivalent to a single bond, that is, a bond in which the pair of electrons occupies the region around the axis between the doubly bonded atoms. We can picture the second bond, which is called a π-bond (e.g., carbon–carbon, carbon–oxygen, carbon–nitrogen, carbon–sulfur, nitrogen–oxygen), by imaging the two bonding *π-electrons* to be present in an "electron cloud" located above and below a plane in which the axes of all other bonds (real or imaginary in the case of nonbonded electron pairs) lay, as in the case of ethene (ethylene):

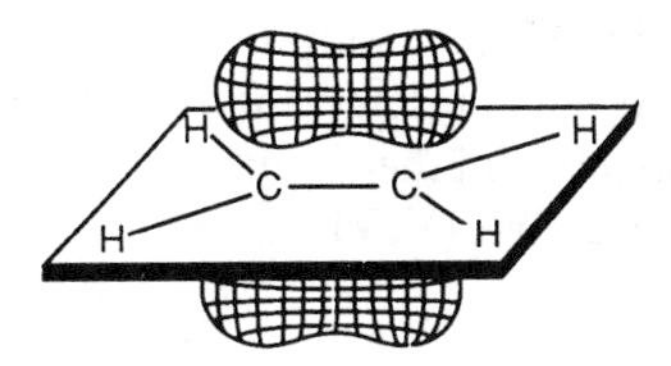

The six atoms closest to a carbon–carbon double bond are in a plane with bond angles of about 120° (see examples given in Fig. 2.3). Rotation about the axis would mean that we would have to break this bond. In triple-bond compounds, as in the case for ethyne (acetylene), there are two π-bond electron clouds which are orthogonal to each other, thus leading to a linear (bond angles = 180°) configuration:

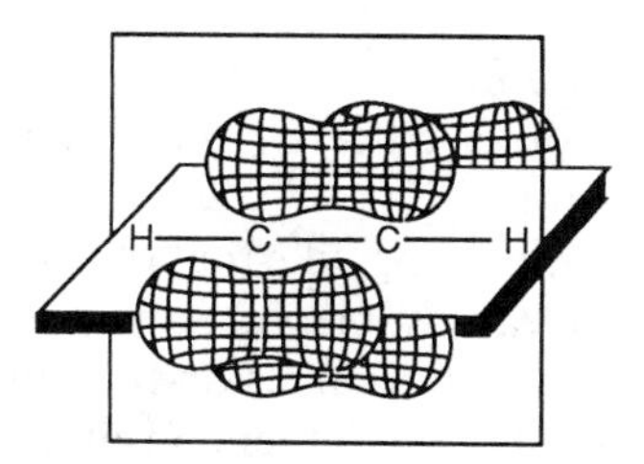

Let us now consider a compound XHC=CHY in which X ≠ Y. In this case, there are two isomers (called geometric isomers), which are distinct (and separable) because we can no longer rotate about the C—C bond.

X(H)C=C(Y)H **IX**   X(H)C=C(H)Y **X**

To distinguish between the two isomers, one commonly uses the terms cis and trans to describe the relative position of two *substituents* (atoms or groups other than hydrogen), each bound to one of the atoms forming the double bond. The term cis is used if the two substituents are on the same side of the double bond, as is the case for X and Y in structure **IX**. In structure **X**, then, X and Y are in the trans (opposite) position. As in the other cases of isomerism, we note that such closely related compounds may exhibit quite different properties. For example, the boiling points of *cis*- and *trans*-1,2-dichloroethylene (**XI, XII**),

Cl(H)C=C(Cl)H — cis — **XI**   Cl(H)C=C(H)Cl — trans — **XII** are 60 and 48°C, respectively. More pronounced differences in properties between cis/trans isomers are observed when interactions between two substituents (e.g., intramolecular hydrogen bonding) occur in the cis but not in the trans form, as is encountered with maleic (**XIII**) and fumaric acid (**XIV**):

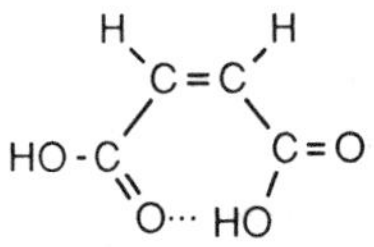

cis

XIII

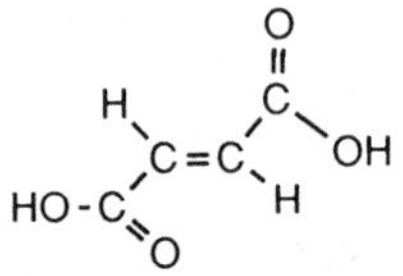

trans

XIV

These two compounds are so different that they have been given different names. For example, their melting points differ by more than 150°C and their aqueous solubilities by more than a factor of 100.

The organization of atoms into a ring, like that of a double bond, also prevents free rotation. Consequently, cis and trans isomers are also possible whenever there is a ring; the cis isomer is the one with two substituents on the same side of the ring (i.e., above or below); the trans isomer exhibits a substituent on either side. An example illustrating cis/trans isomerism in ring systems is given by *cis*- and *trans*-2-methyl cyclohexanol (**XV, XVI**):

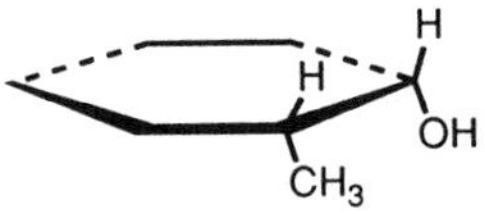

cis

XV

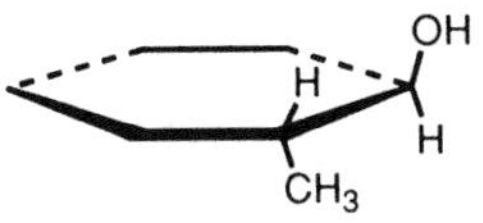

trans

XVI

In ring systems with more than two substituted carbons, more isomers are possible. An example well known to environmental scientists and engineers is 1,2,3,4,5,6-hexachlorocyclohexane. Three of the possible isomers are particularly important from an environmental point of view. The γ-isomer (**XVII**) is the widely used insecticide lindane:

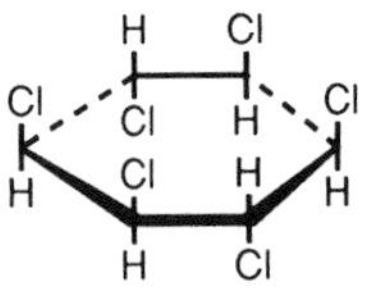

γ-isomer

XVII

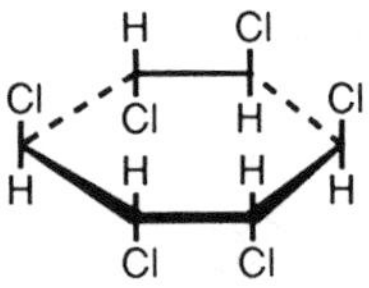

α-isomer

XVIII

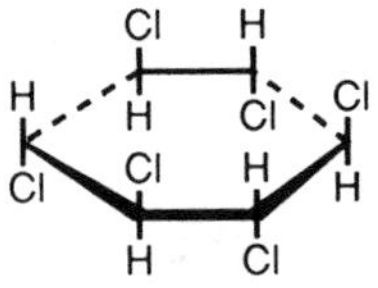

β-isomer

XIX

With the application of lindane, two major contaminants, α- and β-1,2,3,4,5,6-hexachlorocyclohexane (XVIII, XIX), are also introduced into the environment. Problems arise because the β-isomer is very recalcitrant in the environment; that is, it is almost nondegradable under environmental conditions, as compared to the α- and γ-isomers, which are biodegradable (Bachmann et al., 1988).

At this point we should reiterate that the relative positions of atoms in many structures are continuously changing. The term "different conformations" of a molecule is used if two different three-dimensional arrangements in space of the atoms in a molecule are rapidly interconvertible, as is the case if free rotations about sigma bonds are possible. If rotation is not possible, we speak of different *configurations*, which, as we have already discussed, represent isomers that can be separated. Obviously, the conformation(s) with the lowest energy [the most stable form(s)] are the one(s) in which a molecule will preferentially exist. In the case of six-membered rings such as cyclohexane, the most stable conformation is the so-called chair form (**XX**).

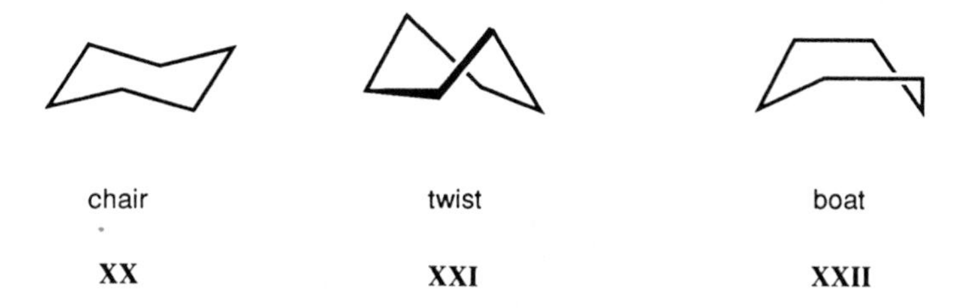

The chair form is more stable than the twist form by about 20 $kJ \cdot mol^{-1}$, and more stable than the boat form by about 25 $kJ \cdot mol^{-1}$ (March, 1985). If we have a closer look at the chair form, we see that six of the bonds linking substitutents to the ring are directed differently than the other six:

XXIII

As can be seen from **XXIII**, six *a*xial bonds are directed upwards or downwards from the "plane" of the ring, while the other six *e*quatorial bonds are more within the "plane." Conversion of one chair form into another converts all axial bonds into equatorial bonds and vice versa. The conversion requires about 40 $kJ \cdot mol^{-1}$ for unsubstituted cyclohexane and is rapid at room temperature. In monosubstituted cyclohexanes, the more stable form is usually the one with the substituent in the equatorial position, because in the axial position the substituent and the axial hydrogens in the 3 and 5 positions begin to "get in each others way." If there is more than one substituent, the situation is more complicated since we have to consider more combinations of substituents which may interact. Often the more stable form is the one with more substituents in the equatorial position. For example, in α-1,2,3,4,5,6-hexachlorocyclohexane (**XVIII**) four chlorines are equatorial (aaeeee), and in the β-isomer all substituents are equatorial. The structural arrangement of the β-isomer also greatly inhibits degradation reactions (the steric arrangement of the chlorine atoms is unfavorable for dechlorination; see Bachmann et al., 1988). Note that at 25°C for each $5.7\,\text{kJ}\cdot\text{mol}^{-1}$ difference in stability (i.e., difference in free energy, ΔG^0, see Chapter 3) between two conformations, the difference in the relative abundance of the molecules in the two forms increases by about a factor of 10. As discussed in Chapter 3, the equilibrium constant K is related to ΔG^0 by $\log K = -\Delta G^0/2.303\,RT$; hence, for $\Delta\Delta G^0 = 5.7\,\text{kJ}\cdot\text{mol}^{-1}$ one obtains $\Delta \log K = 5.7\,\text{kJ}\cdot\text{mol}^{-1}/2.303\cdot 298\cdot 8.31\,\text{J}\,\text{mol}^{-1} \cong 1$.

Finally, with respect to six-membered rings, it should be pointed out that certain substituents cause the ring to be present predominantly in the twist or even boat conformation. An example is compound **XXIV**, in which hydrogen bonding stabilizes the otherwise high-energy form (March, 1985):

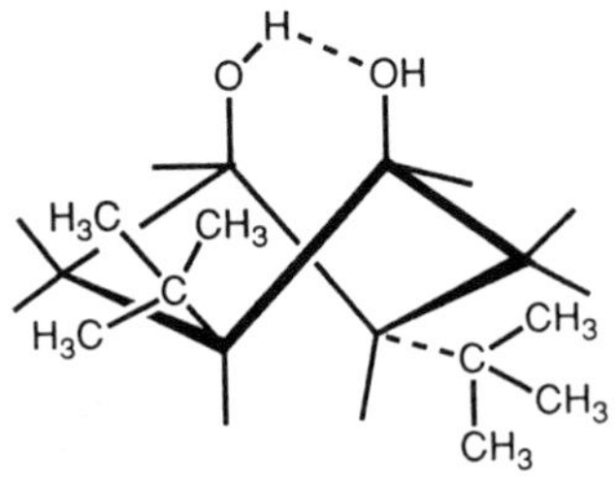

XXIV

Furthermore, in certain polycyclic compounds, the six-membered ring is forced to maintain a boat or twist conformation by bridging structural units, as demonstrated by the insecticide dieldrin (**XXV**).

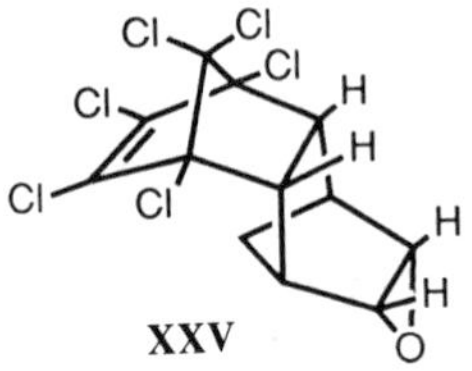

XXV

Delocalized Electrons, Resonance, and Aromaticity

Having gained some insights into the spatial orientation of bonds in chemicals and the consequences for the steric arrangement of the atoms in an organic molecule, we can now proceed to discuss special situations in which electrons move throughout a region covering more than two atoms, The resulting bonds are often referred to as "delocalized chemical bonds." From an energetic point of view, this diminished constraint on the positions of these electrons in the bonds results in their having lower energy and, as a consequence, the molecule exhibits greater stability. For us the most important case of delocalization is encountered in molecules exhibiting multiple *π-bonds* spaced so they can interact with one another. We refer to such a series of π-bonds as conjugated. To effectively interact, π-bonds must be adjacent to each other and the σ-bonds of all atoms involved must *lay in one plane.* In such a conjugated system, we can qualitatively visualize the π-electrons to be smeared over the whole region, as is illustrated below for propenal, also known as acrolein:

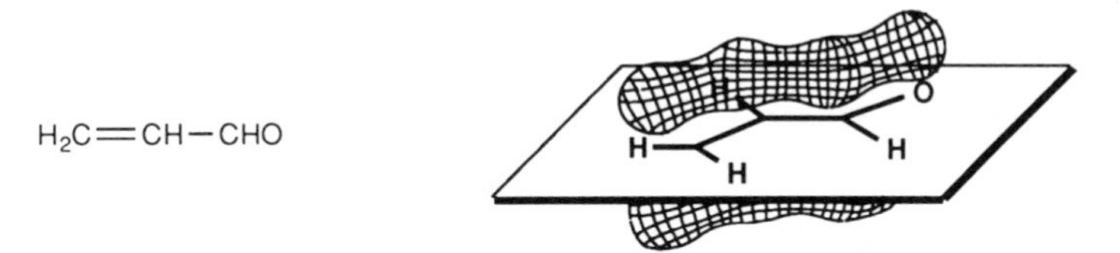

If we try to consider acrolein's structure by indicating the extreme possible positions of these conjugated electrons, we may write

The back and forth arrows are not intended to suggest the three structures are interconvertible, but rather that the location of the four electrons is best thought of as a combination of such extreme possibilities. This freedom in electron positions results in what we call their delocalization, and the visualization of a given molecule by a set of localized structures is called the *resonance* method for representing a structure. The relative contributions of the extremes to the overall resonance structure is determined by their relative stabilities. The stabilizing effect of delocalization is most pronounced in so-called *aromatic systems.* The best-known aromatic system is that of benzene, where we have three conjugated double bonds in a six-membered ring (**XXVI** or **XXVII**):

XXVI **XXVII**

Note again that each of the "static structures" alone does not represent the molecule, but that the molecule is a hybrid of these structures. Thus we sometimes denote the electrons in the conjugated π bonds of benzene with a circle:

XXVIII

In *substituted* benzenes (i.e., benzenes in which hydrogen is substituted by another atom or group of atoms) however, depending on the type and position of the substituents, the different forms may exhibit somewhat different stabilities, and, therefore, their contributions are different.

A quantitative estimate of the stabilization or *resonance energy* of benzene (which cannot be directly measured) may be obtained by determining the heat evolved when hydrogen is added to benzene and to cyclohexene (**XXIX**) to yield cyclohexane (**XXX**):

$$\textbf{XXVIII} + 3H_2 \longrightarrow \textbf{XXX} \qquad \Delta H = -208.6\ \text{kJ}\cdot\text{mol}^{-1}$$

$$\textbf{XXIX} + H_2 \longrightarrow \textbf{XXX} \qquad \Delta H = -120.7\ \text{kJ}\cdot\text{mol}^{-1}$$

If the three double bonds in benzene were identical to the one in cyclohexane, the heat of hydrogenation of benzene would be three times the heat evolved during hydrogenation of cyclohexene. The values given above show that there is a large discrepancy between the "expected" ($-120.7 \times 3 = -362\ \text{kJ}\cdot\text{mol}^{-1}$) and the measured ($-208.6\ \text{kJ}\cdot\text{mol}^{-1}$) ΔH value. Hence, benzene is about $150\ \text{kJ}\cdot\text{mol}^{-1}$ more stable

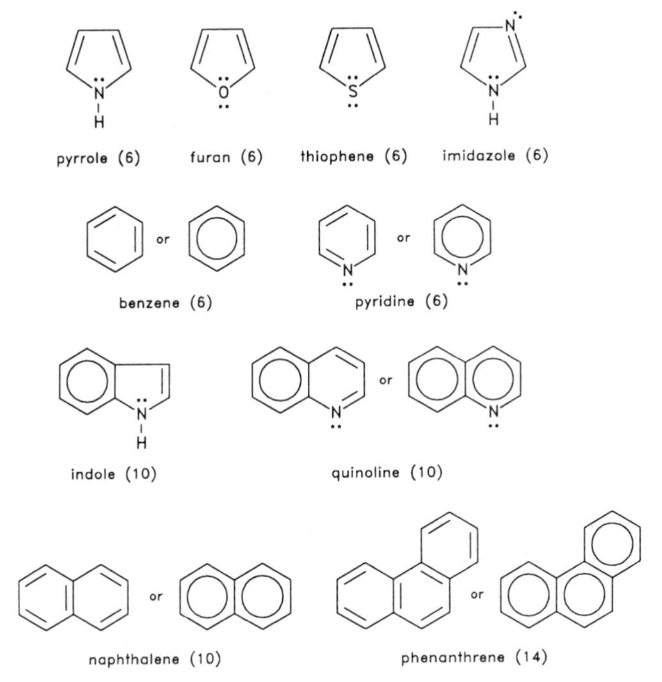

Figure 2.4 Some examples of organic compounds that are aromatic.

than would be expected if there were no resonance interaction between the π-electrons. Large stabilization energies are not only observed in components containing a benzene ring but, in general, in *cyclic* π-bond systems with $4n+2$ (i.e., $6, 10, 14, \ldots$) electrons. In the early days of organic chemistry, it was already recognized that the benzene ring is particularly unreactive compared to acyclic (noncyclic) compounds containing conjugated double bonds. The quality that renders such ring systems especially stable was and still is referred to as *aromaticity*. Some examples of aromatic ring systems are given in Figure 2.4. We note that there are aromatic compounds containing heteroatoms which either contribute one (e.g., pyridine, pyrimidine) or two electrons (e.g., pyrrole, furan, thiophene) to the conjugated π-electron systems. Also note that some polycyclic compounds are referred to as polycyclic *aromatic* compounds, although they are not aromatic throughout their structure in a strict sense (e.g., pyrene with 16 electrons in its π-bond system). Thus, although it is common for such polycyclic aromatic hydrocarbons to be representated by drawing circles in all rings, this is incorrect. However, since this representation is frequently used, we have adopted it throughout this book.

As already indicated for acrolein and for the five-membered heteroaromatic rings (Fig. 2.4), resonance may also be important between nonbonded electrons on a single atom and a π-bond system. For example, an unshared electron pair of oxygen contributes greatly to the stabilization of the carboxylate anion (**XXXI**):

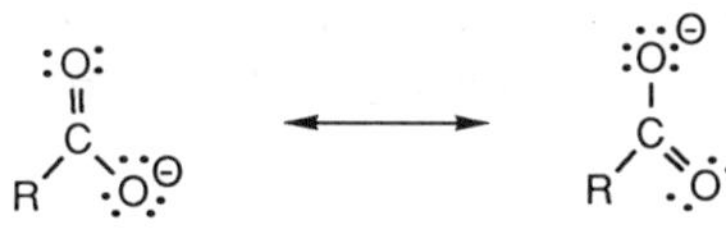

XXXI

Similarly, the two unshared electrons of the nitrogen in aniline (**XXXII**) are in resonance with the aromatic π-electron system:

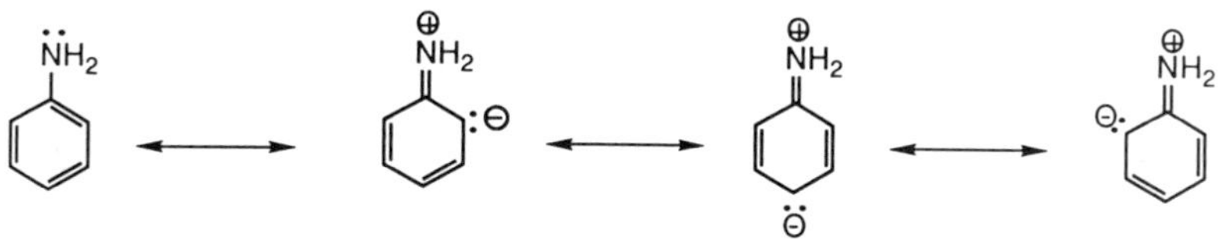

XXXII

As we will see in Chapter 8, the delocalization of the unpaired electron pair in aniline has an important impact on the acid/base properties of anilines as compared to aliphatic amino compounds.

In summary, delocalization of electrons enhances stability, and we can visualize delocalized bonding by using the resonance method. In later chapters we will learn more about the effects of resonance on chemical equilibrium and on the kinetics of chemical reactions of organic compounds.

Multiple Bonding Involving Sulfur and Phosphorus

In general, atoms of the second row of the periodic table do not form very stable double bonds because the overlap of the electrons to form a "regular" π-bond is not optimal, mostly due to the much larger size of the S or P atom. However, there is another type of double bond that is quite common for sulfur and phosphorus, especially when these elements are bound to oxygen. Some examples of such special double bonds are given in Figure 2.5. Note that in each of the examples shown in Figure 2.5 the sulfur or phosphorus atom has (apparently) more than eight valence electrons, and, consequently, more than two and three bonds, respectively. We also note that we can represent these molecules by two resonance structures, but the bond is nevertheless localized. In almost all compounds that have such double bonds, the central atom (S, P) is connected to four atoms (or three atoms and an unshared pair). Since this type of double bond does not significantly change the geometry at the atoms involved (in contrast to the regular π-bond), the bonding at the S or P atom is approximately tetrahedral.

2.3 CLASSIFICATION, NOMENCLATURE, AND EXAMPLES OF ENVIRONMENTAL ORGANIC CHEMICALS

When grouping or classifying environmental organic chemicals, instead of taking a strictly structural approach, it is common practice to use compound categories based

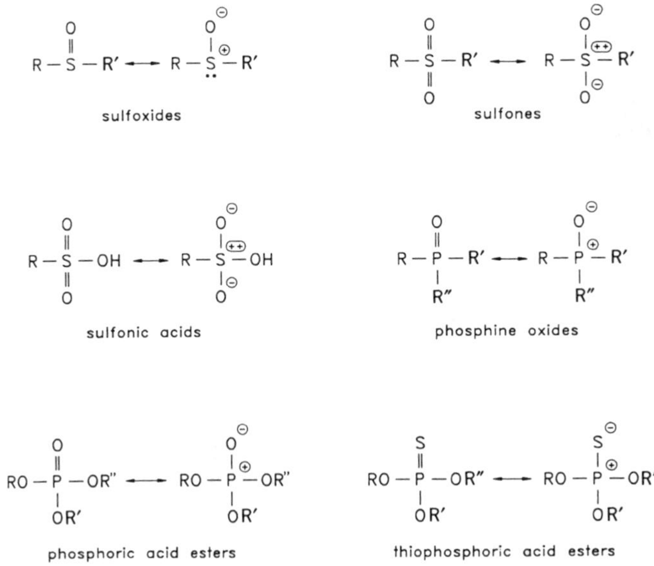

sulfoxides

sulfones

sulfonic acids

phosphine oxides

phosphoric acid esters

thiophosphoric acid esters

Figure 2.5 Examples of molecules exhibiting "special" double bonds betweem sulfur or phosphorus and oxygen. Note that the symbols R, R′, and R″ are used to denote a carbon substituent.

on some physical chemical properties (often reflecting the analytical procedure applied for determining the compounds), and/or based on the source or the use of the chemicals. Terms such as "volatile organic compounds," "hydrophobic compounds," "surfactants", "solvents," "plasticizers," "pesticides" (herbicides, insecticides, fungicides), organic dyes and pigments, or mineral-oil products are very common in the literature. Using such compound categories undoubtedly has its practical value; however, one has to be aware that each of these groups of chemicals usually encompasses compounds of very different structures. Thus the environmental behavior of chemicals within any such group can vary widely.

If we look at organic compounds from a structural point of view, we have already seen that organic molecules are composed of a skeleton of carbon atoms, sheathed in hydrogens, with *heteroatoms* (i.e., O, N, S, P, halogens) or groups of heteroatoms inserted in or attached to that skeleton. Such "sites" are called functional groups, since they are commonly the site of reactivity or function. Hence, the functional groups deserve our special attention if we want to understand or assess a compound's behavior in the environment. For classifying organic chemicals according to structural features, we may then have to base the classification on the type of the carbon skeleton, the type of functional group(s) present, or a combination of both. Consequently, particularly when dealing with a compound exhibiting several functionalities, its assignment to a given compound class may be somewhat arbitrary.

For naming individual organic compounds, there exists a systematic nomenclature that can be found in any organic chemistry textbook. It should be noted, however, that especially in environmental organic chemistry, one frequently uses so-called "common" or "trivial" names instead of a compound's systematic name. Furthermore, quite frequently, the systematic nomenclature is applied incorrectly. For a given compound, one may therefore find a whole series of different names in the literature (see for example synonym listings in the Merck Index), which, unfortunately, may sometimes lead to a certain confusion. Lindane (**XVII**), for example, also goes by the following diverse aliases: γ-HCH, γ-benzene hexachloride, gamma hexachlor, ENT 7796, Aparasin, Aphtiria, γ-BHC, Gammalin, Gamene, Gamiso, Gammexane, Gexane, Jacutin, Kwell, Lindafor, Lindatox, Lorexane, Quellada, Streunex, Tri-6, and Viton. In this book, we shall, however, consider problems encountered with nomenclature to be only of secondary importance as long as we are always aware of the exact structure of the compound with which we are dealing.

The Carbon Skeleton of Organic Compounds

Before we go on discussing some examples of environmentally relevant chemicals, we need to make a few comments about the terms used to describe carbon skeletons encountered in organic molecules. When considering a *hydrocarbon* (i.e., a compound consisting of only C and H) or a *hydrocarbon group* (i.e., a hydrocarbon *substituent*) in a molecule, the only possible "functionalities" are carbon–carbon double and triple bonds. A carbon skeleton is said to be *saturated* if it has no double or triple bond, and *unsaturated* if there is at least one such bond present. Hence, in a hydrocarbon, the term "saturated" indicates that the carbon skeleton contains the maximum number of hydrogen atoms compatible with the requirement that carbon always forms four bonds and hydrogen one. A saturated carbon atom is one that is singly bound to four other separate atoms.

Carbon skeletons exhibiting no ring structures (i.e., only unbranched or branched chains of carbon atoms) are named *aliphatic*, those containing one or several rings are *alicyclic*, or in the presence of an aromatic ring system, *aromatic*. Of course, a compound may exhibit an aliphatic as well as an alicyclic and/or an aromatic entity. The hierarchy of assignment of the compound to one of these three subclasses is commonly aromatic over alicyclic over aliphatic. A saturated aliphatic hydrocarbon is called an *alkane* or a paraffin. If considered as a substituent, it is referred to as an alkyl group. Finally, an unsaturated hydrocarbon exhibiting one or several double bonds is often called an alkene or an olefin. Some examples of common hydrocarbons are given in Figure 2.6.

The most ubiquitous hydrocarbon substituents present in environmental organic chemicals are the alkyl groups. The general formula of an alkyl group is C_nH_{2n+1}—.

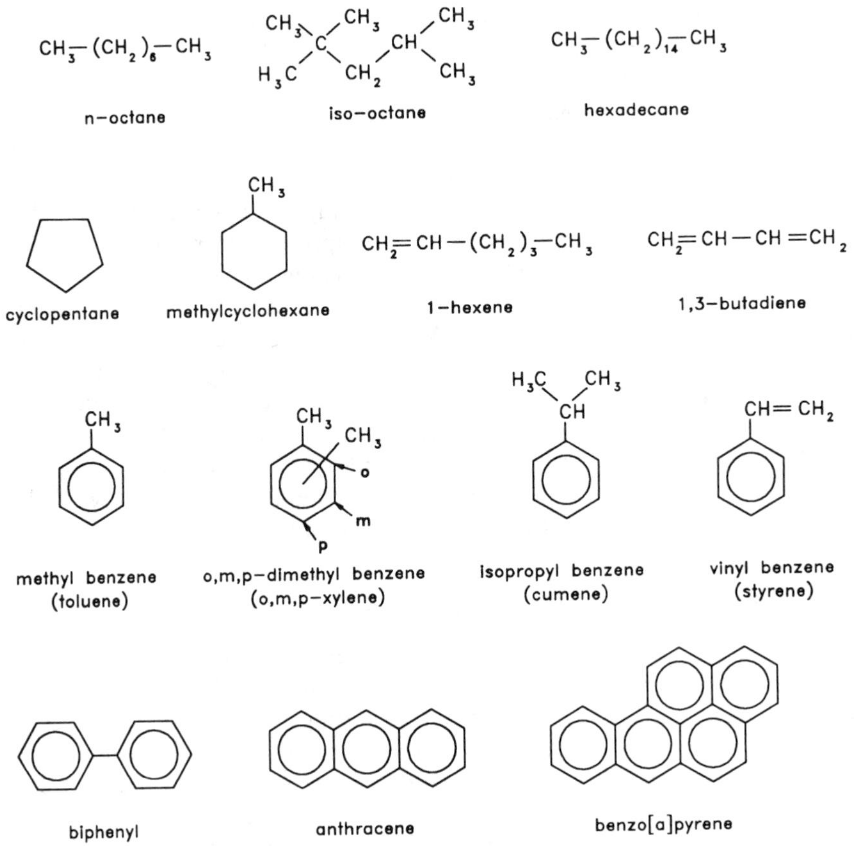

$CH_3-(CH_2)_6-CH_3$
n-octane

iso-octane

$CH_3-(CH_2)_{14}-CH_3$
hexadecane

cyclopentane

methylcyclohexane

$CH_2=CH-(CH_2)_3-CH_3$
1-hexene

$CH_2=CH-CH=CH_2$
1,3-butadiene

methyl benzene
(toluene)

o,m,p-dimethyl benzene
(o,m,p-xylene)

isopropyl benzene
(cumene)

vinyl benzene
(styrene)

biphenyl

anthracene

benzo[a]pyrene

Figure 2.6 Examples of common hydrocarbons.

Examples of common alkyl groups are methyl (CH_3—), ethyl (C_2H_5—), propyl (C_3H_7—), and butyl (C_4H_9—) groups. The prefix *n*, which stands for "normal," is used to denote an unbranched alkyl chain, iso means that there are two methyl groups at the end of an otherwise straight chain, and *neo* is used to denote three methyl groups at the end of the chain. Alkyl groups are further classified according to whether they are primary, secondary, or tertiary. Hence, an alkyl group is referred to as *primary* if the carbon at the point of attachment is bonded to only one other carbon, as *secondary* (s—) if bonded to two other carbons, and as *tertiary* (t—) if bonded to three other carbons:

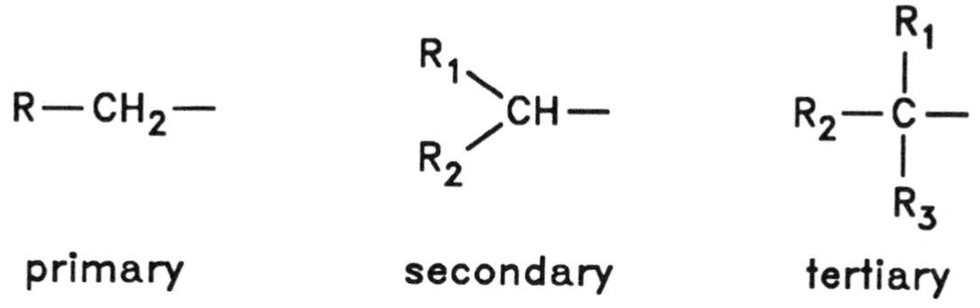

R—CH_2— (primary)

R_1, R_2 >CH— (secondary)

R_2—C(—R_1)(—R_3)— (tertiary)

Note that R stands for a carbon-centered substituent.

Some Common Functional Groups

Organic chemicals sharing common heteroatom moieties are often grouped together in discussing compound properties and reactivities. This is due to the similar behavior of these chemical subunits in the variety of skeletons to which they are attached. Figures 2.5 and 2.7 show some common heteroatom-containing substituents and indicate the trivial compound class names associated with structures containing them. Notably these functional groups change their personalities somewhat when they are attached to alkyl versus aromatic C-containing skeletons; for example, we name a special class of alcohols as phenols when the hydroxyl group is bound to a benzene ring.

Finally, we need to add a brief note concerning the nomenclature in aromatic systems, particularly, in six-numbered rings such as, for example, benzene. Here the terms *ortho-*, *meta-*, and *para-*substitution are often used to express the relative positions of two substituents in a given ring system. Identically, we could refer to those isomers as 1,2- (ortho), 1,3- (meta), or 1,4- (para) disubstituted compounds:

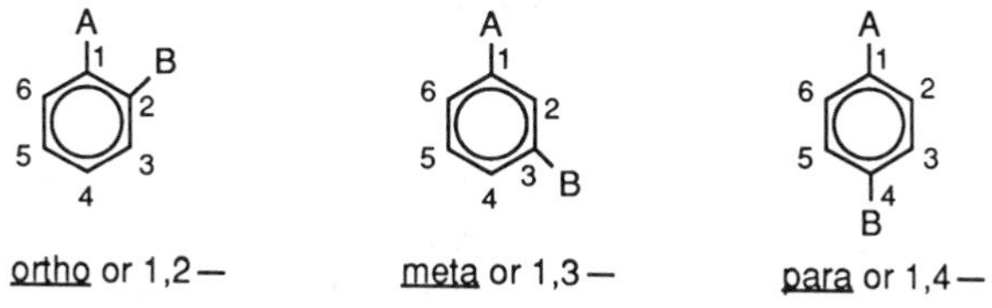

ortho or 1,2—   meta or 1,3—   para or 1,4—

Examples of Environmental Organic Chemicals

Of particular concern among the tens of thousands of existing xenobiotic chemicals are the ones that are introduced into the environment in large quantities. Not only

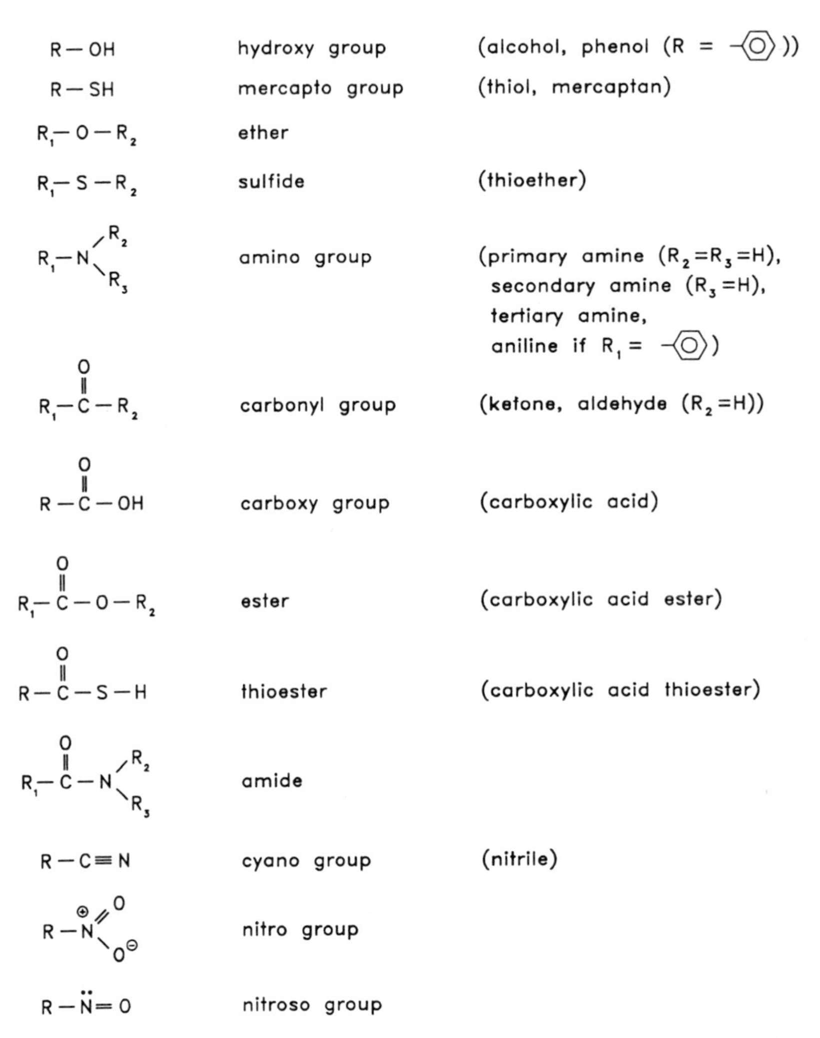

Figure 2.7 Some common heteroatom-containing functional groups present in environmental organic chemicals. Trivial compound class names are used if no other functional group is present in the molecule or to emphasize the particular functional group. If not otherwise indicated, R represents a carbon-centered substituent.

the chemical industry, but all of us, knowingly or not, use and release numerous chemicals in our everyday lives. A few examples illustrate the kinds and quantities of chemicals that are in daily use, and some problems that may be created by such chemicals in the environment.

Let us start with a most familiar case, the pollution of the environment with fossil fuels (i.e., liquid petroleum products and coal) and their combustion products. When we recognize that the global annual production of liquid petroleum products (e.g., gasoline, kerosene, heating oils) is about 3 billion metric tons, it should be no surprise that producing, transporting, processing, storing, using, and disposing of these hydrocarbons poses major problems. Such organic chemicals are released to the atmosphere when we pump gasoline into the tanks of our cars, or are introduced to the surfaces of our streets when these same cars leak crankcase oil. Thus it is not only the spectacular instances such as the wreck of the Exxon Valdez or blow-out of the IXTOC-II offshore oil well through which petroleum hydrocarbons pollute the environment. Furthermore, although they share a common source, the various hydrocarbons in the exceedingly complex mixture that is oil do not necessarily behave the same in the environment. Some constituents are noted for their tendency to vaporize while others clearly prefer to bind to solids; some oil hydrocarbons are extremely unreactive ("methane is the billiard ball of organic chemistry") while others interact beautifully with light; some are quite nontoxic while others are renowned for their carcinogenicity. Thus petroleum hydrocarbons have motivated a great deal of the research in environmental organic chemistry, and the individual components are a good example of why chemical structures must be visualized to predict fates.

Another group of compounds that is with us everywhere are the halogenated methanes, ethanes, and ethenes. Some examples of such C_1- and C_2-*halocarbons* are given in Table 2.5. These compounds are inert, nonflammable, and, depending on the type and number of halogen substituents, exhibit physical properties that render them unique for use as either aerosol propellants, refrigerants, blowing agents for

TABLE 2.5 Examples of Important Industrially Produced C_1- and C_2-Halocarbons[a]

Compound Name(s)	Structural Formula	Approximate Annual World Production (metric tons)	Major Use
Trichlorofluoromethane Dichlorodifluoromethane Chlorodifluoromethane	CCl_3F CCl_2F_2 $CHClF_2$	500,000	Aerosol propellants, refrigerants and blowing agents for plastic foams
Dichloromethane (Methylene chloride)	CH_2Cl_2	500,000	Solvent
Trichloroethene (Trichloroethylene)	$ClHC{=}CCl_2$	600,000	Solvent
Tetrachloroethene (Tetrachloroethylene, Perchloroethylene)	$Cl_2C{=}CCl_2$	1,100,000	Solvent
1,1,1-Trichloroethane	$Cl_3C{-}CH_3$	600,000	Solvent
Methyl bromide	CH_3Br	20,000	Pesticide (fumigant)
1,2-Dibromoethane (Ethylene dibromide)	$BrCH_2{-}CH_2Br$		Pesticide

[a]Data from Russow (1980) and Pearson (1982a).

plastic foams, or solvents for various purposes including dry cleaning and metal degreasing. A few compounds are also used as pesticides. Because of the large quantities released to the environment and their inertness, the C_1- and C_2-halocarbons have become a major environmental concern. It is, for example, estimated that over 85% of the *fluorocarbons* (freons) produced are released to the atmosphere. These compounds are thought to be responsible for depleting the stratospheric ozone layer (Molina and Rowland, 1974; Russow, 1980). The *chlorinated solvents*, dichloromethane, tri- and tetrachloroethene, and 1,1,1-trichloroethane (see Table 2.5) on the other hand, are undoubtedly among the top 10 organic groundwater pollutants. These compounds are quite persistent and mobile in the ground and they may, therefore, lead to the contamination of large groundwater areas. A simple calculation demonstrates the great pollution potential of the chlorinated solvents. In most European countries, the allowed maximum concentration of total chlorinated solvents in drinking water is $25\,\mu g \cdot L^{-1}$. Hence, for example, in Switzerland, where the annual consumption of drinking water is on the order of 2×10^{12} L (more than 70% of which is groundwater), it would take only about 0.2% of the annual consumption of these compounds in Switzerland to spoil the total annual water supply of 6 million people.

The C_1- and C_2-halocarbons are not, however, the only halogenated hydrocarbons that are of great concern. Especially because of their tendency to bioaccumulate, polyhalogenated aromatic hydrocarbons have drawn considerable attention. A classical example involves the *polychlorinated biphenyls* (PCBs, **XXXIII**):

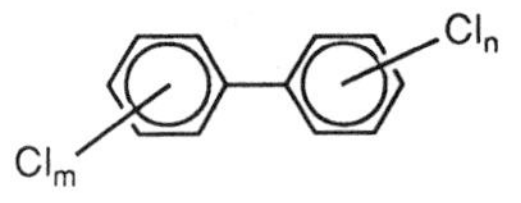

Cl_m Cl_n

XXXIII

To date, more than 1 million metric tons of PCBs have been produced and have been used as capacitor dielectrics, transformer coolants, hydraulic fluids, heat transfer fluids, or plasticizers. Polychlorinated biphenyls are commonly applied as complex mixtures of different *congeners* (i.e., isomers and compounds exhibiting different numbers of chlorine atoms but originating from the same source), and they are lost to the environment mostly from production, storage, and disposal sites. Although their use has been restricted in numerous countries, they are still ubiquitous in the environment. Like DDT, PCBs can be detected everywhere in the world, at the bottom of the ocean as well as in Arctic snow, indicating the powerful environmental transport mechanisms acting on such chemicals (see Fig. 1.2). Figure 2.8 gives the ranges of PCB concentrations found in various environmental samples (Pearson, 1982b). These data illustrate the accumulation of such *hydrophobic* ("water-hating") chemicals in biological tissues.

Another group of high-volume chemicals that we are in daily contact with are the *phthalates* (**XXXIV**), which are diesters of phthalic acid (**XXXV**):

XXXIV **XXXV**

R and R′ denote hydrocarbon groups, in most cases alkyl groups consisting of between 1 and 10 carbon atoms (Giam et al., 1984). The annual world production of phthalates exceeds 1 million metric tons. The phthalates are used mainly as plasticizers, in particular, to make polyvinylchloride (PVC) flexible. They are ubiquitous in the environment, and they are among the most notorious laboratory contaminants encountered when analyzing environmental samples. In aqueous solution, like all *carboxylic acid esters*, phthalates may hydrolyze (react with water) to form the corresponding acid (i.e., phthalic acid) and alcohol(s) (i.e., ROH, R′OH). We discuss this type of reaction in Chapter 12.

From the few examples discussed, we may have already recognized that by varying certain structural moieties (e.g., type and size of carbon skeleton, type and number of halogens), chemicals can be tailor-made to fit a precise purpose. The compounds

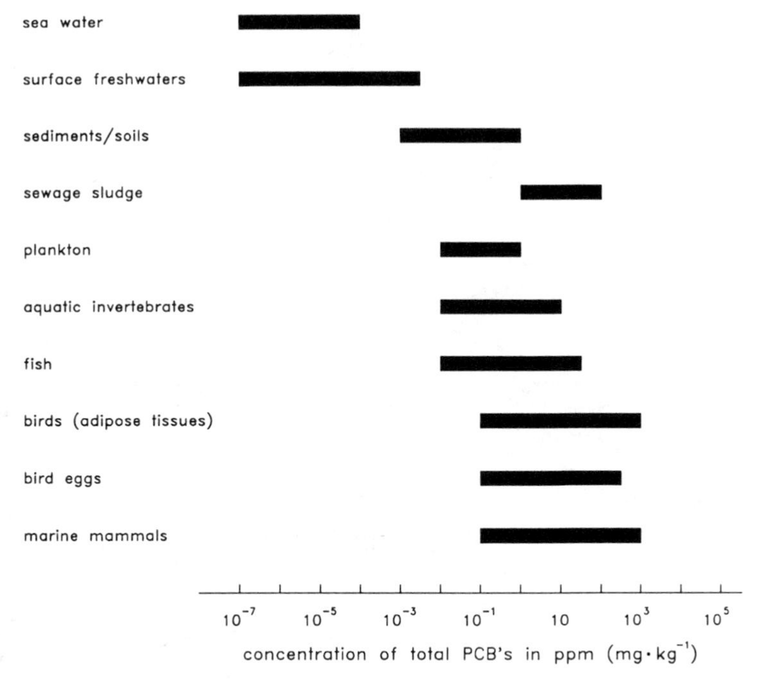

Figure 2.8 Ranges of total PCB concentrations detected in various environmental samples (data from Pearson, 1982b).

considered so far have been mostly hydrophobic in nature, which is the kind of compound we are predominantly concerned with in this book. There are, however, a variety of chemicals that exhibit hydrophilic (liking water) structures or substructures. Important examples are the surfactants, which consist of nonpolar (hydrophobic) and polar (hydrophilic) moieties. As shown in Table 2.6, surfactants are produced by combining a sufficiently large hydrocarbon group with a polar entity that can be either *anionic* (i.e., negatively charged), *cationic* (i.e., positively charged), or *nonionic*

TABLE 2.6 Examples of Commercially Important Surfactants[a]

Common Name of Surfactant Class (Acronym)	General Structure
	Anionic Surfactants
Soaps	$R{-}CH_2{-}COO^{\ominus}Na^{\oplus}$, $R = C_{10\text{-}16}$
Linear alkylbenzene sulfonates (LAS)	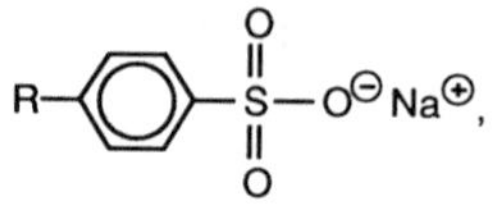 $R{-}C_6H_4{-}SO_2{-}O^{\ominus}Na^{\oplus}$, $R = C_{10\text{-}13}$
Secondary alkyl sulfonates (SAS)	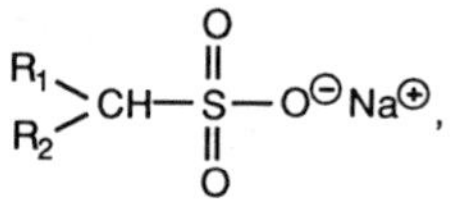 $(R_1)(R_2)CH{-}SO_2{-}O^{\ominus}Na^{\oplus}$, $R_1, R_2 = C_{11\text{-}17}$
Fatty alcohol sulfates (Alkyl sulfates, FAS)	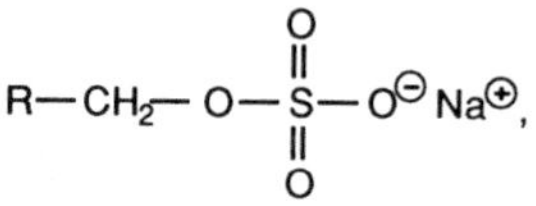 $R{-}CH_2{-}O{-}SO_2{-}O^{\ominus}Na^{\oplus}$, $R = C_{11\text{-}17}$
	Cationic Surfactants
Quaternary ammonium chloride (QAC)	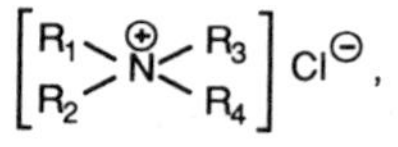 $[R_1R_2N^{\oplus}R_3R_4]Cl^{\ominus}$, $R_1 = R_2 = C_1$, $R_3 = R_4 = C_{16\text{-}18}$
	Nonionic Surfactants
Alkylphenol polyethyleneglycol ethers (APEO)	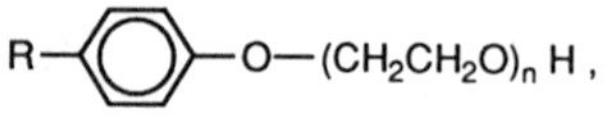 $R{-}C_6H_4{-}O{-}(CH_2CH_2O)_nH$, $R = C_{8\text{-}12}$, $n = 5\text{-}10$
Fatty alcohol polyethyleneglycol ethers (AEO)	$R{-}CH_2{-}O{-}(CH_2CH_2O)_nH$, $R = C_{7\text{-}17}$, $n = 3\text{-}15$

[a]From Piorr (1987).

(i.e., neutral). Note that commercially available surfactants are not uniform substances, but are mixtures of compounds of different carbon chain lengths.

Owing to their *amphiphilic* personality (partly hydrophilic and partly hydrophobic), the surfactants have special properties that render them unique among environmental chemicals. In aqueous solutions they distribute in such a manner that their concentration at the interfaces of water with gases or solids is higher than in the inner regions of the solution. This results in a change of system properties, for example, a lowering of the interfacial tension between water and an adjacent nonaqueous phase, and in a change of wetting properties. Furthermore, inside the solution, on exceeding certain concentrations, surfactants form aggregates, so-called micelles. Hence, surfactants may keep otherwise insoluble compounds in the aqueous phase, and they form, therefore, an important part of any kind of detergent. Surfactants are also widely used as wetting agents, dispersing agents, and emulsifiers in all kinds of consumer products and industrial applications (Piorr, 1987). Because of their direct use in water, a large portion of the annual world production of over 3 million metric tons of surfactants is discharged into municipal and industrial wastewaters. Since these compounds often constitute a significant part of the organic carbon loading of wastewater, their biological degradation behavior is of particular interest. For example, Giger et al. (1984) found that during biological treatment, rather toxic transformation products were formed from nonionic surfactants of the alkylphenol polyethyleneglycol ether type (see Table 2.6 and Fig. 2.9). Some of these products were detected in very high concentrations (grams per kilogram dry matter!) in sewage sludge. In Chapter 14 we discuss biological transformation processes and look at other cases in which, by biologically mediated reactions, compounds of considerable environmental concern were formed.

To round up our tour through the jungle of environmental organic chemicals, we shall have a brief look at those chemicals that, for many environmental scientists and engineers, are the most exotic compounds, the pesticides. In general, pesticides are

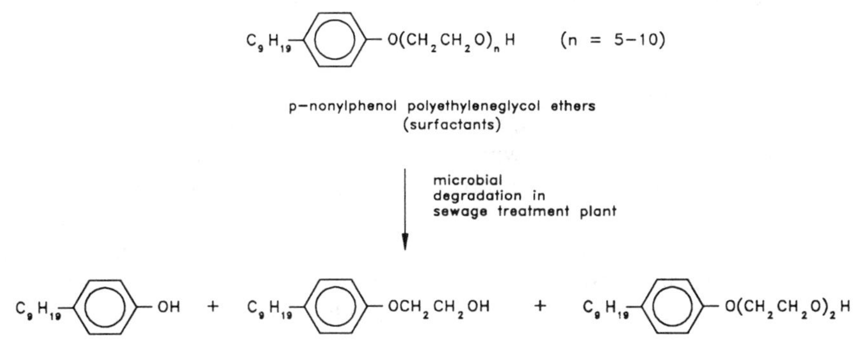

C_9H_{19}–⟨benzene ring⟩–$O(CH_2CH_2O)_nH$ (n = 5–10)

p-nonylphenol polyethyleneglycol ethers
(surfactants)

microbial
degradation in
sewage treatment plant

C_9H_{19}–⟨benzene ring⟩–OH + C_9H_{19}–⟨benzene ring⟩–OCH_2CH_2OH + C_9H_{19}–⟨benzene ring⟩–$O(CH_2CH_2O)_2H$

persistent, ecologically objectionable degradation intermediates

Figure 2.9 Example of a group of widely used chemicals which, as a consequence of biological waste water treatment, are converted into persistent objectionable degradation intermediates (from Giger et al., 1984).

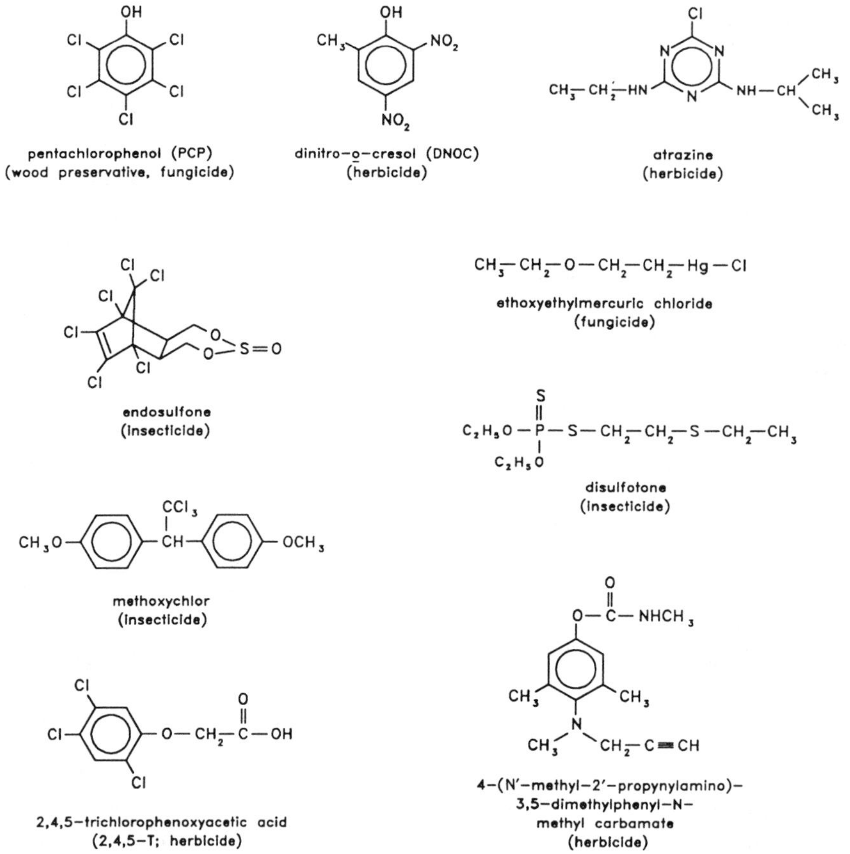

pentachlorophenol (PCP)
(wood preservative, fungicide)

dinitro-o-cresol (DNOC)
(herbicide)

atrazine
(herbicide)

endosulfone
(insecticide)

$CH_3-CH_2-O-CH_2-CH_2-Hg-Cl$

ethoxyethylmercuric chloride
(fungicide)

disulfotone
(insecticide)

methoxychlor
(insecticide)

2,4,5-trichlorophenoxyacetic acid
(2,4,5-T; herbicide)

4-(N'-methyl-2'-propynylamino)-
3,5-dimethylphenyl-N-
methyl carbamate
(herbicide)

Figure 2.10 Examples of pesticides to illustrate the large structural diversity found in this group of environmental chemicals.

designed to have a *specific detrimental biological effect* on one or several target organisms (e.g., plants, insects, fungi), but (ideally) should have little impact on the rest of the living environment in which they are applied. In order to achieve this goal, compounds with very special structures, often exhibiting several functional groups, have to be employed. The present annual world production of pesticides exceeds 2 million metric tons! Figure 2.10 gives a brief illustration of the large structural diversity of these chemicals. Because several functional groups are frequently present in a pesticide molecule, the prediction of the distribution and the reactivity of such compounds in the environment is often more difficult than more "simple" environmental chemicals. Nevertheless, with the general knowledge that we will acquire in the following chapters on how *structural* moieties influence the properties and reactivities of organic compounds, we shall be able to make a reasonable assessment of the behavior of such "complex" chemicals in the environment.

CHAPTER 3

BACKGROUND THERMODYNAMICS

3.1 INTRODUCTION

Before we begin discussing the compounds' properties governing the phase transfer behavior of organic chemicals in the environment, we need to review some of the basic concepts used to describe the relative energy status of molecules in a given system. It is important to realize that it is a molecule's "energy standing in the community" that ultimately controls its ability to move from one phase to another in the environment (e.g., from water to air or from water to solid phases) or to be transformed into other substances. It is appropriate then that this topic be termed *thermodynamics*, which refers to changes "enabled" by the compound's energy content (*therm* from the Greek for "heat" and *dynasthai* for " to be able").

In this chapter we are concerned with *chemical equilibria*; that is, with *reversible* processes. We are interested in the final chemical composition of a given system when no more net changes occur, that is, once the system has reached an energy minimum. We should remember that because of our greater ability to study the energetics (thermodynamics) of systems as opposed to their dynamics, it is usually much easier to describe equilibrium conditions than to deal with *kinetic behavior*. As Morel (1983) so aptly put it, it is very difficult to predict the flight of a feather while its equilibrium position on the floor is a foregone conclusion. In many cases, of course, it is the pathway and the time course of a process that matters most. In the case of chemical and biological transformations in the environment, for example, knowing the final equilibrium condition is often of little help; however, when dealing with phase transfer processes or fast reversible reactions (e.g., proton transfer reactions), thermodynamic considerations are most useful. Furthermore, chemical thermodynamics not only allows a means of evaluating ultimate composition, but also provides information on the direction of spontaneous change, on the energy available from or required for a particular reaction, and in some cases, even on the kinetics of a chemical reaction.

As mentioned earlier, by no means will we provide an in-depth or rigorous treatment of thermodynamics, but rather give only a brief series of descriptive reflections on those thermodynamic concepts most useful to the environmental organic chemist: chemical potential (μ), fugacity (f), activity (a), activity coefficient (γ), Gibbs free energy (G), enthalpy (H), and entropy (S).

3.2 USING THERMODYNAMIC FUNCTIONS TO DESCRIBE MOLECULAR ENERGIES

Chemical Potential

When considering the relative energy status of the molecules of a particular compound in a given environmental system (e.g., benzene in aqueous solution), we can envision the molecules to embody both *internal energies* (i.e., those associated with chemical bonds and bond vibrations, flexations, and rotations) and *external energies* (i.e., whole-molecule translations, orientations, and especially interactions of the molecules with their surroundings). This energy status is dependent on the *temperature*, *pressure*, and *chemical composition* of the system. When we talk about the "energy content" of a given substance, note that we usually are not concerned with the energy status of a single molecule at any given time but, rather, with an average energy status of the entire population of one type of organic molecule (e.g., benzene) in the system. To describe the (average) "energy status" of a compound i mixed in a milieu of substances, Gibbs (1873, 1876) introduced an entity referred to as *total free energy* of this system, which could be expressed as the sum of the contributions from all of the different components present. For this purpose he considered that at *constant* T, P, and *composition*, the Gibbs free energy (which we denote just as free energy) added to the system with each added increment of i should be referred to as the *chemical potential*, μ_i, of component i:

$$\left[\frac{\partial G(\mathrm{kJ})}{\partial n_i(\mathrm{mol})}\right]_{T,P,n_{j\neq i}} \equiv \mu_i(\mathrm{kJ\cdot mol^{-1}}) \tag{3-1}$$

The total free energy then is expressed by

$$G(P, T, n_1, n_2, \ldots n_i) = \sum_i n_i \mu_i \tag{3-2}$$

where n_i is the number of moles of i in the system.

Let us now try to evaluate this important function μ_i. When adding an *incremental number* of molecules of i, free energy is introduced in the form of internal energies of substance i as well as by the interaction of i with other molecules in the system. As more i is added, the composition of the mixture changes and, consequently, subsequent i-type molecules experience an evolving set of interaction energies; thus, μ_i changes as a function of the abundance of i.

Fugacity

Gibbs (1876) recognized that chemical potential could be used to assess the tendency of component i to be transferred from one system to another or to be transformed within a system in a fashion analogous to the use of hydrostatic head potential for identifying the direction of flow between water reservoirs as depicted in Figure 3.1*a*. Figure 3.1*b* shows that equilibrium (no net flow in either direction) is reached when the hydrostatic head potentials of the two reservoirs are equal. Similarly, chemical equilibrium is characterized by equal chemical potentials for all constituents. As with hydrostatic head potential, chemical potential is an *intensive* entity, meaning it is independent of the size of the system (in contrast to the total free energy G, which is an extensive function).

Unfortunately, unlike hydraulic head potentials, there is no way of directly observing chemical potentials. Consequently, the concept of *fugacity* was born. Lewis (1901) reasoned that rather than look into a system and try to quantify all of the chemical potential energies carried by the various components of interest, it would be more practical to assess a molecule's "*urge*" *to escape or flee* that system (hence fugacity from Latin *fugere*, to flee). If one could quantify the *relative tendencies* of molecules to flee various situations, one could simultaneously recognize the *relative* chemical potentials of the compounds of interest in those situations and, based on the differences in their chemical potentials, quantify the direction (higher μ_i to lower μ_i) and extent to which a transfer or transformation process would occur.

Fugacities of Gases

Let us quantify the "fleeing tendency" or fugacity of molecules in a gas (just about the simplest form of a system) in a way we can observe or measure. Imagine a certain number of moles (n_i) of a pure gaseous compound i confined to a volume V, say in a closed beaker, at a specific temperature T. The molecules of the gaseous compound will exert a pressure P_i on the walls of the beaker (a quantity we can feel and measure) as they press upon it seeking to pass (Fig. 3.2*a*). It is not difficult to imagine then that if the gas molecules wish to escape more "insistently" (i.e., a higher chemical potential as a result, for example, of the addition of more i molecules to the beaker), their impact on the walls will increase and we will measure a higher gas pressure. Stating this quantitatively, *relative to some starting chemical potential and pressure*, we see that at constant T, the incremental change in chemical potential of the gaseous compound i may be related to a corresponding change in pressure (deduced from the Gibbs–Duhem equation; see Prausnitz 1969, p. 17):

$$(d\mu_i)_T = \frac{V}{n_i} dP_i \tag{3-3}$$

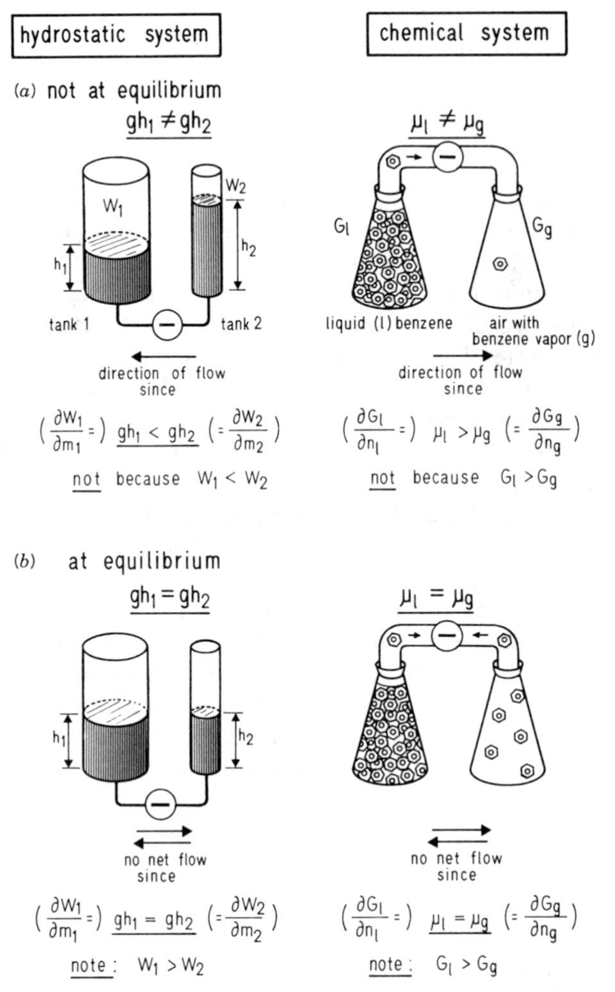

Figure 3.1 Conceptualization of the potential functions in a hydrostatic system and in a simple chemical system. (*a*) In the unequilibrated hydrostatic system, water will flow from reservoir 2 of higher hydrostatic potential ($=g.h_2$, where g is the acceleration due to gravity and h_2 is the observable height of water in the tank) to reservoir 1 of lower hydrostatic potential; total water volumes do not dictate flow. Similarly, benzene molecules move from liquid benzene to the headspace in the nonequilibrated chemical system, not because there are more molecules in the flask containing the liquid, but because the molecules initially exhibit a higher chemical potential in the liquid than in the gas. (*b*) At equilibrium, the hydrostatic system is characterized by equal hydrostatic potentials in both reservoirs (not equal water volumes) and the chemical system reflects equal chemical potentials in both flasks (not equal benzene concentrations).

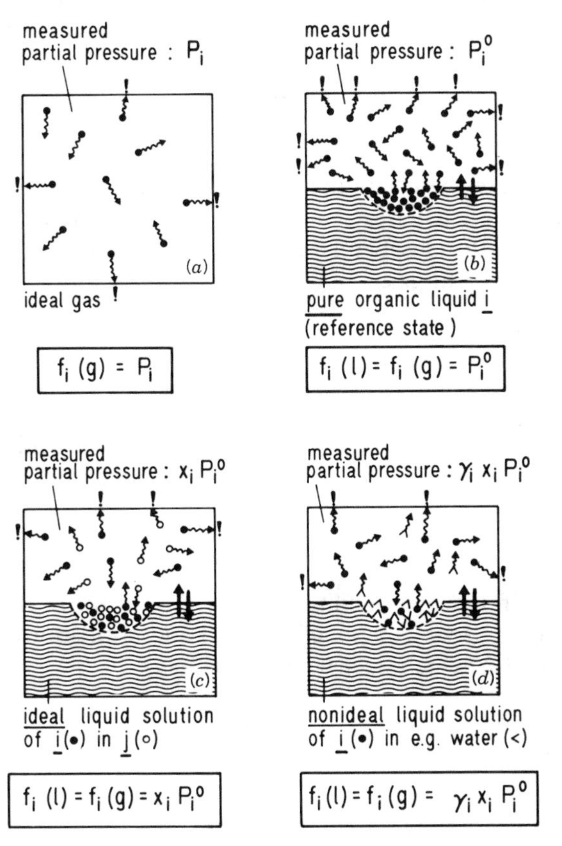

Figure 3.2 Conceptualization of the fugacity of a compound *i* (*a*) in an *ideal* gas; (*b*) in the *pure* liquid compound *i*; (c) in an *ideal* liquid mixture; and (*d*) in a *nonideal* liquid mixture (e.g., in aqueous solution). Note that in (*b*), (*c*), and (*d*), the gas and liquid phases are in equilibrium with one another.

Assuming the compound behaves as an *ideal gas*, we can substitute V/n_i with RT/P_i:

$$(d\mu_i)_T = \frac{RT}{P_i} dP_i \tag{3-4}$$

If we are interested in the chemical potential of an ideal gaseous compound *i* *relative to some standard state* (μ_i^0, P_i^0), at a constant temperature, we can calculate the chemical potential by integration of Eq. 3-4:

$$\int_{\mu_i^0}^{\mu_i} (d\mu_i)_T = RT \int_{P_i^0}^{P_i} \frac{1}{P_i} dP_i \tag{3-5}$$

which yields

$$\mu_i = \mu_i^0 + RT\ln\left[\frac{P_i}{P_i^0}\right] \tag{3-6}$$

We have now related the chemical potential of an ideal gas to pressure, a parameter we can determine. For real gases that experience some molecule–molecule interactions, Lewis invented the closely related parameter *fugacity*, f_i, as a measure of the 'fleeing" tendency of the real gas:

$$\mu_i(\text{real gas}) = \mu_i^0 + RT\ln\left[\frac{f_i}{f_i^0}\right] \tag{3-7}$$

Fugacities of gases then are closely related to their pressures. Indeed, to account for the nonideality of the gas, one can relate these terms by using a fugacity coefficient, θ_i:

$$f_i = \theta_i P_i \tag{3-8}$$

Typically, under environmental conditions (i.e., $P \simeq 1$ atm) θ_i is very nearly equal to 1. In a mixture of gaseous compounds having a total pressure P, P_i is the partial pressure of compound i, which may be expressed as

$$P_i = x_i P \tag{3-9}$$

where x_i is the *mole fraction* of i,

$$x_i = \frac{n_i}{\sum_j n_j} \tag{3-10}$$

where $\sum_j n_j$ is the total number of moles present in the gas, and P is the total pressure; thus, the fugacity of a gas in a mixture is given by

$$\begin{aligned} f_i &= \theta_i x_i P \\ &\simeq P_i \end{aligned} \tag{3-11}$$

Reference States and Standard States

Before we discuss the fugacities of compounds in liquid and solid phases, a few remarks on the choice of *reference states* are necessary. As we have seen in the development of Eq. 3-6, we face one obvious difficulty. Since we cannot compute an absolute value for the chemical potential, we must be content with computing *changes* in the chemical potential as caused by changes in the independent variables of temperature, pressure, and/or composition. This difficulty, however, is apparent rather than fundamental; it is really no more than an inconvenience. It results because the relationship between the chemical potential and the physically measurable quantities are in the form of differential equations which, upon integration, yield only differences. As we will see later, with the choice of an appropriate reference state, it is usually possible to express the energetics of a given process in rather simple terms. Note that in our everyday life, we often choose reference states to express the magnitude of a given entity, for example, the height of a mountain relative to sea level.

When we consider a change in the "energy status" of a compound of interest [e.g., the transfer of organic molecules from the pure liquid phase to the overlying gas phase (vaporization)], we will try to do our energy-change bookkeeping in such a way that we concern ourselves with only those energetic properties of the molecules that undergo change. During the vaporization of liquid benzene, for example, we will not worry about the internal energy content of the benzene molecules themselves, since these molecules maintain the same bonds, and practically the same bond motion, in both the gaseous and liquid states; rather, we will focus on the energy change associated with having benzene molecules in new surroundings. Benzene molecules in gas or liquid phases will, therefore, feel different attractions to their neighboring molecules and will contain different translational energies since in a liquid the molecules are packed fairly tightly together, while in the gas they are almost isolated. This focus on only the changing aspects is the guiding consideration in our choice of *reference states.* For each chemical species of interest, we want to pick a form (a reference form or state) that is closely related to the situation at hand. For instance, it would be silly (though feasible) to consider the energy status of elemental carbon and hydrogen of which the benzene molecule is composed as the reference point with which evaporating benzene should be compared. Instead, we shall be clever and, in this case, choose the "energy status" of pure liquid benzene as a reference state, since liquid benzene includes all of the internal bonding energies common to both the gaseous and liquid forms of the compound.

In the field of environmental organic chemistry, the most common *reference* states used include: (1) the *pure liquid state*, when we are concerned with phase transfer processes; (2) the *infinite dilution state*, when we are dealing with reactions of organic chemicals in solution (e.g., proton transfer reactions in water); and (3) the *elements* in their naturally occurring forms (e.g., C, H_2, O_2, N_2, and Cl_2), when we are interested in reactions in which many bonds are broken and/or formed. Certainly, other reference states may be chosen as convenience dictates, the guiding principle being that one can clearly see how the chemical species considered in a given system is related to this state. Once we have chosen an appropriate reference state, we also must specify the conditions of our reference state; that is, the temperature, pressure, and concentration. These conditions are referred to as *standard conditions* and, together with the reference state, form the *standard state* of a chemical species. We then refer to μ_i^0 in Eqs. 3-6 and 3-7 as the *standard chemical potential*, a value that quantifies the "energy status" under these specific conditions. Since we are most often concerned with the behavior of chemicals in the earth's near-surface ecosystems, 25°C and 1 atm are commonly chosen as standard conditions. In summary, as long as we are unambiguous in our choice of both reference state and standard conditions so that both the starting and final states of a molecular change may be clearly related to these choices, our energy bookkeeping should be fairly straightforward.

Fugacities of Liquids and Solids

Let us now continue with our discussion of how to relate the chemical potential to measurable quantities. We have already seen that the chemical potential of a gaseous compound can be related to pressure. Since substances in both the liquid and solid phases also exert vapor pressures, Lewis reasoned that these pressures likewise reflected the escaping tendencies of these materials from their condensed phases (Fig. 3.2*b*). He thereby extended this logic by defining the fugacities of pure liquids and solids as a function of their vapor pressures, P_i^0 (vapor pressure is discussed in Chapter 4):

$$
\begin{aligned}
f_{i\,\text{pure liquid}} &= \gamma_{i\,\text{pure liquid}} \cdot P_i^0(l) \\
f_{i\,\text{pure solid}} &= \gamma_{i\,\text{pure solid}} \cdot P_i^0(s)
\end{aligned}
\tag{3-12}
$$

where γ_i now accounts for nonideal behavior resulting from molecule–molecule interactions. These activity coefficients are commonly set equal to 1 when we decide to take the pure compound in the phase it naturally assumes under the conditions of interest as the reference state. The molecules are viewed, therefore, as "dissolved" in like molecules, and this condition is assumed to have "ideal" mixing behavior.

If we consider, for example, compound i in a liquid mixture (e.g., in aqueous solution; see Fig. 3.2*d*), we can now relate its fugacity in the mixture to the fugacity of the pure liquid compound by (note that for convenience, we have chosen the *pure liquid compound* as our *reference state*):

$$
\begin{aligned}
f_i &= \gamma_i \cdot x_i \cdot f_{i\,\text{pure liquid}} \\
&= \gamma_i \cdot x_i \cdot P_{i\,\text{pure liquid}}^0
\end{aligned}
\tag{3-13}
$$

where x_i is the mole fraction of i in the mixture (or solution). If the compounds form an ideal mixture (Fig. 3.2*c*), implying that no nonideal behavior results from interactions among unlike molecules, γ_i is equal to 1 and Eq. 3-13 represents the well-known *Raoult's Law*. However, as we will see later when considering solutions of nonpolar organic compounds in polar solvents like water, generally speaking γ_i will be very different from 1.

Activity Coefficients

Analogous to the gaseous phase, we can now express the chemical potential of compound i in a liquid solution by

$$\mu_i = \mu_{i\ \text{pure liquid}}^0 + RT \ln \frac{f_i}{f_{i\ \text{pure liquid}}}$$

or (3-14)

$$\mu_i = \mu_{i\ \text{pure liquid}}^0 + RT \ln \gamma_i x_i$$

The term $f_i/f_{\text{ref}} = \gamma_i x_i = a_i$ is referred to as the *activity* of the compound; that is, a_i is a measure of how active a compound is in a given state (e.g., in aqueous solution) compared to its reference state (e.g., the pure organic liquid at the same T and P). Since γ_i relates a_i, the "apparent concentration" of i, to the real concentration x_i, one refers to γ_i as the *activity coefficient*. It must be emphasized here that activities are relative measures and, therefore, are keyed to the choices of reference state and standard conditions. The numerical value of γ_i will, therefore, depend strongly on the reference state, since molecules of i in different reference states interact differently with their surroundings.

Molecular Scale Influences Resulting in Nonideality

The only task that remains for us is to imagine the molecule–molecule interactions that result in nonideality (and thereby determine the value of γ_i) in an effort to clearly understand all of the terms in chemical potential expressions. From a molecular point of view, one of the easiest ways to conceptualize the factors giving rise to the current free-energy status of a compound (the factors governing γ_i) is through considerations of partial molar *enthalpy* (h_i) and *entropy* (s_i).

In brief, the *enthalpy* of a molecule involves both its attractions or attachments to its surroundings (intermolecular forces) and its internal attractions or bonds (intramolecular forces). The enthalpic contributions may be thought of as the "glue" holding the parts of a molecule to its surroundings. When considering molecular changes then, we must imagine the energy costs or profits resulting from disassembling molecular associations and establishing a new set of intra- and/or intermolecular attachments. In the example of benzene evaporation we used earlier, since no internal bonds are broken or formed, we would be concerned with intermolecular attractions only (i.e., the enthalpy of vaporization) and would use an appropriate reference state (i.e., the pure liquid benzene). On the other hand, if we throw a match into the beaker and burn the benzene to CO_2 and H_2O, clearly intramolecular changes are involved, and a more reasonable reference state choice would be the elements of which benzene is composed. In this case, the corresponding enthalpies of formation of the various molecular species would be of primary interest.

A second contribution to the energy status of a molecule is its entropy (s_i). This property is best imagined as involving the "freedom" or latitude of orientation, configuration, and translation of the molecule. When molecules are forced to be organized or confined, work must be done. As a consequence, energy must be spent in the process. Conversely, the greater the volume in which a molecule can roam (referred to as its free volume), the more ways it can twist and turn, the more freedom the bonding electrons have in moving around in the molecular structure, and the more entropy or "randomness" a substrate has, the lower will be its free-energy content.

3.3 USING THERMODYNAMIC FUNCTIONS TO QUANTIFY MOLECULAR CHANGE PROCESSES

For energy bookkeeping involving molecular changes, we can imagine the populations of molecules having an energy bank account made up of enthalpic and entropic investments (in units of kilojoule per mole rather than dollars). The partial *molar Gibbs free energy* (g_i) (i.e., the chemical potential) is the sum of these contributions:

$$(\mu_i =)\ g_i(\text{J/mol}) = h_i(\text{J/mol}) - T(\text{K}) \cdot s_i(\text{J/mol K}) \tag{3-15}$$

where T is the absolute temperature.

Phase Transfer Processes

In order to see how this energy bookkeeping will be of use, let us now consider the energies of a process that is very important to an environmental organic chemist: the *reversible transfer* of an organic compound A between two phases, say phase 1 and phase 2:

$$\text{A(phase 1)} \rightleftharpoons \text{A(phase 2)} \tag{3-16}$$

which could be, for example, air and water, or perhaps an organic phase and water. As we have already seen (Eq. 3-14), at a given T and P, the energy status of A in each phase is expressed by the corresponding chemical potential (note that we choose the pure liquid A as the reference state):

$$\begin{aligned} \mu_1 &= \mu^0_{\text{pure liquid A}} + RT \ln \gamma_1 x_1 \\ \mu_2 &= \mu^0_{\text{pure liquid A}} + RT \ln \gamma_2 x_2 \end{aligned} \tag{3-17}$$

Recall also that there will be a net flow of compound A from the phase in which A has a higher chemical potential to the phase in which it has a lower chemical potential until the two chemical potentials are equal; that is, until equilibrium is reached. At equilibrium, where $\mu_1 = \mu_2$, we obtain

$$RT \ln \gamma_1 x_1 = RT \ln \gamma_2 x_2 \tag{3-18}$$

Rearrangement of Eq. 3-18 yields:

$$RT \ln \frac{x_1}{x_2} = -(RT \ln \gamma_1 - RT \ln \gamma_2) \tag{3-19}$$

On the left-hand side of Eq. 3-19, we now have an expression that gives us the relative abundance of A (in terms of mole fractions) in the two phases *at equilibrium.* This relative abundance is commonly referred to as the *partition constant* K'_{12} of A between the two phases. The right-hand side of Eq. 3-19 may be interpreted as the energy difference due to the nonideality of the solutions of A in the two phases. Recall that the activity coefficient γ_i is a measure of the additional free energy that a compound is carrying in a nonideal mixture (or solution) as compared with the reference state, which is considered to be the ideal state (e.g., the pure liquid compound). The term $RT \cdot \ln\gamma_i$ is, therefore, commonly referred to as the partial molar *excess free energy*, g_i^e, of the compound in a given phase. Note that the larger g_i^e is, the less comfortable the compound feels in a given phase. We may now reiterate Eq. 3-19 as

$$\ln K'_{12} = -\frac{\Delta g_{12}^e}{RT}$$

or (3-20)

$$K'_{12} = \exp\left[\frac{-\Delta g_{12}^e}{RT}\right]$$

where $K'_{12} = x_1/x_2$ and $\Delta g_{12}^e = RT\ln\gamma_1 - RT\ln\gamma_2$.

This powerful expression, which relates the relative abundance of compound A in two physical states to the energy difference under these two conditions, is the familiar Boltzmann equation. Since the difference in excess free energy can again be seen as a combination of enthalpic and entropic contributions, we have

$$\Delta g_{12}^e = \Delta h_{12}^e - T\Delta s_{12}^e \qquad (3\text{-}21)$$

We now have the means by which we can imagine which molecular scale factors are responsible for causing molecules of A to distribute themselves at equilibrium the way they do! Insight into the specific nature of the forces that direct partitioning processes will allow us to understand why one chemical behaves differently than another and how its fate will vary under new environmental conditions! In summary, when we talk about the free energy of transfer of a given organic compound between two phases, we mean the difference between the *partial molar excess free energies* of the organic compound in the two phases. Note that this energy term is a partial molar entity (or molar in a pure phase system) and therefore has units like joules per mole.

At this point we should direct a few remarks to the problem of how to express abundances of compounds in a given phase. So far, we have used the concept of mole fractions. In environmental organic chemistry, however, the most common way to express concentrations is related to the number of molecules per unit volume, for example, as moles per liter of solution. Although this *molar* concentration scale is sometimes not optimal, it is the most widely used (volumes are, for example, dependent on T and P, whereas masses are not; hence, the use of concentration data normalized per kilogram of seawater is often seen in the oceanographic literature). We can convert mole fractions to molar concentrations by

$$C_i = \frac{x_i(\text{mol/total mol})}{V_{\text{mix}}(\text{L/total mol})} \qquad (3\text{-}22)$$

where C_i is the concentration (moles per liter) of i in a given phase, and V_{mix} is the molar volume of the mixture or solution. When we deal with a mixture of several components, generally we apply *Amagat's Law* as a first approximation; that is, we assume that the compounds mix with no change in volume due to such factors as special packing or molecule:molecule attractions or repulsions:

$$V_{\text{mix}} = \sum_i x_i V_i \tag{3-23}$$

where V_i is the molar volume of the pure compound i. Using molar concentrations, we can then redefine partition constants and use Eq. 3-18 to obtain

$$\begin{aligned} \ln K_{12} &\equiv \ln \frac{C_{i1}}{C_{i2}} = \ln \frac{x_{i1}/V_{\text{mix 1}}}{x_{i2}/V_{\text{mix 2}}} = \ln \frac{\gamma_{i2}/V_{\text{mix 1}}}{\gamma_{i1}/V_{\text{mix 2}}} \\ &= -\frac{\Delta g_{12}^{\text{e}}}{RT} + \ln \frac{V_{\text{mix 2}}}{V_{\text{mix 1}}} \\ &= -\frac{\Delta G_{12}}{RT} \end{aligned} \tag{3-24}$$

where K_{12} no longer has the prime superscript representing the mole fraction basis. Note that ΔG is usually used (unfortunately) to express the difference of two partial molar entities. For *aqueous solutions* of moderately or only sparingly soluble compounds, we can usually neglect the contribution of the organic solute to the molar volume of the mixture, which means we set V_{mix} equal to V_{w}, the molar volume of water ($V_{\text{w}} = 0.018\ \text{L} \cdot \text{mol}^{-1}$ at 25°C). Organic compounds have molar volumes on the order of $0.2\ \text{L} \cdot \text{mol}^{-1}$. The molar volume of an aqueous solution of a compound i is, therefore, given by Eq. 3-23:

$$V_{\text{mix}} \approx 0.2 x_{i\text{w}} + 0.018 x_{H_2O} \tag{3-25}$$

where $x_{i\text{w}}$ is the mole fraction of compound i in aqueous solution. Since $x_{i\text{w}} + x_{H_2O} = 1$, we can write

$$V_{\text{mix, w}} \approx 0.182 x_{i\text{w}} + 0.018 \quad (\text{L} \cdot \text{mol}^{-1}) \tag{3-26}$$

Consequently, V_{mix} will change by less than 2% as long as $x_{i\text{w}} \lesssim 0.002$, which corresponds to about $0.1\ \text{mol} \cdot \text{L}^{-1}$. For nonpolar organic compounds with solubilities below this limit then (nearly all compounds in which we are interested), we will not have to worry about the molar volume change of the aqueous phase due to the organic solute.

Chemical Reactions

Let us now turn to another case of interest, namely, a reversible reaction among chemical species in a given phase, such as in aqueous solution. Let us first consider a very simple reaction in which A reacts with B to give C plus D, such as in the reaction of an organic acid with water which yields the corresponding ("conjugate") base and a solvated proton:

$$A + B \rightleftharpoons C + D \tag{3-27}$$

Instead of considering the energy status of a given chemical species in different phases, here we are concerned with the energy status of several different species (i.e., the reactants and products of a *reversible* reaction) in one given phase, in our case, an aqueous phase. As mentioned earlier, when dealing with reactions in dilute solutions, the most intelligent choice of reference state for solutes is the *infinite dilution state*; one commonly defines the standard state of a solute as one of unit concentration on the scale used (i.e., the *molar* concentration scale) and having the ideal properties of a very dilute solution ($\gamma_i^\infty = 1$). Hence, for each species *at a given P and T* (e.g., 1 atm, 25°C), we may express the chemical potentials as

$$\begin{aligned}
\mu_A &= \mu_A^{0\prime} + RT\ln(\gamma_A'[A]/[A]^0)\\
\mu_B &= \mu_B^{0\prime} + RT\ln(\gamma_B'[B]/[B]^0)\\
\mu_C &= \mu_C^{0\prime} + RT\ln(\gamma_C'[C]/[C]^0)\\
\mu_D &= \mu_D^{0\prime} + RT\ln(\gamma_D'[D]/[D]^0)
\end{aligned} \tag{3-28}$$

where $[i]$ is the actual concentration of the species i, $[i]^0$ is its concentration in the standard state, and the prime superscript is used to denote the infinite dilution reference state (in distinction to the pure organic liquid state). Note that $\mu_i^{0\prime}$ corresponds to the free energy of formation of the species i in aqueous solution, that is, $\mu_i^{0\prime} \equiv \Delta G_f^0[i(\text{aq})]$. Since $[i]^0$ is commonly set to 1 M, we may write (but not forget that every concentration has been divided by 1 M and, therefore, represents a dimensionless number!)

$$\begin{aligned}
\mu_A &= \mu_A^{0\prime} + RT\ln(\gamma_A'[A])\\
\mu_B &= \mu_B^{0\prime} + RT\ln(\gamma_B'[B])\\
\mu_C &= \mu_C^{0\prime} + RT\ln(\gamma_C'[C])\\
\mu_D &= \mu_D^{0\prime} + RT\ln(\gamma_D'[D])
\end{aligned} \tag{3-29}$$

At a given composition [A], [B], [C], [D] of the system (we are not interested in other system components that are not involved in the reaction), it is easy to see that if we convert an incremental number of molecules of A and B to C and D, we cause a change in the total free energy of the system which is given by

$$dG = -\mu_A \cdot |dn_A| - \mu_B \cdot |dn_B| + \mu_C \cdot |dn_C| + \mu_D \cdot |dn_D| \tag{3-30}$$

Note that dn of the educts is negative (the number of educt molecules decreases). In our simple stoichiometric case,

$$|dn_A| = |dn_B| = |dn_C| = |dn_D| = |dn_r|$$

Therefore, we can write

$$dG = (-\mu_A - \mu_B + \mu_C + \mu_D)dn_r \tag{3-31}$$

The quantity $dG/dn_r = -\mu_A - \mu_B + \mu_C + \mu_D$, which is a measure of the free-energy change in the system as the reaction progresses, is called the molar free-energy change of the reaction and is usually (unfelicitously) denoted as ΔG. Note again that we use a capital letter here to express the difference between partial molar entities!

$$\begin{aligned}\Delta G &= -\mu_A - \mu_B + \mu_C + \mu_D \\ &= -\mu_A^{0\prime} - \mu_B^{0\prime} + \mu_C^{0\prime} + \mu_D^{0\prime} + RT\ln\frac{(\gamma'_C[C])(\gamma'_D[D])}{(\gamma'_A[A])(\gamma'_B[B])} \\ &= \Delta G^0 + RT\ln\frac{(C)\cdot(D)}{(A)\cdot(B)}\end{aligned} \tag{3-32}$$

The term ΔG^0 is referred to as the *standard free-energy change of the reaction*, and (i), equal to $\gamma_i[i]$, is the activity of species i. At equilibrium, where $dG/dn_r = \Delta G = 0$, we obtain

$$\ln K \equiv \ln\frac{(C)\cdot(D)}{(A)\cdot(B)} = -\frac{\Delta G^0}{RT} \tag{3-33}$$

K is called the *equilibrium constant* of the reaction. Equation 3-33 is commonly referred to as a *mass law equation.*

Finally, we should consider a more general chemical reaction:

$$\mathrm{aA + bB + \cdots \rightleftharpoons pP + qQ + \cdots} \tag{3-34}$$

where a, b, ..., p, q, ... are the *stoichiometric coefficients* of the reaction; that is, the coefficients that describe the relative number of moles of each reactant consumed or produced by a given reaction. Analogous to the simple case discussed with respect to Eq. 3-30, the molar free-energy change of the reaction is then given by (note that now $|dn_A| = \mathrm{a}|dn_r|$; $|dn_B| = \mathrm{b}|dn_r|$, etc.):

$$\frac{dG}{dn_r} = \Delta G = -a\mu_A - \mathrm{b}\mu_B - \cdots + \mathrm{p}\mu_P + \mathrm{q}\mu_Q + \cdots \tag{3-35}$$

and thus,

$$\ln K \equiv \ln\frac{(P)^{p}\cdot(Q)^{q}\cdots}{(A)^{a}\cdot(B)^{b}\cdots} = \frac{-\Delta G^0}{RT} \tag{3-36}$$

with

$$\Delta G^0 = -\,a\mu_A^{0\prime} - b\mu_B^{0\prime} - \cdots + p\mu_P^{0\prime} + q\mu_Q^{0\prime} + \cdots$$

Note that the standard free-energy change ΔG^0 of a given reaction measures the change in free energy of the system upon infinitesimal progress of the reaction and is considered when all of the reactants are present in their standard states; that is, the corresponding values of $\mu^{0\prime}$ denote the chemical potentials (or free energies of formation) of the various constituents in their reference states under standard conditions of temperature, pressure, and concentration.

3.4 SUMMARY

In conclusion, we can benefit from examining phase transfers or reversible reactions of organic chemicals in the environment using the following approach:

1. Visualize the molecules involved in their surroundings, noting the molecule: molecule interactions and freedom to move, allowing us to understand the enthalpic and entropic contributions to the free-energy status of the molecules relative to a chosen reference state.
2. In light of the process studied, choose both a convenient reference state and standard conditions that incorporate molecular aspects which remain unaltered during the process.
3. Express the energetic status of the species of interest (i.e., the chemical potential) by using fugacity or activity expressions and calculate equilibrium concentrations by balancing chemical potentials.

These are the basics for both understanding and using thermodynamic concepts to judge the environmental behavior of organic chemicals. In the following chapters, in which we deal with various compound properties, we will demonstrate how such measureable quantities can give us important insights into a compound's energy status; that is, its fugacity or activity in various environments.

CHAPTER 4

VAPOR PRESSURE

4.1 INTRODUCTION

The tendency of a chemical to transfer to and from gaseous environmental phases (e.g., the atmosphere, marsh gas bubbles) is determined to a large extent by its vapor pressure. This property is critical for prediction of either the *equilibrium* distribution or the *rates* of exchange to and from natural waters.

The *vapor pressure*, P^0, is defined as the pressure of the vapor of a compound at equilibrium with its pure condensed phase, be it liquid or solid. (We take the vapor pressure of a pure gas to be 1 atm on the surface of the Earth.) According to the Gibbs phase rule (number of degrees of freedom = number of components − number of phases + 2), for a system containing a single chemical distributed between two phases at equilibrium, there is only one degree of freedom. Therefore, if a system temperature is chosen, the vapor pressure of the component in the gaseous phase is fixed. (Note: this forces us to have a gas as one of the two phases). The most familiar vapor pressure/temperature point is the normal boiling point T_b of a compound, which is the temperature at which P^0 is equal to 1 atm. Figure 4.1 shows ranges of vapor pressures at 25°C for some classes of organic chemicals. As can be seen, vapor pressures may differ by many orders of magnitude. These compound-to-compound variations arise from differences in molecule:molecule interactions. One might ask why it is important to know vapor pressures of chemicals with ambient P^0 values as low as 10^{-12} atm [e.g., DDT, PCBs, benzo(a)pyrene] since, intuitively, one would not expect evaporation of such compounds to be a significant environmental process. However, owing to the extremely low solubilities in water of these same chemicals (resulting in their high aqueous phase fugacities), when present in surface waters, these compounds still partition appreciably into the atmosphere. Furthermore, the vapor pressure

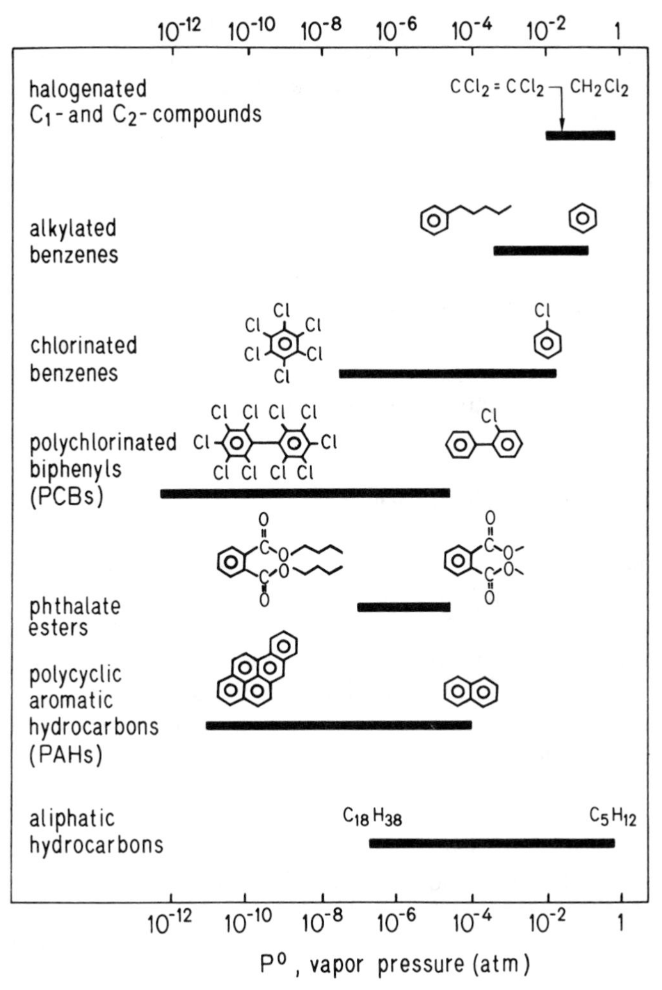

Figure 4.1 Ranges at 25°C in saturation vapor pressure (P°) values for some important classes of organic compounds.

determines the speciation of the compound (gaseous vs. particulate) in the atmosphere (Junge, 1977; Bidleman, 1988; Ligocki and Pankow, 1989). Finally, we note that over the ambient temperature range, the vapor pressure of a given compound may change by more than an order of magnitude; therefore, knowledge of the temperature dependence of vapor pressure is important.

Before we turn to a thermodynamic description of the vapor pressure/temperature relationship, it is illustrative to imagine what the molecules of a substance do to establish an equilibrium vapor pressure. We do this by using a kinetic-molecular description; that is, by picturing the dynamic equilibrium resulting from balancing the rate of evaporation with the rate of condensation. Let us consider a condensed pure compound (either liquid or solid) in equilibrium with its vapor phase (see Fig. 3.2*b*). At a given temperature, a certain number of molecules thermally jostling about in the condensed phase will continuously acquire sufficient energy to overcome the forces of attraction to their neighboring molecules and escape from the condensed phase. Meanwhile in the vapor phase, there will be continuous collisions of some vapor molecules with the surface of the condensed phase. A fraction of the colliding molecules will have so little kinetic energy, or will dissipate their energy upon collision with the condensed surface, that rather than bounce back into the vapor phase, they will be combined into the condensed phase. At a given temperature, these opposing processes of evaporation and condensation reach an equilibrium state that is controlled primarily by molecule–molecule attractions in the condensed phase and is characterized by the abundance of molecules in the vapor above the condensed phase. This gas phase abundance is expressed as the equilibrium vapor pressure.

4.2 THERMODYNAMIC CONSIDERATIONS

Thermodynamically, the phase transitions in a one-component system (i.e., pure compound) can be conveniently diagramed on a pressure–temperature plot. The phase boundaries shown in Figure 4.2 are calculated after noting that at equilibrium the chemical potential (or since we are dealing with a one-compound system, the molar free energy) must be equal in the two phases. If the temperature or pressure of the system is changed but equilibrium is reestablished, the change in chemical potential (or molar free energy) of the two phases must be the same:

$$d\mu_1 = d\mu_2 \tag{4-1}$$

where the subscripts 1 and 2 denote the two phases in equilibrium with one another. The change in chemical potential of the compound in each phase is related to temperature (thermal work) and pressure (mechanical work) (Prausnitz, 1969) by

$$\begin{aligned} d\mu_1 &= -S_1 dT + V_1 dP \\ d\mu_2 &= -S_2 dT + V_2 dP \end{aligned} \tag{4-2}$$

where S_1 and S_2 are the molar entropies, and V_1 and V_2 are the molar volumes of the compound in each of the two phases. Substitution of Eq. 4-2 into Eq. 4-1 and re-arrangement yields

$$\frac{dP}{dT} = \frac{(S_1 - S_2)}{(V_1 - V_2)} = \frac{\Delta S_{12}}{\Delta V_{12}} \tag{4-3}$$

Since at equilibrium $\Delta G_{12} = \mu_1 - \mu_2 = 0$, then ΔS_{12} equals $\Delta H_{12}/T$, and we thereby obtain the mathematical expression from which the boundaries in Fig. 4.2 are derived:

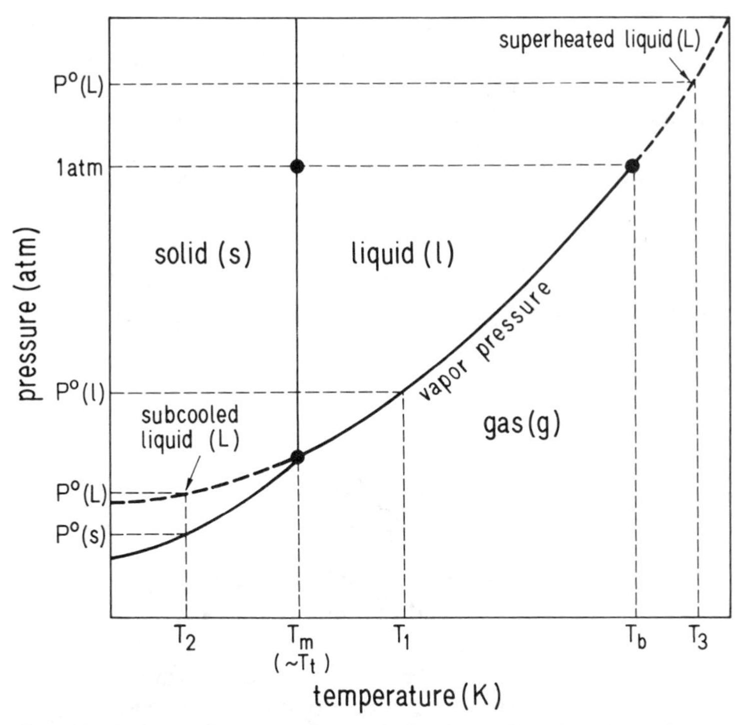

Figure 4.2 Simplified phase diagram of organic liquids and solids. Note that the boundary between the solid and liquid phases has been drawn assuming the chemical's melting point (T_m) equals its triple point (T_t), the temperature–pressure condition where all three phases coexist). In reality, T_m is a little higher than T_t for some compounds, but is a little lower for others.

$$\frac{dP}{dT} = \frac{\Delta H_{12}}{T \cdot \Delta V_{12}} \tag{4-4}$$

where ΔH_{12} is the change in molar enthalpy, and ΔV_{12} is the change in molar volume upon phase transfer. Equation 4-4 is commonly known as the Clapeyron equation. For a more detailed derivation and discussion of this equation and derivation of phase diagrams as shown in Fig. 4.2, the reader is referred to textbooks on physical chemistry or thermodynamics (e.g., Dickerson, 1969, pp. 222–241; Lewis and Randall, 1961, pp. 104–106).

Liquid–Vapor Equilibrium

Here we are especially interested in phase changes that occur at ambient pressures and temperatures and where one of the phases is a gas. With the exception of a few classes of compounds (e.g., low molecular mass hydrocarbons, some halogenated methanes such as freons), the majority of environmentally relevant organic chemicals have boiling points well above ambient temperatures. When these substances are confined, for example, in a reagent bottle at these temperatures, they exist primarily as either liquids or solids. Note that if the bottle is left open, these chemicals slowly but surely evaporate, because both solids and liquids continuously try to maintain some molecules in the adjacent headspace! For the pressure and temperature range of interest here, some simplifying assumptions can be made. First, the change in volume upon evaporation (liquid–gas) or sublimation (solid–gas) may be approximated by the molar volume of the gas ($\Delta V_{12} = V_{\text{gas}} - V_{\text{solid or liquid}} \simeq V_{\text{gas}}$); secondly, the vapor can be assumed to obey the ideal gas law, that is, $V_{\text{gas}} = RT/P^0$ where P^0 denotes the vapor pressure of the compound. The vapor pressure–temperature relationship can then be written as

$$\frac{dP^0}{dT} = \frac{P^0 \Delta H_{12}}{RT^2} \tag{4-5}$$

or

$$\frac{d \ln P^0}{dT} = \frac{\Delta H_{12}}{RT^2} \tag{4-6}$$

where $\Delta H_{12} = \Delta H_{\text{vap}}$ (heat of vaporization) or ΔH_{sub} (heat of sublimation) is the energy required to convert one mole of liquid or solid into a vapor without an increase in temperature. Most vapor pressure–temperature correlation and estimation equations originate from integration of these equations using different assumptions on the temperature dependence of ΔH_{12} and on the compressibility of the vapors. Since we have assumed ideal gas behavior, we have set the compressibility factor equal to 1 (Reid et al., 1977, p. 186). At this point it should also be noted that the presence of external atmospheric pressure (say 1 atm) has a very small effect upon vapor pressures of condensed phases (see Dickerson, 1969, pp. 229–230). In this case, P^0 is not the total pressure of the system but a partial pressure of the compound in the atmosphere. The value of ΔH_{vap} is zero at the critical point T_c (where the vapor and liquid phases cannot be distinguished), rises rapidly at temperatures approaching the normal boiling point, and then rises more slowly at lower temperatures. At temperatures well below the boiling point, H_{vap} is only a very weak function of temperature. Thus, over the range of ambient temperatures, ΔH_{vap} can, as a reasonable approximation, be assumed constant. Integration of Eq. 4-6 in this essentially constant ΔH_{vap} region then yields

$$\ln P^0 = -\frac{B}{T} + A \tag{4-7}$$

where $B = \Delta H_{\text{vap}}/R$. We can interpret Eq. 4-7 in terms of our kinetic–molecular model. As stated before, at a given temperature, only those molecules in the condensed phase that acquire an amount of energy sufficient to overcome the attractive forces of the neighboring molecules can escape. The fraction of molecules in the condensed phase that have energy greater than this amount, and thus the vapor pressure, will be proportional to a Boltzmann-type expression:

$$P^0 \propto e^{-(\text{energy}/RT)} \tag{4-8}$$

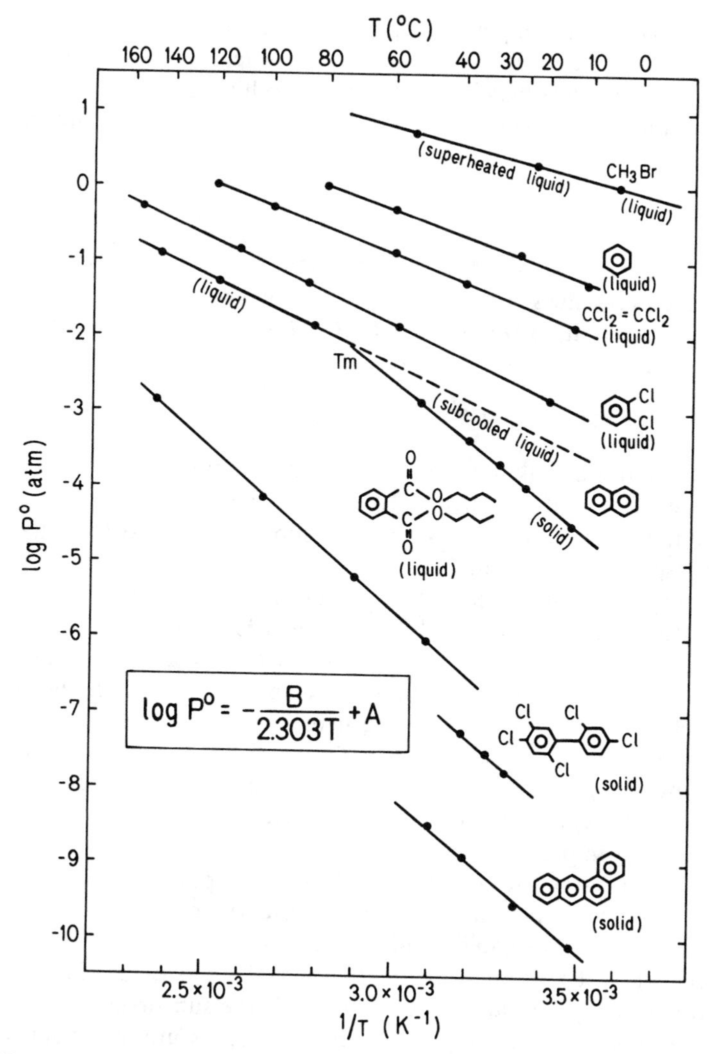

Figure 4.3 Temperature dependence of vapor pressure for some representative compounds.

and

$$\ln P^0 = -\frac{\text{energy}}{RT} + \text{constant} \tag{4-9}$$

which is consistent with Eq. 4-7.

The examples given in Figure 4.3 show that plotting the observed log P^0 versus inverse T (K) over the ambient temperature range yields practically linear relations, as expected from Eq. 4-7. Therefore, over narrow temperature ranges in which there are some vapor pressure data available, Eq. 4-7 can be used to calculate vapor pressures at any other temperature *provided that no other phase change occurs within the temperature range considered.* If the temperature range is enlarged, the fit of experimental data may be improved by modifying Eq. 4-7 by the introduction of a third parameter C, which is used to correct for the temperature dependence of ΔH_{vap}:

$$\ln P^0 = -\frac{B}{T+C} + A \tag{4-10}$$

Equation 4.10, known as the Antoine equation, has been widely used to regress experimental data. Values for A, B, and C have been tabulated for a variety of compounds (Reid et al., 1977, p. 184; *CRC Handbook of Chemistry and Physics* 1985–1986).

Solid–Vapor Equilibrium

If the aggregation state of the condensed phase changes within the temperature range considered, (i.e., if the melting point T_m of the compound lies within this temperature range), the vapor pressure curve shows a break at this phase transition temperature (see naphthalene in Fig. 4.3). Below the melting point, a solid vaporizes without melting; that is, it sublimes. The heat of sublimation ΔH_{sub} may be considered the sum of heat of fusion ΔH_{melt} (which is usually less than a fourth of ΔH_{sub}; Reid et al., 1977, p. 219) and the heat of vaporization ΔH_{vap}(L) of the hypothetical liquid at that temperature. By hypothetical liquid we mean an imaginary liquid that is cooled below its melting point without allowing it to crystallize. This hypothetical state is commonly referred to as the *subcooled liquid state* and we will denote it throughout this book using the symbol (L) (see Fig. 4.2). The concept of the subcooled liquid state, in which molecules continue to be free to move and orient as they do in liquid aggregation, is very important to us since, as discussed in Chapter 3, when considering phase transfer processes, we will commonly use the pure liquid compound as the reference state; hence changes in the chemical potential of compounds that are solids at standard conditions are related to this imaginary condition, meaning that (assuming ideal gas behavior) the vapor pressure P^0(L) of the subcooled liquid becomes the *reference state fugacity*. As can be seen in Fig. 4.2, the subcooled liquid always has a higher P^0 than the corresponding solid. The vapor pressure of the subcooled liquid can be obtained by extrapolation of vapor pressure data above the melting point (e.g., for naphthalene in Fig. 4.3), from the vapor pressure of the solid and the entropy of melting, or, in some cases, determined experimentally (e.g., by gas chromatographic methods). Conversely, the vapor pressure of the solid can be approximated by using the vapor pressure of the subcooled liquid and the entropy of melting (see Section 4.4).

As stated earlier, some organic compounds are gases at ambient conditions. Analogous to the subcooled liquid state, when dealing with solids it is convenient to define a *superheated liquid state* for gases (Fig. 4.2); that is, the liquid is heated above its boiling point at 1 atm without allowing it to boil. Thus, at ambient temperatures, the superheated liquid vapor pressure of a gaseous compound is greater than 1 atm (see example in Fig. 4.3). Consequently, the reference state fugacity of gases is greater than 1 atm when the superheated liquid of such chemicals is chosen as the reference state.

Before we consider a practical approach to estimate vapor pressures and their variations with temperature, it is useful to first consider the intermolecular forces that govern the vapor pressure of a chemical. This will help us in choosing the appropriate empirical factors with which vapor pressure estimates are tuned and will later be useful in understanding a chemical's behavior in phases other than its pure phase.

4.3 MOLECULAR INTERACTIONS GOVERNING VAPOR PRESSURE

Since the extent to which an organic compound escapes to the vapor phase from a liquid solution of like molecules is controlled by the magnitude of ΔH_{vap}, we should consider the underlying molecular factors affecting this thermodynamic property. Molecules in the gas phase at pressures near 1 atm approach one another rather rarely and, consequently, we may assume that these vapor molecules have very little *inter*molecular attraction to one another over the average long distances that separate them. As a result, the change in enthalpy for the process of vaporization chiefly involves disrupting the forces that bind like molecules to one another in the condensed phase. The simplest view of this then is that the more strongly a particular kind of molecule attracts like molecules, the lower its corresponding abundance in the gas phase at equilibrium at a particular temperature.

The forces that attract molecules to one another generally result from the electron-deficient molecular regions being drawn toward the electron-rich counterparts of neighboring molecules. Since average electron abundance in organic molecules varies from position to position on a length scale approximately equal to the bond distances, it should be obvious that the interactions between any two molecules will involve the *summation* of all bond-scale attractions.

Let us examine the four major types of *inter*molecular attractions and their impact on vapor pressure more specifically (Fig. 4.4). First, *all* compounds experience van der Waals forces or dispersive attractions to one another. Even *nonpolar* substances such as alkanes, which exhibit a time-average smooth distribution of electrons throughout their structures, have instantaneous displacements of their electrons such that momentary electron-rich and electron-poor structural regions develop (see Fig. 4.4*a*). This momentary distribution of charges is felt by neighboring alkane molecules whose electrons respond in a complementary fashion. Consequently, there is an instantaneous intermolecular attraction between these regions. In the next moment this attractive interaction shifts elsewhere in the structure. In light of this picture of van der Waals forces, we can perhaps already see why the molecules in a beaker filled with hexane do not instantly "fly apart". Furthermore, we can now understand why larger molecules generally exhibit lower vapor pressures since the summation of van der Waals attractions is directly related to their size. As shown in Fig. 4.5, each additional increment in size for a homologous series of *n*-alkanes causes P^0 to decrease (or the ΔH_{vap} to increase) concomitantly. We can see a similar trend in increasing ΔH_{vap} for each methylene added in the series of alkylated benzenes listed in Table 4.1.

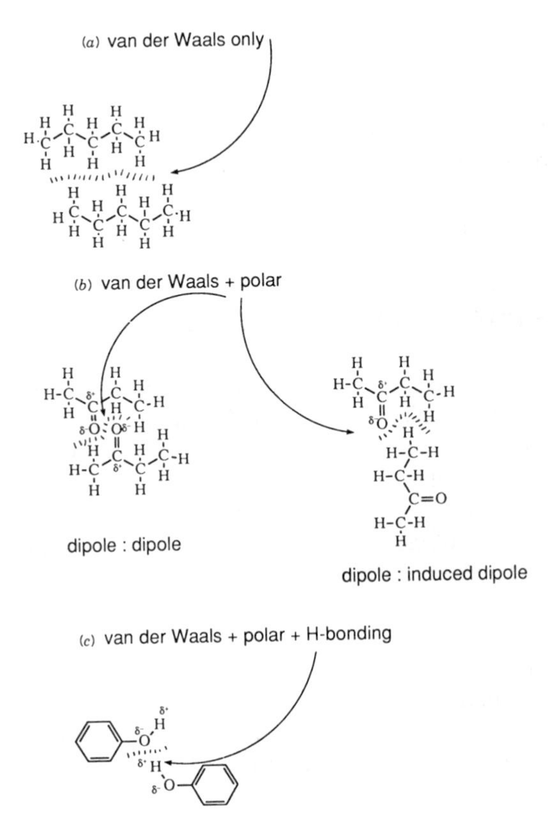

Figure 4.4 Conceptualization of molecular forces and interactions. (*a*) Nonspecific or van der Waals attractions occur between all forms of matter, even nonpolar materials like pentane. (*b*) The presence of unevenly distributed electron densities gives rise to bond-size dipoles which are attractive to other dipoles or which induce electron redistribution in neighbor molecules and thereby establish dipole: induced dipole attractions. (*c*) Hydrogen atoms bonded to oxygen or nitrogen are available to be attracted to the nonbonded electrons of other oxygen or nitrogen atoms, establishing hydrogen bonds.

Owing to differing electron-attracting properties or electronegativities of the various types of atoms included in organic compounds [H, C, S, I < N, Br < Cl < O < F (see Chapter 2)], we may also expect organic structures with different types of atoms bound to one another to exhibit regions either always deficient or always enriched in electrons (Fig. 4.4*b*). This *polar* characteristic results in attractions to a molecule's surroundings in two ways. First, in addition to the van der Waals forces always present, the dipole aligns itself with other dipoles in a head-to-tail fashion resulting in an important dipole:dipole attraction between these molecules. Further, if a dipole is positioned near an evenly charged structural region of an adjacent molecule, electrons in the neighboring molecule will be displaced in response to the approaching polar unit, and an *induced dipole* attraction will result. We can readily see the impact of these polar interactions by examining the ΔH_{vap} for benzenes substituted with various

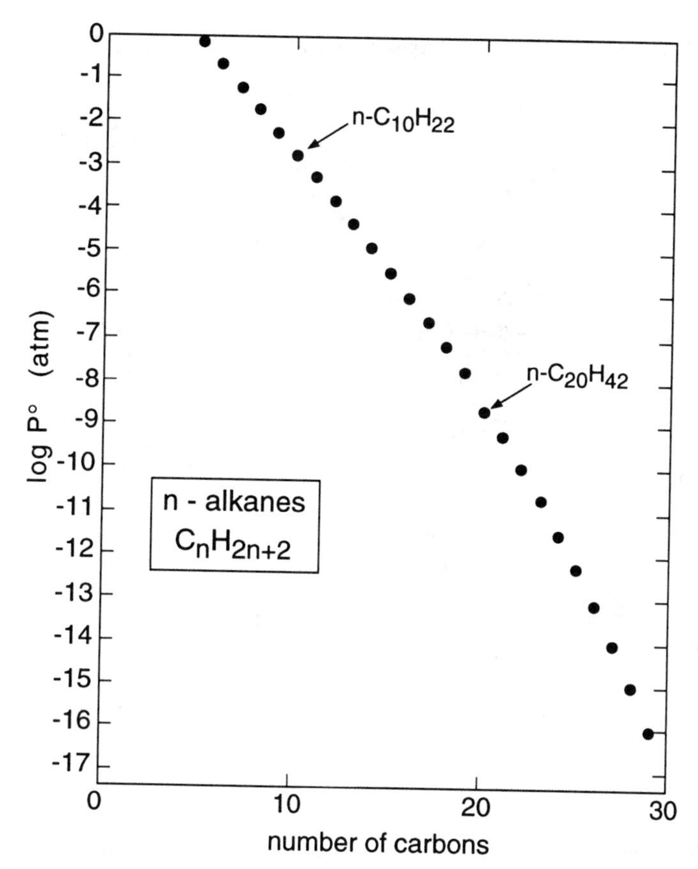

Figure 4.5 Vapor pressure at 25°C of *n*-alkanes as a function of chain length.

groups containing electron-withdrawing heteroatoms (Table 4.1). Replacing a hydrogen with a halogen (F, Cl, Br, I), for example, results in an increase in the heat of vaporization. It should be noted that this trend does not simply follow the electronegativity of the halogens ($F > Cl > Br > I$), since the increasing size of this family and their ability to be polarized (associated with distance between nucleus and outer valence electrons) simultaneously causes the van der Waals and induced dipole attractions to increase. Other polar functional groups such as the nitro ($—NO_2$) group also cause the heat of vaporization to markedly increase. Typically, these polar effects add several kilojoules per mole to the sum of intermolecular attraction energies.

At this point, it is probably good to reiterate that the intermolecular forces are *summed* for all these operating forces on each molecule and result from bond-scale electron displacements. Both of these concepts can be emphasized by considering the heats of vaporization of chlorinated benzenes. Each successive substitution of a chlorine on the ring enhances the heat of vaporization, no matter the substitution position. If overall molecule polarity (reflected by the molecular dipole moment) governed the intermolecular attractions, symmetrical isomers such as p-dichlorobenzene

TABLE 4.1 Variations in Heats of Vaporization at Normal Boiling Points of Substituted Benzenes

Compound	Substituent(s)	T_b(K)	$\Delta H_{vap}(T_b)$ (kJ·mol^{-1})	μ(D)
Benzene	$-H$	353	30.8	0
Methylbenzene (toluene)	$-CH_3$	384	33.2	0.4
Ethylbenzene	$-CH_2CH_3$	409	35.6	0.6
n-Propylbenzene	$-(CH_2)_2CH_3$	432	38.2	
n-Pentylbenzene	$-(CH_2)_4CH_3$	479	41.2	
n-Heptylbenzene	$-(CH_2)_6CH_3$	519	45.2	
n-Nonylbenzene	$-(CH_2)_8CH_3$	555	49.0	
Isopropylbenzene	$-CH(CH_3)_2$	439	37.5	
Vinylbenzene (styrene)	$-CH{=}CH_2$	418	37.0	
Fluorobenzene	$-F$	358	31.2	1.57
Chlorobenzene	$-Cl$	405	36.5	1.73
Bromobenzene	$-Br$	429	37.9	1.71
Iodobenzene	$-I$	462	39.5	1.42
1,2-Dichlorobenzene	2 x Cl	454	40.6	2.5
1,4-Dichlorobenzene	2 x Cl	447	39.7	0
Nitrobenzene	$-NO_2$	484	40.8	4.2
Aminobenzene	$-NH_2$	458	44.5	1.5
Hydroxybenzene	$-OH$	455	40.7	1.5
Benzylalcohol	$-CH_2OH$	478	50.6	1.7
Benzoic acid	$-COOH$	522	50.6	

would show no enhanced attractions arising from polar effects. As shown in Table 4.1, this para isomer has virtually the same heat of vaporization as the ortho compound. There certainly are some intramolecular interactions that are due to the conductivity of the aromatic ring and the "vector nature" of molecular dipoles; therefore, the chlorines at different positions interact at a second order level to cause small differences between positional isomers in their heats of vaporization. Although successive chlorinations do not consistently change ΔH_{vap}, we can readily see how these structural changes will qualitatively affect vapor pressure.

The final intermolecular force we have to consider is *hydrogen bonding* (see also Chapter 2). Because of the substantial differences in electronegativity of hydrogen and oxygen or nitrogen, hydrogens bound to oxygens in alcohols or to nitrogens in amines are particularly electron deficient and "dangle" from the basic structure. Neighboring molecules with substituents containing oxygens or nitrogens and their nonbonding electrons, can engage these available hydrogens and form a hydrogen bond (see Fig. 4.4*c*). As a result, it is not surprising that the substituted benzenes with moieties capable of this special intermolecular interaction (see Table 4.1: —COOH, —OH, $—CH_2OH$, $—NH_2$) show strongly enhanced heats of vaporization over those expected if van der Waals and dipole attractions operated alone. This H-bonding attraction adds about 10 kJ/mol to the total ΔH_{vap}.

In summary, it is clear that structural elements of organic compounds are attracted to their surroundings in a manner that reflects several electronic forces operating simultaneously. These attractions cause molecules to be "bound" to a condensed liquid phase, and the stronger the sum of attractions, the lower the vapor pressure of the compound will be (or the higher its boiling temperature). Thus, as we see in Table 4.1, larger heats of vaporization go hand in hand with higher boiling temperatures.

Trouton's Rule of Constant Entropy of Vaporization

Since the entropy of vaporization at the boiling point, $\Delta S_{vap}(T_b)$, is equal to the ratio $\Delta H_{vap}(T_b)/T_b$, the close relationship between the heat of vaporization and the boiling temperature results in a nearly constant value for this entropy change. If we examine the $\Delta S_{vap}(T)$ of a variety of xenobiotic organic compounds incapable of hydrogen bonding, we obtain values of around 88 $\text{J}\cdot\text{mol}^{-1}\text{K}^{-1}$ (Table 4.2). This result holds for virtually all nonpolar substances and was recognized long ago by Trouton (1884). Later Kistiakowsky (1923) utilized the Clapeyron equation and the ideal gas law to derive

$$\Delta S_{vap}(T_b) = (36.6 + 8.31 \ln T_b) \quad \text{J}\cdot\text{mol}^{-1}\,\text{K}^{-1} \tag{4-11}$$

This expression reflects a weak relationship between the nonpolar compound boiling temperature and the entropy of vaporization, but substantially verifies Trouton's empirical observation. Examination of the $\Delta S_{vap}(T_b)$ for various nonpolar compounds reveals some small differences which are understandable in light of intermolecular forces operating in the liquid phase. For example, elongate molecules such as pentadecane show higher $\Delta S_{vap}(T_b)$ than their corresponding shorter-chain homologues (e.g., pentane). This makes sense since the longer molecules have a tendency to organize in parallel, maximizing van der Waals attractions. This decrease in S_{liq} translates into a larger ΔS_{vap} $(= S_{gas} - S_{liq})$.

For polar organic liquids, especially for hydrogen-bonding liquids such as alcohols and amines, the tendency to orient in the liquid phase, due to these highly directional intermolecular attractions, is greatly increased. We can see the effect of this in the deviation of entropies of vaporization of polar chemicals, like pyridine and nitromethane, or in hydrogen-bonding substances such as methanol, from the predicted entropies

TABLE 4.2 Constancy of Entropy of Vaporization at the Normal Boiling Point for Some Organic Compounds

Compound	Structure	T_b (K)	$\Delta S_{vap}(T_b)(J \cdot mol^{-1} \cdot K^{-1})$ Observed	Kistiakowsky Prediction[a]	Intermolecular Forces[b]
Pentane		309	83.3	84.2	vdW
Pentadecane		544	90.8	88.9	vdW
Benzene		353	87.0	85.4	vdW
Toluene		384	86.6	86.0	vdW
Styrene		418	88.7	86.8	vdW
Naphthalene		491	88.3	88.1	vdW
Phenanthrene		613	90.8	89.9	vdW
Pyridine		389	90.4	86.2	vdW
Methylene chloride	CH_2Cl_2	313	90.0	84.4	vdW + polar
Chloroform	$CHCl_3$	335	88.3	84.9	vdW + polar
Carbon tetrachloride	CCl_4	350	85.8	85.3	vdW + polar
Trichloroethylene	$ClHC{=}CCl_2$	360	87.4	85.5	vdW + polar
p-Dichlorobenzene		447	89.1	87.3	vdW + polar

Acetone		329	88.3	84.8	vdW + polar
Methyl ethyl ketone	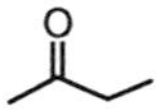	353	88.7	85.4	
Acetonitrile	$CH_3C{\equiv}N$	355	88.3	85.4	
Methyl methacrylate	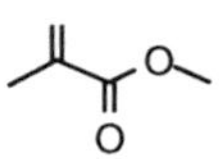	374	97.5	85.8	
Nitromethane	CH_3NO_2	374	90.8	85.8	
Methanol	CH_3OH	338	102	85.0	vdW + polar + H
Ethanol	CH_3CH_2OH	352	110	85.3	
Methyl amine	CH_3NH_2	267	96.7	83.0	
Acetic acid	CH_3COOH	391	60.7	86.2	Dimers in gas phase
Phenol	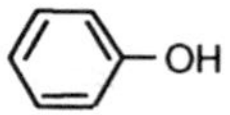	455	89.5	87.5	vdW + polar + H
p-Cresol	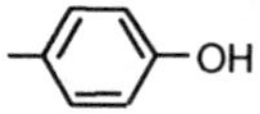	475	90.8	87.8	
Aniline	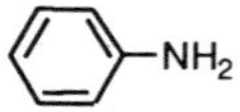	458	97.1	87.5	

[a]See Eq. 4-11.
[b]vdW = van der Waals; polar = polar attractions; H = hydrogen bonding.

based on the Kistiakowsky equation (Table 4.2). Hydrogen bonding is so strong for carboxylic acids (represented in Table 4.2 by acetic acid) that, in addition to organizing in the liquid phase, they form dimers in the gas phase, thereby tremendously lowering S_{gas}; consequently, their ΔS_{vap} is markedly reduced. As a result, these liquids show the greatest deviations from Trouton's Rule of Constant Entropy of Vaporization. Fishtine (1963) has provided a set of empirical correction factors, K_F, which may be applied to the Kistiakowsky estimation of entropy of vaporization as a function of the type of compound (Table 4.3). As expected from intermolecular attractions, polar moieties such as halogens or carbonyl groups lead to very modest adjustments in ΔS_{vap} ($\sim$ 1–8%). Compounds with H-bonding substituents such as amines or alcohols require greater correction factors ($\sim$ 7% and $\sim$ 30%, respectively). Trouton's Rule of constancy for entropies of vaporization when empirically modified for predictable intermolecular interactions, along with the generally applicable integrated Clapeyron expression, establish a highly flexible means of estimating compound vapor pressures as a function of temperature.

4.4 AVAILABILITY OF EXPERIMENTAL VAPOR PRESSURE DATA AND ESTIMATION METHODS

Experimental Data

As illustrated by data given in Fig. 4.1 and in the Appendix, many of the chemicals of environmental concern have very low vapor pressures at ambient temperatures. Unfortunately, vapor pressures are easily measured only if they exceed about 10^{-3} atm ($\approx 10^2$ Pa). In these cases, an isoteniscope (e.g., Thomson, 1959) can be used, and P^0 can be measured directly. Vapor pressure data above 10^{-3} atm are readily available in the literature for many organic compounds (e.g., *CRC Handbook of Chemistry and Physics*). Note, however, that in many cases P^0 values have been determined at elevated temperatures, and ambient values must be extrapolated.

For compounds of very low volatility, that is, compounds with boiling points above about 400°C, experimental data are very scarce and, as emphasized by Mackay et al. (1982), should be treated with caution. For such compounds, vapor pressure measurements are difficult and require that great care be taken to obtain accurate and reproducible results. The methods most widely used are gas saturation (Macknick and Prausnitz, 1979; Sonnefeld et al., 1983) and effusion or vapor balance techniques (Spencer and Cliath, 1983). Both methods are official standard methods accepted by the Organization of Economic Cooperation and Development (OECD, 1981a,b). Results obtained between methods and laboratories vary by as much as a factor of 2 to 3 and, in some cases, by more than an order of magnitude (Spencer and Cliath, 1983). An attractive alternative to these two classical methods is the use of capillary gas chromatography to determine, or at least get a good estimate of, vapor pressures of low volatility compounds (Hamilton, 1980; Bidleman, 1984). Since the method hinges on the selection of an appropriate stationary phase and suitable standard compounds, it is probably best suited for very nonpolar organic substances. Note that for solid compounds, since the molecules are *dissolved* in the stationary phase,

TABLE 4.3 Correction Factors K_F for Calculating $\Delta S_{vap}(T_b)$ for a Variety of Classes of Organic Compounds[a]

Compound Class	Number of Carbons[b] 2	4	6	8	10	Basis for Intermolecular Interactions[c]
Alkanes	1.00	1.00		1.00		
Branched alkanes		0.99		0.99		vdW
Olefins	1.01	1.01		1.01		
Cyclic alkanes		1.00		1.00		
Substituted benzenes			$1 + 2\mu/100$			
Substituted naphthalenes					$1 + \mu/100$	
Alkyl chlorides	1.04	1.03		1.03		
Alkyl bromides	1.03	1.03		1.03		
Incompletely halogenated alkanes	1.05	1.04		1.03		
Completely halogenated alkanes	1.01	1.01		1.01		
Esters	1.14	1.08		1.04		vdW + polar
Ketones		1.07		1.04		
Ethers	1.03	1.02		1.01		
Nitriles	1.05	1.06		1.04		
Nitro compounds	1.07	1.06		1.04		
Sulfides	1.03	1.01		1.01		
Primary amines	1.13	1.11		1.09		
Anilines			1.09			
Alcohols	1.31	1.31		1.28		vdW + polar + H
Cyclo alcohols				1.21		
Phenols			1.15			

[a]After Fishtine, 1963; $\Delta S_{vap}(T_b) = K_F \cdot (36.6 + 8.31 \ln T_b)\ J \cdot mol^{-1} \cdot K^{-1}$.
[b]Includes carbons in functional groups. The variable μ is the molecular dipole moment (in debyes), see *CRC Handbook of Chemistry and Physics* (1985).
[c]vdW = van der Waals forces; polar = polar attractions; H = hydrogen bonding.

the gas chromatographic method yields the vapor pressure of the subcooled liquid [P^0(L)].

In summary, although the literature is replete with vapor pressure values for high to medium volatility compounds (i.e., compounds with $T_b < 400°C$), there is still a paucity of such data for numerous chemicals of environmental importance, especially for compounds with very low vapor pressures. Since knowledge of the vapor pressure is crucial to understanding and describing the environmental behavior of a given organic compound, methods for estimating ambient vapor pressures of organic compounds using more readily available data (e.g., boiling point, structure) must be used.

Vapor Pressure Estimation Methods

Comprehensive and thorough reviews of *vapor pressure estimation methods* are given by Reid et al. (1977), Mackay et al. (1982), and Burkhard et al. (1985). Here we will confine our discussion to a method which can be readily applied since we need to know only the normal boiling, and if the compound is a solid also its melting, point(s). We will first consider liquids and then discuss a procedure to calculate the P^0 of solids from their subcooled liquid vapor pressures.

The starting equation is the Clausius–Clapeyron equation:

$$\frac{d \ln P^0}{dT} = \frac{\Delta H_{vap}(T)}{RT^2} \tag{4-12}$$

Since we would like to estimate the vapor pressure of a compound at ambient temperatures from its normal boiling point (which is usually well above ambient temperatures), we have to account for the temperature dependence of ΔH_{vap} below the boiling point. A first approximation is to assume a linear temperature dependence of ΔH_{vap} over the temperature range considered; that is, to assume a constant heat capacity of vaporization. Thus, if the heat capacity of vaporization $\Delta C_p(T_b)$ (the difference between the vapor and liquid heat capacities) at the normal boiling point is known, ΔH_{vap} can be expressed by

$$\Delta H_{vap}(T) = \Delta H_{vap}(T_b) + \Delta C_p(T_b) \cdot (T - T_b) \tag{4-13}$$

Substitution of Eq. 4-13 into Eq. 4-12 and integration from 1 atm to P^0 and from T_b to T yields

$$\ln P^0 = \frac{\Delta H_{vap}(T_b)}{R}\left(\frac{1}{T_b} - \frac{1}{T}\right) - \frac{\Delta C_p(T_b)}{R}\left(1 - \frac{T_b}{T}\right) - \frac{\Delta C_p(T_b)}{R}\left(\ln \frac{T_b}{T}\right) \tag{4-14}$$

At this point, recall that at the boiling point, $\Delta H_{vap}(T_b) = T_b \Delta S_{vap}(T_b)$; so we may substitute for $\Delta H_{vap}(T_b)$ and combine terms:

$$\ln P^0 = \left(\frac{\Delta S_{vap}(T_b)}{R} - \frac{\Delta C_p(T_b)}{R}\right)\left(1 - \frac{T_b}{T}\right) - \frac{\Delta C_p(T_b)}{R}\left(\ln \frac{T_b}{T}\right) \tag{4-15}$$

Furthermore, for the majority of organic compounds of interest, the ratio of $\Delta C_p(T_b)/\Delta S_{vap}(T_b)$ has a value between -0.6 and -1.0; that is, the heat capacity change upon vaporization of organic liquids is nearly proportional to the entropy change. Smaller nonpolar molecules such as benzene, xylenes, dichlorobenzene, *n*-hexane, and tetrachloroethylene have ratios closer to -0.6. For compounds with very large $\Delta C_p(T_b)$ values, such as long-chain hydrocarbons and many polar compounds, $\Delta C_p(T_b)/\Delta S_{vap}(T_b)$ tends to be closer to -1.0. If $\Delta C_p(T_b)$ is unknown or cannot be estimated (e.g., by one of the methods recommended by Birkett, 1982, or Reid et al., 1977, p. 156), an average value of $-(0.8)\cdot[\Delta S_{vap}(T_b)]$ can be used with reasonable success. Using this information on $\Delta S_{vap}(T_b)$ and $\Delta C_p(T_b)$ in Eq. 4-15, we obtain

$$\ln P^0 = \left(\frac{\Delta S_{vap}(T_b)}{R} + \frac{0.8\Delta S_{vap}(T_b)}{R}\right)\left(1 - \frac{T_b}{T}\right) + \frac{0.8\Delta S_{vap}(T_b)}{R}\left(\ln\frac{T_b}{T}\right)$$

$$= \frac{\Delta S_{vap}(T_b)}{R}\left[(1.8)\left(1 - \frac{T_b}{T}\right) + (0.8)\left(\ln\frac{T_b}{T}\right)\right] \qquad (4\text{-}16)$$

Finally, recalling that the entropy of vaporization is nearly the same for many organic compounds at about $88\ \mathrm{J\cdot mol^{-1}\cdot K^{-1}}$, we have

$$\ln P^0 \simeq 19\left(1 - \frac{T_b}{T}\right) + 8.5\left(\ln\frac{T_b}{T}\right) \quad \text{(atm)} \qquad (4\text{-}17)$$

An even closer estimation (Table 4.4) can be obtained when Kistiakowky's expression, which corrects for strong or weak van der Waals interactions, and Fishtine's correction factors (Table 4.3), which adjust for polar and hydrogen bonding effects, are applied to Eq. 4-16 $[\Delta S_{vap}(T_b) = K_F\ (36.6 + 8.31 \ln T_b)\ \mathrm{J\cdot mol^{-1}\cdot K^{-1}}]$

$$\ln P^0 \approx -K_F(4.4 + \ln T_b)\left[1.8\left(\frac{T_b}{T} - 1\right) - 0.8\ln\left(\frac{T_b}{T}\right)\right] \quad \text{(atm)} \qquad (4\text{-}18)$$

As shown by Mackay et al. (1982), by setting the ratio $\Delta C_p(T_b)/\Delta S_{vap}(T_b)$ equal to -0.8, the potential for misestimation of relatively low boiling compounds (i.e., 100°C) is only about 5%, but the error may exceed a factor of 2 for very high boiling compounds (see also Table 4.5).

Entropy of Melting and the Vapor Pressure of Solids

Below the melting point of a compound, Eqs. 4-17 and 4-18 yield vapor pressures of the subcooled liquid, $P^0(\mathrm{L})$. A useful approximation for relating the corresponding solid vapor pressure $P^0(\mathrm{s})$ to $P^0(\mathrm{L})$ (and vice versa) at a given temperature is given by Prausnitz (1969, p. 390):

$$\ln\frac{P^0(\mathrm{s})}{P^0(\mathrm{L})} = -\frac{\Delta S_{melt}(T_m)}{R}\left(\frac{T_m}{T} - 1\right) \qquad (4\text{-}19)$$

TABLE 4.4 Comparison Between Estimated and Experimental Vapor Pressures at 25°C

			Using Eq. 4-15 and $\Delta S_{vap} = 88$ ($J \cdot mol^{-1} \cdot K^{-1}$)	Using Kistiakowsky and Fishtine approaches					
Compound[a]	T_b (°C)	T_m (°C)	P^0(L)(atm)	ΔS_{vap} ($J \cdot mol^{-1} \cdot K^{-1}$)	K_F	P^0(L) (atm)	Calculated P^0(s)(atm)	Experimental P^0(atm)	Ref.[d]
n-Decane (l)	174.1		2.4×10^{-3}	87.3	1.00	2.4×10^{-3}		2.5×10^{-3}	a
Benzene (l)	80.1		1.3×10^{-1}	85.4	1.00	1.3×10^{-1}		1.2×10^{-1}	a
Tetrachloroethylene (l)	120.8		2.4×10^{-2}	86.3	1.01	2.4×10^{-2}		2.5×10^{-2}	a
Quinoline (l)	237.7		1.3×10^{-4}	88.4	1.02	9.6×10^{-5}		1.4×10^{-4}	a
1-Hexanol (l)	157.0		5.0×10^{-3}	87.0	1.28	1.2×10^{-3}		1.3×10^{-3}	a
Aniline (l)[b]	184.4		2.9×10^{-3}	87.5	1.09	1.7×10^{-3}		1.3×10^{-3}	a
Lindane (s)	323.4	112.5	2.0×10^{-6}	89.7	1.01	1.3×10^{-6}	1.7×10^{-7}	1.0×10^{-6}	b
Phenol (s)[c]	181.9	40.6	1.2×10^{-4}	87.5	1.15	4.2×10^{-4}	2.6×10^{-4}	2.6×10^{-4}	c
Naphthalene (s)	217.9	80.2	3.2×10^{-4}	88.1	1.00	3.0×10^{-4}	8.6×10^{-5}	1.0×10^{-4}	d
Phenanthrene (s)	340.2	99.5	8.7×10^{-7}	89.9	1.00	5.9×10^{-7}	1.1×10^{-7}	1.6×10^{-7}	d
Anthracene (s)	342.0	217.5	8.0×10^{-7}	90.0	1.00	5.4×10^{-7}	6.7×10^{-9}	8.0×10^{-9}	d
Benzanthracene (s)	435.0	162.0	7.0×10^{-9}	91.1	1.00	3.2×10^{-9}	1.4×10^{-10}	2.8×10^{-10}	d

[a] Compounds are in liquid (l) or solid (s) form.
[b] 35°C.
[c] 20°C.
[d] References: (a) *CRC Handbook of Chemistry and Physics.* (b) Spencer and Cliath, 1983. (c) Mackay et al., 1982. (d) Sonnefeld et al., 1983.

TABLE 4.5 Predicted[a] Vapor Pressures (atm) at 25°C for Compounds Exhibiting Different Boiling Points and Ratios of $C = \Delta C_p(T_b)/\Delta S_{vap}(T_b)$ (Dimensionless)

T_b(°C)	$C = 0.6$	$C = 0.8$	$C = 1.0$	Half-Range/ Mean (%)
100	6.3×10^{-2}	5.9×10^{-2}	5.6×10^{-2}	6
200	9.2×10^{-4}	7.0×10^{-4}	5.4×10^{-4}	26
300	8.7×10^{-6}	4.9×10^{-6}	2.7×10^{-6}	53
400	6.0×10^{-8}	2.3×10^{-8}	0.9×10^{-8}	74

[a]From Eq. 4-16; note that for solids $P^0 = P^0(\text{L})$.

where $\Delta S_{melt}(T_m)$ is the entropy of melting at the melting point. Unfortunately, there is no analogous guideline to Trouton's rule for estimating entropies of melting. For nonrigid molecules like aliphatic hydrocarbons, in particular, the entropy of melting may vary by more than a factor of 3 among compounds that have the same number of carbon atoms (e.g., Reid et al., 1977, p. 216). For rigid molecules and their long chain derivatives, however, a reasonable estimate for $\Delta S_{melt}(T_m)$ can be obtained (Yalkowsky, 1979):

$$\Delta S_{melt}(T_m) \simeq [56.5 + 10.5(n - 5)] \quad (\text{J} \cdot \text{mol}^{-1}\,\text{K}^{-1}) \tag{4-20}$$

where n is the total number of flexing chain atoms (exclusive of protons and setting $n = 5$ for $n < 5$) and $R = 8.31\ \text{J} \cdot \text{mol}^{-1} \cdot \text{K}^{-1}$. Substitution of Eq. 4-20 into Eq. 4-19 yields

$$\ln \frac{P^0(\text{s})}{P^0(\text{L})} \simeq -[6.8 + 1.26(n - 5)]\left[\frac{T_m}{T} - 1\right] \tag{4-21}$$

Since many environmentally relevant compounds fall into the category of molecules for which Eq. 4-20 holds, Eq. 4-20 has wide application.

The examples given in Table 4.4 show that the approach presented for estimation of ambient vapor pressures of organic compounds from their boiling points (and melting points for solids) in general yields very acceptable results, even for compounds with very low ambient P^0 values. A slightly different approach to calculating ambient vapor pressures from normal boiling points has been recommended by Grain (1982a), who uses a modification of the Watson correlation (Reid et al., 1977, p. 210) to describe the temperature dependence of ΔH_{vap}; however, this method is more empirical, more cumbersome to apply, and does not yield significantly better results than the method presented here.

If the boiling point of a compound is unknown or cannot be experimentally determined (e.g., because the compound decomposes), it can be estimated with fair to sufficient accuracy from the structure of the compound (Rechsteiner, 1982). Thus, in most cases, it is possible to get an order of magnitude estimate of the ambient vapor pressure of a given compound, even if only the structure of the compound is known.

CHAPTER 5

SOLUBILITY AND ACTIVITY COEFFICIENT IN WATER

5.1 INTRODUCTION

Whether an organic compound "likes" or "dislikes" being surrounded by water molecules is one of the key factors determining its environmental behavior and impact. It is necessary, therefore, that we try to understand the molecular forces and interactions involved when an organic compound dissolves in water. Water is a unique and very complex solvent and, therefore, many of its properties on a molecular level are not yet fully understood (Tanford, 1984); nevertheless, our present knowledge does allow us to picture the dissolution process sufficiently well for our purposes.

A direct measurement of how much an organic compound likes to be present as a solute in water is given by its aqueous solubility [or, more precisely, as we will see by the aqueous solubility of the (subcooled, superheated) liquid compound]. *Aqueous solubility is commonly defined as the abundance of the chemical per unit volume in the aqueous phase when the solution is in equilibrium with the pure compound in its actual aggregation state (gas, liquid, solid) at a specified temperature and pressure (e.g., 25°C, 1 atm)*. We then speak of a saturated solution of the chemical, the concentration of which we will denote as C_w^{sat}.

As may be seen in Fig. 5.1, environmentally relevant organic compounds have aqueous solubilities ranging over several orders of magnitude—from highly soluble compounds [infinitely soluble or completely miscible, e.g., methanol (not shown)] to levels of saturation that are so low that the concentration can scarcely be measured even with the most sophisticated analytical techniques. In the following discussion, we will focus our attention on those compounds that are only moderately or even

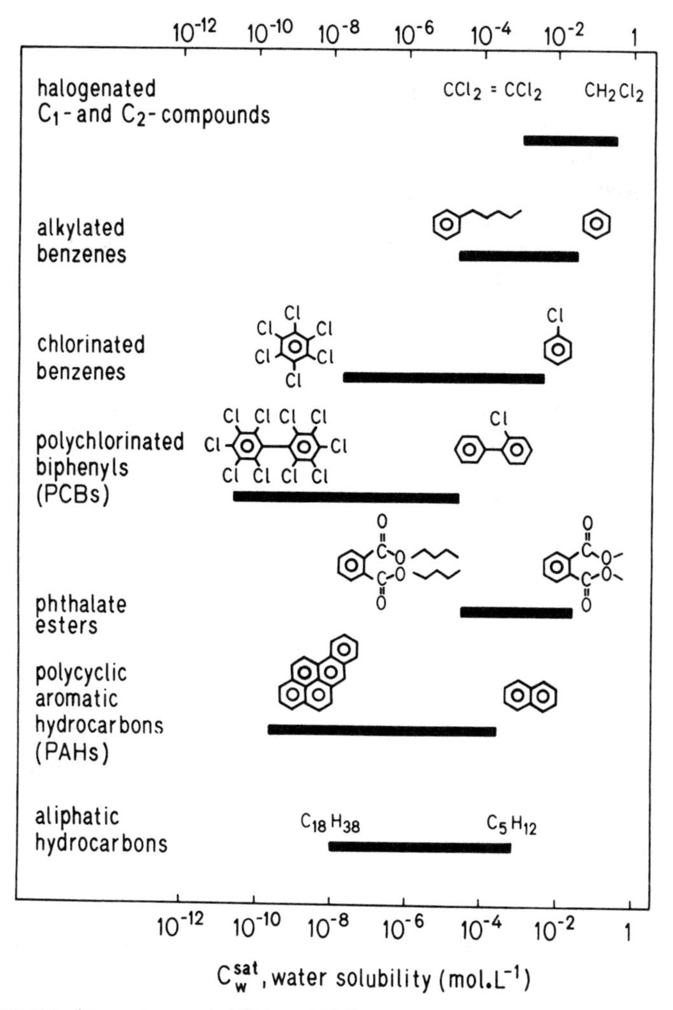

Figure 5.1 Ranges in water solubilities (C_w^{sat}) of some important classes of organic compounds.

sparingly water soluble (i.e., $C_w^{sat} \ll 1\ \text{mol} \cdot \text{L}^{-1}$), since the majority of the chemicals that interest us fall into this category.

5.2 THERMODYNAMIC CONSIDERATIONS

For a quantitative treatment of the dissolution process, we first establish the thermodynamic relationships used to describe the dissolution of an organic compound in water, and then we attempt to understand this process on a molecular level (see Section 5.3). Let us imagine an experiment in which we bring a pure (water immiscible) organic *liquid* (e.g., benzene) in contact with water and ask what will happen. Intuitively, we would say that some organic molecules will leave the organic phase and dissolve in the water, and some water molecules will enter the organic liquid. After some time, so many organic molecules will have entered the aqueous phase that some will begin to return to the organic phase. When the rates of transfer between the two phases are balanced, the system reaches a dynamic equilibrium, and the abundance of organic molecules in the water is the water solubility of that liquid organic compound; similarly, the abundance of water molecules in the organic phase reflects the solubility of water in that organic liquid.

Solubility of Liquids

For a thermodynamic treatment of the dissolution of an organic liquid in water, first recall that the partial molar free energy or the chemical potential of a compound *i* (for simplicity, we will omit the subscript *i*) in a given phase ϕ at a specified temperature and pressure can be expressed by (see Chapter 3)

$$\mu_\phi = \mu^0 + RT \ln \frac{f_\phi}{f^0} = \mu^0 + RT \ln \gamma_\phi x_\phi \tag{5-1}$$

where $\gamma_\phi x_\phi$ is a measure of how "active" the compound is relative to its reference state which, for convenience, we have chosen to be the pure liquid compound at the system temperature and pressure. In our experimental system containing the pure *liquid* organic compound in contact with water, at any instant in time, the chemical potentials of the compound in each of the two phases are

$$\mu_\mathrm{o} = \mu^0 + RT \ln \gamma_\mathrm{o} x_\mathrm{o} \tag{5-2}$$

for the compound in the *organic liquid phase* denoted by subscript o, and

$$\mu_\mathrm{w} = \mu^0 + RT \ln \gamma_\mathrm{w} x_\mathrm{w} \tag{5-3}$$

for the compound in the *water* denoted by subscript w. At a given temperature, pressure, and composition, the difference in the partial (i.e., focusing on the part of the system of concern to us) molar free energy (i.e., the difference between chemical potentials) of the compound between the two phases is given by

$$\Delta G_\mathrm{s} = \mu_\mathrm{w} - \mu_\mathrm{o} = RT \ln \gamma_\mathrm{w} x_\mathrm{w} - RT \ln \gamma_\mathrm{o} x_\mathrm{o} \tag{5-4}$$

We refer to ΔG_s as the molar free energy of *solution*. This free energy difference quantifies the "driving force" for the phase transfer analogously to the way hydrostatic head difference between connected reservoirs dictates the transfer of water (Fig. 3.1). In the beginning of our experiment, μ_o is much larger than μ_w (x_w is near zero); that is, the negative ΔG_s drives a net flux of organic molecules from the organic phase (higher chemical potential) to the aqueous phase (lower chemical potential). This process continues and x_w increases until the chemical potentials become equal in both phases; that is, until equilibrium is reached ($\Delta G_s = 0$; $\gamma_w x_w = \gamma_o x_o$; or $f_w = f_o$).

For the majority of the compounds of interest, we can now make two simplifying assumptions: (1) in the organic phase, the mole fraction of water is small compared with the mole fraction of the compound itself; that is, x_o remains nearly 1 (see examples given in Table 5.1); and (2) the compound shows ideal behavior in its water-saturated liquid phase, that is, $\gamma_o(\text{liquid}) = 1$. The molar free energy of dissolving a liquid organic compound in water is then simply given by

$$\Delta G_s = RT \ln x_w + RT \ln \gamma_w \tag{5-5}$$

where $RT \ln x_w$ is the contribution of the entropy of ideal mixing, and $RT \ln \gamma_w$ is the partial molar *excess* free energy ΔG_s^e of dissolution of the liquid compound in water. Note that since we have assumed that x_o and γ_o are equal to 1, this excess contribution is a direct measure of the nonideality of the solution of the compound in water resulting

TABLE 5.1 Mole Fraction of Some Common Nonpolar Organic Liquids Saturated with Water

Organic Liquid	x_o	Reference
Pentane	0.99952	
Hexane	0.99946	
Heptane	0.99916	Gerrard, 1980
Octane	0.99911	
Benzene	0.9977	
Chlorobenzene	0.9975	
1,2,-Dichlorobenzene	0.9973	
1,2,4-Trichlorobenzene	0.9980	
Trichloroethylene	0.9977	Horvath, 1982
Tetrachloroethylene	0.99913	
Methylene chloride	0.9914	
Chloroform	0.9946	
1,1,1-Trichloroethane	0.9974	
Diethyl ether	0.942	
Butyl acetate	0.89	
Methyl acetate	0.74	Riddick and Bunger, 1970
2-Butanone	0.69	
3-Pentanone	0.89	
Pentanol	0.64	Stephenson et al., 1984
Octanol	0.79	

TABLE 5.2 Aqueous Solution Activity Coefficients γ_w^∞ of Some Sparingly Soluble Organic Compounds in Infinitely Dilute Solutions[a]

Compound	γ_w^∞
Benzene	2.4×10^3
Toluene	1.2×10^4
Naphthalene	1.4×10^5
Phenanthrene	7.4×10^6
Benzo(a)pyrene	2.8×10^9
Methylene chloride	4.2×10^2
Chloroform	8.6×10^2
Carbon tetrachloride	1.0×10^4
1,1,1-Trichloroethane	2.4×10^3
Chlorobenzene	1.9×10^4
1,3-Dichlorobenzene	1.7×10^5
1,2,3,5-Tetrachlorobenzene	1.4×10^7
Pentachlorobenzene	1.2×10^8
Hexachlorobenzene	9.8×10^8
2,4′-Dichlorobiphenyl	5.8×10^7
2,2′,5,5′-Tetrachlorobiphenyl	4.2×10^9
2,2′,4,4′,5,5′-Hexachlorobiphenyl	2.9×10^{11}
Methyl ethyl ketone	3.2×10^1
Diethyl ether	1.6×10^2
Ethyl acetate	1.5×10^2
Octanol	3.7×10^3

[a] After Banerjee, 1985.

from solute–solvent dissimilarities ($\Delta G_s^e = g_w^e$, the partial molar excess free energy of the solute in aqueous solution). For most organic compounds considered here, γ_w is much larger than 1 (see examples given in Table 5.2).

Solubility of Solids and Gases

At this point let us also include those compounds that are solids or gases at ambient conditions. *It is important to realize that organic compounds dissolved in a solvent such as water exist in a liquid state.* Physically, we think of the molecules as free to orient and move as they do when liquids, not as constrained as solids yet not as free as gases; therefore, in order to dissolve solids and gases in water, we must first conceptually convert them into their subcooled or superheated liquid states, respectively. As we discussed in Chapter 4 for solids, the "energy costs" of converting a solid to its subcooled liquid state is given by the ratio of the fugacity of the pure solid to the fugacity of the subcooled liquid (the reference state); that is, $\gamma_o x_o$ in Eq. 5-2 is not equal to 1 (as for a liquid) but to $f(s)/f^0$ which, when assuming ideal gas behavior, is given by $P^0(s)/P^0(L)$. For solids, therefore, we have to include an energy term in Eq. 5-5 that takes into account the energy "cost" of melting the solid solute:

$$\Delta G_s = \underbrace{RT \ln x_w}_{\substack{\Delta G_{mix} \\ \text{(mixing ideal)}}} + \underbrace{RT \ln \gamma_w}_{\substack{\Delta G_s^e \\ \text{(nonideal effects)}}} - \underbrace{RT \ln \frac{P^0(s)}{P^0(L)}}_{\substack{\Delta G_{melt} \\ \text{(melting)}}} \tag{5-6}$$

Similarly for gases, we need to account for the energy required to condense the gas:

$$\Delta G_s = RT \ln x_w + RT \ln \gamma_w - \underbrace{RT \ln \frac{P_i(g)}{P^0(L)}}_{\substack{\Delta G_{cond} \\ \text{(condensation)}}} \tag{5-7}$$

The most common value of the fugacity of pure gas, $P_i(g)$, is arbitrarily chosen to be 1 atm. Solubilities of gases are, therefore, usually reported for 1 atm pressure $[P_i(g) = 1\ \text{atm}]$. Note, however, that there are some exceptions. For example, oxygen solubilities are generally taken with $P_{O_2}(g) = 0.21$ atm, since this is the value appropriate for the Earth's atmosphere at sea level.

Returning to our imagined dissolution experiment, at equilibrium ($\Delta G_s = 0$), we can now express the mole fraction solubility x_w^{sat} of liquids, solids, and gases by

$$\begin{aligned} x_w^{sat} &= \frac{1}{\gamma_w^{sat}} && \text{(for liquids)} \\ x_w^{sat} &= \frac{1}{\gamma_w^{sat}} \cdot \frac{P^0(s)}{P^0(L)} && \text{(for solids)} \\ x_w^{sat} &= \frac{1}{\gamma_w^{sat}} \cdot \frac{1\ \text{atm}}{P^0(L)} && \text{(for gases)} \end{aligned} \tag{5-8}$$

where γ_w^{sat} is the activity coefficient of the organic liquid compound in water *at saturation*. In terms of molar concentrations, the solubility C_w^{sat} is readily obtained by multiplying x_w^{sat} by $1/V_w$ (see Chapter 3):

$$C_w^{sat} \simeq \frac{x_w^{sat}}{V_w} \simeq \frac{x_w^{sat}}{0.018} \quad (\text{mol} \cdot \text{L}^{-1}) \tag{5-9}$$

Equations 5-8 show that the solubility of an organic compound is inversely related to its activity coefficient in aqueous solution. Thus, knowledge of the aqueous solubility of a pure organic *liquid* directly yields information on the activity coefficient of that compound in water, which is one of the key parameters determining the compound's behavior in the environment. Note that solubility data yield information on the activity coefficient at the saturation concentration (γ_w^{sat}). This value is not necessarily identical to the compound's activity coefficient in very dilute solution (γ_w^{∞}) since in concentrated solutions, we cannot *a priori* exclude organic molecule:molecule interactions. We address the problem of concentration dependence of activity coefficients in Section 5.4 and in Chapter 6.

5.3 MOLECULAR INTERPRETATION OF THE DISSOLUTION PROCESS

Let us now consider the molecular factors that govern the excess free energy of solution of a liquid organic compound and, thus, its activity coefficient in water. When we imagine organic solutes mixed in water, we recognize that organic molecules and water molecules differ from one another in two primary ways: (1) they have very dissimilar shapes and sizes, and (2) in general, organic molecules are much less polar than water since they are chiefly constructed from atoms having comparable electronegativities resulting in evenly spaced electronic distribution (see Chapter 2). These dissimilarities between solute and solvent result in various enthalpic and entropic contributions to the excess free energy (ΔG_s^e) of solution. So far, we have expressed these contributions in one lumped parameter, the activity coefficient of the compound in water, γ_w (Eq. 5-5). However, to interpret relationships between molecular structure and solubility (activity coefficient) and evaluate the effects of temperature and chemical composition of the solution on solubility, we must try to understand the molecular factors that contribute to the excess free energy of solution of organic compounds in water.

Enthalpy of Dissolution

A very simple schematic diagram of the transfer of an organic solute from its pure liquid phase into water may help us visualize these molecular factors (Fig. 5.2). Let us first consider the *enthalpic contributions* to the excess free energy of solution, which reflect the changes in interactions between solute:solute, solute:solvent, and solvent: solvent molecules when the liquid organic compound is dissolved in water. (Note that ideal solutions exhibit identical attractions for all combinations of these molecular interactions, and so the excess enthalpy of solution for ideal solutions is zero.) As a first step in our schematic picture, we isolate the organic solute from its pure liquid phase (without leaving a cavity) and create a cavity in the solvent (water) that can accommodate the organic solute. In both cases, we have to "pay" enthalpy ($\Delta H_1 > 0$, $\Delta H_2 > 0$) in order to overcome the molecule:molecule attractions that we have disrupted in the organic phase (o:o) and in the water (w:w), respectively. For the organic molecules, this enthalpy change is equal to the heat of vaporization of the liquid (ΔH_1). The water molecules removed from the interior of the cavity regain their

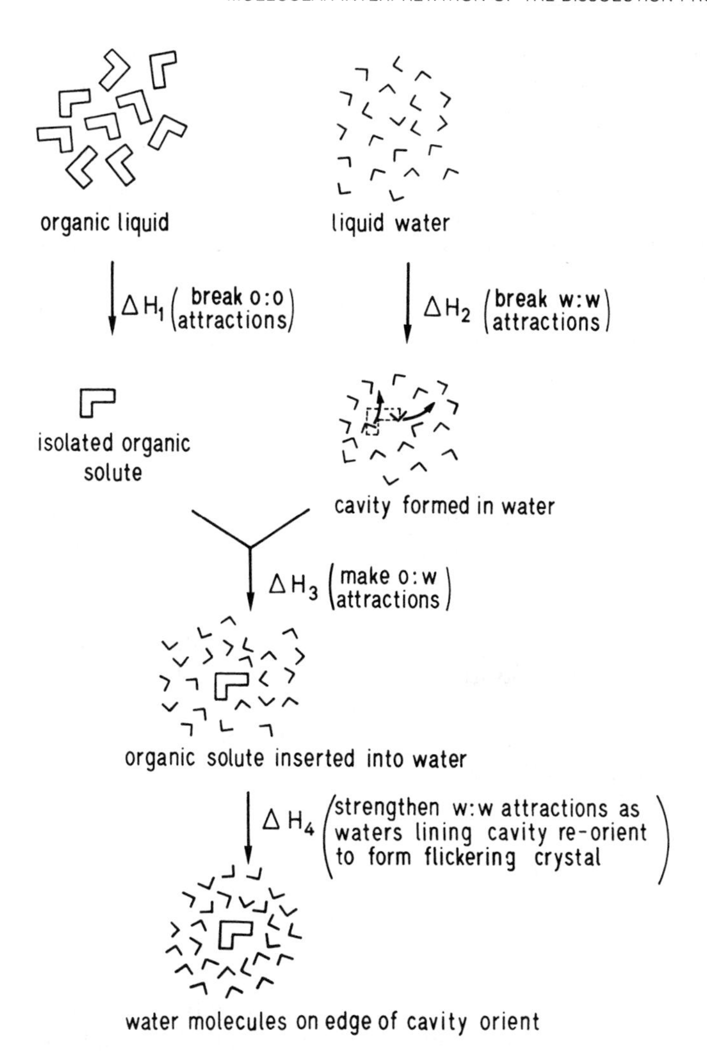

Figure 5.2 Schematic representation of the various enthalpies involved when dissolving a neutral organic molecule in water.

water:water interactions elsewhere in the solution; however, the water molecules lining the cavity now interact with fewer other water molecules. A fraction of their heat of vaporization must then be "invested" (ΔH_2). Depending on the size of the cavity (which is, of course, related to the size of the organic solute), a fairly large number of water molecules will surround one organic molecule. Consequently, this second enthalpy term is very important, especially when dealing with large organic molecules. In the next step we transfer the organic solute into the cavity. In doing so we get back some of the enthalpy spent, since the organic molecule will experience intermolecular attractions to the surrounding water molecules all over its surface, resulting in a negative ΔH_3. (In the case of ideal solutions, this ΔH_3 energy gain precisely balances the two energies "spent," ΔH_1 and ΔH_2.) These forces include, at the very least, van der Waals and induced dipole interactions. In addition, if the organic molecule contains polar functional groups, dipole–dipole and, if appropriate, hydrogen-bonding effects will be felt in the immediate vicinity of these moieties. By examining the solubilities of a homologous series of alcohols, Butler and Ramchandani (1935) suggested that the unit area surface attraction between nonpolar organic molecules and water should be roughly half that which originally existed between the water molecules. The net excess enthalpy change resulting from breaking and forming intermolecular "bonds" upon introduction of our organic solute into the aqueous solution cavity has been referred to as ΔH_{cav} (e.g., Shinoda, 1977):

$$\Delta H_{cav} = \Delta H_1 + \Delta H_2 + \Delta H_3 \tag{5-10}$$

where ΔH_1, $\Delta H_2 > 0$ and $\Delta H_3 < 0$. In addition to these terms, there is an additional enthalpy contribution (ΔH_4) due to "iceberg formation" (Frank and Evans, 1945). This contribution may be visualized as follows: the water molecules immediately surrounding the organic solute have strong polar interactions with nearest-neighbor water molecules only on the side away from the organic molecule. This situation appears to result in a "solidifying of positions and orientations" of these cavity-lining water molecules. The "freezing" effect gives rise to an enthalpy gain ΔH_4 or ΔH_{ice}. Note that this enthalpy contribution, as well as all other enthalpic terms discussed (i.e., ΔH_1, ΔH_2, ΔH_3), are related to the size, or more precisely, to the surface area of the solute molecules.

$$\begin{aligned}\Delta H_s^e &= \Delta H_{cav} + \Delta H_{ice} \\ &\propto \text{molecular size of the solute}\end{aligned} \tag{5-11}$$

This relationship of ΔH_s^e with molecular surface area is illustrated for polycyclic aromatic hydrocarbons and aliphatic alcohols in Figure 5.3. Note that the experimentally determined heats of solution for compounds like naphthalene, fluorene, phenanthrene, anthracene, and pyrene must be corrected for the corresponding heats of melting since these aromatic hydrocarbons are solids at 25°C. For the alcohols, overall negative heats of solution are found resulting from the attraction of the hydroxyl moiety to the water until the aliphatic chain reaches about eight carbons in length. These data strongly support the expectation from theory discussed above that the molecule:molecule attractions which are changed when organic chemicals are dissolved in water become increasingly more unfavorable (ΔH_s^e increases) as the nonpolar surface area of the chemical of interest increases.

Entropy of Dissolution

Compared with ΔH_s^e, it is more difficult to interpret the excess entropic contribution, ΔS_s^e, to the excess free energy of solution. We may group the total excess entropy change

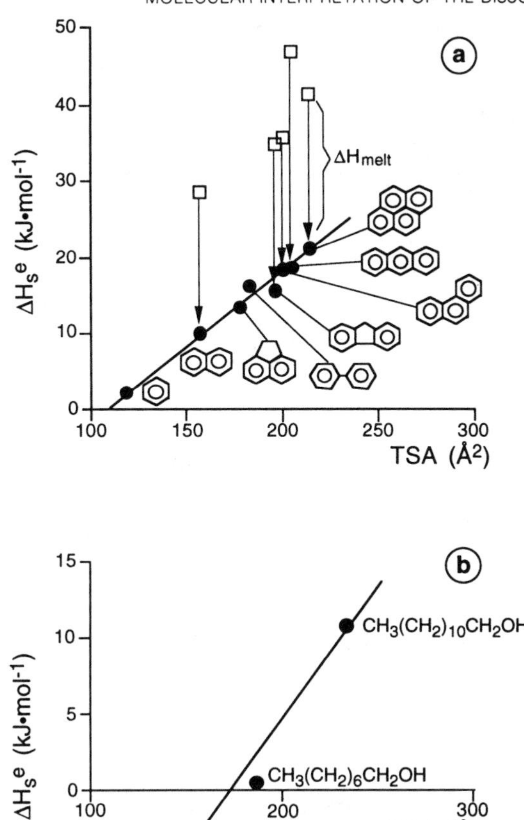

Figure 5.3 Enthalpies of solution of *liquid* organic compounds in water at 25°C versus their total surface areas (TSA): (*a*) a series of nonpolar compounds, (*b*) a series of compounds exhibiting a polar functional group.

into three contributions: (1) an entropy of cavity formation (ΔS_{cav}); (2) an entropy of ice formation around the solute (ΔS_{ice}); and (3) an enhanced randomness (ΔS_{mix}) due to mixing dissimilar molecules. Given our present knowledge of aqueous solutions, it is not clear how ΔS_{cav} and ΔS_{ice} vary from solute to solute.

Since ΔS_{mix} embodies the increased *system randomness* that results from mixing materials, this entropy change will always promote dissolution. When solute and solvent molecules have the same shape and size, this entropy term is simply given by

$$\Delta S_{mix}^{ideal} = -R(n_{solute} \ln x_{solute} + n_{solvent} \ln x_{solvent}) \tag{5-12}$$

and is called the *ideal entropy of solution.* As noted, the free energy derived from this increased system entropy is always gained and, therefore, is not a contributor to the *excess* free energy of dissolution. Organic molecules do not, however, have the same shapes and sizes as water, and the difference or excess entropy of mixing relative to the ideal case (Eq. 5-12) must be considered. Recognizing that the mixing of each large organic solute into water results in the displacement of many water molecules, a better description of the statistical contributions of the solute and solvent molecules is given by their volume fraction, $\tilde{x}$, contribution to solution. The entropy of mixing of a solute and solvent of different molecular sizes (i.e., an organic compound and water) is thus given by

$$\Delta S_{mix}^{real} = -R\,(n_{org} \ln \tilde{x}_{org} + n_{H_2O} \ln \tilde{x}_{H_2O}) \tag{5-13}$$

where $\tilde{x}_{org}$ and $\tilde{x}_{H_2O}$ denote the volume fractions of solute and solvent, respectively. Since we are dealing with dilute solutions, we note that $x_{solvent}$ in Eq. 5-12 and $\tilde{x}_{H_2O}$ in Eq. 5-13 are approximately equal to 1. By assuming that the compounds mix with no change in total volume, we may then write

$$\Delta S_{mix}^{real} = -R\,n_{org} \ln \tilde{x}_{org} = -R\,n_{org} \ln \frac{n_{org} V_{org}}{n_{org} V_{org} + n_{H_2O} V_{H_2O}} \tag{5-14}$$

Furthermore, because $n_{H_2O} \gg n_{org}$, we can approximate ΔS_{mix}^{real} by

$$\Delta S_{mix}^{real} = -R\,n_{org} \ln \frac{n_{org}}{n_{H_2O}} \cdot \frac{V_{org}}{V_{H_2O}}$$

or

$$\begin{aligned} \Delta S_{mix}^{real} &= -R\,n_{org} \ln x_{org} - R\,n_{org} \ln \frac{V_{org}}{V_{H_2O}} \\ &= \quad \Delta S_{mix}^{ideal} \quad + \quad \Delta S_{mix}^{e} \end{aligned} \tag{5-15}$$

Figure 5.4 shows the importance of ΔS_{mix}^{e} on a per mole of organic solute basis for some representative organic compounds. Since the molar volumes of the majority of the organic compounds considered here are on the order of 100–300 $cm^3 \cdot mol^{-1}$, ΔS_{mix}^{e} will cause the ΔS_{mix}^{real} to be less than that calculated for ΔS_{mix}^{ideal} alone by about 4–8 $kJ \cdot mol^{-1}$. Thus, although mixing dissimilar molecules always promotes dissolution, the solubility-enhancing entropic effect is tempered somewhat owing to the larger sizes of many organic solutes than water.

If we examine some of the limited ΔS_s^e data available (Table 5.3), we can get some clues as to the magnitude of this term and how it varies from compound to compound. First, we see that the rigid set of aromatic solutes shown all have nearly the same ΔS_s^e, and this quantity is remarkably like these compounds' entropies of fusion or melting [typically near 56 $J \cdot mol^{-1} \cdot K^{-1}$; see Eq. 4-19]. For the *n*-alkanols we see that compounds with about three to six carbons also have entropies of solution between

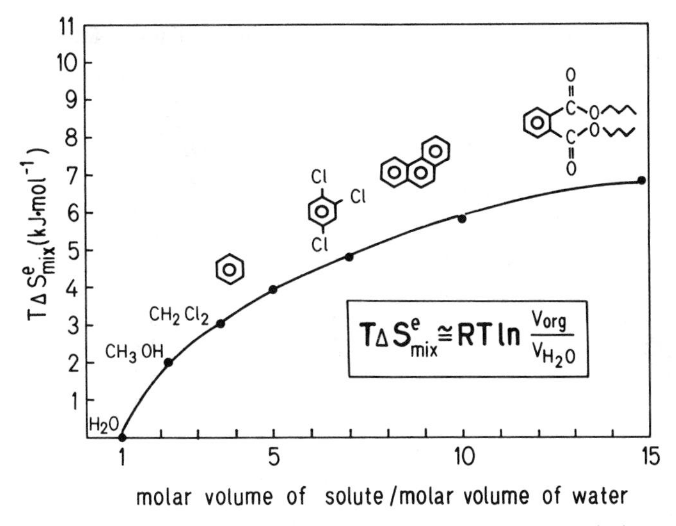

Figure 5.4 Contribution of molecular size to the entropy of dissolution of an organic compound in water.

TABLE 5.3 Entropies of Solution for Liquids (Real or Subcooled) at 25°C[a]

Compound	ΔS^e_s (J·mol^{-1}·K^{-1})
Aromatic Compounds	
Benzene	−58
Naphthalene (L)	−59
Acenaphthene (L)	−52
Fluorene (L)	−62
Phenanthrene (L)	−61
Anthracene (L)	−60
Pyrene (L)	−48
	Mean = −57 ± 5
n-Alcohols	
Methanol	−29
Ethanol	−44
Propanol	−51
Butanol	−65
Pentanol	−71
Hexanol	−68
Octanol	−76
Dodecanol	−92

[a]Entropies estimated from data shown in Table 5-4.

TABLE 5.4 Comparisons of Enthalpic and Entropic Contributions to Excess Free Energies of Solutions at 25°C

Compound	ΔG_s^e $(= -RT \ln x(\mathrm{l, L}))$[a] (kJ·mol^{-1})	ΔH_s^e $(= \Delta H_s - \Delta H_m)$[b] (kJ·mol^{-1})	$-T\Delta S_s^e$ $(= \Delta G_s^e - \Delta H_s^e)$ (kJ·mol^{-1})
Benzene	19.3	2.1	17.2
Naphthalene (L)	27.6	9.9	17.7
Acenaphthene (L)	28.8	13.4	15.4
Fluorene (L)	34.0	15.5	18.5
Phenanthrene (L)	36.3	18.1	18.2
Anthracene (L)	36.4	18.4	18.0
Pyrene (L)	40.6	26.4	14.2
Methanol	1.3	−7.4	8.7
Ethanol	2.4	−10.7	13.1
Propanol	6.2	−8.9	15.1
Butanol	9.6	−9.7	19.3
Pentanol	13.7	−7.4	21.1
Hexanol	16.8	−3.5	20.3
Octanol	23.1	0.5	22.6
Dodecanol	38.0	10.7	27.3

[a] Liquid solubilities are given in the Appendix.
[b] Free energies and enthalpies of solution of aromatic hydrocarbons from May et al. (1983). Enthalpies of fusion of aromatic hydrocarbons from Wauchope and Getzen (1972). Free energies of solution of alcohols from data of Butler et al. (1935) and enthalpies of solution of alcohols from data of Stephenson et al. (1984). Enthalpies of fusion of benzene and alcohols are not subtracted since these are liquids. The products, $-T\Delta S_s^e$, are estimated by difference.

50 and 70 J·mol^{-1}·K^{-1}, but smaller alcohols have lower ΔS_s^e and larger ones are substantially above this range. We can speculate that small alcohols like methanol and ethanol are special because such a large portion of their structure consists of the hydroxyl group, which interfaces very well with the water. On the other hand, very long flexible compounds like octanol and dodecanol appear to be losing more degrees of freedom than either rigid compounds like the aromatic substances or shorter chemicals like butanol or pentanol. These trends are very similar to what is seen for entropies of melting (Yalkowsky, 1979; Eq. 4-19), and suggest that aqueous dissolution of nonpolar compounds or molecules with nonpolar structural parts causes these solutes to lose translational, rotational, and flexing freedoms as if the nonpolar material was frozen.

With this picture of the dissolution process in mind, we can now assemble the terms that contribute to the *excess* free energy of solution in water of the organic compound from its pure liquid phase:

$$\Delta G_s^e = \Delta H_s^e - T\Delta S_s^e \qquad (5\text{-}16)$$

$$= (\Delta H_{cav} + \Delta H_{ice}) - T(\Delta S_{cav} + \Delta S_{ice} + \Delta S_{mix}^e)$$

$$= RT \ln \gamma_w \qquad (5\text{-}17)$$

Now we can clearly see all of the factors lumped into the activity coefficient, γ_w. Both enthalpic interactions and entropic effects govern the magnitude of a compound's aqueous solubility. Table 5.4 summarizes the relative sizes of these contributions for the aromatic hydrocarbons and *n*-alcohols discussed previously. For the alcohols shown, the entropy term dominates. This entropic domination is also seen for the smallest aromatic substances (and other low-molecular-weight nonpolar chemicals not shown). However, the enthalpic costs of dissolution are of comparable importance for compounds like phenanthrene, and these become the greatest contributor to ΔG_s^e for larger nonpolar solutes like pyrene. It is evident from the data shown in Table 5.4 that enthalpic governance would also be the case if we had examined *n*-alcohols much larger than dodecanol. Thus, although for many organic chemicals with small nonpolar structures dissolution in water is currently thought of as limited by entropy costs, this generalization does not apply for substances with large hydrophobic structures.

It is evident that many of the terms in Eq. 5-17 are related to the size of the organic solute (more accurately, the size of the nonpolar portion of the structure). This explains why, for a given class of compounds, quite good correlations are obtained between the

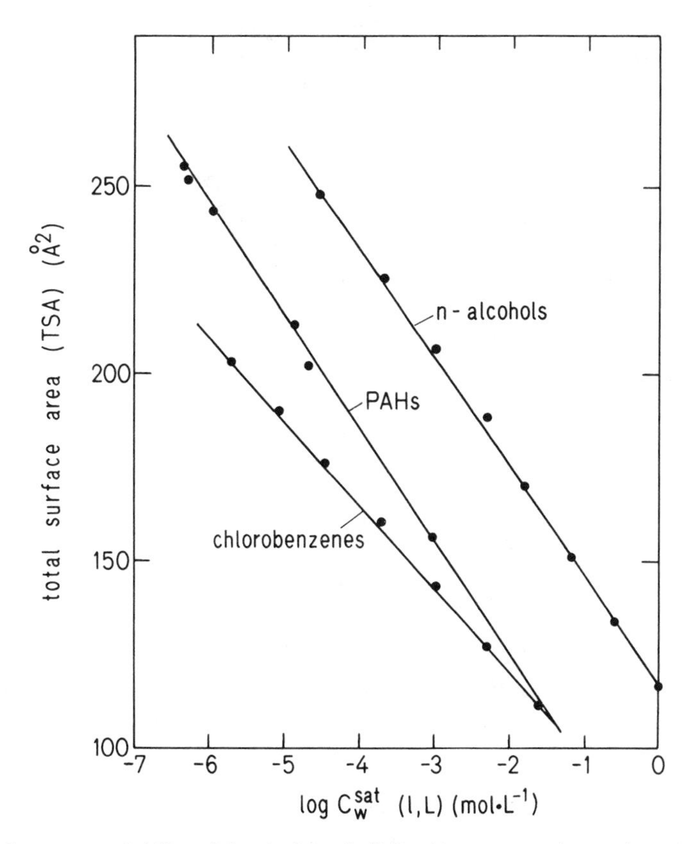

Figure 5.5 Aqueous solubility of the (subcooled) liquid compound as a function of the total surface area (TSA) of the molecule for three different classes of compounds.

activity coefficients in water (or inversely, the aqueous solubilities of liquid organic compounds) and molecular size (McAuliffe, 1966; Hermann, 1972). Figure 5.5 exemplifies these findings for three organic compound classes—alcohols, chlorobenzenes, and polycyclic aromatic hydrocarbons (PAHs). The molecular sizes of the solutes have been quantified in terms of their total surface areas (TSAs) calculated geometrically (Valvani et al., 1976; Yalkowsky et al., 1979; Yalkowsky and Valvani, 1979). As we examine larger and larger molecules within a given compound class, we see that (1) the net free energies of solute transfer become increasingly positive (i.e., disfavoring dissolution), and thus (2) the aqueous solubilities of the organic liquids decrease. Figure 5.5 also illustrates that various compound classes differ from one another because of the specific interactions of the organic molecules with the water molecules. For example, alcohols (which have an —OH moiety capable of hydrogen bonding) show greater solubilities than either the equally sized PAHs or chlorinated benzenes. Obviously this is due to the hydroxyl group's ability to compensate for a certain size of hydrophobic structure (approximately five or six carbons in size). In addition, the degree of change in solubility with increasing molecular size within a given compound class (i.e., the slope of the molecular size versus solubility function), differs with the type of atoms or groups of atoms responsible for the increase in the molecular size. We will return to discussing how various types of atoms in an organic chemical's structure affect that compound's solubility when we treat the problem of organic solvent–water partitioning (Chapter 7).

5.4 EFFECT OF TEMPERATURE AND SOLUTION COMPOSITION ON AQUEOUS SOLUBILITY AND ACTIVITY COEFFICIENT

We have focused on how differences in molecular structure affect the solubilities and activity coefficients of organic compounds dissolved in water. We should now evaluate some important environmental factors that influence the solubility and activity coefficient of a given compound. In the following discussion we consider three such factors: temperature, ionic strength (i.e., dissolved salts), and organic cosolutes. (The pH of the aqueous solution, which is most important for acids and bases, is discussed in Chapter 8.)

Temperature

In the temperature range of natural waters (i.e., ≈ 0–$35°C$), the aqueous solubilities of organic *liquids* (and therefore their activity coefficients) vary only by about a factor of two or less (see examples given in Fig. 5.6 and Table 5.5).

By analogy to the temperature dependence of the vapor pressure (Eq. 4-6), we can deduce the temperature dependence of the mole fraction solubility of a *liquid* compound as:

$$\frac{d \ln x_{\mathrm{w}}^{\mathrm{sat}}}{dT} = \frac{\Delta H_{\mathrm{s}}^{\mathrm{e}}}{RT^2} \qquad (5\text{-}18)$$

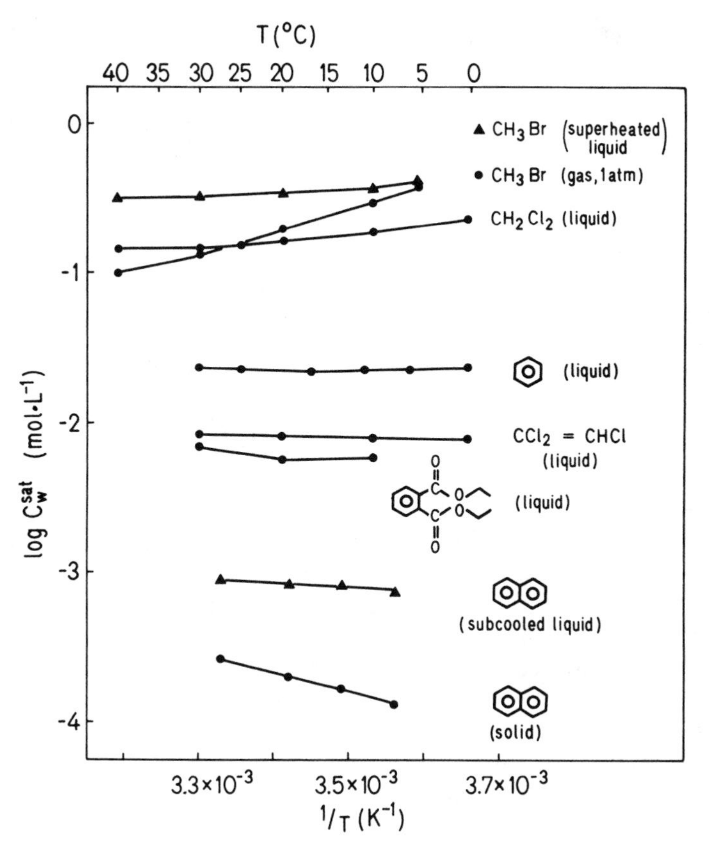

Figure 5.6 Solubility in water as a function of temperature for various compounds.

Assuming a constant ΔH_s^e over a narrow temperature range, and switching to molar units and decadic logarithms, integration of Eq. 5-18 yields:

$$\log C_w^{sat} = -\frac{\Delta H_s^e}{2.303\,RT} + \text{constant} \tag{5-19}$$

Note that when switching to molar units we have also assumed a temperature independent molar volume (V_w) of the solution.

As shown for some *liquid* organic compounds in Fig. 5.6, a plot of $\log C_w^{sat}$ versus $1/T$ yields information on both the sign and magnitude of ΔH_s^e over a narrow temperature range. (Note that the slope of the line is $-\Delta H_s^e/2.303R$.) As is evident in Fig. 5.6 and Table 5.4, within the ambient temperature range, ΔH_s^e is sometimes near zero and therefore may change its sign with increasing temperature. For example, ΔH_s^e for benzene is negative below $\approx 15°C$ and positive above $\approx 20°C$, which means that the solubility of benzene decreases with increasing temperature below $\approx 15°C$ but

TABLE 5.5 Effect of Temperature on the Solubility of Organic Liquids and Solids

Compound (State)	T_1 (°C)	$-\log C_w^{sat}$ ($mol \cdot L^{-1}$)	T_2 (°C)	$-\log C_w^{sat}$ ($mol \cdot L^{-1}$)	$C_w^{sat}(T_2)/C_w^{sat}(T_1)$
1-Pentanol (l)	≈ 0	0.42	30.6	0.64	1.7
1-Heptanol (l)	≈ 0	1.69	30.6	1.88	1.5
Tetrachloroethylene (l)	≈ 0	3.04	30.0	3.04	1.0
1,1,1-Trichloroethane (l)	≈ 0	1.84	30.0	1.96	1.3
1,2-Dichlorobenzene (l)	≈ 0	3.51	30.0	3.16	2.2
Phenanthrene (s)	4.0	5.69	29.9	5.16	3.4
Phenanthrene (L)	4.0	4.67	29.9	4.48	1.5
Anthracene (s)	5.2	7.15	29.3	6.49	4.6
Anthracene (L)	5.2	4.88	29.3	4.64	1.7

increases above $\approx 20°C$. For some liquid compounds (e.g., dichloromethane and superheated bromomethane), solubility decreases with increasing temperature (at ambient temperatures), whereas other compounds show a steady solubility increase with increasing temperature (e.g., trichloroethylene and subcooled naphthalene). The relatively complex ΔH_s^e temperature dependence of liquid organic compounds is a result of the competitive changes with temperature of the various excess enthalpic contributions which we have already discussed (see Fig. 5.2). It seems that for *smaller* molecules at low temperatures, the enthalpy gains from interactions of the organic molecule with water (ΔH_3) and the contribution of iceberg formation (ΔH_4, whose absolute value decreases with increasing temperature) can outcompete the enthalpy costs of "vaporization" of solute and solvent molecules ($\Delta H_1 + \Delta H_2$). On the other hand, for *larger* molecules the enthalpy terms disfavoring solution appear to dominate at all ambient temperatures. In summary, we note that the effect of temperature on the aqueous solubility of *liquid* (liquid, subcooled liquid, superheated liquid) organic compounds and thus on the activity coefficients of organic compounds in water, in general, is rather small (less than about a factor of 2). In the ambient temperature range, therefore, as a first approximation we often neglect the impact of temperature on the activity coefficients of organic compounds in water.

When we are interested in the actual aqueous solubilities of solids and gases, however, the effect of temperature becomes much more important, as is again apparent in the examples given in Fig. 5.6 and Table 5.5. The total enthalpy of solution for gases and solids is given by the sum of the enthalpy of the phase change (i.e., conversion into superheated and subcooled liquid, respectively) and the excess enthalpy of solution:

$$\Delta H_s = \Delta H_\phi + \Delta H_s^e \qquad (5\text{-}20)$$

For solids and gases, we can then write

$$\log C_w^{sat} = -\frac{\Delta H_\phi + \Delta H_s^e}{2.303RT} + \text{constant} \qquad (5\text{-}21)$$

where the constant incorporates entropy and volume terms which are again assumed independent of T. As shown in Table 5.5 and two examples in Fig. 5.6 (compare slopes of subcooled liquid naphthalene versus solid naphthalene or superheated liquid bromomethane versus gaseous bromomethane), the enthalpy of the phase transfer, ΔH_{ϕ}, is usually the dominant factor in determining the sensitivity of the total enthalpy of solution, ΔH_s, to temperature. Consequently, the solubility of solids increases with increasing temperature, since the "cost" of melting solids decreases with increasing temperature. Conversely, the difficulty in condensing gaseous organic compounds increases with increasing temperature; thus, heating an aqueous solution tends to diminish aqueous solubilities of organic gases through this term. These effects can lead to changes in the solubility of solids and gases which may be significant over the ambient temperature range (see Table 5.5).

Dissolved Inorganic Salts

When considering saline environments (e.g., seawater, salt lakes, brines), we have to be concerned with the effects of the dissolved inorganic salt(s) on the aqueous solubilities and activity coefficients of organic compounds. For neutral nonpolar compounds it has been observed that the predominant ionic species found in natural waters (i.e.,

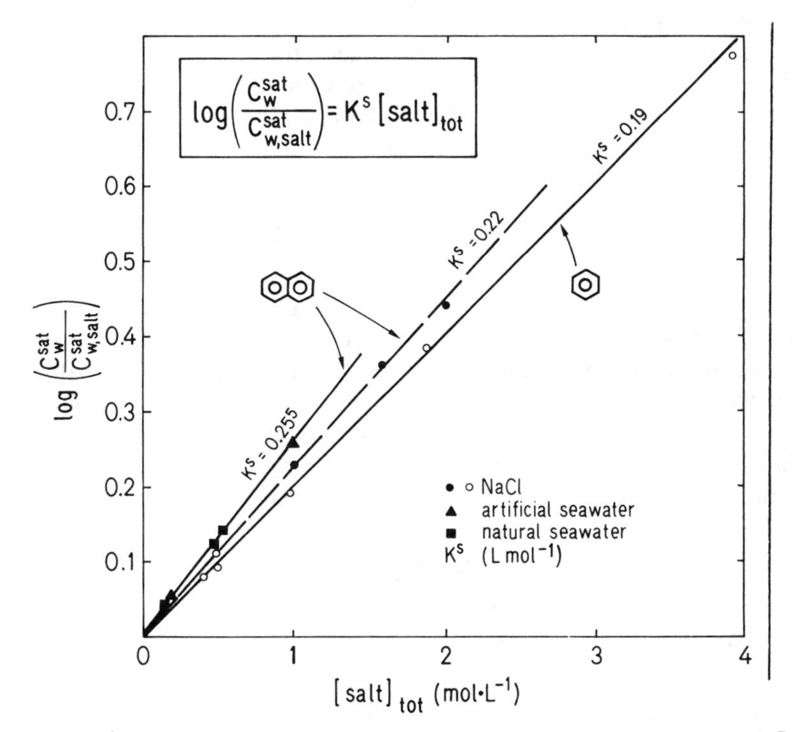

Figure 5.7 Effect of salt concentrations on the aqueous solubility of benzene (McDevit and Long, 1952) and naphthalene (Gordon and Thorne, 1967a).

$Na^+, K^+, Ca^{+2}, Mg^{+2}, Cl^-, SO_4{}^{2-}, HCO_3{}^-$) always decrease the aqueous solubility (increase the activity coefficient). This effect is commonly referred to as "*salting out.*" Setschenow (1889) long ago established an empirical formula relating organic compound solubilities in saline aqueous solutions to those in pure water:

$$\log\left[\frac{C_{\mathrm{w}}^{\mathrm{sat}}}{C_{\mathrm{w,salt}}^{\mathrm{sat}}}\right] = K^{\mathrm{s}}\,[\mathrm{salt}]_{\mathrm{t}} \tag{5-22}$$

where $[\mathrm{salt}]_{\mathrm{t}}$ is the total molar salt concentration and K^{s}, the *Setschenow or salting constant*, is a bulk constant relating the effectiveness of a particular salt or combination of salts to "salt out" a given compound. Figure 5.7 shows that, for nonpolar compounds such as benzene and naphthalene, Eq. 5-22 is valid over a wide range of salt concentrations. Figure 5.7 demonstrates that salt solutions of differing compositions have somewhat different salting-out effects on a given compound (also consult Table 5.6 for Setschenow constants of a variety of compounds in seawater versus sodium chloride solution). We can also see that K^{s} is compound dependent (e.g., naphthalene versus benzene). In seawater ($[\mathrm{salt}]_{\mathrm{t}} \approx 0.5$ M), the solubilities of compounds that interest us are lower by about 10–50% when compared with pure water, depending on the type of compound considered (Table 5.6).

We can qualitatively picture the salting-out phenomenon as follows (McDevit and Long, 1952). The ions produced during dissolution of a salt tightly bind water into hydration shells, which has long been recognized to reduce the volume of aqueous solution even macroscopically by a process known as *electrostriction*. From the point of view of nonpolar organic solute molecules, this tightly bound water is unavailable to "dissolve into." Consequently, as more and more salt is added to an aqueous solution, less and less water remains to create a cavity in which to accommodate an organic solute.

This simple picture also allows us to understand why different inorganic salts exhibit different salting-out effects on a given compound (Table 5.7). Since the degree of hydration, and thus the number of tightly bound water molecules, differs significantly between different ions (compare the effect of CsBr and NaCl on the salting-out of naphthalene in Table 5.7), each ion type uniquely diminishes the available water for dissolving the organic compound. Finally, for solutes that do not specifically interact with the inorganic ions present (i.e., nonpolar organic compounds), the effects of individual salts in salt mixtures are additive, and K^{s} may be expressed as the sum of contributions of the various salts present (Gordon and Thorne, 1967a, b):

$$K^{\mathrm{s}} = \sum_i K_i^{\mathrm{s}} \cdot x_i \tag{5-23}$$

where x_i is the mole fraction and K_i^{s} is the salting constant of salt i in the mixture. Since it is very difficult to quantify the contribution of individual ions, salting constants are only available for combined salts. If, for a given compound, K_i^{s} values are known for the important salts, the salting constant K^{s} for complex mixtures of these salts can be calculated with reasonable accuracy. If one salt compound predominates in

TABLE 5.6 Salting Constants for Some Aromatic Compounds in Seawater (sw), Artificial Seawater (art sw), and Sodium Chloride Solutions (NaCl) at 25°C

		Salting Constant K^s (L·mol^{-1})			Reference[a]		
Compound	TSA (Å^2)	sw	art sw	NaCl	sw	art sw	NaCl
Benzene	110			0.18, 0.19			a, b
Naphthalene	156	0.25, 0.28	0.30	0.21, 0.19, 0.22	d, c	e	a, c, d
Phenanthrene	198	0.25, 0.33	0.39	0.27, 0.29	f, c	e	a, c
Anthracene	202	0.26, 0.35		0.24, 0.25	f, c		a, c
Pyrene	213	0.31, 0.32		0.29, 0.29	c, g		a, c
Chrysene	241			0.34			a
Biphenyl		0.41	0.41	0.26	c	e	c
2,4′-Dichlorobiphenyl		0.3			i		
2,4,4′-Trichlorobiphenyl		0.4			i		
2,3′,4′,5-Tetrachlorobiphenyl		0.2			i		
2,2′,3,4,5′-Pentachlorobiphenyl		0.3			i		
2,2′,3,4,4′,5-Hexachlorobiphenyl		0.3			i		
Toluene (methylbenzene)		0.17	0.28			h	g
Phenol (hydroxybenzene)		0.13		0.12	c		c
4-Aminotoluene		0.19		0.17	c		c
4-Nitrotoluene		0.11		0.14	c		c
Fenuron	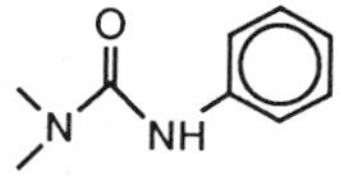			0.23			j
Monuron	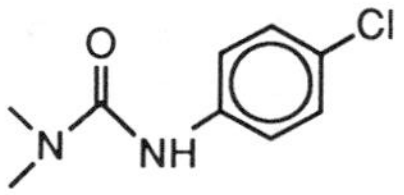			0.24			j

[a] May (1980).
[b] McDevit and Long (1952).
[c] Hashimoto et al. (1984).
[d] Gordon and Thorne (1967b).
[e] Eganhouse and Calder (1976).
[f] Whitehouse (1984).
[g] Rossi and Thomas (1981).
[h] Sutton and Calder (1975).
[i] Brownawell (1986).
[j] van Bladel and Moreale (1974).

TABLE 5.7 Salting Constants for Benzene and Naphthalene at 25°C for Some Important Salts

Salt	Mole fraction of total salt in seawater[a] x_i	Salting Constant	
		K^s (benzene)[b] ($L \cdot mol^{-1}$)	K^s (naphthalene) ($L \cdot mol^{-1}$)
NaCl	0.799	0.19	0.22
$MgCl_2$	0.104		0.30
Na_2SO_4	0.055	0.55	0.70
$CaCl_2$	0.020		0.32
KCl	0.018	0.17	0.19
$NaHCO_3$	0.005		0.32
KBr		0.12	0.13
CsBr			0.01

[a] Gordon and Thorne (1967a, b).
[b] McDevit and Long (1952).

a salt mixture as, for example, NaCl does in seawater, it is often sufficient to use the salting constant for that salt as a surrogate for the whole mixture. For example, the error introduced when using K^s_{NaCl} instead of K^s_{sw} for predicting the effect of salinity on solubility and activity coefficients of organic compounds in seawater is only about 10% (see examples given Table 5.6).

Based on our simple picture of the dissolution process, the introduction of a polar substituent into a molecule should decrease the salting-out effect. This is because the introduction of a polar group generally decreases the hydrophobic surface area, and favorable interactions of the polar group with the ions present in the water are possible. The very few data available on salting effects on polar organic compounds are consistent with this picture. Table 5.6 shows that the measured K^s values for phenol, *p*-aminotoluene, and *p*-nitrotoluene are generally somewhat lower than the values determined for benzene and toluene. In summary, we have seen that the most important dissolved inorganic salts present in natural waters generally decrease the aqueous solubility (or increase the activity coefficient) of neutral organic compounds. At moderate salt concentrations (e.g., in seawater), the effect of salinity on aqueous solubility is usually less than a factor of 2.

Dissolved Organic Solutes and Solvents

Another aspect of solution composition which can affect the solubility (or aqueous activity coefficient) of organic chemicals involves the inclusion of other organic molecules in the water. As depicted in Figure 5.8, such codissolved organic molecules may influence the aqueous cavity surrounding a solute of interest to us, and in so doing, change the energetic costs of forming such a cavity. Three general cases appear to describe the various observations reported. (1) When the other organic molecules are present in relatively large abundances (more than 10% by volume where there is insufficient water to hydrate most of them), these act as *solvent* molecules themselves

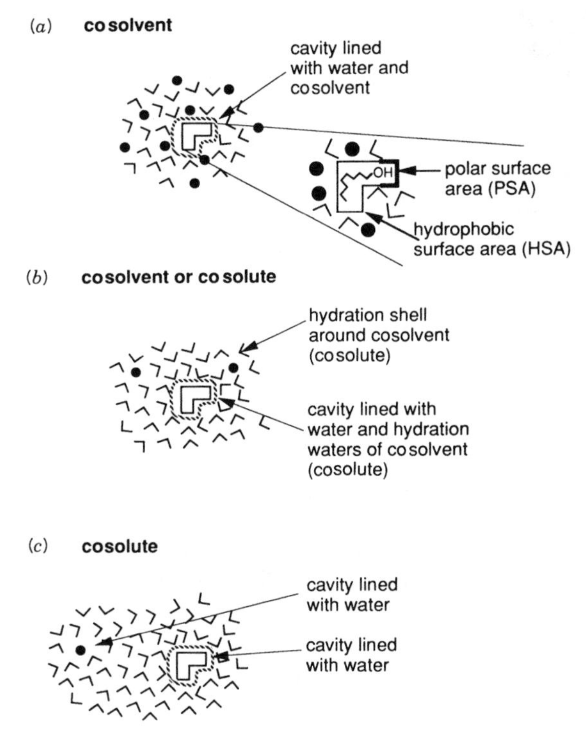

Figure 5.8 Illustrations of how other dissolved organic substances (●) affect the water molecules surrounding an organic compound of interest (⌜⌝).

and partially surround the solute of interest approximately in proportion to their volume fraction in the solution (Yalkowsky et al., 1976). (2) When the other organic compounds are present in somewhat less abundance, these molecules themselves have water-lined cavities surrounding them; and if these hydration shells are somewhat shared by the organic compounds, the overall dissolution cost of the chemical we're considering will be decreased (Banerjee and Yalkowsky, 1988). This situation may be best referred to as an influence of *cosolutes*. This may be the most appropriate picture of the influence of *n*-octanol in water at saturation ($\sim 4.5 \times 10^{-3}$ M or 7×10^{-4} volume fraction), causing a little more hexachlorobenzene (2x) and DDT (3x) to dissolve (Chiou et al., 1982, 1983). (3) Finally, if the organic chemicals are present at low enough levels (less than 10^{-3} volume fraction) that there is a low probability of even their hydration shells overlapping, we can expect no effect on the aqueous activity coefficients or (liquid) solubilities. This is the image we should have for organic compounds that we call "slightly soluble in water" insofar as the molecules of the same kind are too rare to influence one another (Tucker and Christian, 1979; Munz and Roberts, 1986). Similarly, slightly soluble hydrocarbons present in a solution do not appear to enhance the dissolution of other hydrocarbons (e.g., Leinonen and Mackay, 1973).

For the purposes of predicting organic chemical fates in the environment, we are primarily interested in cases where *cosolvents* are present in relatively large proportions (more than 10% by volume). These are the situations where marked changes in nonpolar chemical activity coefficients occur. To estimate the degree of such effects, we can utilize the conceptualization of Yalkowsky et al. (1976). These workers reasoned that the excess free energy of solution of a compound in a water–organic cosolvent mixture should be a linear combination of the compound's excess free energies of solution in each solvent alone:

$$\Delta G^{e}_{s:mix} = (1 - f_c)\,\Delta G^{e}_{s:w} + (f_c)\,\Delta G^{e}_{s:c} \tag{5-24}$$

where

f_c is the volume fraction of the solution consisting of the cosolvent
$\Delta G^{e}_{s:w}$ is the excess free energy of solution in pure water
$\Delta G^{e}_{s:c}$ is the excess free energy of solution in the cosolvent

It is as if part of the organic solute of interest is dissolved in water, while the remainder is dissolved in the organic cosolvent. Recalling that $\Delta G^{e}_{s} = +RT \ln \gamma$, we can also write

$$\ln \gamma_{mix} = (1 - f_c) \ln \gamma_w + (f_c) \ln \gamma_c \tag{5-25}$$

or

$$\ln x^{sat}_{mix} = (1 - f_c) \ln x^{sat}_{w} + (f_c) \ln x^{sat}_{c} \tag{5-26}$$

Yalkowsky and colleagues (1976, and references therein) have reasoned that microscopic-scale situations like that pictured in Fig. 5.8*a* can be thought of much like a macroscopic-scale counterpart of two liquids contacting one another and exhibiting an interfacial surface tension (e.g., Fowkes, 1964). In this case the solute (shown as an octanol molecule in a blowup of Fig. 5.8*a*) may be seen as having both hydrophobic surface area (HSA) and polar surface area (PSA). Each of those microscopic surface area types experiences a different interaction energy when juxtaposed to a polar liquid like water or a relatively nonpolar one like acetone or isopropanol. Thus, Yalkowsky et al. (1976) write for $\Delta G^{e}_{s:w}$:

$$\Delta G^{e}_{s:w} = (\sigma_{h:w})(\text{HSA})(N) + (\sigma_{p:w})(\text{PSA})(N) \tag{5-27}$$

where

$\sigma_{h:w}$ is the interfacial energy (e.g., $J \cdot cm^{-2}$) where the hydrophobic solute contacts water,
HSA is the solute's hydrophobic surface area (cm^2/molecule),
$\sigma_{p:w}$ is the interfacial energy (e.g., $J \cdot cm^{-2}$) where the polar solute contacts water,
PSA is the solute's polar surface area (cm^2/molecule),
N is Avogadro's number (6.02×10^{23} molecules/mol), used to put everything on a per mole basis.

Similarly, for an organic solvent,

$$\Delta G^{e}_{s:c} = (\sigma_{h:c})(\mathrm{HSA})(N) + (\sigma_{p:c})(\mathrm{PSA})(N) \tag{5-28}$$

where the subscript c refers to the energies associated with solute interaction with pure organic cosolvent.

Using the results of Eqs. 5-27 and 5-28 in Eq. 5-26, we find

$$\log x^{\mathrm{sat}}_{\mathrm{mix}} = -\frac{N(1-f_c)}{2.303RT}[(\sigma_{h:w})(\mathrm{HSA}) + (\sigma_{p:w})(\mathrm{PSA})] - \frac{N(f_c)}{2.303RT}[(\sigma_{h:c})(\mathrm{HSA}) + (\sigma_{p:c})(\mathrm{PSA})] \tag{5-29}$$

Rearranging we see that

$$\log x^{\mathrm{sat}}_{\mathrm{mix}} = -\left[\frac{N(\sigma_{h:w})(\mathrm{HSA})}{2.303\,RT} + \frac{N(\sigma_{p:w})(\mathrm{PSA})}{2.303\,RT}\right] + \left[\frac{f_c N(\sigma_{h:w}-\sigma_{h:c})(\mathrm{HSA})}{2.303\,RT} + \frac{f_c N(\sigma_{p:w}-\sigma_{p:c})(\mathrm{PSA})}{2.303\,RT}\right] \tag{5-30}$$

The first term in brackets on the right-hand side of Eq. 5-30 is equal to $\log x^{\mathrm{sat}}_{w}$. For chemicals that consist of mainly nonpolar structural parts so that (PSA) is small, we can approximate

$$\log x^{\mathrm{sat}}_{\mathrm{mix}} = \log x^{\mathrm{sat}}_{w} + \frac{f_c N(\sigma_{h:w}-\sigma_{h:c})(\mathrm{HSA})}{2.303RT} \tag{5-31}$$

This result is reasonably well supported by experimental observations. For example, we see that for a series of aromatic hydrocarbons of varying HSA dissolved in a particular cosolvent–water mix (e.g., 40% methanol, 60% water in Fig. 5.9*a*), enhancement in solubility [i.e., $\log(x^{\mathrm{sat}}_{\mathrm{mix}}/x^{\mathrm{sat}}_{w})$] increases exponentially in proportion to HSA. In this case, benzene with its surface area of 110 Å^2 has about five times greater solubility than in pure water alone, and pyrene (210 Å^2) dissolved in 40:60 methanol:water at more than 50 times its pure water solubility. If we focus our attention on a single organic solute (e.g., the *n*-octyl ester of *p*-aminobenzoate in Fig. 5.9*b*), we also see its log solubility is enhanced linearly as the fraction of cosolvent in the solution mixture is increased.

The term, $(\sigma_{h:w}-\sigma_{h:c})$, remains the most difficult component of Eq. 5-31 to evaluate. Yalkowsky et al. (1976) found this difference was related to the difference in macroscopic liquid:liquid interfacial tensions when the reference hydrophobic liquid was a hydrocarbon-like tetradecane. That is, water overlaying liquid tetradecane is found to exhibit an interfacial energy ($\sigma_{t:w}$) of about 52 erg/cm^2, while a less polar liquid like propylene glycol contacts tetradecane with a lower interfacial energy ($\sigma_{t:p}$) of about

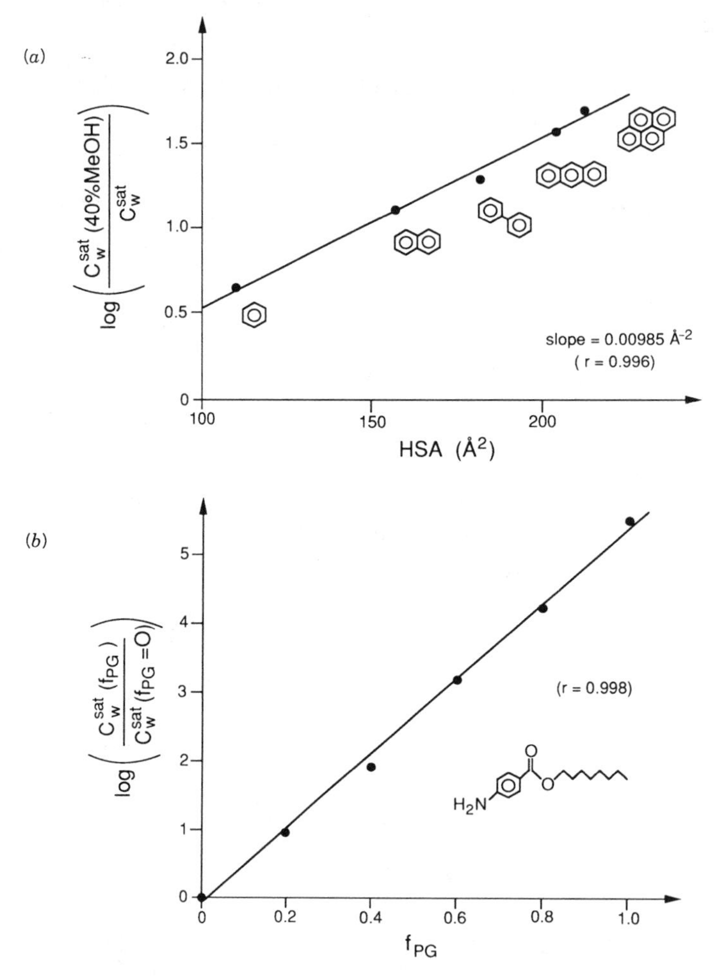

Figure 5.9 Illustration of cosolvent-enhanced solubility as a function of (*a*) the hydrophobic surface area (HSA in Å^2) of a set of aromatic solutes (data from Morris et al., 1988) and (*b*) the fraction of propylene glycol in aqueous solutions of *n*-octyl-*p*-amino-benzoate (data from Yalkowsky et al., 1976).

13 erg/cm^2. Such polar liquid:nonpolar liquid interfacial tensions also strongly correlate with polar liquid:air surface tensions which have been measured (see Table 5.8). This is because air, like a nonpolar liquid, offers virtually no specific across-interface attractive forces to the polar liquid on the other side. Nonpolar liquids provide only about 20 erg/cm^2 of van der Waals attraction; thus the polar liquid:air tensions correlate well with the polar liquid:tetradecane interfacial tensions, except that they are offset by about 20 erg/cm^2. Generally, these surface energies are large when dissimilar materials like polar and nonpolar fluids are contacting one another and

TABLE 5.8 Estimated Solubility Enhancement of Pyrene (HSA = 212 Å^2) for Solutions Containing 10% Cosolvent at 20°C[a]

Cosolvent	Structure	Liquid:Air Surface Tensions at 20°C[d] (erg/cm^2)	$\sigma_{t:c}$ (erg/cm^2)	$\sigma_{t:w} - \sigma_{t:c}$ (erg/cm^2)	C_{mix}^{sat}/C_w^{sat}
Water	H_2O	73	52	0	1
Glycerin	HOCH$_2$CH(OH)CH$_2$OH	63	35	17	1.6[b]
Formamide	HC(=O)NH$_2$	58	31	21	1.7
Ethylene glycol	HOCH$_2$CH$_2$OH	48	19	33	2.3[b], 1.9[c]
Propylene glycol	CH$_3$CH(OH)CH$_2$OH	—	13	39	2.8[b], 2.5[c]
Isopropanol	(CH$_3$)$_2$CHOH	22	< 10[e]	~52	3.9[b], 3.3[c]
Methanol	CH_3OH	23	< 10[e]	~52	3.9[b], 2.8[c]
Ethanol	CH$_3$CH$_2$OH	23	< 10[e]	~52	3.9[b], 3.3[c]
Acetonitrile	$CH_3{-}C{\equiv}N$	29	< 10[e]	~52	3.9[b], 3.6[c]
Acetone	(CH$_3$)$_2$C=O	24	< 10[e]	~52	3.9[b], 3.6[c]

[a] Also shown are liquid:air surface tensions, liquid:tetradecane surface tensions ($\sigma_{t:c}$), and an estimated value of the difference ($\sigma_{t:w} - \sigma_{t:c}$) used in Eq. 5-31. Note 1 erg = 10^{-7} J.

[b] Results using the equation of Yalkowsky et al. (1976):

$$\log(C_{mix}^{sat}/C_w^{sat}) = 0.5(\sigma_{t:w} - \sigma_{t:c})(N \cdot \text{HSA} \cdot f_c)/2.303RT$$

where $0.5(\sigma_{t:w} - \sigma_{t:c})$ is the estimated change in solute:solvent interaction energy, N is Avogadro's number, HSA is the solute's hydrophobic surface area, f_c is the volume fraction cosolvent, R is the gas constant, and T is the temperature.

[c] Calculated using results of Morris et al. (1988):

$$\log(C_{mix}^{sat}/C_w^{sat}) = (a \log K_{ow} + b)(f_c)$$

where a and b are coefficients unique to a particular solvent, K_{ow} is the solute's octanol–water partition coefficient (see Ch. 7), and f_c is the volume fraction of the cosolvent.

[d] *CRC Handbook of Chemistry and Physics* (1985).

[e] Cosolvent miscible with hydrocarbons so no macroscopic interface can be formed and tension determined.

decrease as the difference in polarity is diminished. By extension we can imagine these "tensions" also correspond to cavity costs around an organic solute dissolved in a water–cosolvent mix; that is, $(\sigma_{t:w} - \sigma_{t:c})$ is proportional to $(\sigma_{h:w} - \sigma_{h:c})$. Where water touches the organic solute's surface, there is a large energy cost; where cosolvent molecules lie near the solute, the microscopic interfacial energy is greatly reduced.

To utilize the macroscopic polar liquid:tetradecane interfacial tensions in Eq. 5-31, Yalkowsky et al. (1976) have suggested that a small correction should be made for using a flat interfacial region resulting in a case where the solvent cavity is curved around the solute. When solute surface areas are calculated as the geometric exterior of only the solute itself, this curvature correction factor appears to be about 0.5. Using these results, Eq. 5-31 becomes

$$\log x_{\text{mix}}^{\text{sat}} = \log x_{\text{w}}^{\text{sat}} + \frac{0.5\, f_{\text{c}} N(\sigma_{\text{t:w}} - \sigma_{\text{t:c}})(\text{HSA})}{2.303RT} \tag{5-32}$$

where $\sigma_{t:w}$ and $\sigma_{t:c}$ are the macroscopic liquid:liquid surface energies (erg/cm^2) of the organic cosolvent $(\sigma_{t:c})$ and water $(\sigma_{t:w})$ against tetradecane (or like nonpolar liquid).

We are now in a position to estimate the importance of various cosolvents to dissolution of organic chemicals. For example, the polycyclic aromatic hydrocarbon, pyrene, can be calculated to have a total surface area of 212 Å^2 ($= 212 \times 10^{-16}$ cm^2);

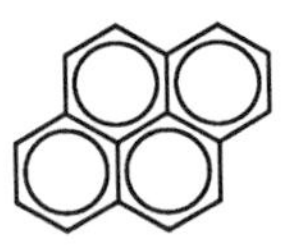

pyrene

and since this compound is entirely hydrophobic over its structure, we can use this result in Eq. 5-32 to predict the influence of a variety of cosolvents on its solubility (Table 5.8). Generally, we estimate that a small proportion of cosolvent, such as 10% by volume, increases pyrene's solubility between 2 and 4 times. Greater proportions of cosolvent increase pyrene's solubility exponentially. Lower-volume fractions of cosolvent have a rapidly diminishing effect as their low abundance allows the individual cosolvent molecules themselves to be hydrated, and they can only play the much less influential role of a cosolute. Morris et al. (1988) developed another approach to predicting the influence of cosolvents which relies on knowledge of octanol–water partition coefficients (discussed in Chapter 7); some estimates of pyrene's enhanced solubility in several 10% cosolvent:water mixtures using this approach are also shown in Table 5.8 for comparison.

A final observation on the use of macroscopic surface tensions to estimate microscopic energies is probably worth making. Other factors such as temperature and salt change the surface tensions of aqueous solutions. If we consider salts rather than

TABLE 5.9 Predicted versus Observed Setschenow Constants for Three Aromatic Hydrocarbons in Aqueous Solutions of 1:1 Salts[a]

				Benzene		Toluene		Naphthalene	
Salt, T(°C)	$\sigma_{\text{air:saltwater}}$ (erg/cm^2)	[Salt] (M)	$-\frac{\Delta\sigma}{[\text{Salt}]}$(J/Å^2·M)	$K^{s,\text{pred}}$	$K^{s,\text{obs}}$	$K^{s,\text{pred}}$	$K^{s,\text{obs}}$	$K^{s,\text{pred}}$	$K^{s,\text{obs}}$
$HClO_4$, 25	69.01	25.92	-0.114×10^{-23}	−0.013	−0.041[b]				
LiCl, 25	78.10	3.82	1.61×10^{-23}	0.186	0.141[b]	0.213	0.191[d]		
NaOH, 18	101.05	13.998	2.00×10^{-23}	0.238	0.256[b]				
NaCl, 20	82.55	5.99	1.64×10^{-23}	0.193	0.195[b]	0.223	0.267[d] 0.208[e]	0.274	0.220[c]
NaBr, 20	76.55	2.903	1.309×10^{-23}	0.154	0.155[b]				
$NaNO_3$, 20	80.25	6.998	1.072×10^{-23}	0.127	0.199				
KCl, 20	78.75	4.40	1.36×10^{-23}	0.160	0.166[b]	0.185	0.205[d]	0.227	0.186[c]

[a]Surface Tension Data from *CRC Handbook of Chemistry and Physics* (1985). Using Eq. 5-35 and assuming $c = 1$.
[b]McDevit and Long, 1952.
[c]Gordon and Thorne, 1967a.
[d]Sada et al., 1975.
[e]Garbarini and Lion, 1985.
[f]$\sigma_{\text{air:water}}$ (18°C) = 73.05 erg/cm^2; $\sigma_{\text{air:water}}$ (20°C) = 72.75 erg/cm^2; $\sigma_{\text{air:water}}$ (25°C) = 71.97 erg/cm^2.

cosolvents, we might rewrite Eq. 5-32 as

$$\log x_{\mathrm{sw}}^{\mathrm{sat}} = \log x_{\mathrm{w}}^{\mathrm{sat}} + \frac{c \cdot f_{\mathrm{s}} \cdot N(\sigma_{\mathrm{t:w}} - \sigma_{\mathrm{t:sw}})(\mathrm{HSA})}{2.303RT} \tag{5-33}$$

where

$x_{\mathrm{sw}}^{\mathrm{sat}}$ is the solubility in a salty solution,
c is the curvature correction factor for salty solution cavities,
f_{s} is the fraction of approach to a solution filled with hydrated salt ions (i.e., fraction of approach to a saturated solution for very soluble salts),
$\sigma_{\mathrm{t:sw}}$ is the tetradecane:salt water macroscopic interfacial surface energy.

This expression is strikingly reminiscent of the empirical Setschenow equation if

$$(K^{\mathrm{s}}) \cdot [\mathrm{salt}] = -\frac{c \cdot f_{\mathrm{s}} \cdot N(\sigma_{\mathrm{t:w}} - \sigma_{\mathrm{t:sw}})(\mathrm{HSA})}{2.303RT} \tag{5-34}$$

In practice we do not find values of $\sigma_{\mathrm{t:sw}}$, but rather those reflecting surface tensions against air ($\sigma_{\mathrm{air:sw}}$). If we assume $f_{\mathrm{s}} \cdot (\sigma_{\mathrm{t:w}} - \sigma_{\mathrm{t:sw}}) = (\sigma_{\mathrm{air:w}} - \sigma_{\mathrm{air:sw}})$ at a given value of [salt], we can use available data to estimate K^{s} via

$$K^{\mathrm{s}} = -\frac{(c)(N)(\sigma_{\mathrm{air:water}} - \sigma_{\mathrm{air:saltwater}})(\mathrm{HSA})}{2.303 \cdot R \cdot T \cdot [\mathrm{salt}]} \tag{5-35}$$

Table 5.9 compares some measured Setschenow constants with those predicted for benzene, toluene, and naphthalene in 1:1 salt solutions keeping $c = 1$. Clearly salts like $HClO_4$ which are observed to "salt-in" benzene are predicted to do so as their presence in aqueous solutions lowers the solution's surface tension. Also, Eq. 5-35 predicts the increase in K^{s} for larger and larger hydrophobic solutes as we have seen before (recall Table 5.6). Thus, although estimated absolute values may diverge from measurements, all of the functional trends carried in Eq. 5-35 appear to be well represented.

5.5 DISSOLUTION OF ORGANIC COMPOUNDS IN WATER FROM ORGANIC LIQUID MIXTURES

Often we are concerned with situations where liquid *mixtures* of organic chemicals are dissolving in water. Examples of such mixtures are gasoline, other petroleum products, or PCBs in transformer oils. In these cases, we need to rethink Eq. 5-2 so that it describes the chemical potential of the chemical of interest in an organic mixture:

$$\mu_{\mathrm{org\,mix}} = \mu^0 + RT \ln \gamma_{\mathrm{org\,mix}} x_{\mathrm{org\,mix}} \tag{5-36}$$

As before, we know that $\mu_w = \mu_{org\,mix}$ at dissolution equilibrium, or we may recognize by combining Eq. 5-3 and 5-36:

$$x_w = \gamma_{org\,mix}\, x_{org\,mix}\, \gamma_w^{-1} \tag{5-37}$$

This result differs from the solubility of a pure organic compound, x_w^{sat}, in three important ways. First, it is possible that the organic chemical of concern does not form an ideal solution in the liquid organic mixture. The more dissimilar (shape, size, polarity) the chemical is from the average nature of the organic mix, the greater will be this nonideality. This will cause the factor $\gamma_{org\,mix}$ in Eq. 5-37 to become greater than 1. For example, benzene dissolved in hexane exhibits $\gamma_{org\,mix}$ (benzene) ~ 2. Often for want of information on mixtures of interest, however, we can only assume $\gamma_{org\,mix}$ is somewhere near 1 for the typical compounds in the organic liquid.

The second difference from pure compound dissolution involves accounting for the dilution of the chemical by the other substances of the organic mixture using $x_{org\,mix}$. Since mixtures are often made of numerous and unidentified compounds, it is often impossible to determine the concentration in the mix of the particular substance of concern on a mole fraction basis. Thus one often assumes an average molecular weight of the materials in the mix and uses that to convert concentrations measured on a weight basis to those on a mole fraction one. For example, if naphthalene was found to occur at 5g per 100g of coal tar (a mix of hydrocarbons produced in the oxygen-free pyrolysis of coal), we would need the average molecular weight of compounds in the tar to calculate how many moles 100g of tar represented. For a particular tar, Picel et al. (1988) have estimated this to be 150g/mol. Thus

$$x_{org\,mix}(\text{naphthalene}) = \frac{(5\text{ g})/(128\text{ g/mol})}{(100\text{ g})/(150\text{ g/mol})} = 0.059 \tag{5-38}$$

Obviously, for chemicals that have nearly the same molecular weight as the average chemicals in the mixture, the weight percent abundance will not differ greatly from the mole fraction.

The third very important distinction regarding dissolution of organic chemicals from a liquid mixture, as opposed to from the pure substance, involves the last term in Eq. 5-37, γ_w^{-1}. This term reflects the ability of *liquid* organic molecules to dissolve in water; notably terms that account for phase change costs of solids and gases dissolving in water [i.e., $P(s)/P^\circ(L)$ and 1 atm/$P^\circ(L)$, respectively, in Eq. 5-8] are not required. This distinction arises because all the molecules in a liquid mixture are present in a liquid state. Thus naphthalene dissolved in coal tar liquid is *present as liquid naphthalene molecules* (i.e., not exhibiting restriction due to crystallization) even though pure naphthalene would be a solid at ambient temperatures [T_m(naphthalene) = 80.6°C]. This is also the case for PCB congeners in transformer oils, although as pure substances many would exist as solids at 0–35°C. We may understand the absence of the terms accounting for phase transformation costs on dissolution by remembering that such melting or condensation is not needed to exchange molecules from one liquid mixture to another.

Let us conclude this discussion by estimating the dissolution in water of 1,2,4,5-tetrachlorobenzene ($T_m = 140°C$) mixed in chlorobenzene ($T_m = -45.6°C$) at 20°C. First we *assume* $\gamma_{\text{org mix}}$(tetrachlorobenzene) is about 1 since these organic compounds are fairly similar in type. Next we estimate γ_w(tetrachlorobenzene) from knowledge of its pure compound solubility and its melting temperature:

$$\gamma_w^{-1} = x_w^{\text{sat}} \cdot \frac{P°(\text{L})}{P°(\text{s})}$$

$$= x_w^{\text{sat}} \cdot \exp\left(\frac{+\Delta S_m}{R}\left[\frac{T_m}{T} - 1\right]\right) \tag{5-39}$$

using Eq. 4-19. Inserting these results in Eq. 5-37 we have

$$x_w(\text{tetrachlorobenzene}) \approx x_{\text{org mix}} \cdot x_w^{\text{sat}} \cdot \exp\left(\frac{+\Delta S_m}{R}\left[\frac{T_m}{T} - 1\right]\right) \tag{5-40}$$

Since tetrachlorobenzene has a rigid structure, ΔS_m is probably about 56.5 J/(mol·K), and the exponential expression can be calculated as 16. Banerjee's (1984) data suggest a pure compound solubility for this tetrachlorobenzene isomer of 3.9×10^{-8} mole fraction. Taken together, we may now estimate the concentration of tetrachlorobenzene in water at 20°C if this polychlorinated compound occurred at 0.023 mole fraction in the mixture with chlorobenzene:

$$\begin{aligned} x_w(\text{tetrachlorobenzene}) &\approx (0.023)(3.9 \times 10^{-8})(16) \\ &\approx 1.4 \times 10^{-8} \text{ mole fraction} \\ &\approx 7.9 \times 10^{-7} \text{ M} \end{aligned} \tag{5-41}$$

Banerjee (1984) observed 1.4×10^{-6} M when he tested this mixture against water in the laboratory. The small discrepancy could be due to our assumption that $\gamma_{\text{org mix}} = 1$ (a reasonable value of 1.8 would cause the prediction to match the observations), and/or because our presumed value for ΔS_m was too low.

Note that as the mole fraction of tetrachlorobenzene in the organic liquid mix goes up, it appears that the concentrations of this substance in water could exceed its pure component solubility! However, this does not happen because when such water supersaturation occurs, solid crystals of tetrachlorobenzene form and all phases in the system are "buffered" at levels in equilibrium with this pure solid phase. Put another way, $x_{\text{org mix}}$ could not exceed about 0.1 mole fraction as tetrachlorobenzene precipitation even from the organic solvent prevented this from going any higher. Again this corresponds well to the observations of Banerjee (1984), who noted solid precipitates in his experimental setup soon after exceeding this compound's pure solubility.

5.6 AVAILABILITY OF SOLUBILITY DATA; ESTIMATION OF AQUEOUS SOLUBILITY AND ACTIVITY COEFFICIENT

Owing to the analytical limitations of the past, many organic substances have acquired the reputation of "being insoluble in water" (e.g., *CRC Handbook of Chemistry and Physics*, 1985); however, all organic compounds are soluble in water to some extent! Numerous approaches have been utilized to assess aqueous solubilities, and it is now possible to determine these values for compounds dissolved in water at roughly parts per trillion levels ($\approx$ picomolar) and above.

Before we talk about the available data, we should discuss some of the more common experimental approaches that have been used to determine aqueous solubilities of hydrophobic compounds. For compounds containing chromophores (i.e., moieties that absorb visible or ultraviolet light), water samples have been exposed to excess pure organic material, and the eventual saturated concentrations have been quantified by spectroscopic analysis of either the water solution itself (e.g., Bohon and Claussen, 1951; Alexander, 1959) or an organic solvent extract of the saturated aqueous solution (e.g., Yalkowsky et al., 1979). With the eventual widespread availability of gas chromatography, aqueous solutions saturated with the compounds of interest were quantitatively analyzed for their organic solute contents by direct aqueous injection of the saturated solution (e.g., McAuliffe, 1966) or by gas chromatographic analysis of solvent extracts (e.g., Dexter and Pavlou, 1978). A still more recent advance is the use of liquid chromatographic columns containing immobile solids coated with the organic compound of interest and through which volumes of water are slowly passed which generate saturated solutions (e.g., May, 1980; May et al., 1978a; Whitehouse, 1984). Subsequently, the aqueous effluent is assessed for organic solute contents by collecting the solute on a reverse-phase "extractor column" and then flushing this concentrate into a high performance liquid chromatograph equipped with UV or fluorescence detectors. With each application of increasingly sensitive analytical techniques, the solubilities of increasingly less soluble materials are being acquired. These activities have resulted in solubility compilations for various classes of compounds: (1) hydrocarbons (McAuliffe, 1966, 1971; Sutton and Calder, 1974); (2) alkyl benzenes (Bohon and Claussen, 1951; Sutton and Calder, 1975; Tewari et al., 1982); (3) polycyclic aromatic hydrocarbons (Eganhouse and Calder, 1976; Mackay and Shiu, 1977; May et al., 1978a, b; May, 1980; May et al., 1983; Whitehouse, 1984); (4) polychlorinated biphenyls (Dexter and Pavlou, 1978; Mackay et al., 1980a, b; Miller et al., 1984); (5) halobenzenes (Yalkowsky et al., 1979; Banerjee, 1984; Miller et al., 1984); (6) halosolvents (McGovern, 1943; Yalkowsky et al., 1979); (7) alcohols (Amidon et al., 1974; Tewari et al., 1982); and (8) ketones and esters (Tewari et al., 1982).

If the aqueous solubility of an organic compound is unknown, there are two main approaches available for estimating this property. The first utilizes known solubilities of structurally related compounds and the general observation that within a class of organic solutes, solubility decreases with increasing solute size. As we have seen in Section 5.3, this often-observed empirical relationship appears to succeed since the predominant free-energy contributions to the dissolution process are proportional to molecular size. For example, referring again to Fig. 5.5, we can estimate the solubilities of other PAHs based on the correlation shown. Thus, for fluorene, whose TSA is 194 $Å^2$, we can interpolate and predict the subcooled liquid solubility to be 4.5×10^{-5} M. The observed value is 6.1×10^{-5} M, so there is some error; but a reasonable estimate can be quickly made.

A second common approach used to estimate aqueous solubilities involves methods to predict the aqueous activity coefficients of hydrophobic solutes. As we shall see in Chapter 7, as a first approximation, γ_w can be related to the octanol–water partition coefficient, K_{ow}, for the solute of interest, and thereby estimated. Using this estimate of the aqueous activity coefficient together with the knowledge of melting or condensation free-energy "costs" for solids or gases, we can predict aqueous solubilities using Eq. 5-8.

Finally, we recognize γ_w as an integrated factor reflecting the contributions of many interactions between solute molecule subparts and water molecules; therefore, we may use estimation techniques to tally these "structural contributions" to nonideal solution behavior of organic solutes. This is the approach of activity coefficient estimation techniques such as UNIFAC (Fredenslund et al., 1975; Gmehling et al., 1982; Arbuckle, 1983; Banerjee, 1985) and ASOG (Pierotti et al., 1959), both of which have been described by Grain (1982b). Of course, all of these methods apply to the solute in its liquid state only; thus, we must keep in mind the "energy costs" involved in phase transitions from gaseous or solid states to corresponding hypothetical liquid states.

CHAPTER 6

AIR–WATER PARTITIONING: THE HENRY'S LAW CONSTANT

6.1 INTRODUCTION

Having gained some insight into the energy status of molecules in the gaseous state as well as in aqueous solution, it follows that our next question should concern how a chemical will be distributed between the gas phase (i.e., air) and water when the two phases are in contact. Transfer between the atmosphere and bodies of water is one of the key processes affecting the transport of many organic compounds in the environment. In this chapter we will discuss the *equilibrium partitioning* of organic chemicals between the gas phase and an aqueous solution. By partitioning we mean the subdivision of a population of molecules of a given compound between any two phases, in this case the gas and solution phases, determined by the compound's relative compatibility with each medium. For neutral compounds at *dilute solute concentrations in pure water*, the air–water distribution ratio is referred to as the *Henry's Law constant* K_H. For real aqueous solutions (i.e., solutions that contain many other chemical species), we use the term "air–water distribution ratio" which, for practical purposes, we approximate by the Henry's Law constant. As discussed in Chapter 10, this air–water partition ratio characterizes not only the air–water equilibrium distribution of a compound, but also plays a role in the rate expressions describing air–water systems which are evolving toward equilibrium. The Henry's Law constant K_H may be thought of as simply the ratio of a compound's abundance in the gas phase to that in the aqueous phase at equilibrium (see Fig. 6.1*a*). As discussed previously, we commonly express the abundance of a chemical in the gas phase as its partial pressure, P_i, and its abundance in the aqueous solution as a molar

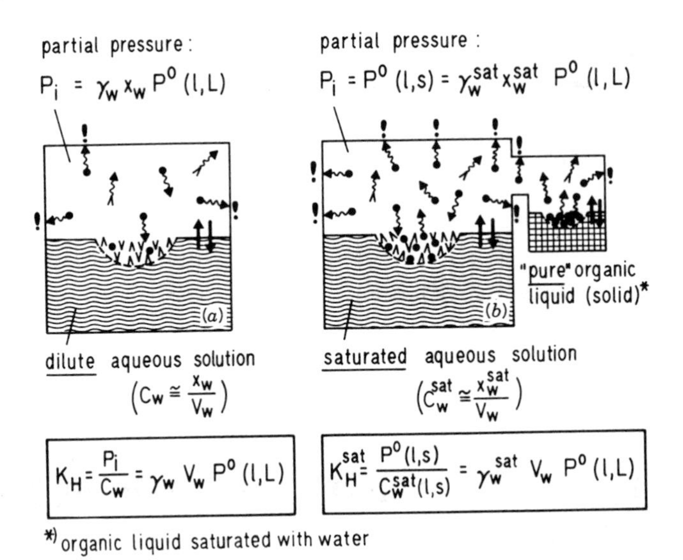

Figure 6.1 Equilibrium partitioning of a compound between a gas phase (air) and water. (*a*) System at dilute concentrations. (*b*) System at saturated concentrations.

concentration, C_w:

$$K_H = \frac{P_i}{C_w} (\text{atm} \cdot \text{L} \cdot \text{mol}^{-1}) \tag{6-1}$$

The units of K_H are, of course, dependent on our choice of measures. If we express the abundance of a compound in air as moles per liter of air (C_a), we obtain the so-called *dimensionless* Henry's Law constant:

$$K'_H = \frac{C_a}{C_w} (\text{mol} \cdot \text{L}_a^{-1} \cdot \text{mol}^{-1}\,\text{L}_w) \tag{6-2}$$

K'_H and K_H may be related to one another by applying the ideal gas law for converting partial pressure in atmospheres to moles per liter of air [$P_i = (n_i/V)RT$]:

$$K'_H = \frac{K_H}{RT} \tag{6-3}$$

All of the other constants (e.g., Bunsen coefficients) used to relate the distribution of a chemical between air and water are convertible to either K_H or K'_H. It does not really matter then in what form the air–water partition ratios are given as long as we pay careful attention to how they are defined (as air–water or water–air ratios), the temperature at which they are given, and the units of concentration that are used.

No matter which form is used, the air–water or water–air partition constant quantifies the relative escaping tendency (fugacity) of a compound existing as vapor molecules as opposed to being dissolved in water. Figure 6.2 exhibits the ranges of K_H values for a few compound classes. From our previous discussions we would predict that compounds with high vapor pressures (i.e., low fugacity in the gas phase) and high activity coefficient in water (i.e., high fugacity in aqueous solution) should partition appreciably from water into air (i.e., high K_H values). Furthermore, we can intuitively understand why water–air transfer of polar compounds (low P^o, low γ_w) is not very important, but that air–water transfer (i.e., scavenging from the atmosphere by rain) will play a role for these compounds. We cannot, however, automatically assume that either process is insignificant for nonpolar high boiling compounds like

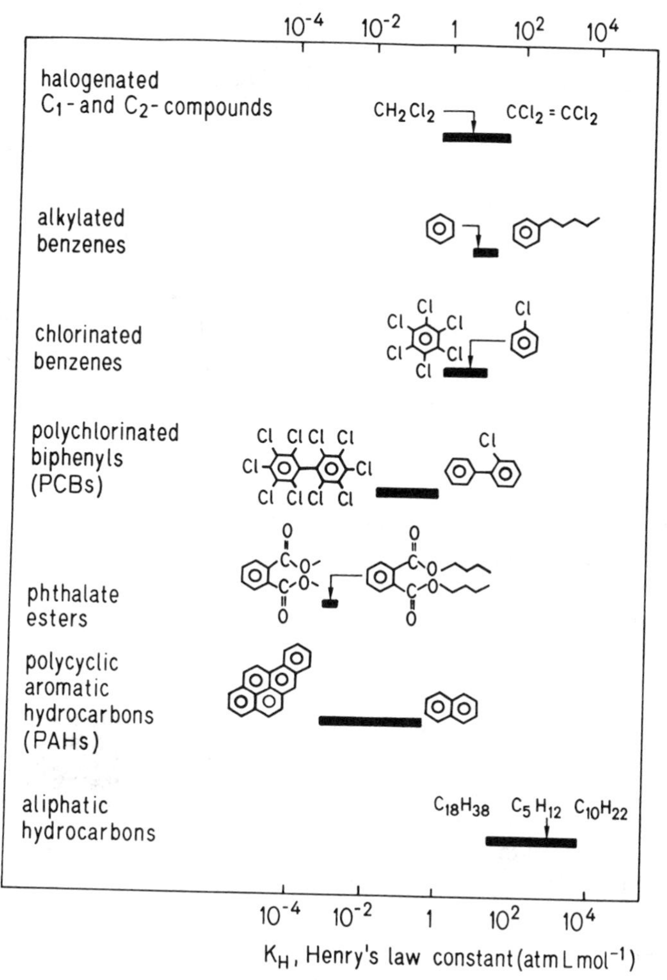

Figure 6.2 Ranges in Henry's Law constants (K_H) for some important classes of organic compounds.

PCBs or PAHs, since these compounds exhibit high fugacities in both phases. Let us now consider a more quantitative treatment of air–water partitioning.

6.2 THERMODYNAMIC CONSIDERATIONS: EFFECTS OF CONCENTRATION, TEMPERATURE, SALT, AND COSOLUTES

In Chapter 3 we defined the fugacities of a compound in an *ideal* gas and in aqueous solution as (see Fig. 3.2):

$$\begin{aligned} f_g &= P_i \\ f_w &= \gamma_w x_w P^o(\mathrm{l, L}) \end{aligned} \tag{6-4}$$

Note that we always use the pure liquid compound i as the reference state. Imagine a simple experiment in which we now bring a dilute "solution" of the compound in air (in which the concentration of compound i is given by its initial partial pressure, P_i^{initial}) in contact with pure water. In the beginning of our experiment, there will be a net flux of i out of the gas phase into the aqueous phase due to the much higher fugacity (or chemical potential) of the compound in the gas phase. (Note that x_w is equal to zero at the beginning of the experiment). Eventually reaching equilibrium, when there are equal chemical potentials or fugacities in both phases, there is no longer a net flux between the two phases. The partial pressure (i.e., the fugacity) of the compound P_i in the gas phase above the aqueous solution will be diminished relative to P_i^{initial} and will now be equal to the fugacity of the compound in the water:

$$P_i = \gamma_w x_w P^o(\mathrm{l, L}) \tag{6-5}$$

If we express the concentration of the compound in water in mole per liter ($C_w \simeq x_w/V_w$), the equilibrium partition constant (the Henry's Law constant, see Eq. 6-1) is then given by (see Fig. 6.1*a*)

$$K_H \equiv \frac{P_i}{C_w} = \frac{\gamma_w x_w P^o(\mathrm{l, L})}{x_w/V_w} = \gamma_w V_w P^o(\mathrm{l, L}) \tag{6-6}$$

Equation 6-6 shows that, as expected, K_H is directly proportional to both the activity coefficient of the compound in water (γ_w) and the vapor pressure of the pure organic liquid [$P^o(\mathrm{l, L})$]. Note that if $P^o(\mathrm{l, L})$ is known, measurements of K_H at dilute concentrations allow an approximate experimental determination of γ_w^∞, the activity coefficient of the compound in aqueous solution at infinite dilution, where we assume that the organic molecules do not "feel" one another.

Effect of Concentration

The problem that worries us next is the concentration dependence of the air–water partition ratio; that is, the question of how K_H changes when we increase the concentration of the organic compound in the system until saturation is reached. Let us imagine a second experiment (Fig. 6.1*b*) in which we connect the pure organic compound via the gas phase with an aqueous phase and then allow the system to equilibrate. Recall from Chapter 4 that at equilibrium the fugacity of the compound in the gas phase will be equivalent to the vapor pressure P^o of the compound (assuming ideal gas behavior) and since $f_g = f_w$, we obtain

$$P^o = \gamma_w x_w P^o(\mathrm{l, L}) \tag{6-7}$$

If we rearrange Eq. 6-7, we recognize that, just as if the pure compound is in direct contact with the water, we now have a saturated aqueous solution of the compound (and a water-saturated organic phase, see Chapter 5):

$$x_w^{sat} = \frac{1}{\gamma_w^{sat}} \cdot \frac{P^o}{P^o(\mathrm{l, L})} \tag{6-8}$$

Note that $P^o = P^o(\mathrm{l, L})$ for a liquid compound, and commonly $P^o \equiv 1$ atm for a gaseous compound. *Thus, the equilibrium partial pressure in air above the saturated aqueous solution of a compound is equal to the vapor pressure of the pure compound at the same temperature.* The air–water distribution ratio, K_H^{sat}, is then

$$K_H^{sat} \equiv \frac{P^o}{C_w^{sat}} = \gamma_w^{sat} V_w P^o(\mathrm{l, L}) \tag{6-9}$$

In Eq. 6-9 we assumed that the molar volume of the aqueous solution is not appreciably altered by the organic solute; that is, we take V_w as the molar volume of pure water (see Chapter 3). Since we also assume ideal gas behavior, our question about the concentration dependence of K_H, or simply how different K_H^{sat} is from K_H at infinite dilution, comes down to a question of how much the aqueous solution activity coefficient γ_w changes as a function of the solute concentration.

In Chapter 5 we envisioned γ_w as a quantification of the partial molar excess free energy of a compound completely surrounded by water molecules as compared with the compound mixed in a liquid of other structurally identical organic molecules. This led us to consider the free energies of cavity and "ice" formation arising from intermolecular interactions of the water molecules surrounding the organic solute. However, if any significant proportion of the organic molecules position themselves adjacent to one another at elevated concentrations, it is easy to see how the intermolecular interactions of cavity and "iceberg" formation will change and cause the activity coefficient to vary. This is, of course, observed with very highly soluble compounds (Prausnitz, 1969); however, our focus here is on compounds that are moderately or even only slightly soluble in water. If for such compounds any solute: solute interaction is important, it is reasonable to say that some of the organic compound in the water will exist as a "dimer." As a consequence, at higher and higher "monomer" concentrations, an increasingly greater proportion of the organic solute is no longer isolated in the water (i.e., completely surrounded by water molecules), and one can predict a decline in γ_w. This, in turn, results in a corresponding decrease in the measured air–water ratio or the "apparent" K_H value. Tucker and Christian (1979) have obtained some experimental evidence for this effect using aqueous benzene solutions. The difference found in air–water ratios between very dilute and saturated solutions (solubility of benzene $0.023\,\text{mol}\cdot\text{L}^{-1}$) was less than 4%. Furthermore, in the few other studies in which air–water ratios have been directly determined, no significant effects of solute concentration on K_H have been observed (e.g., Mackay et al., 1979, for benzene and toluene; Lincoff and Gossett, 1984, Munz and Roberts, 1986, for chlorinated solvents; and Przyjazny et al., 1983, for organosulfur compounds). Prausnitz (1969) suggests a limit of 3 mole % (or $\approx 1.5\,\text{mol}\cdot\text{L}^{-1}$) as the dissolved concentration at which solute:solute interactions can no longer be neglected. Thus, for compounds that are slightly or even moderately soluble in water, with reasonable accuracy we can *approximate* K_H *by* K_H^{sat}; *that is, by the ratio of the compound's vapor pressure and its aqueous solubility.*

Effect of Temperature

Since we have extensively discussed the temperature dependence of P^o and C_w^{sat} in the previous two chapters, the close correlation of K_H with vapor pressure and aqueous solubility immediately enables us to evaluate the temperature dependence of the Henry's Law constant. Recall that over small temperature ranges, we can express the temperature dependence of $P^o(\text{l}, \text{L})$ and γ_w^{sat} $(= 1/x_w^{sat})$ for a liquid compound:

$$\ln P^o = -\frac{\Delta H_{vap}}{R}\cdot\frac{1}{T} + \text{constant} \tag{6-10a}$$

$$\ln \gamma_w^{sat} = +\frac{\Delta H_s^e}{R}\cdot\frac{1}{T} + \text{constant} \tag{6-10b}$$

Neglecting the effect of temperature on V_w, substitution of Eqs. 6-10a and 6-10b into Eq. 6-6 yields an expression for the temperature dependence of K_H^{sat} $(\simeq K_H)$:

$$\ln K_H^{sat} = -\left[\frac{\Delta H_{vap} - \Delta H_s^e}{R}\right]\frac{1}{T} + \text{constant} \tag{6-11}$$

The same result is obtained for gaseous and solid compounds. Note that for solids, $\Delta H_{vap} = \Delta H_{sub} - \Delta H_{melt}$ and $\Delta H_s^e = \Delta H_s - \Delta H_{melt}$. Therefore, since the energy costs of melting the solids cancel, the ΔH_{Henry} for solids is also given by $\Delta H_{vap} - \Delta H_s^e$. Our knowledge of ΔH_{vap} and ΔH_s^e now enables us to make a few statements about the temperature dependence of K_H. As we have seen in Chapter 4 and as indicated by the examples in Table 6.1, the heat of vaporization of a (subcooled, superheated) liquid organic compound is always quite positive and increases with both increasing molecular size and increasing polarity of the compound. The excess heat of solution is generally much smaller, but also increases with molecular size. For some small

TABLE 6.1 Heats of Air–Water Exchange as a Function of Compound Size

Compound	TSA (Å^2)	ΔH_{vap} (kJ·mol^{-1})	ΔH_s^e (kJ·mol^{-1})	ΔH_{Henry}[a] (kJ·mol^{-1})
	Predicted values for aromatic compounds versus TSA			
Benzene	110	31	2	29
Naphthalene(L)	156	43	10	33
Phenanthrene(L)	198	56	18	38
Anthracene(L)	202	56	18	38

Compound	$\bar{V}$ (cm^3·mol^{-1})	T_b (°C)	ΔH_{Henry} (kJ·mol^{-1})
	Observed values for thioesters versus molar volume $\bar{V}$		
Dimethyl sulfide	73	37	0.20
Diethyl sulfide	108	92	0.24
Dipropyl sulfide	141	142	0.22
Diisopropyl sulfide	145	120	0.25

[a] Predicted from ΔH_{vap} minus ΔH_s^e.

and/or polar compounds, ΔH_s^e may be close to zero or may even be slightly negative. Combining our knowledge of these trends in ΔH_{vap} and ΔH_s^e, we realize that ΔH_{Henry} typically will be positive but less than ΔH_{vap}. From these observations, we can conclude that the effect of temperature on the Henry's Law constant at its maximum is of a similar magnitude as the temperature effect on the vapor pressure of the liquid compound (e.g., trichloroethylene in Fig. 6.3). In most cases, however, the effect will be smaller. Furthermore, since ΔH_{Henry} and ΔH_s^e both increase with molecular size, the increase in the heat of air–water partitioning will be more gradual as we look at larger and larger hydrophobic molecules within a single compound class. This type of result is shown in Table 6.1 for several thioesters (Przyjazny et al., 1983). Despite a large range in their size, these compounds have nearly the same ΔH_{Henry} values.

Effect of Salt

From our previous discussions, we can also readily deduce the effect of dissolved salts on air–water partitioning. Since dissolved salts have an effect on γ_w (as discussed for γ_w^{sat} in Chapter 5), but not on the fugacity of the compound in the gas phase, we expect air–water partition constants to vary proportionally (typically directly) with changes in γ_w. Note that at very high salt concentrations, we also have to account for the effect of the dissolved salts on the molar volume of the solution; in many cases, however (e.g., seawater), we may neglect this effect. As mentioned in Chapter 5, seawater solubilities are usually smaller, but within a factor of 2 of distilled water

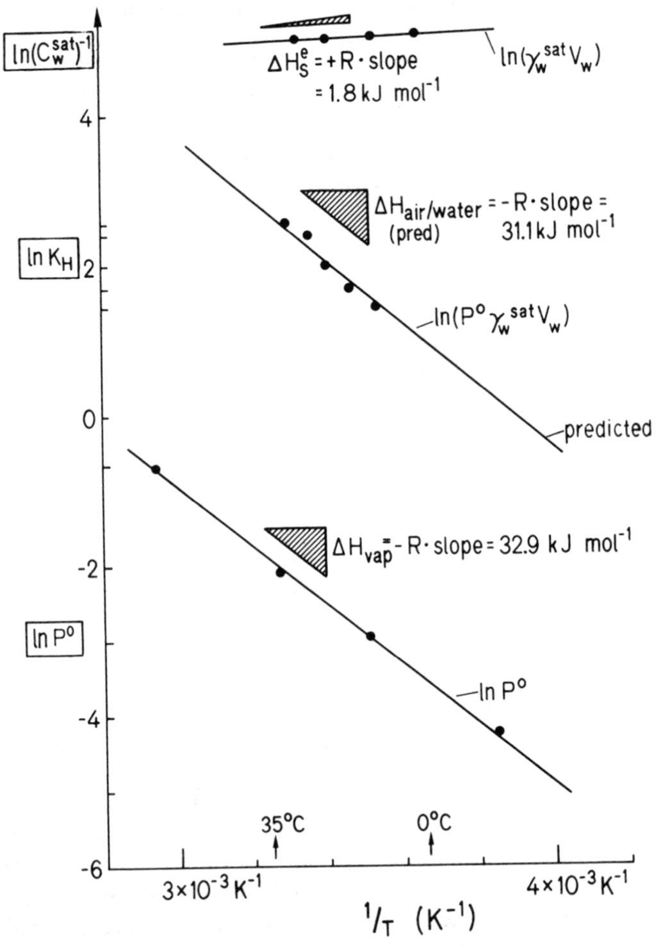

Figure 6.3 Comparison of the observed (shown as data points) temperature dependence of the Henry's Law constant for trichloroethylene with that predicted from the temperature variations of vapor pressure and solubility.

values. The examples given in Table 6.2 show that air–water ratios determined for seawater are generally enhanced to the same degree.

6.3 EXPERIMENTAL DETERMINATION OF AIR–WATER PARTITION RATIOS

By determining the air–water partition ratio, we can gain direct information on the activity ($\gamma_w \cdot x_w$) of a specific compound in a given aqueous phase. Thus, it is of interest not only to know the Henry's Law constant of the compound (i.e., its air–water partition ratio with pure water), but also to determine deviations from K_H, which

TABLE 6.2 Observed Effect of Salt on "Dimensionless" Henry's Law Constants ($mol \cdot L_a^{-1}/mol \cdot L_w^{-1}$) at 25°C

Compound	K'_H (Distilled Water)	K'_H (Seawater)	$K'_H(sw)/K'_H(dw)$	Reference[a]
CCl_3F (F-11)	3.6	5.0	1.4	a
CCl_4	0.98	1.5	1.5	a
CH_3CCl_3	0.53	0.94	1.8	a
Hexachlorobenzene	0.054	0.07	1.3	b
2,4′-Dichlorobiphenyl	0.00713	0.079, 0.0102	1.4	b
2,4,4′-Trichlorobiphenyl	0.00595	0.00885	1.5	b
2,5,3′,4′-Tetrachlorobiphenyl	0.00357	0.00461	1.3	b
Dimethyl sulfide	0.075	0.089	1.2	c
Thiophene	0.095	0.11	1.2	c

[a](a) Hunter-Smith et al., 1983. (b) Brownawell, 1986. (c) Przyjazny et al., 1983.

may be caused by other water constituents present in "real" samples, for example, dissolved organic matter or salts. It is worthwhile, therefore, to spend some time discussing the methodological aspects of determining air–water partition ratios.

Techniques for measuring air–water partition ratios involve determining chemical concentrations in air and water which have been mixed for a long enough time to establish a partitioning equilibrium. A now classical method was introduced by McAuliffe (1971), who used a syringe containing an aqueous solution of toluene (C_w) to which an equal volume of nitrogen gas was added. After vigorously shaking the syringe to establish gas–water equilibrium of the toluene, the nitrogen gas was expelled by advancing the plunger. The toluene content of the gas was then determined by gas chromatographic (GC) analysis. Repetition of the N_2 addition, equilibration, expulsion of the vapor phase, and GC analysis yielded a series of gas samples containing progressively less toluene, since the original concentration of toluene in solution was reduced stepwise. The fraction of toluene remaining each time in the water phase (w) was

$$\text{fraction in water} = \frac{C_w V_w}{C_g V_g + C_w V_w} \tag{6-12}$$

where C_g is the concentration in the gas ($\text{mol} \cdot L_g^{-1}$), V_g is the volume of gas (L_g), C_w is the concentration in the water ($\text{mol} \cdot L_w^{-1}$), and V_w is the volume of water (L_w). Using a dimensionless gas–water partition ratio, $D_{gw} = C_g/C_w$, we find that this fraction is a function of V_g, V_w, and D_{gw} only:

$$\text{fraction in water} = \frac{V_w}{D_{gw} V_g + V_w} \tag{6-13}$$

For pure water, note that $D_{gw} = K'_H$. Consequently, the concentration of toluene measured in the gas phase after the nth equilibration is given by

$$C_{g,n} = D_{gw} \cdot (\text{fraction in water})^n \cdot C_{w,0} \tag{6-14}$$

or using Eq. 6-12 in 6-14 and taking the logarithms of both sides

$$\log C_{g,n} = n \cdot \log \left[\frac{V_w}{D_{gw} V_g + V_w} \right] + \log(C_{w,0} D_{gw}) \tag{6-15}$$

A plot of the log toluene concentration measured in the N_2 versus the number of successive equilibrations has a slope of $\log[(V_w)/(D_{gw} V_g + V_w)]$ and an intercept of $\log(C_{w,0} D_{gw})$ (see Fig. 6.4). Since the volumes V_g and V_w are known precisely, McAuliffe was able to deduce D_{gw} and thereby determine the Henry's Law constant for toluene.

Mackay and his colleagues (1979) extended this approach by using a stripping apparatus. Gas bubbles continuously produced near the bottom of the vessel slowly rise to the surface of the solution and during transit become equilibrated with respect

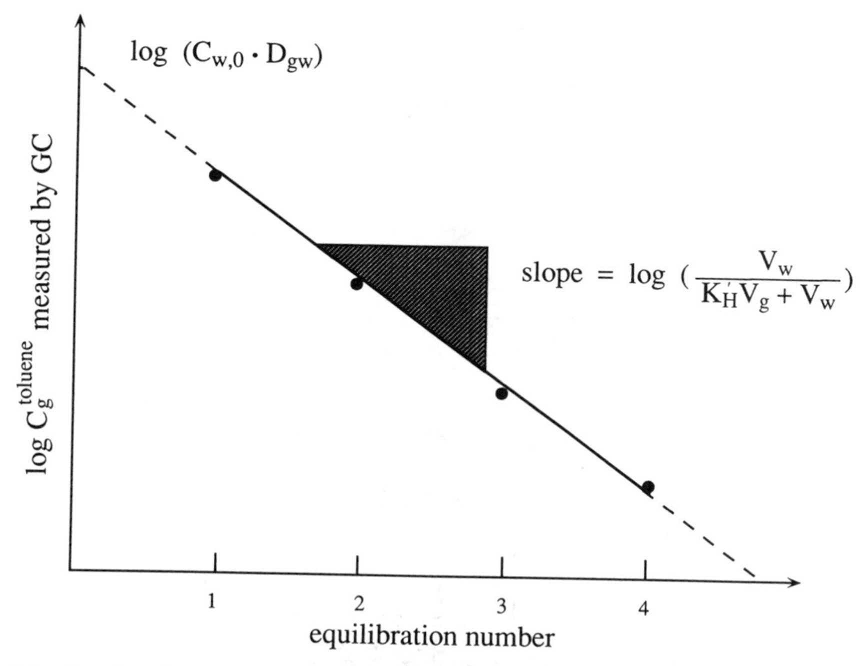

Figure 6.4 Results of successive air–water equilibrations used to calculate Henry's Law constants.

to gas–water partitioning. Consequently, the solution phase concentration diminishes according to

$$C_w = C_w^o \exp[(-D_{gw}V_g^*/V_w)t] \tag{6-16}$$

where V_g^* now quantifies the volume flow per unit time and where t is the accumulated time of stripping. By monitoring the solution phase concentrations (by using UV absorbance in the case of Mackay et al., 1979), one can fit the slope of the log C_w versus t curve to deduce D_{gw}, given known values of V_g^* and V_w. This experimental design requires that the velocity of the rising bubbles be sufficiently small enough and the height of the water column be sufficiently great enough to establish an air–water partitioning equilibrium. Lincoff and Gossett (1984) have shown that we must be very careful with this assumption. Using batch equilibrium measurements with varying ratios of V_g to V_w and assuming air–water equilibrium, they found somewhat larger Henry's Law constants ($\approx 10\%$ for chlorinated solvents) than those previously obtained from stripping experiments.

All of these approaches require accurate knowledge of gas and water volumes; furthermore, only those compounds with suitable D_{gw} (K'_H) values (near ≈ 0.1 mol$\cdot L_g^{-1}$/mol$\cdot L_w^{-1}$) may be investigated using convenient volumes of gases and solutions. As a result, very few experimental K_H determinations have been made for compounds that are not very volatile (i.e., very low P^o) and/or have a low γ_w (high C_w^{sat}).

TABLE 6.3 Comparison of Measured and Estimated (from P^o and C_w^{sat}) Henry's Law Constants

Compound (State at 25°C)	K'_H (Experimental) $\left(\frac{mol \cdot L_a^{-1}}{mol \cdot L_w^{-1}}\right)$	Ref	P° $(mol \cdot L_a^{-1})$	C_w^{sat} $(mol \cdot L_w^{-1})$	K'_H (Estimated) $\left(\frac{mol \cdot L_a^{-1}}{mol \cdot L_w^{-1}}\right)$	Deviation (%)
Dichloromethane	0.119	a	2.4×10^{-2}	2.3×10^{-1}	0.10	+19
Trichloromethane	0.176, 0.160	a, b	1.1×10^{-2}	6.5×10^{-2}	0.17	+3, −6
Tribromomethane	0.0246	b	3.0×10^{-4}	1.2×10^{-2}	0.025	−2
1,1,1-Trichloroethane	0.702	a	6.8×10^{-3}	8.5×10^{-3}	0.80	−14
Trichloroethylene	0.423	a	4.0×10^{-3}	9.1×10^{-3}	0.44	−4
Tetrachloroethylene	0.727	a	1.0×10^{-3}	9.1×10^{-4}	1.1	−51
Trichlorofluoromethane (F-11)(g)	3.72	c	4.1×10^{-2}	1.1×10^{-2}	3.7	0
Dichlorodifluoromethane (F-12)(g)	13.5	c	4.1×10^{-2}	3.0×10^{-3}	13.7	0
Benzene	0.229	d	5.1×10^{-3}	2.3×10^{-2}	0.22	+4
Toluene	0.271	d	1.6×10^{-3}	5.6×10^{-3}	0.29	−7
Ethylbenzene	0.347	d	5.1×10^{-4}	1.6×10^{-3}	0.32	+8
Naphthalene(s)	0.0197	d	4.3×10^{-6}	2.5×10^{-4}	0.017	+14
1-Methylnaphthalene	0.0106	e				

Biphenyl	0.0167, 0.0123	d, e	4.1×10^{-7}	4.5×10^{-5}	0.0091	+ 45, + 26
Fluorene	0.00407	e	3.2×10^{-8}	1.1×10^{-5}	0.0029	+ 29
Anthracene(s)	0.0294	e	3.2×10^{-10}	3.5×10^{-7}	0.00091	+ 97
Phenanthrene(s)	0.00161, 0.00147	d, e	6.6×10^{-9}	6.3×10^{-6}	0.0010	+ 38, + 32
Pyrene(s)	0.000444	e	2.5×10^{-10}	6.8×10^{-7}	0.00037	+ 17
Chlorobenzene	0.154, 0.127	d, e	6.5×10^{-4}	4.5×10^{-3}	0.14	+ 9, − 10
Bromobenzene	0.0996	e	2.3×10^{-4}	2.3×10^{-3}	0.10	− 0.4
1,2-Dichlorobenzene	0.0779	e	8.0×10^{-5}	6.3×10^{-4}	0.13, 0.082	− 67, + 5
				9.8×10^{-4}		
1,4-Dichlorobenzene(s)	0.0968	e	3.7×10^{-5}	4.1×10^{-4}	0.090	+ 7
1,2,3,5-Tetrachlorobenzene(s)	0.0641	e	4.0×10^{-6}	1.5×10^{-5}	0.26	− 306
Hexachlorobenzene(s)	0.054 (23°C)	f	1.3×10^{-9}	2.0×10^{-8}	0.065	− 20
2,2′,5,5′-Tetrachlorobiphenyl(s)	0.038 (23°C)	f	1.0×10^{-9}	8.7×10^{-8}	0.011	+ 71

[a] Lincoff and Gossett, 1984.
[b] Nicholson et al., 1984.
[c] Wisegarver and Cline, 1985.
[d] Mackay et al., 1979.
[e] Mackay and Shiu, 1981.
[f] Atlas et al., 1982.

TABLE 6.4 Comparison of Henry's Law Constants Estimated by Structural Unit Contributions to Other Values Measured or Estimated (by P^o and C_w^{sat}) at 25°C[a]

Bond	Contribution	Bond	Contribution
	Contributions to the Logarithms of K'_H		
C—H[b]	+0.11	C_{ar}—Br	−0.21
C—F	+0.50	C_{ar}—NO_2[c]	−1.83
C—Cl	−0.30	C_{ar}—O	+0.74
C—Br	−0.87	C_{ar}—S	−0.53
C—I	−1.03	C_{ar}—CO[d]	−1.14
C—CN[c]	−3.28	C_{ar}=C_{ar}[e]	−0.33
C—NO_2[c]	−3.10	C_{ar}=N_{ar}[e]	−1.64
C—O	−1.00	C_d—H	+0.15[f]
C—S	−1.11	C_d—Cl	−0.16[f]
C—N	−1.35	C_d—C_d	−0.48[f]
C—C	−0.04	C_d—CO[d]	−2.24[f]
C—CO[d]	−1.78	C_t—H[g]	−0.00
C—C_d	−0.15[f]	CO—H[d]	−1.19
C—C_t[g]	−0.64	CO—O[d]	−0.28
C—C_{ar}	−0.11	O—H	−3.21
C_{ar}—H	+0.21	S—H	−0.23
C_{ar}—Cl	+0.14	N—H	−1.34

[a]After Hine and Mookerjee, 1975.
[b]C without a subscript refers to a carbon atom bound by single bonds to four other atoms except in CN.
[c]The cyano and nitro groups are treated as univalent atoms.
[d]The CO group is treated as a divalent atom.
[e]The double bond is the $\sigma + \pi$ bond in an aromatic ring.
[f]This contribution includes one-fourth the contribution of the carbon–carbon double bond(s).
[g]This contribution includes one-half the contribution of the carbon–carbon triple bond.

Example Compound	Contributions
1. Bromodichloromethane	1 (C—H) + 1 (C—Br) + 2 (C—Cl)
	1 (+0.11) + 1 (−0.87) + 2 (−0.30)
	$\log(K'_H) = -1.36$
	$K'_{H\,estim} = 0.044 \dfrac{mol \cdot L_a^{-1}}{mol \cdot L_w^{-1}}$; $K'_{H\,obs} = 0.085$ Nicholson et al., 1984
2. Phenol	$6(C_{ar}{=}C_{ar}) + 5(C_{ar}{-}H) + 1(C_{ar}{-}O) + 1(O{-}H)$
	6(−0.33) + 5(+0.21) + 1(+0.74) + 1(−3.21)
	$\log(K'_H) = -3.40$
	$K'_{H\,estim} = 0.00040 \dfrac{mol \cdot L_a^{-1}}{mol \cdot L_w^{-1}}$; $K'_{H\,calc'd} = 0.00041$ from P° and soly

6.4 AVAILABILITY OF K_H DATA—ESTIMATION METHODS

Mackay and Shiu (1981) reviewed the literature (through 1978) for Henry's Law constants of organic compounds of environmental interest. Nearly all of the data is based on estimation techniques since, as noted before, for many compounds, experimental measurements of equilibrated air and water volumes are difficult to make. The estimation method most commonly used involves the calculation of K_H values from vapor pressure and solubility data. Table 6.3 compares some measured Henry's Law constants with values estimated by calculation. For lower molecular weight compounds with relatively easily measured vapor and solution concentrations, the calculated values differ from experimental ones by about 10% or less; those values showing the greatest divergence (tetrachloroethylene, anthracene, phenanthrene, and 1,2,3,5-tetrachlorobenzene) suggest that one or another of the measured parameters (P^o, C_w^{sat}, or K_H) is in error! Alternatively, one must conclude that there is a problem with the theory!

Hine and Mookerjee (1975) have proposed a K_H estimation method based on structural contributions. Its underlying idea is that each subunit of organic compounds (e.g., all C—H bonds) has a substantially constant effect on air–water partitioning regardless of the substance in which it occurs. Viewed in this way, the Henry's Law constant may be calculated for any compound by simply summing the contributions of each of its structural parts. Hine and Mookerjee have collected a large number of Henry's Law constants and by setting up linear algebraic equations for each structure following the form

$$\log K_H = a_1(\text{subunits of type } a_1) + a_2(\text{subunits of type } a_2) + \cdots$$

have solved for the best fits of the coefficients a_1, a_2 and so on. Table 6.4 shows the results of such calculations when the organic molecules are structurally subdivided by bond type. Most of the symbols are self-explanatory; for example, C—H is a singly bonded carbon–hydrogen subunit; C_{ar}—Cl is a chlorine bound to an aromatic carbon; and C—C_d is a carbon bound to a vinylic carbon. Some groups, such as the carbonyl group (C=O), are treated as a single "atom." Just looking at the signs and values of the bond contribution, we readily see that units such as C—H bonds tend to encourage molecules to partition into the air, while other units like O—H groups strongly induce molecules to remain associated with the water. These tendencies correspond to expected behaviors deduced qualitatively from considerations of intermolecular interactions with water. At the bottom of Table 6.4, two sample calculations are performed, and the results are compared with K'_H values obtained by other means. This simple bond contribution approach is usually accurate to within a factor of 2 or 3. One major drawback, however, is that it does not account for special intermolecular interactions that may be unique to the molecule in which a particular bond type occurs. Hine and Mookerjee (1975) demonstrated that some such errors can be alleviated by considering larger subunits [e.g., CH_3 rather than 3·(C—H)] *as the currency of contributions.* However, when there are multiple polar units in a single organic structure, intramolecular interactions, in particular, may become very complex, and the estimated Henry's Law constants are correspondingly less accurate.

CHAPTER 7

ORGANIC SOLVENT–WATER PARTITIONING: THE OCTANOL–WATER PARTITION CONSTANT

7.1 INTRODUCTION

So far, we have dealt with cases of equilibrium partitioning of organic compounds between two phases in which in at least one of the phases considered (i.e., gas phase, pure organic liquid) the compound showed approximately ideal behavior. In our next case, we discuss the partitioning behavior of organic compounds between two *immiscible* liquids—water and a liquid organic compound that is not identical to the pure liquid compound. In this organic phase we cannot expect ideal solution behavior a priori.

As we will see in later chapters, the distribution of nonpolar organic compounds between water and natural solids (e.g., soils, sediments, and suspended particles) or organisms can in many cases be viewed as a partitioning process between the aqueous phase and the bulk organic matter present in natural solids or in biota. As early as 1900, investigators studying the uptake of nonpolar drugs by organisms discovered that they could use water-immiscible organic solvents like *n*-octanol as a surrogate for organisms or parts of organisms insofar as accumulation of these pharmaceutically important organic molecules from the water was concerned (Meyer, 1899; Overton, 1899). Although the extent of uptake from water into these solvents was not identical to that into organisms, it was directly proportional; that is, within a series of compounds, higher accumulation into an organism corresponded to more favorable partitioning into the organic solvent. More recently, environmental chemists have found similar correlations with soil humus and other naturally occurring organic phases.

These correlations exist because the same molecular factors controlling the distribution of compounds between water-immiscible organic solvents and water also determine environmental partitioning from water into natural organic phases. Thus, by directing our attention in this chapter to organic solvent–water partitioning, we will: (1) identify the molecular factors controlling this process; (2) gain insight into how these factors differ from compound to compound; and (3) assemble the quantitative basis for predicting the partitioning of organic compounds between natural organic matter and water.

In a fashion analogous to air–water partitioning, we can envision the molecules of a given compound partitioning between the organic phase and the aqueous phase. This partitioning process is determined by the relative fugacity of the compound in each phase and at equilibrium may be described by a "dimensionless" equilibrium constant:

$$K_{\mathrm{sw}} = \frac{C_{\mathrm{s}}}{C_{\mathrm{w}}} (\mathrm{mol} \cdot \mathrm{L}_{\mathrm{s}}^{-1}\, \mathrm{mol}^{-1} \cdot \mathrm{L}_{\mathrm{w}}) \tag{7-1}$$

where C_s is the concentration of the compound in the organic phase (here, the organic solvent), and C_w is the concentration in the water.

7.2 THERMODYNAMIC CONSIDERATIONS AND MOLECULAR INTERPRETATION OF SOLVENT–WATER PARTITIONING

To understand the factors that influence the partitioning behavior of neutral organic solutes between an organic solvent and water, let us again imagine a simple experiment. We add water and a water-immiscible organic solvent to a flask, shake it vigorously, and then wait for the two phases to separate. From our discussion in Section 5.2, we realize that even before we add the compound for which we want to assess the partitioning behavior between the two phases, some important things have already happened in our system, insofar as at equilibrium, the two phases will be mutually saturated with one another (Fig. 7.1*a*). When we introduce our compound into the system, it will find a somewhat different molecular environment in each of the liquid phases than it would encounter if we had dissolved it in pure organic solvent or in pure water. For example, if we use *n*-octanol as the organic solvent, at equilibrium there will be roughly one water molecule for every four octanol molecules in the organic phase (see Table 5.1). Conversely, as pointed out in Section 5.2, owing to the difficulty of dissolving organic molecules in water, there will only be about eight octanol molecules for every 100,000 water molecules in the aqueous phase. In certain cases, therefore, we must be prepared to include the effects of this crossover insofar as intermolecular interactions of solute molecules with different solvent molecules are concerned. Furthermore, for some organic solvents, the presence of appreciable amounts of water will have an effect on the molar volume, V_s, of the organic phase. At 25°C, water-saturated *n*-octanol, for example, has a molar volume of $0.12\,\mathrm{L \cdot mol^{-1}}$ compared with $0.16\,\mathrm{L \cdot mol^{-1}}$ for pure *n*-octanol. On the other hand, if we choose a very nonpolar solvent such as *n*-hexane, we can neglect the effect of the water on the

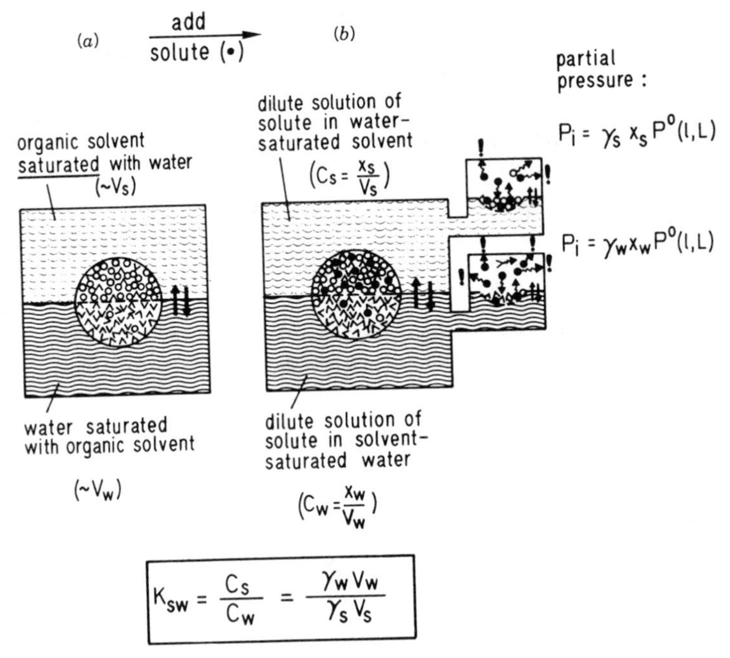

Figure 7.1 Equilibrium partitioning of a compound between an immiscible organic liquid and water. (*a*) Mutually saturated organic and aqueous phases. (*b*) System after solute is added.

molar volume (the mole fraction of water in *n*-hexane is less than 0.001; see Table 5.1). Similarly we can approximate the molar volume of octanol-saturated water by the molar volume of pure water.

Let us now get back to our experiment. We add a known amount of our compound into the aqueous phase (initially $x_w \gg x_s$). Because of the much higher chemical potential of the compound in the aqueous phase, there will be a net flux of the compound from the aqueous phase to the organic phase until equilibrium is reached; that is, as we recall from Chapter 3, until the fugacity of the compound is equal in both phases (see Fig. 7.1*b*). As indicated in Fig. 7.1, we could monitor the fugacity of the compound in each phase by determining its partial pressure in the headspace above each solution.

At equilibrium ($f_s = f_w$), we obtain (see Fig. 7.1)

$$\gamma_s x_s = \gamma_w x_w \tag{7-2}$$

or, in terms of molar concentrations,

$$\gamma_s C_s V_s = \gamma_w C_w V_w \tag{7-3}$$

Our partition constant K_{sw} (Eq. 7-1), which we can determine by measuring the equilibrium concentration of the compound in each phase, is therefore related to the activity coefficients of the compound in each phase by

$$K_{sw} = \frac{C_s}{C_w} = \frac{\gamma_w}{\gamma_s} \cdot \frac{V_w}{V_s} \tag{7-4}$$

In the previous chapters (especially Chapter 5), we have interpreted the quantity ($RT \ln \gamma$) to be the partial molar excess free energy (g^e) of the compound in a given phase; that is, as a measure of the nonideality of a solution of the chemical in a phase of dissimilar substances compared with a solution in a liquid of identical molecules. Thus, our partition constant, K_{sw}, in essence expresses the difference between the nonideality of the solution of the compound in the organic solvent and in the aqueous phase:

$$\ln K_{sw} = \ln \gamma_w - \ln \gamma_s + \ln \frac{V_w}{V_s} = -\frac{\Delta G^e_{sw}}{RT} + \text{constant} \tag{7-5}$$

For nonpolar compounds, we already know that the activity coefficient in water, γ_w, can be quite large (i.e., 10^2–10^{11}; see Table 5.2) owing to the high free-energy costs of what we have referred to as cavity and iceberg formation (see Fig. 5.2). In contrast, in organic phases, in particular for nonpolar solutes in nonpolar solvents, we would intuitively expect very small γ_s values (even close to 1), since the compound molecules should experience very similar interactions to those in their pure liquid phase (where we have set $\gamma_{\text{pure liquid}} = 1$). Consequently, we expect the magnitude of K_{sw} of nonpolar solutes to be dominated primarily by γ_w. It should come as no surprise that we will find increasing K_{sw} values with decreasing (liquid) water solubilities. We would also expect that compounds with polar functional groups, which favor the aqueous phase and simultaneously disfavor very nonpolar organic phases (e.g., *n*-hexane), to have very low K_{sw} values in such solvent–water systems. But let us look at this question in a more quantitative fashion. To this end, we make two assumptions:

1. First, we assume that the activity coefficient of the partitioning chemical is independent of its concentration in the aqueous phase. In Chapter 6 we indicated that this assumption is reasonable for a pure aqueous phase for most compounds that are of interest to us. In other words, even at saturation concentrations (γ_w^{sat}), the likelihood of two solute molecules positioning themselves near one another is so small that we may neglect such interactions. It seems reasonable then to extend this

TABLE 7.1 *n*-Octanol–Water and *n*-Hexane–Water Partition Constants at 25°C for Some Organic Compounds[a]

Partitioning Compound	K_{ow} $\left[\frac{(\text{mol}\cdot\text{L octanol}^{-1})}{(\text{mol}\cdot\text{L water}^{-1})}\right]$	K_{hw} $\left[\frac{(\text{mol}\cdot\text{L hexane}^{-1})}{(\text{mol}\cdot\text{L water}^{-1})}\right]$	$C_w^{sat}(\text{l, L})$ $(\text{mol}\cdot\text{L water}^{-1})$	γ_o	γ_h
n-Hexane	13,000	52,000	1.5×10^{-4}	4.4	1.0
Benzene	130	170	2.3×10^{-2}	2.7	2.0
Toluene	490	560	5.6×10^{-3}	3.0	2.4
Chlorobenzene	830	810	4.5×10^{-3}	2.2	2.1
Naphthalene	2,300	2,400	8.7×10^{-4}	4.2	3.7
Benzaldehyde	30	13	3.1×10^{-2}	8.9	19.0
Nitrobenzene	68	29	1.7×10^{-2}	7.3	16.0
1-Hexanol	34	2.8	1.3×10^{-1}	1.9	21.0
Aniline	7.9	0.8	3.9×10^{-1}	2.7	25.0
Phenol	28	0.1	8.9×10^{-1}	0.3[b]	61.0
Water	0.04	0.00005	5.5×10^{1}	3.6	2600

[a]Also shown are the liquid aqueous solubilities and the organic phase activity coefficients calculated using Eq. 7-10 or 7-11. (Data taken from Hansch and Leo, 1979.)

[b]Value probably incorrect since $\gamma_w^{sat} \neq \gamma_w^{\infty}$ because of intermolecular interactions of the phenol species at saturation concentrations.

assumption to an aqueous solution that contains only very small amounts of organic solvent.

2. The second assumption we make is that the organic solvent molecules present in the water do not affect the activity coefficient of the compound in the aqueous phase; hence, we do not worry about the interactions of the organic solvent molecules with our solute—interactions that would lower the compound's activity coefficient in the aqueous phase (see Chapter 5). For certain solvents (i.e., solvents with appreciable water solubility like *n*-octanol), the effect of the solvent molecules on γ_w may be of some significance for very hydrophobic compounds.

Using our two assumptions, we can now approximate the activity coefficient of the compound at dilute concentrations in the *solvent-saturated* aqueous phase by γ_w^{sat}, the activity coefficient of the compound *at saturation* in *pure* water. As we have seen in Chapter 5, we can deduce γ_w^{sat} from the aqueous solubility $C_w(\mathrm{l, L})$ of the liquid organic compound. Substituting $\gamma_w V_w$ with $\gamma_w^{sat} V_w = [C_w^{sat}(\mathrm{l, L})]^{-1}$ in Eq. 7-4 yields

$$K_{sw} = \frac{1}{C_w^{sat}(\mathrm{l, L})} \cdot \frac{1}{\gamma_s} \cdot \frac{1}{V_s} \tag{7-6}$$

Table 7.1 contains experimentally determined partition constants for a series of organic compounds of different polarities in two common organic solvent–water systems, together with the water solubilities of the liquid compounds. As we have intuitively anticipated, nonpolar compounds (e.g., *n*-hexane) with low water solubilities (i.e., high γ_w^{sat} values) strongly favor the organic phase. Furthermore, from the few examples of nonpolar compounds given in Table 7.1 (compounds 1–5), we already recognize that such compounds feel equally comfortable in the very nonpolar solvent *n*-hexane, and in the much more polar solvent, *n*-octanol. We also see that the activity coefficients of the compounds in these solvents are less than 5. Since water-saturated *n*-octanol ($V_s = 0.12\,\mathrm{mol \cdot L^{-1}}$) and other (water-saturated) solvents have similar molar volumes, it is not surprising that for a given *nonpolar compound*, partition constants in various nonpolar or moderately polar organic solvent–water systems are often of similar magnitude.

For more polar substances, particularly compounds with oxygen or nitrogen functional groups, the ability of the solvent molecules to accommodate polar groups is important. As the examples given in Table 7.1 show, there are significant differences between the K_{ow} and K_{hw} values, mostly due to the inability of the hexane molecules to develop polar interactions with the polar group, in particular hydrogen bonds. This reduced solvent:solute attraction is expressed by a larger activity coefficient, γ_h. In contrast, the amphiphilic (i.e., part lipophilic and part hydrophilic) solvent *n*-octanol is quite versatile and can accommodate rather indiscriminantly a diverse group of organic solutes. It is interesting to note that benzene and water have very similar activity coefficients in octanol, while in hexane the coefficients differ by more than three orders of magnitude (Table 7.1). Apparently, the *n*-octanol solvent molecules use their nonpolar end to interface diplomatically with nonpolar molecules (or parts of molecules), but turn around and use their polar alcohol moieties to provide

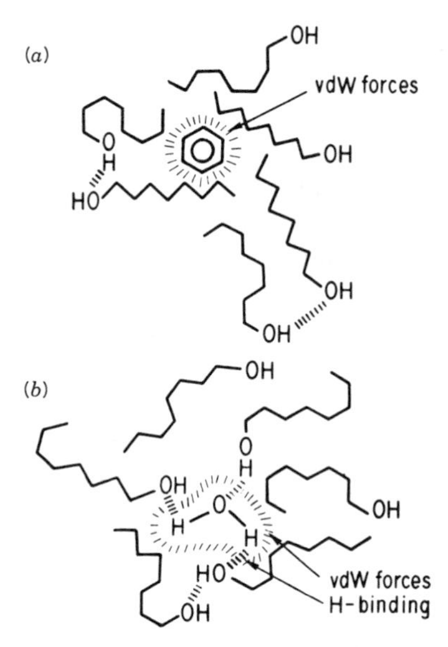

Figure 7.2 Schematic representation of solvent:solute interactions showing the versatility of amphiphilic solvents such as *n*-octanol. (*a*) Benzene solute molecule surrounded by octanol solvent molecules. Solute:solvent interaction chiefly of van der Waals type like those when a benzene molecule is in benzene liquid. (*b*) Water solute molecule surrounded by octanol solvent molecules. Solute:solvent interaction chiefly of hydrogen-bonding and dipole–dipole types like those when a water molecule is in water.

favorable intermolecular interactions with polar groups (see Fig. 7.2). In summary, the data in Table 7.1 suggest that the major factor that determines the magnitude of the partition constant of a nonpolar or moderately polar organic compound between an organic solvent and water is the incompatibility of the compound with water and that the nature of the organic solvent is generally of secondary importance. We shall now explore these factors in more detail for a particular organic solvent–water system, the *n*-octanol–water system, which has been investigated extensively.

7.3 THE *n*-OCTANOL–WATER PARTITION CONSTANT (K_{ow}) OF NEUTRAL ORGANIC COMPOUNDS

Partly because of the choices of early workers such as Overton (1899) and Meyer (1899) and the special amphiphilic nature of C_4 to C_{10} alcohols, but mostly because *n*-octanol is a reasonable surrogate for many kinds of environmental and physiological organic matter, *n*-octanol has become the most popular reference phase for assessing the organic phase–water partitioning behavior of organic solutes. Figure 7.3 shows the ranges of K_{ow} values for some common xenobiotic compound classes. In light of the previous discussion, it is not surprising that, as we have already seen with the water

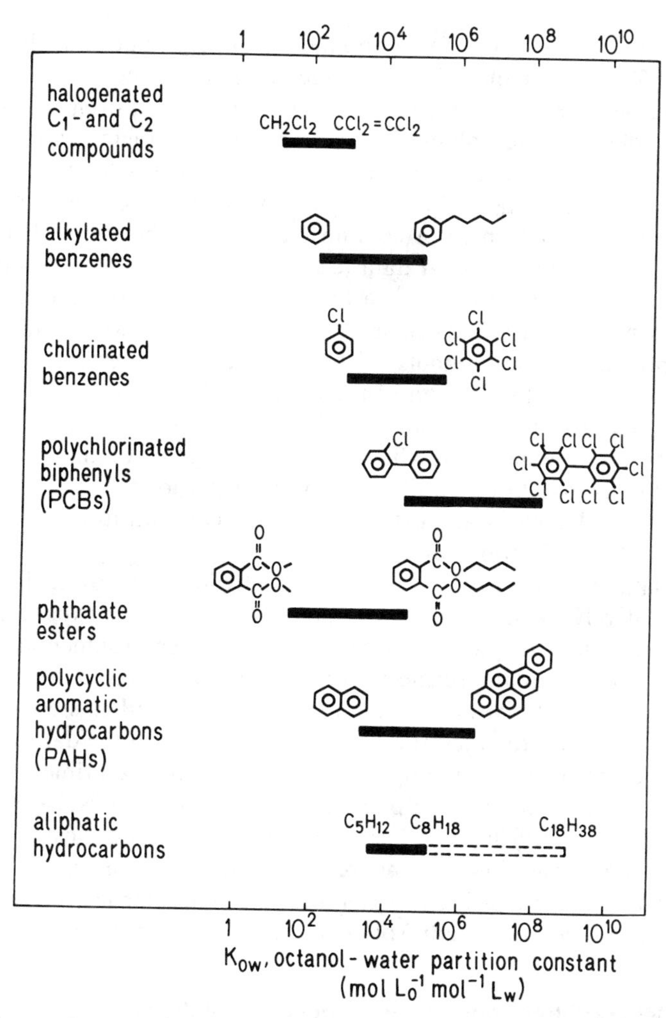

Figure 7.3 Ranges in octanol–water partition constants (K_{ow}) for some important classes of organic compounds

solubilities (Fig. 5.1), these constants may vary over many orders of magnitude within a given compound class.

Availability of Experimental Octanol–Water Partition Constants

Before we continue our discussion about the factors that determine the K_{ow} value of a compound, a few comments on the availability and accuracy of experimental octanol–water partition constants are necessary. Traditionally, direct measurements of K_{ow} have been carried out by using conventional "shake flask" methods (Leo et al., 1971; OECD, 1981a), not unlike our simple experiment in Section 7.2. This experimental approach is restricted, however, to compounds with K_{ow} values less than about 10^5 (i.e., $\log K_{ow} < 5$), since for more hydrophobic compounds the concentration in the aqueous phase becomes too low to be accurately measured even when using very small octanol volume:water volume ratios. Consequently, there is a large data base for compounds of low-to-medium hydrophobicity (e.g., Hansch and Leo, 1979), but reliable measurements of K_{ow} values of highly hydrophobic compounds are scarce. Only recently, the use of "generator columns" coupled with solid sorbent cartridges has allowed determination of such data (e.g., Tewari et al., 1982; Woodburn et al., 1984). Briefly, large volumes of octanol-saturated water (up to 10 L) are passed through small columns packed with beds of inert supports coated with octanol solutions (typically 10 mL) of the compound(s) of interest. As the water passes through the column, an equilibrium distribution of the compound(s) is(are) established between the octanol and water. By retrieving the chemical of interest with a solid sorbent cartridge from large volumes of the effluent water leaving the column, enough material may be accumulated to allow quantification of the water load. This result, along with knowledge of the volume of water extracted and the concentration of the compound in the octanol, ultimately provides the K_{ow} value.

Since K_{ow} data are commonly measured by investigators evaluating the partitioning behavior of relatively small groups of compounds, much of the experimental data is scattered throughout the pharmaceutical, analytical, environmental engineering, chemical engineering, aquatic sciences, and toxicology literature. A compilation of K_{ow} values for representatives of some important compound classes is given in the Appendix, together with the literature references from which the data were taken. It should be noted that there is still appreciable scatter in the experimental K_{ow} values, especially for very hydrophobic compounds. Furthermore, in most cases, the concentration dependence of K_{ow} has not been evaluated, which is critical for compounds that tend to self-associate in either phase. Nevertheless, the available experimental data allow us to gain interesting insights into the molecular factors that determine the extent to which a neutral organic solute partitions between water and *n*-octanol.

Octanol–Water Partition Constant and Aqueous Solubility

In Fig. 7.4 *experimental* K_{ow} values of numerous compounds representing a variety of different compound classes have been plotted versus their liquid aqueous solubilities. The $C_w^{sat}(L)$ values of the solids have been calculated from *experimental* solubilities and melting-point data using Eqs. 5-8 and 4-19.

Let us first compare the experimental data in Fig. 7.4 with the solid lines, which represent the logarithmic form of Eq. 7-6:

$$\log K_{ow} = -\log C_w^{sat}(l, L) - \log \gamma_o - \log V_o \tag{7-7}$$

where the activity coefficient in octanol (γ_o) has been set equal to 1 (heavy line) and to 10, 100, and 1000. Recall that when expressing K_{ow} by Eq. 7-7, we assume that the activity coefficient of a compound is independent of concentration in both phases, and that the influence of the octanol in the water on the γ_w is negligible; that is, γ_w

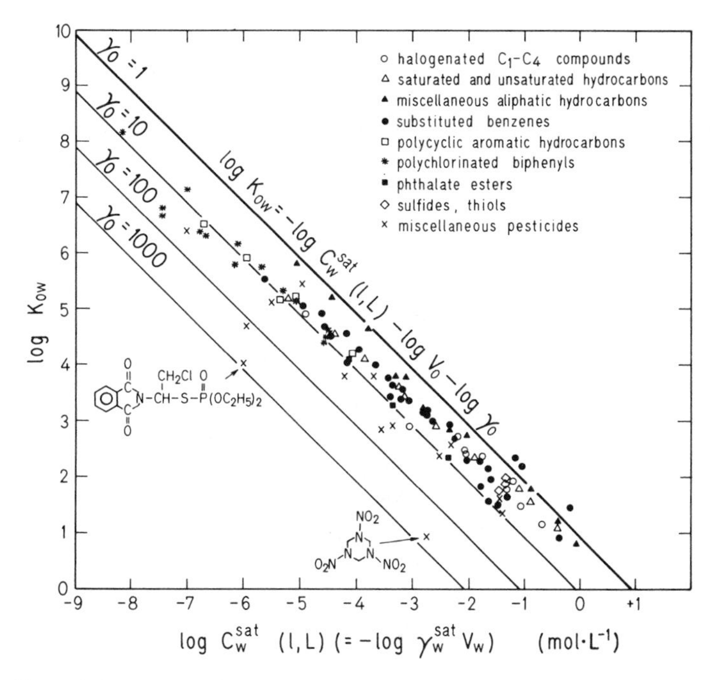

Figure 7.4 Plot of log octanol–water partition constants versus log (subcooled) liquid aqueous solubilities for a variety of organic compound classes. Diagonal lines show the results using Eq. 7-7 with various values for γ_o.

in octanol-saturated water is equal to $[C_w^{sat}(l, L)V_w]^{-1}$. Furthermore, we assume that the concentrations of the compounds used in the experiments are low enough so that the solutes have no impact on the molar volume of the water-saturated octanol phase (V_o), which we have set equal to $0.12\,L\cdot mol^{-1}$. With all of these assumptions, we now interpret the difference between the solid line and a given data point in Fig. 7.4 as the logarithm of the activity coefficient of that compound in the water-saturated octanol phase. Note that all of the γ_o values in Table 7.1 have been derived in this way.

From the experimental data in Fig. 7.4, we see that for most (liquid) compounds exhibiting water solubilities of greater than about $10^{-6}\,mol\cdot L^{-1}$, the calculated values for γ_o are between 1 and 10, indicating nearly ideal solution behavior of these compounds in water-saturated octanol as compared with their solution behavior in water (where activity coefficients are much larger). Again, we stress the point that for such chemicals, which primarily include nonpolar compounds and compounds with only one polar functional group, their incompatibility with water (i.e., their hydrophobicity) is by far the most important factor in determining their octanol–water partitioning behavior. This dislike for water also dominates their partitioning behavior between natural organic phases and water.

The few data available for highly hydrophobic compounds (i.e., PCBs, PAHs) indicate that, with increasing size of the molecules (i.e, decreasing liquid aqueous solubility), the incompatibility of a compound with water-saturated octanol and hence γ_o seems to increase. However, as Chiou et al. (1982, 1983) have demonstrated for hexachlorobenzene (HCB) and 1,1,1,-trichloro-2,2′-bis(*p*-chlorophenyl)ethane (DDT), part of the deviation in the experimental data from the line of ideal solution behavior in octanol ($\gamma_o = 1$ line in Fig. 7.4) is due to the influence of even the limited amount of octanol in the water on the activity coefficients of such highly hydrophobic compounds in the aqueous phase. Chiou et al. (1982, 1983) found that the solubilities of HCB and DDT were 1.9 (HCB) and 2.8 (DDT) times higher in octanol-saturated water as compared with pure water, demonstrating the increasing effect of the organic cosolute (octanol) with increasing hydrophobicity of the compound. Taking the effect of the octanol molecules dissolved in the aqueous phase into account, the calculated "experimental" values for γ_o (rearranging Eq. 7-7) would be 6 and 13, respectively. These values compare reasonably well with the γ_o^{sat} values of 5 for HCB and 8 for DDT, as estimated by Chiou et al. from measurements of the solubilities of these compounds in water-saturated octanol and using knowledge of their melting points (i.e., as in Eqs. 5-8 and 4-19). Nevertheless, the conclusion that with increasing molecular size the incompatibility of nonpolar compounds with octanol increases and results in higher γ_o values remains valid.

When dealing with compounds with a variety of polar functional groups, the incompatibility of the molecule with the water-saturated octanol phase may become a significant factor in determining the partitioning behavior of the compound. As illustrated by several polar pesticides (indicated by the symbol "x" in Fig. 7.4), such solutes apparently can no longer be as completely accommodated by the octanol (and water) molecules as nicely as shown for ''simple'' solutes in Fig. 7.2. One extreme case in the set of data shown in Fig. 7.4 is the pesticide RDX (hexahydro-1,3,5-trinitro-1,3,5-triazine) which has a calculated γ_o of about 1000. From these findings we may conclude that when dealing with multifunctional polar compounds γ_o may become quite large. For such compounds, one has to be very careful when estimating K_{ow} values from liquid aqueous solubility data and vice versa, as discussed below.

7.4 APPROACHES USING CHEMICAL STRUCTURE TO RELATE AND ESTIMATE PARTITION CONSTANTS

Linear Free-Energy Relationships (LFERs)

Since the partitioning of organic compounds between organic phases and water is of interest in many fields of research, considerable efforts have been and are being made to develop the means to estimate partition constants from chemical structure (structure–property relationships) and to relate partition constants between different organic phase–water systems (linear free-energy relationships, LFERs). We have already used a structure–property relationship in Chapter 6 to estimate Henry's Law constants by summing the contributions of the structural parts from which organic molecules are assembled. Possibly, the most extensively developed structure–property relationship is that for estimating the K_{ow} of a compound (Hansch and Leo, 1979).

In this section we explore this particular correlation in depth, not only to provide a very useful tool for estimating K_{ow} values, but also to serve as an example of the "chemical thinking" that goes into such approaches. This exercise demonstrates that the behavior of organic molecules may be quantified because their subparts interact both with one another and with their surroundings in consistently characteristic ways. Additionally, we explore the utility of LFERs when attempting to estimate the free energy of transfer of an organic solute between *n*-octanol and an aqueous phase based on some knowledge of the free energy of partitioning the same solute between a different organic phase and water. Linear free-energy relationships are not only extremely useful tools for evaluating and relating partition constants, but also form the basis from which other equilibria (e.g., acid–base equilibrium constants) and kinetic data (e.g., hydrolysis rate constants) can be organized and used to estimate properties of compounds.

Let us return to solvent–water partition constants. Recall that the equilibrium partition constant of an organic compound for a given solvent–water system is related to the difference between the partial molar excess free energy of the compound in the organic solvent and that in the aqueous phase, $\Delta G^{e}_{sw} = g^{e}_{s} - g^{e}_{w}$ (see Eq. 7-5):

$$\log K_{sw} = -\frac{\Delta G^{e}_{sw}}{2.303RT} + \log\frac{\bar{V}}{\bar{V}_s} \tag{7-8}$$

We will refer to $\Delta G_{sw} = \Delta G^{e}_{sw} - 2.303RT \log \bar{V}_w/\bar{V}_s$ as the "free energy of transfer" in the particular solvent–water system. To better understand how structure–property rela-

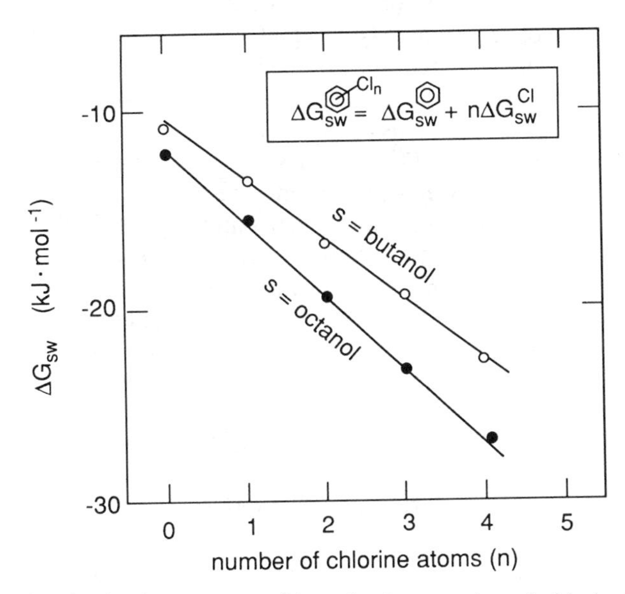

Figure 7.5 Variation in the free energy of transfer for a series of chlorinated benzenes in the *n*-butanol–water and *n*-octanol–water systems.

tionships work in any one solvent–water system or how LFERs work in relating partition constants in different solvent–water systems, we will first consider a specific example—the partitioning of a series of chlorobenzenes between octanol and water and between butanol and water (Westall, 1983). Data for ΔG_{sw} are plotted in Fig. 7.5 for both solvent–water systems as a function of the number of chlorine substituents on the benzene ring. As may be seen, in both solvent–water systems for this series of compounds, ΔG_{sw} decreases regularly for each chlorine added. Therefore, there is a clear and consistent contribution to the free energy of transfer made for each incremental change in its structure. This type of observation leads us to expect that we can predict ΔG_{sw} from the structural parts of the compound of interest, hence, the basis for a structure–property relationship. For both solvent–water systems (where s = *o*ctanol and *b*utanol, respectively), we can then write

$$\begin{aligned} \Delta G_{ow}(\phi Cl_n) &= \Delta G_{ow}(\phi) + n \cdot \Delta G_{ow}(Cl) \\ \Delta G_{bw}(\phi Cl_n) &= \Delta G_{bw}(\phi) + n \cdot \Delta G_{bw}(Cl) \end{aligned} \tag{7-9}$$

where $\Delta G_{sw}(\phi)$ corresponds to the free-energy transfer of the unsubstituted benzene ring, and $\Delta G_{sw}(Cl)$ is the incremental change in ΔG_{sw} when substituting one hydrogen atom with a chlorine atom. Each of these equations states that the free energy of transfer is related to a linear combination of the contributions of the structural parts.

Figure 7.5 also shows that the effect of exchanging a $-H$ for a $-Cl$ is less pronounced for partitioning into butanol than into octanol. Since water-saturated butanol (9.4 moles H_2O per liter solution) is a significantly more polar solvent than water-saturated octanol (2.3 moles H_2O per liter solution), we can understand that the incremental free-energy change contributed by the hydrophobic chlorine is smaller in the butanol–water system than in octanol–water partitioning and that the free energy of the benzene nucleus transfer into butanol is somewhat less negative than that into octanol from water. Despite these partitioning system differences, the information from one system yields insight into the other.

To understand both why and under what conditions LFERs relating partition constants in different nonaqueous phase–water systems exist, we first develop a generalization of Eq. 7-9. To this end, we imagine a hypothetical substituent x to which we assign the free-energy term ΔG^x_{sw}. Given a set of compounds *which undergo the same types of intermolecular interactions with both organic solvents*, we can now relate ΔG^i_{sw} for compound i in both solvent–water systems (ow, sw) to the variable n:

$$\begin{aligned} \Delta G^i_{ow} &= \Delta G^A_{ow} + n\Delta G^x_{ow} \\ \Delta G^i_{sw} &= \Delta G^A_{sw} + n\Delta G^x_{sw} \end{aligned} \tag{7-10}$$

where ΔG^A_{sw} refers to the free energy of transfer of the central structure (the "stem" of the compound) on which the moiety x is attached. If we choose octanol–water as the reference system and define A so that $\Delta G^A_{ow} = 0$, we have $n = \Delta G^i_{ow}/\Delta G^x_{ow}$, and by substitution of n we obtain

$$\Delta G^{i}_{\mathrm{sw}} = \frac{\Delta G^{x}_{\mathrm{sw}}}{\Delta G^{x}_{\mathrm{ow}}} \cdot \Delta G^{i}_{\mathrm{ow}} + \Delta G^{\mathrm{A}}_{\mathrm{sw}} \tag{7-11}$$

or in terms of partition constants (see Eq. 7-8),

$$\log K_{\mathrm{sw}} = a \log K_{\mathrm{ow}} + b \tag{7-12}$$

where a is proportional to $\Delta G^{x}_{\mathrm{sw}}/\Delta G^{x}_{\mathrm{ow}}$, and b is proportional to $\Delta G^{\mathrm{A}}_{\mathrm{sw}}$.

Given a particular set of organic compounds whose partition constants are known in both solvent–water systems, parameters a and b may be obtained by linear regression. Such relationships are important for estimating the partitioning behavior of compounds between natural nonaqueous phases and water. However, the LFER (i.e., the values of a and b) found for one set of compounds will not necessarily be similar to those found for another set. Furthermore, as already mentioned and assumed when deriving Eq. 7-12, the goodness of correlation will critically depend on whether the compounds used to establish the LFER have the same types of interactions with the two nonaqueous solvents; that is, that the structural dissimilarities of the compounds are similarly reflected in their partial molar excess free energies (i.e., in their activity coefficients) in the two solvents. This is usually the case when dealing with nonpolar compounds where, as we have already seen, the activity coefficients in organic solvents are small and are primarily influenced by the size of the solute and not by specific interactions of the solute molecules with the organic solvent. Hence, for *nonpolar* organic compounds, fairly good correlations may be expected and are obtained. This is illustrated in Fig. 7.6 for two organic solvents with very different polarities: hexane–water and octanol–water. As discussed earlier and also shown in Fig. 7.6, polar compounds, particularly compounds with functional groups that may interact with solvent molecules by *hydrogen bonding*, may exhibit very different activity coefficients (and hence, very different partition constants) in two solvents where one solvent (i.e., octanol) may interact with the solute by hydrogen bonding, while the other solvent (i.e., hexane) may not. We can easily understand why very poor LFERs are found in such cases. For more detailed discussions of this topic see Leo et al. (1971) and Campbell et al. (1983).

Before we discuss some applications of such LFERs, it is useful to reflect on the meaning of the slope a in the LFER shown in Eq. 7-12. To do so, we return to our initial example—the chlorinated benzenes. In this example, a is given by

$$a = \frac{\Delta G^{\mathrm{Cl}}_{\mathrm{bw}}}{\Delta G^{\mathrm{Cl}}_{\mathrm{ow}}} \tag{7-13}$$

that is, by the ratio of the slopes of the two lines in Fig. 7.5. A value of $a = 1$ means that substitution of a hydrogen atom on the benzene ring by a chlorine atom has the same effect on the partial molar excess free energies (and hence on the activity

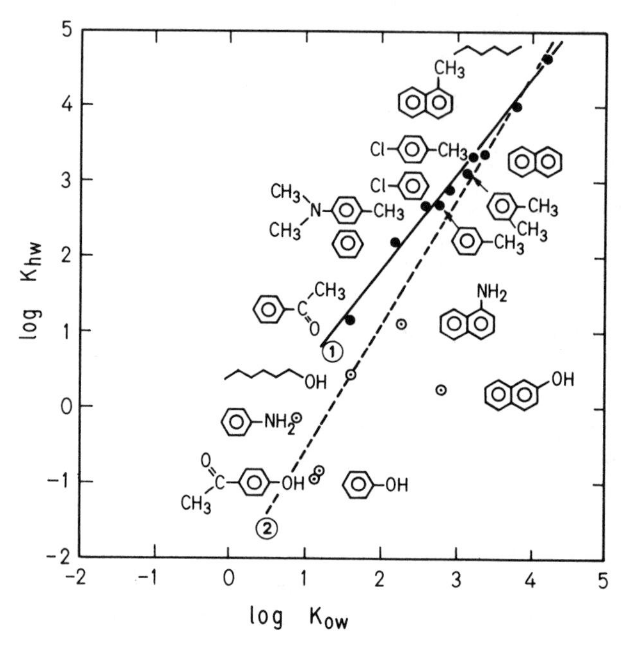

Figure 7.6 Relationship between log hexane–water partition constants and log octanol–water partition constants for organic compounds with and without polar functional groups (data from Hansch and Leo, 1979). The regression, obtained for the nonpolar compounds, yields (line 1) $\log K_{nw} = 1.28 \cdot \log K_{ow} - 0.76$, $r^2 = 0.97$; inclusion of the polar compounds yields (line 2) $\log K_{hw} = 1.66 \cdot \log K_{ow} - 2.26$, $r^2 = 0.79$.

coefficients) of the solute in both organic phases. (Recall that ΔG_{sw} is proportional to $g_s^e - g_w^e$; see Eq. 7-8). A value of $a < 1$, as in our example ($\Delta G_{bw}^{Cl}/\Delta G_{ow}^{Cl} = 0.72$), means that the chlorine substitution renders the molecule more incompatible with the first solvent (e.g., butanol) as compared with the reference solvent octanol; that is, $\Delta\gamma_b > \Delta\gamma_o$ (change of activity coefficient in butanol and octanol, respectively).

From the previous discussion, we can summarize and conclude that for nonpolar organic compounds in particular, LFERs as expressed by Eq. 7-12 can be successfully used to estimate the K_{ow} value of a particular compound if its partition constant is known for another solvent–water system (or vice versa). Generally, the estimated value will be better the more closely related the structure of the compound is to those compounds for which parameters a and b in the LFER have been determined and the more similar the two solvents are with respect to their ability to undergo hydrogen bonding in particular.

K_{ow} from Aqueous Solubility

We now come back to the LFER between K_{ow} and the aqueous solubility of a liquid compound. Inspection of the data in Fig. 7.4 shows that we can easily find sets of compounds for which a LFER of the form:

$$\log K_{ow} = a \log \frac{1}{C_w^{sat}(\mathrm{l, L})} + b = -a \log C_w^{sat}(\mathrm{l, L}) + b \qquad (7\text{-}14)$$

can be established. In this particular case, one of the organic "solvents" (the pure organic liquid) exhibits special properties; it always forms an ideal mixture with the solute (i.e., $g_s^e = 0$ or $\gamma_s = 1$), and the mole fraction of the solute is equal to 1. Thus, the goodness of the LFER shown in Eq. 7-14 and the values of the regression parameters a and b depend to a great extent on the variability of the activity coefficients in octanol, γ_o, within the set of compounds chosen to establish the LFER, and also, of course, on the accuracy of the experimental data used. Thus, as shown by the examples given in Table 7.2, excellent relationships are found for specific classes of "simple" nonpolar or moderately polar organic compounds, including the alkanes, polycyclic aromatic hydrocarbons, alkylated and halogenated benzenes, and phthalates. Mixing nonpolar with polar compounds and/or with compounds of very different molecular sizes generally results in poorer correlations. This is demonstrated by the substituted benzenes where inclusion of nine additional polar compounds in the set of 23 nonpolar alkyl- and chlorobenzenes substantially decreases the goodness of the fit. Furthermore, as illustrated by the last compound set in Table 7.2, for a very diverse set of chemicals such as the pesticides, very poor correlations may be found. We have already seen that these compounds contain a variety of very different polar groups and, therefore, do not share the same nonideal interactions in octanol. In contrast, a more uniform set of polar compounds like the aliphatic alcohols show much better LFERs. Finally, it should be noted that not all of the scatter in the data found when establishing a relationship between K_{ow} and $C_w^{sat}(\mathrm{l, L})$ is due to real chemical factors, because experimental errors in both solubility and K_{ow} determina-

TABLE 7.2 Linear Free-Energy Relationships Between Octanol–Water Partition Constants and (Liquid) Aqueous Solubilities for Various Sets of Compounds

			$\log K_{ow} = -a \log C_w^{sat}(\mathrm{l, L}) + b$	
Set of Compounds	n	R^2	$a(\pm\sigma)$	$b(\pm\sigma)$
Alkanes	16	0.91	0.81	−0.20
Polycyclic aromatic hydrocarbons	8	0.99	0.87(±0.03)	0.68(±0.16)
Substituted benzenes				
Only nonpolar substituents	23	0.98	0.86(±0.03)	0.75(±0.09)
Including polar substituents	32	0.86	0.72(±0.05)	1.18(±0.16)
Phthalates	5	1.00	1.06(±0.03)	−0.22(±0.09)
PCBs	14	0.92	0.85(±0.07)	0.78(±0.47)
Alcohols	41	0.94	0.90	0.83
Miscellaneous pesticides	14	0.81	0.84(±0.12)	0.12(±0.49)

tion are still large enough to contribute significantly, particularly when dealing with highly hydrophobic compounds. The data for PCBs, for example, are sufficiently uncertain as to yield substantially different regressions depending on the data chosen.

In summary, as many publications on this subject suggest (e.g., Hansch et al., 1968; Chiou et al., 1977, 1982; Banerjee et al., 1980; Mackay et al., 1980a; Lyman, 1982a; Miller et al., 1984), LFERs are very useful tools for estimating K_{ow} values from aqueous solubilities (of the liquid organic compound) and *vice versa*, if an appropriate relationship is used (i.e., derived from similar types of compounds, same range of hydrophobicity). For a *nonpolar* organic compound a less specific LFER combining various classes of compounds (e.g., see last example in Table 7.2) may yield satisfactory results, since γ_o is usually small and does not vary much between these compounds; however, it should be emphasized that more caution has to be taken when dealing with *polar* compounds, in particular those with several polar functional groups and/or groups that may undergo hydrogen bonding such as $-OH$ and $-NH_2$.

K_{ow} from Chromatographic Data

Another frequently applied approach for estimating the K_{ow} of a given compound is based on the retention behavior of the compound in a *chromatographic system.* [high performance liquid chromatography (HPLC) or thin-layer chromatography (TLC); e.g., see Fujita et al., 1964; Tomlinson, 1975; Veith et al., 1979; Butte et al., 1981; Eadsforth and Moser, 1983]. Here, the organic solute is transported in a *polar mobile phase* (e.g., water and water–methanol) through a porous stationary phase which exhibits *hydrophobic properties* as typically used in *reversed-phase chromatography.* The reversed-phase solid media generally consists of C_{10}–C_{18} *n*-alkanes covalently bound to a silica support (e.g., silica beads). As such, these hydrocarbons form an oily coating and, as the compounds of interest move through the system, they partition between this oily surface and the polar mobile phase according to their compatibility (or incompatibility) with the two phases. In principle, we have the same situation as discussed earlier when treating organic solvent–water systems, the only differences being that the compound is *ad*sorbed at a surface rather than totally *ab*sorbed (dissolved) in a liquid and that the "aqueous" phase frequently contains large amounts of an organic solvent such as methanol. Nevertheless, for nonpolar compounds in which the hydrophobicity (or solvophobicity when dealing with methanol–water mixtures) of the solute primarily determines its partitioning behavior, good correlations between K_{ow} and the *s*tationary phase/*m*obile phase partition constant K_{sm} of a given compound are obtained. Since, in a given chromatographic system, the travel time or *retention time t* of a solute is directly proportional to K_{sm}, a LFER of the following form is obtained:

$$\log K_{ow} = a \log t + b \tag{7-15}$$

Figure 7.7 shows the very good correlation obtained for a series of aromatic hydrocarbons spanning a K_{ow} range of four orders of magnitude. To compare different chromatographic systems, however, it is more useful to use the *relative retention time*

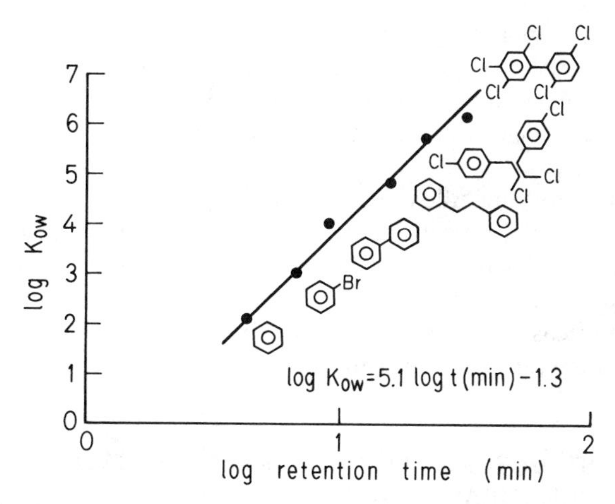

Figure 7.7 Relation between log octanol–water partition constants and log retention times on a reversed-phase liquid chromatography system for a series of nonpolar organic compounds (data from Veith et al., 1979).

(also called the *capacity factor*), which is the retention of the compound relative to a nonretained chemical species, such as a very polar organic compound or an inorganic species such as nitrate. Equation 7-15 is then written as

$$\log K_{\mathrm{ow}} = a \log \frac{t - t^0}{t^0} + b' \tag{7-16}$$

where t^0 is the travel time of the nonretained species in the system. It should be pointed out that the coefficients a and b or b' in Eqs. 7-15 and 7-16 must be determined using appropriate reference compounds for *each* chromatographic system. With respect to the choice of reference compounds (type, range of hydrophobicity) and the goodness of the LFER, in principle the same conclusions as drawn earlier for organic solvent–water systems are valid. The situation, however, may sometimes be more complicated by surface phenomena such as steric or geometric factors. Since the interaction of solutes with nonpolar surfaces occurs via an adsorption mechanism, the critical question concerns how well the contact is between organic solute and the surface. If, owing to steric hindrance, molecules are forced to project away from the surface, they maintain a greater surface area in contact with the aqueous medium. As a consequence, these compounds do not obtain as great a benefit of "water:solute cavity relief" as their planar counterparts and, therefore, have comparably lower retention. In organic solvent–water partitioning or absorption systems, this geometric factor is usually not as important, since solvent molecules may adjust themselves in space to accommodate this steric effect. Rapaport and Eisenreich (1984) investigated HPLC techniques for estimating the K_{ow} values of PCBs. They found good correlations only for congeners without chlorine substitutions on the carbon atoms adjacent to the carbon atom that is connected to the other ring. Such substitution causes the rings to twist out of coplanarity. Thus, when using chromatographic properties that reflect adsorptive interactions of molecules with a stationary phase, steric factors and/or specific chemical interactions (e.g., reactions of polar groups with the support material) may have to be evaluated when establishing or using LFER-relating retention data with K_{ow}.

In summary we may conclude that, when appropriate, the use of chromatographic systems for evaluating the partitioning behavior of organic compounds between nonaqueous phases and water (e.g., octanol–water) offers several advantages. Once a chromatographic system is set up and calibrated, several compounds may be investigated at once. The measurements are fast and, above all, accurate compound quantitation (which is a prerequisite when using solvent–water systems) is not required.

7.5 CHEMICAL STRUCTURE AND PARTITIONING BEHAVIOR—ESTIMATION OF K_{ow} FROM STRUCTURAL GROUP CONTRIBUTIONS

A very intriguing approach used to estimate partition constants is to try to express the free energy of transfer of a compound in a given nonaqueous phase–water system as the sum of free-energy contributions of the different parts of the molecule. Since the thousands of organic chemicals with which we have to deal are all constructed from a fairly small and, therefore, manageable number of pieces (atoms, functional groups, etc.), such structure–property relationships allow us, with a minimum of effort, to estimate a specific property of a compound, particularly if that property is known for a structurally related compound.

In Chapters 5 and 6 we briefly mentioned some approaches that have been taken to estimate activity coefficients and Henry's Law constants from chemical structures. Although it is not our purpose in this book to provide "cookbook recipes" for the estimation of the compound properties from chemical structure, it is both useful and instructive to have a closer look at one example. We choose the octanol–water partition constant because, to date, estimation methods for K_{ow} from chemical structure are among the best developed and most widely used structure–property relationships (Leo et al., 1971; Nys and Rekker, 1973; Rekker, 1977; Hansch and Leo, 1979; Lyman, 1982a). Furthermore, the following discussion will allow us to deepen our insights into the molecular forces and interactions that determine the excess free energy of a compound in both aqueous and organic solutions, and thereby understand its partitioning behavior as well.

We begin by assuming that the fairly small number of subunits, from which all organic compounds are assembled, each exhibits fairly reproducible behavior in every chemical in which it occurs. Thus, for the case of transfer of any particular compound between octanol and water, we assume that the total free energy of transfer is simply the sum of the free energies carried by individual atoms or groups of atoms of which the chemical is composed. Consequently, we have

$$\log K_{ow} = \sum_i f_i + \sum_j F_j \qquad (7\text{-}17)$$

where f_i values quantify the contributions arising from each building block i in the particular chemical, and F_j values account for any special *intra*molecular interaction, j between these fundamental pieces which cause them to act a little special in the particular chemical. By experience with K_{ow} values from a variety of chemicals, investigators have been able to assign average values for contributions to $\log K_{ow}$ for each of the relatively manageable number of structural parts used to assemble the menagerie of organic compounds. Using these quantities, we can now estimate K_{ow} for other compounds based on the structural pieces from which they are constructed. Let us now examine these ideas in more detail.

Fragment Constants

Workers such as Fujita et al. (1964) have noted that specific structural units such as a methyl group ($—CH_3$) always increase a compound's K_{ow} by about the same amount when they are added to the substance instead of a hydrogen. For example,

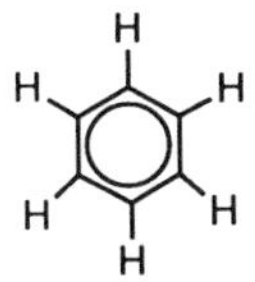

benzene
$\log K_{ow} = 2.13$

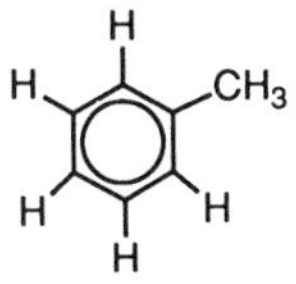

toluene (or methyl benzene)
$\log K_{ow} = 2.69 \quad \Delta = 0.56$

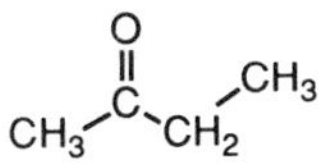

methyl ethyl ketone
(or 2-butanone)
$\log K_{ow} = 0.29$

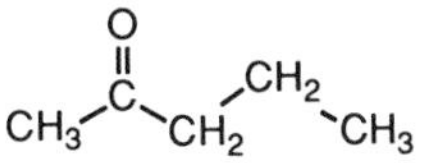

2-pentanone
Log $K_{ow} = 0.79 \quad \Delta = 0.50$

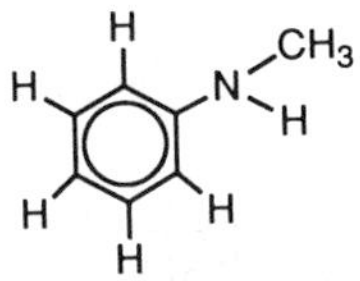

N-methyl aniline
$\log K_{ow} = 1.66$

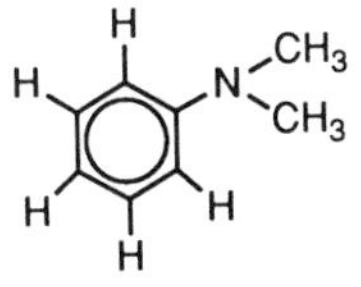

N,*N*-dimethyl aniline
$\log K_{ow} = 2.31 \quad \Delta = 0.55$

On average, we see that for many organic compounds this methyl *structural* piece seems to carry with it a *contribution* to the $\log K_{ow}$ of about half a log unit (i.e., a factor of 3 in K_{ow}). (The small variations from compound to compound result from intramolecular effects caused by differences in the remainder of the structures in each case.) Similarly consistent contributions have been observed for a variety of other structural substituents. It is not too difficult to imagine taking this observation one step further and postulating that the entire compound structure can be disassembled to its unit parts and that each of these structural building blocks, according to its attributes

TABLE 7.3 Some Common Fragment Constants (f) Useful for log K_{ow} Estimation[a]

Fragment	f	f^{ϕ}	$f^{\phi\phi}$
—H	0.23	0.23	
—C—	0.20	0.20	
C aromatic	0.13		
C aromatic between rings	0.23		
—F	−0.38	0.37	
—Cl	0.06	0.94	
—Br	0.20	1.09	
—I	0.59	1.35	
—O—	−1.82	−0.61	0.53
—OH	−1.64	−0.44	
—N<	−2.18	−0.93	
—NH—	−2.15	−1.03	−0.09
—NH_2	−1.54	−1.00	
—NO_2	−1.16	−0.03	
Ketone —C(=O)—	−1.90	−1.09	−0.50
Carboxylate —COO^-	−5.19	−4.13	
Carboxylic acid —COOH	−1.11	−0.03	
Ester—COO—	−1.49	−0.56	−0.09

[a]From Hansch and Leo, 1979. Superscript ϕ indicates constant for substituents bonded to aromatic carbons, and when it is used twice it refers to fragments bound to aromatic carbon on both sides. Fragment constants for halogens bonded to isolated double bonds are given by $1/2\ (f + f^{\phi})$ and those for polar substituents containing N or O are given by $2/3 f + 1/3 f^{\phi}$.

(e.g., size, polarity), contributes a consistent portion to the resulting K_{ow}. Such building blocks have been termed fragments (Rekker, 1977; Hansch and Leo, 1979), and the quantities that we empirically find they contribute to a chemical's log K_{ow} are called "fragment constants." Some typical values are shown in Table 7.3; additional fragment constants for other substituents can be found in Hansch and Leo (1979).

Nonpolar and Polar Fragments

A brief discussion of the relative quantities contributed to log K_{ow} by the various fragments is useful in light of our concepts concerning solute:solvent intermolecular interactions, solubilities in liquids, and octanol–water partitioning. First, we see that molecular pieces made of carbons and hydrogens have positive fragment constants and, therefore, always cause an increase in log K_{ow}. These are the organic building blocks that interact with their surroundings chiefly via van der Waals forces and, thus, readily dissolve into octanol but disrupt the polar environment of liquid water.

On the other extreme, we see that functional groups containing nitrogen and oxygen always exhibit negative fragment constants; therefore, their presence in compound structures diminishes $\log K_{ow}$. Again, this general result is not surprising in light of the ability of these moieties to be involved in polar and hydrogen-bonding interactions. Thus, these fragments work to pull the rest of the molecule in which they are contained into the polar aqueous phase. Interestingly, the halogens (F, Cl, Br, and I) seem to straddle the polar-to-nonpolar world. For example, the small and highly electronegative fluorine exhibits a negative contribution to $\log K_{ow}$, whereas the bulkier and relatively low electronegative bromine and iodine substituents significantly increase compound hydrophobicity.

Fragments Attached to Double Bonds and Aromatic Rings

A special set of fragment constants is used when a polar substituent, X, is attached to a carbon double-bonded to other parts of the molecule:

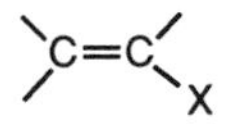

This is because the π electron cloud, which forms the second bond to the carbon, can be deflected enough by polar moieties to spread the uneven time-averaged electron distribution over a greater molecular area. For example, a nitro group attached to a carbon in nitromethane establishes a dipole moment that is confined to only about two bond lengths; on the other hand, a nitro group attached to a benzene carbon finds the dipole it establishes spread throughout the molecule by resonance positions of the π electron cloud (see also Chapter 2):

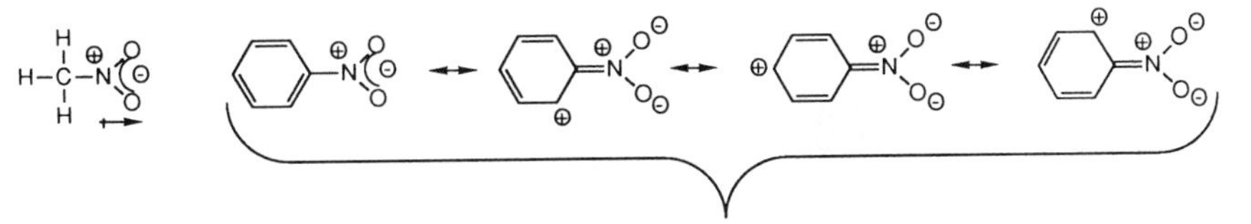

$d = 3.46$ debyes

$f_{NO_2} = -1.16$

$d = 4.22$ debyes

$f^{\phi}_{NO_2} = -0.03$

In the case of nitrobenzene, which has a dipole moment similar in strength to nitromethane, its polar character is extended over about five bond lengths. Since the polar solute:solvent molecule interactions that enhance aqueous solubilities are important for dipole:dipole attractions if they are similar in length to individual water molecules, this spreading of the polar nature of the nitro group in nitrobenzene reduces the forces drawing it into water. Consequently, the fragment constants for this and other polar moieties bonded to unsaturated carbons must be adjusted to a more positive value.

As demonstrated by the fragment constants f^{ϕ}, shown in Table 7.3, substituents attached to aromatic systems with their extensive π electron systems typically make contributions to the log K_{ow} estimate that are about one full log unit more positive than corresponding constants for the same substituents attached to aliphatic carbons (i.e., K_{ow} increases by about 10x). In the event that moieties (e.g., —NH— or —O—) are linked to two aromatic systems (e.g., C_6H_5—NH—C_6H_5), the fragment constant ($f^{\phi\phi}$) becomes even more positive.

Since the π-electron system of olefins or alkenes is usually not as extended as in aromatic systems, this effect is not as strong for fragments attached to isolated double bonds (referred to as vinyllic). Hansch and Leo (1979), therefore, recommend using $(f + f^{\phi})/2$ for vinyllic halogens and $(2/3f + 1/3f^{\phi})$ for vinyllic polar groups containing N or O. Note that nonpolar fragments such as —H or —C show no difference in attachment to an unsaturated carbon as compared with a saturated carbon.

In summary, the fragment constants for molecular parts which are used to build the nonpolar regions of molecules are positive, reflecting the unfavorable interactions these molecular surfaces have in aqueous solution. On the other hand, fragment constants for N- and O-containing moieties that both contribute dipoles and the ability to hydrogen bond are typically negative, consistent with their water-solubilizing nature. Perhaps surprisingly, the halogens usually enhance compound hydrophobicity, as indicated by their positive fragment constants.

We are now in a position to estimate log K_{ow} for some simple organic chemicals. For example, nitromethane (CH_3—NO_2) is composed of a carbon, three hydrogens, and a nitro substituent; thus, we calculate

$$\log K_{ow} = f_C + 3f_H + f_{NO_2} = (0.20) + 3(0.23) + (-1.16) = -0.27$$

The observed result is log $K_{ow} = -0.34$ (Hansch and Leo, 1979). For a slightly more complex molecule like nitrobenzene (C_6H_5—NO_2), we can use literature information on the K_{ow} of benzene together with these fragment constants:

$$\begin{aligned} \log K_{ow}(\text{nitrobenzene}) &= \log K_{ow}(\text{benzene}) - f_H^{\phi} + f_{NO_2}^{\phi} \\ &= 2.13 - (0.23) + (-0.03) \\ &= 1.87 \end{aligned}$$

The observed result is log $K_{ow} = 1.84$ (Hansch and Leo, 1979). As long as the molecular fragments in which we are interested do not participate in any important intramolecular interactions, this adding and subtracting of fragment constants can yield quite good log K_{ow} estimates.

Quantification of Intramolecular Interactions

As the assemblages of atomic parts become more complex in organic molecules, such as in many pesticides, the molecular pieces begin to exert important influences on one another through steric, electronic, and resonance *intra*molecular interactions. In turn, these interactions affect γ_w and γ_o and, thus, the octanol–water partitioning. Hansch and Leo (1979) have begun the process of quantifying the impacts of these *intra*molecular Factors, designated as F values. Usually, the sign and relative importance of the F constants can be rationalized easily in terms of their effect on aqueous solution incompatibility, the dominant characteristic causing $\log K_{ow}$ to vary from compound to compound. Table 7.4 reproduces some common intramolecular interaction F values; a larger compilation is found in Hansch and Leo (1979).

Generally, *intramolecular* interactions fall into two categories: geometric or electronic. As we shall see, geometric factors usually involve organizing the parts of a molecule so as to affect aqueous cavity formation. These factors always make the compound less hydrophobic than we would estimate simply by adding up the fragment constants appropriate for the molecular parts present. Consequently, as a rule, geometric factors decrease K_{ow}. On the other hand, electronic factors always involve intramolecular interactions of polar pieces of molecules. When such polar groups are sufficiently near one another in the structure, they neutralize one another's polarity somewhat by pulling the electrons in opposite directions. Consequently, dipole interactions of the molecular parts with water molecules are diminished. Hence, electronic factors always increase K_{ow}.

Let us briefly describe these factors to see how they are applied and what underlying chemical principles they reflect. A more detailed discussion can be found in the monograph by Hansch and Leo (1979).

Geometric Factors

First, examine the geometric factors listed under unsaturation in Table 7.4. When we form multiple-bonded carbons, the new units are somewhat smaller in size (e.g., molar volumes of hexane, 1-hexene, and 1-hexyne are 126.7, 125, and 114.8 cm^3/mol, respectively). Further, the electrons in these multiple bonds are somewhat easier to polarize. As a result, molecular pieces exhibiting unsaturation are pushed less hard from water than their saturated counterparts (smaller cavity needed, stronger induced dipole attraction to water). Consequently, when double or triple bonds appear in organic chemical structures, we must correct the sum of fragment contributions ($\sum_i f_i$) by an appropriate negative F (see Table 7.4). Thus, in order to estimate $\log K_{ow}$ of 1-hexene from hexane,

$$\begin{aligned}\log K_{ow}(\text{1-hexene}) &= \log K_{ow}(\text{hexane}) - 2f_H + F_{\parallel}\\ &= 4.11 - 2(0.23) + (-0.09)\\ &= 3.56\end{aligned}$$

The observed result is $\log K_{ow} = 3.39$ (Tewari et al., 1982). Note that the "double bonds" in aromatic systems do not require this unsaturation factor because this effect

Table 7.4 Some Common Intramolecular Interaction Factors Useful for log K_{ow} Estimation[a]

Structural Feature	Symbol	Influence on Aqueous Solubility and K_{ow}	F Value
		Geometric Effects	
Unsaturation			
Double bond	$F_{\|\|}$	greater polarizability, smaller size	-0.09[b]
Triple bond	$F_{\|\|\|}$	inc. soly ⇒ dec. K_{ow}	-0.50[b]
Skeletal arrangement			
Long-chain flexing	F_{ch}	upsets "flickering ice" cavity	$(n-1)(-0.12)$
Ring flexing	F_r	formation, inc. soly ⇒ dec. K_{ow}	$(n-1)(-0.09)$
Nonpolar chain branch	$F_{br\ nonpolar}$	dec. molecular size	(-0.13)
Polar chain branch	$F_{br\ polar}$	inc. soly ⇒ dec. K_{ow}	(-0.22)
		Electronic Effects	
Nearby polyhalogenation	$F_{polyhalo}$	Opposing nearby dipoles	
2 on same C		diminish polarity, dec.	0.60
3 on same C		soly ⇒ inc. K_{ow}	1.59
4 on same C			2.88
2 on adjacent single-bonded C			0.28
3 on adjacent single-bonded C			0.56
4 on adjacent single-bonded C			0.84
5 on adjacent single-bonded C			1.12
6 on adjacent single-bonded C			1.40
Nearby polar groups	$F_{nearby\ polar\ groups}$	Opposing nearby dipoles diminish polarity, dec. soly ⇒ inc. K_{ow}	

	In Chain	In Alicyclic Ring	In Aromatic Ring
On same C	$-0.42(f_1+f_2)$		
On adjacent C's	$-0.26(f_1+f_2)$	$-0.32(f_1+f_2)$	$-0.16(f_1+f_2)$
On C's separated by one C	$-0.10(f_1+f_2)$	$-0.20(f_1+f_2)$	$-0.08(f_1+f_2)$

Structural Feature	Symbol	Influence on Aqueous Solubility and K_{ow}	F Value
Intramolecular hydrogen bonding			
With —OH	$F_{H\text{-bond with oxy.}}$	Ties up moiety diminishing	$+1.0$
With —NH |	$F_{H\text{-bond with nit.}}$	hydrogen bonding with water, dec. soly ⇒ inc. K_{ow}	$+0.6$

[a] From Hansch and Leo (1979); n refers to the number of bonds, soly means aqueous solubility, inc. means increase, dec. means decrease.

[b] Unlike Hansch and Leo (1979), we have subtracted the effect of removing H's from these factors.

has already been incorporated in the fragment constants for carbons in the aromatic rings:

$$\begin{aligned}\log K_{ow}(\text{benzene}) &= 6 f_C + 6 f_H^{\phi}\ (no\ F_{\|}) \\ &= 6(0.13) + 6(0.23) \\ &= 2.16\end{aligned}$$

The observed result is $\log K_{ow} = 2.13$.

Flexing

A second geometric factor that we must consider involves the arrangement of carbons in chains. When the skeleton of molecules is made up of simply bonded atoms (except H), we recognize that they may wriggle, twist, and bend with respect to one another. This flexing, as it is called by Hansch and Leo (1979), causes the shape of the surface of molecules to change continuously. Therefore, when we insert these chain substances into water, the water molecules lining the cavity never fully orient and "freeze." As

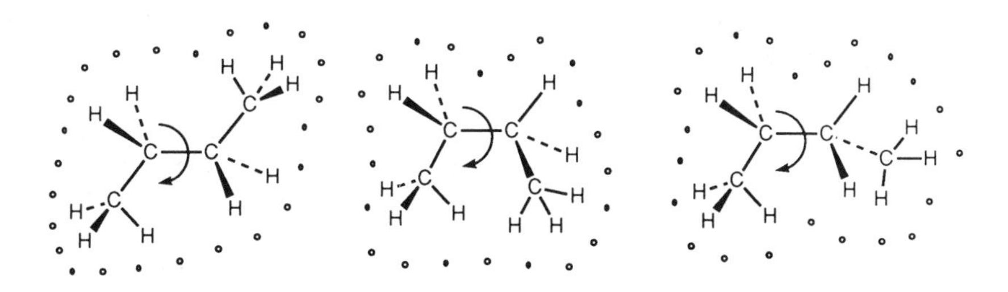

Twisting in n-butane

a result, the excess free-energy cost of dissolution is not as great as it is for counterpart compounds that are more rigid.

The flexing factor is applied as a multiple of the number of bonds in a chain. Since a chain of one bond such as ethane is still rigid, we do not count the first bond in any given chain. Thus, the chain-flexing factor is given as $F_{ch} = (n-1)(-0.12)$, where n is the number of bonds. For hexane, n is 5:

Hexane

so we would calculate

$$\begin{aligned}\log K_{ow} &= 6f_C + 14f_H + (5-1)(F_{ch}) \\ &= 6(0.20) + 14(0.23) + (4)(-0.12) \\ &= 3.94\end{aligned}$$

$$\log K_{ow}\text{observed} = 4.11 \text{ (Tewari et al., 1982)}$$

Alicyclic ring systems can also flex, so a similar Factor reducing compound K_{ow} should also be applied.

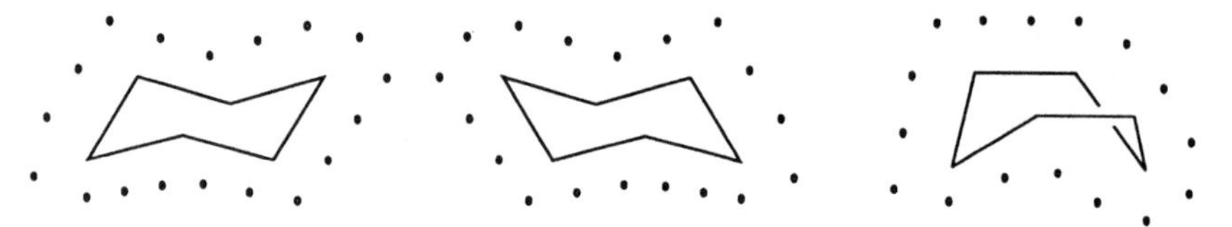

Molecular Flexing in Cyclohexane

Owing to limitations imposed by the ring nature of these chains, the factor is less per bond (-0.09). For cyclohexane, we would apply $F_r = (6-1)(-0.09) = -0.45$. Interestingly, these ring-flexing factors are nearly as great or greater than those in straight chains of the same carbon number. Mathematically this arises because rings exhibit one more bond per carbon than do chains. Maybe a good chemical rationalization for the relative hydrophobicity of rings is that they exhibit reduced molecular surface area because of their folded nature (e.g., *n*-hexane is 142 Å^2/molecule versus cyclohexane at 121 Å^2/molecule). The chain-flexing factor thus also reflects the reduced aqueous cavity size of rings. Note that aromatic rings are rigid and, therefore, no correction for chain flexing is applied; the reduced molecular size arising from the ring geometry is also incorporated into the lower $f_{C_{ar}}$ constant.

Branching

In a fashion similar to rings reducing compound size, so too does branching since the compound more closely approaches sphericity and thereby diminishes its molecular surface area (and concomitantly increases aqueous solubility). The total surface area calculated for 3-methyl pentane is 137.5 Å^2/molecule, whereas that of the corresponding straight chain compound as noted above is 142 Å^2/molecule. Thus, compounds exhibiting a branch in their structure require a geometric correction factor which reduces their hydrophobicity ($F_{br} = -0.13 \log K_{ow}$ units). This effect is even more important for polar substituents located at branching points from aliphatic chains ($F_{br\ polar} = -0.22 \log K_{ow}$ units). Not only do such compounds (e.g., secondary alcohols) exhibit lower total molecular surface area, but the centrally located polar substituent helps accomodate a larger portion of the nonpolar molecular surface to any surrounding water environment. Aqueous solubilities of these polar-branched

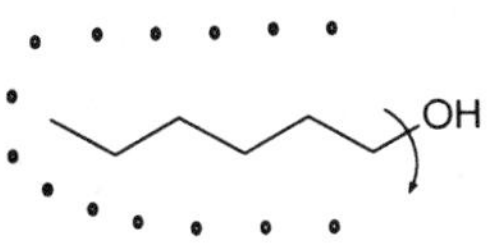

1-Hexanol
152 Å^2/molecule
$C_w^{sat} = 0.6\,mol \cdot L^{-1}$

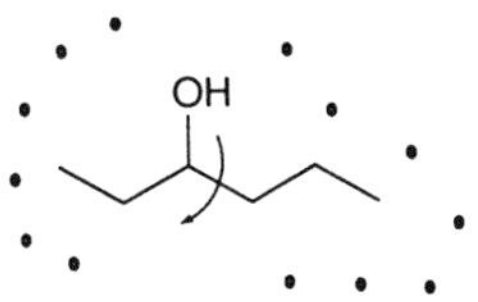

3-Hexanol
150.1 Å/molecule
$C_w^{sat} = 1.6\,mol \cdot L^{-1}$

compounds then are greater (as is shown above for 3-hexanol versus 1-hexanol), and their K_{ow} values must be reduced to reflect this fact.

Having described these geometric factors, we can now estimate $\log K_{ow}$ for any monofunctional organic chemical (i.e., those without polar intramolecular interactions). For a compound such as aminocyclohexane, we must first think about the pieces of the structure with which we are working:

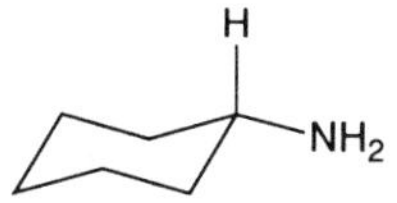

We tally the structural pieces and their contributions to $\log K_{ow}$:

$$6f_C + 11f_H + 1f_{NH_2} = 6 \cdot (0.20) + 11 \cdot (0.23) + 1 \cdot (-1.54) = 2.19$$

Next we consider any special intramolecular interactions arising from the way in which the pieces of this molecule are put together. Since we have a chain (in the ring, but not in the branch), we must correct for flexing: $F_r = (6 - 1)(-0.09) = -0.45$. We also have a polar functional group branching from the ring which will solubilize that region of the molecule in water, so a factor $F_{br\ polar} = (-0.22)$ should be used. Summing these factors with the fragment constants:

$$\log K_{ow}(\text{aminocyclohexane}) = 2.19 - 0.45 - 0.22$$
$$= 1.52$$

The observed result is $\log K_{ow} = 1.49$ (Hansch and Leo, 1979)

Electronic Factors

As soon as the molecules of interest contain multiple polar substituents that may interact with one another electronically, we must also add electronic factors to our octanol–water partition constant calculations. Since these polar substituents always involve moieties that strongly attract electrons to themselves, it is easy to see that if two or more of these polar molecular pieces are located close to one another in an organic compound structure, they will pull electrons toward themselves but against one another, thereby reducing one anothers' polarity. Consequently, when these compounds are in aqueous solution, the strength of polar intermolecular attractions to adjacent water molecules will be diminished, and compound hydrophobicity will increase. Hansch and Leo (1979) have organized such electronic intramolecular interactions into four groups: (a) those involving only halogens (thus practically limited to changes in dipole:dipole interactions); (b) those involving halogens interacting with hydrogen-bonding moieties (thus reflecting both diminished dipoles and H-bonding); (c) those involving interactions between two hydrogen-bonding groups; and (d) special cases in which the appropriate geometry combines with suitable substituents so as to allow *intra*molecular hydrogen bonding. In general, the closer the polar substituents are to one another, the stronger (i.e., more positive F value) will be the effect on compound hydrophobicity (Table 7.4). In all of these cases, the intramolecular interactions reduce the ability of the molecule to have intermolecular polar interactions with water. These factors, therefore, cause aqueous activity coefficients, and concomitantly K_{ow} values, to increase. Three brief sample calculations may best serve to demonstrate the use of these electronic factors.

Polyhalogenation

First, let us consider more than one electron-attracting halogen in a chemical, for example,

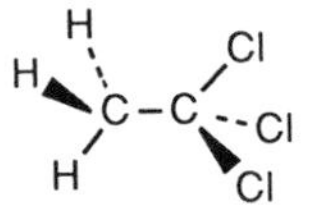

1,1,1-Trichloroethane (also called methyl chloroform)

We sum the fragment constants: $2f_C + 3f_H + 3f_{Cl}$, then look for intramolecular interactions due to geometry and electronic effects. In this case, we have

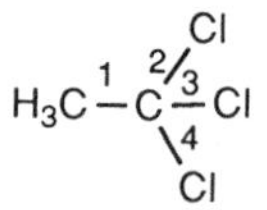

four flexing single bonds $[(4-1)\cdot F_{ch}]$ and three polar substituents pulling electrons against one another. Since the three chlorines are on the same carbon, we apply (from Table 7.4) $F_{polyhalo} = +1.59$ to reflect the overall diminished polarity; therefore,

$$\begin{aligned}\log K_{ow} &= 2f_C + 3f_H + 3f_{Cl} + (4-1)F_{ch} + F_{polyhalo}\\ &= 2(0.20) + 3(0.23) + 3(0.06) + 3(-0.12) + (1.59)\\ &= 2.50\end{aligned}$$

The observed $\log K_{ow}$ is 2.49 (Hansch and Leo, 1979).

Note that although a F_{ch} factor is applied to the C—Cl bonds, no branching correction is made. This factor is included in the polyhalogenation factor. Also, it should be noted that the more halogens involved on either the same carbon or adjacent carbons, the greater is the incremental contribution to compound hydrophobicity. This polar correction factor is not effective when adjacent carbons are linked by a double bond; hence, tetrachloroethylene exhibits no polyhalogenation factor between the carbons:

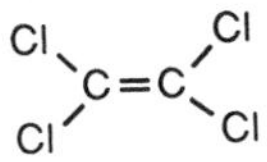

$$\begin{aligned}\log K_{ow}(\text{tetrachloroethylene}) &= 2f_C + 4([f_{Cl} + f^{\phi}_{Cl}]/2) + (5-1)F_{ch} + F_{\parallel} \\ &\quad + 2(F_{\text{two halo on same C}}) \\ &= 2(0.20) + 4([0.06 + 0.94]/2) + 4(-0.12) \\ &\quad + (-0.09) + 2(0.6)\end{aligned}$$

$$\log K_{ow}(\text{estimated}) = 3.03$$

$$\log K_{ow}(\text{observed}) = 2.88$$

We can now understand how other polar functional groups positioned closely enough together in the same molecule will somewhat neutralize one another's polarity. Additionally, if these groups contain oxygen or nitrogen (and, therefore, are capable of important hydrogen-bonding attractions with water molecules), it is not surprising that neighboring polar substituents will reduce this molecule:molecule interaction mechanism. As the nonbonded electrons or the hydrogens are electronically drawn more tightly to the organic molecule, the strength of the hydrogen bonds formed with nearby water molecules is reduced. Thus, we expect the intermolecular forces drawing such chemicals to water to be diminished (they would also decline in the octanol which is capable of hydrogen bonding, but somewhat less so per unit volume), and consequently a factor reflecting the increased hydrophobicity should be applied in $\log K_{ow}$ calculations. Hansch and Leo (1979) have made this factor a function of the hydrophobicity of the unit moieties involved (i.e., use the fragment constants of the specific groups to reflect their polarity), the distance separating these polar substituents, and whether they occur in chains or rings (i.e., less able to align themselves).

An estimate of $\log K_{ow}$ for dimethyl phthalate exemplifies the use of these factors. Usually the nonbonded electrons of oxygens, such as those in the phthalate ester, will be involved in hydrogen bonding. This attractive interaction to water would be reduced by nearby polar groups. First, we list the pieces of the molecule:

$$6f_{C_{ar}} + 10f_H + 2f^{\phi}_{COO} + 2f_C$$

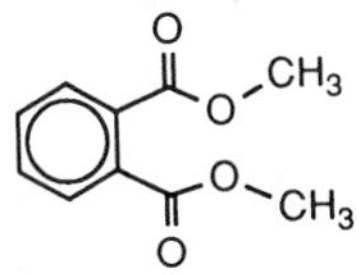

Note the —COO— group is treated as one building block. Then we look for geometric factors such as chains or branches: $2 \cdot (2-1) \cdot F_{ch}$. Finally, we consider polar intramolecular interactions. Since oxygen atoms are relatively electronegative, the two —C(=O)O— groups pull electrons toward themselves; however, owing to their proximity these polar groups neutralize one another somewhat, so we apply a factor appropriate for aromatic rings: $-0.16(f^{\phi}_{\text{COO}} + f^{\phi}_{\text{COO}})$. Altogether we have

$$\begin{aligned}\log K_{\text{ow}}(\text{estimated}) &= 6(0.13) + 10(0.23) + 2(-0.56) + 2(0.20) \\ &\quad + 2(-0.12) + (-0.16)(-0.56 - 0.56) \\ &= 2.30 \\ \log K_{\text{ow}}(\text{observed}) &= 1.53\end{aligned}$$

We see that since polar hydrogen-bonding fragment constants are always negative, the mathematical formulation reflecting their interaction always results in a positive contribution to $\log K_{\text{ow}}$. Hansch and Leo (1979) provide a special listing of factors for halogens near hydrogen-bonding substituents; generally halogens located on a carbon adjacent to a polar functional group increase the compound hydrophobicity by about one $\log K_{\text{ow}}$ unit.

Intramolecular Hydrogen Bonding

For some special compounds containing two groups capable of hydrogen bonding with one another (and, therefore, reducing the tendency to hydrogen bond with water) and having these groups spaced just so they can reach around and be near, we must also provide a correction to $\log K_{\text{ow}}$ estimation calculations. When oxygen is donating the hydrogen (i.e., hydroxyl and carboxyl groups), this factor is one full $\log K_{\text{ow}}$ unit; when nitrogen provides the hydrogen (i.e., amines), we use $+0.6 \log K_{\text{ow}}$ units. A chemical for which this effect is important is 2-ethyl-1,3-hexanediol in which one hydroxy oxygen serves as the nonbonded electron donor, while the other supplies the hydrogen. The $\log K_{\text{ow}}$ estimation would include:

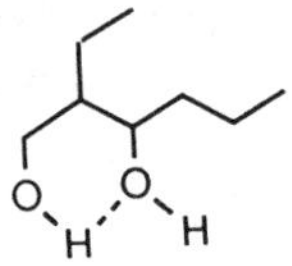

$$\begin{aligned}\log K_{\text{ow}} &= 8f_{\text{C}} + 16f_{\text{H}} + 2f_{\text{OH}} + (9-1)F_{\text{ch}} + F_{\text{br polar}} + F_{\text{br nonpolar}} \\ &\quad + F_{\text{nearby polar groups}} + F_{\text{H-bond with oxy}} \\ \log K_{\text{ow}} &= 8(0.20) + 16(0.23) + 2(-1.64) + 8(-0.12) + (-0.22) \\ &\quad + (-0.13) + (-0.10)(-1.64 - 1.64) + (1.0) \\ &= 2.02\end{aligned}$$

$$\log K_{\text{ow}}(\text{observed}) = 3.22$$

(This poor match between estimate and observation suggests that the reported log K_{ow} is suspect, especially in light of the log K_{ow} discussed next.) A diol (i.e., a compound containing two hydroxyl groups) which is incapable of hydrogen bonding such as 1,8-octanediol would not contain intramolecular hydrogen bonding and proximate polarity correction factors, so its estimated log K_{ow} is only

$$8f_C + 16f_H + 2f_{OH} + (9-1)F_b$$
$$8(0.2) + 16(0.23) + 2(-1.64) + (9-1)\cdot(-0.12) = 1.04$$

The observed log K_{ow} for this compound is between 0.96 and 1.07 (Hansch and Leo, 1979).

Estimation of K_{ow} from Structurally Related Compounds

One important estimation shortcut that we have used before and which becomes increasingly effective as the complexity of organic structures increases is to employ known log K_{ow} values for closely related compounds. All we need to do is supply fragment constants and intramolecular interaction factors as necessary to account for structural differences. Thus, the estimation expression becomes:

$$\log K_{ow}(\text{new}) = \log K_{ow}(\text{old}) - \sum_{\text{removed}} f + \sum_{\text{added}} f - \sum_{\text{removed}} F + \sum_{\text{added}} F$$

Generally, this approach is much more accurate than building up a log K_{ow} estimate from fundamental fragments because the most important contributions are accurately included. A sample calculation may help serve to illustrate this idea. Suppose we are interested in estimating log K_{ow} for methoxychlor. Notice that the structure is closely related to that of DDT, a substance whose log K_{ow} is already known. Thus,

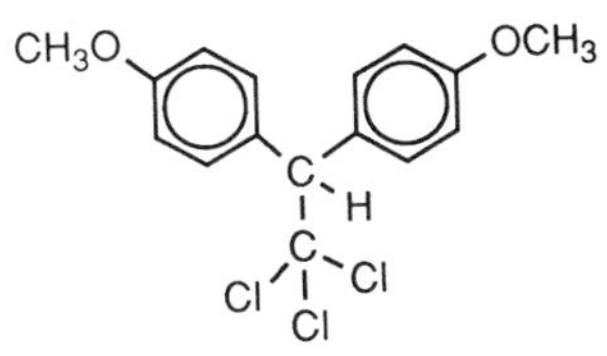

methoxychlor 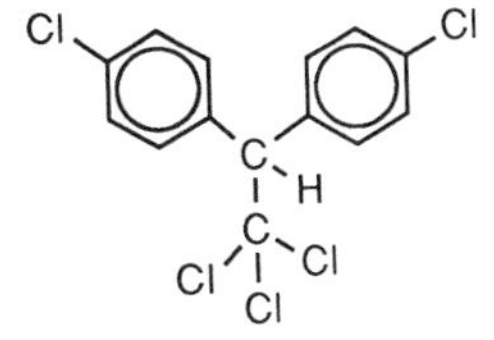 DDT

$$\log K_{ow}(\text{methoxychlor}) = \log K_{ow}(\text{DDT}) - 2f^{\phi}_{Cl} + 2f^{\phi}_{O} + 2f_C + 6f_H + 2(2-1)(F_{ch})$$
$$= 6.36 - 2(0.94) + 2(-0.61) + 2(0.2) + 6(0.23) + 2(-0.12)$$
$$= 4.80$$

The observed result is log $K_{ow} = 4.20$ (Veith et al., 1979). Obviously, simple molecular structure changes such as —OCH_3 for —Cl can be quickly seen to reduce compound hydrophobicity. (Such insights are used by chemical manufacturers to adjust chemical properties to suit specific purposes.)

In summary, the use of fragment constants and structural factors in calculating octanol–water partition constants from no more information than compound structure is a very attractive approach. In light of the LFERs between K_{ow} and aqueous solubility, it is clear that such a physical–chemical property estimate can be used to deduce other information about the substance. Because of the large number of fragment constants and structural factors available, log K_{ow} estimates from structural contributions are often satisfactory (within 1 log K_{ow} unit), particularly when dealing with simple structures. Eventually, computerized versions of the method will become widely available (Chou and Jurs, 1979). Some caution should be exercised when large, complex, or polyfunctional compounds are examined by structural contribution approaches, since our data and experience with these structural arrangements is limited.

CHAPTER 8

ORGANIC ACIDS AND BASES: ACIDITY CONSTANT AND PARTITIONING BEHAVIOR

8.1 INTRODUCTION—ACIDITY CONSTANT

Until now, we have confined our discussions about compound properties to *neutral* chemical species. Some important environmental organic chemicals may, however, undergo proton transfer reactions resulting in the formation of *charged* species (e.g., cations or anions: see examples given in Tables 8.1 and 8.2). These charged species have *very different properties and reactivities as compared to their neutral counterparts*; thus, it is important to know whether or not and to what extent the molecules of an organic compound may form ions in a given environmental system.

A proton transfer can occur only if an *acid* (HA), that is, a proton *donor* (Brønsted and Pedersen 1924), reacts with a *base* (B), that is, a *proton acceptor*, since isolated protons are quite unstable species:

$$\begin{array}{l} HA \rightleftharpoons A^- + H^+ \\ H^+ + B \rightleftharpoons BH^+ \\ \hline HA + B \rightleftharpoons BH^+ + A^- \end{array} \tag{8-1}$$

Note that A^- is called the conjugate base of HA, and BH^+ the conjugate acid of B. Proton transfer reactions as described by Eq. 8-1 are usually *very fast and reversible*. It makes sense then that we treat such reactions as *equilibrium processes*, and that we are interested in the equilibrium distribution of the species involved in the reaction. In this chapter we confine our discussion to proton transfer reactions in *aqueous solution*, although in some cases, such reactions may also be important in nonaqueous media.

Our major concern will be the speciation of an organic acid or base (neutral vs. ionic species) in water under given conditions.

Organic Acids

Let us first consider the reaction of an organic acid (HA) with water which, in reaction 8-1, plays the role of the base:

$$HA + H_2O \rightleftharpoons H_3O^+ + A^- \tag{8-2}$$

From Chapter 3 we recall that for describing the energetics of a chemical reaction in dilute solution, it is convenient to use the infinite dilution state as a reference state for the solutes. For the solvent water, which in this case takes part in the reaction, the choice of the pure liquid as the reference state is, of course, more appropriate. Using the thermodynamic convention (e.g., Stumm and Morgan, 1981):

$$H^+ + H_2O \rightleftharpoons H_3O^+; \qquad K = 1 \quad \text{or} \quad \Delta G^0 = 0 \tag{8-3}$$

we can rewrite Eq. 8-2 as

$$HA \rightleftharpoons H^+ + A^- \tag{8-4}$$

At equilibrium, we then obtain (see Eq. 3-33)

$$\ln K_a = \ln \frac{(\gamma'_{H^+}[H^+])(\gamma'_{A^-}[A^-]}{(\gamma'_{HA}[HA])} = -\frac{\Delta G^0}{RT} \tag{8-5}$$

with $\Delta G^0 = -\mu^{0'}_{HA} + \mu^{0'}_{A^-}$ (since $\mu^{0'}_{H^+} = 0$ by convention; recall that primes indicate the infinite dilution reference state). Note that $\mu^{0'}_{HA}$ and $\mu^{0'}_{A^-}$ are the free energies of formation in aqueous solution of HA and A^-, respectively. The equilibrium constant

$$K_a = \frac{(\gamma'_{H^+}[H^+])(\gamma'_{A^-}[A^-])}{(\gamma'_{HA}[HA])} \tag{8-6}$$

is commonly referred to as the *acidity constant* or *acid dissociation constant* of the compound.

When determining K_a values of organic acids, one generally uses techniques by which the hydrogen ion activity $[pH = -\log(\gamma'_{H^+}[H^+])]$ is measured, while HA and A^- are determined as molar concentrations. Thus, most acidity constants reported in the literature are so-called "mixed acidity constants" which are operationally defined for a given aqueous medium (e.g., 0.05–0.1 M salt solution):

$$K_a^* \equiv (\gamma'_{H^+}[H^+])\frac{[A^-]}{[HA]} = K_a \frac{\gamma'_{HA}}{\gamma'_{A^-}} \tag{8-7}$$

Note that for small compound concentrations in aqueous media of moderate to low ionic strength, γ'_{A^-} and γ'_{HA} are close to 1, and hence $K_a^* \simeq K_a$. Using the common chemical shorthand of $pX = -\log X$, we obtain

$$\log\frac{[A^-]}{[HA]} = \log K_a - \log(\gamma'_{H^+}[H^+]) = pH - pK_a \tag{8-8}$$

Equation 8-8 allows us now to visualize the meaning of the *acidity constant* for a given organic compound. We can see that the pK_a is a measure of the *strength of an organic acid* relative to the acid–base pair: H_3O^+/H_2O. For example, it tells us at which hydrogen ion activity (expressed by the pH) our organic acid is present in equal parts in the dissociated (A^-) and nondissociated (HA) forms:

$$[A^-] = [HA] \quad \text{at} \quad pH = pK_a \tag{8-9}$$

If the pK_a of an organic acid is very low, (i.e., $pK_a \simeq 0$–3), we speak of a *strong organic acid*. A strong acid has a high tendency to deprotonate even in an aqueous solution of high H^+ activity (low pH). Examples of strong organic acids are trichloroacetic acid and 2,4,6-trinitrophenol (see Table 8.1). Consequently, at ambient pH values (i.e., pH = 4–10), such acids will be present in natural waters predominantly in their dissociated form; that is, as anions. The other examples given in Table 8.1 show that organic acids of environmental concern cover a broad pK_a range. Logically, *weaker acids* are those with higher pK_a values. Hence, very weak acids (i.e., $pK_a \gtrsim 12$) will be present in natural waters always in their nondissociated form. Many important organic acids, however, have pK_a values between 4 and 10. In these cases, exact knowledge of the pK_a value is necessary since, as already pointed out, the environmental behavior of the dissociated form of the molecule is very different from that of the nondissociated form.

Organic Bases

By analogy with the acids, we can define a basicity constant for the reaction of an organic base with water:

$$B + H_2O \rightleftharpoons OH^- + BH^+$$

$$K_b = \frac{(\gamma'_{OH^-}[OH^-])(\gamma'_{BH^+}[BH^+])}{(\gamma'_B[B])} \tag{8-10}$$

Here the reaction of a neutral base with water results in the formation of a cation. To compare acids and bases on a uniform scale, it is convenient to use the acidity constant of the conjugate acid, BH^+, as a measure of the base strength:

$$BH^+ \rightleftharpoons H^+ + B$$

$$K_a = \frac{(\gamma_{H^+}[H^+])(\gamma'_B[B])}{(\gamma'_{BH^+}[BH^+])} \tag{8-11}$$

TABLE 8.1 Examples of Neutral Organic Acids

Name	Structure of Acid	pK_a[a] (at 20–25°C)	Fraction[b] in Acid Form at pH 7 (α_a)
2,4,6-Trinitrophenol	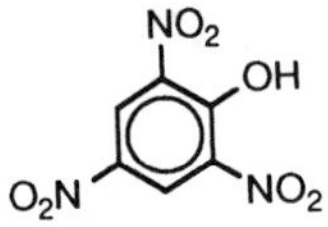	0.38	≪ 0.001
Trichloroacetic acid	Cl_3CCOOH	0.70	≪ 0.001
Chloroacetic acid	ClH_2CCOOH	2.85	≪ 0.001
Acetic acid	H_3CCOOH	4.75	0.006
Pentachlorophenol	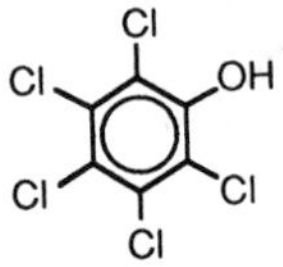	4.75	0.006
2,4,6-Trichlorophenol	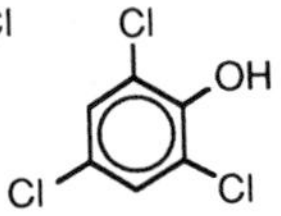	6.13	0.112
2-Nitrophenol	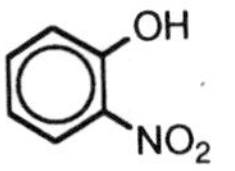	7.17	0.597
4-Chlorophenol		9.18	0.993
β-Naphthol	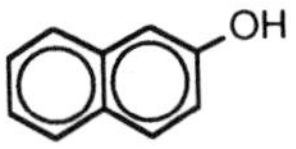	9.51	0.997
Phenol	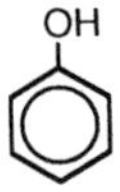	9.82	0.998
Methyl mercaptan	CH_3SH	10.7[c]	≫ 0.999
Aliphatic alcohols	$R{-}CH_2OH$	> 14.0	≫ 0.999

[a] From *CRC Handbook of Chemistry and Physics* (1985).
[b] Calculated using Eq. 8-16.
[c] From Reid (1958).

TABLE 8.2 Examples of Neutral Organic Bases

Name	Structure of Base	pK_a[a] ($= pK_{BH^+}$) (at 20–25°C)	Fraction[b] in *Base* Form at pH 7 ($1 - \alpha_a$)
Acetamide	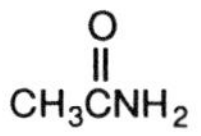 $CH_3C(=O)NH_2$	0.63	$\gg 0.999$
4-Nitroaniline	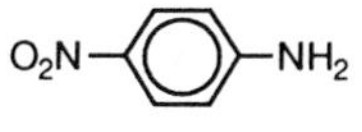 $O_2N-C_6H_4-NH_2$	1.00	$\gg 0.999$
3-Chloropyridine	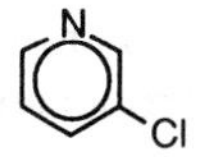	2.84	> 0.999
4-Chloroaniline		4.15	0.999
Aniline	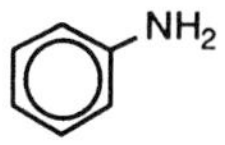	4.63	0.996
Isoquinoline	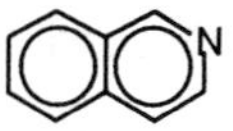	5.25	0.983
Pyridine	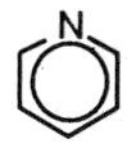	5.42	0.974
2,4-Dimethylpyridine	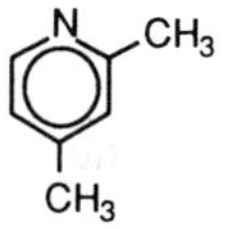	6.95	0.529
Imidazol	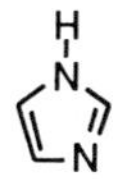	6.95	0.529
2,4-Dimethylimidazol	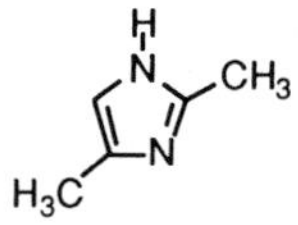	8.36	0.042
Trimethylamine	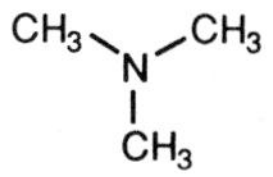 $(CH_3)_3N$	9.81	0.002
Methylamine	CH_3-NH_2	10.66	< 0.001
Pyrrolidine	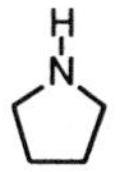	11.27	$\ll 0.001$

[a]From *CRC Handbook of Chemistry and Physics* (1985).
[b]Calculated using Eq. 8-16.

K_b and K_a are quantitatively interrelated by the *autodissociation constant of water* (*ion product of water*), K_w:

$$K_w = K_a \cdot K_b = (\gamma'_{H^+}[H^+])(\gamma'_{OH^-}[OH^-]) = 1.01 \times 10^{-14} \quad (8\text{-}12)$$

at 25°C for pure water. Note that K_w is strongly temperature dependent (see Section 8.3). Using our pX nomenclature:

$$pK_a = pK_w - pK_b \quad (8\text{-}13)$$

From Eq. 8-13 it follows that the stronger an acid is (low pK_a), the weaker the basicity of its conjugate base (high pK_b), while the stronger the base (low pK_b), the weaker its conjugate acid (high pK_a). Thus, a neutral base with a pK_b value < 3 (i.e., the pK_a of the conjugate acid > 11!) will be present in water predominantly as a cation at ambient pH values. Some examples of important organic bases are shown in Table 8.2.

Overview of Acid and Base Functional Groups

Figure 8.1 gives the range of pK_a values for some important functional groups that have either proton donor or proton acceptor properties. As already pointed out, we are primarily interested in compounds having pK_a values in the range of 3 to 11; therefore, the most important functional groups we have to consider include hydroxyl groups bound to aromatic rings (e.g., phenolic compounds), aliphatic and aromatic carboxyl groups, aliphatic and aromatic amino groups, nitrogen atoms incorporated in aromatic compounds (e.g., pyridines), and aliphatic or aromatic thiols (see examples given in Tables 8.1 and 8.2). The range in pK_a values shown for a given functional group in Figure 8.1 may vary by many units because of the structural characteristics of the remainder of the molecule. Depending on the type and number of substituent groups on the aromatic ring, for example, the pK_a values for substituted phenols may differ by almost 10 units (Table 8.1). It is necessary, therefore, that we make an effort to understand the effects of various structural entities on the acid or base properties of a given functional group. But before doing so, a few more general remarks about organic acids and bases are needed.

So far, we have dealt with organic acids and bases that possess only one acid or base group in the pK_a range of interest. There are, however, compounds with more than one acid or base function. An example of a "two-protic acid" is given in Figure 8.2*a*. In such cases, it is possible that a molecule is present in aqueous solution as a doubly charged anion. Similarly, as illustrated by 1,2-diaminopropane in Figure 8.2*b*, a "two-protic" base may form doubly charged cations. A very interesting case involves those compounds that have both acidic and basic functions, such as amino acids and the hydroxy-isoquinoline shown in Figure 8.2*c*. Here it is not always possible to unambiguously specify a proton transfer reaction in terms of the actual chemical species involved. In the case of a simple amino acid, for example, proton transfer may occur by two different pathways:

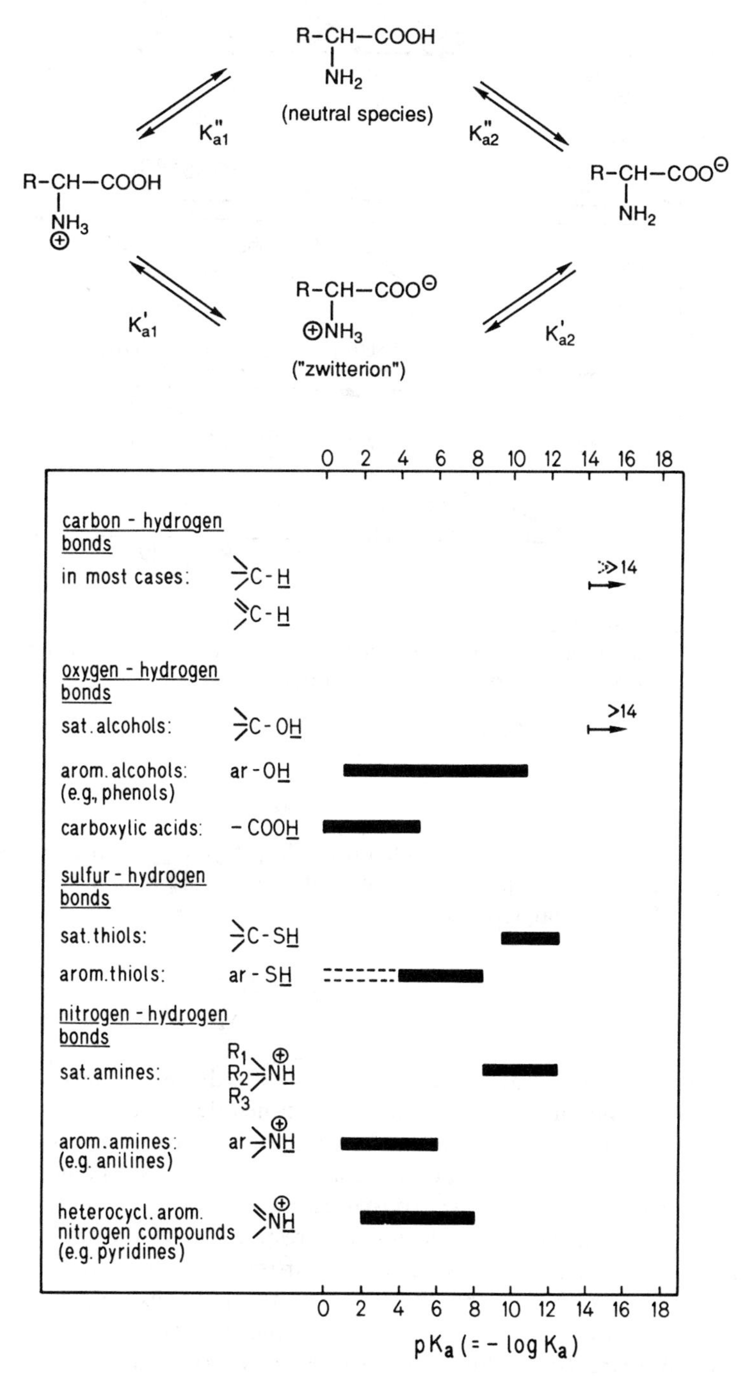

Figure 8.1 Typical ranges of aqueous acidity constants (expressed as pK_a) for hydrogen bound to various positions in organic molecules.

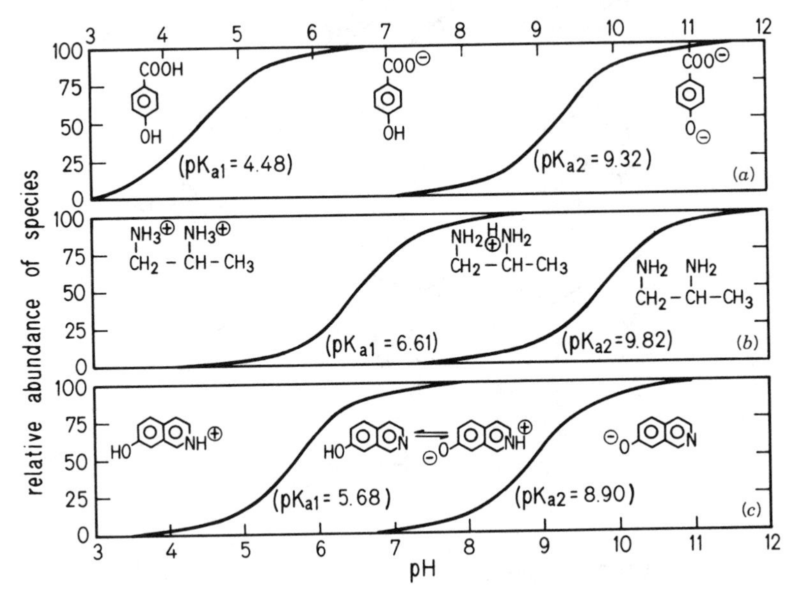

Figure 8.2 Relative abundances of the conjugated acid–base species as a function of pH for some compounds exhibiting more than one acid or base moiety: (*a*) 4-hydroxy benzoic acid, (*b*) 1,2-diaminopropane, and (*c*) 7-hydroxyisoquinoline.

Although four microscopic acidity constants ($K'_{a1}, K''_{a1}, K'_{a2}, K''_{a2}$) may be defined, only two apparent (macroscopic) acidity constants, K_{a1}, and K_{a2}, may be determined experimentally (Fleck, 1966, Chapter 5):

$$K_{a1} = K'_{a1} + K''_{a1}; \qquad K_{a2} = \frac{K'_{a2} K''_{a2}}{K'_{a2} + K''_{a2}} \tag{8-14}$$

By comparing the magnitude of K_{a1} and K_{a2} with the K_a values of structurally comparable acid functions, one can, however, conclude whether or not *zwitterion* formation is important. In the case of the amino acids with $pK_{a1} = 2–3$ and $pK_{a2} = 9–11$, zwitterion formation is very likely, since pK_{a1} is very similar to that of a carboxylic acid carrying an electron-withdrawing α-substituent (compare with chloroacetic acid in Table 8.1), and the pK_{a2} corresponds to that of an aliphatic amine (see examples in Table 8.2). In contrast, for 7-hydroxy-quinoline (Fig. 8.2*c*), zwitterion formation is very unlikely, since the pK_{a1} corresponds to that exhibited by the nitrogen in isoquinoline (Table 8.2), and the pK_{a1} is more typical of monosubstituted β-naphthols (compare β-naphthol, Table 8.1). Finally, note that for compounds such as amino acids and hydroxy-isoquinolines, a pH value exists at which the *average net charge* of all species present is zero. This pH value is called the *isoelectric pH* and is given by

$$pH_{isoelectric} = \tfrac{1}{2}(pK_{a1} + pK_{a2}) \tag{8-15}$$

Speciation in Natural Waters

Given the pK_a of an organic acid or base, we can now ask to what extent this compound is ionized in a natural water; that is, what are the relative abundances of the neutral versus the charged species? The pH of a natural water is primarily determined by various *inorganic* acids and bases (e.g., H_2CO_2, HCO_3^-, $CO_3^=$) which are usually present at much higher concentrations than the compounds that interest us (Stumm and Morgan, 1981, Chapters 3 and 9). These acids and bases act as hydrogen ion buffers, meaning that the addition of a very small quantity of acid or base will not detectably change the pH. We can easily visualize this buffering effect by the following simple example. Let us assume that a hypothetical acid–base pair ($pK_a = 7.00$) is present at equal concentrations in one liter of water, say 10^{-3} mol·L^{-1}. According to Eq. 8-8, the pH of this aqueous solution will then be

$$pH = 7.00 + \log \frac{10^{-3}\,\text{mol}\cdot\text{L}^{-1}}{10^{-3}\,\text{mol}\cdot\text{L}^{-1}} = 7.00$$

If we now add 10^{-5} moles of a *strong* organic acid, (i.e., we add 10^{-5} moles H^+), for example, 2,4,6-trinitrophenol (which would correspond to a total concentration of this compound of 10^{-5} mol·L^{-1} or 2 mg L^{-1}), the pH would change by less than 0.01 units:

$$pH = 7.00 + \log \frac{0.99 \times 10^{-3}\,\text{mol}\cdot\text{L}^{-1}}{1.01 \times 10^{-3}\,\text{mol}\cdot\text{L}^{-1}} = 6.991$$

As a first approximation then, we may assume that adding a "trace" organic acid or base (where trace < 0.1 mM) to a natural water will, in most cases, not significantly affect the pH of the water.

For a given pH, we may now express the fraction of our organic acid (denoted as HA, the same holds for BH^+) present in the acid form in the water, α_a, by

$$\alpha_a = \frac{[HA]}{[HA] + [A^-]} = \frac{1}{1 + \dfrac{[A^-]}{[HA]}} = \frac{1}{1 + 10^{(pH - pK_a)}} \tag{8-16}$$

Tables 8.1 and 8.2 give calculated α_a values for various acids and bases in water at pH 7. It should be reemphasized that the neutral and ionic "forms" of a given neutral acid (base) behave very differently in the environment. Depending on the process considered, either the neutral or ionic species may be the dominant factor in the compound's "reactivity," even if the relative abundance of that species is very low.

8.2 CHEMICAL STRUCTURE AND ACIDITY CONSTANT

Let us now consider the structural features that determine the acidity constant of a given functional group. First, recall from Chapter 3 that the acid–base equilibrium constant of a compound is related to the standard chemical potentials of the conjugate acid–base species; that is, to the partial molar free energies of these species in their standard state (e.g., infinite dilution state, 1 M, 25°C, 1 atm); see Eq. 8-5:

$$\ln K_a = -\frac{\Delta G^0}{RT}$$

$$= \frac{1}{RT}(\mu^{0'}_{HA(BH^+)} - \mu^{0'}_{A^-(B)}) \qquad (8\text{-}17)$$

$$pK_a = -\frac{1}{2.303\,RT}(\mu^{0'}_{HA(BH^+)} - \mu^{0'}_{A^-(B)})$$

Hence, the pK_a is determined by the difference between the standard chemical potential of the acid ($\mu^{0'}_{HA}$) and its conjugate base ($\mu^{0'}_{A^-}$) in aqueous solution. Therefore, when comparing acidity constants of compounds exhibiting a specific acid (or base) functional group, the question is how much the rest of the molecule favors (decreases μ^0) or disfavors (increases μ^0) the ionic versus the neutral form of the compound in aqueous solution.

Inductive Effects

Let us first consider a simple example, the influence of a chloro-substituent on the pK_a of butyric acid:

	$CH_3CH_2CH_2COOH$	$CH_2(Cl)CH_2CH_2COOH$	$CH_3CH(Cl)CHCOOH$	$CH_3CH_2CH(Cl)COOH$
pK_a	4.81	4.52	4.05	2.86

In this example, we see that if we substitute a hydrogen atom by chlorine, which is much more electronegative than hydrogen (see Chapter 2), the pK_a of the carboxyl group decreases. Furthermore, the closer the electron-withdrawing chlorine substituent is to the carboxyl group, the stronger is its effect on pK_a. We can intuitively explain these findings by realizing that any group that will have an electron-withdrawing effect on the carboxyl group (or any other acid function) will help to accommodate a negative charge and increase the stability of the ionized form. In the case of an organic base, an electron-withdrawing substituent will, of course, destabilize the acidic form (the cation) and, therefore, also lower the pK_a. This effect is called a

TABLE 8.3 Inductive and Resonance Effects of Some Common Substituents[a]

Effect[b]	Substituents
	Inductive
+I	O^-, NH^-, alkyl
−I	SO_2R, NH_3^+, NO_2, CN, F, Cl, Br, COOR, I, COR, OR, SR, phenyl, NR_2
	Resonance
+R	F, Cl, Br, I, OH, OR, NH_2, NR_2, NHCOR, O^-, NH^-
−R	NO_2, CN, CO_2R, $CONH_2$, phenyl, COR, SO_2R

[a]From Clark and Perrin (1964).
[b]A plus sign means that the effect increases the pK_a; a minus sign means that the effect decreases the pK_a.

negative inductive effect (− I). Table 8.3 shows that most functional groups with which we are concerned have inductive electron-withdrawing (− I) effects, and only a few electron-donating (+ I) effects such as, for example, alkyl groups:

	CH_3COOH	CH_3CH_2COOH
pK_a	4.75	4.87

As illustrated by the chlorobutyric acids discussed above, in *saturated* molecules inductive effects usually fall off quite rapidly with distance.

Delocalization Effects

In unsaturated chemicals, such as aromatic or olefinic compounds (i.e., compounds with "mobile" π-electrons; see Chapter 2), the inductive effect of a substituent may be felt over larger distances (i.e., more bonds). In such systems, however, another effect, the *delocalization of electrons*, may be of even greater importance. In Chapter 2, we learned that the delocalization of electrons (i.e., the "smearing" of π-electrons over several bonds) may significantly increase the stability of an organic species. In the case of an organic acid, delocalization of the negative charge may, therefore, lead to a considerable decrease in the pK_a of a given functional group, as one can see from comparing the pK_a of an aliphatic alcohol with that of phenol (Fig. 8.3*a*). Analogously, by stabilizing the neutral species, the delocalization of the free electrons of an amino group has a very significant effect on the pK_a of the conjugated ammonium ion (see Fig. 8.3*b*).

In the next step, we introduce a substituent on the aromatic ring which, through the aromatic π-electron system, may develop shared electrons (i.e., through "resonance" or "conjugation") with the acid or base function (e.g., the —OH or —NH_2 group). For example, the much lower pK_a value of *para*-nitrophenol as compared with *meta*-nitrophenol may be attributed to additional resonance stabilization of the (*a*) $R-CH_2-OH \rightleftharpoons R-CH_2-O^{\ominus}$ $pK_a > 14$

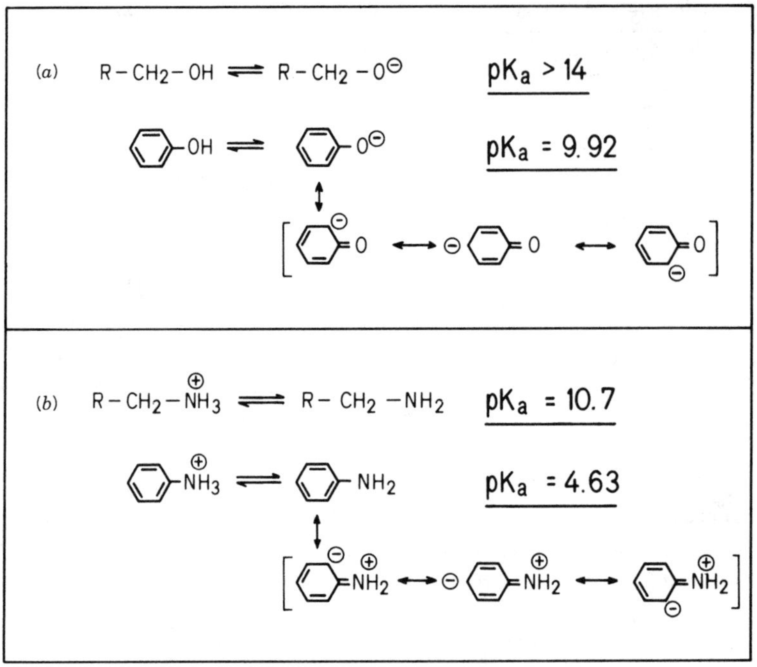

$C_6H_5-OH \rightleftharpoons C_6H_5-O^{\ominus}$ $pK_a = 9.92$

(*b*) $R-CH_2-\overset{\oplus}{N}H_3 \rightleftharpoons R-CH_2-NH_2$ $pK_a = 10.7$

$C_6H_5-\overset{\oplus}{N}H_3 \rightleftharpoons C_6H_5-NH_2$ $pK_a = 4.63$

Figure 8.3 Effect of delocalization on the pK_a of $-OH$ and $-NH_3^+$.

anionic species by the *para*-positioned nitro group (see Fig. 8.4). In the meta position, only the electron-withdrawing inductive effect of the nitro group is felt by the —OH group. Other substituents that increase acidity (i.e., that lower the pK_a, "$-$ R" effect) are listed in Table 8.3. All of these substituents can help to accommodate electrons. On the other hand, substituents with heteroatoms having nonbonding electrons that may be in resonance with the π-electron system, have an *electron-donating resonance effect* (+ R, see examples given in Table 8.3), and, will therefore, decrease acidity (i.e., increase pK_a). Note that many groups that have a negative inductive effect ($-$ I) at the same time have a positive resonance effect (+ R). The overall impact of such substituents depends critically on their location in the molecule. In monoaromatic molecules, for example, resonance in the meta position is negligible, but will be significant in both the ortho and para positions.

Proximity Effects

Another important group of effects are *proximity effects*; that is, effects arising from the influence of substituents that are physically close to the acid or base function under consideration. Here, two interactions are important: *intra*molecular (within the same molecule) *hydrogen bonding* and *steric effects*. An example of the effect of intramolecular hydrogen bonding is given in Figure 8.5*a*. The stabilization of the carboxylate anion by

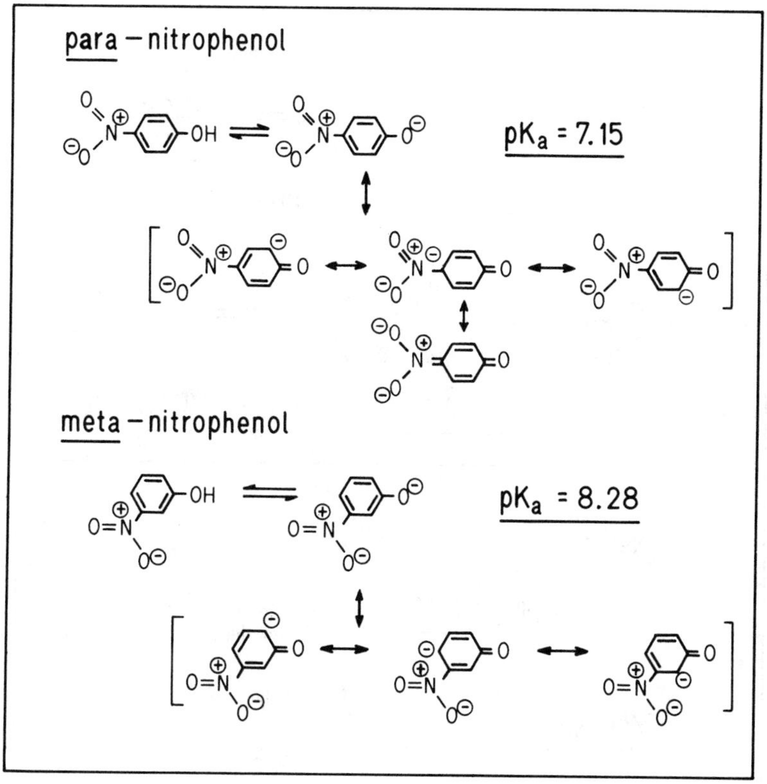

para – nitrophenol

$pK_a = 7.15$

meta – nitrophenol

$pK_a = 8.28$

Figure 8.4 Influence of the position of a nitro substituent on the pK_a of a phenolic hydrogen.

the hydroxyl hydrogen in ortho-hydroxy-benzoic acid (salicylic acid) leads to a much lower pK_{a1} value and to a much higher pK_{a2} value compared with *para*-hydroxy-benzoic acid in which no intramolecular hydrogen bonding is possible.

In some cases, steric effects may have a measurable impact on the pK_a of a given acid or base function. This involves steric constraints that inhibit optimium solvation of the ionic species by the water molecules (and thus increase the pK_a), or hinder the resonance of the electrons of a given acid or base group with other parts of the molecule by causing these groups to twist with respect to one another and to avoid coplanarity. For example, the large difference found between the pK_a of *N*,*N*-dimethylaniline and of *N*,*N*-diethylaniline (Fig. 8.5*b*) is partially due to the larger ethyl substituents that limit the orientation of the free electrons of the nitrogen atom and, thus, their resonance with the π-electrons of the aromatic ring.

In summary, the most important factors influencing the pK_a of a given acid or base function are inductive, resonance, and steric effects. The impact of a substituent on the pK_a depends critically on where the substituent is located in the molecule relative to the acid or base group. In one place, a given substituent may have only one of the mentioned effects, while in another location, all effects may play a role. It is quite

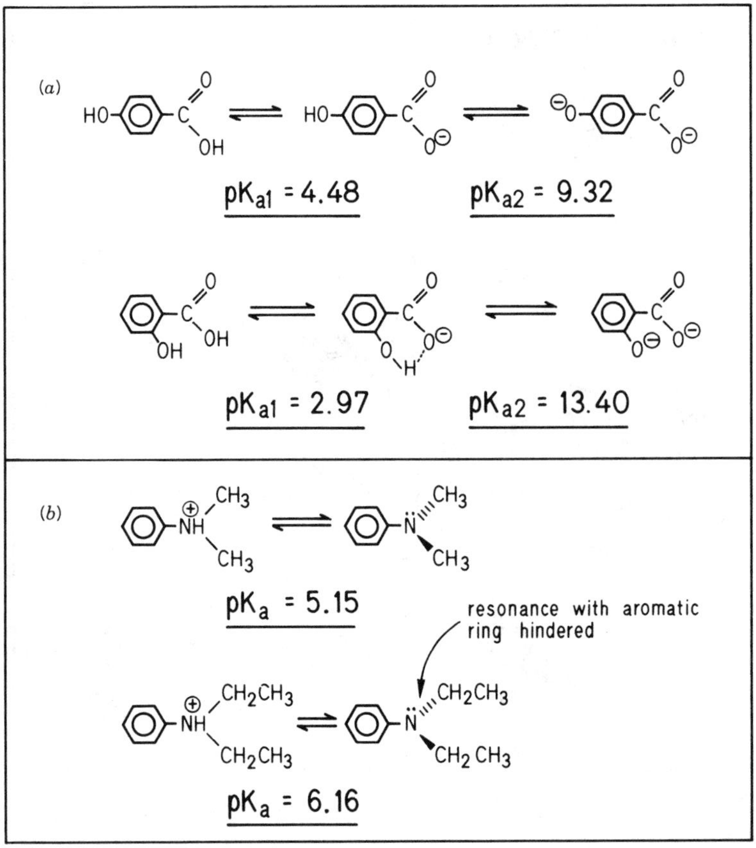

Figure 8.5 Examples of proximity effects on acidity constants: (*a*) hydrogen bonding and (*b*) steric interactions.

difficult, therefore, to establish simple general rules for *quantifying* the effect(s) of structural entities on the pK_a of an acid or base function. Nevertheless, in certain restricted cases, a quantification of the effects of substituents on the pK_a value is possible by using LFERs. In the next section, we discuss one example of such an approach, the *Hammett correlation* for aromatic compounds. First, however, a few comments on the availability of experimental pK_a values are necessary.

8.3 AVAILABILITY OF EXPERIMENTAL pK_a VALUES AND pK_a ESTIMATION METHODS

Experimental pK_a Values

Acidity has long been recognized as a very important property of some organic compounds. Experimental methods for determining acidity constants are well established, and there is quite a large data base of pK_a values of organic acids and bases (e.g., Kortüm et al., 1961: Perrin, 1972; Serjeant and Dempsey, 1979; *CRC Handbook of Chemistry and Physics*, 1985). The most common procedures discussed by Kortüm et al. (1961) include titration, determination of the concentration ratio of acid–base pairs at various pH values using conductance methods, electrometric methods, and spectrophotometric methods. It should again be noted that pK_a values reported in the literature are often "mixed acidity constants" (see Section 8.1), that are commonly measured at 20 or 25°C, and at a given ionic strength (e.g., 0.05–0.1 M salt solution). Depending on the type of measurement and the conditions chosen, therefore, pK_a values reported in the literature for a given organic acid or base may vary by as much as 0.3 pK_a units. Depending on the type of acid or base, the effect of temperature may be more or less pronounced. The pK_a values of carboxylic acids, for example, are rather insensitive to changes in temperature. With increasing pK_a value, however, the enthalpy of reaction, and thus the effect of temperature, increases. For example, the pK_a of 2-nitrophenol is 7.40 at 5°C and 7.20 at 25°C, that is, a decrease of 0.01 pK_a units is observed with one degree increase in temperature (Schwarzenbach et al., 1988). Nevertheless, as a first-approximation, reported pK_a values of 20 or 25°C can also be used at lower temperatures, since often the uncertainty of the measurement is of a similar magnitude as the temperature effect on pK_a. [Note, however, the autodissociation constant of water (Eq. 8-12) is strongly temperature dependent: pK_w(5°C) = 14.74; pK_w(25°C) = 13.99; pK_w(45°C) = 13.40]. Finally, for practical purposes, one can usually neglect the effect of ionic strength on pK_a up to about 0.1 M.

Estimation of pK_a Values: The Hammett Correlation

In Chapter 7 we used LFERs to quantify the effects of structural entities on the partitioning behavior of organic compounds. In an analogous way, LFERs can be used to *quantitatively* evaluate the influence of structural moieties on the pK_a of a given acid or base function, particularly if only electronic effects are important.

About 50 years ago, Hammett (1940) recognized that for *substituted benzoic acids* (see Fig. 8.6), the effect of substituents in either the meta or para position on the free-energy change of dissociation of the carboxyl group could be expressed as the sum of the free-energy change of the dissociation of the unsubstituted compound, ΔG_H, and the contributions of the various substituents; ΔG_i^0:

$$\Delta G^0 = \Delta G_H^0 + \sum_i \Delta G_i^0 \tag{8-18}$$

To express the effect of substituent i on pK_a, Hammett introduced a constant, σ_i, that is defined as

$$\sigma_i = \frac{-\Delta G_i^0}{2.303\,RT} \tag{8-19}$$

Since σ_i differs for meta and para substitutions, there are two sets of σ values, σ_{meta}

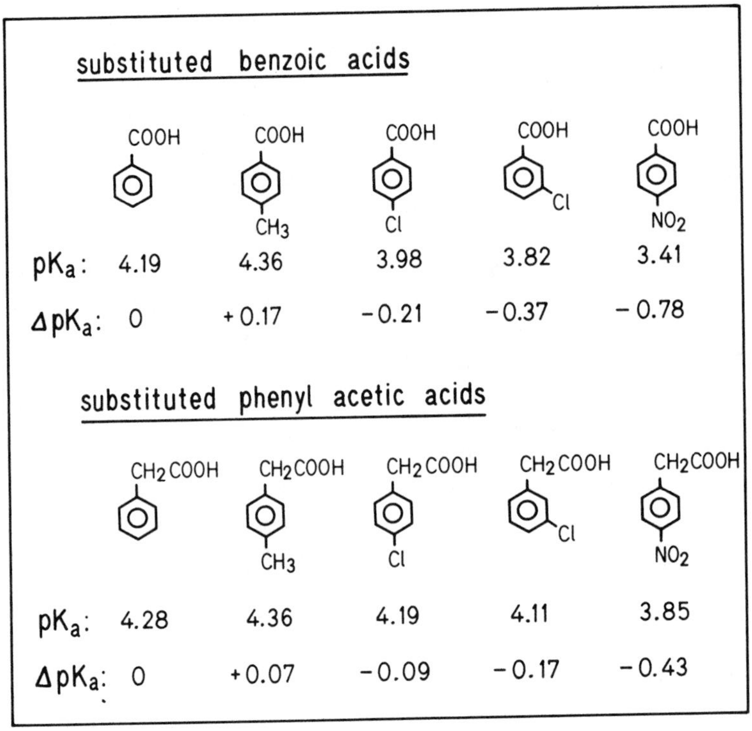

substituted benzoic acids

	COOH	COOH, CH_3	COOH, Cl	COOH, Cl	COOH, NO_2
pK_a:	4.19	4.36	3.98	3.82	3.41
ΔpK_a:	0	+0.17	−0.21	−0.37	−0.78

substituted phenyl acetic acids

	CH_2COOH	CH_2COOH, CH_3	CH_2COOH, Cl	CH_2COOH, Cl	CH_2COOH, NO_2
pK_a:	4.28	4.36	4.19	4.11	3.85
ΔpK_a:	0	+0.07	−0.09	−0.17	−0.43

Figure 8.6 Effect of ring substituents on the pK_a of benzoic acid and phenyl acetic acid.

and σ_{para}. *Ortho* substitution is excluded since, as we have already seen, proximity effects, which are difficult to separate from electronic factors, may play an important role. Since $\Delta G^0 = -2.303\,RT \log K_a$, we may write Eq. 8-18 in terms of acidity constants:

$$\log \frac{K_a}{K_{aH}} = \sum_i \sigma_i \qquad \text{or} \qquad pK_a = pK_{aH} - \sum_i \sigma_i \tag{8-20}$$

Table 8.4 lists σ_{meta} and σ_{para} values for some common substituent groups. Note that these σ values are a quantititative measurement of the effect of a given substituent on the pK_a of *benzoic acid*. As we would expect from our previous discussion, the sign of the σ value reflects the net electron-withdrawing (negative sign) or electron-donating (positive sign) character of a given substituent in either the meta or para position. For example, we see that $-NO_2$ and $-C{\equiv}N$ are strongly electron-withdrawing in both positions, while the electron-providing groups, $-NH_2$ or $-N(CH_3)_2$, are strongly electron-donating in the para position, but show a much weaker effect in the meta position. The differences between σ_{meta} and σ_{para} of a given substituent are due to the difference in importance between the inductive and resonance effects which, as we mentioned earlier, may have opposite signs (see Table 8.3).

TABLE 8.4 Hammett Constants for Some Common Substituents[a]

Substituent	σ_{meta}	σ_{para}	σ^{-}_{para}	$\sigma^{phenols}_{ortho}$	$\sigma^{anilines}_{ortho}$
$-H$	0.00	0.00	0.00	0.00	0.00
$-CH_3$	−0.07	−0.17		−0.13	0.10
$-CH_2CH_3$	−0.07	−0.15			
$-CH_2CH_2CH_2CH_3$	−0.07	−0.16		−0.18	
$-C_6H_5$	0.06	0.01			
$-CHO$	0.36	0.22	1.13	0.75	
$-CH_2OH$	0.08	0.08		0.04	
$-COCH_3$	0.38	0.50	0.87		
$-COOCH_3$	0.32	0.39	0.64		
$-CONH_2$	0.28	0.36	0.63	0.72	
$-CN$	0.56	0.66	0.89		
$-OH$	0.10	−0.37			−0.09
$-OCH_3$	0.07	−0.24	−0.20	0.00	0.02
$-NH_2$	−0.04	−0.66			0.00
$-N(CH_3)_2$	−0.05	−0.83			
$-NO_2$	0.71	0.78	1.25	1.24	1.72
$-F$	0.34	0.06	−0.02	0.54	0.47
$-Cl$	0.37	0.23		0.68	0.67
$-Br$	0.39	0.23	0.26	0.70	0.71
$-I$	0.35	0.18		0.63	0.70

[a]Values taken from Williams (1984) and Lowry and Schueller-Richardson (1981). Apparent σ values for *ortho* substitution in phenols and anilines from Barlin and Perrin (1966) and Clark and Perrin (1964).

Let us now examine the effects of the *same* substituents on the pK_a of another group of acids, the substituted phenyl acetic acids. As we might have anticipated, Figure 8.6 shows that each of the various substituents exerts the same relative effect as in their benzoic counterparts; however, in the case of phenyl acetic acid, the greater separation between substituent and reaction site makes the impact less pronounced than in the benzoic acid. Plotting pK_{aH}−pK_a values for meta- and para-substituted phenyl acetic acids versus $\sum_i \sigma_i$ values results in a straight line with a slope ρ of less than 1 (Fig. 8.7). In this case, introduction of a substituent on the aromatic ring has only about half the

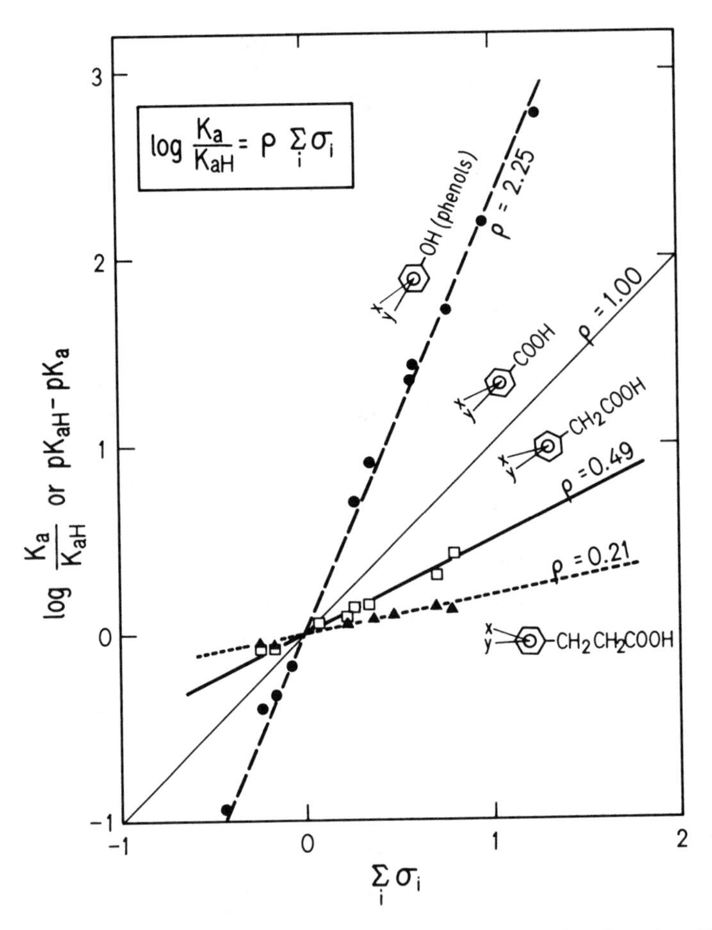

Figure 8.7 Hammett plots for *meta*- and *para*-substituted phenols, phenylacetic acids, and 3-phenylpropionic acids (data from Serjeant and Dempsey, 1979).

effect on the pK_a as compared with the effect of the same substituent on the pK_a of benzoic acid. Thus, ρ is a measure of how sensitive the dissociation reaction is to substitution as compared with substituted benzoic acid and is commonly referred to as the *susceptibility factor* that relates one set of reactions to another. If we consider another group of acids, the substituted β-phenyl propionic acids, where the substituents are located at even greater distances from the carboxyl group, yet smaller ρ values are expected and found ($\rho = 0.21$, Fig. 8.7).

If we express these findings in energetic terms, we obtain

$$\Delta G^{0'} = \Delta G_{\mathrm{H}}^{0'} - \rho\, 2.303\, RT \sum_i \sigma_i \tag{8-21}$$

where the primes on the ΔG^0 values indicate that they apply to β-phenyl propionic acids. By substituting $-2.303\,RT\sum_i \sigma_i$ with $\Delta G^0 - \Delta G_H^0$ (Eqs. 8-18 and 8-19) and rearranging Eq. 8-21, we realize that we have applied a linear free-energy relationship to relate pK_a values between two different groups of acids in a very similar way to that used to relate organic solvent–water partition constants in different solvent–water systems (see Section 7.4):

$$\Delta G^{0'} = \rho(\Delta G^0 - \Delta G_H^0) + \Delta G_H^{0'} \tag{8-22}$$

The classical form of the *Hammett Equation* is Eq. 8-21, expressed in terms of equilibrium constants (i.e., acidity constants):

$$\log K_a = \log K_{aH} + \rho \sum_i \sigma_i \qquad \text{or} \qquad pK_a = pK_{aH} - \rho \sum_i \sigma_i \tag{8-23}$$

Note that we have omitted the prime notation in Eq. 8-23. Examples of some Hammett relationships for ionization constants are given in Table 8.5.

In many cases, the simple approach of using σ_{meta} and σ_{para} values is applied with reasonable success. One should be aware, however, that good correlations are not always obtained, simply implying that in those cases one has not incorporated all of the molecular interactions into the LFER that play a role in that particular system. This is usually encountered when substituents exhibit a more complex interaction with the reaction center or when substituents interact with one another.

A simple case where the general σ constants in Table 8.4 do not succeed in correlating acidity constants is when the acid or base function is in *direct resonance* with the substituent. This may occur in substituted phenols and anilines. For example, owing to resonance (see Fig. 8.4), a para nitro group decreases the pK_a of phenol much more than would be predicted from the σ_{para} constant obtained from the dissociation of *p*-nitrobenzoic acid. In such "resonance" cases (another example would be the anilines), a special set of σ values (denoted as σ_{para}^-) has been derived (Table 8.4) to try to account for both inductive and resonance effects. If these values are employed, good correlations are obtained, as shown for meta- and para-substituted phenols in Fig. 8.7. Note that for compound classes such as phenols, anilines, and pyridines where the acid (base) function is in resonance with the aromatic ring, the ρ values obtained are significantly greater than 1 (Table 8.5); that is, the electronic effect of the substituents is greater than in the case of benzoic acid.

We have stated earlier that because of proximity effects, no generally applicable σ values may be derived for ortho substitution. Nevertheless, one can determine a set of *apparent* σ_{ortho} values for a specific type of reaction, as for example, for the dissociation of substituted phenols. Table 8.4 gives such apparent σ_{ortho} constants for estimating pK_a values of substituted phenols and anilines. Figure 8.8 shows that such values can be applied with reasonable success even for many polysubstituted phenols. Of course, in cases of multiple substitution, substituents may interact with one another, thereby resulting in larger deviations of experimental from predicted pK_a values.

It should also be noted that ortho substituents that allow hydrogen bonding with the acid or base functionality, may also have an influence on the susceptibility factor

TABLE 8.5 Hammett Relationships for Ionization Constants of Various Acids[a]

Acid	ρ	pK_{aH} (Unsubstituted Ring)
X,Y-C_6H_3—COOH	1.00 (by definition)	4.20
X,Y-C_6H_3—CH_2—COOH	0.49	4.30
X,Y-C_6H_3—CH_2—CH_2—COOH	0.21	4.55
X,Y-C_6H_3—O—CH_2—COOH	0.30	3.17
X,Y-C_6H_3—OH	2.25	9.92
X,Y-C_6H_3—$NH_3^{\oplus}$	2.89	4.63
X,Y-C_6H_3—CH_2—$NH_3^{\oplus}$	1.06	9.39
X,Y-C_5H_3N—$NH^{\oplus}$	5.90	5.25

[a] $pK_a = pK_{aH} - \rho \sum_i \sigma_i$. Data from Williams (1984).

ρ. For example, as compared to $\rho = 2.25$ for meta- and para-substituted phenols (Fig. 8.8), a ρ value of 2.59 has been found for a series of substituted 2-nitrophenols (Schwarzenbach et al., 1988). This difference probably results because a nitro substituent ortho to the phenolic group may stabilize the nondisassociated form by intramolecular hydrogen bonding:

R, X-substituted 2-nitrophenol: O—H ··· $O^{\ominus}$—$N^{\oplus}$(=O)

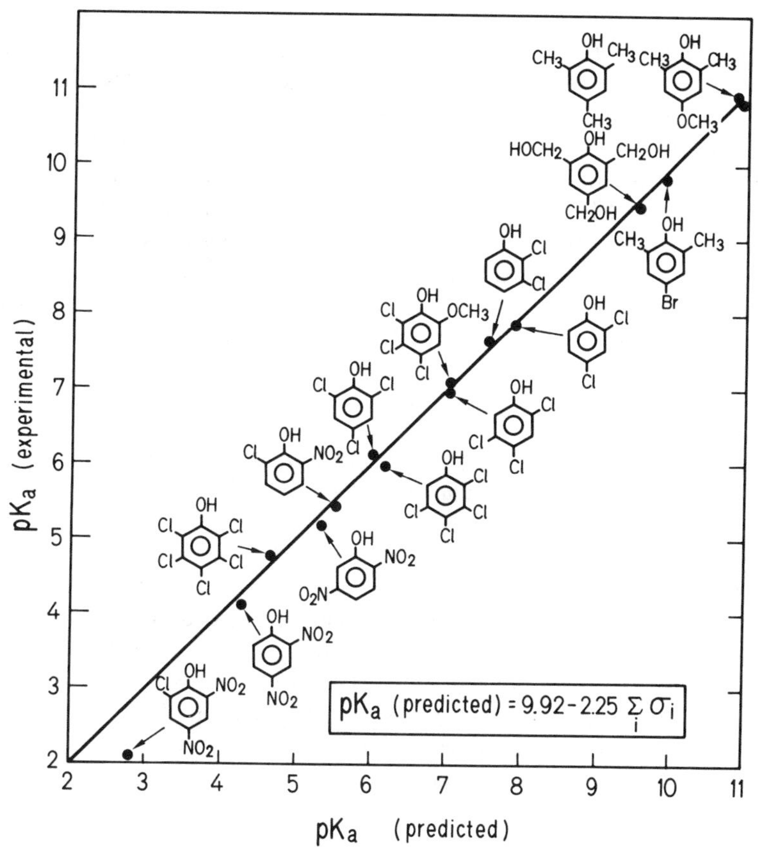

Figure 8.8 Comparison of predicted and experimental pK_a values of polysubstituted phenols using the σ values given in Table 8.4 for phenolic compounds.

An electron-withdrawing substituent on the ring will now weaken this hydrogen bond, which means that, in addition to its stabilizing effect on the anionic form, the substituent will destabilize the nondissociated form (the opposite effect will be observed for an electron-donating substituent). Thus, the effect of a substituent on the pK_a of the phenolic group will be enhanced by the presence of a nitro group in the ortho position.

In our discussion of the Hammett correlation, we have confined ourselves mostly to benzene derivatives. Of course, a similar approach can be taken for other aromatic systems, such as for the derivatives of polycyclic aromatic hydrocarbons and heterocyclic aromatic compounds. For a discussion of such applications, we refer to papers by Clark and Perrin (1964), Barlin and Perrin (1966), and Perrin (1980). Using the Hammett equation as a starting point, a variety of refinements using more sophisticated sets of constants have also been suggested. The interested reader can find a treatment of these approaches, as well as compilations of substituent constants, in various textbooks (e.g., Hine, 1975; Lowry and Schueller-Richardson, 1981; Williams, 1984) and in data collections (e.g., Hansch and Leo, 1979; Harris and Hayes, 1982). These references also give an overview of parallel approaches, such as the *Taft correlation* developed to predict pK_a values in *aliphatic and alicyclic* systems.

In summary, in this section we have discussed the electronic and steric effects of structural moieties on the pK_a value of acid and base functions in organic molecules. We have seen how LFERs can be used to quantitatively describe these electronic effects. At this point, it is important to realize that we have used such LFERs to evaluate the relative stability and, hence, the relative energy status of organic species in aqueous solution (e.g., anionic vs. neutral species). It should come as no surprise then that we will find similar relationships when dealing with chemical reactions other than proton transfer processes.

8.4 COMMENTS ON AQUEOUS SOLUBILITY AND PARTITIONING BEHAVIOR OF ORGANIC ACIDS AND BASES

Since the water solubility of the ionic form (salt) of an organic acid or base is generally several orders of magnitude higher than the solubility of the neutral species, $C_w^{sat}(HA)$, the total concentration of the compound (nondissociated and dissociated forms) at saturation, $C_{w,tot}^{sat}$, is strongly pH-dependent. As is illustrated schematically for an organic acid, HA, in Figure 8.9, at low pH, the saturation concentration is given by the solubility of the neutral compound. At higher pH values, $C_{w,tot}^{sat}$ is determined by the fraction in the nondissociated form, α_a (Eq. 8-16):

$$C_{w,tot}^{sat} = \frac{C_w^{sat}(HA)}{\alpha_a} \tag{8-24}$$

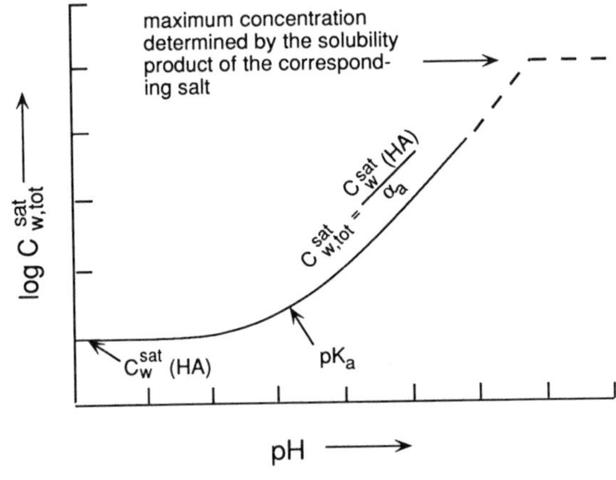

Figure 8.9 Schematic representation of the solubility of an organic acid (HA) as a function of pH.

Eq. 8-24 is valid, of course, only up to the solubility product(s) of the salt(s) of the ionized organic species (which is dependent on the type of counterion(s) present). Unfortunately, solubility data of organic salts are extremely scarce. Note that for an organic base, $C^{sat}_{w,tot}$ is given by

$$C^{sat}_{w,tot}(B) = \frac{C^{sat}_{w}(B)}{1-\alpha_a} \tag{8-25}$$

and that the corresponding relation would be symmetrical to the one shown for HA in Figure 8.9 (i.e., higher total concentrations at lower pH where the ionic form dominates).

When considering the air–water equilibrium partitioning of an organic acid or base, we may, in general, assume that the ionized species will not be present in the gas phase. The air–water distribution ratio of an organic acid, $D_{aw}(HA, A^-)$ (note that we speak of a ratio and not of a partition constant since we are dealing with more than one species), is then given by

$$\begin{aligned} D_{aw}(HA, A^-) &= \frac{[HA]_a}{[HA]_w + [A^-]_w} = \frac{[HA]_w}{[HA]_w + [A^-]_w} \cdot \frac{[HA]_a}{[HA]_w} \\ &= \alpha_a K'_H \\ &= \alpha_a \cdot \frac{K_H}{RT} \end{aligned} \tag{8-26}$$

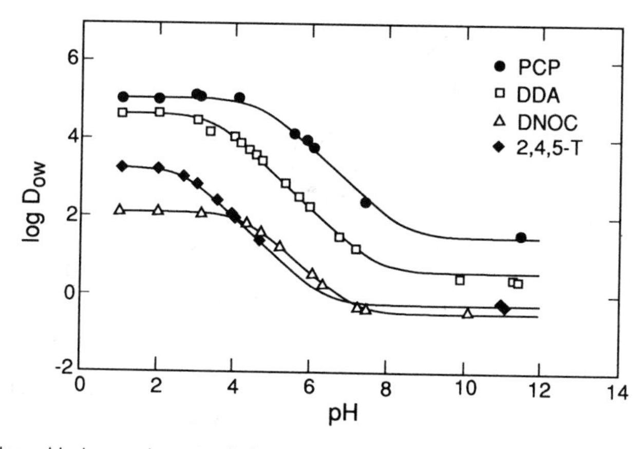

Figure 8.10 The pH dependence of the *n*-octanol–water distribution ratio of pentachlorophenol (PCP, $pK_a = 4.75$), 4-chloro-α-(4-chlorophenyl) benzene acetic acid (DDA, $pK_a = 3.66$), 2-methyl-4,6-dinitrophenol (DNOC, $pK_a = 4.46$), and 2,4,5-trichlorophenoxy acetic acid (2,4,5-T, $pK_a = 2.83$). From Jafvert et al., 1990.

By analogy, for an organic base, we obtain

$$D_{aw}(B, BH^+) = (1 - \alpha_a)K'_H$$
$$= (1 - \alpha_a)\frac{K_H}{RT} \qquad (8\text{-}27)$$

where K_H is the Henry's Law constant of the neutral acid or base species, respectively.

In contrast to air–water partitioning, the situation may be a little bit more complicated when dealing with organic solvent–water partitioning of organic acids and bases. As an example, Figure 8.10 shows the pH dependence of the *n*-octanol–water partition ratios, $D_{ow}(HA, A^-)$, of 4 pesticides exhibiting an acid function:

$$D_{ow}(HA, A^-) = \frac{[HA]_{o,tot}}{[HA]_w + [A^-]_w} \qquad (8\text{-}28)$$

where $[HA]_{o,tot}$ is the total concentration of HA in octanol. Since in octanol, not only the nondissociated acid but also ion pairs (with inorganic counterions) as well as ionic organic species may be present (Jafvert et al., 1990), D_{ow} of an acid may have a significant value even at high pH, particularly when dealing with hydrophobic acids. For pentachlorophenol (PCP, $pK_a = 4.75$), for example, at pH 12 (virtually all PCP present

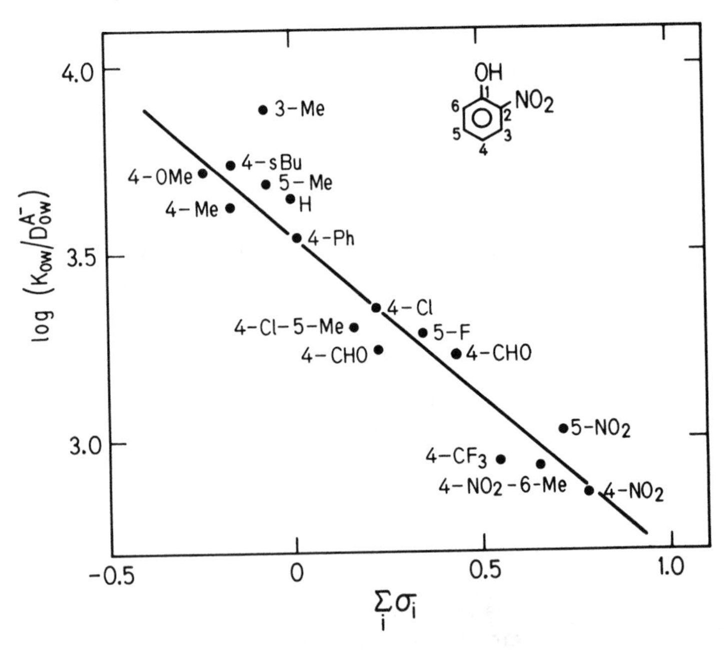

Figure 8.11 Plot of $\log[K_{ow}(HA)/D_{ow}(A^-)]$ versus $\Sigma\sigma_i$ (Hammett substituent constants) for a series of substituted 2-nitrophenols (from Schwarzenbach et al., 1988).

as phenolate in the aqueous phase) and 0.1 M KCl, a $D_{ow}(A^-)$ value of about 100 has been determined (Fig. 8.10). Note that the partitioning of the ionic species depends on the type and concentration of the counterions present in the aqueous phase. Hence, for calculating the organic phase–water equilibrium distribution ratio of an organic acid or base, a variety of species in both phases have to be considered (for details, see Jafvert et al., 1990).

As is illustrated by Figure 8.11 for a series of substituted 2-nitrophenols, the relative magnitude of the $K_{ow}(HA)$ value of the nondissociated species compared to the distribution ratio of the ionic species, $D_{ow}(A^-)$ (determined at high pH) depends on the type of substituent. Substituents with electron withdrawing properties (positive σ constants) tend to decrease the $K_{ow}(HA)/D_{ow}(A^-)$ ratio. This can be explained by the fact that such substituents tend to "smear" the charge over the molecule, thus rendering it more compatible with the organic phase and less compatible with the aqueous phase. Anionic species that exhibit both electron-withdrawing and hydrophobic moieties partition, therefore, relatively well into organic phases. As a consequence, such species may exert a toxic effect on organisms by acting as "uncoupling agents;" that is, they degrade proton gradients that organisms build up across membranes for energy storage. To act as uncouplers, the anionic species have to diffuse through organic phases (i.e., membranes). For more details on this important process see, for example, Benz and McLaughlin (1983).

From Figures 8.10 and 8.11, it can be seen that for organic acids (and similarly for organic bases, Johnson and Westall, 1990), the K_{ow} of the neutral species is more than two orders of magnitude larger than the corresponding $D_{ow}(A^-)$ value of the ionic species. Hence, at pH $\lesssim (pK_a + 2)$ for acids and pH $\gtrsim (pK_a - 2)$ for bases, the neutral species is the dominant species in determining the octanol–water distribution ratio of the compound. At these pH values, by analogy to the air–water distribution ratio (Eqs. 8-26 and 8-27), $D_{ow}(HA, A^-)$ may be expressed by

$$D_{ow}(HA, A^-) \cong \alpha_a \cdot K_{ow}(HA) \tag{8-29}$$

and

$$D_{ow}(B, BH^+) \cong (1 - \alpha_a) \cdot K_{ow}(B) \tag{8-30}$$

With this, we conclude our discussion on the equilibrium partitioning of organic compounds between a well-defined nonaqueous phase (i.e., air, organic solvent) and water. The knowledge that we have acquired in this and the previous chapters forms the basis for our discussions of phase transfer processes of organic compounds in the (macroscopic) environment (Chapters 10 and 11).

CHAPTER 9

DIFFUSION

9.1 THE GRADIENT-FLUX LAW: OFFSPRING OF RANDOMNESS AT THE MOLECULAR LEVEL

We all have an intuitive perception for the direction in which spontaneous mixing processes proceed. In a fluid for instance, spatial variations of concentrations, pressure, and temperature are steadily diminished, provided that no external mechanism keeps them alive. The erosion of spatial unevenness in the distribution of mass, heat, and other properties is a manifestation of the second law of thermodynamics stating that, in the absence of an external energy source, entropy in a system is always growing until equilibrium is attained.

As an example let us assume that at some initial time, t_0, a drop of dye is added to one end of a narrow tube filled with water (Fig. 9.1). As we know, the dye will eventually be spread evenly along the tube. When the dye concentration has become constant everywhere in the tube, net fluxes of dye across any section of the tube will be zero. If this were not true, the concentration on, say, the right side of any dividing interface would increase at the expense of the left side's concentration. Such behavior would violate both our intuitive understanding of mixing as well as the second law of thermodynamics (i.e., randomness is always increasing in the absence of some organizing energy input). However, as long as the dye distribution is not homogeneous, net fluxes will proceed across any interface such that transport is directed from the higher to the lower concentration.

This behavior of the dye concentration, C, implies that we may expect the net flux to be proportional to the concentration difference, $\Delta C = C_{\text{right}} - C_{\text{left}}$, across the imaginary wall:

$$\text{Flux along } x \equiv F_x = \text{constant} \cdot \Delta C \qquad [\mathrm{M \cdot L^{-2} \cdot T^{-1}}] \tag{9-1}$$

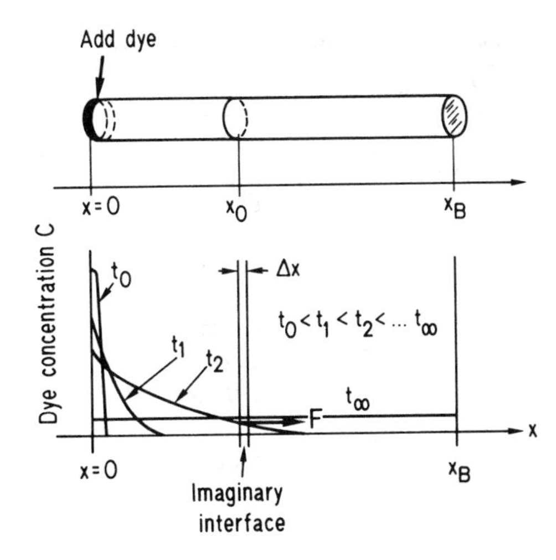

Figure 9.1 Qualitative view of dye diffusion along a tube with impermeable walls at $x=0$ and x_B. The dye is added to the tube at time t_0 at the end $x=0$.

(Note that in this and most of the following discussion, equations will be presented which are independent of the choice of specific units and thus can be characterized by a combination of the three basic dimensions: M = mass, L = length, and T = time as indicated in the brackets.) We indicate the direction of the flux by the sign of the flux. If the flux is directed in the positive x direction (to the right, see Fig. 9.1), F_x is positive; transport to the left means F_x is negative. A similar convention can be introduced for the sign of ΔC: if the concentration on the right-hand side of the interface is larger, ΔC is positive, otherwise it is negative. Since the transport must occur from large to small concentrations, we conclude that the constant factor appearing in Eq. 9-1 must be a negative number.

Natura non facit saltus—nature does not make jumps—is an axiom of classical physics. Thus, the concentration drop ΔC across the imaginary wall approaches zero if the "wall thickness," Δx, tends to zero. What remains as a measure of the concentration "slope" at the interface is the derivative of C along x: dC/dx $[\mathrm{M \cdot L^{-3} \cdot L^{-1}}]$. Thus, we get

$$F_x = -D\frac{dC}{dx} \qquad [\mathrm{M \cdot L^{-2} \cdot T^{-1}}] \tag{9-2}$$

where F is the mass flux per unit (cross-sectional) area per unit time, and D, a positive number, has the dimension length squared per time. For reasons explained below, D is called the molecular diffusion coefficient; it is a property of the moving chemical and the ambient medium. The standard metric unit for D is square meters per second but in most handbooks square centimeters per second is used. If the latter units are adopted, the appropriate units for concentration C and flux F are mole per cubic centimeter and mole per square centimeters per second, respectively.

TABLE 9.1 Physical Processes Obeying the "Gradient-Flux Law"[a]

Law	Equation	Definition of Variables	
First Fick's law for molecular diffusion	$F = -D\frac{\partial C}{\partial x}$	$F\ (\mathrm{mol \cdot m^{-2} \cdot s^{-1}})$	Mass flux
		$C\ (\mathrm{mol \cdot m^{-3}})$	Concentration
		$D\ (\mathrm{m^2 \cdot s^{-1}})$	Molecular diffusion coefficient
Conduction of heat (Fourier)	$F_{th} = -\kappa\frac{\partial T}{\partial x}$	$F_{th}\ (\mathrm{W \cdot m^{-2}})$	Heat flux
		$T(\mathrm{K})$	Temperature
		$\kappa(\mathrm{W \cdot m^{-2} \cdot K^{-1}})$	Thermal conductivity
Flow of fluid through porous medium (Darcy's Law)	$q = -K\frac{\partial h}{\partial x}$	$q\ (\mathrm{m \cdot s^{-1}})$	Velocity of fluid
		$\mathrm{h}\ (\mathrm{m})$	Hydraulic head (or pressure change along flow path x)
		$K\ (\mathrm{m \cdot s^{-1}})$	Hydraulic conductivity of medium
Electric conductivity[b] (Ohm's Law)	$j = +k\frac{\partial V}{\partial x}$	$j\ (\mathrm{A \cdot m^{-2}})$	Electric current per area
		$V\ (\mathrm{V})$	Electric field
		$k\ (\Omega^{-1} \cdot \mathrm{m^{-1}})$	Electric conductivity

[a]The partial derivatives ($\partial/\partial x$) are used to point out that the property variables (C, T, etc.) generally depend on time and space and that the spatial derivatives are calculated by keeping the other (spatial and temporal) coordinates constant.

[b]The positive sign results from the special sign convention used for electric currents and fields.

Equation 9-2 is known as Fick's first law of diffusion. In fact, its mathematical form has a much broader application: any physical law that combines some property *flux* with a corresponding property *gradient* (e.g., dC/dx) is called a "gradient flux law." Equation 9-2 is just one example, others are listed in Table 9.1. What do these laws have in common and what is the physical meaning behind them? The name "molecular diffusion" rightly suggests a linkage to the molecular nature of chemicals; but this is just one side of the coin. Another aspect refers to the randomness of the processes which are responsible for the flux. This property permits us to use the gradient-flux scheme for other processes which are of random nature but do not necessarily act on the molecular level. Finally, randomness is linked to the definition of entropy, thus the entropy-producing nature of gradient-flux mechanisms does not come as a surprise.

Let us demonstrate the alleged role of randomness in the gradient-flux law for the case of mass flux (Eq. 9-2) by using a simple model (Csanady, 1973). We consider an infinite linear array of discrete boxes (Fig. 9.2). At time $t = 0$, N particles (e.g., 256) are in box 0; all other boxes are empty. At equally spaced times $t_1 = \Delta t$, $t_2 = 2\Delta t, \ldots,$ $t_n = n\Delta t$, the particles in any box m move with equal probability but randomly to the two neighboring boxes $(m - 1)$ and $(m + 1)$, respectively. Generally, this means about

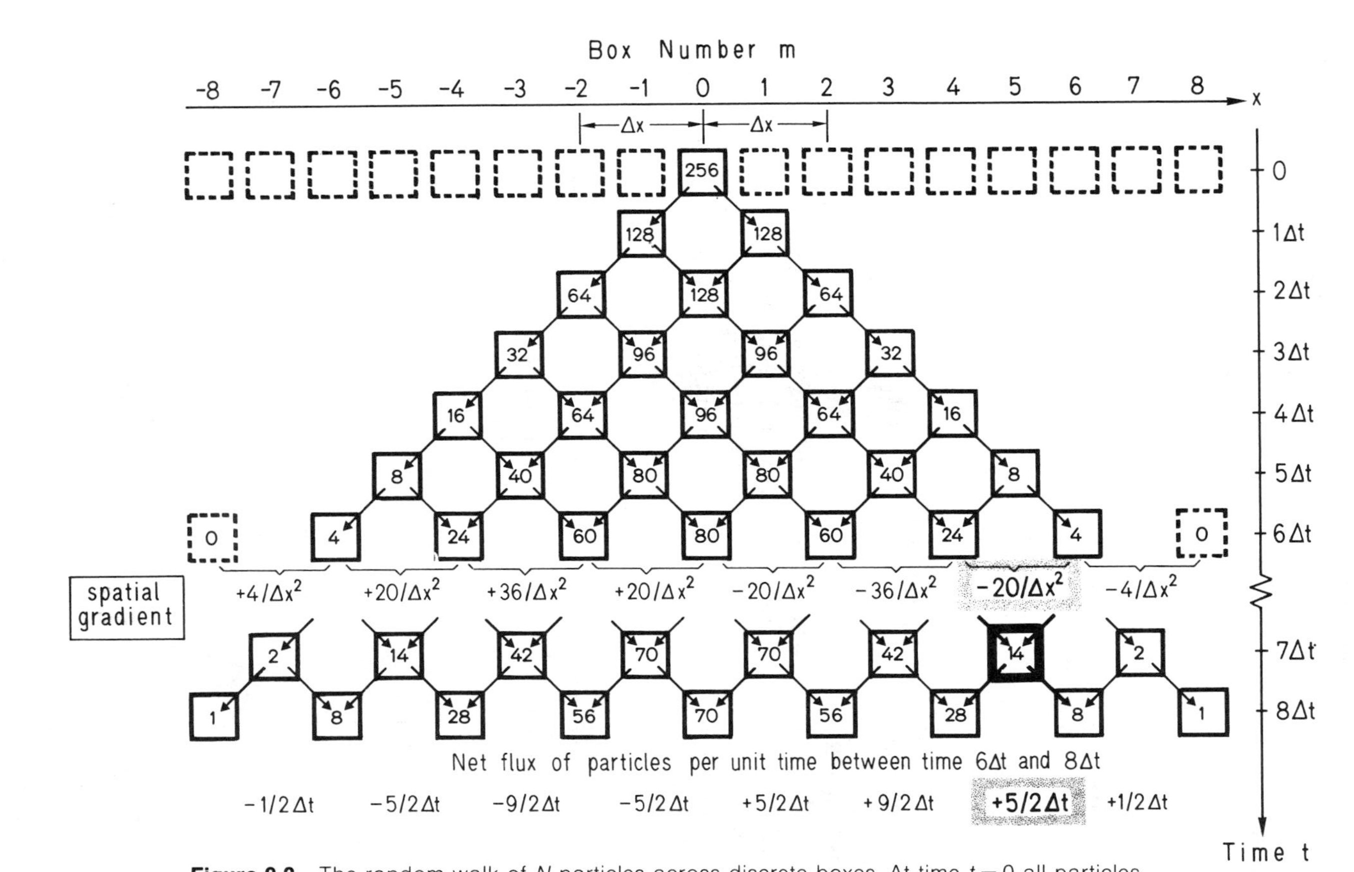

Figure 9.2 The random walk of N particles across discrete boxes. At time $t = 0$ all particles are in the box $m = 0$. After every time step Δt the particles move with equal probability to the two adjacent boxes. A single experiment leads to an unknown particle distribution $N(n, m)$ after n time steps. If the experiment is repeated many times, the average occupation numbers $\bar{N}(n, m)$ approach the Bernoulli coefficients $p(n, m)$ exemplified in the figure for $N = 256$. See also Figure 9.3.

half the particles move to the left while the rest move to the right. After n time steps (say $n = 8$), we find certain numbers of particles $N(8, m)$ in the 9 boxes with even numbers $m = -8, -6, \ldots, +6, +8$. For an individual experiment the exact values of $N(n, m)$ are unknown; yet if the experiment is repeated many times, the average occupation numbers, $\bar{N}(n, m)$, approach

$$\bar{N}(n, m) = N \cdot \frac{1}{2^n} \cdot \frac{n!}{[\frac{1}{2}(n-m)]!\,[\frac{1}{2}(n+m)]!} = N \cdot p(n, m) \tag{9-3}$$

[$n! = 1 \cdot 2 \cdot 3 \cdot \cdots \cdot (n-1) \cdot n$ and is referred to as n factorial]. The $p(n, m)$ are the so-called Bernoulli coefficients. Note that $m = -n, \ldots, +n$ runs through even or odd numbers only, if n is even or odd, respectively. In most mathematical textbooks they are written in terms of the two integers n and $k = \frac{1}{2}(n + m)$ as

$$p(n, k) = \frac{1}{2^n} \frac{n!}{k!(n-k)!}$$

where k runs from 0 to n.

For large n, that is, for a great number of time steps, $p(n, m)$ becomes increasingly cumbersome to calculate. If plotted for a fixed n as a function of m, the Bernoulli coefficients clearly remind us of a normal distribution (Fig. 9.3). Indeed, many years ago two French mathematicians, DeMoivre and Laplace, showed that $p(n, m)$ can be

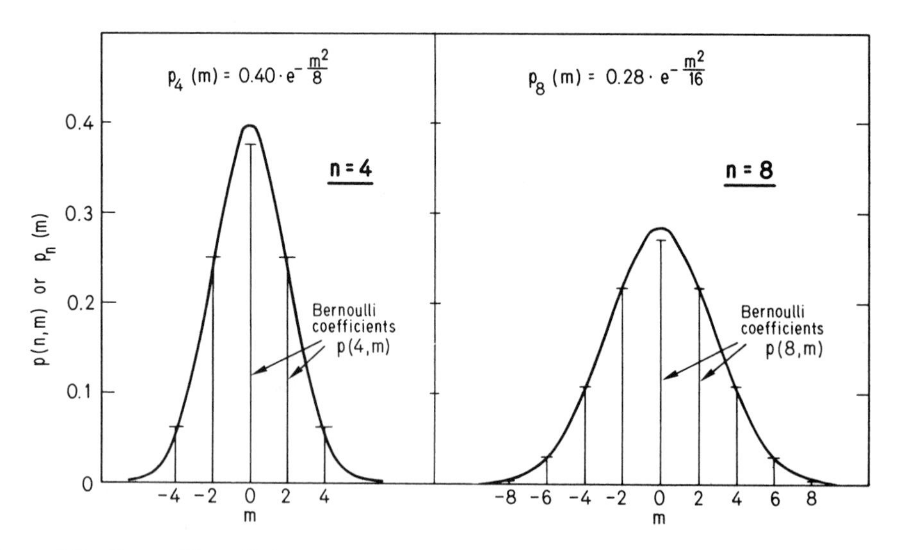

Figure 9.3 Bernoulli coefficients $p(n, m)$ for $n = 4$ and $n = 8$ compared to the corresponding normal density approximation by DeMoivre and Laplace $p_n(m)$, Eq. 9-4. Note that in most mathematical handbooks $p(n, k)$ is listed where $k = 1/2\ (n + m)$.

approximated by the continuous distribution (Feller, 1957)

$$p_n(m) = \left(\frac{2}{\pi n}\right)^{1/2} \exp\left\{-\frac{(m/2)^2}{n/2}\right\} \tag{9-4}$$

In this expression, n and m are no longer restricted to integer values. Eq. 9-4 describes a normal distribution along the x axis ($x = m/2$) centered at $x = 0$ with standard deviation $\sigma = \sqrt{n}/2$ (see Table 9.2). For large n the approximation becomes very accurate.

The pathway of an individual particle in the system is called a random walk. At any moment the particle has equal chance to move to either side. Thus, the direction of movement at time step n does not depend on the side from which the particle had entered the box during the preceding time step ($n - 1$). The particle does not carry any memory from one step to the next; it is completely randomized at every time step. The movement itself is characterized by the jumping distance $\Delta x/2$ (the distance between adjacent boxes) and the mean jumping velocity $v^* = \Delta x/2\Delta t$.

The random walk model can be applied to the Brownian motion of molecules: $\Delta x/2$ describes the mean free path used in the ideal gas theory, v^* quantifies the mean thermal velocity. But what about the gradient flux, Eq. 9-2? Let us consider the net flux of particles across box 5 during two time steps, for example $7\Delta t$ and $8\Delta t$ (Fig. 9.2).

For instance, at time step $7\Delta t$, 12 particles are entering box 5 from the left and 2 particles from the right. (Of course, all these numbers are mean values of many individual particle flux experiments!) In the following time step, 7 particles are leaving the box in either direction. Thus, the net flux across the boundary between boxes 4 and 5 for both steps $7\Delta t$ and $8\Delta t$ combined is 12 to the right minus 7 returning to the left equaling a net flow of 5 particles to the right (positive sign). The flux per unit time follows after division by the total time interval, $2\Delta t$:

$$F = \frac{5}{2\Delta t} \tag{9-5}$$

To express the particle flux in terms of the gradient flux, Eq. 9-2, a concentration value has to be constructed from the discrete box occupation numbers $\bar{N}(n, m)$. Since for a given time t_n every second box is filled only, we divide $\bar{N}(n, m)$ by the distance between two adjacent occupied boxes, Δx, to calculate the continuous particle concentration:

$$C(n, m) = \frac{\bar{N}(n, m)}{\Delta x} \quad \text{(particles per unit length)} \tag{9-6}$$

The spatial concentration gradient dC/dx follows from $C(n, m)$ after division of the concentration difference between neighboring boxes, ΔC, by the distance between occupied boxes, Δx:

$$\frac{dC}{dx} = \frac{\Delta C}{\Delta x} = \frac{\Delta \bar{N}}{(\Delta x)^2} \tag{9-7}$$

For the flux of particles across the boundary between boxes 4 and 5 we examined above, we can now see that the concentration gradient driving this transport at time $6\Delta t$ was $(4-24)/(\Delta x)^2 = [-20/(\Delta x)^2]$. We are now ready to calculate the diffusion coefficient D of the gradient flux, Eq. 9-2, using the Eqs. 9-5 and 9-7:

$$D = -\frac{F}{dC/dx} = -\frac{5/(2\Delta t)}{-20/(\Delta x)^2} = \frac{1}{8}\frac{(\Delta x)^2}{\Delta t} = \frac{1}{2}(\Delta x/2)\, v^* \qquad [\mathrm{L}^2 \cdot \mathrm{T}^{-1}] \qquad (9\text{-}8)$$

It turns out that the diffusion coefficient is one half of the product of "jumping distance" $\Delta x/2$ and "jumping velocity" $v^* = \Delta x/(2\Delta t)$. By studying Fig. 9-2, you may convince yourself that Eq. 9-8 also holds for the other locations along the x axis at all other times.

Why do we perform this lengthy derivation for a common concept like molecular diffusion? First, because the insight which we got, especially through Eq. 9-8, enables us to extend the formalism to other processes based on randomness, though not necessarily on the molecular level. The description of turbulent mixing by the so-called turbulent diffusion coefficient makes use of this possibility (see Section 9-4). Second, the effect of diffusion can be characterized by a mixing distance that depends on diffusion time t. Remember that we found the function $p_n(m)$, Eq. 9-4, to be a convenient approximation for the particle distribution among the boxes after time $t = n \cdot \Delta t$. The "spread" or standard deviation of $p_n(m)$, expressed in terms of the length scale Δx, is $\sigma_x = (\sqrt{n}/2)\Delta x$. Noting $n = t/\Delta t$ and using Eq. 9-8, we find

$$\text{measure of diffusion distance} = \sigma_x = (2Dt)^{1/2} \qquad (9\text{-}9)$$

This law, independently found by Einstein and the Polish physicist Smoluchowski when they studied the Brownian motion of small particles, became an important clue in the proof of the real existence of atoms as discrete pieces of mass moving around randomly.

9.2 FICK'S SECOND LAW

In the preceding section the general principles behind the "gradient-flux law" and its implications have been demonstrated. In this section we use this law together with the general principles of mass conservation to derive Fick's second law. The essential idea can be stated in the following way:

> The change with time of a conservative property inside a given volumetric element, ΔV, is equal to the algebraic sum of the fluxes across all the boundaries separating ΔV from the outside world.

Two expressions have to be explained first. A "conservative property" is a property that does not have *in situ* sources or sinks. For instance, thermal energy per unit volume (usually called temperature!) is conservative if neither heat-producing or -consuming reactions occur within the volume nor radiation processes carry energy to or from this test volume. The concentration of a chemical *element* is usually conservative (unless it appears or disappears by radioactive decay) but not necessarily the concentration of a specific chemical species or *compound* in which this element occurs.

The second term to be explained is "algebraic sum." It means summation of quantities (which can have a positive or negative sign) in such a way that the sign determines whether the numbers are actually added or subtracted. Thus, the algebraic sum of nonzero numbers can be zero. We have given a sign to the mass flux introduced in Eq. 9-2 to indicate the direction of the flux along the x axis. Therefore, the algebraic sum of several fluxes (e.g., those portrayed in Fig. 9.2) is the net flux along the x axis.

We will demonstrate the consequences of the general law for one dimension and later extend the result to three dimensions. The simplest choice for ΔV is a rectangular element with edges along the axis of a Cartesian coordinate system and lengths Δx, Δy, and Δz (Fig. 9.4*a*). The three-dimensional concentration field of some property shall be $C(x, y, z, t)$. Let us first consider the concentration variation along a straight line which is parallel to the x axis, that is, a line with constant coordinates y and z. For simplicity we chose $y = z = 0$, so that our line is just the x axis (Fig. 9.4*b*). We

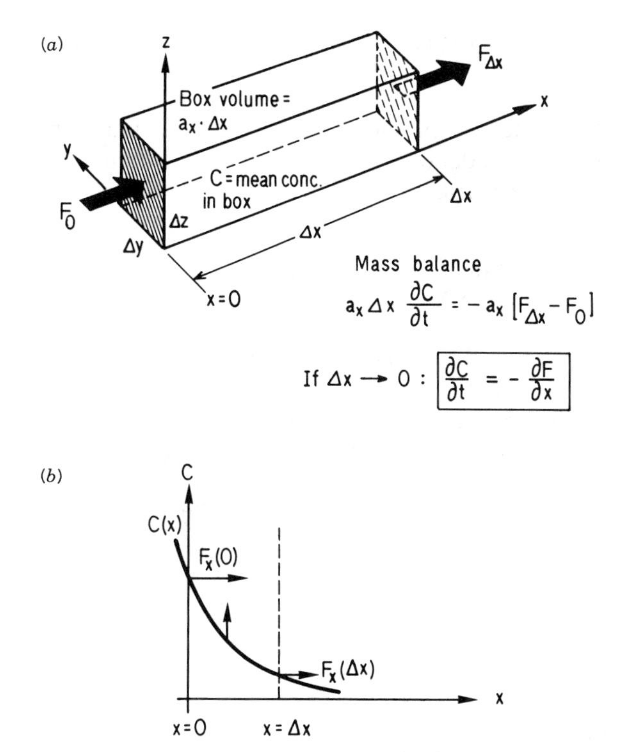

Figure 9.4 (*a*) Rectangular test volume of size $\Delta x \cdot \Delta y \cdot \Delta z$ and mass flux along the x axis. (*b*) For the concentration distribution along the x axis, $C(x)$, the diffusional transport causes $C(x)$ to increase between $x = 0$ and Δx.

calculate the diffusive fluxes through the walls of ΔV, which are perpendicular to the x axis. Both walls have the area $a_x = \Delta y \cdot \Delta z$ and are positioned at $x = 0$ and $x = \Delta x$, respectively.

The mass flux at any point of the system is a vector (i.e., a directed quantity) with components (F_x, F_y, F_z). Only the x component (F_x) can produce a flux across the mentioned walls of ΔV; the other flux components are parallel to these walls. We call the respective fluxes per unit area and time $F_x(0)$ and $F_x(\Delta x)$. Note that the index x indicates the direction of the flux, the argument in the parentheses gives the position of the flux on the x axis.

The mass balance for the concentration C in the constant volumetric element ΔV is

$$\frac{d}{dt}(\Delta V C) = \Delta V \frac{dC}{dt} = \underset{\substack{\text{flux in} \\ \text{from left}}}{a_x F_x(0)} - \underset{\substack{\text{flux out} \\ \text{to right}}}{a_x F_x(\Delta x)} \qquad [\mathrm{M \cdot T^{-1}}] \tag{9-10}$$

Division by ΔV $(\Delta V = a_x \Delta x)$ yields

$$\frac{dC}{dt} = \frac{a_x}{\Delta V}[F_x(0) - F_x(\Delta x)] = \frac{1}{\Delta x}[F_x(0) - F_x(\Delta x)] \qquad [\mathrm{M \cdot L^{-3} \cdot T^{-1}}] \tag{9-11}$$

If the size of the volume element along the x axis, Δx, is made smaller and smaller, both the numerator and the denominator of the far right-hand side of Eq. 9-11 tend to zero, but not the quotient itself. In fact,

$$\lim_{\Delta x \to 0} \frac{1}{\Delta x}[F_x(0) - F_x(\Delta x)] = -\frac{dF_x}{dx} \tag{9-12}$$

is the negative spatial derivative of the flux component F_x calculated at $x = 0$.

Before we insert this result into Eq. 9-11, we have to make a slight adjustment of our notation which often puzzles the nonmathematician. Obviously, we now have two kinds of derivatives, one with respect to time (dC/dt) and another with respect to space (dF_x/dx) along the x axis. To indicate that each derivative is to be evaluated by keeping all *other* variables constant, the symbol for *partial differentiation*, ∂, has to be used. Thus, from Eqs. 9-11 and 9-12 follows

$$\left.\frac{\partial C}{\partial t}\right|_{x = \text{constant}} = -\left.\frac{\partial F_x}{\partial x}\right|_{t = \text{constant}} \tag{9-13}$$

This equation states that the temporal concentration change is equal to the negative spatial gradient of the flux along the x axis.

Note that we have not yet made use of the actual definition of this flux, for instance Eq. 9-2. In fact, Eq. 9-13 is generally valid for all kinds of transport processes. For the special case of diffusion, we get from Eq. 9-2:

$$\frac{\partial C}{\partial t} = -\frac{\partial}{\partial x}\left(-D\frac{\partial C}{\partial x}\right) \tag{9-14}$$

which for a spatially constant diffusion coefficient D turns into the well-known *Fick's second law*:

$$\frac{\partial C}{\partial t} = D\frac{\partial^2 C}{\partial x^2} \qquad [\mathrm{M{\cdot}L^{-3}{\cdot}T^{-1}}] \tag{9-15}$$

It is easy to generalize the considerations given above to include the other spatial dimensions. By introducing the fluxes along the y and z axes, F_y and F_z, respectively, we can directly add the corresponding effects to Eq. 9-13:

$$\frac{\partial C}{\partial t} = -\frac{\partial F_x}{\partial x} - \frac{\partial F_y}{\partial y} - \frac{\partial F_z}{\partial z} \qquad [\mathrm{M{\cdot}L^{-3}{\cdot}T^{-1}}] \tag{9-16a}$$

The special "diagonal" combination of the derivatives of a vector $\mathbf{F} = (F_x, F_y, F_z)$ is called the divergence (div) of **F**, often also written as $\nabla\mathbf{F}$. Thus,

$$\frac{\partial C}{\partial t} = -\operatorname{div}\mathbf{F} \tag{9-16b}$$

(Note that bold letters are used to indicate vectors.) In fact, Eq. 9-16b explains why the name divergence has been chosen by mathematicians: div **F** describes the "parting" of the flux.

From Eq. 9-16a it is only a small step to the three-dimensional version of Fick's second law:

$$\frac{\partial C}{\partial t} = \frac{\partial}{\partial x}\left(D_x\frac{\partial C}{\partial x}\right) + \frac{\partial}{\partial y}\left(D_y\frac{\partial C}{\partial y}\right) + \frac{\partial}{\partial z}\left(D_z\frac{\partial C}{\partial z}\right) \tag{9-17}$$

where D_x, D_y, and D_z are the respective diffusion coefficients along the three axes. For $D_x = D_y = D_z = D =$ a constant:

$$\frac{\partial C}{\partial t} = D\left(\frac{\partial^2 C}{\partial x^2} + \frac{\partial^2 C}{\partial y^2} + \frac{\partial^2 C}{\partial z^2}\right) \tag{9-18}$$

where the sum of the second derivatives is the so-called Laplace operator.

All the equations related to Fick's second law (Eqs. 9-13 to 9-18) are linear partial differential equations. Generally, their solutions are exponential functions (or integrals of exponentials such as the error function). They strongly depend on the shape of the volume within which diffusion occurs and on the conditions imposed on C or its spatial derivative at these boundaries. In most cases, solutions for real systems are only found by numerical approximations. For simple boundary conditions, the mathematical techniques for the solution of the diffusion equation (such as the Laplace transformation) are extensively discussed in Crank (1975) and Carslaw and Jaeger (1959).

As an illustration we take the case of one-dimensional diffusion along the unbounded x axis of a concentration patch of total mass M which at time $t=0$ is concentrated at the point $x=0$. [A function which is infinite at $x=0$ and zero otherwise, but has a finite integral M along the x axis, is called a delta function $\delta(x)$.] The concentration distribution

$$C(x,t)=\frac{M}{2(\pi Dt)^{1/2}}\exp\left(-\frac{x^2}{4Dt}\right) \tag{9-19}$$

is a solution of the one-dimensional diffusion Eq. 9-15 with initial conditions as stated

TABLE 9.2 The Normal (Gaussian) Distribution Function

Definition	$p(x)=\frac{1}{(2\pi)^{1/2}\sigma}\exp\left[-\frac{x^2}{2\sigma^2}\right]$ σ: standard deviation
Normalization	$\int_{-\infty}^{\infty} p(x)\,dx=1$
Mean value	$\bar{x}=\int_{-\infty}^{\infty} xp(x)\,dx=0$
Variance	$\overline{x^2}=\int_{-\infty}^{\infty} x^2p(x)\,dx=\sigma^2$

Relative size $p(x)/p(0)$:

For $x=\pm\sigma\ : p(\sigma)/p(0) =\exp(-0.5)=0.607$
$x=\pm 2\sigma : p(2\sigma)/p(0)=\exp(-2) =0.135$
$x=\pm 3\sigma : p(3\sigma)/p(0)=\exp(-4.5)=0.011$

Integral within boundaries $\pm x$

$$\int_{-\sigma}^{\sigma} p(x)\,dx=0.683$$

$$\int_{-2\sigma}^{2\sigma} p(x)\,dx=0.954$$

above. You may convince yourself by calculating the derivatives $(\partial C/\partial t)$ and $(\partial^2 C/\partial x^2)$ and inserting them into Eq. 9-15. Strictly speaking, the solution given above is not defined for $t = 0$. Yet we can approach the value $C(x, t = 0)$ from the shape of $C(x, t)$ for small t and the limit $t \to 0$. This is quite easy to do since Eq. 9-19 is the well-known Gaussian (normal) distribution (Table 9.2) with standard deviation

$$\sigma = (2Dt)^{1/2} \tag{9-20}$$

and constant integral

$$\int_{-\infty}^{\infty} C(x, t)\, dx = M \tag{9-21}$$

The shape of $C(x, t)$ for different times t is shown in Figure 9.5. The limit $t \to 0$ of the Gaussian function can be used to define the δ function since the value at $x = 0$ approaches infinity while Eq. 9-21 remains valid.

Note the resemblance between Eq. 9-19 and the particle distribution among discrete boxes, Eq. 9-4. In fact, the "diffusion distance" equations, Eqs. 9-9 and 9-20 are identical. Of course, this is not a mere coincidence, but demonstrates that our simple discrete model (Fig. 9.2) effectively portrays the process of diffusion due to molecular random walk. The development of the concentration patch into a sequence of normal

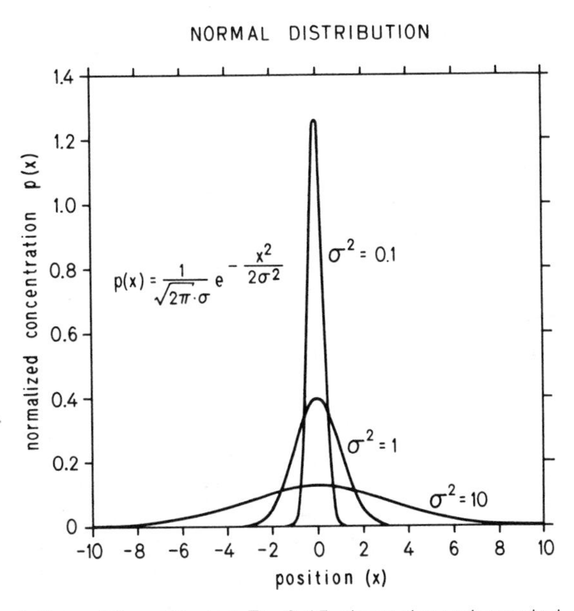

Figure 9.5 The solution of the diffusion Eq. 9-15 along the unbounded x axis (see Eq. 9-19) is a Gaussian distribution with standard deviation σ increasing as $(2Dt)^{1/2}$ (D is the diffusion coefficient).

distributions $p(x)$ with increasing standard deviation σ allows us to assign a more precise physical meaning to the "measure of diffusion distance" introduced in Eq. 9-9.

From Table 9.2 we can learn that the mean value x of the distribution (i.e., the mean position of the "particles" within the patch) keeps its position at $x = 0$. The variance of the distribution σ increases as $t^{1/2}$. Thus, one interpretation of the "diffusion distance" would be the mean quadratic dislocation of the particles from their original position.

Another way to interpret Figure 9.5 is by considering just half of the distribution. We do this by looking at the concentration "front" as it moves away from the origin toward the right or left, respectively. Since the distribution drops continuously from its maximum at the center to zero at $|x| \rightarrow \infty$, we have to define somewhat arbitrarily the position of the "front." For instance, we may choose the front at the position where $p(x)$ drops below 60% of its maximum value at $x = 0$. From Table 9.2 it follows that this relative concentration level is reached at about $x = \pm\sigma$. Alternatively, at $x = \pm 2\sigma$ and $\pm 3\sigma$ the relative concentration is 13.5 and 1.1% of the maximum, respectively.

Finally, we can define the diffusion distance by those $\pm x$ values within which a certain fraction of the total mass M, say 95%, should be located. No matter what percentage we choose, the corresponding boundaries $\pm x$ are always linearly related to σ. For example, if we want to have 95% of the total mass within the boundaries, then $x = \pm 1.4\sigma$ (see Table 9.2). Thus, in all the cases $\sigma = (2Dt)^{1/2}$ serves as the key parameter to describe the diffusion distance.

Having discussed these mathematical fundamentals, we now realize that, to actually calculate diffusive fluxes and diffusion distances, we need information on the molecular diffusivity of the chemical of interest to determine diffusive fluxes.

9.3 MOLECULAR DIFFUSIVITIES

Diffusivities reflect the action of random walk of molecules moving by Brownian motion through any medium of interest. This section deals with the estimation of diffusivities of organic substances in both air and aqueous solution. In light of the foregoing discussion, we can readily see that larger chemicals progress more slowly because the mean velocity (v^*) of their thermal motion is reduced and their increased cross-sectional area reduces their mean free path, that is, their ability to slip through a crowd of other molecules. As indicated in Figure 9.6, diffusivities of molecules in air are usually on the order of $0.1\ cm^2 \cdot s^{-1}$. For a small molecule like water, diffusivities are near $0.3\ cm^2 \cdot s^{-1}$ and go down to about $0.07\ cm^2 \cdot s^{-1}$ for organic molecules of molecular mass near $100\ g \cdot mol^{-1}$ (thus varying only within about a factor of 4). Additionally, we see from Figure 9.6*a* that diffusion coefficients decrease approximately as the two-thirds power of molar volume, indicative of the importance of molecule cross-sectional area (cross-sectional area of a *sphere* $= \pi r^2 = \pi(3\bar{V}/4\pi)^{2/3} = 1.21\bar{V}^{2/3}$). Finally, as shown in Figure 9.6*b*, the chemical's molecular mass may also serve as a useful measure of molecule size, and thus we see molecular masses are inversely correlated with gas phase diffusivities.

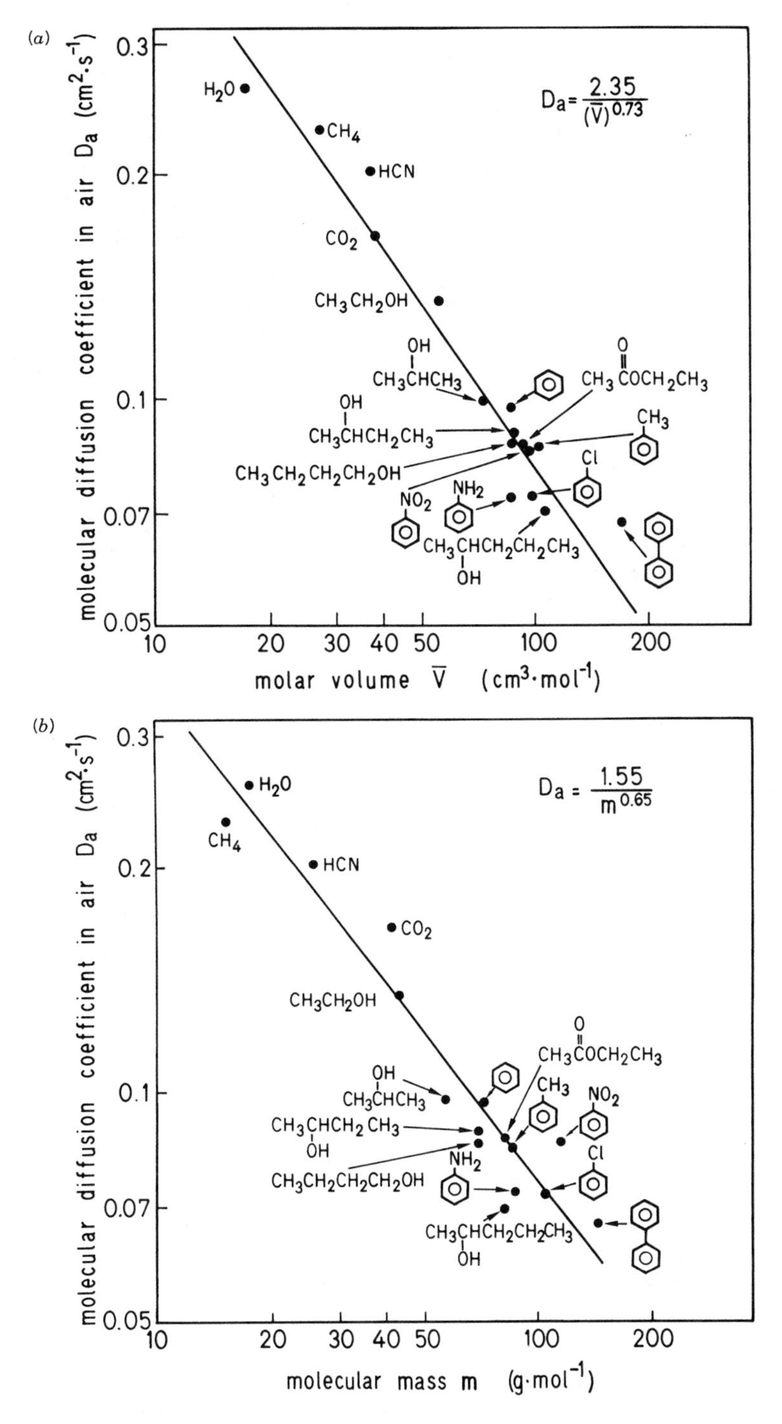

Figure 9.6 Molecular diffusion coefficients D_a at 25°C of molecules in air plotted as a function of: (*a*) their liquid molar volumes $\bar{V}$ (calculated as ratio of molecular weight to liquid density), and (*b*) their molecular mass. Data from references reviewed by Fuller et al. (1966).

Similar relationships between molecule size and diffusivities in aqueous solution can be seen in Figure 9.7. Here we see diffusivities ranging from about 3×10^{-5} for small molecules to about $0.5 \times 10^{-5}\ cm^2 \cdot s^{-1}$ for those of molecular mass near 300 $g \cdot mol^{-1}$. Once again, the data indicate the importance of the cross-sectional area of the molecular structure. Inverse correlations between molecular masses of compounds and their diffusivities in water are also apparent (Fig. 9.7*b*), but one additional point should be made. Note how substances like Radon-222 (Rn) deviate from this correlation. This is because the liquid density of radon is quite large ($\sim 4.4\ g \cdot cm^{-3}$) relative to the other substances (usually $\sim 1\ gm \cdot cm^{-3}$). Thus, radon molecular mass overestimates

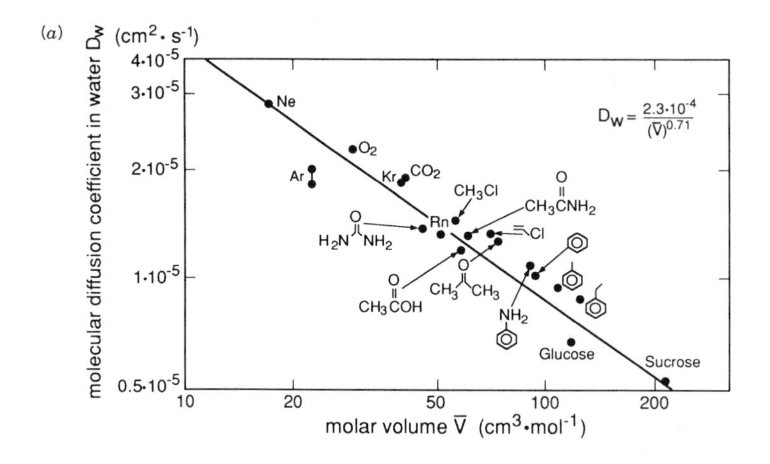

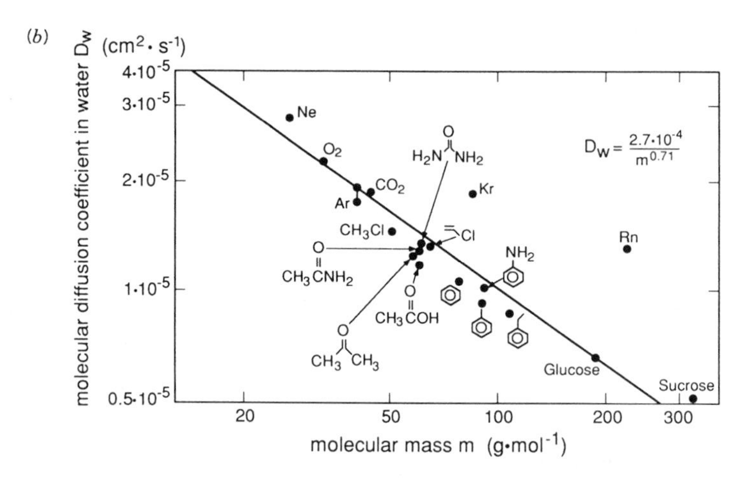

Figure 9.7 Molecular diffusion coefficients D_w at 25° C of molecules in water plotted as a function of: (*a*) their *liquid* molar volumes $\overline{V}$ (calculated as ratio of molecular weight to liquid density), and (*b*) their molecular mass. Data from review by Hayduk and Laudie (1974).

the relative size of this important geochemical tracer. If diffusivity is related to molar volume, instead of molecular mass, radon behaves as most other substances (Fig. 9-7*a*).

Other factors beside molecular size affect diffusivity. Media exhibiting more crowding or viscosity will inhibit the Brownian movements of molecules. Hence air, which is much less densely packed than water, reflects much higher diffusivities for molecules of interest than for the same chemical in aqueous solution (usually less by about 10^4 times). Finally, certain conditions such as the temperature can affect molecular diffusivity. Elevated temperatures result in more vigorous Brownian motion and consequently more rapid "random walk;" and heated media are less densely packed so that "percolation" of chemicals through them is facilitated.

A variety of semiempirical approaches, which make use of our physicochemical understanding of diffusion, are available to quantify chemical diffusivities in both gas and solution phases and to explain the correlations found in Figures 9.6 and 9.7. The method of Fuller et al. (1966) is a good example, illustrating how such an approach works for estimating diffusivities (D_a) of organic molecules in air:

$$D_a = 10^{-3} \frac{T^{1.75}[(1/m_{air}) + (1/m)]^{1/2}}{P[\bar{V}_{air}^{1/3} + \bar{V}^{1/3}]^2} \qquad (cm^2 \cdot s^{-1}) \qquad (9\text{-}22)$$

where

T is the absolute temperature (K),
m_{air} is the average molecular mass of air (28.97 $g \cdot mol^{-1}$),
m is the organic chemical molecular mass ($g \cdot mol^{-1}$),
P is the gas phase pressure (atm),
$\bar{V}_{air}$ is the average molar volume of the gases in air ($\sim$ 20.1 $cm^3 \cdot mol^{-1}$), and
$\bar{V}$ is the molar volume of the chemical of interest ($cm^3 \cdot mol^{-1}$).

Thus, we see that the critical component of such equations is the molecular size, here reflected by terms in both the numerator and denominator. There are various ways to estimate $\bar{V}$. As we did for Figures 9.6 and 9.7, we can simply divide the chemical molecular mass by its *liquid* density. Alternatively, we can sum the "size" of the atoms making up the chemical's structure. Table 9.3 shows some "diffusion sizes" of various atoms deduced by regression of available diffusion data. From such a listing we could estimate the size of benzene:

$$\begin{aligned} \bar{V}(\text{benzene}) &= 6(\text{C}) + 6(\text{H}) + \text{ring} \\ &= 6(16.5) + 6(2.0) - 20.2 \qquad (9\text{-}23) \\ &= 90.8\ cm^3 \cdot mol^{-1} \end{aligned}$$

Using benzene's molecular mass (78 $g \cdot mol^{-1}$) and liquid density (0.88 $g \cdot cm^{-3}$), we

TABLE 9.3 Estimation of Diffusion Volumes of Organic Molecules (Fuller et al., 1966)

Element	Volume Contribution ($cm^3 \cdot mol^{-1}$)
C	16.5
H	2.0
O	5.5
N	5.7
Cl	19.5
S	17.0
Rings	−20.2

calculate $\bar{V}$(benzene) to be $89\,cm^3 \cdot mol^{-1}$. Both approaches yield similar results (Table 9.4). Incorporating such size estimates with benzene molecular mass into Eq. 9-22 we determine:

$$D_a(\text{benzene}, 25°\text{C}, 1\text{ atm}) = \frac{(10^{-3})(298)^{1.75}\left(\frac{1}{29}+\frac{1}{78}\right)^{0.5}}{(1)[(20.1)^{1/3}+(90)^{1/3}]^2} = 0.09\,cm^2 \cdot s^{-1} \tag{9-24}$$

The experimental result is $0.096\,cm^2 \cdot s^{-1}$. Fuller et al. (1966) found that such diffusivity estimates match observations to within 10%.

A related technique involves adjusting diffusivities known for one chemical to approximate values for related structures, recognizing that diffusivities vary approximately inversely with molecular size. To the extent that organic chemical molecular masses also reflect molecular size, we can use this readily available parameter to adjust known diffusivities for one chemical to predict new diffusivities for similar molecules. Thus, to estimate the diffusivity of toluene in air, we can use knowledge about benzene:

$$\frac{D_a(\text{toluene})}{D_a(\text{benzene})} = \frac{\text{Diffusion size of benzene}}{\text{Diffusion size of toluene}} \simeq \left[\frac{m(\text{benzene})}{m(\text{toluene})}\right]^{1/2} \tag{9-25}$$

Thus, we estimate $D_a(\text{toluene}) \simeq (0.096\,cm^2 \cdot s^{-1}) \cdot [78/92]^{1/2} \simeq 0.088\,cm^2 \cdot s^{-1}$. The experimental value is $0.086\,cm^2 \cdot s^{-1}$ (Gilliland, 1934).

Similar semiempirical methods can be utilized to obtain solution phase diffusivities of organic molecules. For example, Othmer and Thakar (1953) provide a simple

TABLE 9.4 Comparison of Two Methods to Calculate the Molar Volume $\bar{V}$ of Organic Compounds

Compound	$\bar{V}\,(m/\rho_{\text{liquid}})^a$ ($cm^3 \cdot mol^{-1}$)	Volume by $\sum$ contributions[b] ($cm^3 \cdot mol^{-1}$)
Benzene	$\frac{78.1}{0.879} = 88.9$	6(C) + 6(H) − ring = 90.8
Trichloroethylene	$\frac{131.4}{1.456} = 90.2$	2(C) + 1(H) + 3(Cl) = 93.5
Ethyl acetate	$\frac{88.1}{0.901} = 97.8$	4(C) + 8(H) + 2(O) = 93.0
Trimethylamine	$\frac{59.1}{0.662} = 89.3$	3(C) + 9(H) + 1(N) = 73.2
Dimethyl amine	$\frac{45.1}{0.680} = 66.3$	2(C) + 7(H) + 1(N) = 52.7
Dimethyl sulfide	$\frac{62.1}{0.846} = 73.4$	2(C) + 6(H) + 1(S) = 62.0

[a]Method used in Figures 9.6 and 9.7; m = organic molecular mass ($g \cdot mol^{-1}$); ρ_{liquid} = *liquid* density of compound ($g \cdot cm^{-3}$).
[b]From Fuller et al. (1966), data shown in Table 9.3.

expression (coefficients modified slightly by Hayduk and Laudie, 1974):

$$D_w = \frac{13.26 \times 10^{-5}}{\mu^{1.14} \cdot (\bar{V})^{0.589}} \qquad (cm^2 \cdot s^{-1}) \tag{9-26}$$

where

μ is the solution viscosity in centipoise ($10^{-2}\, g \cdot cm^{-1} \cdot s^{-1}$) at the temperature of interest, and
$\bar{V}$ is the molar volume of the chemical ($cm^3 \cdot mol^{-1}$).

In this expression, the primary environmental determinant of solution diffusivities is the viscosity of water. This factor inhibits diffusive transport, and hence it is in the denominator. Since water viscosity decreases by about a factor of 2 over the ambient temperature range (0–30°C corresponds to water viscosities of 1.519–0.719 centipoise), D_w only increases by about 50% over the same range.

As an example, we estimate the solution phase diffusivity of trichloroethylene at 25°C using atomic contributions. First, we estimate this molecule's size using information shown in Table 9.3:

$$\begin{aligned} \bar{V}(\text{trichloroethylene}) &= 2(C) + 1(H) + 3(Cl) \\ &= 2(16.5) + 1(2.0) + 3(19.5) \\ &= 93.5\ cm^3 \cdot mol^{-1} \end{aligned} \tag{9-27}$$

For comparison, we note that $m/\rho_{\text{liquid}} = 90.3\ \text{cm}^3\cdot\text{mol}^{-1}$. Using this result in Eq. 9-26 we obtain

$$\begin{aligned} D_w(\text{trichloroethylene}) &= \frac{13.26 \times 10^{-5}}{(0.894)^{1.14}(93.5)^{0.589}}\ \text{cm}^2\cdot\text{s}^{-1} \\ &= 1.04 \times 10^{-5}\ \text{cm}^2\cdot\text{s}^{-1} \end{aligned} \tag{9-28}$$

Generally, this approach also yields results that are correct to within 10%.

In light of our intuition, as well as the semiempirical equation given above, it should come as no surprise that solution phase diffusivities can also be estimated by comparison to known results for related compounds. Given the Othmer and Thaker equation (Eq. 9-26), one expects

$$\frac{D_w(\text{unknown})}{D_w(\text{known})} \simeq \left(\frac{\text{diffusion volume}_{\text{known}}}{\text{diffusion volume}_{\text{unknown}}}\right)^{0.589} \tag{9-29}$$

and this compares favorably with what we saw in the data plotted in Figure 9.7.

Again, molecular masses are widely used as relative indices of molecular size, and a square-root functionality is used for simplicity:

$$\frac{D_w(\text{unknown})}{D_w(\text{known})} \simeq \left(\frac{m_{\text{known}}}{m_{\text{unknown}}}\right)^{0.5} \tag{9-30}$$

Figure 9.7*b* indicates that within a class of molecules like benzene derivatives, the inverse relationship of diffusivity to molecular mass is clear.

In summary, reasonably accurate means exist for estimating gas phase and solution diffusivities of organic chemicals of interest. Generally, only some knowledge of chemical size is needed (e.g., diffusion volume, or more approximately, molecular mass) to obtain estimates within 10% of measured results.

9.4 TURBULENT DIFFUSION

Molecular diffusion is important mainly on the microscopic scale. It brings reactants into contact with each other and transports chemicals across boundaries, for example, into a living cell, onto a particle surface, or across the air–water interface (see Chapter 10). Yet on a macroscopic scale—that of rivers, lakes, or subsurface aquifers—molecular diffusion is extremely slow. From Eq. 9-9 we can calculate the characteristic transport time t_d to diffuse the distance L:

$$t_d = \frac{L^2}{2D} \tag{9-31}$$

TABLE 9.5 Characteristic Molecular Diffusion Times Taking $D_w = 1 \cdot 10^{-5}\,cm^2 \cdot s^{-1}$ in Water and $D_a = 0.1\,cm^2 \cdot s^{-1}$ in Air

L	Time in water	Time in air
0.1 μm	$5 \cdot 10^{-6}$ s	$5 \cdot 10^{-10}$ s
1 μm	$5 \cdot 10^{-4}$ s	$5 \cdot 10^{-8}$ s
10 μm	0.05 s	$5 \cdot 10^{-6}$ s
100 μm	5s	5.10^{-4}s
1 mm	~ 8 min	0.05 s
1 cm	~ 14h	5 s
10 cm	~ 58 days	~ 8 min
1 m	~ 16 y	~ 14 h
10m	~ 1600 y	~ 58 days

As shown in Table 9.5, diffusion occurs very quickly (seconds or less) in water over distances less than 100 μm and in air across less than 1 cm; however, to diffuse as far as a meter requires quite a long time! Take, for instance, a chemical in water with a diffusion coefficient $D = 10^{-5}\,cm^2 \cdot s^{-1}$: for $L = 10^2$ cm (1 m), t_d becomes 16 years. Such large mixing times are certainly not realistic! What is wrong with Eq. 9-31?

The equation is not wrong, rather its application to distances over which transport does not primarily occur by molecular diffusion is inappropriate. Over large distances transport is performed by the motion of the fluid itself, by advection. Only at very short distances, where because of viscosity the fluid motion cannot reach, does transport by molecular diffusion become relevant. Such "niches of no motion" exist, for instance, in the pore space of sediments and at the various interfaces.

Transport by advection is described by the vector $\mathbf{F}_{ad}$ which is parallel to the velocity vector of fluid motion $\mathbf{v}$:

$$\mathbf{F}_{ad} = C \cdot \mathbf{v} \qquad [M \cdot L^{-2} \cdot T^{-1}] \tag{9-32}$$

This expression is analogous to Fick's first law, Eq. 9-2. Consequently, we can also apply Eq. 9-16 to calculate the local concentration changes due to advection:

$$\left(\frac{\partial C}{\partial t}\right)_{ad} = -\operatorname{div} \mathbf{F}_{ad} = -\left(v_x \frac{\partial C}{\partial x} + v_y \frac{\partial C}{\partial y} + v_z \frac{\partial C}{\partial z}\right) \qquad [M \cdot L^{-3} \cdot T^{-1}] \tag{9-33}$$

whereby we have made use of the *continuity equation* which is valid for incompressible fluids:

$$\operatorname{div} \mathbf{v} = \left(\frac{\partial v_x}{\partial x} + \frac{\partial v_y}{\partial y} + \frac{\partial v_z}{\partial z}\right) = 0 \tag{9-34}$$

(In fact, for our purpose, water can be assumed to be incompressible!) For the case of a constant current velocity v, transport time due to advection is given by

$$t_{ad} = \frac{L}{v} \tag{9-35}$$

Let us use Eqs. 9-31 and 9-35 to calculate the distance L_{crit} for which molecular diffusion and advection play equal roles in chemical transport:

$$\underbrace{\frac{L^2}{2D}}_{\text{diffusion time}} = \underbrace{\frac{L}{v}}_{\text{advection time}} \rightarrow L_{crit} = \frac{2D}{v} \tag{9.36}$$

Over distances larger than L_{crit}, transport by advection is faster and thus more important than transport by diffusion. In open waters (lakes, rivers, and oceans), typical advection velocities lie between 1 and 10^2 cm·s^{-1}. For a molecule with $D_w = 10^{-5}$ cm^2·s^{-1}, the critical distance is of the order 10^{-7}–10^{-5} cm; but even for the tiny advection velocity of 10^{-2} cm·s^{-1}, the critical length L_{crit} is still only 2×10^{-3} cm. Thus, in the presence of fluid motion—as small as it may be—transport by molecular diffusion is surpassed by advective transport after only very short distances. In air (typical values $D_a = 0.1$ cm^2·s^{-1} and $v = 10$ cm·s^{-1}), the critical distance is about 0.02 cm.

In spite of the simplicity of the advective transport (Eq. 9-32), dealing with transport by currents is extremely complicated. This is because currents are usually turbulent.

It is difficult to define turbulence. Intuitively, we associate it with the fine structure of the fluid motion, as opposed to the flow pattern of the large-scale currents. Although it is not possible to describe in an exact way the distribution in space and time of this small-scale motion, we can still characterize it in terms of some statistical parameters such as the variance of the current velocity at some fixed location. This is very similar to motion at the molecular level. It is not possible to describe the movement of some "individual" molecule (we cannot even distinguish individual molecules), but groups of molecules still obey certain characteristic laws since the random individual behavior sums to yield average motion in response to macroscopic forcings.

In light of this analogy, we anticipate that the effect of turbulence may be dealt with in a similar manner as the random motion of molecules for which the gradient-flux law of diffusion has been developed. Let us start with the simple model shown in Figure 9.8. The concentration of a compound along the x axis, $C(x)$, is influenced by turbulent velocity fluctuations producing occasional water exchange across a plane at x_0 perpendicular to the x axis. Owing to the continuity of water flow, any water transport in the positive x direction has to be compensated for by a corresponding flow in the negative x direction. In a simplified way we can thus visualize the effect of turbulence as occasional exchange events of water volumes Q_{ex} over distance L_x across the interface x_0 (see Fig. 9.8). A single exchange event causes a net transport of the compound equal to $(C_1 - C_2)\, Q_{ex}$. The concentration difference $(C_1 - C_2)$ can be

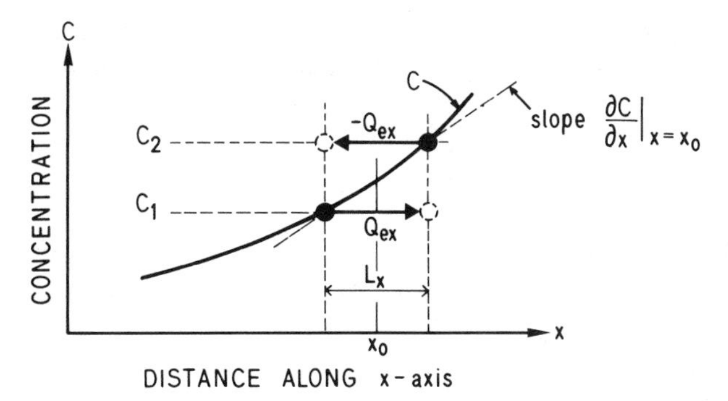

Figure 9.8 Exchange model for mass transport by turbulent diffusion. C is mean concentration along the x axis. Turbulence causes "exchange events" of water (Q_{ex}) over distance L_x. The resulting net mass transport is proportional to the mean gradient $\partial C/\partial x$ (see Eq. 9-37).

approximated by the linear gradient of the concentration curve $(-L_x \cdot \partial C/\partial x)$. Summation over all exchange events across some area Δa within a given time interval Δt yields the mean turbulent flux of the compound per unit area and time:

$$F_{\text{turb},x} = -\frac{\sum L_x Q_x}{\Delta a \cdot \Delta t} \cdot \frac{\partial C}{\partial x} = -E_x \frac{\partial C}{\partial x} \qquad [\text{M} \cdot \text{L}^{-2} \cdot \text{T}^{-1}] \tag{9-37}$$

The coefficient E_x is called the turbulent (or eddy) diffusion coefficient; it has the same dimensions (length squared per time) as the molecular diffusion coefficient. The index x indicates the coordinate axis along which the transport occurs. Note that the new diffusion coefficient can be interpreted as the product of a mean transport distance $\bar{L}_x$ times a mean velocity $\bar{v} = \bar{Q}_{ex}/(\Delta a \cdot \Delta t)$, as found in the random walk model for particles, Eq. 9-8.

Though Eq. 9-37 may be adequate as a qualitative model for turbulent diffusion, it is certainly not suitable to quantify E_x or to relate it to measurable quantities such as current velocities. A different model going back to Reynolds (1894) and Schmidt (1917) casts more light on the concept of turbulent diffusion. The central idea is to distinguish turbulence from mean motion by isolating the temporal (or spatial) fine structure of any field variable f, such as current velocity or concentration. This is done by separating f into a mean value $\bar{f}$ and a residual fluctuation f':

$$f(t) = \bar{f} + f'(t) \tag{9-38}$$

Here $\bar{f}$ is the time average of the variable measured at some fixed point over the time interval s:

$$\bar{f} = \frac{1}{s} \int_{t-s/2}^{t+s/2} f(t')\,dt' \tag{9-39}$$

Obviously, the residual $f'(t)$ varies in time because sometimes $f(t)$ is a little bigger than $\bar{f}$ and at other moments it is less. Note that the average of all deviations of the instantaneous value $f(t)$ from its mean $\bar{f}$ is zero:

$$\overline{f'} = \overline{(f - \bar{f})} = \bar{f} - \bar{f} = 0 \tag{9-40}$$

The fluctuation model, Eq. 9-38, can be applied to the description of advective transport, Eq. 9-32. For simplicity, we restrict ourselves to the x component:

$$F_{\mathrm{ad},x} = Cv_x = (\bar{C} + C')(\bar{v}_x + v'_x) = \bar{C}\bar{v}_x + C'v'_x + C'\bar{v}_x + \bar{C}v'_x \tag{9-41}$$

If the time average is taken on both sides of Eq. 9-41, the last two terms of the right-hand side become zero because $\overline{C'}$ and $\overline{v'_x}$ are zero (after Eq. 9-40). Thus,

$$\overline{F_{\mathrm{ad},x}} = \bar{C}\bar{v}_x + \overline{C'v'_x} \tag{9-42}$$

and similar for the other two flux components ($\overline{F_{\mathrm{ad},y}}, \overline{F_{\mathrm{ad},z}}$). This means that the average advective mass flux $\overline{F_{\mathrm{ad},x}}$ consists of two contributions: (1) the product of the mean concentration $\bar{C}$ and the mean velocity $\bar{v}_x$ and (2) the "turbulent flux":

$$F_{\mathrm{turb},x} = \overline{C'v'_x} \tag{9-43}$$

This flux is the mean value of the product of two fluctuating quantities, the concentration and the velocity. Unlike the last two terms of the far right-hand side of Eq. 9-41 which contain only one fluctuation quantity, the product of two fluctuations does not generally disappear when the mean is taken over time. This is because there may exist some correlation between the fluctuating part of the two quantities. For instance, consider the concentration profile shown in Figure 9.8. It is quite possible that if the velocity v_x is a little more than average to the right ($v'_x > 0$), then the corresponding water parcel carries with it a concentration which is below average ($C' < 0$) since in our example concentration is generally smaller at small x values. Similarly, we expect situations with $v'_x < 0$ to be associated with $C' > 0$. Thus, the two means of the fluctuations "co-vary." That is why mathematicians call the right-hand side of Eq. 9-43 "covariance."

Consequently, $F_{\mathrm{turb},x}$ would only be zero either in the absence of turbulence ($v'_x = 0$) or if the concentration distribution were completely homogeneous ($\partial C/\partial x = 0$, thus $C' = 0$). On the other hand, we expect C' (and thus $F_{\mathrm{turb},x}$) to increase with increasing mean gradient ($\partial C/\partial x$). We may even assume a linear relationship between $F_{\mathrm{turb},x}$ and the mean gradient: $F_{\mathrm{turb},x} = \text{constant} \cdot (\partial C/\partial x)$.

This brings us back to an expression which we derived for the exchange model (Fig. 9.8), that is, to Eq. 9-37. In fact, we can interpret the concentration gradient in Eq. 9-37 as a mean value; thus we get similar expressions for $F_{\mathrm{turb},x}$ in both models:

$$F_{\mathrm{turb},x} = \overline{C'v'_x} = -E_x \frac{\partial \bar{C}}{\partial x} \tag{9-44}$$

The empirical coefficient E_x, formally defined by

$$E_x = -\frac{\overline{C'v'_x}}{\partial\bar{C}/\partial x} \tag{9-45}$$

should only depend on the fluid motion (the turbulence structure of the fluid, to be more precise), and not on the substance described by the concentration C. This is an important difference from the case of molecular diffusion which is specific to the physicochemical properties of the substance and the medium. Since the intensity of turbulence must strongly depend on the forces (wind, solar radiation, river flow, etc.) driving the currents, the coefficients of turbulent diffusion constantly vary in space and time. Of course, we cannot tabulate them in chemical or physical handbooks as we do their molecular counterparts.

Eq. 9-45 shows one way to determine E values, though not an easy one. For the case of temperature T (the "concentration of heat"), modern instrumentation allows us to determine directly the velocity and temperature fluctuations v' and T', but the correlations $\overline{v'T'}$ are very small and difficult to measure with any significance (Gregg, 1987). For the case of vertical turbulent heat flux, $\overline{v'_zT'}$, a method introduced by Osborn and Cox (1972) is frequently used. It is based on the assumption that turbulent overturns against a mean temperature gradient produce temperature fluctuations that are balanced by local diffusive smoothing.

Then the vertical eddy diffusivity can be expressed as

$$E_z = \kappa \cdot \overline{(\nabla T')^2} / \overline{(\partial T/\partial z)}^2 \tag{9-46}$$

where $(\nabla T')$ is the fluctuation of the vertical temperature gradient,

$$(\nabla T') = \frac{\partial T}{\partial z} - \overline{\frac{\partial T}{\partial z}} \tag{9-47}$$

and κ is the molecular diffusivity of heat in water. Equation 9-46 is widely used in physical oceanography; it allows us to determine E_z from temperature measurements alone. Other methods are discussed at the end of this section.

The decomposition of turbulent motion into mean and random fluctuations which resulted in the convenient separation of the flux, Eq. 9-42, leaves us a serious problem of ambiguity. It concerns the question of how to choose the averaging interval s introduced in Eq. 9-39. In a schematic manner we can visualize turbulence to consist of eddies of different sizes. Their velocities overlap to yield the turbulent velocity field. When these eddies are passing by a fixed point, they cause fluctuation in the local velocity. We expect that some relationship should exist between the spatial dimension of those eddies and the typical frequencies of velocity fluctuations produced by them. Small eddies would be connected to high frequencies and large eddies to low frequencies. Consequently, the choice of the averaging time s determines which eddies appear in the mean advective transport term and which ones appear in the fluctuating part (and thus are interpreted as turbulence).

Let us make this point clearer by the following hypothetical experiment. At some initial time t_0 a droplet of dye is put on the surface of a turbulent fluid (Fig. 9.9). At some later time t_1 the large-scale fluid motion has moved the dye patch to a new location which can be characterized by the position of the center of mass of the patch. In addition, the patch has grown in size because of the small (turbulent) eddies, more precisely those eddies with size similar to or smaller than the patch size. The patch size can be quantified by the variance σ^2 of the dye concentration in space about its center of mass.

With increasing time, the growth of the patch will continue at an increasing rate since larger and larger eddies will contribute to the spreading while the mean motion becomes more and more restricted to the very large scales. Eventually, the dye patch extends over the whole size of the water body. Then no significant mean motion is possible any more, and all mixing has become turbulent.

A relation between the diffusion coefficient and the growth of the variance σ^2 can be derived from Eq. 9-20 if we note that eddy diffusion behaves mathematically like molecular diffusion:

$$\sigma^2 = 2Et$$
$$\frac{\partial \sigma^2}{\partial t} = 2E \qquad (9\text{-}48)$$

We have replaced D by E to indicate the turbulent nature of eddy diffusion. As mentioned before, the growth of σ^2 becomes faster with increasing σ^2. Thus, we

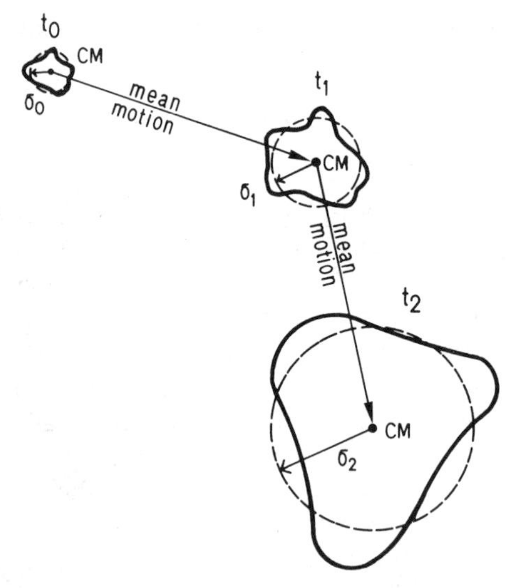

Figure 9.9 Growth and movement of a tracer cloud due to turbulent currents. While the mean current velocity moves the patch (represented by its center of mass, *CM*) as a whole, the turbulent part of the current increases the size of the cloud (represented by the variance, σ_i).

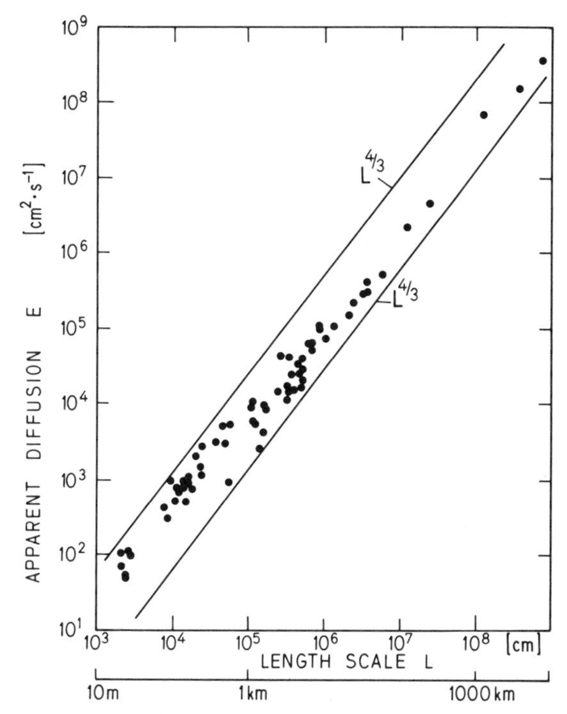

Figure 9.10 Apparent turbulent diffusivity E in the ocean versus length scale $L = \sigma$. Redrawn after Okubo (1971).

conclude that the eddy diffusion coefficient E must grow with increasing patch size σ^2. This point has been confirmed for the ocean by Okubo (1971) who, using different oceanic diffusion studies, has shown that the apparent diffusion coefficient E (i.e., the temporal change of σ^2 with time) grows with the length scale of diffusion L as $L^{4/3}$ (Fig. 9.10). The "4/3 law" is valid over horizontal length scales ranging from 10 m to 1000 km. It is a consequence of the so-called turbulence spectrum, that is, the typical distribution of the kinetic energy into eddies of different size.

From Figure 9.10 we can learn that in the surface water of the ocean horizontal diffusivities vary by several orders of magnitude, that is, between 10^2 and $10^8\ \text{cm}^2\cdot\text{s}^{-1}$. Even the lowest values are still about 10^7 times larger than a typical molecular diffusion coefficient ($\sim 10^{-5}\ \text{cm}^2\cdot\text{s}^{-1}$) and 10^5 times larger than molecular diffusion of heat ($\sim 10^{-3}\ \text{cm}^2\cdot\text{s}^{-1}$). Similar values are found in lakes. Since the shear stress of the wind acting at the water surface is one of the most important sources of turbulent movement, diffusivities in the deeper parts of oceans and lakes are usually smaller by one to two orders of magnitude (Table 9.6).

While molecular diffusivity is commonly independent of direction (isotropic—to use the correct expression), turbulent diffusivity in the horizontal direction is usually much larger than vertical diffusion. One reason is the scale factor. In the troposphere (the lower part of the atmosphere) and in surface waters, the available vertical distances

TABLE 9.6 Typical Turbulent Diffusivity in the Atmosphere, Oceans, and Lakes

Location	Diffusivity ($cm^2 \cdot s^{-1}$)[a]
Atmosphere	
Vertical	10^4–10^5
Note: horizontal transport mainly by advection (wind)	—
Ocean	
Vertical, mixed layer[b]	0.1–10^4
Vertical, deep sea	1–10^2
Horizontal (scale dependent, see Fig. 9.10)	10^2–10^8
Lakes	
Vertical, mixed layer[b]	0.1–10^4
Vertical, deep waters	10^{-3}–10^{-1}
Horizontal	10^1–10^7

[a] $1\,cm^2 \cdot s^{-1} = 8.64\,m^2 \cdot d^{-1}$.
[b] Maximum numbers for storm conditions.

are limited to about 10 km or less, while horizontal scales are much larger. Yet, there is another and often more important factor which distinguishes vertical from horizontal diffusion, that is, the influence of vertical density gradients. In contrast to the troposphere, which is vertically well mixed during most of time (only inversion situations are an exception), water bodies are usually vertically stratified, meaning that water density is increasing with depth. The strength of the stratification could be expressed by the vertical density gradient of the fluid, $(d\rho/dz)$. Yet, the rather peculiar units which such a quantity would have ($g \cdot cm^{-4}$), do not really help us understand its relationship to the classical concept of stability used in mechanics, for instance. There, a stable system can always be characterized by a restoring force which brings the system back to its original state every time a perturbation drives it away from the stability point. Examples are the pendulum or mass hanging on a spring. The "restoration capacity" (that is, the stability) can be described by the time needed to move the system back: *small* restoring times would then indicate *large* stabilities. Conversely, we get a direct correlation if we consider the inverse of the restoring time (the restoring rate or frequency): *large* restoring frequencies mean *large* stabilities.

This stability frequency is most commonly used to describe the stability of fluid systems. In the case of a vertically stratified water column, the appropriate quantity is called the Brunt–Väisälä frequency; it is defined by

$$N = \left(\frac{g}{\rho}\frac{d\rho}{dz}\right)^{1/2} \qquad [T^{-1}] \tag{9-49}$$

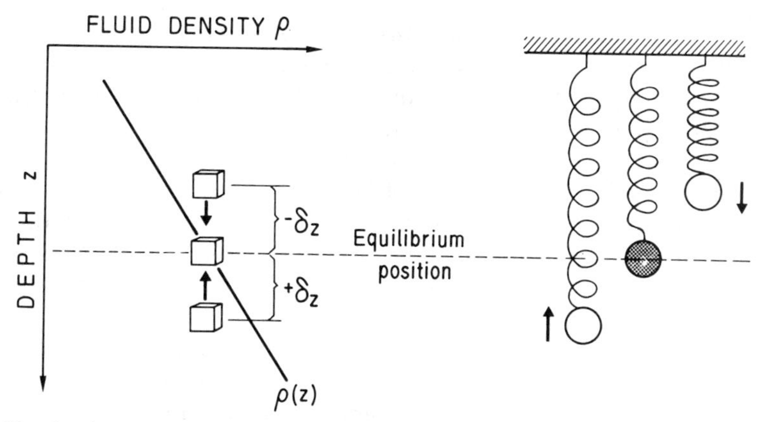

Figure 9.11 Analogy between stability oscillation in a stratified water column and the motion of a body hanging on a spring (linear oscillator). For small vertical displacements, $\pm\delta_z$, the restoring forces are proportional to δ_z and to the force constant (density gradient or spring constant, respectively) leading to an oscillation with a fixed frequency.

where $d\rho/dz$ is the vertical gradient of the water density ρ, and g is the acceleration of gravity ($981\,\text{cm}\cdot\text{s}^{-2}$). In this expression the vertical coordinate z is increasing downward.

The meaning of N can be understood by looking at a small water parcel which moves vertically within a stratified water column without exchanging heat or solutes with its new environment, that is, without changing its density (Fig. 9.11). If the water parcel is displaced upward, its own density is larger than the density of the surrounding fluid. Therefore, the parcel experiences a "restoring" force downward which is proportional to the vertical excursion δz and the vertical density gradient $d\rho/dz$. The reverse happens if the parcel moves below its equilibrium depth. Thus, the so-called buoyancy forces act on the water parcel in an analogous way as gravity acts on a sphere hanging on a spring (the so-called linear oscillator); both situations result in an oscillation around the equilibrium point. The stronger the spring (or the density gradient), the larger will be the frequency of oscillations about the equilibrium point. Therefore, large stability frequencies indicate stronger stratification.

In a stratified fluid, the turbulent motion is concentrated along the planes of constant density (which are usually the horizontal surfaces), while across these planes (in the vertical direction) the turbulent motion is suppressed. Therefore, the eddies are not fully developed three-dimensional (isotropic) features, but flat horizontal pancakes. Since the eddy diffusion coefficient is directly related to the turbulent velocities (Eq. 9-45), the vertical eddy diffusivity should decrease with increasing density gradient or increasing N. In fact, Welander (1968), based on theoretical considerations on the nature of turbulence, postulated a relationship of the form

$$E_z = a(N^2)^{-q} \tag{9-50}$$

where the parameter a depends on the overall level of kinetic energy input and the parameter q on the mechanism that transforms this energy into turbulent motion. Welander distinguished between two extreme cases: (1) $q = 0.5$ for shear-generated turbulence (i.e., turbulence produced by the friction between waters flowing a different speeds), and (2) $q = 1$ for turbulence generated by energy cascading from the large-scale fluid motion (such as tidal motion) down to the small-scale turbulent motion.

Before we can evaluate the validity of Welander's equation we have to discuss how turbulent diffusion coefficients are actually measured. As mentioned earlier, the relationship between E and turbulent fluctuations, such as Eq. 9-45 or the expression by Osborn and Cox (Eq. 9-46) could be used in principle. Yet the measurement of turbulent fluctuations is not trivial. In addition, such measurements only yield information on the instantaneous and often highly variable size of E. In contrast, the dispersion of organic compounds in natural systems is usually a long-term process influenced by the action of turbulence integrated over some space and time interval (days, months, or years).

Information on average diffusivities can best be gained by visually observing the transport of so-called *tracers*, that is, natural or artificial compounds (with known sources and sinks) which can be employed to view turbulence. Since turbulent diffusion, compared to molecular diffusion, is usually very large, the distribution of tracers can often be analyzed without taking molecular processes into account. The data set assembled by Okubo (Fig. 9.10) was determined mainly from the dispersion of tracer patches using Eq. 9-48.

Vertical eddy diffusivity in closed basins (lakes, estuaries) can also be determined from the temporal change of heat in the water column below a fixed depth (Fig. 9.12). Let us assume that in a lake the temperature varies only in the vertical, not in the horizontal. This means that at a fixed depth, for instance at 10 m, the water temperature is constant across the lake. (In fact, this assumption is not true for an instantaneous view of the lake since the surfaces of constant temperature move up and down like waves at the water surface; yet for the time-averaged temperatures the picture of a horizontally homogeneous temperature field is reasonable.) Consequently, if we obtain a vertical temperature profile taken at time t_i we can calculate the heat content of

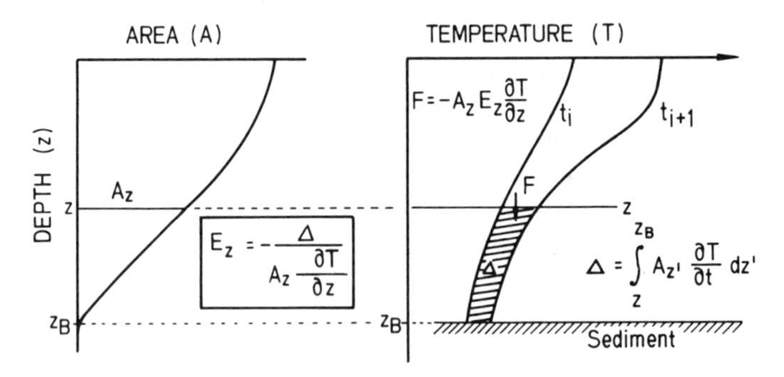

Figure 9.12 Comparison of vertical temperature profiles measured at consecutive times t_i and t_{i+1} can be used to determine the vertical turbulent diffusivity E_z. From Imboden et al. (1979).

the whole lake or part of it, for instance, of the zone between depth z and the lake bottom at z_B:

$$\text{heat content} = \int_z^{z_B} A_{z'} T(z') dz' \tag{9-51}$$

where z' denotes the integration variable. The lake area at depth z' ($A_{z'}$) appears because each temperature value $T(z')$ has to be weighted with the corresponding area in this depth.

Owing to vertical (turbulent) diffusion, heat is transported from regions of warm water to adjacent colder layers. Mathematically this appears as a heat flux in the direction opposite to the vertical temperature gradient (remember the Fick's first law, Eq. 9-2). Thus, at a later time t_{i+1} we expect to find warmer water between z and z_B. One way to write the change of the heat content, Δ, with time is

$$\begin{aligned} \Delta \equiv \frac{\partial}{\partial t}\begin{pmatrix}\text{heat} \\ \text{content}\end{pmatrix} &= \frac{\partial}{\partial t}\int_z^{z_B} A_{z'} T(z') dz' \\ &= \int_z^{z_B} A_{z'} \frac{\partial T(z')}{\partial t} dz' \end{aligned} \tag{9-52}$$

(Note that $A_{z'}$ does not change with time and that the derivative can be taken into the integral.)

Another approach to describe the changing heat content involves writing it in terms of the total Fickian flux (use Eq. 9-2, replace D by E and multiply with lake area at depth z, A_z):

$$\Delta = -A_z E_z \frac{\partial T}{\partial z} \tag{9-53}$$

Rearranging Eq. 9-53 and substituting for Δ from Eq. 9-52 yields the expression for E_z:

$$E_z = -\frac{\Delta}{A_z(\partial T/\partial z)} = -\frac{\int_z^{z_B} A_{z'}[\partial T(z')/\partial t]\, dz'}{A_z(\partial T/\partial z)} \tag{9-54}$$

The E_z values calculated by this method represent the average action of turbulence during the time interval t_i to t_{i+1}.

Radioactive tracers are also used to measure turbulent diffusion. As an example, the radioactive isotope radon-222 (^{222}Rn, half-life 3.8 days) is mentioned. Radon-222, a noble gas, is the decay product of radium-226, a natural component of sediments and soil. At the bottom of lakes and oceans, radon diffuses from the sediments to the overlying water where it is transported upward by turbulence. Broecker (1965) used vertical radon profiles in the deep sea by analyzing the steady-state solution of

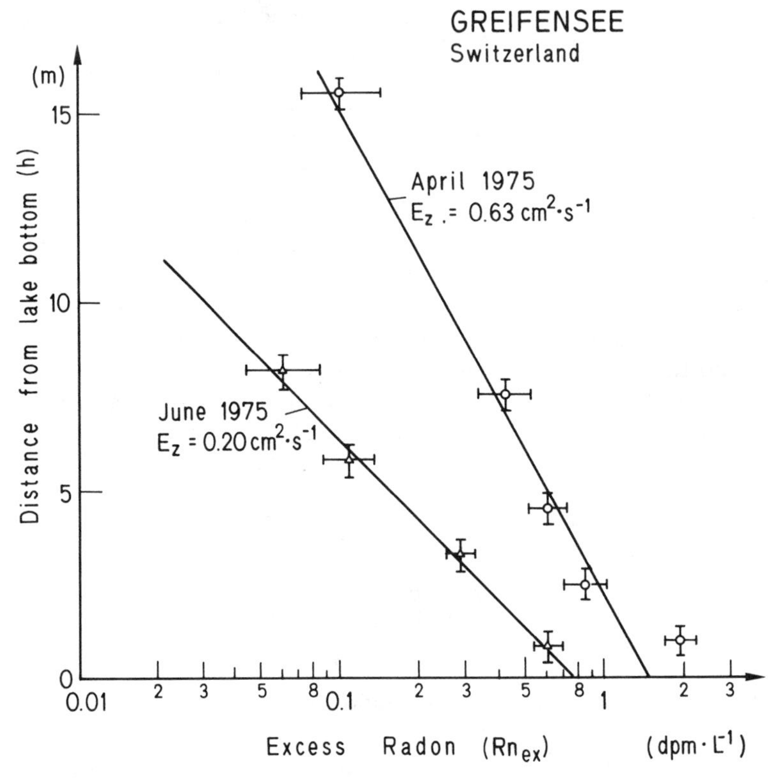

Figure 9.13 Vertical profile of dissolved excess radon-222 activity in the bottom waters of Greifensee (Switzerland) serves to compute vertical turbulent diffusivity E_z. Data from Imboden and Emerson (1978).

the one-dimensional diffusion–decay equation:

$$\frac{\partial \mathrm{Rn_{ex}}}{\partial t} = E_z \frac{\partial^2 \mathrm{Rn_{ex}}}{\partial h^2} - \lambda_{\mathrm{Rn}} \mathrm{Rn_{ex}} = 0 \tag{9-55}$$

where $\mathrm{Rn_{ex}}$ is the so-called excess radon-222 activity (i.e., the radon-222 activity exceeding the activity of its parent isotope radium-226) and $\lambda_{\mathrm{Rn}} = 0.181\, d^{-1}$ is the decay constant of radon-222. In Eq. 9-55, horizontal transport is disregarded relative to vertical transport, and the eddy diffusivity E_z is assumed to be independent of height above the sediment surface h.

Because of the relatively short half-life of radon-222, the vertical radon distribution approaches steady-state within about a week, once the turbulent conditions have remained constant long enough. For $\partial \mathrm{Rn_{ex}}/\partial t = 0$, the solution of Eq. 9-55 is

$$\mathrm{Rn_{ex}}(h) = \mathrm{Rn_{ex}}(h=0) \exp\left[-\left(\frac{\lambda_{\mathrm{Rn}}}{E_z}\right)^{1/2} h\right] \tag{9-56}$$

Taking the natural logarithm yields

$$\ln \mathrm{Rn_{ex}}(h) = \ln \mathrm{Rn_{ex}}(h=0) - \left(\frac{\lambda_{\mathrm{Rn}}}{E_z}\right)^{1/2} h \tag{9-57}$$

Thus, a plot of $\ln \mathrm{Rn_{ex}}(h)$ versus h should yield a straight line with slope, $-(\lambda_{\mathrm{Rn}}/E_z)^{1/2}$. Since λ_{Rn} is known, E_z can be determined from the slope.

Figure 9.13 gives an example from Greifensee, a small Swiss lake (area about 8 km^2, maximum depth 32 m). Though there are limits to the one-dimensional interpretation of the data brought about by lateral transport from the sides (Imboden and Emerson, 1978), the method can still be useful under certain conditions to yield insight into the vertical mixing regime at the bottom of lakes and oceans. We see that E_z in this small lake was between 0.1 and 1 cm$^2\cdot$s^{-1} on these two occasions, thereby indicating vertical Rn-222 transport far greater than that explainable by molecular diffusion.

Let us now come back to the relationship between vertical diffusivity and density stratification, Eq. 9-50. Jassby and Powell (1975) determined N and E_z from water temperature (the method shown in Fig. 9.12) in Crater Lake (California). They found E_z to vary like N^{-1} in most cases. Based on Welander's theory this would mean $q = 0.5$; that is, turbulence is mainly produced by local shear. Since then many other measurements have been made in lakes and oceans in which a variety of relationships between N and E_z were found. A good summary was given by Gargett and Holloway (1984). Two examples from Swiss lakes (Wüest, 1987) are shown in Figure 9.14. An overview of typical environmental turbulent diffusivities is presented in Table 9.6. The values show that turbulent diffusion is generally larger in the horizontal than in the vertical direction. Atmospheric diffusion is typically greater than that in water bodies. This is consistent with greater fluid velocities in air. Also turbulent diffusivities in the

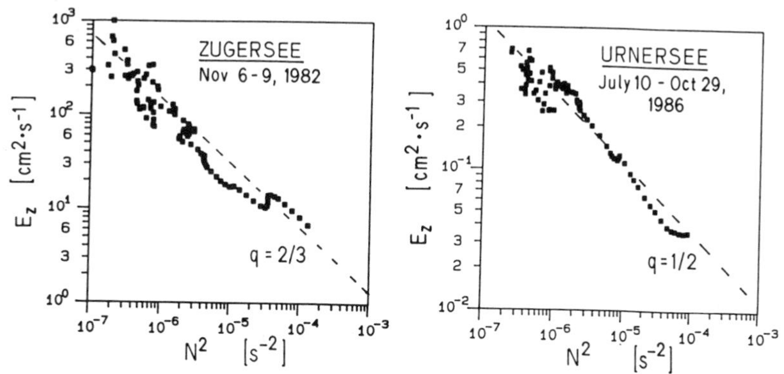

Figure 9.14 Vertical turbulent diffusivity E_z versus square of stability frequency N^2 in two Swiss lakes. For Zugersee (maximum depth 198 m) the values are calculated for an extreme storm of about two days duration. The data refer to the depth interval between 10 and 70 m; they show a mixture between turbulence production by local shear and large-scale motion. For Urnersee (maximum depth 196 m), a basin of Lake Lucerne, the data refer to 10–100 m depth and indicate shear-produced turbulence. From Wuest (1987).

ocean are on average larger than in lakes. Lakes are less exposed to wind, especially if the lakes are small or surrounded by steep terrain. Thus mixing energies are transmitted less effectively to the surfaces of lakes. Additionally, lakes rarely have long-lasting current systems comparable to those in the oceans (e.g., the Gulf Stream). As a result of these factors, deep portions of lakes are usually less turbulent than shallower intervals.

Let us summarize this section on turbulent diffusion in the environment by the following statements:

1. The action of turbulence on the transport of heat and chemicals in the atmosphere and water bodies can be approximated by the gradient-flux law, that is, by the two Fickian equations where the molecular diffusivities D are replaced by turbulent eddy diffusivities E.

2. Turbulent diffusivities are usually much larger than molecular diffusion coefficients (1 or more orders of magnitude). Therefore, molecular diffusion is only important in those regions where turbulence is absent, for example, at boundaries or in the pore space of sediments.

3. Horizontal turbulent diffusivity is scale-dependent, that is, the coefficients grow with increasing "patch" or system size.

4. Because of vertical density stratification, which frequently occurs in oceans and lakes below the mixed layer, vertical diffusivity is commonly much smaller than horizontal diffusivity. For a given energy input, vertical diffusivity is inversely related to the strength of the density gradient.

CHAPTER 10

THE GAS–LIQUID INTERFACE: AIR–WATER EXCHANGE

10.1 INTRODUCTION

In the previous chapters we learned how chemical structures determine the partitioning behavior of organic compounds between different phases (e.g., air and liquid phases, organic solvents, and water) in closed systems. With this background, we are now prepared to consider the natural environment, and the problems related to the transfer of organic chemicals between different environmental compartments. We start out by considering the transfer of compounds between gaseous phases (e.g., air) and condensed phases (e.g., dissolved in water). In Chapter 11 we discuss transfer processes between aqueous solutions and solid phases.

In most environmental situations we are concerned with the transfer of chemicals between gaseous phases and condensed phases that are not at equilibrium. This is due to our interest in times shortly after "catastrophic inputs" of chemicals to the environment and/or to cases involving compounds dispersed in huge environmental compartments (e.g., the atmosphere or Lake Superior) which do not quickly transmit materials from throughout their bulk interior to adjacent phases. Exceptions where equilibrium treatments may be appropriate include exchange of compounds between the air and raindrops (Slinn et al., 1978; Leuenberger et al., 1985), between water-saturated peat and included gas bubbles (Army, 1987), or between headspace gases accumulated over enclosed water samples (Pankow, 1986). In such instances, the large surface area-to-volume ratio of the rain or marsh bubbles facilitates overall exchange, and the associated contact times between phases may be long enough to assure attainment of equilibrium partitioning. We can then simply apply Henry's Law (see Chapter 6), together with knowledge of gas and solution volumes, to quantify such chemical transfers.

To understand, for example, the impact of air–water exchange on the concentration of a compound in a natural water body not at equilibrium with the compound's concentration in the air, we need to develop kinetic models for describing the rate at which the transfer occurs. Hence, in contrast to equilibrium considerations (where we are only interested in the final composition of a system, and where we neither need to know by which path or mechanism nor at which rate a process occurs), we here need to try to understand and describe the physical processes, often referred to as *mass transfer processes*, which dictate the rate of exchange of molecules between these compartments. In this chapter, our major focus will be on cases in which organic compounds are being transported between the atmosphere and natural waters such as oceans, lakes, and streams. However, many of the same principles such as diffusion through very narrow quiescent intervals or equilibrium partitioning across phase interfaces are also applicable to quantify, for example, volatilization of substances from soils to air (Thomas, 1982) or from open drums of toxic liquid wastes. In our discussion we place particular emphasis on the properties of organic chemicals that influence their rate of air–water transfers, as well as on the environmental factors that play an important role in this process.

It is probably obvious to most of us that exchange between condensed and gaseous phases is an important phenomenon controlling the fate of many low-boiling chemicals such as gasoline hydrocarbons or organic solvents like trichloroethylene. For example, after a tank load of chloroform (trichloromethane) was leaked into the Mississippi River, Neely et al. (1976) interpreted the continuously diminished chloroform concentrations downstream as reflecting losses due to mass transfer to the air (volatilization). Additionally, a variety of investigations have also implicated air–water exchange as important to the environmental movements of not-so-volatile chemicals like PCBs (Harvey and Steinhauer, 1973; Murphy et al., 1983; Eisenreich and Looney, 1983). As we suggested in our discussions of air–water equilibrium partitioning (Chapter 6), even compounds with high boiling points may partition substantially into air from water if the aqueous solubilities of these substances are commensurately low. Finally, air–water exchange is obviously important for various gases such as O_2 or H_2S which potentially interact chemically or via microbial mediation with organic compounds. Often the extent of interactions of these inorganic volatiles with organic chemicals is limited by the rate at which such "environmental reagents" are transported into or out of the place where the organic substance of interest is present. Thus our discussions of the air–water exchange transport process will apply to quite a broad suite of environmentally relevant species.

In the following, we try to imagine and develop physical pictures of the air–water boundary, especially insofar as it controls mass transfer. We follow this visualization with the development of the mathematical framework needed to quantify this process. Since our physical models indicate that molecular diffusivities in both air and water are critical traits needed to describe air–water mass transfer, we use the information developed in Chapter 9 where we have related these compound properties to chemical structure. Then we discuss how environmental factors like wind or water currents influence the mass transfer between air and water. After examining how chemical reactions may play a role in air–water exchange, we briefly consider the importance of special effects like the presence of surface films to air–water exchange.

10.2 VISUALIZATION OF THE MECHANISMS INVOLVED IN AIR–WATER EXCHANGE

To describe organic chemical transport into water from the overlying air or vice versa, we must first imagine the transport mechanisms governing the speed of molecule movements along their travel route. Referring to Figure 10.1, we see that in a somewhat simplified picture, the world at the contact of the atmosphere and a surface water can be discerned to consist of four layers arranged in series: (1) turbulent air, (2) a quiescent skin of air about 1 mm thick, (3) a quiescent skin of water on the order of 0.1 mm across, and finally (4) the well-mixed bulk of water below the interface region. It is at the interface between layers (2) and (3) that the air molecules and the molecules in the water below actually contact one another. The visualization of the important intervals of interest neglects situations in which special conditions arise which cause us to include in our picture physical entities like bubbles, aerosol droplets, or oily surface films. All

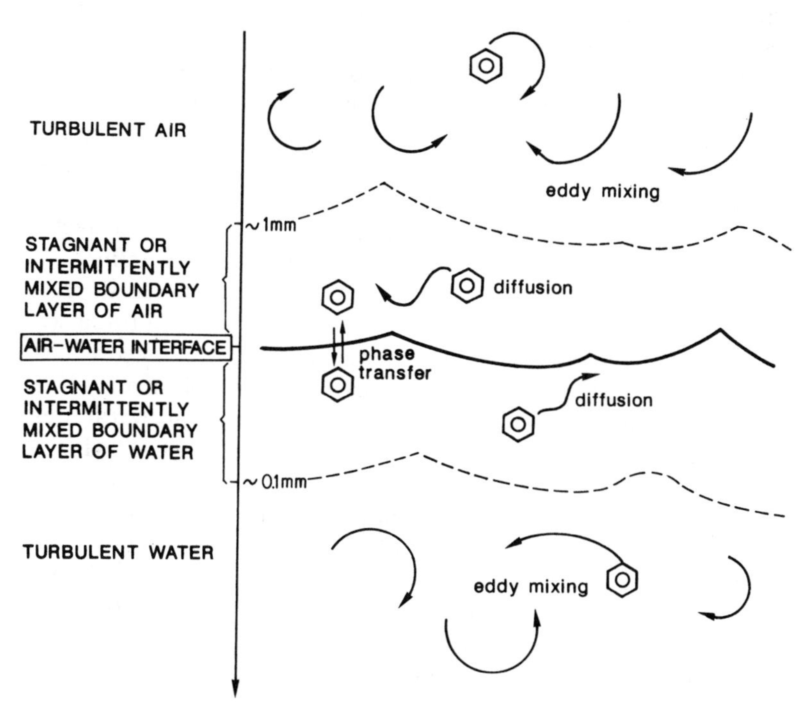

Figure 10.1 Depiction of the physical processes responsible for the movement of chemicals through four zones spanning an "intact" air–water interface (i.e., no bubbles or aerosols).

of these special entities could certainly affect the overall rate of air–water transport if they are sufficiently present, but for now our attention will remain on the four intervals and the interface depicted in Figure 10.1, which are always present.

Generally, we recognize that bulk fluid movements in the air of region (1) or in the water of layer (4) are sufficient to generate turbulent motions (see Section 9.4). Such swirling carries organic vapor molecules in the air or dissolved organic molecules in the water very quickly to adjacent regions in these bulk fluids and acts to homogenize chemical concentrations throughout. Consequently, we anticipate the vertical transport of organic chemicals is fast through regions (1) and (4) shown in Figure 10.1 and does not generally play a role in limiting air–water exchange of volatile chemicals.

Thus, our focus narrows to the processes operating in the remaining two intervals very near the contact of the air and water fluids and to the actual contact zone between the two phases. Although we do not fully understand the exact nature of the air and water movements in the "films" or "boundary layers" adjacent to the interface, we do have some critical insights. First, owing to the increasingly viscous nature of fluids on smaller and smaller length scales, the energy required to drive eddies that are "tighter" than about 100 μm in water or about 1 mm in air becomes less and less available. Additionally, when wind rushes across the surface of the water, a certain frictional force opposing airflow is applied where the air and water touch, causing the wind speed to decrease dramatically in a thin zone just above the water. This reduced movement leads to diminished eddy mixing in this thin layer. Together, the interfacial shearing and the absence of turbulence below a certain range of sizes seem to produce "boundary layers" of air and water which are different from the bulk volumes in that they are stagnant or only intermittently mixed. These layers need not be of the exact same character everywhere in time and space, but exhibit a kind of time- and space-averaged thickness or frequency of mixing with the bulk fluids that dictates their average ability to mediate the transport of organic molecules passing through them.

From this understanding of the fluid movements (or more correctly, lack of movements) in zones 2 and 3 of Figure 10.1, two important "extremes" in physical simplifications of the situation have evolved. First, one can envision the case of water and air turbulence which is not strong enough to stir the thin air and water layers adjacent to the interface at all. In this extreme, we imagine the presence of a "completely stagnant boundary layer" of air and water on each side (Whitman, 1923; Liss and Slater, 1974). On the other hand, we can imagine a somewhat more boisterous situation at the interface (possibly more likely with higher winds or currents) in which the eddies of the bulk air or water actually force their way on occasion up to the air–water contact plane displacing parcels of air or water of similar size (Higbie, 1935; Danckwerts, 1951). Since turbulence in air diminishes rapidly below lengths of less than *a millimeter* and in water below distances of less than about *100 μm*, we visualize the stagnant or renewed boundary layers as being about these thicknesses. Now in either case, to move organic molecules across such subeddy sized boundary regions, we can no longer rely on swirling fluid motions, but instead must anticipate that molecular diffusion predominates.

Finally, we need to understand the exchange of molecules right at the actual molecule-scale interface between regions 2 and 3. Typically this contact boundary is treated as gas in equilibrium with the adjacent surface water. In other words, whatever the speed of transport in air or in water of molecules to or from the boundary may be, the molecular equilibrium at the interface itself is established fast enough to compensate immediately for concentration changes in the bulk volumes. From the kinetic theory of motions of gas molecules (e.g., Atkins, 1978), we know the mean absolute velocity component of the molecules along a fixed coordinate axis $\overline{|u_i|}$ to be

$$\overline{|u_i|} = \left(\frac{2RT}{\pi \cdot M}\right)^{1/2} \qquad (\mathrm{cm \cdot s^{-1}}) \qquad (10\text{-}1)$$

where

R is the gas constant ($8.31 \times 10^7\ \mathrm{g \cdot cm^2 \cdot K^{-1} \cdot s^{-2} \cdot mol^{-1}}$),
T is the absolute temperature (K), and
M is the chemical's molecular mass ($\mathrm{g \cdot mol^{-1}}$).

If we choose the axis of motion to be perpendicular to the air–water interface and consider only the molecules with velocities *toward* the interface, the mean transfer velocity from air to water is

$$u_{\mathrm{a/w}} = \frac{\overline{|u_i|}}{2} = \left(\frac{RT}{2\pi M}\right)^{1/2} \qquad (10\text{-}2)$$

For $T = 283$ K (10°C) and a compound with $M = 100\ \mathrm{g \cdot mol^{-1}}$, $u_{\mathrm{a/w}}$ is about 6000 $\mathrm{cm \cdot s^{-1}}$ or $5.3 \times 10^6\ \mathrm{m \cdot d^{-1}}$. Even if only a small fraction of the colliding molecules, say 1 out of 1000, penetrates into the aqueous solution, the transfer velocity is still several kilometers per day. As we will see, this is much faster than the transfer velocities across the two boundary layers in the adjacent bulk phases. Therefore, equilibrium at the interface is not a limiting step in the overall gas exchange.

10.3 AIR–WATER EXCHANGE MODEL FORMULATIONS

In light of our physical picture of the movement of molecules from bulk air to bulk water, or vice versa, we recognize that the overall rate of chemical transfer will generally be controlled by the speed of passage through the two boundary layers. Thus we need a quantitative treatment focusing on these interfacial "films" to calculate the rate of air–water exchange. Due to the various simplified pictures people make for these layers, different mathematical models have evolved—but as we will see, the critical modeling parameters of each one are not experimentally observable and therefore do not permit the identification of which modeling view is more appropriate to describe the actual physical situation. In other words, we cannot measure the average thickness of any boundary layer, nor can we measure the rate at which microscopic eddies renew the interfacial layers. Consequently, all available *a priori* models retain fitting parameters which, when appropriately adjusted, yield reasonable volatilization flux estimates. To demonstrate, we develop the now classical "stagnant film" and "film renewal" models below.

Stagnant Two-Film Model

First we begin with the conceptually simpler "stagnant boundary theory" (Whitman, 1923; Liss and Slater, 1974). This model formulation envisions an unstirred or stagnant condition in both a water layer and an air layer adjacent to the interface and is thus often referred to as the two-film model (Fig. 10.2). In this case, the rate of movement of molecules through these two stagnant transport bottlenecks *arranged in series* is controlled by the chemical's ability to diffuse across them. Molecules in the aqueous solution and as vapors in air are continuously changing places with one another due to their Brownian motions. As a result, there will be a net movement of molecules of a given kind from all positions of high concentrations to those sites where they are relatively lacking. For our case, where chemical concentrations are assumed the same everywhere horizontally, we only need to be concerned with net transport in the vertical, that is, between the air and water.

From Fick's first law (Eq. 9-2), we conclude that the concentration gradient, dC/dz, across a layer of thickness z_a (e.g., the stagnant boundary layer of air, Fig. 10.2) must eventually become constant, provided that two conditions hold: (1) the chemical does not undergo any reaction within the layer, and (2) the concentrations at the boundaries of the layer (e.g., C_a and $C_{a/w}$) are kept constant long enough that the concentration profile reaches a steady state. The validity of these conditions is discussed later. Let us, at first, take them for granted. Then the flux of molecules, F_a, across the layer is simply

$$F_a = -D_a \frac{dC}{dz} = -D_a \frac{C_a - C_{a/w}}{z_a} \qquad [\mathrm{M \cdot L^{-2} \cdot T^{-1}}] \qquad (10\text{-}3)$$

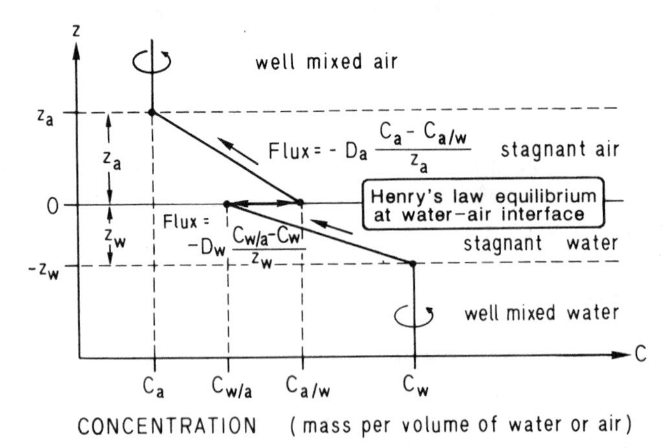

Figure 10.2 Stagnant boundary layer model for air–water exchange of chemicals. Note that the vertical coordinate z is defined as positive upwards and that $z = 0$ at the level of air–water contact.

because the gradient is just the concentration jump across the layer divided by the layer thickness. Note that according to the sign convention introduced in Eq. 9-1 and the orientation of the z axis, the flux F_a is positive if directed from the water surface upward. Since the molecular diffusion coefficient in air, D_a, is constant across the layer, Eq. 10-3 causes the flux F_a to be constant at every point within the layer, as it should be under steady-state conditions. The flux F_w, across the adjacent stagnant boundary layer of water, can be described in a similar way (see Fig. 10.2)

$$F_w = -D_w \frac{(C_{w/a} - C_w)}{z_w} \tag{10-4}$$

where D_w is the chemical's molecular diffusion coefficient in water (at the temperature of the water) and z_w is the thickness of the stagnant water film. Note that utilizing the same argument as for the air film allows us to assume a constant concentration gradient across the water film.

To deduce the overall rate of chemical transfer from the bulk water to the atmosphere, we must evaluate the combined influence of both boundary layers acting in series. We do this by recognizing that when volatilization is finally acting to deliver a steady stream of molecules from the water to the air (or vice versa), the number of molecules passing through each boundary film per unit area per unit time must be the same. As analogized in Figure 10.3, a steady state occurs when one has the same number of any units of interest (e.g., cars, molecules) passing each stage in a series (road sections or environmental compartments) in each increment of time. Notably, sections that are "harder" to pass usually play an important role in governing the time needed to execute the entire course. The *equilibrium* condition contrasts one of *steady state* in that two equilibrated compartments exhibit a balance of units coming and going between them.

Returning to our volatilization problem, when air–water exchange is proceeding at steady state, the flux of chemical through every layer in series must be the same. Hence $F_a = F_w = F$ where

$$F = -D_w \frac{(C_{w/a} - C_w)}{z_w} = -D_a \frac{(C_a - C_{a/w})}{z_a} \tag{10-5}$$

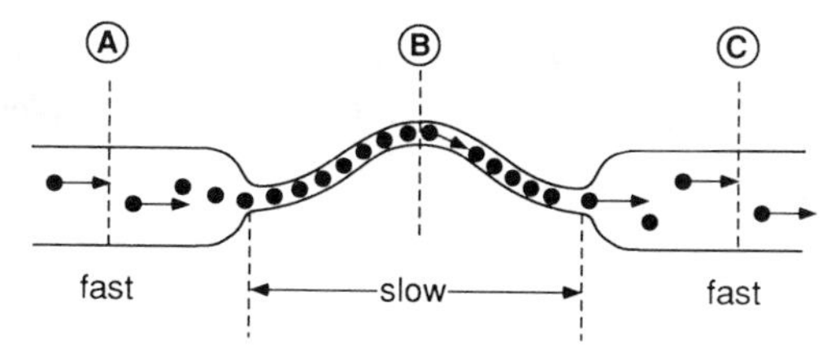

Figure 10.3 Visualization of a steady-state flux. The concentration is large in those compartments with slow particle motion.

Since we assumed that the layer of air molecules immediately above the interface was always equilibrated with the layer of water molecules immediately below, we can relate $C_{w/a}$ and $C_{a/w}$ with the compound's Henry's Law constant (using K'_H with units of $(\text{mol}\cdot L_a^{-1})\cdot(\text{mol}\cdot L_w^{-1})^{-1}$ for convenience):

$$K'_H = \frac{C_{a/w}}{C_{w/a}} \tag{10-6}$$

Substituting in Eq. 10-5, we can solve for $C_{w/a}$ in terms of bulk media concentrations C_a and C_w which we can measure:

$$\frac{D_w(C_w - C_{w/a})}{z_w} = \frac{D_a(K'_H C_{w/a} - C_a)}{z_a} \tag{10-7}$$

$$C_{w/a} = \frac{(D_w/z_w)C_w + (D_a/z_a)C_a}{(D_w/z_w) + (D_a K'_H/z_a)} \tag{10-8}$$

Substituting this somewhat complex result in Eq. 10-4, we arrive at a quantitative description of the chemical flux through the two transport bottlenecks in series:

$$F = \left(\frac{1}{(z_w/D_w) + (z_a/(D_a K'_H))}\right)\left(C_w - \frac{C_a}{K'_H}\right) \tag{10-9}$$

The flux F is positive (i.e., directed from the water into the air) if the bulk water concentration C_w is larger than C_a/K'_H, the water concentration in equilibrium with the bulk air concentration C_a. Since the relevant mechanisms, diffusive transport and Henry's Law partitioning, are reversible phenomena, Eq. 10-9 also quantifies transfers in the opposite direction (air-to-water), that is, negative fluxes. They occur if C_w is smaller than the equilibrium concentration C_a/K'_H. Also, when the concentrations in the water and air are at equilibrium ($K'_H = C_a/C_w$), we see that the flux is zero.

Let us now focus on the first parenthetical term in the flux expression—the so-called overall or total mass transfer velocity, v_{tot}. This term reflects the combined effects of requiring the organic molecules to diffuse through two stagnant layers in series. A simple dimensional analysis shows that this term indeed has the dimensions of a velocity $[L\cdot T^{-1}]$. Choosing specific units, for example, moles, centimeters, and seconds:

$$\underset{(\text{mol}\cdot\text{cm}^{-2}\cdot\text{s}^{-1})}{F} = \underset{(\text{cm}\cdot\text{s}^{-1})}{v_{tot}} \cdot \underset{(\text{mol}\cdot\text{cm}^{-3})}{\left(C_w - \frac{C_a}{K'_H}\right)} \tag{10-10}$$

The mass transfer coefficient v_{tot} can be expressed in terms of the "partial" transfer velocities defined for the two stagnant layers:

$$v_w \equiv \frac{D_w}{z_w}, \quad v_a \equiv \frac{D_a}{z_a} \qquad [L \cdot T^{-1}] \tag{10-11}$$

Combining these definitions yields

$$\frac{1}{v_{tot}} = \frac{1}{v_w} + \frac{1}{v_a K'_H} \tag{10-12}$$

There exists an obvious analogy between the boundary layer theory and the theory of electrical conductance. The mass transfer velocity corresponds to the conductivity, and thus the inverse of the transfer velocity can be interpreted as a transfer resistance. In this way, Eq. 10-12 means that the combined resistance of the two boundary layers is equal to the sum of the resistances of the individual layers. (By the way, this is also true for the case of three and more boundary layers!) The analogy is given by the total electrical resistance of two resistors arrayed in series. If one resistance is much smaller than the other, it has only little influence on the total resistance. Likewise, if one transfer resistance, say v_w^{-1}, is much larger than the other $(v_a K'_H)^{-1}$, then $v_{tot}^{-1} \sim v_w^{-1}$. Expressed for transfer velocities instead, we conclude that the small transfer velocity dominates the overall velocity. Thus, we have two extreme situations:

(1) Transfer dominated by water boundary layer:

$$v_w \ll v_a K'_H \Rightarrow v_{tot} \sim v_w \tag{10-13a}$$

(2) Transfer dominated by air boundary layer:

$$v_w \gg v_a K'_H \Rightarrow v_{tot} \sim v_a K'_H \tag{10-13b}$$

Of course, there are situations in which neither the air nor the water boundary layer is the sole controlling factor; then Eq. 10-12 has to be employed to combine the partial transfer velocities.

Because of their units, the v_i are frequently called "piston velocities," reflecting the rate at which an imaginary piston would sweep through the vertical direction moving molecules ahead of it. Yet, the discussion above demonstrates that this picture can be misleading since the piston moves at different speed through the different layers. Thus in speaking of overall volatilization rates using piston velocities, we should be careful to specify which layer we mean.

From Eq. 10-9 we see what properties of the chemical and of the aquatic environment are responsible for controlling the rate of volatilization. Notably, a chemical's diffusivities in air and in water and its Henry's Law constant are the personality traits (dictated by its structure!) that govern its air–water exchange rate. The poorly understood environmental characteristics, z_w and z_a, are the most obvious factors dictating how volatilization varies from time to time and place to place. Although these film thicknesses are somewhat fictitious constructs insofar as they do not accurately portray the real physics at the interface, they do allow a certain degree of rationalization of how volatilization rates vary. Thus, since we anticipate that higher winds cause greater shearing stress on the air–water boundary, we expect that this forcing causes z_a and z_w to diminish in thickness. Our mathematical model indicates that such decreases in stagnant layer thicknesses will cause greater volatilization rates—which of course is what we observe in the laboratory and field tests of wind effects on air–water exchange (see Section 10.4). We also note that environmental factors, such as water or air temperature and water ionic strength, will influence volatilization through the Henry's Law constant term.

Finally, we should address the question of whether the assumptions introduced to derive Eq. 10-9 are reasonable. The first condition refers to the possible *in situ* production or decay of the chemical under consideration. Since no organic compound is really conservative in the strict sense, it is more accurate to require that the compound not react in the typical time needed for the compound to diffuse through the boundary layers. From the diffusion distance derived in Chapter 9, Eq. 9-9, we can calculate the time needed for an average molecule to cross, for instance, the water boundary layer to be of the order

$$\tau_w \simeq \frac{z_w^2}{D_w} = \frac{z_w}{v_w} \tag{10-14}$$

and likewise for the air boundary layer:

$$\tau_a \simeq \frac{z_a^2}{D_a} = \frac{z_a}{v_a} \tag{10-15}$$

where a factor of 2 has been dropped. Note that this equation yields another interpretation of the piston velocity as the mean velocity at which the molecules move across the boundary.

As discussed in Section 10.4, typical ranges for z_w and z_a are 5×10^{-3} to 5×10^{-2} cm and 0.1–1 cm, respectively; and typical diffusivities D_w and D_a are 10^{-5} cm$^2 \cdot$s^{-1} and 0.1 cm$^2 \cdot$s^{-1}, respectively. Thus, the resulting total diffusion times (τ_w and τ_a) are typically on the order of seconds. If chemicals are produced or decay in times longer than about seconds, then we can justifiably neglect the reaction during air–water exchange. This sets one limit to the simple stagnant boundary layer model; modifications necessary for fast-reacting species are presented in Section 10.5.

The second requirement for the model is that the boundary layer configuration should not change at a pace that would make it impossible for the concentration profile to attain steady state. As before, time-to-steady-state can be estimated from the diffusion time τ_w and τ_a, Eqs. 10-14 and 10-15. If the surface is turned over at a rate that could make it impossible for the concentration profile across the boundary layers to reach steady state, an alternative formulation for the flux of molecules across the air–water interface is needed. The extreme version of this concept is the so-called surface renewal model which is now discussed in greater detail.

Surface Renewal Model

The surface renewal model formulation of the air–water exchange process attempts to reflect a continual turnover of air and water "parcels" with their associated organic chemical load at the air–water interface (Danckwerts, 1951). This modeling approach couples the rate of delivery of new bulk media "packets" to the interface, a consequence of the level of turbulence in these media, with the local molecular diffusive exchange out of, or into, these air or water elements. Typically, one assumes: (1) on arrival at the interface, the concentration of chemical at all points in immediate contact with the other phase is equilibrated with the other side, and (2) the packets do not stay at the phase boundary long enough for diffusive losses or gains to translate far enough to affect the local chemical concentration on the side of the packet away from the boundary. Thus the diffusion expression takes on a time-varying form in which initially after arrival of a packet containing a chemical of interest at the interface, the flux is very large, but steadily slows down as the subregion nearest the phase boundary becomes depleted of diffusing substance (i.e., lowering dC/dz). This situation is shown in a simplified way in Figure 10.4. At time zero, a new interface is created because of some turnover event. Henry's Law equilibrium at the contact zone is achieved without time delay:

$$C_{a/w} = K'_H C_{w/a} \tag{10-6}$$

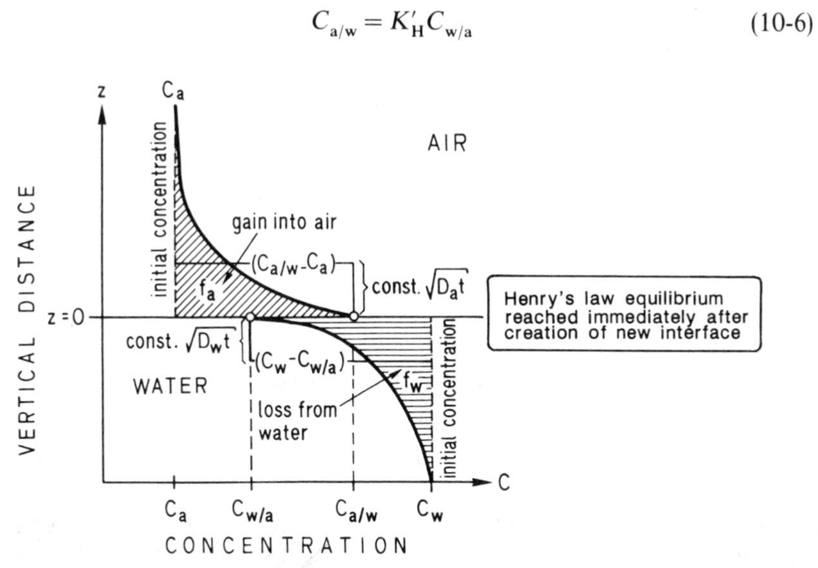

Figure 10.4 Surface renewal model for air–water exchange of chemicals. At time t, after the creation of a new interface, a total mass f_w of the chemical has diffused out of the water leading to a gain f_a of chemical in the air. Mass conservation requires that $f_a = f_w$. The equilibrium at the interface is given by Henry's law: $C_{a/w} = K'_H C_{w/a}$. Both f_a and f_w are approximated by the piston flow rectangles, $(C_{a/w} - C_a)\cdot\sqrt{D_a t}$ and $(C_w - C_{w/a})\cdot\sqrt{D_w t}$, respectively.

whereas at time $t = 0$ the concentration away from the interface still corresponds to the bulk concentrations in water and air, C_w and C_a, respectively. Owing to the sudden change of the boundary concentrations, the chemical starts to move out of the water into the air. At time t a total amount of chemical, f_w (in Fig. 10.4 represented by the hatched area), has left the water. It is equal to the new amount of chemical, f_a, which appeared on the air side of the interface:

$$f_a = f_w \qquad [\mathrm{M \cdot L^{-2}}] \tag{10-16}$$

The exact shape of the areas f_a and f_w could be calculated from the appropriate solution of the time-dependent second Fickian equation, Eq. 9-15 (e.g., Crank, 1975, or Carslaw and Jaeger, 1959). Since the model will inevitably contain quantities that cannot be directly measured (e.g., the mean contact time of interfaces, t), it is not so important to calculate this shape as long as we know how it depends on the diffusivities D_a and D_w. From the Einstein–Smoluchowsky equation (Eq. 9-9), we know the diffusion penetration depth to be proportional to $\sqrt{Dt}$ (the factor $\sqrt{2}$ can be built into the unknown proportionality). Thus, we can approximate f_a and f_w by the product of ΔC and the penetration depth (the rectangles shown in Fig. 10.4):

$$f_a = \text{constant} \cdot (C_{a/w} - C_a)\sqrt{D_a t} \tag{10-17a}$$

$$f_w = \text{constant} \cdot (C_w - C_{w/a})\sqrt{D_w t} \tag{10-17b}$$

The constants take care of the details of the concentration profile across the boundary and are equal in both expressions, since they both originate from the same time-dependent solution of the diffusion equation. Using Eq. 10-17 and the equilibrium relation 10-6 allows us to eliminate the unknown interface concentrations $C_{a/w}$ and $C_{w/a}$, for instance,

$$C_{w/a} = \frac{C_w\sqrt{D_w} + C_a\sqrt{D_a}}{\sqrt{D_w} + K'_H\sqrt{D_a}} \tag{10-18}$$

We insert this result into Eq. 10-17b:

$$f_w = \text{constant} \cdot \frac{(C_w - C_a/K'_H)}{\left(\dfrac{1}{\sqrt{D_w t}}\right) + \left(\dfrac{1}{K'_H\sqrt{D_a t}}\right)} = f_a \tag{10-19}$$

Note that $f_a = f_w$ is the total mass exchanged per unit contact area during an event of duration t. Thus, division by t yields the exchange flux per unit time. Instead of the contact time t, we can introduce the surface renewal rate

$$r = \frac{1}{t} \qquad [\mathrm{T^{-1}}] \tag{10-20}$$

representing the mean frequency of creating contact surfaces. Finally, the exchange flux per unit time and area is

$$F = \frac{f_w}{t} = r \cdot t_w = \text{constant} \cdot \left(\frac{1}{(1/\sqrt{rD_w}) + (1/(K'_H\sqrt{rD_a}))} \right) \left(C_w - \frac{C_a}{K'_H} \right) \quad (10\text{-}21)$$

Since the unknown surface renewal rate r is only an idealized model parameter to be experimentally determined, we can set the constant equal to 1. We can even decouple the surface renewal processes on either side of the interface by introduction of two separate renewal rates, r_a and r_w, reflecting the turnover of the air layer and the water film, respectively.

Note that Eq. 10-21 strongly resembles the flux calculated from the stagnant two-film model, Eq. 10-9. Thus, we may express F by a total transfer velocity:

$$v_{tot} = \left(\frac{1}{\sqrt{r_w D_w}} + \frac{1}{K'_H \sqrt{r_a D_a}} \right)^{-1} \quad (10\text{-}22)$$

which, in turn, can be put together from the "partial" transfer velocities:

$$v_w = \sqrt{r_w D_w}, \qquad v_a = \sqrt{r_a D_a} \quad (10\text{-}23)$$

Thus, Eq. 10-22 turns out to have the same mathematical form as Eq. 10-12, derived for the stagnant film model. The essential difference between the models lies in the definition of the transfer velocities, Eqs. 10-11 and 10-23. Both contain empirical constants, two film thicknesses for the stagnant film model, and two surface renewal rates for the surface renewal model. Most importantly, the two models differ in the way in which the flux depends on the diffusivities, proportional to D in the former, proportional to $\sqrt{D}$ in the latter. Various attempts to discern this functional relationship (e.g., Torgersen et al., 1982, who looked at relative volatilization rates of helium and radon in lakes; or Ledwell, 1984, who looked at mass transfer coefficients exhibited by N_2O, CH_4, and He in a laboratory wind-wave tunnel) have yielded disparate results. Possibly relatively quiescent conditions (such as those in lakes) yield results more in line with "stagnant boundary physics," whereas more turbulent conditions (as in flumes or streams) correspond more closely to the "renewal" picture. Generally it was found that F depends on D^{α} where α varies between 0.5 (the result of the renewal model) and 1 (the stagnant film model). The physical meaning behind the result could be that the exposure of surfaces is often too long to allow the assumption of diffusion into infinite fluid or air volumes (as assumed for the renewal model), but too short to create steady-state fluxes across stagnant films. In any case, an important point to recognize in these mass transfer coefficient formulations is that the values of z_w, z_a, r_w, and r_a cannot be independently assessed. Thus these parameters, though imagined to have physical meaning, remain fitting coefficients which vary in rational ways as winds and water turbulence change from place to place or time to time.

10.4 IMPACT OF WIND AND WATER CURRENTS ON v_w AND v_a

Let us now turn our attention to the experimental information on gas exchange. A variety of laboratory and field measurements of air–water exchange of chemicals have been performed in an attempt to understand this process and quantify its mass transfer coefficients (v_w and v_a). Although such research has gone on for many years, we still do not know how to predict these coefficients based solely on properties of the chemical and of the environment. Since chemical diffusivities and Henry's Law constants are reasonably well known (even insofar as they change with temperature, etc.), this problem primarily boils down to our inability to describe accurately the physics of a temporally and spatially variable air–water interface.

As a consequence, it is not trivial to extrapolate air–water transfer velocities gained from laboratory experiments to the field. Certainly, at a fixed wind speed the structure of the water surface (distortion by waves, formation of bubbles and spray) is not the same in a laboratory tank as on the ocean or on a lake, and so the gas transfer may be different. Furthermore, the physical properties of the boundary layers should depend on the size of the wind shear stress at the water surface whereas the quantity usually determined in the experiments is the wind speed measured at some distance above the water surface. For a comparison, a model is needed to transform wind speeds from different heights to a reference level (usually 10 m above the water surface; see Mackay and Yeun, 1983). Using a so-called boundary roughness height of 0.03 cm, the wind speed u_z (measured at distance z, in meters, from the water surface) is related to u_{10} by

$$u_z = \left(\frac{\ln z + 8.1}{10.4} \right) u_{10} \qquad \text{(10-24)}$$

Nonetheless, certain systematic studies have provided various means to correlate "film thicknesses" or "renewal rates" with measurable conditions such as wind speed or current velocity and water depth. Thus although our understanding is imperfect, useful predictions of air–water exchange rates can be made.

Transfer Velocity in Air (v_a)

Probably the best place to start is by considering the rate of evaporation of water itself from a body of water. We quickly recognize that water molecules exist in abundance throughout any "stagnant" or "intermittently mixed" water layer below the air–water interface; and consequently we can reasonably assume that the flux of water itself is a special case in which $v_w \to \infty$ or $v_{tot} = v_a K'_H$. Then Eqs. 10-10 and 10-21 simplify to

$$\text{Flux(water)} = v_a (K'_H C_w - C_a) \qquad (\text{mol}\cdot\text{cm}^{-2}\cdot\text{s}^{-1}) \qquad \text{(10-25)}$$

where $v_a = (D_a/z_a)$ or $\sqrt{D_a r_a}$ is measured in $\text{cm}\cdot\text{s}^{-1}$, C_w is in $\text{mol}\cdot\text{cm}^{-3}$, and K'_H is the Henry's Law constant of water [e.g., at 25°C, $K'_H = 2.3 \times 10^{-5}\ (\text{mol}\cdot\text{L}_a^{-1})\cdot(\text{mol}\cdot\text{L}_w^{-1})^{-1}$].

Thus, by measuring the relative humidity,

$$h = \frac{C_a}{K'_H C_w} = \frac{C_a}{C_a^{sat}} \tag{10-26}$$

and the evaporation flux of water to the atmosphere, we can directly obtain v_a (or z_a or r_a):

$$v_a = \frac{\text{Flux(water)}}{(K'_H C_w - C_a)} = \frac{\text{Flux(water)}}{(C_a^{sat} - hC_a^{sat})} = \frac{\text{Flux(water)}}{C_a^{sat}(1-h)} \tag{10-27}$$

Hydrologists have long recognized the relationship of v_a to environmental conditions such as wind speed (Dalton, 1802; Fitzgerald, 1886; Rohwer, 1931; Sverdrup et al., 1942). In short-term laboratory experiments (Liss, 1973; Penman, 1948; Münnich et al., 1978; Mackay and Yeun, 1983), increasing wind velocities distinctly enhance this phase

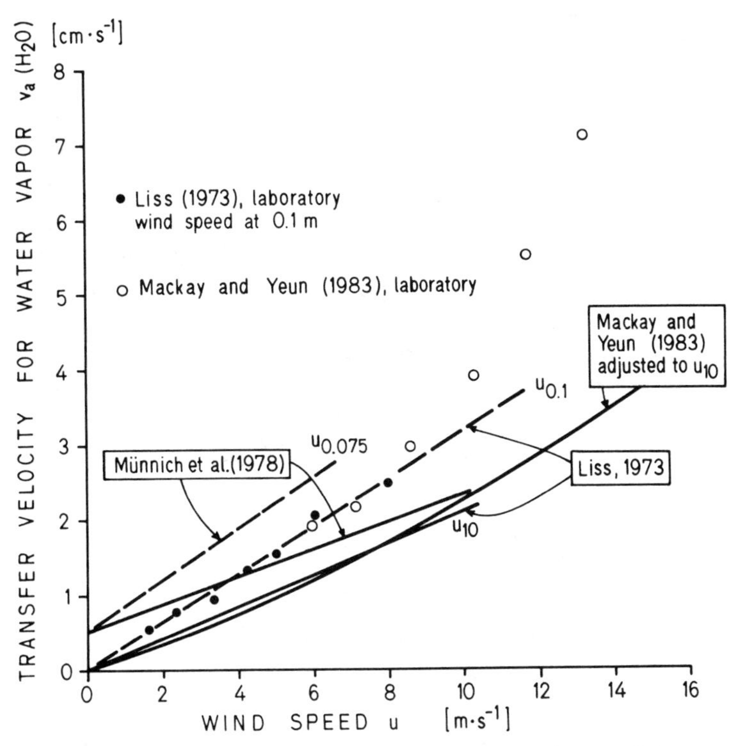

Figure 10.5 Impact of wind speed u_z on the air-side mass transfer coefficient v_a, as observed by the evaporation of water in laboratory experiments and as predicted using various correlation expressions (see Table 10.1). Note that the wind speeds refer to different heights above the water surface. This explains the higher v_a values for wind measurements made at small height z. In Table 10.1, values are also adjusted to wind speeds at 10 m using Eq. 10-24.

transfer coefficient approximately linearly (Fig. 10.5), although as winds exceed $10\,m\cdot s^{-1}$ this coefficient may increase more quickly, possibly due to aerosol ejection and a nonlinear increase of the wind shear due to wave breaking (see data by Mackay and Yeun, 1983, in Fig. 10.5 and Table 10.1). It is also interesting to note that in some of the resulting expressions at zero wind speed, v_a clearly remains different from zero (Table 10.1, examples 2, 3, and 5), while in others v_a becomes very small or zero (Table 10.1, examples 1, 4, and 6). Münnich et al. (1978) relate the zero-wind transfer velocity to convective motion in the air because saturated air at the water surface may be up to 1% lighter than the dry air above.

The data suggest that v_a is typically in the range of 0.3 to $3\,cm\cdot s^{-1}$ (or $10–100\,m\cdot h^{-1}$). If we consider physical interpretations of this range of magnitudes of v_a, in light of the stagnant film model of volatilization, it corresponds to "stagnant film thicknesses" of 0.1–1 cm (assuming $D_a(H_2O) = 0.26\,cm^2\cdot s^{-1}$). On the other hand, these phase transfer coefficients would imply a surface renewal rate of the air layer adjacent to the water of between 0.4 and 40 times per second. No matter what the exact physics is, the lumped parameter $v_a(H_2O)$ has been repeatedly seen to be a direct function of wind speed. Thus, empirical correlations obtained by oceanographers, hydrologists, and engineers look similar (Table 10.1). Perhaps a suitable approximation deduced from all these investigations is

$$v_a(H_2O) \sim 0.2\,u_{10}(m\cdot s^{-1}) + 0.3 \qquad (cm\cdot s^{-1}) \tag{10-28}$$

where u_{10} is the wind speed measured 10 m above the water surface and the intercept value of $0.3\,cm\cdot s^{-1}$ reflects the minimum value of v_a typically observed corresponding to a stagnant air layer of about 1 cm. Note that via the influence of D_a, v_a is also temperature dependent. In the following approximations we will, however, neglect this effect.

To demonstrate the feasibility of extrapolating such laboratory results to the field, we can compare the range of observed water fluxes from natural waters to predicted results. Thus for evaporation of water from a lake of temperature 20°C and overlying air of typical relative humidity 80% (i.e., $C_a = 0.8\,K'_H C_w$), we could estimate the water-to-air flux of water as

$$\begin{aligned} \text{Flux} &= v_a(C_w K'_H - C_a) \\ &= v_a(C_w K'_H - 0.8\,C_w K'_H) \\ &\sim (0.3 \text{ to } 3\,cm\cdot s^{-1})(1.0\,g\cdot cm^{-3})(2.3\times 10^{-5})(0.2) \\ &\sim 10^{-6} \text{ to } 10^{-5}\,g\cdot cm^{-2}\cdot s^{-1} \end{aligned}$$

which corresponds to an evaporation rate of about $0.1–1\,cm\cdot d^{-1}$. This is in accordance with observed evaporative loss rates of water from lakes and the ocean (Miller, 1977; Sverdrup et al., 1942).

Given such an estimate of v_a for water, we are now in an excellent position to predict v_a for other volatilizing chemicals of interest. Recalling the two extreme model

TABLE 10.1 Empirical Relationships Between Wind Velocity and the Transfer Velocity at the Air–Side Boundary Deduced from Observations of Water Evaporation Rates[a]

Source	Data Type	Original Data	v_a related to u_{10}
(1) Sverdrup et al. (1942)	Field	$v_a = 0.16 u_6$	$v_a = 0.15 u_{10}$
(2) Penman (1948)	Evaporation pans	$v_a = 0.59 + 0.27 u_2$	$v_a = 0.59 + 0.23 u_{10}$
(3) Rohwer (1931)	Evaporation pans	$v_a = 0.35 + 0.12 u_{10}$	$v_a = 0.35 + 0.12 u_{10}$
(4) Liss (1973)	Wind–water tunnel	$v_a = 0.005 + 0.32 u_{0.1}$	$v_a = 0.005 + 0.21 u_{10}$
(5) Münnich et al. (1978)	Circular wind–water tank	$v_a = 0.5 + 0.35 u_{0.075}$	$v_a = 0.5 + 0.185 u_{10}$
(6) Mackay and Yeun (1983)	Wind–wave tank	$v_a = 0.065(6.1 + 0.63 u_{10})^{0.5} u_{10}$	$v_a = 0.065(6.1 + 0.63 u_{10})^{0.5} u_{10}$

[a] u_z = wind ($m \cdot s^{-1}$) measured at height z (m) above the water surface; v_a = air–side transfer velocity ($cm \cdot s^{-1}$).

expressions, Eqs. 10-11 and 10-23, and the experimental evidence cited earlier, we know that v_a varies with molecular diffusivity in air as $(D_a)^\alpha$ where α lies between 0.5 (surface renewal model) and 1 (stagnant film model). Therefore, we can use the value measured for water vapor, $v_a(H_2O)$, reflecting the typical physical condition of the air above the water body (windiness) to calculate v_a for some other chemical:

$$v_a(\text{compound}) = v_a(H_2O)\left[\frac{D_a(\text{compound})}{D_a(H_2O)}\right]^\alpha \qquad (10\text{-}29)$$

where $0.5 \leqslant \alpha \leqslant 1$. Experimental information concerning the size of α is discussed later.

Transfer Velocity in Water (v_w)

Having established the magnitude of v_a, we can now focus our attention on the size of v_w. Early workers pursued laboratory observations of extremely volatile gases such as O_2 or CO_2 to this end (e.g., Kanwisher, 1963; Liss, 1973), and more recent geochemistry investigations have expanded the data base to include natural radiotracers such as ^3He or ^{222}Rn (Emerson et al., 1973; Broecker and Peng, 1974; Peng et al., 1979; Torgersen et al., 1982; Jähne et al., 1984) and artificial chemical tracers such as SF_6 (Wanninkhof et al., 1987). Additionally, many laboratory experiments on mixed or wind-blown waters containing organic chemicals have now been reported (e.g., Dilling et al., 1975, 1977; Smith et al., 1980; Mackay and Yeun, 1983). By and large the early investigations revealed that typical values of v_w lie between 5×10^{-4} and 5×10^{-3} $cm \cdot s^{-1}$, or about 100–1000 times less than v_a (Fig. 10.6). Since gas exchange of molecular oxygen (O_2) only depends on v_w (and not on v_a since K'_H is so large), O_2 is commonly used as the reference substance for v_w as water vapor is used for v_a.

Unlike the air-side mass transfer velocity, wind shear stress or wind speed are not the only important parameters controlling the magnitude of v_w. As demonstrated by Jähne et al. (1984), v_w is strongly affected by the wave field, which itself depends in a complicated way on wind stress, wind fetch, surface contamination (affecting the surface tension of the water), and the water currents. Thus, it is not surprising that field and laboratory data show such a large disparity in the results. Generally, laboratory experiments tend to overestimate gas exchange rates occurring under natural conditions.

At wind speeds above about $10\ m \cdot s^{-1}$, that is, above the onset of wave breaking and formation of air bubbles (Blanchard and Woodcock, 1957; Monahan, 1971; Kolovayev, 1976; Johnson and Cooke, 1979; Wu, 1981), transfer velocities determined from natural systems are possibly distorted by an additional effect, by the so-called wind pumping. In this situation, bubbles injected deep below the water surface experience pressures in excess of atmospheric, and as a result dissolve more of the gases than required for equilibrium of the water surface, leading to supersaturation of O_2, N_2, and CO_2 of up to 15% (Smith and Jones, 1985). Yet the importance of this phenomenon is not fully understood; Thorpe (1984) estimates O_2 supersaturation does not exceed 3% at wind speeds of about $12–14\ m \cdot s^{-1}$, and Craig and Hayward (1987) conclude that 72–86%

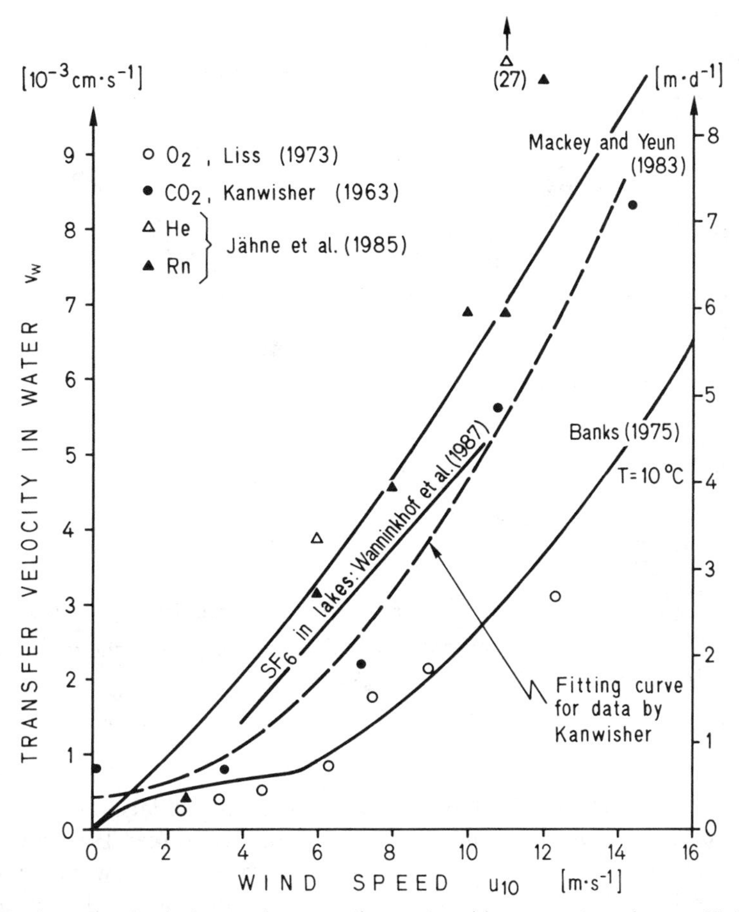

Figure 10.6 Impact of wind speed u_z on the water-side mass transfer coefficient v_w, as measured by experiments with various volatiles and as predicted using some reported correlations (see Table 10.2). Wind speeds are adjusted to a value measured 10 m above water surface, u_{10}, with Eq. 10-24.

of the O_2 supersaturation measured in ocean surface waters is due to photosynthesis rather than bubble trapping. Whatever the final resolution of this problem may be, it is clear that gas exchange rates measured from natural gases close to the saturation equilibrium may be misleading when applied to volatile organic compounds far from their water–air exchange equilibrium.

If we now turn our attention to flowing waters, we have to be aware that the physics of the boundary layer is influenced by both motion from the air side (wind) and the water side (water currents). For example, it has long been recognized that in rapidly flowing water, such as in shallow streams, where interactions of the current with the bottom topography (rocks and logs, rapidly varying depths) are the primary source of turbulence, we must utilize knowledge of the nature of this water-side fluid physics to

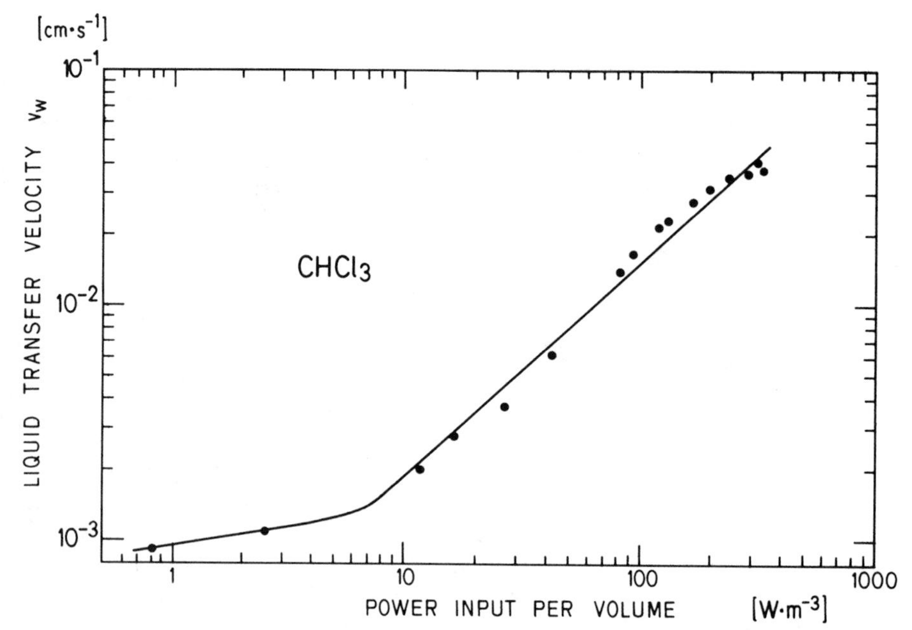

Figure 10.7 Variation of liquid film mass transfer coefficient v_w of chloroform with water stirring intensity. From Roberts and Dändliker (1983).

estimate v_w (O'Connor and Dobbins, 1958). Thus stirring in laboratory experiments (Fig. 10.7) clearly promotes volatilization (Roberts and Dändliker, 1983). In most other cases, where the water is fairly deep or the current is relatively slow (i.e., slow flowing rivers, lakes, estuaries, oceans), it appears that the action of wind friction on the air–water interface probably dictates the magnitude of v_w (e.g., Jirka and Brutsaert, 1984). Thus we tend to use two kinds of empirical relationships to estimate v_w—one based on current speed and depth, the other on wind velocity.

As is the case for $v_a(H_2O)$, numerous predictive equations are available to estimate $v_w(O_2)$ in streams. O'Connor and Dobbins (1958) argued that the more rapidly flowing is the water in a shallow stream, the more tumultuous should be the mixing, and hence the greater the surface replacement rate r_w (or more eroded the stagnant boundary layer z_w). Further, these investigators noted that the shallower is the stream, in general, the greater is the transfer of eddies in the water flow from the irregular stream bottom to the water surface. These notions led O'Connor and Dobbins to propose a relation of r_w to water velocity u_w and water depth d_w:

$$r_w \simeq \frac{u_w}{d_w} \tag{10-30}$$

Given this approximation, we can readily estimate v_w for any chemical of interest in a turbulent stream:

$$v_w \simeq (D_w r_w)^{1/2}$$
$$\simeq \left(\frac{D_w u_w}{d_w}\right)^{1/2} \tag{10-31}$$

Thus for a stream of depth 50 cm and water velocity 1 m·s^{-1}, one would estimate v_w for a solute like benzene ($D_w \simeq 1.02 \times 10^{-5}$ cm^2·s^{-1}) to be

$$\left(\frac{1.02 \times 10^{-5}\,\text{cm}^2\cdot\text{s}^{-1}\cdot 100\,\text{cm}\cdot\text{s}^{-1}}{50\,\text{cm}}\right)^{1/2} = 4.5 \times 10^{-3}\,\text{cm}\cdot\text{s}^{-1} \qquad (\simeq 16\,\text{cm}\cdot\text{h}^{-1})$$

From this empirical formulation, we can judge when current-generated turbulence must be included in our thinking. Typically v_w is observed to be greater than about 5×10^{-4} cm·s^{-1}. Thus we can evaluate current velocities for various water depths necessary for the fluid motion to contribute significantly:

$$5 \times 10^{-4}\,\text{cm}\cdot\text{s}^{-1} < \left(\frac{D_w \cdot u_w}{d_w}\right)^{1/2}$$

and if D_w is about 10^{-5} cm^2·s^{-1}, then the critical current velocity-to-depth ratio is $(u_w/d_w) > 0.03$ s^{-1}. For water of depth 10 cm, u_w need only be greater than 0.3 cm·s^{-1}; and for water of 10 m, u_w must be about 30 cm·s^{-1} or more for current interactions with the river bottom to be important to gas exchange at the surface.

Various field studies have been done to investigate volatilization of organic chemicals from streams and rivers (O'Connor and Dobbins, 1958; Neely et al., 1976; Rathbun and Tai, 1982, 1983; Schwarzenbach, 1983; Duran and Hemond, 1984; Wilcock, 1984; Genereux, 1991). Figure 10.8 shows some field results for chemicals (added purposely as a tracer or inadvertently) in streams and rivers in which the decline in concentration can be used to quantify the rates of volatilization. Generally the transfer coefficients predicted by the relation of O'Connor and Dobbins (Eq. 10-31) yield results within a factor of 2 or 3 of that which is observed, but are biased to an *underprediction.* Such agreement appears more than adequate when we recall that we have neglected wind effects and aeration effects of waterfalls, and recognize that average values of current velocities and depths are used (themselves varying by factors of 2 over the reaches of interest).

In slowly flowing waters (lakes, oceans), the motion of the air dictates the physical nature of the air–water interface. By shearing the water surface, wind acts to transfer energy into the water body and the associated result is to enhance the rate of molecular transfer out of aqueous solution (i.e., either visualized as thinning the stagnant water boundary layer or increasing the rate of renewal of surface water packets from below.) Numerous laboratory and field studies (Fig. 10.6) have sought to establish predictive equations relating v_w and wind speed, as summarized in Table 10.2. For typical winds of 2–10 m·s^{-1}, such equations yield v_w estimates of 5×10^{-4} to 5×10^{-3} cm·s^{-1}

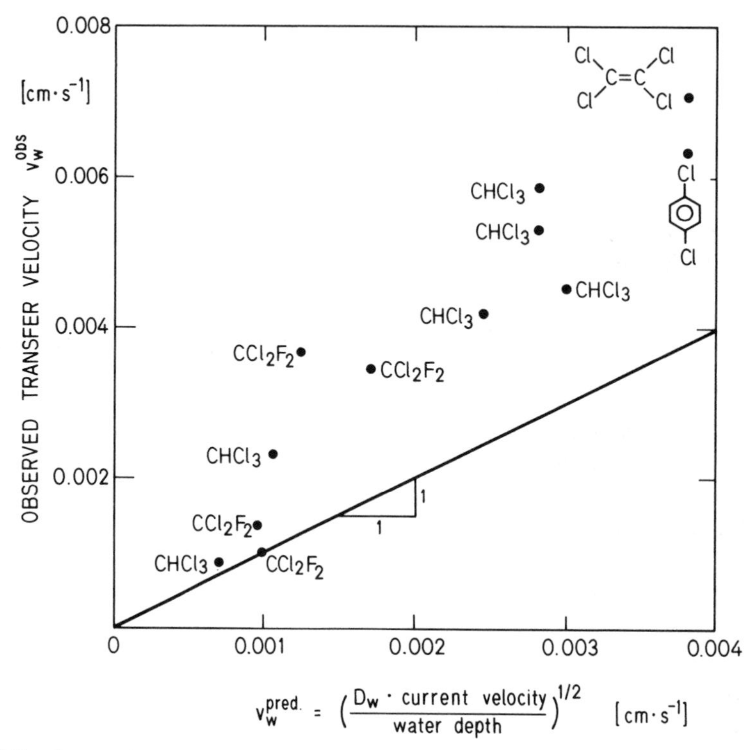

Figure 10.8 Comparison of observed liquid film mass transfer coefficients v_w^{obs} for organic compounds in streams and rivers versus values predicted by the relation (Eq. 10-31) of O'Connor and Dobbins (1958). $CHCl_3$ data from Neely et al. (1976) and from Wilcock (1984); CCl_2F_2 data from Duran and Hemond (1984); $Cl_2C{=}CCl_2$ and Cl–⟨benzene⟩–Cl data from Schwarzenbach (1983).

($2–20\,\text{cm}\cdot\text{h}^{-1}$). We suggest a quadratic relation in which a no-wind condition still yields a minimum v_w:

$$v_w(O_2) = 4 \times 10^{-4} + 4 \times 10^{-5}\, u_{10}^2 \qquad (\text{cm}\cdot\text{s}^{-1}) \tag{10-32}$$

where u_{10} is the wind speed in meters per second measured 10 m above the water surface. This range of values for v_w implies a "stagnant water film" of about 20–200 μm thickness or a renewal rate of approximately 0.03–3 times per second.

This result allows us to estimate transfer coefficients for other chemicals besides those for which they were originally derived. By analogy to Eq. 10-29, written for the air-side transfer velocity, v_a, we can write

$$v_w(\text{compound}) = v_w(O_2)\left[\frac{D_w(\text{compound})}{D_w(O_2)}\right]^{\beta} \tag{10-33}$$

Following Mackay and Yeun (1983), the exponents (α and β) are not the same for the

TABLE 10.2 Empirical Relationships Between Wind Velocity and Transfer Velocity at the Water-Side Boundary Layer[a]

Source	Data Type	Original Data	v_w related to u_{10}
(1) Kanwisher (1963)	CO_2, wind–water tunnel	$v_w = (4.1 + 1.33u_{0.1}^2)\cdot 10^{-4}$	$v_w = (4.1 + 0.417u_{10}^2)\cdot 10^{-4}$
(2) Banks (1975)	O_2, lakes		If $u_{10} < 5.5\,m\cdot s^{-1}$: $v_w = 1.024^{(T_w - 10°C)}(4.2\cdot 10^{-6}u_{10}^{0.5})$ If $u_{10} > 5.5\,m\cdot s^{-1}$: $v_w = 1.024^{(T_w - 10°C)}(0.32\cdot 10^{-6}u_{10}^2)$
(3) Liss (1973)	Laboratory tank, O_2	For $u_{0.1}$	See Figure 10.6
(4) Mackay and Yeun (1983)	Laboratory, various organic solutes		$v_w = 1.75\cdot 10^{-4}(6.1 + 0.63u_{10})^{0.5}u_{10}$ (for O_2)
(5) Wanninkhof et al. (1987)	SF_6 in lakes		$v_w = (-8.9 + 5.8u_{10})\cdot 10^{-4}$

[a] u_z = wind speed ($m\cdot s^{-1}$) measured at height z(m) above the water surface; v_w = transfer velocity ($cm\cdot s^{-1}$). T_w is the water temperature (°C).

TABLE 10.3 Importance of Air and Water Boundary Layers for Controlling the Total Air–Water Transfer Velocity v_{tot} for a Variety of Organic Compounds at Two Wind Speeds[a]

Compound	$K'_H = K_H/RT$ (25°C)	D_a[b] ($cm^2 \cdot s^{-1}$)	D_w[c] ($cm^2 \cdot s^{-1}$)
Methane	27	0.28	3.0×10^{-5}
Trichlorofluoromethane	5.3	0.094	1.0×10^{-5}
Octadecane	1.1	0.069	0.74×10^{-5}
Tetrachloroethylene	0.73	0.086	0.92×10^{-5}
Benzene	0.23	0.12	1.3×10^{-5}
1,2,4-Trichlorobenzene	0.11	0.086	0.88×10^{-5}
Naphthalene	0.018	0.097	1.0×10^{-5}
2,2′,4,4′,5,5′-Hexachlorobiphenyl	3×10^{-3}	0.058	0.63×10^{-5}
1-Hexanol	5×10^{-4}	0.11	1.2×10^{-5}
Benzo(a)pyrene	5×10^{-5}	0.069	0.75×10^{-5}
Phenol	1.7×10^{-5}	0.11	1.2×10^{-5}

[a]See also Figure 10.9. $v'_a \simeq (D_a/D_a^{H_2O})^{0.67}\,(K_H/RT)\,(0.2u_{10}+0.3)$ and $v_w \simeq (D_w/D_w^{O_2})^{0.57}\,(4 \cdot 10^{-5} u_{10}^2 + 4 \cdot 10^{-4})$ where u_{10} is in meters per second.
[b]D_a is estimated from $(0.26\,cm^2 \cdot s^{-1})$ [mw (H_2O)/mw (chemical)]$^{0.5}$ after Eq. 9-25.
[c]D_w estimated from $(2.1 \cdot 10^{-5}\,cm^2 \cdot s^{-1})$ [mw (O_2)/mw (chemical)]$^{0.5}$ after Eq. 9-30.
[d]$R_{a/w} = v_w/v'_a$: resistance ratio air–water boundary layer, after Eq. 10-36.

air and water boundary layers. From laboratory data, these investigators conclude:

$$\alpha = 0.67 \text{ (air)} \quad \text{and} \quad \beta = 0.5 \text{ (water)}$$

Holmén and Liss (1984) summarize results for β, mostly from laboratory experiments. They extract

$$\beta = 0.57 \pm 0.15$$

to be the best estimate. As concluded for the absolute values for v_a and v_w, their dependence on the molecular diffusivities of the chemicals may be different between laboratory and natural conditions and also depend on other factors characterizing the structure and shape of the air–water interface.

Total Transfer Velocity

An important consequence of having a knowledge of the order of magnitude of v_w and v_a is that we can now make a quick estimation regarding the relative importance of the two boundary layers to the overall transfer velocity v_{tot}. Remember that for both models, the stagnant two-film model and the surface renewal model, the total velocity follows from the partial velocities according to the relation (see Eq. 10-12)

$$\frac{1}{v_{tot}} = \frac{1}{v_w} + \frac{1}{v'_a} \tag{10-34}$$

TABLE 10.3 (*cont'd*)

$u_{10} = 1\ m \cdot s^{-1}$				$u_{10} = 20\ m \cdot s^{-1}$			
v'_a ($cm \cdot s^{-1}$)	v_w ($cm \cdot s^{-1}$)	v_{tot} ($cm \cdot s^{-1}$)	$R_{a/w}$ [d]	v'_a ($cm \cdot s^{-1}$)	v_w ($cm \cdot s^{-1}$)	v_{tot} ($cm \cdot s^{-1}$)	$R_{a/w}$ [d]
1.4×10^{1}	5.4×10^{-4}	5.4×10^{-4}	4×10^{-5}	1.2×10^{2}	2.0×10^{-2}	2.0×10^{-2}	2×10^{-4}
1.3×10^{0}	2.9×10^{-4}	2.9×10^{-4}	2×10^{-4}	1.1×10^{1}	1.1×10^{-2}	1.1×10^{-2}	1×10^{-3}
2.2×10^{-1}	2.4×10^{-4}	2.4×10^{-4}	1×10^{-3}	1.9	9.0×10^{-3}	9.0×10^{-3}	5×10^{-3}
1.7×10^{-1}	2.7×10^{-4}	2.7×10^{-4}	1.6×10^{-3}	1.5	1.0×10^{-2}	1.0×10^{-2}	7×10^{-3}
6.7×10^{-2}	3.3×10^{-4}	3.3×10^{-4}	5×10^{-3}	5.8×10^{-1}	1.2×10^{-2}	1.2×10^{-2}	2×10^{-2}
2.6×10^{-2}	2.7×10^{-4}	2.7×10^{-4}	1×10^{-2}	2.2×10^{-1}	1.0×10^{-2}	9.6×10^{-3}	5×10^{-2}
4.5×10^{-3}	2.9×10^{-4}	2.7×10^{-4}	0.06	3.9×10^{-2}	1.1×10^{-2}	8.6×10^{-3}	0.3
5.2×10^{-4}	2.2×10^{-4}	1.5×10^{-4}	0.4	4.5×10^{-3}	8.3×10^{-3}	2.9×10^{-3}	2
1.3×10^{-4}	3.2×10^{-4}	9.2×10^{-5}	2.4	1.1×10^{-3}	1.2×10^{-2}	1.0×10^{-3}	11
1.0×10^{-5}	2.4×10^{-4}	9.6×10^{-6}	24	8.6×10^{-5}	9.1×10^{-3}	8.5×10^{-5}	100
4.7×10^{-6}	3.2×10^{-4}	4.6×10^{-6}	68	4.0×10^{-5}	1.2×10^{-2}	4.0×10^{-5}	300

where we have introduced the compound-specific air velocity

$$v'_a = v_a \frac{K_H}{RT} = v_a K'_H \tag{10-35}$$

We have interpreted the inverse transfer velocities as layer resistances. Thus, the resistance ratio

$$R_{a/w} = \frac{(1/v'_a)}{(1/v_w)} = v_w / v'_a \tag{10-36}$$

measures the relative importance of the air film resistance compared to the water film resistance. The overall transfer rate is primarily controlled by the water film if $R_{a/w} < 0.1$, and it is air-film controlled if $R_{a/w} > 10$. Values of $R_{a/w}$ in between indicate that both boundary layers are contributing significantly to the size of v_{tot}.

Let us now make a specific example using a set of chemical compounds with Henry's Law constants extending from 4×10^{-4} to $660\ L \cdot atm \cdot mol^{-1}$ (Table 10.3) and by choosing two extreme wind velocities, $u_{10} = 1\ m \cdot s^{-1}$ and $20\ m \cdot s^{-1}$, respectively. From Eqs. 10-28 and 10-32 we can estimate the transfer velocities:

For $u_{10} = 1\ m \cdot s^{-1}$: $v_a(H_2O) = 0.5\ cm \cdot s^{-1}$, $v_w(O_2) = 4.4 \times 10^{-4}\ cm \cdot s^{-1}$

For $u_{10} = 20\ m \cdot s^{-1}$: $v_a(H_2O) = 4.3\ cm \cdot s^{-1}$, $v_w(O_2) = 1.6 \times 10^{-2}\ cm \cdot s^{-1}$

The molecular diffusion coefficients of the compounds listed in Table 10.3 are generally smaller than the diffusivities of our reference substances, water vapor for the

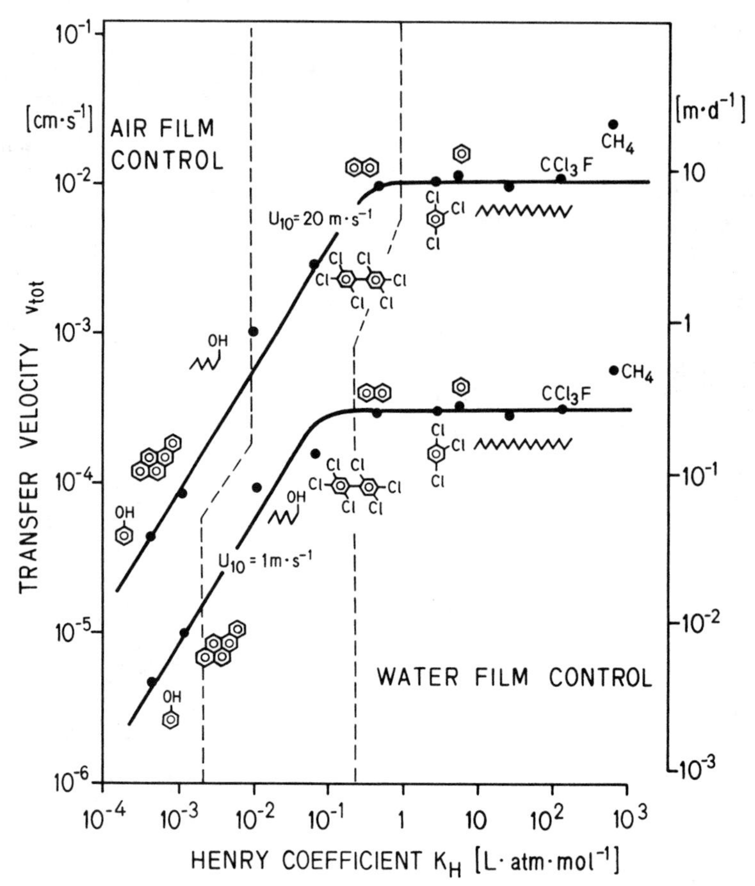

Figure 10.9 Total air–water transfer velocity v_{tot} as a function of Henry's coefficient K'_H for two different wind speeds, u_{10} (see Table 10.3). The lines are calculated for typical diffusion coefficients ($D_a = 0.09\,cm^2 \cdot s^{-1}$, $D_w = 1.0 \times 10^{-5}\,cm^2 \cdot s^{-1}$).

air, dissolved oxygen for the water. For our purpose it is sufficient to estimate the compound diffusivities in air using Eq. 9-25 and the known diffusivity of water in air of $0.26\,cm^2 \cdot s^{-1}$ (25°C). Using this result with our empirical prediction of v_a (Eq. 10-29), we can calculate transfer velocities for numerous chemicals diffusing across the air film (Table 10.3). Similarly, we use Eq. 9-30, the known diffusivity of O_2 in water of $2.1 \times 10^{-5}\,cm^2 \cdot s^{-1}$ (25°C), and Eq. 10-33 to adjust v_w for compounds of interest to us (Table 10.3) at any wind condition of interest. The resulting total transfer velocities for the different compounds are listed in Table 10.3 for the two wind speeds.

The ratio v_w/v'_a, the $R_{a/w}$ ratio introduced in Eq. 10-36, shows that for $u_{10} = 1\,m \cdot s^{-1}$, relatively small and nonpolar compounds are water-film controlled and only large (e.g., benzo(a)pyrene) or polar (e.g., phenol) chemicals are air-film controlled. For the larger wind speed, the region of air-film control slightly shifts to larger K_H values (Fig. 10.9) owing to the faster increase of the water transfer velocity v_w with increasing wind. These results lead to the expectation that compounds like gasoline hydrocarbons and nonpolar solvents (relatively high K_H values) can be modeled with only liquid-film control; however, fluxes of volatile and polar compounds like acetone or relative nonvolatile and nonpolar compounds like PCBs or PAHs are likely to be governed by air-side dynamics, or a combination of resistances. A single value of K_H allowing these modeling simplifications cannot be chosen because the ratio of v_w to v'_a is not a constant across environments. Typically, this ratio approaches 0.001 when wind dominates air–water interfacial physics, but can increase to about 0.01 at very high wind speed or when mixing from below is important.

10.5 INFLUENCE OF CHEMICAL REACTIONS ON AIR–WATER EXCHANGE RATES

In certain situations, a chemical of interest may be involved in a rapid reversible transformation in the water phase. Such a reaction would clearly affect the local concentration of the substance, and as a result will alter the overall rate of diffusive transport (which, of course, is concentration gradient dependent.) Thus, to estimate accurately the rate of air–water exchange of these reactive volatiles, we must account for this reactivity effect.

Typically, the rate of chemical reaction determines whether air–water exchange is influenced by the reaction or not. Let us consider the situation in a single boundary layer (water or air) by distinguishing three cases:

1. The reaction is slow relative to the time needed to transport the chemical across the boundary layer. In this case we can reasonably assume that such a transformation has no significant impact on the molecules during the time they spend diffusing through the (liquid or air) film bottleneck. Thus, the expressions derived in Section 10.3 remain valid.
2. The reaction is extremely fast compared to the transport time. In this case we should include the newly formed species in our thinking, but we can do it in a simplified way by assuming, throughout the boundary layer, immediate equilibrium between the species linked by the fast reaction.
3. Reaction and transport rates are of the same order of magnitude. This situation requires a more detailed analysis of the fluxes and the concentration profiles within the boundary layer.

Before we discuss the new cases (2 and 3) in greater depth, we should have approximate values for the typical transport times against which the reaction rates are to be analyzed. Table 10.4 summarizes the situation for both layers, air and water, for the stagnant film as well as the surface renewal model. Thus, upper limits for the time to cross the surface layer are about 10 s and 50 s for the air and water side of the interface, respectively. For periods of rapid gas exchange, these times may become significantly smaller and drop below 0.1 s.

TABLE 10.4 Characteristics of the Air–Water Interface and Typical Transport Times for Organic Molecules Traversing the Air or Water Boundary Layer

	Air	Water
Thickness of boundary layer (cm)	0.1–1	$2 \cdot 10^{-3}$–$2 \cdot 10^{-2}$
Surface renewal rate (s^{-1})	0.4–40	0.03–3
Typical molecular diffusion coefficient for organic compound ($cm^2 \cdot s^{-1}$)	$7 \cdot 10^{-2}$	$8 \cdot 10^{-6}$
Transport time[a] for stagnant film model (s)	0.1–10	0.5–50
Transport time[b] for surface renewal model (s)	0.025–2.5	0.1–10

[a]Transport time $= z^2/D$; see Eqs. 10-14 and 10-15.
[b]Transport time $= r_a^{-1}$ or r_w^{-1}.

Since case 1 does not need any further discussion (the equations remain as they are), we now address case 2 in which the reaction is extremely fast compared to time of transport across the boundary. This may occur with proton exchange as illustrated in Figure 10.10 for the case of a volatile acid (e.g., acetic acid). At all depths in the aqueous solution the acid is assumed to be in instantaneous equilibrium with its conjugate base (see Chapter 8):

$$K_a = \frac{H^+ \cdot A^-}{HA} \qquad (10\text{-}5)$$

Note that for simplicity, we use the symbols HA (H^+, A^-, of the chemical species) to denote the corresponding activities or concentrations. However, since the conjugate base is charged, it has virtually no ability to leave the polar solvent and enter the

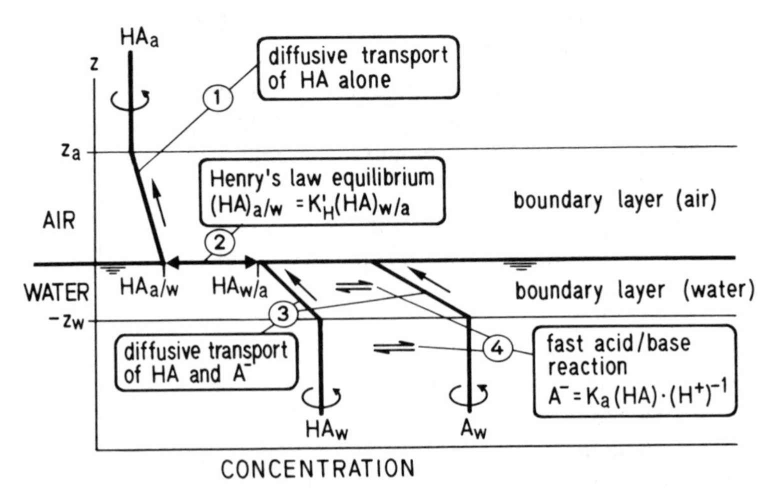

Figure 10.10 Air–water exchange of an organic acid with $pK_a < pH$. Note A^- is not volatile; proton exchange is very fast; $(HA)_{total} = HA + A^- = HA\ (1 + K_a/H^+)$.

air. Thus only protonated neutral molecules diffuse into the air and exhibit a Henry's Law equilibrium with the molecules in the water just below the air–water boundary:

$$K'_{\mathrm{H}} = \frac{\mathrm{HA}_{\mathrm{a/w}}}{\mathrm{HA}_{\mathrm{w/a}}} \tag{10-6}$$

We thus write flux expressions based on the same species concentration gradient in the air, but a combination of diffusing gradients in the water:

$$F_{\mathrm{a}} = \frac{D_{\mathrm{a}}(\mathrm{HA}_{\mathrm{a/w}} - \mathrm{HA}_{\mathrm{a}})}{z_{\mathrm{a}}} \tag{10-3}$$

$$F_{\mathrm{w}} = \frac{D_{\mathrm{w}}(\mathrm{HA}_{\mathrm{w}} - \mathrm{HA}_{\mathrm{w/a}})}{z_{\mathrm{w}}} + \frac{D_{\mathrm{w}}(\mathrm{A}^-_{\mathrm{w}} - \mathrm{A}^-_{\mathrm{w/a}})}{z_{\mathrm{w}}} \tag{10-37}$$

Typically it is reasonable to assume that the diffusivities of HA and A^- are the same since these species differ so little in size. Recalling the acid–base equilibrium relationship between HA and A^-, we rewrite

$$\begin{aligned} F_{\mathrm{w}} &= \frac{D_{\mathrm{w}}(\mathrm{HA}_{\mathrm{w}} - \mathrm{HA}_{\mathrm{w/a}})}{z_{\mathrm{w}}} + \frac{D_{\mathrm{w}}\left(\dfrac{K_{\mathrm{a}}\mathrm{HA}_{\mathrm{w}}}{\mathrm{H}^+} - \dfrac{K_{\mathrm{a}}\mathrm{HA}_{\mathrm{w/a}}}{\mathrm{H}^+}\right)}{z_{\mathrm{w}}} \\ &= \frac{D_{\mathrm{w}}}{z_{\mathrm{w}}}\left(1 + \frac{K_{\mathrm{a}}}{\mathrm{H}^+}\right)(\mathrm{HA}_{\mathrm{w}} - \mathrm{HA}_{\mathrm{w/a}}) \end{aligned} \tag{10-38}$$

assuming the solution pH and temperature are the same throughout the boundary region. Effectively we have obtained the same flux expression as before (Eq. 10-4) for nonreactive volatiles modified by a term: $(1 + K_{\mathrm{a}}/\mathrm{H}^+)$. Obviously the more acidic our volatile chemical of interest (increasing K_{a}) or the more basic our aqueous solution (decreasing H^+) the greater will be this additional term as diffusion transport through the water film becomes more and more mediated by the chemical's conjugate base. Thus it appears we have an enhanced diffusivity D'_{w}:

$$D'_{\mathrm{w}} = D_{\mathrm{w}}\left(1 + \frac{K_{\mathrm{a}}}{\mathrm{H}^+}\right) \tag{10-39}$$

At steady state we find (by equating the flux expressions):

$$F = \left(\frac{1}{\dfrac{z_{\mathrm{a}}}{K'_{\mathrm{H}}D_{\mathrm{a}}} + \dfrac{z_{\mathrm{w}}}{D'_{\mathrm{w}}}}\right)\left(\mathrm{HA}_{\mathrm{w}} - \frac{\mathrm{HA}_{\mathrm{a}}}{K'_{\mathrm{H}}}\right) \tag{10-40}$$

If we derive the flux expression from the surface renewal conceptualization, a parallel result arises:

$$F = \left[\frac{1}{K'_H \sqrt{D_a r_a}} + \frac{1}{\sqrt{D'_w r_w}} \right]^{-1} \left(HA_w - \frac{HA_a}{K'_H} \right) \tag{10-41}$$

Two important lessons can be gained by these derived results. First, the conductance through the surface water film of such highly reactive compounds is greatly increased whenever much of the compound is ionized. This is reflected by the adjusted diffusivity, D'_w. Additionally, we note that it is the concentration difference of the volatile species only (e.g., CH_3COOH, but not CH_3COO^-) which drives the air–water exchange process. In this vein, we must be careful that Henry's Law constants which reflect this nonionic species equilibrium alone are utilized, that is, recalling Eq. 8-26

$$D_{aw} = \left(\frac{(CH_3COOH)_a}{(CH_3COOH)_w + (CH_3COO^-)_w} \right) \neq K'_H$$

Interestingly, most compounds that undergo rapid acid–base reactions, such that significant diffusivity enhancement could be anticipated at the pH's of natural waters (pH 4 to 10), do so because they have fairly polar structural moieties. As a result these chemicals also exhibit fairly low Henry's Law constants, and virtually all of the overall resistance to air–water exchange is in the stagnant or intermittently mixed gas film. Thus apparent enhancement of aqueous diffusivity has a negligible effect on the overall transfer coefficient. This may seem a somewhat counterintuitive result, especially for the case of a volatile acid dissolving into water from overlying air ($HA_w = 0$), where one would expect the ionization reaction to "pull" the acid into the solution (e.g., as in trapping CH_3COOH with KOH solutions). Although the final equilibrium condition relating acetic acid in the gas versus in the basic solution strongly favors the bulk of the volatile acid residing in the aqueous solution, the rate at which this transfer occurs is controlled by passage of the nonionic species through the air boundary adjacent to the water.

Although our definition of the enhanced diffusivity (Eq. 10-39) is based on the specific acid–base equilibrium given in Eq. 8-1, the result can easily be generalized to every fast interconversion between any two species and expressed by

$$Y = K \cdot X \tag{10-42}$$

(K = equilibrium constant). If only the species X can cross the air–water interface, then the Henry's Law constant reflects the equilibrium between the X species only,

$$K_H = \frac{X_{a/w}}{X_{w/a}} \tag{10-43}$$

Equation 10-39 becomes

$$D'_w = D_w(1 + K) \tag{10-44}$$

and Eqs. 10-40 and 10-41 remain valid. In fact, Eq. 10-39 just appears as a special case of Eq. 10-44 with $K = K_a/(H^+)$.

Case 3 involves situations where reactions occur over time periods similar to those required for diffusive transfer across the water boundary layer (about 5 s, see Table 10.4). A reaction of this type is the hydration of formaldehyde. Many aldehydes react substantially with water in a reversible fashion to yield a diol (two alcohols), see Bell and McDougall, 1960):

$$\underset{\text{an aldehyde}}{R-\overset{\overset{\displaystyle O}{\|}}{C}-H} + H_2O \underset{\text{dehydration}}{\overset{\text{hydration}}{\rightleftharpoons}} \underset{\text{a diol}}{R-\underset{\underset{\displaystyle OH}{|}}{\overset{\overset{\displaystyle OH}{|}}{C}}-H}$$

Formaldehyde (R = H) occurs more than 99% as the diol when it is dissolved in water and acetaldehyde ($R = CH_3$) is about half hydrated at ambient temperatures. As long as the concentration of water is constant ($[H_2O] = 55\,\text{mol}\cdot L^{-1}$), we can express the hydration–dehydration processes as (pseudo) first-order reactions (see Section 12.2). For example, for formaldehyde we have

$$\begin{aligned}\frac{d}{dt}[CH_2O] &= -k_1[H_2O][CH_2O] + k_2[CH_2(OH)_2] \\ &= -k'_1[CH_2O] + k_2[CH_2(OH)_2]\end{aligned} \tag{10-45}$$

The pseudo-first-order rate constant for this hydration, k'_1, is about $10\,s^{-1}$ and its dehydration rate k_2 is on the order of $5 \times 10^{-3}\,s^{-1}$ (Table 10.5). Consequently we can readily see that this compound undergoes reversible transformations between the hydrated and dehydrated species over times which are shorter than that necessary to diffuse across a stagnant water boundary layer, but not short enough to guarantee equilibrium between the two species (Table 10.4). This latter point may be obvious for dehydrated formaldehyde entering the water from the air and reacting with a half-life on the order of 0.07 sec ($\ln 2/k'_1$), but may not be so obvious for a dehydration process necessary to facilitate transfer from aqueous solution to the overlying air since dehydration occurs with a half-life of about 140 s ($\ln 2/k_2$).

It is worth noting that the sum of the two rates ($k'_1 + k_2$), in a "back-and-forth" process, dictates the time necessary to reach equilibrium. We can illustrate this by a general reaction:

$$A + \cdots \underset{k_2}{\overset{k_1}{\rightleftharpoons}} B + \cdots \tag{10-46}$$

TABLE 10.5 Pseudo-First-Order Hydration Rates and Equilibrium Constants Reported for Some Aldehydes at Neutral pH[a]

Aldehyde		Hydration k'_1 (s^{-1})	Dehydration k_2 (s^{-1})	$K'(25°C) = k'_1/k_2$	K_H ($L \cdot atm \cdot mol^{-1}$)
Formaldehyde	HCHO	10	$5 \cdot 10^{-3}$	2000	0.40
Acetaldehyde	CH_3CHO	10	10	1, 1.4	0.21
Chloral (trichloroacetaldehyde)	Cl_3CCHO	—	—	25,000 to 28,000	0.071
Acrolein	$CH_2{=}CH{-}CHO$	—	—	17	—

[a]Data from Bell and Evans, 1966; Bell and McDougall, 1960; Schecker and Schulz, 1969; Betterton and Hoffman, 1988. See Eq. 10-45.

where the points indicate that on both sides of the reaction other chemicals may be involved without affecting the rate of transformation. For instance, in the formaldehyde hydration reaction, A means CH_2O and the points on the left-hand side refer to H_2O. On the right-hand side of Eq. 10-46, B stands for $CH_2(OH)_2$ and no other species are involved. The temporal change of the species concentrations A and B are

$$\frac{dA}{dt} = -k_1 A + k_2 B$$
$$\frac{dB}{dt} = k_1 A - k_2 B \tag{10-47}$$

Note that $dA/dt = -dB/dt$, that is $(A + B) = \text{constant}$. The steady-state condition is reached if dA/dt and dB/dt are zero, that is,

$$B_{eq} = \frac{k_1}{k_2} A_{eq} = K A_{eq} \qquad \text{where } K = k_1/k_2 \tag{10-48}$$

This is the familiar equilibrium condition of a reversible reaction. We will now derive an equation which describes the transition of the system from any initial condition (A_0, B_0) to the steady state (A_{eq}, B_{eq}). The problem we encounter with Eq. 10-47 is the crosswise relationship between A and B. The time derivative of A also depends on B, and the derivative of B also depends on A. Such a system of coupled differential equations is often solved by introducing some specific new variables which lead to decoupled equations. There are recipes for finding the new variables, but they lie beyond the scope of this discussion. Instead of a long theoretical explanation, let us just guess a little bit. For instance, we may choose an expression that measures the deviation of the A/B ratio from its equilibrium $(k_1 A = k_2 B)$ by defining

$$y(t) = k_1 A(t) - k_2 B(t) \tag{10-49}$$

Note that at equilibrium, y becomes zero. The dynamic equation

$$\frac{dy}{dt} = k_1 \frac{dA}{dt} - k_2 \frac{dB}{dt} \tag{10-50}$$

after some algebraic rearrangement becomes

$$\frac{dy}{dt} = -(k_1 + k_2)(k_1 A - k_2 B) = -(k_1 + k_2)y \tag{10-51}$$

which has the solution

$$y(t) = y_0 e^{-(k_1 + k_2)t} \tag{10-52}$$

since, indeed the derivative of y only depends on y and not on the single concentrations A and B. Equation 10-52 shows that the sum of the two rate constants $(k_1 + k_2)$ determines how fast $y(t)$ drops to zero. The half life of y, $\ln 2/(k_1 + k_2)$, can still be small, even if one k_i is very small, as long as the other rate is large. This explains the fast equilibration of the hydration–dehydration reaction of formaldehyde, in spite of the small dehydration constant k_2.

In the environment, chemical reactions such as Eq. 10-46 are always accompanied by mixing and transport. For instance, if we consider the reaction 10-46 to occur in the stagnant water film at the air–water interface, both parts of Eq. 10-47 should also include a term expressing transport by molecular diffusion (see Eq. 9-15):

$$\frac{\partial A}{\partial t} = D_A \frac{\partial^2 A}{\partial z^2} - k_1 A + k_2 B$$
$$\frac{\partial B}{\partial t} = D_B \frac{\partial^2 B}{\partial z^2} + k_1 A - k_2 B \qquad (10\text{-}53)$$

where D_A and D_B are the molecular diffusion coefficients for species A and B, respectively. Note that the notation of partial derivatives is used (see Section 9.2) because two variables, t and z (z coordinate perpendicular to the interface) appear in Eq. 10-53.

To calculate the transfer rate across the water surface we are looking for the steady-state solution ($\partial A/\partial t = \partial B/\partial t = 0$) of Eq. 10-53. Let us assume that species A is in equilibrium with the atmospheric concentration at the water surface (see Fig. 10.11):

$$A_{w/a} = \frac{A_{a/w}}{K'_H} \qquad (10\text{-}54)$$

In contrast, species B is not able to escape from the water to the air. As a result, the slope of the B profile becomes zero at the interface:

$$\frac{\partial B}{\partial z} \text{(at the water side of the interface)} = 0 \qquad (10\text{-}55)$$

since any spatial gradient would mean transport by molecular diffusion from or to the boundary for which there is no counterpart (reaction, gas transfer) at the interface. At the other side of the stagnant film, the concentrations A and B are assumed to be at equilibrium (Eq. 10-48):

$$B_w = K \cdot A_w \qquad (10\text{-}56)$$

Though in principle the steady-state solution of Eq. 10-53 together with the boundary conditions Eqs. 10-54 to 10-56 can be derived by well-known techniques, we shall spare the reader the tedious derivation. Instead, we prefer to discuss the qualitative aspects of the concentration of species A and B across the stagnant film. In Figure 10.11, we

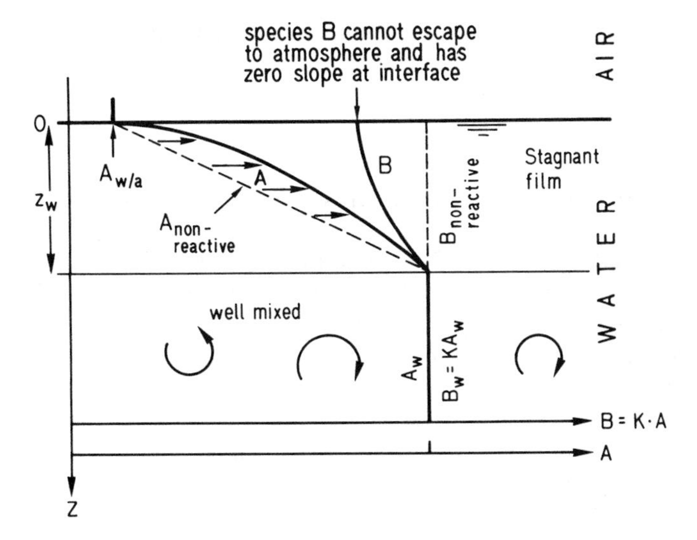

Figure 10.11 Air–water exchange of a volatile chemical species A interconverting to a non–volatile species B in a time comparable to the diffusion time of A and B across the water film of thickness z_w. Arrow between $A_{non-reactive}$ and A concentration curve indicates the change of A due to production from B. Note that scales of A and B concentration are normalized to a situation with equilibrium constant $K = 1$.

have drawn the concentrations A and B with different scales in such a way that at equilibrium (Eq. 10-48) the concentrations seem to be equal. As we have discussed in Section 10.3, for a nonreactive species the concentration profile of the volatile species A across the stagnant film is a straight line connecting the boundary values $A_{w/a}$ and A_w. If A can be reversibly transformed into a species B, at the lower boundary the two species are in equilibrium while at the upper boundary (the air–water interface) A is below its equilibrium value with B because of the constant loss to the atmosphere. In contrast, B does not "feel" the interface since it cannot escape into the air. As a consequence of the disequilibrium, there is a steady flux of species B to species A (indicated by arrows in Fig. 10.11). Thus, the process of diffusive transport of species A toward the interface has to take care of a growing number of molecules the closer we get to the interface. Since the diffusion coefficient across the boundary layer is constant, the only way to increase the flux of species A is by increasing the spatial gradients, that is, by making the concentration curve deform upward relative to the straight line of the nonreactive species (see Fig. 10.11).

As a consequence of Fick's first law, the relative flux enhancement ψ, that is, the flux ratio between a reactive and nonreactive species, is given by the ratio of the concentration slopes of those species at the interface. For the situation described by Eq. 10-53 and the conditions 10-54 to 10-56, the flux enhancement is given by the expression

$$\psi = \frac{\text{Flux reactive species}}{\text{Flux nonreactive species}} = \frac{K+1}{1+(K/q)\cdot\tanh q} \qquad (10\text{-}57)$$

with K defined in Eq. 10-48 and the nondimensional parameter

$$q^2 = \frac{(z_w^2/D_w)}{(k_r^{-1})} = \frac{k_r z_w^2}{D_w} \qquad (10\text{-}58)$$

is the ratio between the time of diffusion across the boundary layer (z_w^2/D_w) and the time of reaction $k_r^{-1} = (k_1 + k_2)^{-1}$.

In order to discuss expression 10-57, let us first find out whether cases 1 and 2, which were discussed earlier, are included. For case 1 (reaction slow compared to diffusion across the boundary layer), we put $q = 0$ and get [with $(\tanh q)/q \to 1$ for $q \to 0$] $\psi = 1$ in accordance with the conclusion made earlier that a slow reaction does not influence the gas exchange. For case 2 (reaction fast compared to diffusion), we put $q = \infty$. With $\tanh(\infty) = 1$, we get $\psi = K + 1$ in accordance with Eq. 10-44. The cases in between are shown in Figure 10.12. As expected, the flux enhancement increases with reaction–diffusion parameter q and reaches a steady-state value at $(K + 1)$. Thus, large equilibrium constants K in combination with fast reaction rates or thick boundary layers favor the occurrence of flux enhancement.

As an example let us again use the hydration of formaldehyde. From Table 10.5

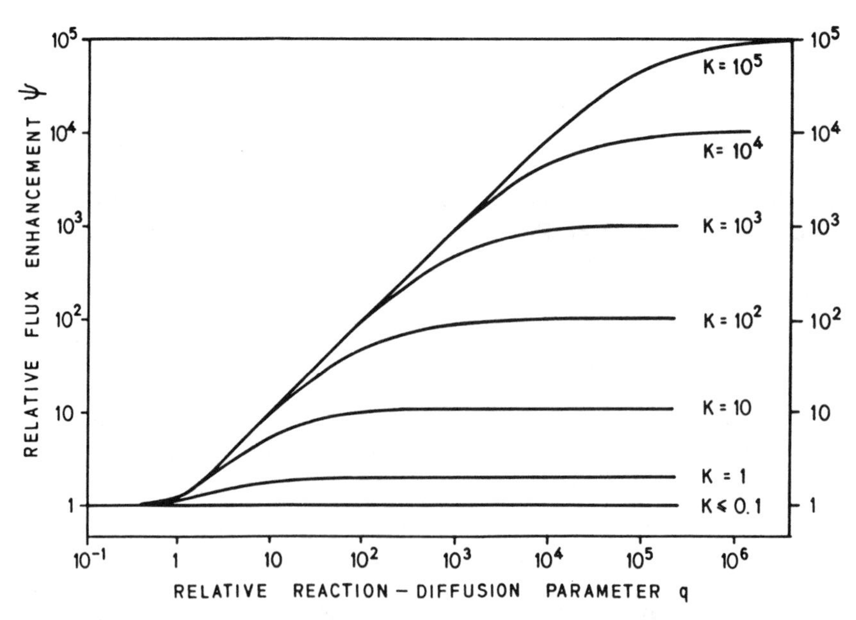

Figure 10.12 Flux enhancement ψ (Eq. 10-57), that is, the ratio of gas exchange fluxes of a reactive and non-reactive species, respectively, as a function of equilibrium constant, K (Eq. 10-56) and the nondimensional parameter $q = k_r \cdot z_w^2 \cdot D_w^{-1}$ (Eq. 10-58). k_r = first-order reaction rate, D_w = molecular diffusivity of species A and B, z_w = thickness of water-side boundary layer.

and assuming $D_w \sim 10^{-5}\ cm^2 \cdot s^{-1}$ we get

$$K = 2000 \qquad k_r \sim k_1' = 10\ s^{-1}$$

thus $q = 10^3 \cdot z_w$ (z_w measured in centimeters). As stated in connection with Eq. 10-32, z_w typically varies between 200 μm (0.02 cm) for no-wind conditions and 20 μm (0.002 cm) for a wind speed of $10\ m \cdot s^{-1}$. Thus, q lies between 20 (no wind) and 2 (wind speed $10\ m \cdot s^{-1}$) and the flux enhancement decreases from 20 (no wind) to 2 at $10\ m \cdot s^{-1}$. This example demonstrates the importance of relative flux enhancement during calm conditions. However, since under natural conditions the air–water exchange rate averaged over some time (days, weeks) is often dominated by the "events" of high wind speed for which flux enhancement is generally smaller, the mean flux enhancement for such a period is closer to the high wind value than to the value for calm conditions.

10.6 SURFACE FILMS

Oily liquids or amphiphilic substances (referred to as *surface active*) often accumulate at air–water interfaces. As a result, the rate of exchange of volatile chemicals through this zone may be slowed. It appears that two types of effects are discernible: (1) the addition of another stagnant film in the series through which a volatile chemical must diffuse, and/or (2) the damping of mixing and turbulence, thereby lengthening diffusion lengths or slowing renewal rates.

In their flume studies of reaeration, Downing and Truesdale (1955) clearly demonstrated that oily films could act as an additional diffusion barrier (Fig. 10.13). First, the figure shows that such an additional resistance to transport does not become important until the oil has a thickness which is comparable to that of the "stagnant water layer." Such oily layers would be very unusual in nature, and would likely be limited to instances immediately following events like oil spills. We can fit such data quite well using the principles described in Section 10.3, only now considering *three* layers in series (Note, we take the flux to be positive if directed from water toward air.):

$$\text{Flux}_{\substack{\text{overall}\\ \text{w/o/a}}} = \text{Flux}_{\substack{\text{boundary}\\ \text{water layer}}} = \text{Flux}_{\text{oil layer}} = \text{Flux}_{\substack{\text{boundary}\\ \text{air layer}}} \qquad (10\text{-}59)$$

Utilizing v_o to reflect the piston velocity of a volatile substance through the oily film, we write

$$v_w(C_w - C_{w/o}) = v_o(C_{o/w} - C_{o/a}) = v_a(C_{a/o} - C_a) \qquad (10\text{-}60)$$

where

$C_{w/o}$ is the concentration in the water immediately adjacent to the oil,
$C_{o/w}$ is the concentration in the oil immediately adjacent to the water,
$C_{o/a}$ is the concentration in the oil immediately adjacent to the air, and
$C_{a/o}$ is the concentration in the air immediately adjacent to the oil.

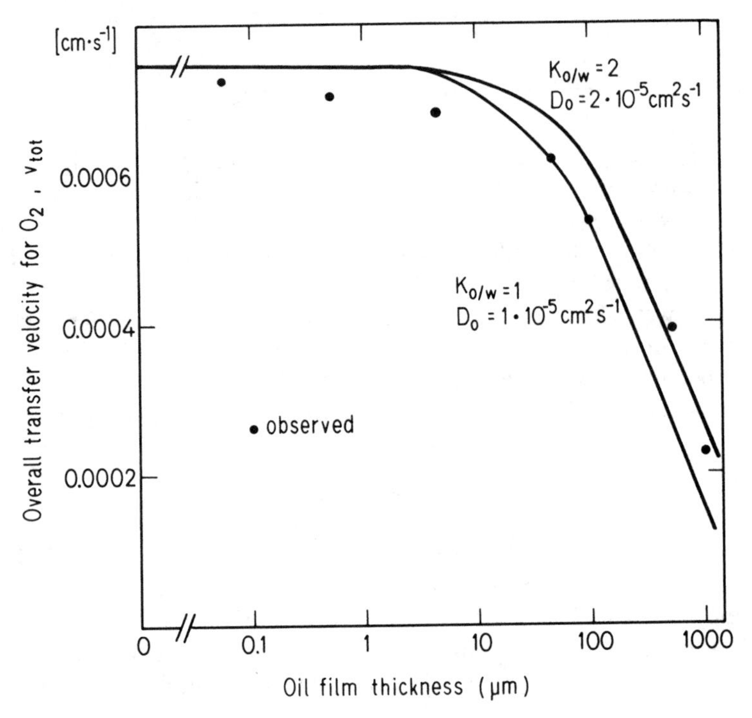

Figure 10.13 Decreasing reaeration rates for water covered with increasingly thick films of oil. Comparison between film model as calculated using Eq. 10-64 (lines) and measurements (dots). $K_{o/w}$ = equilibrium distribution constant of O_2 between oil and water; D_o = molecular diffusivity of O_2 in oil. From Downing and Truesdale (1955).

As before, we assume that the molecular scale layers on either side of the air–oil and oil–water interfaces are at equilibrium so that we may apply equilibrium relationships

$$K_{a/o} = C_a/C_o \tag{10-61}$$

$$K_{o/w} = C_o/C_w \tag{10-62}$$

where C_o is the concentration of the chemical in the oil. Since the overall equilibrium between the concentrations in air and water, respectively (Eq. 10-6), is not altered by the oil film, we have

$$C_a/C_w = K_{a/o} \cdot K_{o/w} = K'_H \tag{10-63}$$

Thus, $K_{a/o}$ can be replaced by $K'_H/K_{o/w}$. By algebraic manipulations, we may eliminate the unmeasurable boundary concentrations $C_{a/o}$, $C_{o/a}$, $C_{o/w}$, and $C_{w/o}$ from the flux expressions in Eq. (10-60) and deduce

$$\text{Flux}_{w/o/a} = \left(\frac{1}{K'_H v_a} + \frac{1}{K_{o/w} v_o} + \frac{1}{v_w}\right)^{-1} \left(C_w - \frac{C_a}{K'_H}\right) = v_{tot} \cdot \left(C_w - \frac{C_a}{K'_H}\right) \tag{10-64}$$

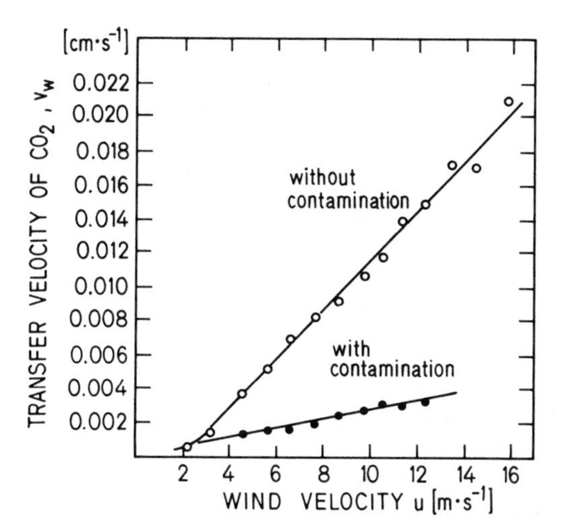

Figure 10.14 Effect of wind velocity on exchange rate of CO_2 with and without contaminating the tank water with a monolayer of oleyl alcohol. Lines indicate linear approximation of experimental data. From Broecker et al. (1978).

Once again our overall flux expression consists of two terms: the first reflecting the combined effects of three transport resistances arranged in series and the second quantifying the overall concentration gradient from bulk air to bulk water. Note that the transfer velocities are multiplied with the "concentration enhancement" relative to the water, that is, with $K_{o/w}$ for the oil film and K'_H for the air film. For chemicals like oxygen which exhibit high K'_H, we know that we can neglect the term $1/K'_H v_a$. Oxygen exhibits a $K_{o/w}$ value of about 2, depending on the oil. If the diffusivity of oxygen in the oil is 2×10^{-5} $cm^2 \cdot s^{-1}$ and v_0 can be estimated as D_o/z_o, then one predicts the trend in v_{tot} as a function of z_o as shown in Figure 10.13. Generally such a model adequately predicts the rapid fall off in v_{tot} as the oil film thickness increases from 10 to 100 μm. Had the volatilizing chemical of interest exhibited a large $K_{o/w}$ (see Chapter 7), as would most organic volatiles, then we could neglect the oil film as an important transport resistance (i.e., $1/K_{o/w}v_o \ll 1/v_w$, unless of course the diffusivity of the volatilizing compound through the oil was especially low (such as for highly viscous oils). The final point to note in Figure 10.13 is that the overall rate of gas exchange decreases, albeit slowly, even before film thicknesses are large enough that we can reasonably attribute diminished transport to an additional film resistance. This effect is probably indicative of the dampening of surface layer mixing, the second mechanism by which surface films inhibit gas exchange.

It has long been realized that the addition of minute amounts of oils to the sea surface calms the waves and yields a "glassy" slick (Plinius and Benjamin Franklin have been quoted in this regard!). Such action appears to require a continuous monolayer of substances like oleyl alcohol or oleic acid, corresponding to films of approximately nanometer thickness and coverages of about 100 ng of organic chemical per square centimeter (Garrett, 1967; Jarvis et al., 1967). By extracting organic matter from seawater, Jarvis et al. (1967) have shown that natural organic materials can also coat a water–air interface and damp capillary waves. If such surface-calming reflects diminished mixing in the water just below the surface, we would anticipate that water–air exchange will become more and more limited by the ability of chemicals to diffuse rather than be carried by microscopic eddies; thus piston velocities should decrease. This is exactly the result seen by Broecker et al. (1978) for exchange of CO_2 as a function of wind speed (Fig. 10.14). At very low winds ($\leqslant 2\,m \cdot s^{-1}$), the importance of a monolayer of oleyl alcohol is negligible, since such conditions are insufficient to ruffle a water in any case. However, above the wind velocity at which capillary waves begin to form (2–$3\,m \cdot s^{-1}$), the oleyl alcohol was seen to both "prohibit" surface roughening (up to speeds of $12\,m \cdot s^{-1}$) and concomitantly to reduce CO_2 gas exchange relative to the "clean surface" case by as much as a factor of 4 at winds of $10\,m \cdot s^{-1}$. Interestingly, much of the inconsistency in data sets for air–water exchange may be due to surface contamination in the experiments (Asher and Pankow, 1986). The net result is that if sufficient organic material has accumulated on a water surface to cause it to have a "slick appearance," the gas exchange rates (v_w^{contam}) would be better modeled with a much weaker dependency on wind speed than exhibited in the relationship derived in Fig. 10.6 and Eq. 10-32.

CHAPTER 11

SORPTION: SOLID–AQUEOUS SOLUTION EXCHANGE

11.1 INTRODUCTION

The process in which chemicals become associated with solid phases is generally referred to as *sorption* (either *ad*sorption onto a two-dimensional surface, or *ab*sorption into a three-dimensional matrix). This phase transfer process may involve interacting either vapor molecules or dissolved molecules with adjacent solid phases.

Sorption is extremely important because it may dramatically affect the fate and impact of chemicals in the environment. Such importance is readily understood if we recognize that structurally identical molecules behave very differently if they are surrounded by water molecules and ions as opposed to clinging onto the exterior of solids or being buried within a solid matrix (Fig. 11.1). Clearly, the environmental movements of water-borne molecules must differ from that fraction of the same kind of molecules carried by particles that settle. Additionally, only the dissolved molecules are available to collide with the interfaces leading to other environmental compartments such as air; and thus these phase transfers, for practical purposes, are limited to the dissolved species of a chemical. Finally, the chemical milieu of the solution and solid worlds differ greatly. For example, the thin layer of water surrounding silicate surfaces is typically "more acidic" than bulk water, and thus reactions involving protons or hydroxide ions proceed at different rates for sorbed molecules which are otherwise structurally identical to dissolved molecules. It is possible that molecules located within particles are substantially shaded from incident light; therefore, these molecules may not get involved with direct photochemical processes or short-lived reactive species such as $\cdot OH$ (see Chapter 13). Finally insofar as molecular transfer into microorganisms is frequently a prerequisite to a substance's biodegradation, it

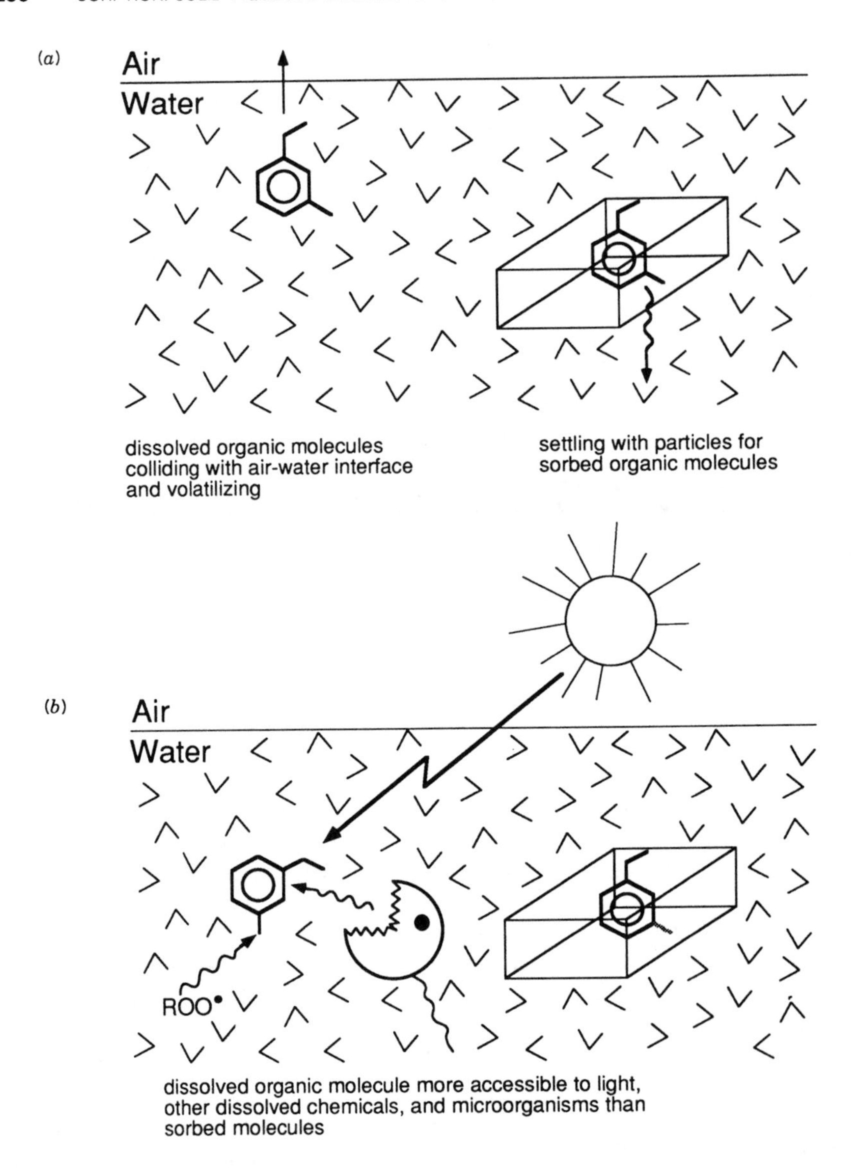

Figure 11.1 Some processes in which sorbed species behave differently than dissolved molecules of the same substance. (*a*) For example, dissolved species may undergo air–water exchange while sorbed species may sediment. (*b*) Also, dissolved species may react at different rates as compared to their sorbed counterparts.

should be recognized that the greater ease of chemical movement from solution versus from within solids to bacteria generally causes the biological decomposition of the sorbed form of the chemical to be slower than its dissolved counterpart. Hence, we must understand solid–solution exchange phenomena before we can quantify virtually any other process affecting the fate of chemicals in the environment.

Unfortunately, sorption is not always a single simple process (Westfall, 1987). Rather, some combination of interactions may be responsible for governing the association of any particular chemical (called a *sorbate*) with any particular solid (called a *sorbent*). Figure 11.2 illustrates this point for 4-chloroaniline (4-chloroaminobenzene). First, chiefly because of unfavorable free-energy costs of remaining in aqueous solution, such an organic substance may escape the water by penetrating

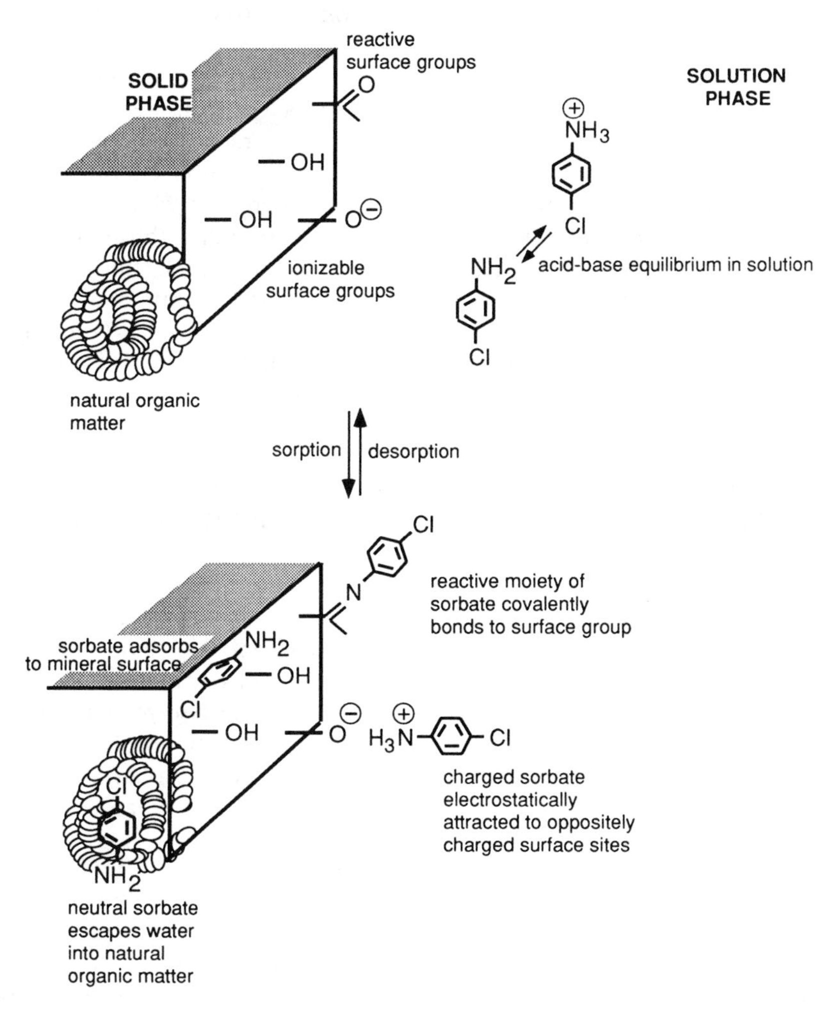

Figure 11.2 Some sorbent–sorbate interactions possibly controlling the association of a chemical with particles.

natural organic matter in the particulate phase. Additionally, such a molecule may displace water molecules from the region near the mineral surface to some extent and thereby be associated with the surface via van der Waals, dipole–dipole, and other weak intermolecular forces. These two mechanisms of sorption are general and will operate for any organic chemical and any natural solid. Additionally, if the sorbate is ionizable in the aqueous solution, then attraction to specific surface sites exhibiting the opposite charge will promote sorption of the ionic species. Finally, should the sorbate and sorbent exhibit mutually reactive moieties (e.g., in Fig. 11-2 a carbonyl group on the sorbent and an amino group on the sorbate), some portion of the chemical may actually become bonded to the solid. All of these interaction mechanisms will operate simultaneously, and the combination that dominates the overall solution–solid distribution will depend on the structural properties of the organic chemical and solid medium of interest.

In this chapter we try to visualize the sets of molecular interactions involved in each of the sorption processes. This means that we have to consider van der Waals and dipole–dipole interactions, H-bonding, ionic interactions between charged species, and specific bonding of reactive moieties and solid surface atoms or groups. With such pictures in our minds, we will seek to understand what makes various sorption mechanisms important under various circumstances. Establishing the critical molecular properties and solid characteristics will enable us to understand why certain predictive approaches may be applied. Ultimately, we should gain some feeling for what structural features of a chemical and what characteristics of solids (and solutions) are important to sorption interactions. Finally, we conclude by examining the factors limiting the rate of approach to sorption equilibrium.

11.2 QUANTIFYING THE RELATIVE ABUNDANCES OF DISSOLVED AND SORBED SPECIES: THE SOLID–WATER DISTRIBUTION RATIO K_d

When we are interested in assessing the *equilibrium* proportion of a particular chemical's presence in association with solids for any particular volume of an aquatic environment, we begin by considering how the total sorbate concentration associated with the sorbent, C_s ($\text{mol} \cdot \text{kg}^{-1}$), depends on the total chemical concentration in the solution, C_w ($\text{mol} \cdot \text{L}^{-1}$). Such a relationship is commonly referred to as a *sorption isotherm*. The term isotherm is used to indicate that one is considering sorption at a constant temperature. Depending on the dominating mechanism(s), sorption isotherms may exhibit different shapes (Fig. 11.3). Experimentally determined isotherms can commonly be fit with a relationship of the form

$$C_s = K \cdot C_w^n \tag{11-1}$$

This equation is known as the Freundlich isotherm; K is referred to as the Freundlich constant; and n is a measure of the nonlinearity involved. Case I in Figure 11.3 ($n < 1$) reflects the situation in which at higher and higher sorbate concentrations, it becomes more and more difficult to sorb additional molecules. This may occur in cases where

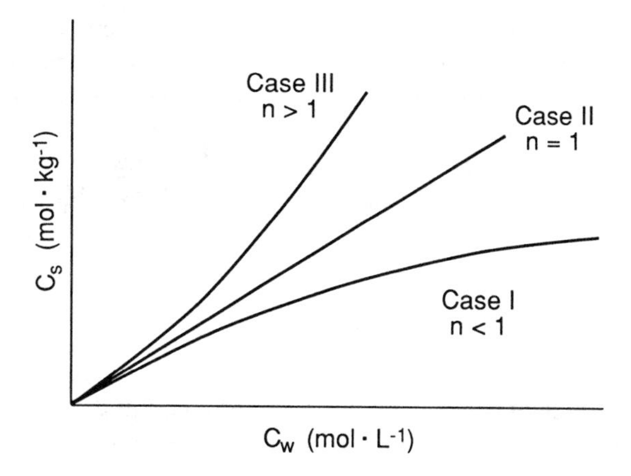

Figure 11.3 Three types of observed relationships between concentrations of a chemical in the sorbed state, C_s, and the dissolved state, C_w. All can be fit with a relationship of the form $C_s = K \cdot C_w^n$ where K and n are constants.

specific binding sites become filled or remaining sites are less attractive to the sorbate molecules. Case III ($n > 1$) describes a contrasting situation in which previously sorbed molecules lead to a modification of the surface which favors further sorption. Such effects have been seen for surface active compounds like alkyl benzene sulfonates (see Table 2.6), where the sorbent becomes coated and increasingly exhibits a nonpolar nature. Finally, Case II ($n = 1$) reflects those situations in which the attractiveness of the solid for the sorbates remains the same for all levels of C_s. This is the so-called *linear* isotherm case; one should also realize that over narrow ranges in C_w, particularly at low concentrations, both Cases I and III appear to be linear.

Let us now consider a case in which we know the *ratio* of a substance's total equilibrium concentrations in the sorbed phase and in the solution. We denote this distribution ratio with K_d:

$$K_d = \frac{C_s}{C_w} \qquad \frac{(\text{mol} \cdot \text{kg}^{-1})}{(\text{mol} \cdot \text{L}^{-1})} \tag{11-2}$$

The value of this ratio may only apply at the given solute concentration (i.e., n in Eq. 11-1 may not equal 1). In fact, inserting Eq. 11-1 into Eq. 11-2 yields

$$K_d = K \cdot C_w^{n-1} \tag{11-2a}$$

Often, one assumes that K_d is constant over some concentration range. From Eq. 11-2a we can differentiate K_d with respect to C_w, rearrange the result, and find

$$\frac{dK_d}{K_d} = (n-1)\frac{dC_w}{C_w} \tag{11-2b}$$

so this assumption about the constancy of K_d is equivalent to presuming: (a) the overall sorption process is either described by a linear isotherm ($n = 1$ in Eq. 11-1), or (b) the relative concentration variation, dC_w/C_w, is sufficiently small to guarantee that the relative K_d variation, dK_d/K_d, is also small.

Armed with such a K_d parameter for a case of interest, we may evaluate what fraction of a compound is in the water solution, f_w, in a volume containing both solids and water (but only these phases):

$$f_w = \frac{C_w \cdot V_w}{C_w V_w + C_s M_s} \tag{11-3}$$

where V_w is the volume of water (L) in total volume V_{tot}, and M_s is the mass of solids (kg) present in that same total volume. Now if we substitute the product $K_d \cdot C_w$ from Eq. 11-2 for C_s in Eq. 11-3, we have

$$\begin{aligned} f_w &= \frac{C_w V_w}{C_w V_w + K_d C_w M_s} \\ &= \frac{V_w}{V_w + K_d M_s} \end{aligned} \tag{11-4}$$

Finally, noting that we frequently refer to the quotient M_s/V_w as the solid-to-water phase ratio r_{sw} (e.g., $kg \cdot L^{-1}$) in the environmental compartment of interest, we may describe the fraction of chemical in solution as a simple function of K_d and this ratio:

$$\begin{aligned} f_w &= \frac{1}{1 + (M_s/V_w)K_d} \\ &= \frac{1}{1 + r_{sw} \cdot K_d} \end{aligned} \tag{11-5}$$

Such an expression clearly indicates that for substances exhibiting a great affinity for solids (hence a large value of K_d) or in situations having large amounts of solids per volume of water (large value of r_{sw}), we predict that correspondingly small fractions of the chemical remain dissolved in the water. Note the fraction associated with solids, f_s, must be given by $(1 - f_w)$ since we assume that no other phases are present (e.g., air, other immiscible liquids).

The fraction of the total volume V_{tot} that is not occupied by solids, the so-called porosity ϕ, is often used instead of r_{sw} to characterize the solid–water phase ratio in a given system. In the absence of any gas phase (as is the case in water-saturated soil and usually in sediments and the open water column), ϕ is equal to

$$\phi = \frac{V_w}{V_{tot}} = \frac{V_w}{V_w + V_s} \tag{11-6}$$

where V_s, the volume occupied by particles, can be expressed by M_s/ρ_s, (where ρ_s is the density of the solids). Thus

$$\phi = \frac{V_w}{V_w + M_s/\rho_s} = \frac{1}{1 + r_{sw}/\rho_s} \tag{11-7}$$

Solving for r_{sw} yields the corresponding relation

$$r_{sw} = \rho_s \frac{1 - \phi}{\phi} \tag{11-8}$$

It is a matter of convenience whether r_{sw} or ϕ is used.

The application of such solution- versus solid-associated speciation information may be illustrated by considering an organic chemical, say 1,4-dimethylbenzene (DMB), in a lake and in flowing groundwater. In the lake, the solid–water ratio is given by the suspended solids concentration (since $V_w \approx V_{tot}$), which is typically near $10^{-6}\,kg \cdot L^{-1}$. From experience we may know that the K_d value for DMB in this case is $1\ L \cdot kg^{-1}$; therefore we can see that virtually all of this compound is in the dissolved form in the lake:

$$f_w = \frac{1}{1 + 10^{-6} \cdot 1} \approx 1$$

In the groundwater situation, ρ_s for aquifer solids is about $2.5\ kg \cdot L^{-1}$ (e.g., quartz density is 2.65); ϕ is often between 0.2 and 0.4. If our particular groundwater situation has ϕ of 0.2, r_{sw} is $10\,kg \cdot L^{-1}$. Hence, we predict that the fraction of DMB in solution, assuming again a K_d of $1\ L \cdot kg^{-1}$, is drastically lower than in the lake:

$$f_w = \frac{1}{1 + 10.1} \simeq 0.09$$

So we deduce that only one DMB molecule out of 11 will be in the moving groundwater at any instant (Fig. 11.4). This result has implications for the fate of the DMB in that subsurface environment. If DMB sorptive exchange between the aquifer solids and the water is fast relative to the groundwater flow and if sorption is reversible, we can conclude that the whole population of DMB molecules moves at one eleventh the rate of the water. The phenomenon of diminished transport speed relative to the water seepage velocity is referred to as retardation; and it is quantified using a retardation factor which is simply f_w^{-1}.

Many situations require us to know something about the distribution of a chemical between a solution and solids. Our task then is to see how we can get K_d values suited for the cases that concern us. As we will see, these K_d values are determined by the structures of the sorbates as well as the composition of the aqueous phase and the sorbents.

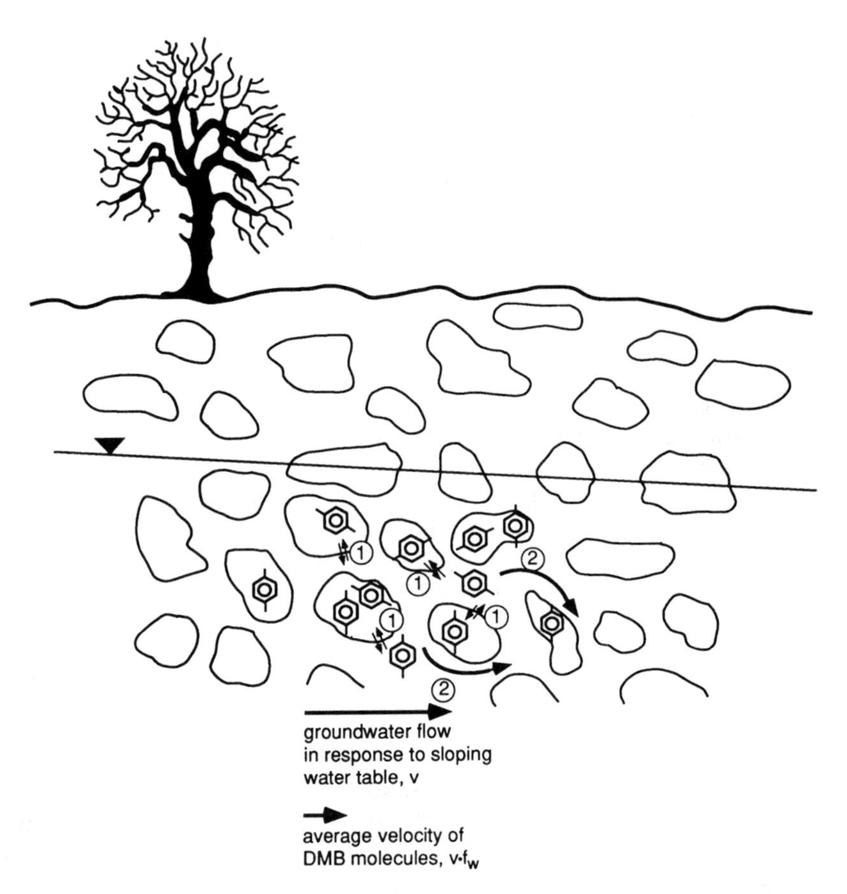

Figure 11.4 Illustration of the retardation of 1,4-dimethylbenzene (DMB) transport in groundwater due to: (1) reversible sorptive exchange between water and solids, and (2) limiting transport of DMB to that fraction remaining in the flowing water. As dissolved molecules move ahead they become sorbed and stopped, while molecules sorbed at the rear return to the water and catch up. Thus, overall transport of DMB is slower than that of the water itself.

11.3 THE COMPLEX NATURE OF K_d's

The prediction of K_d for any particular combination of organic chemical and solids in the environment can be difficult, but fortunately many situations appear reducible to fairly simple limiting cases. We begin by emphasizing that the way we defined K_d means that we may have lumped together many chemical species in each phase. For example, referring again to Figure 11.2, we recognize that the total concentration of 4-chloroaniline in the sorbed phase combines the contributions of molecules in many different sorbed forms. Even the solution in this case contains both a neutral and a charged species of this chemical. Thus the distribution ratio for this case would have to be written

$$K_d = \frac{C_{om} \cdot f_{om} + C_{min} \cdot A + C_{ie} \cdot \sigma_{ie} \cdot A + C_{rxn} \cdot \sigma_{rxn} \cdot A}{C_{w,neut} + C_{w,ion}} \qquad (11\text{-}9)$$

where

C_{om} is the concentration of sorbate associated with the natural organic matter ($mol \cdot kg^{-1}$ om),

f_{om} is the weight fraction of solid which is natural organic matter (kg om $\cdot kg^{-1}$ solid),

C_{min} is the concentration of sorbate associated with the mineral surface ($mol \cdot m^{-2}$),

A is the area of mineral surface per mass of solid [$m^2 \cdot kg^{-1}$ solid),

C_{ie} is the concentration of ionized sorbate drawn toward positions of opposite charge on the solid surface ($mol \cdot mol^{-1}$ surface charges),

σ_{ie} is the net concentration of suitably charged sites on the solid surface (mol surface charges$\cdot m^{-2}$),

C_{rxn} is the concentration of sorbate bonded in a reversible reaction to the solid ($mol \cdot mol^{-1}$ rxn sites),

σ_{rxn} is the concentration of reactive sites on the solid surface (mol rxn sites$\cdot m^{-2}$),

$C_{w,neut}$ is the concentration of uncharged chemical in solution ($mol \cdot L^{-1}$), and

$C_{w,ion}$ is the concentration of the charged chemical in solution ($mol \cdot L^{-1}$).

It is possible that some of the terms in Eq. 11-9 also deserve further subdivision. For example, $C_{min} \cdot A$ may reflect a linear combination of the interactions of several mineral surfaces present in a particular soil or sediment with a single sorbate. Thus, a soil consisting of montmorillonite, kaolinite, iron oxide, and quartz mineral components may actually have $C_{min} \cdot A = C_{mont} \cdot a \cdot A + C_{kao} \cdot b \cdot A + C_{iron\ ox} \cdot c \cdot A + C_{quartz} \cdot d \cdot A$ where the parameters a, b, c, and d are the area fractions exhibited by each mineral type. Similarly, $C_{rxn} \cdot \sigma_{rxn} \cdot A$ may reflect bonding to several different kinds of moieties, each with its own reactivity with the sorbate (e.g., chloroaniline). For now, we will work from the simplified expression which is Eq. 11-9, primarily because there are little data available allowing rational subdivisions of soil or sediment differentially sorbing organic chemicals beyond that reflected in this equation.

It is very important to realize that only particular combinations of species in the numerator and denominator of complex K_d expressions like that of Eq. 11-9 are involved in any one exchange process. For example, in the case of 4-chloroaniline

$$(\text{4-}ClC_6H_4NH_2)_{water} \rightleftharpoons (\text{4-}ClC_6H_4NH_2)_{om} \qquad (11\text{-}10)$$

reflects the molecular interchange between the uncharged chloroaniline species dissolved in water and the species in the particulate natural organic phase with the same molecular structure. Similarly, the combination

$$(Cl{-}C_6H_4{-}NH_2)_{\text{water}} \rightleftharpoons (Cl{-}C_6H_4{-}NH_2)_{\text{min surf}} \qquad (11\text{-}11)$$

would indicate the exchange of uncharged chloroaniline molecules from aqueous solution to the available mineral surfaces. Each of these exchanges is characterized by a unique free energy difference reflecting the equilibria shown as Eqs. 11-10 and 11-11. Similarly, the exchange of

$$(Cl{-}C_6H_4{-}NH_2)_{\text{water}} \rightleftharpoons (Cl{-}C_6H_4{-}N{=})_{\text{rxn site}} \qquad (11\text{-}12)$$

should be considered if it is the neutral sorbate which can react with components of the solid. Note that such specific binding to a particular solid phase moeity may prevent rapid desorption, and therefore such sorbate–solid associations may cause part or all of the sorption process to appear irreversible on some timescale of interest.

So far we have considered sorptive interactions in which the neutral chloroaniline species was involved. In contrast, it is the charged chloroaniline species that is important in the ion exchange process:

$$(Cl{-}C_6H_4{-}NH_3^{\oplus})_{\text{water}} \rightleftharpoons (Cl{-}C_6H_4{-}NH_3^{\oplus})_{\text{ion exchange site}} \qquad (11\text{-}13)$$

Again, we emphasize this solution–solid exchange has to be described using the appropriate equilibrium expression relating corresponding species in each phase. The influence of each sorption mechanism on the overall K_d is weighted by the availability of the respective sorbent property (i.e., f_{om}, A, σ_{ie}, σ_{rxn}, or A) in the total solid. By combining information on the individual equilibria (e.g., Eqs. 11-10 through Eq. 11-13) with these sorbent properties, we can develop versions of the complex K_d expression (Eq. 11-9) which take into account the structure of the chemical we are considering. In the following sections, we discuss these individual equilibrium relationships.

11.4 SORPTION OF NEUTRAL ORGANIC CHEMICALS TO SOILS AND SEDIMENTS

Role of the Natural Organic Matter

We begin by treating the case of *neutral organic chemicals* distributing themselves between an aqueous solution and natural solids. Our major emphasis involves nonpolar compounds constructed primarily from carbon, hydrogen and halogen atoms such as those shown below:

Naphthalene

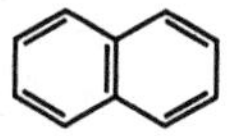

Isooctane

Tetrachloroethylene

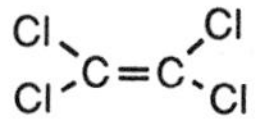

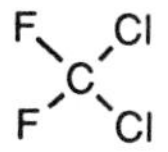

$Cl_2C{=}CCl_2$

CF_2Cl_2

Freon-12

2,4,4′-Tribromobiphenyl

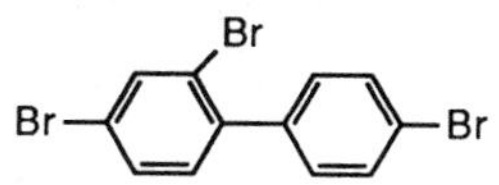

However, much of the following treatment also applies for more polar, but still neutral, chemicals like

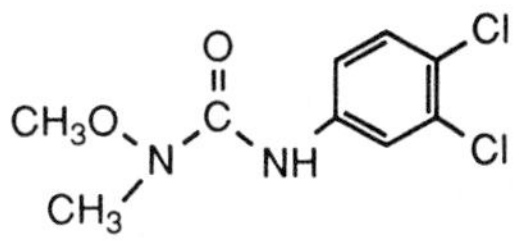

Linuron
(a phenyl urea herbicide)

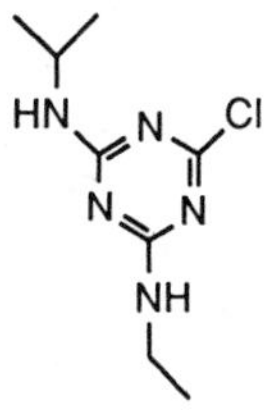

Atrazine
(a chloro-s-triazine herbicide)

As we have seen in discussing the aqueous solubilities of nonpolar compounds (Chapter 5), these molecules do not "enthusiastically" dissolve in water. This incompatibility principally arises because water molecules change their overall H-bonding to their surroundings when they are forced to interface with such nonpolar solutes or nonpolar structural subunits of compounds like linuron and atrazine.

In like manner, most natural minerals are polar and expose a combination of hydroxy- and oxy-moieties to their exterior. Naturally, then, these polar surfaces strongly favor interactions which allow them to form hydrogen bonds—such as with liquid water:

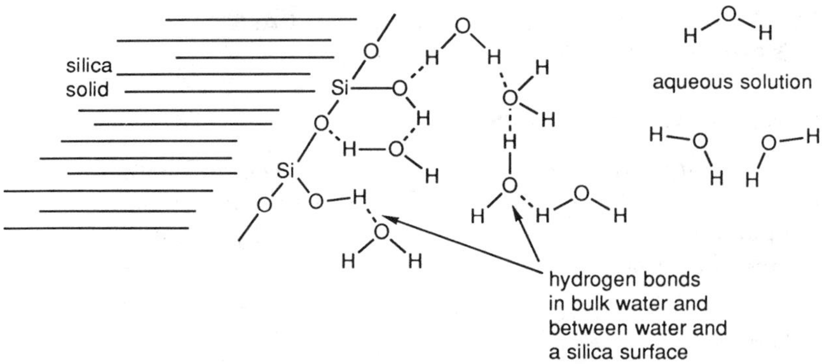

silica solid

aqueous solution

hydrogen bonds in bulk water and between water and a silica surface

As a result, replacing the water molecules at such a mineral surface by nonpolar organic compounds is unfavorable from an energetic point of view, despite the sorption-favoring activity coefficients of these solutes in the water.

On the other hand, penetration of neutral organic chemicals into any natural organic matter included in the solid phase does not require displacement of tightly bound water molecules. This solid organic phase material may include recognizable biopolymers like proteins, lignin, and cellulose, but also a menagerie of macromolecules from the partial degradation and crosslinking of organic residues remaining from organisms or photochemical reactions. Naturally, the structure of such altered materials will depend on the ingredients supplied by the particular organisms living in or near the water or soil and will tend to be somewhat randomized depending on what particular substituents have been altered and where crosslinking has happened to occur. For example, soil scientists (Schnitzer and Khan, 1972; Stevenson, 1976) have deduced that the recalcitrant remains of woody terrestrial plants make up a major portion of the natural organic matter in soils. Such materials also make up an important fraction of organic matter suspended in freshwaters (Liao et al., 1982). Similarly, marine chemists believe that the natural organic matter, suspended in the oceans at sites far from land, consists of altered biomolecules such as amino acids, sugars, and triglycerides that have been linked together (Hedges, 1977; Stuermer and Payne, 1976; Harvey et al., 1983). At intermediate locales, such as large lakes and estuaries, the natural organic material in sediments and suspended in water appears to derive from a variable mixture of terrestrial organism and planktonic organism remains (Hedges and Parker, 1976; Sigleo et al., 1982; Hedges et al., 1984; Thurman, 1985). These altered complex organic substances are typically referred to as *humic substances* if they are soluble or extractable in aqueous base, and *humin or kerogen* if they are not. The humic substances are further subdivided into *fulvic acids* if they are soluble in both acidic and basic solutions and *humic acids* if they are not soluble in acidic conditions but are soluble at high pHs. Such materials are predominantly made of carbon (about 40–50% by weight), but some have nearly as many oxygens as carbons included in their structures (Table 11.1). Thus, they are not as polar as water, especially since they can only be involved in H-bonding at limited points on their structures (e.g., carboxy, phenoxy, hydroxy, and carbonyl substituents); but they are not as nonpolar as hydrocarbons or chlorinated hydrocarbons either, since they

TABLE 11.1 Properties of Natural Organic Matter Relevant to Sorption

	Mole Ratio				Moles H-bonding Substituents per kg					
	C	H	N	O	ether	keto	hydroxy	carboxy	Molecular Mass Range (amu)	Reference
Proteins	10	15	4	2					$1 \times 10^4 - 1 \times 10^5$ (cytochrome c to serum γ-globulin)	5
Cellulose (*linear* polymer)	10	16	0	8	1.2	0	1.9	0	$3 \times 10^5 - 4 \times 10^6$ (cotton)	2
Lignin	10	11	0.1	3						2
Fulvic acids										
Dissolved (river)	10	12	0.2	6		2	9	6	1×10^3	3
Soils	10	8	0.1	7	0.1	3.1	6.9	9.1		1
Humic acids										
Dissolved (river)	10	11	0.3	6				4.3	$2 \times 10^3 – 5 \times 10^3$	3
Sedimentary (lake)	10	13	0.9	5						4
Soils	10	12	0.6	4	0.3	4.4	4.9	4.5	$1 \times 10^3 – 2 \times 10^5$	1,4
Humin	10	19	0.5	11						2

References: 1. Khan, 1980. 2. Garbarini and Lion, 1986. 3. Thurman, 1985. 4. Schnitzer and Khan, 1972. 5. Oser, 1965.

do have oxygens and a few nitrogens in their structures. Such natural organic matter occurs in a very broad spectrum of molecular sizes from the smallest fulvic acids of about 1000 amu (corresponding to spheres of about 2 nm diameter) to the huge complexes of solid kerogen. The important point is that when these natural organic constituents are associated with particles (and even when they remain suspended as nonsettling particles called colloids in aqueous solution), they offer a relatively nonpolar environment into which a hydrophobic compound may escape without undue competition with water.

In light of this discussion, we may not be too surprised to find that nonreactive, neutral chemicals show greater solid–water distribution ratios for soils or sediments that contain high amounts of natural organic matter, as illustrated for pyrene in Figure 11.5. Additionally, we see in this figure that when f_{om} approaches zero, K_d (pyrene) is very small. (Note: f_{om} is often determined by measurement of organic carbon (oc); as a result, it is often discussed using a parameter called f_{oc}, or fraction organic carbon, with units kg oc·kg^{-1} solid. As shown by the materials listed in Table 11.1, natural organic matter is typically made up of about half carbon, so f_{om} approximately equals $2 \cdot f_{oc}$.)

Consequently, we conclude that we only need to consider sorption of neutral nonpolar organic compounds to the natural organic matter as long as f_{om} is "significant". A more precise definition of "significant" will emerge below as we consider competing sorption mechanisms. For now, we realize that in this type of case, $C_s \approx C_{om} \cdot f_{om}$. Since chemicals like pyrene cannot ionize, C_w is equal to $C_{w,neut}$. Thus the composite

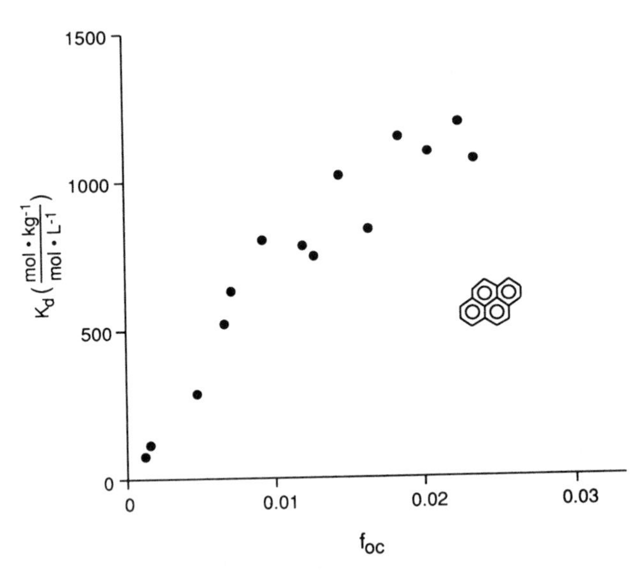

Figure 11.5 Observed increase in solid–water distribution ratios for a hydrophobic compound, pyrene, as a function of the organic matter content of the solid (measured as organic carbon, $f_{oc} \simeq 1/2\, f_{om}$) in a variety of soils and sediments. Data from Means et al. (1980).

K_d (Eq. 11-9) simplifies to

$$K_d = \frac{C_{om} \cdot f_{om}}{C_{w,neut}} \tag{11-14}$$

The Organic Matter—Water Partition Coefficient, K_{om}

To ascertain what controls the ratio of C_{om} to $C_{w,neut}$ for various situations, we consider briefly the molecular environment of a neutral organic molecule sorbed by natural organic matter. We picture this natural organic material to exist in large part as organic chains coiled into globular units, much like globular proteins, and to occur in somewhat isolated patches coating mineral solids (Fig. 11.6). Such coiling is brought about because the natural organic matter minimizes the hydrophobic surface area it exposes to the aqueous solution. Because of the "porous" nature of such flexible macromolecules, nonpolar sorbates can physically penetrate between the chains and find themselves "dissolved" in the nonaqueous medium (Fig. 11.6). Note that such sorbent associations are not limited to the interfacial area where the water solvent contacts the natural organic sorbent. Indeed, since the sorbate goes *into* the "volume" of the sorbent, this process may be referred to as an *ab*sorption (from Latin, *ab*, away from, and *sorbere*, to suck). The entry of nonpolar sorbates into the natural organic matter amounts to mixing these components as in solutions.

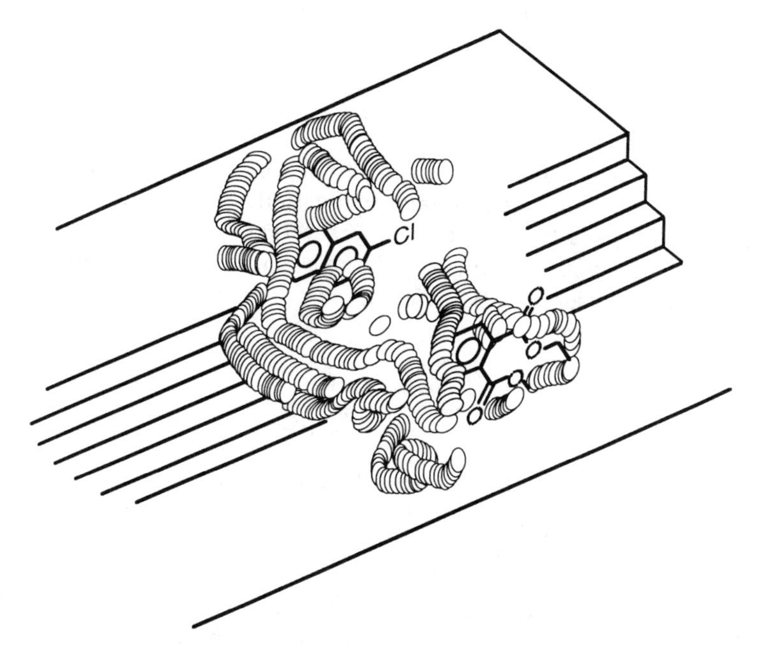

Figure 11.6 Conceptualization of nonpolar organic sorbates (here, 2-chloronaphthalene and diethylphthalate) associated with natural organic matter in a solid phase.

Thus we can imagine the organic sorbate is distributing itself between two immiscible solutions, and we may define a partition coefficient which reflects the ratio of these concentrations:

$$K_{\text{om}} = \frac{C_{\text{om}} \ (\text{mol}\cdot\text{kg}_{\text{om}}^{-1})}{C_{\text{w,neut}} \ (\text{mol}\cdot\text{L}_{\text{water}}^{-1})} \tag{11-15}$$

which is analogous to what we defined for organic solvent–water partitioning (Eq. 7-1). In defining K_{om} this way, we have led to a new expression for the solid–water distribution ratio of such neutral, nonpolar chemicals:

$$K_{\text{d}} = \frac{f_{\text{om}}\cdot C_{\text{om}}}{C_{\text{w,neut}}} \tag{11-14}$$

$$= f_{\text{om}}\cdot K_{\text{om}} \tag{11-16}$$

This formulation will prove to be very convenient because it is broken into two parts: f_{om}, which is a property of the soil or sediment sorbent, and K_{om}, which is primarily a characteristic of the organic chemical sorbate.

Given our conceptualization of hydrophobic neutral chemical absorption as two solutions competing for the sorbate, we can deduce several important results. First, considering C_{om} and $C_{\text{w,neut}}$ to be solution concentrations, we may rewrite the equilibrium partition coefficient expression (Eq. 11-15):

$$K_{\text{om}} = \frac{x_{\text{om}}\cdot \bar{V}_{\text{om}}^{-1}\cdot \rho_{\text{om}}^{-1}}{x_{\text{w}}\cdot \bar{V}_{\text{w}}^{-1}} \tag{11-17}$$

where x_{om} (mol sorbate$\cdot$mol^{-1} om) and x_{w} (mol sorbate$\cdot$mol^{-1} water) refer to the mole fraction concentrations in the organic matter and aqueous solutions, respectively; $\bar{V}_{\text{om}}$ (L om$\cdot$mol^{-1} om) and $\bar{V}_{\text{w}}$ (L water$\cdot$mol^{-1} water) are the molar volumes of the two phases, and ρ_{om} (kg om$\cdot$L^{-1} om) is the organic matter density. Choosing the pure liquid of the sorbate to serve as the reference state, we may use the relation $x = \gamma^{-1}$ to rewrite Eq. 11-17:

$$K_{\text{om}} = \frac{\gamma_{\text{w}}\cdot \bar{V}_{\text{w}}}{\gamma_{\text{om}}\cdot \bar{V}_{\text{om}}\cdot \rho_{\text{om}}} \tag{11-18}$$

where γ_{om} is the activity coefficient reflecting the incompatibility of the sorbate associated with the natural organic matter as compared to in solution with a liquid of itself. Again, this result is analogous to what we had before for partitioning between organic solvents and water (see Eq. 7-4).

Now, for a given chemical sorbate and a particular soil or sediment, all the terms on the right-hand side of Eq. 11-18 may be virtually constant for all concentrations of the hydrophobic sorbate in water. As we saw before, γ_{w} is practically independent of C_{w} for neutral nonpolar compounds (see Section 6.2); and since $\bar{V}_{\text{w}}$, $\bar{V}_{\text{om}}$, and ρ_{om} are properties of the solution and solid phase which are generally unaffected by the

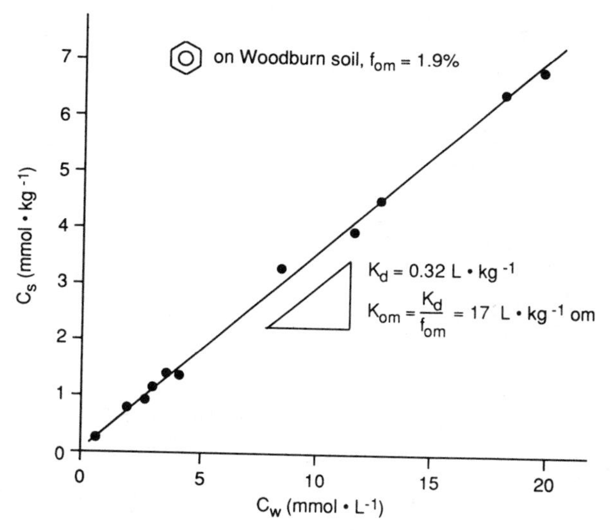

Figure 11.7 Linear variation of sorbed benzene concentration C_s with aqueous benzene concentration C_w up to 85% of benzene's solubility in water. Data from Chiou et al. (1983).

presence of submillimolar levels of neutral nonpolar compounds, these inputs to Eq. 11-18 may reasonably by presumed constant. Finally, the activity coefficient of a sorbate in natural organic matter may not be too much above 1 if this material can be as accommodating as a solvent like octanol (recall Table 7.1); and therefore it should not change much if more and more organic sorbate is mixed into this particular phase. Thus, one can expect K_{om} to remain constant for a particular sorbate–sorbent combination at all concentrations of the compound in the water. This means that the solid–water distribution ratio K_d must also remain invariant (i.e., a linear isotherm), since the product $f_{om} \cdot K_{om}$ does not change. This is seen experimentally, as shown in Figure 11.7 for benzene absorption on a soil. Here we see the concentration of benzene associated with this particular soil increased linearly as the levels of this compound in the water were increased. The slope of this graph yields the solid–water distribution ratio for this case:

$$K_d(\text{benzene, Woodburn soil}) = \frac{C_s}{C_w} = 0.32 \frac{\text{mol} \cdot \text{kg}^{-1}\ \text{soil}}{\text{mol} \cdot \text{L}^{-1}\ \text{water}} \tag{11-19}$$

And realizing that the natural organic matter of the soil ($f_{om} = 0.019$) was predominately responsible for benzene sorption, we may also calculate

$$\begin{aligned} K_{om}(\text{benzene}) &= \frac{K_d(\text{benzene, Woodburn soil})}{f_{om}(\text{Woodburn soil})} \\ &= 17 \frac{\text{mol} \cdot \text{kg}^{-1}\ \text{om}}{\text{mol} \cdot \text{L}^{-1}\ \text{water}} \end{aligned} \tag{11-20}$$

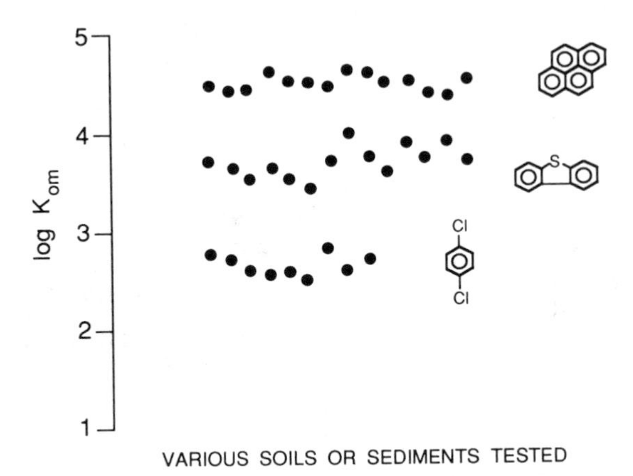

Figure 11.8 Similarity of log K_{om} values observed for particular nonpolar organic chemicals absorbing to a wide variety of soils and sediments. Data from Means et al. (1980), Hassett et al. (1980), Schwarzenbach and Westall (1981), and Chiou et al. (1983).

When one examines the K_{om} values for a single neutral nonpolar chemical sorbing to a variety of soils and sediments, one typically finds almost the same K_{om} result. Figure 11.8 shows such constancy of K_{om} for three chemicals: 1,4-dichlorobenzene, dibenzothiophene, and pyrene, sorbing to many soils and sediments. This near constancy of K_{om} values for a nonpolar organic chemical can be interpreted by considering Eq. 11-18. First, it is obvious that the numerator, $\gamma_w \cdot \bar{V}_w$, does not change for different sorbents. If the composition of natural organic matter does not vary much from the point of view of acting as a solvent for neutral nonpolar chemicals, then for a variety of sediments and soils, the product $\gamma_{om} \cdot \bar{V}_{om} \cdot \rho_{om}$ may also be similar. Much of the variability shown in Figure 11.8 is due to substantially different analytical techniques used by various investigators, but some is probably caused by differences in the polarity of the natural organic matter investigated (Garbarini and Lion, 1986; Chiou et al., 1987; Gauthier et al., 1987). But even including such variability, for any one chemical it seems log K_{om} is a constant within ± 0.3 log K_{om} units ($\pm$ a factor of 2 in K_{om}).

Linear Free Energy Relationships to Estimate K_{om}'s Now we may consider what factor(s) governs the magnitude of the K_{om} for a particular neutral nonpolar sorbate. For a series of compounds absorbing into the natural organic matter of a single soil or sediment (hence $\bar{V}_w$, $\bar{V}_{om}$ and ρ_{om} in Eq. 11-18 held constant), we need to evaluate the sizes of their γ_w's and γ_{om}'s. This is very similar to the situation we discussed before regarding organic solvent–water partitioning (Chapter 7). Since the organic sorbates are much more similar to the natural organic matter than to the water in terms of polarity and ability to interact with their surroundings via intermolecular attractions, we expect the γ_w's will be much larger than γ_{om}'s, and that the γ_{om}'s will not be much greater than 1. Once again, this is analogous to what we encountered in the case of organic chemical partitioning between octanol and water (Table 7.1). Since the γ_{om}'s are near 1, they will not vary much from organic chemical to chemical; in contrast, we know that various neutral organic chemicals exhibit γ_w's that vary by orders of magnitude (see Chapter 5). Consequently, differences, in K_{om} from chemical to chemical primarily arise from corresponding differences in their γ_w's.

A major result of realizing the strong relationship of K_{om}'s and γ_w's for a series of organic sorbates is that we deduce that the free energy driving this absorption is determined largely by the excess free energy of aqueous solution (i.e., $\Delta G_{\text{sorption to om}} \sim \Delta G_s^e$). This situation is therefore well suited to a free energy relationship between $\Delta G_{\text{absorption}}$ and ΔG_s^e. Indeed, plots of log K_{om} ($= -\Delta G_{\text{absorption}}/2.303\ RT$) versus log C_w^{sat} (l, L) ($= -\Delta G_s^e/2.303RT - \log \bar{V}_w$) for various sets of organic compounds show strong inverse correlations of the form (Table 11.2)

$$\log K_{om} = -a \cdot \log C_w^{sat}(\text{l, L}) + b \tag{11-21}$$

where K_{om} is given in units of (mol·kg^{-1} om/mol·L^{-1} water) and C_w^{sat} has units of (mol·L^{-1} water). For different sets of sorbates, different values of the coefficients a and b are seen and this undoubtedly reflects systematic variations in the compatibility of compound classes with the natural organic matter. Nonetheless, an overall strong inverse correlation of log K_{om} with log C_w^{sat}(l, L) exists as shown in Figure 11.9.

Since K_{om} and C_w^{sat}(l, L) properties of sorbates are related, it follows that any other

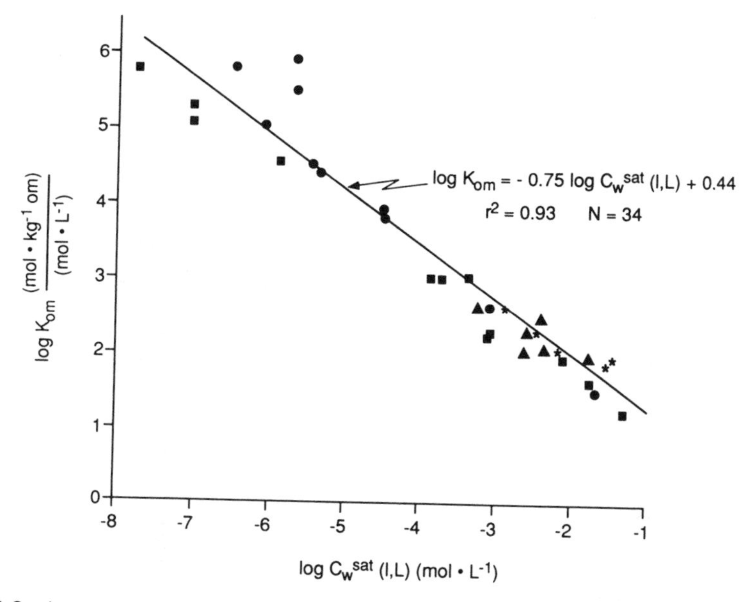

Figure 11.9 Inverse relationship of log K_{om} and log (liquid) aqueous solubility of neutral organic compounds: (●) aromatic hydrocarbons, (■) chlorinated hydrocarbons, (▲) chloro-S-triazines, and (*) phenyl ureas (data compiled by Karickhoff, 1981). See Table 11.2 for correlations for individual compound classes.

TABLE 11.2 LFERs Between $\log K_{om}$ and $\log C_w^{sat}(l, L)$ or $\log K_{ow}$[a]

		$\log K_{om} = -a \cdot \log C_w^{sat}(l, L) + b$ $(L \cdot kg^{-1}\ om)(mol \cdot L^{-1})$			(r^2)	Compound Class
Eq. 11-21a	$a =$	0.93	$b =$	−0.17	(0.92)	Aromatic hydrocarbons
Eq. 11-21b		0.70		+0.35	(0.99)	Chlorinated hydrocarbons
Eq. 11-21c		0.41		+1.20	(0.54)	Chloro-*S*-triazines
Eq. 11-21d		0.56		+0.97	(0.94)	Phenyl ureas
		$\log K_{om} = c \cdot \log K_{ow} + d$ $(L \cdot kg^{-1}\ om)(L_{water} \cdot L_{octanol}^{-1})$				
Eq. 11-22a	$c =$	1.01	$d =$	−0.72	(0.99)	Aromatic hydrocarbons
Eq. 11-22b		0.88		−0.27	(0.97)	Chlorinated hydrocarbons
Eq. 11-22c		0.37		+1.15	(0.93)	Chloro-*S*-triazines
Eq. 11-22d		1.12		+0.15	(0.93)	Phenyl ureas
Eq. 11-22e		0.81		−0.25	(0.98)	Chlorophenols

[a] Data from Karickhoff, 1981 and Schellenberg et al., 1984.

chemical property which is related to $C_w^{sat}(l, L)$ must also correlate with K_{om}. For example, log K_{om} should be directly related to log K_{ow} [inversely proportional to log $C_w^{sat}(l, L)$, see Chapter 7]. Table 11.2 also shows the values of the slope and intercept of relationships for compound classes of the form

$$\log K_{om}\left[\frac{\text{mol}\cdot\text{kg}^{-1}\text{ om}}{\text{mol}\cdot\text{L}^{-1}\text{ water}}\right] = c\cdot\log K_{ow}\left[\frac{\text{mol}\cdot\text{L}^{-1}\text{ octanol}}{\text{mol}\cdot\text{L}^{-1}\text{ water}}\right] + d \qquad (11\text{-}22)$$

Again, the values of this equation's coefficients vary between compound classes, but the overall trend is strong (Fig. 11.10).

Such linear free-energy relationships (Eqs. 11-21 and 11-22) provide one of the best means with which we can use well-known physical chemical properties of neutral chemicals [e.g., $C_w^{sat}(l, L)$ and K_{ow}] to estimate the magnitude of K_{om} for a compound of interest. Obviously, one obtains a somewhat more accurate estimate if one uses a correlation equation based on a set of chemicals from the sorbate's own compound class; however, without this kind of information, a reasonable K_{om} estimate can be obtained from expressions averaging results from many compound classes (like those illustrated in Figs. 11.9 and 11.10):

$$\log K_{om}\left[\frac{\text{mol}\cdot\text{kg}^{-1}\text{ om}}{\text{mol}\cdot\text{L}^{-1}\text{ water}}\right] \simeq -0.75\log C_w^{sat}(l, L)(\text{mol}\cdot\text{L}^{-1}\text{ water}) + 0.44 \qquad (11\text{-}23)$$

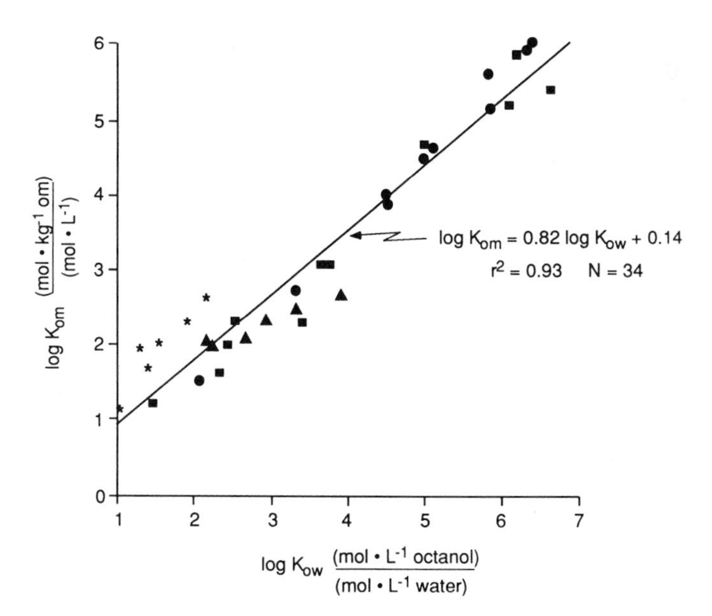

Figure 11.10 Relationship of log K_{om} and log K_{ow} for a series of neutral organic compounds: (●) aromatic hydrocarbons, (■) chlorinated hydrocarbons, (▲) chloro-*S*-triazines, and (*) phenyl ureas (data compiled by Karickhoff, 1981). See Table 11.2 for correlations of each compound class.

or

$$\log K_{\mathrm{om}}\left[\frac{\mathrm{mol \cdot kg^{-1}\ om}}{\mathrm{mol \cdot L^{-1}\ water}}\right] \simeq +0.82 \log K_{\mathrm{ow}}\left[\frac{\mathrm{mol \cdot L^{-1}\ octanol}}{\mathrm{mol \cdot L^{-1}\ water}}\right] + 0.14 \qquad (11\text{-}24)$$

Reversibility and Competitive Effects of Neutral Compound Sorption to Natural Organic Matter Our image of neutral chemical interactions with the natural organic matter of particles suggests that such sorption must be *reversible*. We see that the only "bonds" that are made between the sorbate and the natural organic sorbent are through relatively weak intermolecular attractions. Thus, to desorb these nonpolar chemicals, only small activation energies are needed to separate these compounds from their points of solid attachment. This reversibility has been demonstrated experimentally for nonpolar compounds like polycyclic aromatic hydrocarbons and nonpolar insecticides (Karickhoff et al., 1979), PCB congeners (Gschwend and Wu, 1985), and solvents like tetrachloroethylene and toluene (Garbarini and Lion, 1985). Additionally, many nonpolar chemicals can partition between water and the available volume of natural organic matter at the same time and without getting in each others' way (Karickhoff et al., 1979; Chiou et al., 1983). We refer to this as *noncompetitive* sorption. As was the case for organic solvent–water partitioning, this implies that so little nonpolar chemical enters the organic phase that the overall properties of the natural organic matter do not change and the resultant intermolecular interactions within the medium remain essentially constant.

Effect of Temperature on K_{om} Since we envision neutral compound absorption to be like a solvent–water partitioning process, we can immediately deduce the likely effect of conditions such as temperature on K_{om}. First, as we recall from our discussions of aqueous solubility, over a narrow temperature range, the logarithm of a chemical's aqueous activity coefficient varies according to (see Section 5.4)

$$\ln \gamma_{\mathrm{w}} = \frac{\Delta H_{\mathrm{s}}^{\mathrm{e}}}{RT} + \text{constant} \qquad (11\text{-}25)$$

Thus, the variation in γ_{w} with temperature will depend on the magnitude of $\Delta H_{\mathrm{s}}^{\mathrm{e}}$. Typically, small nonpolar solutes like benzene exhibit small values of $\Delta H_{\mathrm{s}}^{\mathrm{e}}$, but within any one family of compounds (e.g., PAHs or alcohols), we saw that $\Delta H_{\mathrm{s}}^{\mathrm{e}}$ steadily increased as the nonpolar molecular surface area of contact with the water solvent increased. A similar expression should apply for γ_{om}.

$$\ln \gamma_{\mathrm{om}} = \frac{\Delta H_{\mathrm{s,om}}^{\mathrm{e}}}{RT} + \text{constant} \qquad (11\text{-}26)$$

where $\Delta H_{\mathrm{s,om}}^{\mathrm{e}}$ is the excess enthalpy of the sorbate's mixing with natural organic matter. If the natural organic matter can approximately mimic the van der Waals attractions over the molecular surface that the sorbate enjoyed in a pure liquid of itself, then the change in enthalpy of dissolution into natural matter from the pure organic liquid will be quite small. This result, in turn, implies that γ_{om} is not particularly sensitive to temperature.

Since $\bar{V}_w$ and $\bar{V}_{om}$ vary only slightly with temperature, the effect of temperature on K_{om} values will chiefly arise from its impact on γ_w. Although very little data is available to check this conclusion, the literature does support it. First, Chiou et al. (1979) reported very small heats of absorption for organic chemicals frequently used as solvents, consistent with their small heats of solution in water. Also, Wu and Gschwend (1986) observed ΔH_{abs} of 1,2,3,4-tetrachlorobenzene to be about $-15\,kJ \cdot mol^{-1}$. Although the heat of solution of this compound has not been directly measured, it can be estimated using Eq. 5-18, the known (liquid) solubility of this compound ($10^{-4.20}$ M or 1.14×10^{-6} mol fraction), and an estimate of ΔS_s^e of $-56\,J \cdot mol^{-1}\,K^{-1}$ (see Table 5-3). Thus,

$$\begin{aligned} \Delta H_s^e &\simeq -RT \ln x_w + T\Delta S_s^e \\ &= -(8.31\,J \cdot mol^{-1}\,K^{-1})(293\,K)(\ln 1.14 \times 10^{-6}) \\ &\quad + (293\,K)(-56\,J \cdot mol^{-1}\,K^{-1}) \\ &= 33\,kJ \cdot mol^{-1} - 16\,kJ \cdot mol^{-1} \\ &= +17\,kJ \cdot mol^{-1} \end{aligned} \tag{11-27}$$

For the process removing the solute from the sorbent, this enthalpy would be negative in approximate agreement with the observed enthalpy of sorption ($-15\,kJ \cdot mol^{-1}$). Consequently, it appears reasonable to expect K_{om} to vary in an equal, but opposite, manner as the neutral chemical's liquid solubility in water. For large nonpolar compounds like pyrene exhibiting $\Delta H_s^e \simeq +25\,kJ \cdot mol^{-1}$, we then anticipate an *exothermic* $\Delta H_{abs} \approx -25\,kJ \cdot mol^{-1}$. Such enthalpy effects imply pyrene's sorption to natural particles should decrease as we increase temperature:

$$\begin{aligned} \frac{K_{om}(T_1 = 293\,K)}{K_{om}(T_2 = 283\,K)} &\simeq \frac{\gamma_w(293\,K)}{\gamma_w(283\,K)} \\ &\simeq \exp\left[\frac{\Delta H_s^e}{R}\left(\frac{1}{293} - \frac{1}{283}\right)\right] \\ &\simeq 0.7 \end{aligned} \tag{11-28}$$

That is, warming a solid–water suspension by 10°C results in an estimated 30% diminished absorption coefficient as a consequence of correspondingly decreased activity coefficient for pyrene in water.

Effect of Dissolved Salts on K_{om} Since solvated inorganic ions would not be expected to interfere or compete with the penetration of nonpolar organic compounds into natural organic matter, one can reasonably assume that salts affect the values of K_{om} primarily through the γ_w term also. As we saw in Chapter 5, a quantitative description of the impact of dissolved salts on aqueous solubilities (and hence γ_w's) was provided by the Setschenow relationship; a transformation of this formula (Eq. 5-22) yields

$$C_{\mathrm{w,salt}}^{\mathrm{sat}} = C_{\mathrm{w}}^{\mathrm{sat}} \cdot 10^{-K^{\mathrm{s}}[\mathrm{salt}]} \tag{11-29}$$

Since K_{om} values are predictable from a chemical's $C_{\mathrm{w}}^{\mathrm{sat}}(1, \mathrm{L})$, we may substitute Eq. 11-29 into Eq. 11-21 to characterize how K_{om} changes in the presence of dissolved salts:

$$\begin{aligned} \log K_{\mathrm{om,salt}} &= -a \cdot \log (C_{\mathrm{w,salt}}^{\mathrm{sat}}) + b \\ &= -a \cdot \log (C_{\mathrm{w}}^{\mathrm{sat}} \cdot 10^{-K^{\mathrm{s}}[\mathrm{salt}]}) + b \\ &= -a \cdot \log C_{\mathrm{w}}^{\mathrm{sat}} - a \cdot (-K^{\mathrm{s}}) \cdot [\mathrm{salt}] + b \\ &= \log K_{\mathrm{om}} + a \cdot K^{\mathrm{s}} \cdot [\mathrm{salt}] \end{aligned} \tag{11-30}$$

where $K_{\mathrm{om,salt}}$ is the organic matter–water partition coefficient when the aqueous solution contains dissolved salts. The second term on the right-hand side of Eq. 11-30 quantifies the magnitude of change we expect for individual cases. As we saw in Table 11.2, a values are typically between 0.5 and 1 for various compound classes, and from Table 5.6 we recall K^{s} was about 0.1–0.4 depending on the chemical of interest. Thus we may expect $\log K_{\mathrm{om,salt}}$ in a salty solution like seawater ([salt] $\simeq$ 0.6 M) to be higher than $\log K_{\mathrm{om}}$ by about $\{(0.5 \text{ to } 1) \cdot (0.1 \text{ to } 0.4) \cdot [0.6]\}$ or 0.03–0.24 log units. This amounts to increasing $K_{\mathrm{om,salt}}$ over K_{om} by only several percent for relatively small and polar neutral organic compounds (i.e., a nearer to 0.5 and K^{s} nearer to 0.1) and by a factor approaching 2 for large, very nonpolar sorbates (i.e., a close to 1 and K^{s} around 0.4).

As was the case for the effects of temperature on K_{om}, there is not much data to verify these results. Karickhoff et al. (1979) found that pyrene sorption coefficients increased by 15% in going from pure water to sodium chloride solution concentration of 0.34 M. We may compare the effectiveness of Eq. 11-30 by estimating what we would expect for this case. First, for aromatic hydrocarbons Eq. 11-21a suggests a value of a to be 0.93; from Table 5.6 we see that a suitable value for K^{s}(pyrene, NaCl) would be 0.29. Thus we may estimate for this case:

$$\begin{aligned} \log K_{\mathrm{om,salt}}(\text{pyrene}, 0.34\ \mathrm{M\ NaCl}) &= \log K_{\mathrm{om}}(\text{pyrene}) \\ &\quad + (0.93)(0.29)(0.34) \\ &= \log K_{\mathrm{om}}(\text{pyrene}) + 0.09 \end{aligned} \tag{11-31}$$

This predicts that the organic matter–water partition coefficient of pyrene with 0.34 M NaCl would be about 24% higher than the corresponding case without dissolved salt, a result not too different than that which was observed. This discrepancy (15% observed versus 24% predicted) may indicate other factors are at play, such as salts causing changes in the coiling of the natural organic matter and concomitant changes in its solvency for neutral organic sorbates (i.e., changing γ_{om}). However, the prediction of Eq. 11-31 looks reasonable to a very good first approximation, and one probably rightly concludes that typical levels of dissolved salts do not cause major changes in K_{om} values.

Influence of Cosolvents on K_{om} Sometimes we may be concerned with the sorption of neutral organic compounds to soils from solutions in which a substantial concentration of organic solvent is also present. This could occur, for example, at a groundwater site where water-miscible liquids like acetone or methanol were codisposed with other organic chemicals. In such a case, the tendency of the organic chemicals to sorb to the soil's organic matter might be changed by the organic cosolvent's effects on γ_{om} and γ_w. The inclusion of substantial amounts of the cosolvent into the soil's natural organic matter would influence γ_{om}, but since these activity coefficients are near 1 to start with, such effects are probably not too big. However, as we discussed in Chapter 5, the presence of cosolvents can greatly change a chemical's aqueous solubility. Using our knowledge of that effect, we are already in a position to estimate the importance of cosolvent to K_{om} values.

First, we recall Eq. 5-32 and convert to a molar basis to describe the influence of a cosolvent on a sorbate's solubility:

$$\log C_{mix}^{sat} = \log C_w^{sat} + \log \frac{\bar{V}_w}{\bar{V}_{mix}} + \frac{0.5 f_c N(\sigma_{t:w} - \sigma_{t:c})(\text{HSA})}{2.303\, RT} \tag{11-32}$$

where C_{mix}^{sat} is the solubility in an aqueous solution containing an organic cosolvent ($\text{mol} \cdot \text{L}^{-1}$ mixed solution), $\bar{V}_{mix}$ is the molar volume of the mixed solution (L mixed solution$\cdot\text{mol}^{-1}$ total solvents), and the other terms remain as defined previously (see Section 5.4).

Using this result in our relationship between K_{om} and C_w^{sat} (Eq. 11-21), we arrive at:

$$\begin{aligned}\log K_{om,mix} &= -a \cdot \log(C_{mix}^{sat}) + b \\ &= -a \cdot \left(\log C_w^{sat} + \log \frac{\bar{V}_w}{\bar{V}_{mix}} + \frac{0.5 f_c N(\sigma_{t:w} - \sigma_{t:c})(\text{HSA})}{2.303\, RT} \right) + b \\ &= \log K_{om} - a \cdot \left[\log \frac{\bar{V}_w}{\bar{V}_{mix}} + \frac{0.5 f_c N(\sigma_{t:w} - \sigma_{t:c})(\text{HSA})}{2.303\, RT} \right]\end{aligned} \tag{11-33}$$

This equation first suggests that the natural organic matter–water partition coefficients will change because of differences in the solution's molar volume from that of water. Also, the third term on the right-hand side of Eq. 11-33 indicates that the attractiveness of the cosolvent–water mixture for maintaining the sorbate in solution will lower $\log K_{om,mix}$ relative to $\log K_{om}$.

A brief calculation exemplifies the magnitude of these effects. Suppose we were interested in the sorption of anthracene (HSA = 202 Å^2) from an aqueous solution containing 10% methanol by volume. First, we could calculate that $\bar{V}_{mix} \approx 0.0185\ \text{L} \cdot \text{mol}^{-1}$ total by assuming the methanol and water mix with no change in total volume. Thus the term $\log(\bar{V}_w/\bar{V}_{mix})$ is quite small (≈ 0.01). Next, to estimate the importance of the second term in brackets of Eq. 11-33, we apply Eq. 11-33 with $(\sigma_{t:w} - \sigma_{t:c})$ taken as $52 \times 10^{-7}\ \mathrm{J/cm^2}$:

$$a \cdot \left(\frac{0.5 f_c N(\sigma_{t:w} - \sigma_{t:c})(\mathrm{HSA})}{2.303\,RT} \right) =$$

$$(0.93) \cdot \frac{0.5 \cdot 0.1 \cdot 6.02 \times 10^{23} \dfrac{\text{molecules}}{\text{mol}} \cdot \left(52 \times 10^{-7} \dfrac{\mathrm{J}}{\mathrm{cm^2}} \right) \left(202 \times 10^{-16} \dfrac{\mathrm{cm^2}}{\text{molecule}} \right)}{2.303 \cdot 8.31\,(\mathrm{J/mol \cdot K}) \cdot 293\,\mathrm{K}}$$

$$= 0.52 \tag{11-34}$$

This result implies that $K_{om,mix}$(pyrene, 10% methanol) is lower than the corresponding K_{om}(pyrene) by a factor of $10^{-0.52}$ or 0.30. Experiments by Nkedi-Kizza et al. (1985) observed anthracene's K_{om} to decrease by a factor of 0.4 under these conditions. These investigators demonstrated that this approach for estimating the influence of cosolvents on K_{om} appears to work well for cosolvent fractions less than 25%.

Sorption of Neutral Chemicals to Organic Colloids Until now, we have not been concerned about the size of the particles with which the nonpolar chemicals became associated. This factor is generally not critical unless we are interested in sorption kinetics (see Section 11.7) or when we are concerned with separating the particles from suspensions. Such a separation becomes important if we are trying to measure contaminant concentrations in the dissolved versus sorbed phases or if we are interested in predicting how a chemical might be carried by moving water. In these cases, one often finds that the smallest of particles, called *colloids*, are not separable from the water. Thus, chemicals of interest to us would occur in the fluid both as truly dissolved species and as molecules associated with very small particles or macromolecules.

Colloids are microparticles or macromolecules that are small enough to move primarily by Brownian motion, as opposed to gravitational settling, and are large enough to provide a microscopic "environment" into or onto which molecules of interest to us can escape the aqueous solution. This means colloids range from about a few nanometers to around a few micrometers in dimension. Based on this perception, organic colloids in natural waters would include humic substances and proteins, viruses and nonmotile bacteria, and organic coatings on very small inorganic particles (e.g., aluminosilicates, iron oxides).

A variety of approaches have been used to demonstrate that colloidal organic matter may serve to absorb nonpolar organic chemicals much like the organic materials associated with larger soil and sediment particles. For example, Chiou et al. (1986) have shown that the "apparent solubility" of compounds like DDT or PCB congeners is increased in solutions containing humic substances over that seen for pure water alone (Fig. 11.11). The interpretation of these data is that the true aqueous solubility is not affected, but rather the additional nonpolar chemical molecules are associated

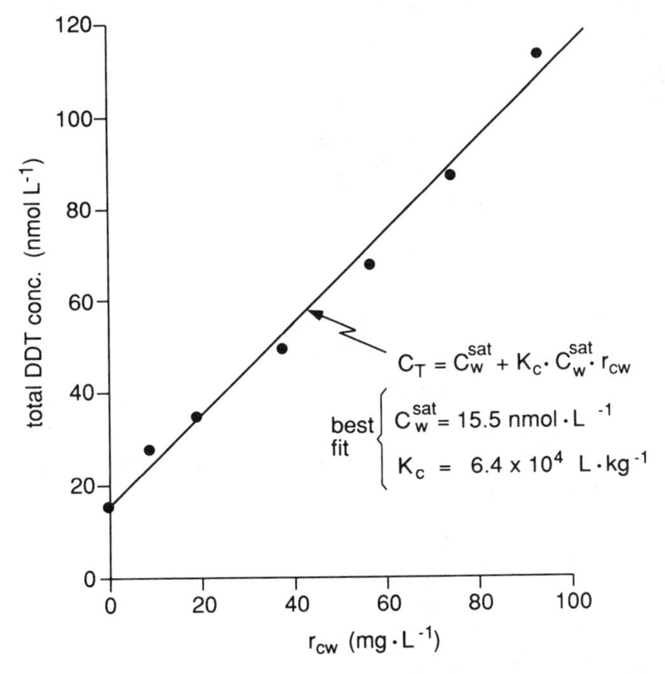

Figure 11.11 Increase in total DDT concentration in aqueous solution as a function of concentration of humic colloids, r_{cw} (data of Chiou et al., 1986). Interpreting the total DDT as a sum of truly dissolved DDT (C_w^{sat}) and DDT bound to colloids ($K_c C_w^{sat} r_{cw}$) fits the data very well.

with the humic colloids. Similarly, Carter and Suffet (1982) have shown that DDT associates with humic materials retained within dialysis tubing.

Much as we saw before for large sediment and soil particles, the tendency for different organic chemicals to associate with organic colloids is directly related to the aqueous activity coefficient of the chemicals. This should not be too surprising since it is largely the same types of organic matter, whether attached to large mineral grains or dispersed as colloidal suspensions, which are acting as absorbents. Some indications exist which show that there is some decrease in colloidal sorbent effectiveness, especially as the organic matter becomes more polar as reflected in C/O ratios (Chiou et al., 1986; Garbarini and Lion, 1986; Gauthier et al., 1987; Chin and Weber, 1989). Nonetheless, K_{om} values for colloidal organic matter generally appear similar to these reported for larger particulate organic matter.

The association of nonpolar chemicals with colloidal organic matter results in several very important effects on the behavior of those compounds. First, the organic molecules in a colloidal suspension exhibit a tendency to leave the suspension and move into adjacent gaseous phases only in proportion to the truly dissolved fraction of molecules (i.e., proportional to their fugacity). This has been seen in a dynamic system such as that employed by Mackay et al. (1979), where bubbled solutions transmitted the chemicals to the headspace at diminished rates when colloid association became important. Similarly, Brownawell (1986) observed diminished PCB congener vapor pressures over seawater suspensions of colloids compared to seawater solutions devoid of organic colloids. It is clear from these data that nonpolar chemicals distribute themselves between truly dissolved and colloid-bound states, and that the dissolved species dictate equilibria with other phases. Consequently, when we are interested in colloid-containing solutions that exchange chemicals into the air, we must recognize that the air–water distribution ratio (D_{aw}) describing the concentrations in each phase has the form

$$D_{aw} = \frac{C_a}{C_w + C_c \cdot r_{cw}} \tag{11-35}$$

where C_c is the concentration of chemical associated with colloids per mass of colloid (mol·kg^{-1} colloids) and r_{cw} is the mass of colloids per volume of solution (kg colloids·L^{-1}). If we define a sorption coefficient for the colloids analogous to that for larger particles (Eq. 11-1),

$$K_c = \frac{C_c}{C_w} \tag{11-36}$$

we may combine Eqs. 11-35 and 11-36 to get

$$\begin{aligned} D_{aw} &= \frac{C_a}{C_w + K_c C_w r_{cw}} \\ &= K_H \cdot (1 + K_c r_{cw})^{-1} \end{aligned} \tag{11-37}$$

where K_H is the "true" Henry's Law constant reflecting the equilibrium between the truly dissolved species and its vapor molecules. We should note that the term $(1 + K_c r_{cw})^{-1}$ is simply the fraction of molecules remaining dissolved in this case. Thus the air–water distribution ratio is simply the true K_H multiplied by the fraction of molecules participating in air–water exchange. This result indicates that when colloids are quite abundant in a solution (high r_{cw}) or the chemical–colloid combination of interest exhibits a large K_c, we should expect air–water partitioning to diverge substantially from the simple situation quantified by K_H alone.

Another important ramification of such colloid binding of nonpolar chemicals is that this process diminishes the tendency of such substances to bioaccumulate in aquatic organisms. Leversee et al. (1983) showed this for the uptake of polycyclic aromatic hydrocarbons by *Daphnia*, a freshwater zooplankter. Also, colloid association changes the light-processing characteristics of nonpolar chemicals. For example, Gauthier et al. (1986, 1987) and Backhus and Gschwend (1990) have shown how association of aromatic hydrocarbons with organic colloids diminishes their ability to fluoresce. Clearly, organic chemicals behave differently in many ways when associated with colloids than when they are dissolved in water.

A final major consequence of nonpolar chemical sorption by organic colloids involves efforts to distinguish experimentally the abundances in sorbed and dissolved forms. In batch sorption tests, when we try to separate water from all particles in a soil or sediment suspension, it is practically impossible not to include some colloids with the water. Such colloids go through typical filters and are difficult to sediment, even with intensive centrifugation. As a result, experimental attempts to assess solid–water distribution coefficients virtually always measure colloid-bound molecules with the dissolved ones in the aqueous separate. For that matter, to some extent some water (and associated solute molecules) always remains with the solid separates. As a result, the subsequent processing of measured chemical concentrations for the purpose of K_d calculations should recognize the possibility of such cross-over phases:

$$K_d^{\text{apparent}} = \frac{C_s + C_w V}{C_w + C_c r_{cw}} \qquad (11\text{-}38)$$

where

C_s is the sorbate concentration on the separated particles ($mol \cdot kg^{-1}$),
C_w is the solute concentration in the water ($mol \cdot L^{-1}$),
V is the volume of water left with the separated particles ($L \cdot kg^{-1}$),
C_c is the sorbate concentration on the colloids ($mol \cdot kg^{-1}$), and
r_{cw} is the colloid mass remaining with the bulk of the water ($kg \cdot L^{-1}$).

If the solid–water distribution coefficient, defined in Eq. 11-1 as $K_d = C_s \cdot C_w^{-1}$, applies for the separated particles, and we again use a second coefficient, $K_c = C_c \cdot C_w^{-1}$, which applies for the colloidal particles that are not separated from the water, then we may rewrite Eq. 11-38 as

$$K_d^{\text{apparent}} = \frac{K_d + V}{1 + K_c r_{cw}} \qquad (11\text{-}39)$$

This expression indicates that the apparent solid–water distribution coefficient will only equal the "true" one (as defined by Eq. 11-1) if $V \ll K_d$ and if $K_c r_{cw} \ll 1$. For compounds that do not sorb extensively (low K_d), this result suggests that experimental data may be erroneously high (Fig. 11.12). For chemicals that do tend to sorb (high K_c) and in situations where colloidal phases are substantial (high r_{cw}), batch observations of solid–water partitioning will indicate results lower than K_d (Fig. 11.12). These phase separation difficulties are probably one of the major explanations for the so-called "solids concentration effect" in which K_d appears to decrease with greater and greater loads of total solids (and colloids) in batch sorption systems. Most likely, there is no change in the relative abundances of sorbing nonpolar molecules as the proportions of water and particles are varied (Gschwend and Wu, 1985).

An interesting spinoff of this recognition of colloids in laboratory experiment aqueous phases is the realization that such microparticles and macromolecules undoubtedly exist in natural waters too. As a consequence, one must be concerned

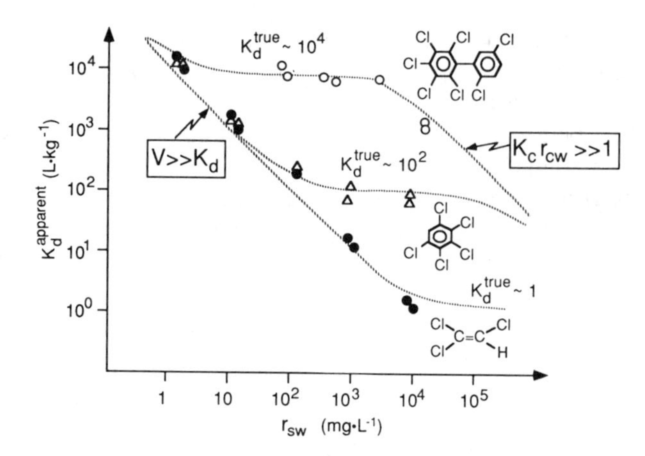

Figure 11.12 Effects of incomplete phase separations on apparent K_d's of three nonpolar sorbates studied as a function of solids-to-water ratio, r_{sw}, using sediment from the Missouri River with $f_{oc} = 0.72\%$ (data from Gschwend and Wu, 1985, and Gschwend, Brownawell, and Wu, unpublished). Model lines through data are for Eq. 11-39, assuming $V = 0.2\,\text{mL}\cdot\text{g}^{-1}$ and $r_{cw} = 0.02\ r_{sw}$. Only the horizontal portions of the curves reflect "true" K_d's.

that some portion of the molecules of interest moving with fluid flows are attached to colloids (e.g., Enfield et al., 1989). There is evidence that hydrophobic chemicals like PCBs exist in the surface waters of the Great Lakes, not just as dissolved molecules, but also in association with colloidal phases (Baker et al., 1986). Thus to predict the fate of such chemicals, we may need to consider transport both as dissolved and sorbed-to-colloids forms, and we may need to evaluate transformations remembering that colloid-bound molecules are not free to participate like truly dissolved ones.

11.5 SORPTION OF NEUTRAL ORGANIC CHEMICALS TO POLAR MINERAL SURFACES

In some environments, solids in aquatic systems do not include important amounts of natural organic matter. Consequently, association of organic solutes, especially hydrophobic ones, with mineral surfaces may become significant. This situation exists in groundwater environments where the organic fraction of the aquifer solids is very small (Schwarzenbach and Westall, 1981; Banerjee et al., 1985; Piwoni and Banerjee, 1989). Additionally, clay walls and liners are often used to isolate organic wastes buried below ground; and we may be interested in not only the impact of this low-permeability material on the subsurface hydraulics, but also its suitability for binding organic pollutants and preventing their passage offsite (Boyd et al., 1988). Also, sorption to mineral surfaces may be important when dealing with surface-catalyzed transformations (Ulrich and Stone, 1989). Finally, laboratory glass surfaces and sampling vessels may sorb hydrophobic compounds from aqueous solutions, confusing subsequent data interpretation. Thus, some consideration of the magnitude of the distribution ratio of such compounds between the mineral surface and water and the factors affecting this ratio are in order.

Several investigators have reported "binding" of neutral organic compounds to mineral surfaces (Table 11.3). Although somewhat limited and typically normalized to the mass rather than the surface area of sorbent, these data suggest the following generalizations. First, coarser particles (e.g., silica sand) exhibit less binding than corresponding finer particles made of the same material (e.g., porous silica adsorption of lindane). This is presumably due to the influence of the solid surface area which obviously differs between fine and coarse particles of the same material. Thus, values of sorption coefficients for minerals ($K_{min} = C_{min}/C_w$) may be more useful if they are normalized to the solid's surface area rather than its mass (i.e., K_{min} in units like moles per square meter per moles per liter). The second tendency we see is that for any one sorbent, binding increases within a series of sorbates as a function of their aqueous activity coefficients. Therefore, to predict mineral binding as a function of sorbate properties, it appears that, as for absorption into natural organic matter, we should use characteristics related to γ_w.

Another clue as to the nature of nonionic compound binding to mineral surfaces comes from studies on the effect of temperature on this process. Increasing the system temperature resulted in diminished sorption in three studies (Mills and Biggar, 1969b; Boucher and Lee, 1972; and van Bladel and Moreale, 1974), indicating that the overall process in these cases was exothermic. Since the dissolution of such nonpolar compounds in water is generally an endothermic process (recall Table 5.4), we may reasonably anticipate that some of this energy yield on mineral sorption came from the removal of those chemicals from aqueous solution. Indeed, in all three studies noted above, after accounting for solution enthalpies, the remaining steps in mineral binding of nonpolar sorbates proved to be energetically neutral or even slightly endothermic. From these results it appears that strong molecule:surface interactions are not involved. Unless some moiety is added which can specifically interact with atoms on a solid's surface, the differential extent of mineral adsorption from one chemical to another may be due chiefly to the variations in their aqueous activity coefficients.

A brief description of the chemical nature of mineral surfaces may help us interpret these neutral–chemical sorption phenomena. Most major minerals expose a surface to the exterior which consists of hydroxyls protruding into the medium from a checkboard plane of electron-deficient atoms (e.g., Si, Al, Fe) and electron-rich ligands (e.g., oxygen, carbonate) (Fig. 11.13*a*). Like water molecules, these surface hydroxyls and ligands prefer to form hydrogen bonds with the molecules adjacent to the mineral surface. We can use data on pure liquid:silica solid attractions reported by Fowkes (1964) to understand the strength of such molecule:surface interactions (Fig. 11.13*b*). While all sorbates are attracted to the surface by van der Waals or dispersive forces, as functional groups capable of dipole:dipole interaction and H-bonding are added, incrementally stronger attractions per unit surface area of silica are observed. Interestingly, from the point of view of the particular series of sorbates shown in Figure 11.13*b*, the attraction energy per molecule remains nearly constant. The relatively

TABLE 11.3 Observed Values of K_{min} for Nonionic Organic Compounds and Inorganic Solids

Adsorbent[a]	Sorbate	$\log \gamma_w^{sat}$	K_{min} $[mol \cdot kg^{-1} \cdot (mol \cdot L^{-1})^{-1}]$	Reference[b]
Porous silica				
20°C, 500 m^2/g	Chlorobenzene	4.09	4	1
	1,4-Dichlorobenzene	4.85	6	
	1,2,4-Trichlorobenzene	5.39	8	
	1,2,4,5-Tetrachlorobenzene	6.09	12	
20°C, 890 m^2/g	γ-1,2,3,4,5,6-Hexachloro-cyclohexane (lindane)	5.45	7	2
	β-1,2,3,4,5,6-Hexachloro-cyclohexane	4.99	2	
20°C, 400 m^2/g	Tetrachloroethylene	4.78	13	3
23–27°C, 200 m^2/g	Nitrotoluene	4.04	5	4
23–27°C, 250 m^2/g	1,4-Dichlorobenzene	4.85	3.4	
23–27°C, 1.6 m^2/g	Pentachlorobenzene	6.69	26	
Silica sand				
5°C	γ-1,2,3,4,5,6-Hexachloro-cyclohexane (lindane)	5.45	0.1	5
	Dieldrin	6.72	2	
	Aldrin	7.20	1.5	6
Silica beads				
22°C, 0.6 m^2/g	Pyrene	7.09	0.7	7
γ-Alumina				
20°C, 120 m^2/g	Chlorobenzene	4.09	0.6	1
	1,4-Dichlorobenzene	4.85	0.9	
	1,2,4-Trichlorobenzene	5.39	1.5	
	1,2,4,5-Tetrachlorobenzene	6.09	2.2	

Kaolinite				
20°C, 12 m^2/g	Chlorobenzene	4.09	0.6	1
	1,4-Dichlorobenzene	4.85	1.1	
	1,2,4-Trichlorobenzene	5.39	2.4	
	1,2,4,5-Tetrachlorobenzene	6.09	4.9	
Unknown	Aldrin	7.20	3	6
22°C, 12 m^2/g	1,2,4-Trichlorobenzene	5.39	0.3	7
	Pyrene	7.09	50	
	Perylene	8.1	1.3×10^5	
	Methyl perylene	8.7	2.0×10^5	
Montmorillonite	Aldrin	7.20	3	6
Ca–Montmorillonite				
20°C	γ-1,2,3,4,5,6-Hexachloro-cyclohexane (lindane)	5.45	3	2
	β-1,2,3,4,5,6-Hexachloro-cyclohexane	4.99	4	
26.5°C, 760 m^2/g	Fenuron	2.32	8	8
	Monuron	3.13	13	
Na–montmorillonite				
25°C, 610 m^2/g	Diuron	4.16	23	9
	Fenuron	2.32	14	
	Monuron	3.13	24	
26.5°C, 760 m^2/g	Fenuron	2.32	8	8
	Monuron	3.13	13	
800 m^2/g	Phenol	1.7	0.2	10
	Ethanol	~0	0.2	
	1,4-Dioxane	~0	0.5	
	Urea	~0	0.6	

[a] Temperature and surface area, if known.

[b] 1. Schwarzenbach and Westall, 1981. 2. Mills and Biggar, 1969a. 3. Estes et al., 1988. 4. Szecsody and Bales, 1989. 5. Boucher and Lee, 1972. 6. Yaron et al., 1967. 7. Backhus, 1990. 8. van Bladel and Moreale, 1974. 9. Bailey et al., 1968. 10. Zhang et al., 1990.

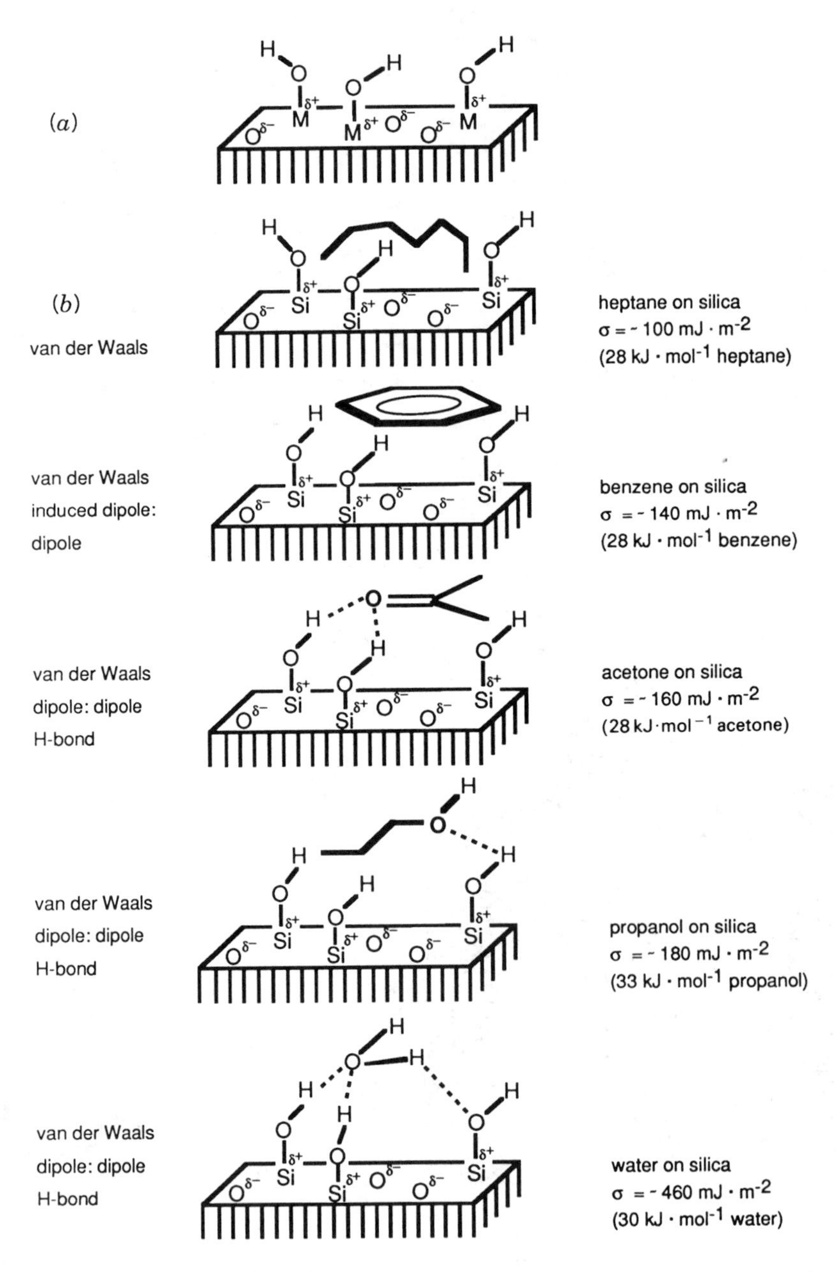

Figure 11.13 (*a*) A schematic view of a mineral surface exhibiting loci of partial positive charges where metal atoms occur (M), partial negative charges where linking anions occur (shown as oxygen atoms), and hydroxyl substituents extending to the exterior. (*b*) Interactions of organic chemicals wetting silica surfaces and a measure of the interaction strengths derived from surface tension data (Fowkes, 1964).

small water molecules make up what they lack in dispersive attraction by using H-bonding. Thus, one anticipates that water is extremely favored in competitions with nonionic organic sorbates for positions on such minerals surfaces, and the energy change on adsorbing such organic chemicals directly to the solid must also reflect the desorption of water from the same area.

The strong interactions of water molecules with solid surfaces causes the water molecules near the surface to be oriented. This organizing effect seems to extend for several layers of water out toward the bulk medium (Drost-Hansen, 1969; Etzler and White, 1987). Thus, a complete image of organic chemical "binding" to minerals from aqueous solutions, especially for nonpolar compounds (i.e., the ones least able to displace adsorbed water), may need to include the advantage gained upon transfer of these compounds into the ordered water layer immediately adjacent to solid surfaces. The volume of this so-called "vicinal water" per mass of sorbent would be directly related to the intraparticle porosity and surface area and consequently should be greater for porous silica ($\sim 0.5\,\mathrm{mL/g}$) than for quartzite sand ($< 0.01\,\mathrm{mL/g}$) and greater for expandable montmorillonite ($\sim 1\,\mathrm{mL/g}$) than for the two-layer clay kaolinite ($< 0.02\,\mathrm{mL/g}$). Such volumes may approach a milliliter per gram in highly porous solids (Ogram et al., 1985; Mikhail et al., 1968a, b) and may reflect the water's behavior for nanometers away from the solid surface. In view of this concept of vicinal water, nonpolar "binding" to minerals may involve not only the exchange of organic sorbates with water molecules at the surface, but also the partitioning between relatively disorganized bulk water and this special volume of ordered water near the solid's surface.

In light of these observations, we may be able to find linear free-energy relationships which are suitable for estimating new K_{min} values. If the tendency of sorbates to escape aqueous solutions is an important factor, as it appears from the very limited available data, we expect the free energy of sorption of neutral organic compounds on minerals to be inversely related to the free energy of aqueous dissolution of those same (liquid) chemicals. This expectation is supported by the available data (Fig. 11.14) when we examine $\log K_{\mathrm{min}}$ (proportional to the sorption free energy) versus $\log \gamma_{\mathrm{w}}$ (reflecting the free energy of dissolution). For both silica and kaolinite sorbents, a correlation is seen of the form

$$\log K_{\mathrm{min}}(\mathrm{L \cdot m^{-2}}) \simeq a \cdot \log \gamma_{\mathrm{w}} + b \tag{11-40}$$

where a and b are fitted coefficients. One recent study found that $a \simeq 1.4$ and $b \simeq -11$ for a small set of organic sorbates interacting with silica and kaolinite (Backhus, 1990). Of course, one should not necessarily infer that this expression applies for other mineral types (e.g., carbonates, iron oxides). Although the data are very few and exhibit substantial scatter, this result does suggest it may be feasible to predict mineral surface binding coefficient from properties like $\gamma_{\mathrm{w}}^{\mathrm{sat}}$, $C_{\mathrm{w}}^{\mathrm{sat}}(\mathrm{l, L})$ or K_{ow} of the sorbates.

Based on this limited understanding of the magnitudes of K_{min} values, one may be tempted to try to predict the conditions under which nonpolar chemical sorption to mineral surfaces dominates association with natural organic matter (i.e.,

(*a*)

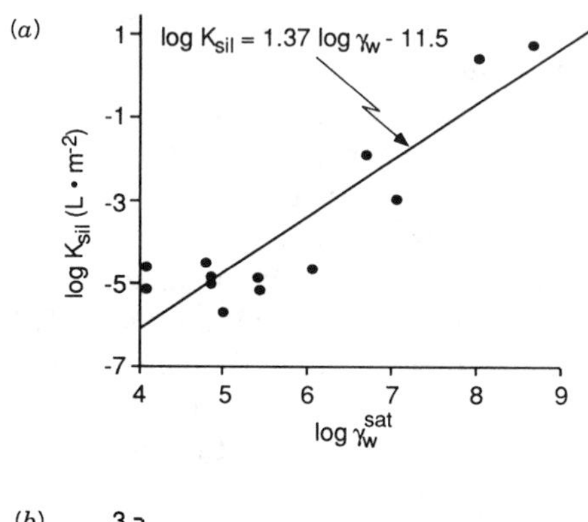

(*b*)

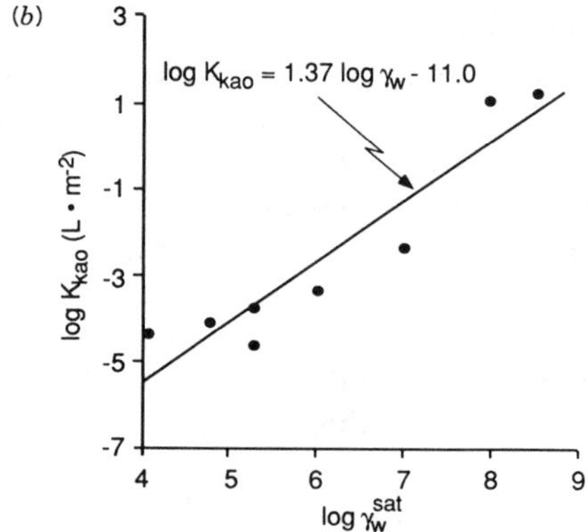

Figure 11.14 Relationship of log K_{min} and log γ_w as illustrated by the data of Yaron et al. (1967), Mills and Biggar (1969a), Schwarzenbach and Westall (1981), Estes et al. (1988), Szecsody and Bales (1989), and Backhus (1990): (*a*) for silica sorbents and (*b*) for kaolinite sorbents.

$A \cdot K_{min} > f_{om} \cdot K_{om}$). However, since sorption of nonpolar sorbates to mineral surfaces is an adsorption phenomenon, it is probably subject to competitive effects arising from the presence of other sorbates. Included in this competition for space on mineral surfaces is the natural organic matter itself. These natural materials are known to attach to kaolinite, iron oxides, and other minerals, especially when the inorganic surfaces are positively charged (Tipping, 1981; Davis, 1982). As a consequence, the available mineral surface area may be inversely related to f_{om}. Thus, we must know not only when $A \cdot K_{min} > f_{om} \cdot K_{om}$, but also how A varies with f_{om} in the system.

Some empirical approaches have been suggested to indicate when mineral surface sorption of nonpolar compounds starts to become important. First, sorption data appear to fit the pattern exhibited by pyrene in Fig. 11.5 as long as $f_{om} > 0.002$ to 0.004 (Schwarzenbach and Westall, 1981; Banerjee et al., 1985). Below this cutoff, the influences of mineral surfaces start to be felt. On the other hand, experiments to test for the extent of sorption before and after soil organic carbon oxidation indicate that this natural organic matter continues to play an important role even to virtually unmeasurable levels (Karickhoff, 1984; Lion et al., 1990). Unfortunately, these "thresholds" appear to vary as a function of the sorbate's hydrophobicity (Karickhoff, 1984; Banerjee et al., 1985). Competition between sorption to natural organic matter and to mineral surfaces is probably also a function of other soil or sediment properties, such as the proportion of swelling clays present with their very large specific surface areas (Karickhoff, 1984). Thus, the ratio of A to f_{om} is undoubtedly important. In sum, at f_{om} less than about 0.002 we should begin to consider sorption to mineral surfaces as contributing to the total picture, but this additional sorption mechanism may or may not become dominating depending on the properties of the sorbate and solid mixture of interest.

11.6 ADSORPTION OF IONIZABLE ORGANIC CHEMICALS FROM AQUEOUS SOLUTIONS

Influences of Charged Moieties Present in Organic Compounds on K_d

Now we begin to consider organic species that exhibit at least one ionic group in their structure (e.g., $—COO^-$, $—NH_3^+$, $—SO_3^-$). Much of the work on this topic has been performed by investigators interested in surfactants, since the inclusion of a charged moiety on an otherwise nonpolar chemical skeleton renders the resultant compound *amphiphilic* (part liking water, part disliking water) and capable of participating in many interesting interfacial phenomena. Also, a good deal of progress has been made by researchers studying the chemistry of inorganic surfaces (e.g., metal oxides) and how these minerals are affected by organic *ligands* (compounds that bind metals; from the Latin *ligare*: to bind).

When organic chemicals include structural components that are ionized, for example,

3-Dodecylbenzene sulfonate (anionic detergent)

Hexadecyl trimethyl ammonium (cationic detergent, disinfectant)

Pentachlorophenolate (wood preservative)

a variety of new effects become important insofar as the interactions of these sorbates with solid surfaces are concerned. This is largely due to two phenomena that we have not considered for neutral sorbates: (1) the electrostatic interactions of charged

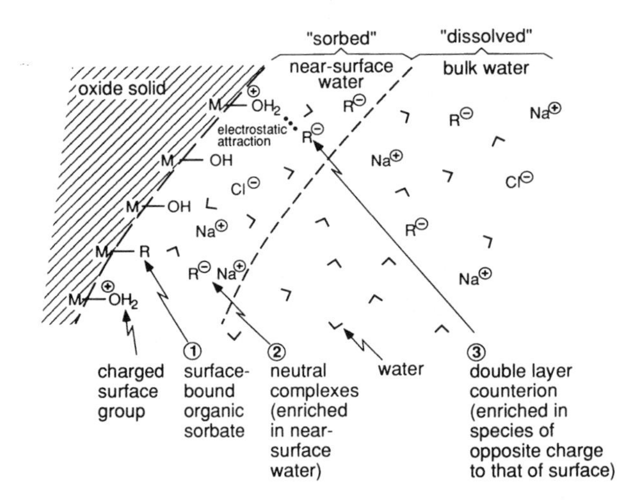

Figure 11.15 A positively charged oxide particle in water attracts anionic species including organic ones (e.g., R^-) to the near-surface water. Some of these anionic species may also react with the surface, displacing other ligands (e.g., H_2O or OH^-), to form surface-bound sorbate. M in the solid refers to atoms like Si, Al, or Fe.

molecules with charged sites on the sorbent (i.e., giving rise to the product $C_{ie} \cdot \sigma_{ie} \cdot A$ in the numerator of Eq. 11-9), and (2) exchange reactions with ligands previously bound to the solid (i.e., contributing to the terms $C_{rxn} \cdot \sigma_{rxn} \cdot A$ is Eq. 11-9).

First, owing to the ubiquitous phenomenon of ionizable surface groups on wet particles, virtually every solid presents a charged surface to the aqueous solution. If this surface charge is of opposite sign (e.g., positive) to that exhibited by an organic functional group (e.g., negative), then there will be an electrostatic attraction (ΔG_{elect}) between the organic sorbate in the bulk solution and the particle surface (Fig. 11.15). Such organic ions will accumulate in the thin film of water surrounding the particle as part of the population of charges in solution balancing the charges on the solid surface. Conversely, organic molecules with charges of like sign as the surface will be repulsed from the near-surface water. These electrostatic effects act similarly for all charged sorbates.

The second interaction involves chemical bonding of the organic compound to the surface or to some component of the solid phase. As depicted in Fig. 11.15, this may involve displacing some ligand previously bound to the surface (e.g., OH^- in Fig. 11.15). Alternatively, such a reaction could look more like the condensation reaction shown in Figure 11.2. Such a surface reaction involves another free energy change (ΔG_{rxn}). *We must note that a surface reaction forms a second sorbed species which is distinct from like-structured organic ions dissolved in the near-surface water* (e.g., $M - R \neq M - OH_2^+ \ldots R^-$).

Finally, we will find that some charged organic compounds include a sufficiently large hydrophobic portion in their structure that transfer into near-surface (i.e., vicinal) water is favored (i.e., due to a negative $\Delta G_{hydrophob}$) even in the absence of surface

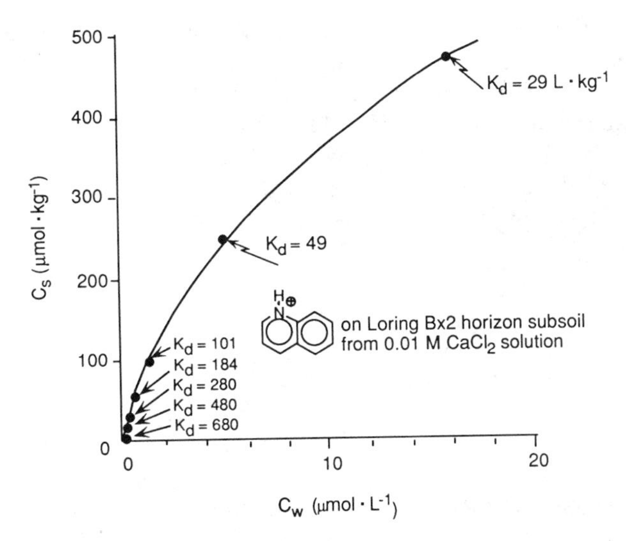

Figure 11.16 Nonlinear sorption isotherm exhibited by quinoline on a subsoil of $f_{om} \simeq 0.48\%$, surface area $\simeq 30.5\ m^2/g$, and cation exchange capacity of 8.4 mmol/100g (data from Zachara et al., 1986).

charges or reactive surface sites. Such sorption requires the cotransfer of a counterion (e.g., Na^+ with R^- in Fig. 11.15) to maintain local electroneutrality.

Experiments typically show that the isotherms of charged organic sorbates are decidedly nonlinear (Fig. 11.16). Said another way, the solid–water distribution ratio changes markedly as a function of the sorbate's own dissolved concentration. The extent of solid association of charged organic compounds also varies as a function of factors like solution pH, since this property governs both the presence of charges on mineral surfaces and the fraction of sorbate in an ionized form. Solution ionic strength and ionic composition also affect the sorption of charged organic chemicals, especially if inorganic ions compete with organic ones for binding sites. Obviously, the mineral composition of the sorbent is a key factor. Not surprisingly, all of these factors make the *a priori* estimation of charged organic chemical sorption to natural soils and sediments much more difficult and complex than for neutral compounds where everything is usually reducible to a few key factors (e.g., γ_w and f_{om}). In the following sections, we examine in more detail the nature of the interactions of charged molecules with charged surfaces and discuss how we might estimate the extent of such (ad)sorption.

Influence on K_d of Organic Compound Speciation in Solution

Before we discuss the sorbate–sorbent interactions that control charged organic species sorption, let us briefly examine the impact of solution pH on the sorption of substances that can be protonated or deprotonated to varying extents depending on their pK_a (recall Chapter 8). As a result of rapid acid–base interconversion, the solution may contain two (or more) organic species, a neutral one, and its conjugate acid or base. Each of these solution species will be involved in its own sorptive exchange phenomena as we noted before (e.g., Eqs. 11-10 through 11-13). Thus, we must begin analyzing ionizable chemical sorption by realizing what fraction of that chemical occurs in its charged, as opposed to its neutral form. For example, if we consider sorption of phenolic compounds like 2,4,5-trichlorophenol ($pK_a = 6.94$):

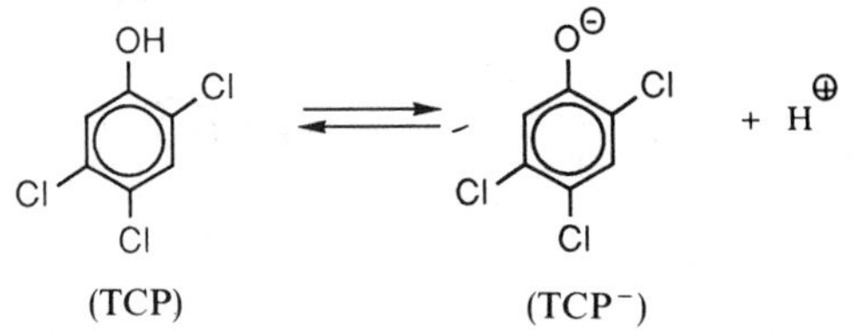

(TCP) $\rightleftharpoons$ (TCP⁻) + $H^{\oplus}$ (11-41)

we may calculate that the fraction of this compound in the neutral form at any pH is (after Eq. 8-16):

$$\alpha_a = \frac{1}{1 + K_a/[H^+]} \tag{11-42}$$

Similarly, the proportion present as the charged species is

$$(1 - \alpha_a) = \frac{K_a/[H^+]}{1 + K_a/[H^+]} \tag{11-43}$$

Now *we proceed by evaluating the solid interactions of each of these dissolved species separately*. In this case, we anticipate that the neutral species will sorb to natural solids containing natural organic matter just like any other nonionic organic chemical (recall Eq. 11-16):

$$K_d(\text{TCP}) = \frac{[\text{TCP}]_{om}}{[\text{TCP}]_w} \cdot f_{om} \tag{11-44}$$

where $[\text{TCP}]_{om}$ is the concentration of the nonionized trichlorophenol in the solid organic matter, and $[\text{TCP}]_w$ is the concentration of the nonionized trichlorophenol in the water. We may similarly define for the anionic phenolate species:

$$K_d(\text{TCP}^-) = \frac{[\text{TCP}^-]_{om}}{[\text{TCP}^-]_w} \cdot f_{om} \tag{11-45}$$

where, in this case, we assume the major sorption mechanism for the phenolate is dissolution into natural organic matter. Next, we may reasonably expect that the presence of a charge on the deprotonated phenol species will cause this molecule to be much more water-soluble than its conjugate acid. Therefore, such ionized phenol molecules exhibit correspondingly less tendency to go from solution into particulate organic matter:

$$K_{om}(TCP^-) \ll K_{om}(TCP) \tag{11-46}$$

We would expect $K_{om}(TCP^-)$ to be about three orders of magnitude less than $K_{om}(TCP)$ since the K_{ow} of phenolate is about that much less than the K_{ow} of the corresponding phenol (see Chapter 8 or fragment constants in Lyman et al., 1982). If the phenolate species does not sorb by another major mechanism besides dissolving in the natural organic matter, and this appears to be true for many natural solids which are primarily negatively charged at natural water pH's, then we may estimate the total sorption of the trichlorophenol using a distribution ratio:

$$K_d(\text{TCP plus TCP}^-) = \frac{([TCP]_{om} + [TCP^-]_{om}) \cdot f_{om}}{[TCP]_w + [TCP^-]_w} \tag{11-47}$$

where we have written K_d as the ratio of all major sorbed species divided by the sum of all the forms in the water (TCP_w and TCP_w^-). Rearranging and substituting with the K_a and K_{om} expressions defined above, we deduce

$$K_d(\text{TCP plus TCP}^-) = \frac{\left[\dfrac{[TCP]_{om}}{[TCP]_w} + \dfrac{[TCP^-]_{om}}{[TCP]_w}\right] \cdot f_{om}}{1 + \dfrac{[TCP^-]_w}{[TCP]_w}} \tag{11-48}$$

$$= \frac{\left[\dfrac{[TCP]_{om}}{[TCP]_w} + \dfrac{[TCP^-]_{om} \cdot K_a}{[TCP^-]_w[H^+]}\right] \cdot f_{om}}{1 + \dfrac{K_a}{[H^+]}} \tag{11-49}$$

$$= \frac{[K_{om}(TCP) + K_{om}(TCP^-) \cdot K_a/[H^+]] \cdot f_{om}}{1 + \dfrac{K_a}{[H^+]}} \tag{11-50}$$

As long as we are interested in cases where the solution pH is less than about two units above the acid's pK_a (i.e., $K_a/[H^+] \ll 1{,}000$), then we may neglect the involvement of TCP^- with the particulate organic matter:

$$K_d(\text{TCP plus TCP}^-) \simeq \frac{K_{om}(TCP) \cdot f_{om}}{1 + \dfrac{K_a}{[H^+]}} \tag{11-51}$$

This result is equivalent to

$$K_d(\text{TCP plus TCP}^-) = \alpha_a \cdot K_d(\text{TCP}) \tag{11-52}$$

Equation 11-52 implies that sorption of this trichlorophenol isomer will depend not only on the tendency of the ionized phenol to associate with particulate organic matter [K_d(TCP)], but also on the fraction (α_a) of nondissociated species in aqueous solution at a given pH. The accuracy of this result is illustrated in Figure 11.17. Using a lake sediment of f_{om} near 19%, Schellenberg and co-workers (1984) measured the solid–water distribution coefficient of 2,4,5-trichlorophenol at several solution pH's. They saw diminished sorption at higher pH, and this trend in K_d closely mirrored the change in TCP present in its protonated form. Further, the K_d found at pH's well below TCP's pK_a matched the result these investigations expected for nonionized TCP sorption to the particulate organic matter.

When the solution pH is such that the anionic species finally becomes much more abundant than its neutral counterpart in solution, we cannot neglect the sorption of TCP^- (Jafvert, 1990; Jafvert et al., 1990). Unfortunately $K_{om}(TCP^-)$ is not constant at all solution pH's and ionic conditions. In some part this is due to the variation in solid charging as solution pH changes, causing differential electrostatic repulsions at varying pH. Various ions (e.g., Ca^{2+} versus Na^+) also affect the value of K_{om}(anions) owing to the differential formation of ion pairs [e.g., $Ca^{2+}(\text{phenolate}^-)_2$].

Thus in cases where an organic chemical may be present as more than one species in solution, we must consider the sorptive interactions of each of these species. We

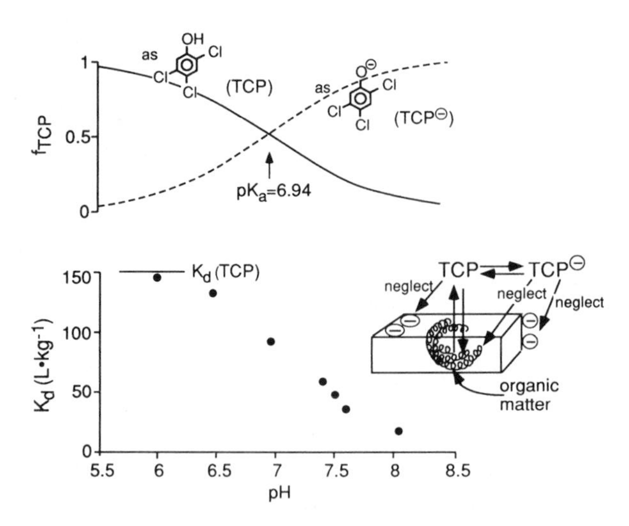

Figure 11.17 2,4,5-Trichlorophenol ionization (upper) and sorption to a lake sediment (lower) as a function of solution pH (data from Schellenberg et al., 1984). Inset indicates the importance of TCP interaction with particulate organic matter, in this case, relative to other sorption processes for TCP or TCP^-.

may often find that the overall solid–water distribution coefficient can be estimated from information on the dominant sorption interaction and the fraction of molecules available to participate in that interaction. In the following sections we focus on the processes governing the sorption of ionized organic species, thereby enabling us to determine when sorbed species besides nonionized molecules in natural organic matter become dominant.

Sorption of Charged Organic Species

Let us continue our discussion by evaluating the sorption of the *charged organic species themselves*. We first examine cases where specific chemical reactions (i.e., bond breaking and making) with the solid are *not* important to K_d. For example, for now we consider "monodentate" compounds like benzoate or unreactive ones such as dodecyl-pyridinium:

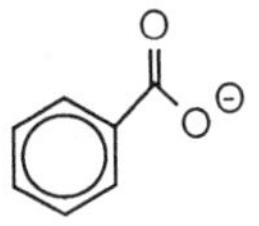

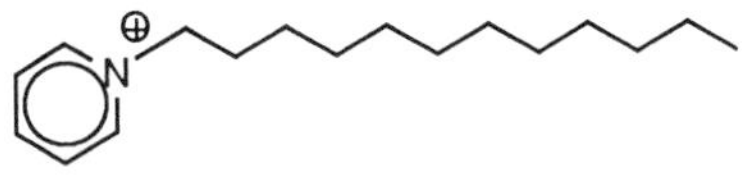

but not bidentate ones like salicylate or compounds like aniline with its reactive amine moiety:

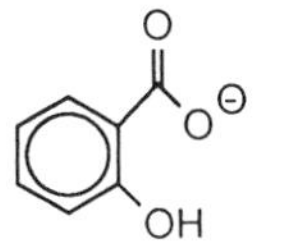

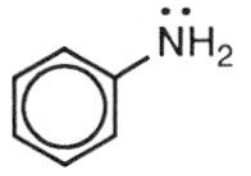

We make this distinction because "monodentate" compounds (i.e., those with a "single tooth" with which to bond to atoms on the surface of solids) seem to sorb more as counterions dissolved in the near-surface thin film of water than as surface-bound species. The "polydentate" compounds like salicylate form, not only the near-surface counterion species, but also nonnegligible quantities of surface-bound sorbates (Yost et al., 1990). Quaternary ammonium compounds like dodecyl pyridinium do not retain any nonbonded electrons with which to bond to components of the solid phase, in contrast to aniline which can react with carbonyl moieties included in particulate organic matter (Hsu and Bartha, 1976). Before we can treat surface-bound forms, it is useful to know how to handle electrostatic influences in isolation.

Surface Charges on Solids in Water To evaluate the importance of charged organic compound interchange between the bulk aqueous solution and the thin layer of water surrounding a charged particle (Fig. 11.18), we need to know how many charges are on the surfaces of solids. The adjacent surficial layer of water (i.e., Region II in Fig. 11.18) contains an excess of ions, called counterions (e.g., Na^+ in Fig. 11.18), that carry charge equal in magnitude and opposite in sign to that exhibited by the particle surface (i.e., Region I in Fig. 11.18). The thickness of this ion-rich water layer, which is sometimes called the *diffuse double layer*, varies inversely with the ionic strength

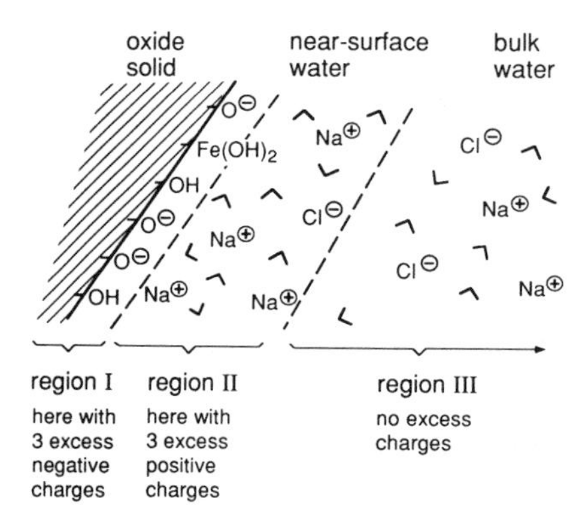

Figure 11.18 A schematic view of the accumulation of counterions in a narrow solution volume adjacent to charged particle surfaces. Region I depicts the charged surface layer formed by reactions of surface groups with acids or bases (e.g., $M—OH + OH^- \rightleftharpoons M—O^- + H_2O$) and with certain strongly sorbed ions [e.g., $M—O^- + Fe(OH)_2{}^+ \rightleftharpoons M—O—Fe(OH)_2$]. Region II is the diffuse double layer in which counterions (e.g., Na^+) accumulate in excess over bulk solution concentrations to match the net charge on the solid surface. Region III is the aqueous solution outside the diffuse double layer where electroneutrality is achieved by the sum of cation charges equaling the sum of anion charges.

of the solution. The e^{-1} characteristic width is given by $0.28 \times I^{-0.5}$ nm, where I is the solution's ionic strength in molar units (Morel, 1983). For typical ionic compositions of natural waters (10^{-3}–0.5 M), this means that most [i.e., $(1 - e^{-1}) \cdot 100\% = 63\%$] of the counterions are packed into a layer of water between 0.3 and 10 nm thick, and nearly all [i.e., $(1 - e^{-3}) \cdot 100\% = 95\%$] are within 1–30 nm of the surface. It is worth noting that this range is very similar to the 1–10 nm range believed to reflect ordered vicinal water (Drost-Hansen, 1969); thus, this is indeed a very special microscopic water environment.

For charged species that do not react with the surface, we recognize that the extent of their accumulation in this layer of water is caused at least in part by electrostatic attractions or repulsions (ΔG_{elect}). A major factor in governing these attractions is the abundance of charges on the particle's surface. Several kinds of surfaces can be envisioned; here we especially consider: (1) oxides or oxyhydroxides, (2) aluminosilicates or clay minerals, and (3) natural organic matter.

For natural solids that are *oxides or oxyhydroxides* (e.g., quartz, SiO_2; goethite, FeOOH; alumina, Al_2O_3), their water-wet surface is covered by hydroxyl groups (recall Fig. 11.13). These hydroxyl moieties can undergo proton-exchange reactions with the aqueous solution much like dissolved acids:

$$\text{surface}{\equiv}M{-}OH_2^+ \rightleftharpoons \text{surface}{\equiv}M{-}OH + H^+ \qquad (11\text{-}53)$$

$$\text{surface}{\equiv}M{-}OH \rightleftharpoons \text{surface}{\equiv}M{-}O^- + H^+ \qquad (11\text{-}54)$$

where M refers to an atom like Si, Fe, or Al at the particle surface, and $\equiv$ refers to the attachments of that atom to the solid. We may define acid–base equilibrium constants for those reactions:

$$K_{a1} = \frac{[\equiv MOH][H^+]}{(\equiv MOH_2^+]} \qquad (11\text{-}55)$$

$$K_{a2} = \frac{[\equiv MO^-][H^+]}{[\equiv MOH]} \qquad (11\text{-}56)$$

These surface acid equilibrium constants differ from their solution counterparts in that they reflect both an *intrinsic* reactivity of the particular O—H bond and an electrostatic free energy of moving H^+ to and from a charged surface:

$$K_{a1} = K_{a1}^{\text{int}} \cdot e^{F\psi/RT} \qquad (11\text{-}57)$$

$$K_{a2} = K_{a2}^{\text{int}} \cdot e^{F\psi/RT} \qquad (11\text{-}58)$$

where

F is the Faraday constant ($96{,}485\,C \cdot mol^{-1}$),
ψ is the surface potential relative to the bulk solution (V or $J \cdot C^{-1}$),
R is the molar gas constant ($8.31\,J \cdot mol^{-1}\,K^{-1}$), and
T is the absolute temperature (K).

At higher and higher pH's, ψ becomes less and less positive as reaction 11-53 proceeds to the right and more and more negative as reaction 11-54 continues to the right. This variation in surface charge buildup makes it increasingly more difficult to move H^+ away from an oxide surface as solution pH is increased. The magnitude of this effect is calculated with the exponential terms in Eqs. 11-57 and 11-58.

It is also possible for some other inorganic species (e.g., Fe^{+3} or $PO_4{}^{-3}$) to bond with the surface [e.g., $Fe(OH)_2{}^+$ bound in Region I of Fig. 11.18); in such a case, these inorganic ions along with H^+ and OH^- are responsible for establishing the extent of charging on the solid surface. The combination of ions responsible for this charge formation are called "potential (ψ) determining." For now, we neglect specific inorganic adsorption other than H^+ and OH^- and their effects on surface charge (e.g., see Dzombak and Morel, 1990, for examples of specific sorption and its associated impact on surface charge for hydrous ferric oxide).

For the case at hand, it is easy to see that the abundance of $\equiv MOH_2{}^+$ and $\equiv MO^-$ species on the solid surface control the surface's charge. The concentration of this charge, σ_{ie} (mol charges$\cdot m^{-2}$) can be estimated (neglecting other specifically sorbed species):

$$\sigma_{ie} = [\equiv MOH_2{}^+] - [\equiv MO^-] \qquad (11\text{-}59)$$

where the surface species concentrations are given in units of mole per meter squared of exposed surface. When these two surface species are present in equal concentration, the surface exhibits zero net charge (also $\psi = 0$); we call the solution pH that establishes this condition, the pH of zero point of charge, or pH_{zpc}. This pH_{zpc} can be calculated if we know the intrinsic acidities of $\equiv MOH_2^+$ and $\equiv MOH$:

$$[\equiv MOH_2^+] = [\equiv MO^-] \qquad \text{at } pH_{zpc} \tag{11-60}$$

Substituting from Eqs. 11-55 to 11-58 and recalling $\psi_{zpc} = 0$, we have

$$[\equiv OH][H^+]_{zpc}(K_{a1}^{int})^{-1} = [\equiv OH][H^+]_{zpc}^{-1}(K_{a2}) \tag{11-61}$$

Simplifying Eq. 11-61 allows us to relate pH_{zpc} and the intrinsic acidities of the surface:

$$[H^+]_{zpc}^2 = K_{a1}^{int} K_{a2}^{int} \tag{11-62}$$

$$[H^+]_{zpc} = (K_{a1}^{int} K_{a2}^{int})^{0.5} \tag{11-63}$$

$$\log [H^+]_{zpc} = 0.5(\log K_{a1}^{int} + \log K_{a2}^{int}) \tag{11-64}$$

$$pH_{zpc} = 0.5(pK_{a1}^{int} + pK_{a2}^{int}) \tag{11-65}$$

Equation 11-65 shows that an oxyhydroxide's pH_{zpc} is midway between the intrinsic pK_a's of its surface groups. Now, when the aqueous solution pH is below the pH_{zpc}, we have the condition $[\equiv OH_2^+] > [\equiv O^-]$, and the solid exhibits a net positive surface charge. Conversely, when we are above the solid's pH_{zpc}, then $[\equiv O^-] > [\equiv OH_2^+]$, the surface is negatively charged, and it becomes increasingly so at higher pH.

Our task now is to estimate the concentration of surface charge for cases that interest us as a function of solid and solution properties. Table 11.4 shows how this can be done by solving for the abundances of the important surface species, $\equiv OH_2^+$ and $\equiv O^-$, using two sets of information: (1) knowledge of the intrinsic acidities for the oxide of interest, and (2) the feedback relationship of surface potential on surface charge density. Figure 11.19 illustrates the results of such calculations on charge density at pH's below and above an oxide's pH_{zpc} for aqueous solutions of $I = 0.001$ and 0.5 M. These results may be understood with a specific example. If we were interested in iron oxide with a pK_{a1}^{int} of 7 (Table 11.5) and a surface hydroxyl density of $2 \times 10^{-6}\, mol \cdot m^{-2}$, we could estimate that this solid would have about $2 \times 10^{-7}\, mol \cdot m^{-2}$ of positive charges on its surface at pH 6 in freshwater of $I = 10^{-3}$ M (i.e., ~10% as $M \equiv OH_2^+$ in Fig. 11.19). In salty water of $I = 0.5$ M and pH = 6, the same solid would have a little more than $1 \times 10^{-6}\, mol \cdot m^{-2}$ of positive charges (i.e., ~50% as $M \equiv OH_2^+$ in Fig. 11.19). If the solution pH was 7 instead of 6, the surface charge density would decrease by about a factor of 2. It would not be until pH was increased to above 8 [$pH_{zpc} = 0.5(7 + 9) = 8$] that this solid would start to show a net negative surface charge. Table 11.5 shows some other important oxide minerals common in aquatic environments, their typical surface areas, and their

TABLE 11.4 Estimating the Surface Charge σ_{ie} of Oxides When H^+ and OH^- Are the Potential-Determining Ions

$$\sigma_{ie}(\text{mol charges}\cdot m^{-2}) = [\equiv MOH_2{}^+] - [\equiv MO^-] \quad (11\text{-}59)$$

where

$[\equiv MOH_2{}^+]$ is the concentration of protonated surface sites ($mol\cdot m^{-2}$), and
$[\equiv MO^-]$ is the concentration of deprotonated surface sites ($mol\cdot m^{-2}$).

When $[\equiv MOH_2{}^+] > [\equiv MO^-]$, σ_{ie} is positive, indicating a net positive surface charge; conversely, when the $[\equiv MO^-] > [\equiv MOH_2{}^+]$, σ_{ie} is negative and its absolute value reflects the concentration of negative sites.

Using acidity relationships (Eqs. 11-55 through 11-58)

$$\sigma_{ie} = [\equiv MOH][H^+]K_{a1}^{-1} - [\equiv MOH][H^+]^{-1}K_{a2} \quad (11\text{-}66)$$

$$\sigma_{ie} = [\equiv MOH][H^+](K_{a1}^{int})^{-1}e^{-F\psi/RT} - [\equiv MOH][H^+]^{-1}K_{a2}^{int}e^{F\psi/RT} \quad (11\text{-}67)$$

where

K_{a1}^{int} and K_{a2}^{int} are the intrinsic equilibrium constants quantifying the extent of proton exchange in the absence of charging the surface,
ψ is the surface potential ($V = J\cdot C^{-1}$),
R is the molar gas constant ($8.31\ J\cdot deg\ K^{-1}\ mol^{-1}$),
T is the absolute temperature (K), and
F is the Faraday constant ($96{,}485\ C\cdot mol^{-1}$).

The surface charge and surface potential are related (Stumm and Morgan, 1981):

$$\psi = \frac{2RT}{zF}\sinh^{-1}\left[\left[\frac{\pi F^2 10^{-3}}{2\varepsilon RTI}\right]^{0.5}\sigma_{ie}\right] \quad (11\text{-}68)$$

where

z is the valence of ions in the background electrolyte (e.g., NaCl, $z = 1$),
ε is the dielectric constant of water ($7.2 \times 10^{-10}\ C\cdot V^{-1}m^{-1}$ at 25°C), and
I is the solution ionic strength ($mol\cdot L^{-1}$; 10^{-3} is needed to convert L to m^3).

Substituting Eq. 11-68 into 11-67 yields:

$$\sigma_{ie} = [\equiv MOH][H^+](K_{a1}^{int})^{-1}\exp\left(-\frac{2}{z}\sinh^{-1}\left[\left[\frac{\pi F^2 10^{-3}}{2\varepsilon RTI}\right]^{0.5}\sigma_{ie}\right]\right)$$
$$- [\equiv MOH][H^+]^{-1}(K_{a2}^{int})\exp\left(+\frac{2}{z}\sinh^{-1}\left[\left[\frac{\pi F^2 10^{-3}}{2\varepsilon RTI}\right]^{0.5}\sigma_{ie}\right]\right) \quad (11\text{-}69)$$

(Continued)

TABLE 11.4 *(Continued)*

Well below the pH_{zpc} ($[\equiv MOH_2{}^+] \gg [\equiv MO^{-1}]$), Eq. 11-69 simplifies to:

$$\sigma_{ie} = [\equiv MOH][H^+](K_{a1}^{int})^{-1}\exp\left(-\frac{2}{z}\sinh^{-1}\left[\left[\frac{\pi F^2 10^{-3}}{2\varepsilon RTI}\right]^{0.5}\sigma_{ie}\right]\right) \quad (11\text{-}70a)$$

or

$$\log\sigma_{ie} + \frac{2}{2.303z}\sinh^{-1}\left[\left[\frac{\pi F^2 10^{-3}}{2\varepsilon RTI}\right]^{0.5}\sigma_{ie}\right] - \log\left[[\equiv MOH_0] - |\sigma_{ie}|\right] = -pH + pK_{a1}^{int} \quad (11\text{-}70b)$$

At $T = 25°C$, $I = 10^{-3}$ M, a 1:1 background electrolyte ($z = 1$), and with an initial hydroxyl site density ($\equiv MOH_0$) of 2×10^{-6} mole/m^2, this becomes:

$$\log\sigma_{ie} + 0.868\sinh^{-1}(9.05\times10^{7}\sigma_{ie}) - \log(2\times10^{-6} - |\sigma_{ie}|) = -pH + pK_{a1}^{int} \quad (11\text{-}71)$$

which can be solved for σ_{ie} at pH's below the solid's pK_{a1}^{int}.

At $T = 25°C$, $I = 0.5$ M, a 1:1 background electrolyte, and with an initial hydroxyl site density of 2×10^{-6} mole/m^2, this becomes:

$$\log\sigma_{ie} + 0.868\sinh^{-1}(4.05\times10^{6}\sigma_{ie}) - \log(2\times10^{-6} - |\sigma_{ie}|) = -pH + pK_{a1}^{int} \quad (11\text{-}72)$$

Conversely, well above pH_{zpc} ($[\equiv MO^-] \gg [\equiv MOH_2{}^+]$), we have from 11-67:

$$\log\sigma_{ie} + \left(\frac{2}{2.303z}\sinh^{-1}\left[\left[\frac{\pi F^2 10^{-3}}{2\varepsilon RTI}\right]^{0.5}\sigma_{ie}\right]\right) + \log\left[[\equiv MOH_0] - |\sigma_{ie}|\right] = -pH + pK_{a2}^{int} \quad (11\text{-}73)$$

At $T = 25°C$, $I = 0.5$ M, a 1:1 background electrolyte, and with an initial hydroxyl site density of 2×10^{-6} mole/m^2, this becomes:

$$\log(-\sigma_{ie}) - 0.868\sinh^{-1}(9.05\times10^{7}\sigma_{ie}) - \log(2\times10^{-6} - |\sigma_{ie}|) = pH - pK_{a2}^{int} \quad (11\text{-}74)$$

In the cases where pH is between pK_{a1}^{int} and pK_{a2}^{int}, both parts of the right-hand side of Eq. 11-69 must be evaluated.

In the special case near the oxide's pH_{zpc} where ψ is small (absolute value less than 25 mV) and with a low ionic strength, Eq. 11-68 becomes:

$$\psi = \frac{2.303\,RT}{F}(pH_{zpc} - pH) \quad (11\text{-}75)$$

and we may estimate the solid's surface charge density:

$$\sigma_{ie} = [6.40\times10^{-10}(T\cdot I\cdot 10^{3})^{0.5}]\sinh[1.15\,z(pH_{zpc} - pH)] \quad (11\text{-}76)$$

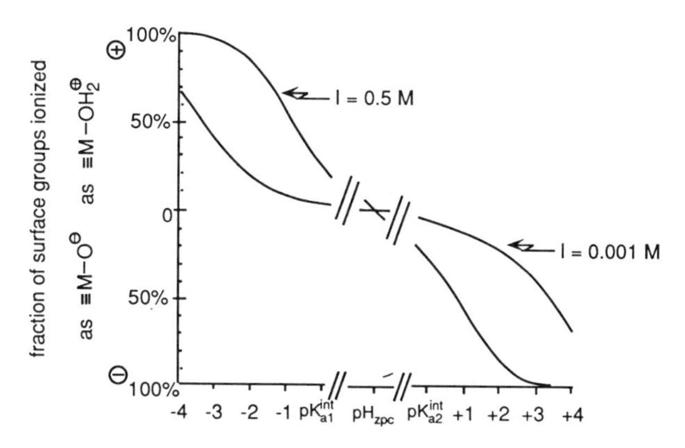

Figure 11.19 Variation of surface charge on a solid oxide (25° C) due to protonation and deprotonation of surface hydroxyls (at 2×10^{-6} mol·m^{-2}) as a function of solution pH for (a) $I = 0.5$ M of a dissolved salt with both the cation and the anion exhibiting one charge (1:1 electrolyte) and (b) $I = 0.001$ M 1:1 electrolyte. Note the breaks in the curves between unspecified values of pK_{a1}^{int} and pK_{a2}^{int}.

intrinsic acidity constants. Typically, surface charge densities in the range of 10^{-6} to 10^{-8} mol·m^{-2} are seen for oxides, and this implies that 10^{-6}–10^{-8} moles of counterions, including some charged organic molecules, will accumulate opposite each meter squared of surface area due to electrostatic attractions. This property of surfaces is sometimes referred to as the oxide's *c*ation *e*xchange *c*apacity (CEC) or *a*nion *e*xchange *c*apacity (AEC). Note again, this treatment neglects the influence of specifically sorbed ions which would neutralize some of this surface charge [e.g., $Fe(OH)_2^+$ bound in Region I of Fig. 11.18].

Clay minerals present a different case with regard to assessing their surface charge. These mixed aluminum oxides and silicon oxides (thus aluminosilicates) expose two kinds of surface to the external media, and therefore the same particles may exhibit both a CEC and an AEC at the same time (Table 11.5). First, the edges of these flakey-shaped minerals are somewhat like aluminum oxides in their behavior and respond to pH changes in solution much like pure aluminum oxides (e.g., pH_{zpc} of kaolinite edge ~7, Williams and Williams, 1978). The consequent anion exchange capacity observed empirically for clays is near 1×10^{-2} mole/100 g for a wide variety of clays (Grim, 1968), but this value changes with solution pH and ionic strength. On the other hand, the faces of these platey particles have a "siloxane" structure (—Si—O—Si—) which does not leave any free hydroxyl groups (—Si—OH) to participate in proton exchange reactions with the bulk solution. Instead, the faces exhibit a charge due to cation substitutions for the aluminum or silicon atoms within the internal structure. These "isomorphic" substitutions often involve cations of lower total positive charge (e.g., Al^{+3} for Si^{+4} or Mg^{+2} for Al^{+3}). The result is a fixed and permanent charge deficiency that looks like a negative surface charge to the surrounding solution. Empirical measures of this negative surface charge or CEC are

TABLE 11.5 Sorbent Properties of "Pure Solids" Commonly Present in Aquatic Environments[a]

Category		Specific Surface Area	CEC	AEC				
Sorbent	Compositions	($m^2 \cdot g^{-1}$)	($mol \cdot m^{-2}$)	($mol \cdot m^{-2}$)	pK_{a1}^{int}	pK_{a2}^{int}	pH_{zpc}	Ref.[b]
Oxides								
Quartz	SiO_2	0.14	9×10^{-8}*		(−3)	7	2.0	1, 2
Amorphous silica	SiO_2	500	9×10^{-8}*		(−3)	7	2.0	1, 2
Goethite	α-FeOOH	46		2×10^{-8}*	6	9	7.5	2
Amorphous iron oxide	$Fe(OH)_3$	600		5×10^{-8}*	7	9	8	2, 4, 5
Alumina	Al_2O_3	15		8×10^{-8}*	7	10	8.5	2
Gibbsite	$Al(OH)_3$	120	2×10^{-8}*		5	8	6.5	2,6
Aluminosilicates								
Na–montmorillonite	$Na_3Al_7Si_{11}O_{30}(OH)_6$	600–800	0.9 to 2×10^{-6}	3 to 4×10^{-7}			2.5	6,7
Kaolinite	$Al_2Si_2O_5(OH)_4$	12	0.2 to 1×10^{-5}	0.6 to 2×10^{-5}			4.6	
Illite	$KAl_3Si_3O_{10}(OH)_2$	65–100	1 to 6×10^{-6}	3×10^{-7}				
Organic								
Humus	$C_{10}H_{12}N_{0.4}O_6$	1	1 to 10×10^{-3}					8, 9
Carbonate								
Calcite	$CaCO_3$	1		9×10^{-6}			8–9.5	10, 11

[a]Calculated CEC and AEC values (*) assume solution pH = 7, ionic strength of 10^{-3} M, $T = 293$ K, solid-site density of 2×10^{-6} $mol \cdot m^{-2}$, and use of Eq. 11-69 or 11-76. Intrinsic acidity constants are rounded off to the nearest unit.

[b]References: 1. Parks (1965). 2. Schindler and Stumm (1987). 3. Mikhail et al. (1968a, b). 4. Tipping (1981). 5. Dzombak and Morel (1990). 6. Davis (1982). 7. Grim (1968) 8. Chiou et al. (1990). 9. Khan (1980). 10. Zullig and Morse (1988). 11. Somasundaran and Agar (1967).

made by assessing the maximum concentrations of weakly bound cations such as ammonium, NH_4^+, that can be sorbed. Table 11.5 shows the results of such cation exchange capacity tests on three common clays, montmorillonite, illite, and kaolinite. Expandable three-layer clays like montmorillonite exhibit the highest CEC's near 1×10^{-1} mol/100 g (Grim, 1968) or 1.4×10^{-6} moles of charged sites per meter squared (assuming a specific surface area of 700 m^2/g). On the other extreme, two-layer kaolinite clays exhibit the lowest CEC's of about 1×10^{-2} mol/100 g (Grim, 1968). This is chiefly due to their greatly reduced specific surface areas compared to the expandable three-layer clays, since per unit area these kaolinites actually have greater charge density, $\sim 10^{-5}$ mol/m^2.

Particulate natural organic matter may also contribute to the assemblage of charged sites of solids in water. This is mostly due to ionization reactions of carboxyl groups (—COOH), and at higher pH values, phenolic groups (aromatic ring —OH), which have been found at about 1–10 mM per gram of natural organic matter. Depending on the surrounding molecular environment, the carboxyl moieties exhibit pK_a's ranging from about 3 to 6. Consequently, the extent of charge buildup in the organic portion of natural particles will vary as a function of pH.

Still other solid phases, like carbonates, are common in nature, and these materials also exhibit surface charging. Realizing there will almost always be charges on particle surfaces submerged in water, we can now examine their impact with regard to sorbing ionized organic chemicals from solution.

Sorption of Organic Counterions to Single Solids by Ion Exchange Mechanisms

Let us now proceed to examine the quantities of organic ions that associate through ion exchange with a single type of solid surface in aqueous solution. For example, we consider the case of adding ethyl ammonium chloride ($CH_3CH_2NH_3Cl$; pK_a of the alkyl ammonium ~ 10) to a 10^{-2} M NaCl solution at pH 6 which also contains suspended montmorillonite particles. We do not need to consider ammonium ion ($CH_3CH_2NH_3^+$) bond formation with the clay surface, nor must we be concerned with neutral amine absorption into particulate organic matter in this case, since we have chosen solids which lack this phase. Consequently, the solid–water distribution coefficient involves only

$$K_d(\text{ethyl amine plus ethyl ammonium ions}) = \frac{[\text{ethyl ammonium ions near clay surface}]}{[\text{ethyl ammonium ions in solution}]} \quad (11\text{-}77)$$

at pH's (i.e., < 8) where we can assume that the positively charged species is much more abundant than its neutral conjugate base both in solution and near the clay surface. Before we added any alkyl ammonium salt, the negative surface charges of the clay were balanced by an excess of hydrated sodium cations relative to hydrated chloride ions accumulated in the thin film of water surrounding the particles (Fig. 11.18). When we add a small quantity of the alkyl ammonium salt to the suspension (insufficient to change the ionic strength), an ion exchange reaction occurs

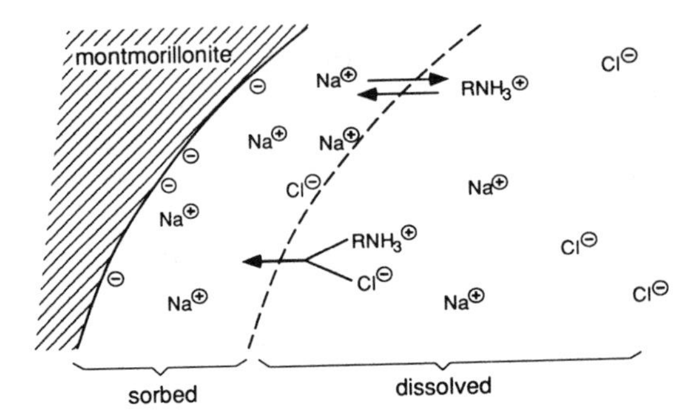

Figure 11.20 Illustration of two types of transfer enabling organic ammonium ions to collect near a negatively charged solid surface: exchange with Na^+ or cotransfer with Cl^-.

(Fig. 11.20):

$$RNH_3{}^+ + \text{Na:surface} \rightleftharpoons RNH_3\text{:surface} + Na^+ \tag{11-78}$$

resulting in some of the ammonium ions exchanging with Na^+ counterions in Region II near the solid's surface (the colon here indicates association without bond formation and $R = CH_3CH_2$). We can write an equilibrium constant expression for this ion exchange reaction:

$$K_{ie} = \frac{[RNH_3\text{:surface}][Na^+]}{[RNH_3{}^+][\text{Na:surface}]} \tag{11-79}$$

where we have used concentrations in place of activities. (A complete equilibrium expression would include activity coefficients for the charged species; however, since this equation involves monovalent ions both in the bulk medium and near the surface, such activity coefficients virtually cancel one another out in the equilibrium quotient.) K_{ie} reflects the preference of organic sorbate relative to the competing sorbates, in this case Na^+. Recognizing that the accumulaton of counterion cations around the negatively charged particles must be a constant equal to the product of the surface charge density σ_{ie} ($\text{mol}\cdot\text{m}^{-2}$) and the specific particle surface area A ($\text{m}^2\cdot\text{kg}^{-1}$), we can write

$$\sigma_{ie}\cdot A = [RNH_3\text{:surface}] + [\text{Na:surface}] \tag{11-80}$$

where $[RNH_3\text{:surface}]$ and $[\text{Na:surface}]$ have units of moles per kilogram. We can now eliminate the term in $[\text{Na:surface}]$ from the equilibrium constant expression 11-79:

$$K_{ie} = \frac{[RNH_3\text{:surface}][Na^+]}{[RNH_3^+](\sigma_{ie}\cdot A - [RNH_3\text{:surface}])} \tag{11-81}$$

Rearranging this expression, we find

$$[\mathrm{RNH_3:surface}] = \frac{(\sigma_{\mathrm{ie}} \cdot A)(K_{\mathrm{ie}})[\mathrm{RNH_3}^+]}{[\mathrm{Na}^+] + (K_{\mathrm{ie}})[\mathrm{RNH_3}^+]} \tag{11-82}$$

or more generally,

$$\begin{bmatrix}\text{organic ion}\\ \text{at surface}\end{bmatrix} = \frac{\sigma_{\mathrm{ie}} \cdot A \cdot K_{\mathrm{ie}}[\text{organic ion in solution}]}{[\text{competing ion}] + K_{\mathrm{ie}}[\text{organic ion in solution}]} \tag{11-83}$$

In terms of the particular example at hand, Eq. 11-82 says that the concentration of bound alkyl ammonium ion varies as we change the concentration of the dissolved species (Fig. 11-21*a*). At low organic ion concentrations ($[\mathrm{Na}^+] \gg K_{\mathrm{ie}}[\mathrm{RNH_3}^+]$), the bound-to-dissolved *ratio* is almost constant:

$$K_{\mathrm{d}} \simeq \begin{pmatrix}\text{total net}\\ \text{surface charge}\end{pmatrix} \cdot \begin{pmatrix}\text{equilibrium}\\ \text{constant}\end{pmatrix} \cdot \begin{pmatrix}\text{competing cation}\\ \text{concentration}\end{pmatrix}^{-1} \tag{11-84}$$

On the other hand, at high levels of $\mathrm{RNH_3}^+$, the bound counterion concentrations asymptotically approach a constant value set by the total surface charge density (in this case, the cation exchange capacity of the clay). At these elevated charged organic sorbate levels, the bound-versus-dissolved distribution ratio actually declines, and the isotherm is hyperbolic (Fig. 11-21*a*). An isotherm having this shape is generally referred to as a *Langmuir isotherm*:

$$[\text{Sorbed}] = \frac{[\text{maximum sorbed}]\, K_{\mathrm{Langmuir}}[\text{dissolved}]}{1 + K_{\mathrm{Langmuir}}[\text{dissolved}]} \tag{11-85}$$

where in this case $[\text{maximum sorbed}] = \sigma_{\mathrm{ie}} \cdot A$, and $K_{\mathrm{Langmuir}} = K_{\mathrm{ie}}/[\text{competing ion}]$.

We can confirm the direct relationship (expected from Eq. 11-84) of the bound-to-dissolved distribution ratio with the surface charge density for low concentrations of organic sorbate ions by examining the results of Fuerstenau and Wakamatsu (1975). In this case, alumina was used as a positively charged sorbent ($\mathrm{pH_{zpc}} \simeq 8.5$) and dodecylsulfonate anions were exchanged with background chloride ions for ion exchange sites on this solid as a function of solution pH. (Alkyl sulfonates, $\mathrm{R{-}SO_3}^-$, do not appear to participate in substantial ligand exchange reactions with alumina.) Based on Eq. 11-69, we calculate how the alumina's surface charge density σ should have decreased with increasing pH (Fig. 11-22*a*). Using these results, we anticipate that $K_{\mathrm{d}}(\text{dodecylsulfonate}) \cdot A^{-1}$ should be lowest at high pH when σ is lowest, and this is exactly what these workers observed (Fig. 11-22*b*). The direct correlation of these observed $(K_{\mathrm{d}} \cdot A^{-1})$'s with the calculated σ_{ie}'s yields a slope which, when adjusted by the competing chloride concentration, suggests K_{ie} is between 20 and 50. This implies the dodecylsulfonate was accumulated in the diffuse double layer surrounding the alumina relative to its bulk solution concentration more than an order of

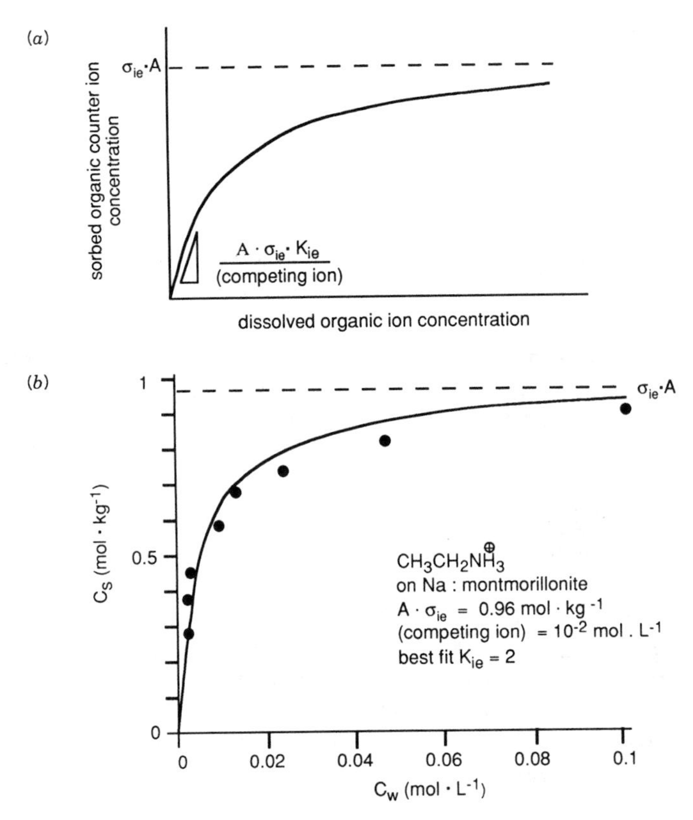

Figure 11.21 (*a*) Schematized Langmuir isotherm showing the variation of sorbed organic counterions (i.e., those within the near surface water layer) as dissolved (or bulk solution) organic ion concentration changes: $\sigma_{ie} \cdot A$ is the total surface charge density; K_{ie} is the exchange reaction equilibrium constant; and (competing ion) is the concentration of competing counterions. (*b*) For the specific case of ethyl ammonium sorption to montmorillonite, a Langmuir isotherm with best fit $K_{ie} = 2$ matches the experimental data well (data from Cowan and White, 1958).

magnitude more preferentially than the inorganic chloride adsorbate ($\Delta G_{ads} \simeq 7$–10 kJ·mol^{-1}). These findings clearly illustrate the direct effect of the particle's surface charge density on the extent of adsorption of organic counterions. We should note that at the point where [organic ion] $\simeq$ [competing ion], further increases in [organic ion] cause substantial changes in the solution ionic strength. Were we considering an oxide surface, such an important change in the solution would result in a concomitant slow increase in the surface charge (recall Eq. 11-68 in Table 11.4). Obviously, this effect on σ_{ie} would feed back into the extent of sorption.

Now let's examine the effectiveness of such a modeling approach for ethyl

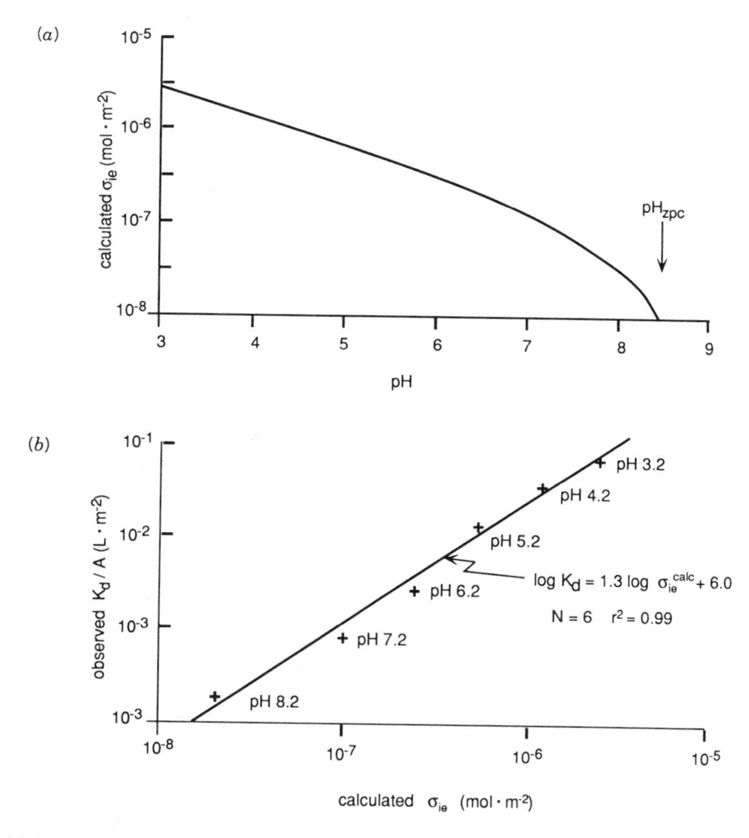

Figure 11.22 (*a*) Expected variation of the surface charge density σ_{ie} of alumina as a function of pH ($I = 2 \times 10^{-3}$ M, $T = 25°$ C, $[\equiv \text{MOH}]_0 = 5 \times 10^{-6}\,\text{mol}\cdot\text{m}^{-2}$). (*b*) Direct relationship between observed sorption of dodecyl sulfonate on alumina (Fuerstenau and Wakamatsu, 1975) and calculated σ_{ie}.

ammonium sorption to a Na–montmorillonite (Cowan and White, 1958). These workers measured the CEC ($=\sigma_{\text{ie}}\cdot A$) of their clay sorbent to be nearly $1\,\text{mol}\cdot\text{kg}^{-1}$ (recall that this surface charge is not sensitive to solution ionic strength like the oxides, since it arises principally from isomorphic substitutions). Working with $[\text{Na}^+] = 10^{-2}$ M, the observed isotherm data can be nicely fit with $K_{\text{ie}} = 2$ (Fig. 11-21*b*). Since $K_{\text{ie}} = 1$ would imply no preference between the sodium and the alkyl ammonium ions, this fit value of K_{ie} indicates only a little selection of the organic cation over the sodium ion, presumably because of the hydrophobicity of the ethyl substituent. We also deduce that for ethyl ammonium concentrations less than about 10^{-2} M (i.e., less than the Na^+ concentration) we have a constant K_d(ethyl ammonium ion) of about 200 $(\text{mol}\cdot\text{kg}^{-1})(\text{mol}\cdot\text{L}^{-1})^{-1}$ $[=\sigma_{\text{ie}}\cdot A\cdot K_{\text{ie}}(\text{competing ion})^{-1} = \text{CEC}\cdot K_{\text{ie}}\cdot [\text{Na}^+]^{-1}]$. Obviously, solutions of lower competing cation concentrations would allow the sorbed-to-dissolved ratio to be even greater. Given the ionic strength of 10^{-2} M, we recognize the characteristic length of the diffuse double layer is about 3 nm; together with an estimate of this montmorillonite's surface area ($\sim 700\,m^2/g$), we can calculate that this distribution ratio corresponds to about 60 ($mol \cdot L^{-1}$ double layer water)$\cdot$($mol \cdot L^{-1}$ bulk water)$^{-1}$. Clearly, the electrostatic attraction of the negatively charged clay faces is concentrating cations like ethyl ammonium ions in the water near the particles. In addition, as also indicated by the dodecylsulfonate sorption to alumina, we recognize that the rest of the organic structure (R) plays a role in determining the magnitude of K_{ie}.

Effects of Sorbate Hydrophobicity Thus our problem involves the question of how various hydrophobic portions of charged organic sorbates influence their sorption. Presumably the sorbate's chemical structure determines the preference of the sorbate for the near-particle water region versus the bulk solution. Cowan and White (1958) investigated the sorption of a series of alkyl ammonium ions to the same Na–montmorillonite (Fig. 11-23). A very interesting pattern emerged: *the longer the alkyl chain, the steeper was the initial isotherm slope.* Exactly parallel results have been seen for sorption of other amphiphiles, for example, negatively charged *n*-alkyl benzene sulfonates binding to positively charged alumina particles (Somasundaran et al., 1984). Further, *above about eight carbons in the chain, the extent of binding*

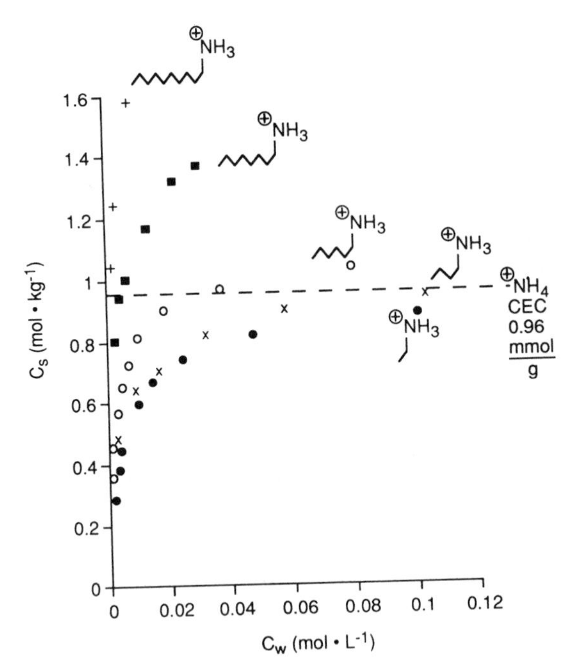

Figure 11.23 Adsorption isotherms for a series of alkyl ammonium compounds on sodium montmorillonite (adapted from Cowan and White, 1958). The horizontal dashed line indicates the cation exchange capacity of the clay.

could far exceed the clay's cation exchange capacity. Using quartz as a sorbent, Somasundaran et al. (1964) showed that chain lengths of about 10 carbons or more in alkyl ammonium ions could bind to the particles enough to reverse the particles' surface charge (i.e., convert negatively charged particles to positively charged ones by including counterions within what we called Region I in Fig. 11.18). These effects are undoubtedly due to the increasing hydrophobicity of the longer and longer chains. By favoring chemical partitioning to the near surface (e.g., into near surface water as we saw in Section 11.5) from the bulk solution, hydrophobic forces augment the electrostatic forces and thereby enhance the tendency of the sorbates to collect near the particle surface (Somasundaran et al., 1984). Thus, we anticipate little differences in sorption for organic chemicals due to moieties of like charge (e.g., $—COO^-$ vs. $—SO_3^-$), since the electrostatic attraction to a surface is fairly nonselective; but we do expect substantial variations between sorbates if they differ in the hydrophobicity of their nonpolar parts (analogous to Eq. 11-40 for neutral sorbates associating with minerals).

To provide an estimate of K_{ie} suitable for use in Eq. 11-83, let us try to isolate the contribution of the sorbate's hydrophobicity using some available data. First, recalling the ion exchange equilibrium expression,

$$K_{ie} = \frac{[RNH_3\text{:surface}][Na^+]}{[RNH_3^+][Na\text{:surface}]} \tag{11-79}$$

we see that this can be considered to consist of two individual partitioning processes, one for RNH_3^+ and the other for Na^+. Each of these species finds itself distributed between the bulk solution and the near-surface environment according to the respective free energy differences. For the sodium ion, this free energy difference consists of only an electrostatic contribution. The solid–water ratio is therefore given by

$$\begin{aligned} \frac{[Na\text{:surface}]}{[Na^+]} &= \exp(-\Delta G_{elect}/RT) \\ &= \exp(-zF\psi/RT) \end{aligned} \tag{11-86}$$

In contrast, the alkyl ammonium ions are driven to the surface, not only by electrostatic forcing, but also by a hydrophobic effect:

$$\begin{aligned} \frac{[RNH_3\text{:surface}]}{[RNH_3^+]} &= \exp[(-\Delta G_{elect} - \Delta G_{hydrophob})/RT] \\ &= \exp[(-zF\psi - \Delta G_{hydrophob})/RT] \end{aligned} \tag{11-87}$$

Here we assume that the Na^+ and the RNH_3^+ ions experience the same electrostatic attractions to the surface, although we know that factors like differences in solvation have some effect. Substituting Eqs. 11-86 and 11-87 into Eq. 11-79 yields

$$\begin{aligned} K_{ie} &= \exp[(-zF\psi - \Delta G_{\text{hydrophob}})/RT]/\exp(-zF\psi/RT) \\ &= \exp(-\Delta G_{\text{hydrophob}}/RT) \end{aligned} \quad (11\text{-}88)$$

Equation 11-88 indicates that the preference in ion exchange for organic ions over inorganic ones of the same valency (i.e., $z = 1$ for both sorbates) is mostly due to factors directing the organic sorbate to escape solution in bulk water. If we could evaluate this $\Delta G_{\text{hydrophob}}$ as a function of chemical structure, we could estimate K_{ie} for various organic sorbates competing with inorganic ions that do not bond with the surface.

Since this hydrophobic effect appears to regularly increase with the size of the nonpolar part of the chemical structure (Cowan and White, 1958; Somasundaran et al., 1984), we may reasonably propose this energy term is composed of contributions from each of the nonpolar parts of the structure. Consequently, we expect for the alkyl ammonium ions studied by Cowan and White (1958),

$$\Delta G_{\text{hydrophob}} = m \cdot \Delta G_{\text{—CH}_2\text{—}} \quad (11\text{-}89)$$

where m is the number of methylene (—CH_2—) groups in each sorbate's alkyl chain, and $\Delta G_{\text{—CH}_2\text{—}}$ is the hydrophobic contribution made by each methylene driving these sorbates into the diffuse double layer–vicinal water layer.

Thus, the total free energies directing these organic ions to distribute between the montmorillonite surface region and the bulk water would be

$$\begin{aligned} \Delta G_{\text{ads}} &= -RT\ln(K_d \text{ of } RNH_3^+) = -RT\ln\left(\frac{[RNH_3\text{:surface}]}{[RNH_3^+]}\right) \\ &= zF\psi + m \cdot \Delta G_{\text{—CH}_2\text{—}} \end{aligned} \quad (11\text{-}90)$$

Figure 11-24 shows the variation in ΔG_{ads} for these alkyl ammonium ions (when these organic sorbates are present at levels much less than Na^+) as a function of the number of methylenes in the alkyl chains. The least-squares correlation line through the data yields

$$\Delta G_{\text{ads}} = -10.9 - m \cdot 0.75 \qquad (\text{kJ}\cdot\text{mol}^{-1}) \quad (11\text{-}91)$$

This result implies that the alkyl ammonium ions experienced an electrostatic attraction to the clay surface corresponding to

$$zF\psi \approx -10.9\,\text{kJ}\cdot\text{mol}^{-1} \quad (11\text{-}92)$$

or that $\psi \simeq -0.11$ V. Also, we see $\Delta G_{\text{—CH}_2\text{—}} = -0.75\,\text{kJ}\cdot\text{mol}^{-1}$. Examination of the variation in aqueous solubilities for compound classes like alkanes or alcohols

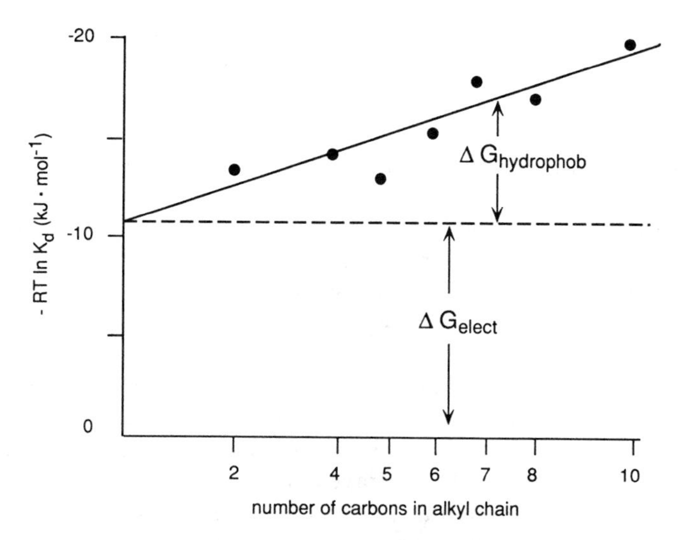

Figure 11.24 Variation in observed ion exchange free energy change ($-RT\ln K_d$) for a series of alkyl ammonium ions associating with a sodium montmorillonite. All K_d's taken at low organic ion concentrations. The least-squares fit line yields an intercept of $-10.9\,\text{kJ}\cdot\text{mol}^{-1}$ and a slope of $0.75\,\text{kJ}\cdot\text{mol}^{-1}$ per methylene group (data from Cowan and White, 1958).

(Chapter 5) as a function of additional methylene groups reveals that ΔG_s^e changes by almost $4\,\text{kJ}\cdot\text{mol}^{-1}$ for each increase in chain length. Thus, the ΔG_{-CH_2-} contributing to ΔG_{ads} in Eq. 11-91 corresponds to "a relief" of about 20% of the excess free energy of aqueous solution per methylene group. Somasundaran et al. (1984) noted that inclusion of the phenyl group in alkyl aryl sulfonates increases the ion exchange sorption tendency of these amphiphiles to a degree corresponding to lengthening the alkyl chain by 3–4 methylene groups. This is consistent with increasing the nonpolar structure's hydrophobicity to the same extent [i.e., $\Delta\log K_{ow}$(phenyl) $\simeq$ 1.68 and $\Delta\log K_{ow}$(3–4 methylenes] $\simeq$ 1.59 to 2.12). Thus, we may be justified in estimating the hydrophobic contribution to K_{ie} for charged organic sorbates as a fraction, say 20%, of the excess free energy of aqueous solution in the corresponding hydrocarbon:

$$K_{ie} = \exp(-\Delta G_{\text{hydrophob}}/RT) \tag{11-88}$$

$$\simeq \exp(+0.2\Delta G_s^e/RT) \tag{11-93}$$

$$\simeq \exp\left(+0.2\cdot\left(RT\ln\frac{55.3}{C_w^{sat}(\text{l, L})}\right)\bigg/RT\right) \tag{11-94}$$

$$\simeq 2.2[C_w^{sat}(\text{l, L})]^{-0.2} \tag{11-95}$$

The data of Cowan and White (1958) yield the empirical result

$$K_{\mathrm{ie}} = 1.1(C_{\mathrm{w}}^{\mathrm{sat}})^{-0.19} \tag{11-96}$$

using the solubilities of the corresponding alkanes. Such expressions predict that K_{ie} of decyl amine to be 19 (Eq. 11-96) or 45 (Eq. 11-95), since $10^{-6.57}$ M is the liquid solubility of decane. Cowan and White (1958) observed K_{ie}(decyl amine) to be 36. There is little doubt that hydrophobic phenomena are playing a role in determining the extent of amphiphilic sorption; however, a great deal more work is necessary before approaches such as Eq. 11-95 or 11-96 are proven to be robust.

Multisite, Multimechanism Sorption Because the sorbed concentration can exceed $\sigma_{\mathrm{ie}} \cdot A$ (Fig. 11-23), and the observation that real-world sorbents are not well fit with a single Langmuir isotherm (Fig. 11-25), we must consider other sorptive mechanisms in addition to the ion exchange one portrayed in Eq. 11-78. For example, recognizing the need to maintain electroneutrality near the solid's surface, a second exchange process has been postulated (Brownawell et al., 1990), which for our alkyl ammonium case would look like

$$RNH_3^+ + Cl^- \rightleftharpoons RNH_3{:}Cl \quad \text{ion pair formation in solution} \tag{11-97a}$$

$$RNH_3{:}Cl + \text{surface} \rightleftharpoons RNH_3{:}Cl{:}\text{surface} \quad \text{sorption} \tag{11-97b}$$

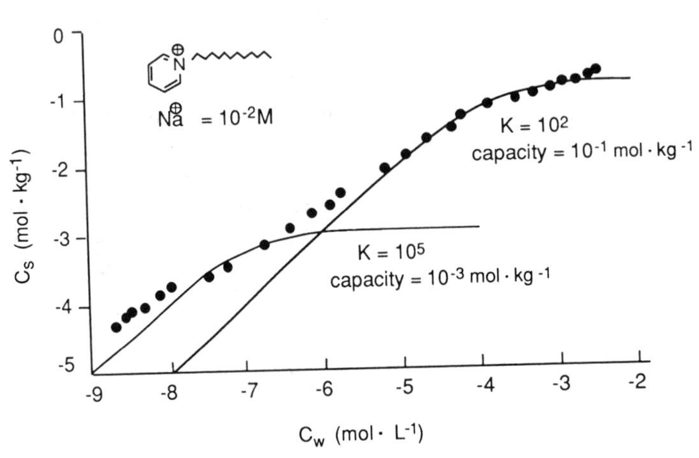

Figure 11.25 Observed sorption of dodecylpyridinium on a soil (EPA—12) exhibiting an overall cation exchange capacity of 0.135 mol·kg^{-1}. Two Langmuir isotherms are placed on the data to illustrate how different portions of the observed isotherm may reflect the influence of different materials in the complex soil sorbent or possibly different mechanisms (data from Brownawell et al., 1990).

We note especially that *the surface need not be charged to act as a sorbent in this process*. The physical picture associated with Eqs. 11-97a and 11-97b is that the organic ion with a companion (inorganic) counterion form an ion pair and moves together into the layer of water near the particle surface (Fig. 11-20). Presumably this transfer occurs because the hydrophobic portion of the organic ion has some desire to escape the bulk water and move into the near surface water. The equilibrium expression for this *n*on-*i*on *e*xchange process is

$$K_{\text{nie}} = \frac{[\text{RNH}_3\text{:Cl:surface}]}{[\text{RNH}_3^{\,+}][\text{Cl}^-][\text{surface}]} \tag{11-98}$$

where we include the activity coefficients applicable to charged species in K_{nie}. Similar to the ion exchange mechanism, this electroneutral sorption is ultimately limited by the capacity of the solid surface available to sorb the amphiphile. Thus we may write

$$\begin{aligned} \text{Total capacity} &= [\text{RNH}_3\text{:Cl:surface}] + [\text{surface}] \\ &= A/a \end{aligned} \tag{11-99}$$

where a is the surface area "covered" by a mole of sorbate molecules ($\text{m}^2 \cdot \text{mol}^{-1}$) and A is the total available surface area of the solids ($\text{m}^2 \cdot \text{kg}^{-1}$).

Substituting Eq. 11-99 into Eq. 11-98, we have

$$K_{\text{nie}} = \frac{[\text{RNH}_3\text{:Cl:surface}]}{[\text{RNH}_3^{\,+}][\text{Cl}^-](A/a - [\text{RNH}_3\text{:Cl:surface}])} \tag{11-100}$$

which, upon rearrangement, yields a second Langmuir isotherm:

$$[\text{RNH}_3\text{:Cl:surface}] = \frac{A/a \cdot K_{\text{nie}} \cdot [\text{RNH}_3^{\,+}]}{[\text{Cl}^-]^{-1} + K_{\text{nie}}[\text{RNH}_3^{\,+}]} \tag{11-101}$$

For this mechanism, the maximum sorbed concentration is given by A/a (mole per kilogram) and K_{Langmuir} is equal to $K_{\text{nie}} \cdot$[companion counterion].

Now we can see how the sorbed concentration can exceed the cation (or anion) exchange capacity for amphiphilic sorbates if we assume K_{nie} values (Fig. 11.26). The total sorption process involves both ion exchange and non-ion exchange mechanisms operating simultaneously (i.e., the cations shown in Eqs. 11-78 and 11.97; Fig. 11.20). The solid–water distribution coefficient can be deduced by summing the two contributions:

$$\begin{aligned} [\text{RNH}_3\,\text{sorbed}] &= [\text{RNH}_3\text{:surface}] + [\text{RNH}_3\text{:Cl:surface}] \\ &= \frac{\sigma_{\text{ie}} \cdot A \cdot K_{\text{ie}}[\text{RNH}_3^{\,+}]}{[\text{Na}^+] + K_{\text{ie}}[\text{RNH}_3^{\,+}]} + \frac{A \cdot a^{-1} \cdot K_{\text{nie}} + [\text{RNH}_3^{\,+}]}{[\text{Cl}^-] + K_{\text{nie}}[\text{RNH}_3^{\,+}]} \end{aligned} \tag{11-102}$$

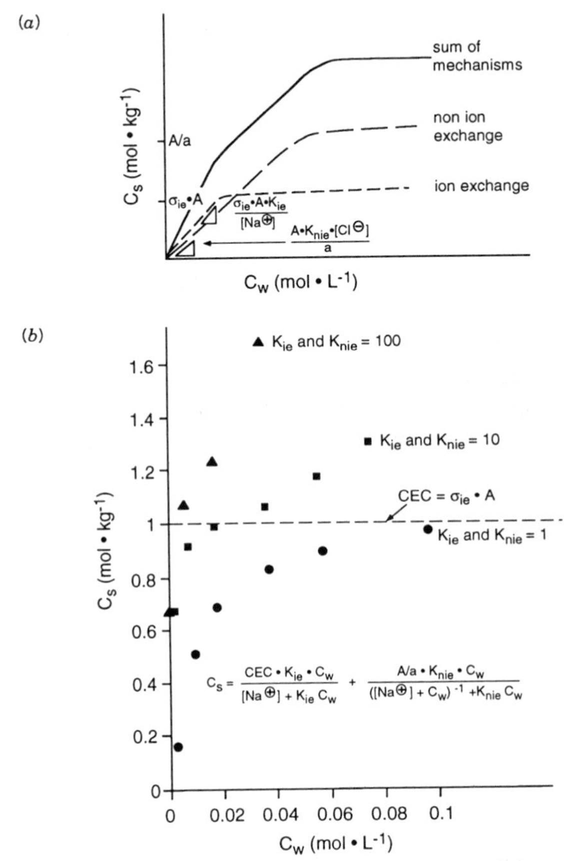

Figure 11.26 (*a*) Schematized isotherm for combination of ion exchange and nonion exchange mechanisms acting together: assumes $A/a > A \cdot \sigma_{ie}$ and $K_{ie} \cdot [Na^+]^{-1} > K_{nie} \cdot [Cl^-]$. (*b*) Calculated data points for specific cases where $K_{nie} = K_{ie}, \sigma_{ie} \cdot A = 1$ mol·kg^{-1}, $A/a =$ 4 mol·kg^{-1}, $[Na^+] = 10^{-2}$ M, and $[Cl^-] = [Na^+] + [R^+]$. Note that $C_w = [RNH_3{}^+]$ and $C_s = [RNH_3$: surface$] + [RNH_3$:Cl:surface$]$.

The combined isotherm appears as a somewhat "nonsharp" hyperbola, exhibiting a linear portion at sufficiently low dissolved sorbate concentrations and a plateau at suitably high levels (Fig. 11.26*a*). At intermediate concentrations, a gradual transition results from the differential maximization of the two sorption mechanisms. Depending on the magnitudes of K_{ie} and K_{nie}, various isotherms ranging from those looking like Langmuir (when one mechanism dominates) to Freundlich relations (when both mechanisms contribute) can be found (Fig. 11.26*b*). Obviously, the shape of the isotherm an experimentalist would see depends on the range of dissolved concentrations utilized. Nonetheless, the calculated isotherm points shown in Fig. 11.26*b* look remarkably like the measured data seen by Cowan and White (1958) for the several alkyl ammoniums they tested on a single sorbent (Fig. 11.23*b*).

Recent work (Brownawell et al., 1990) performed using soils and subsoils and examining dissolved concentrations spanning several orders of magnitude, suggests the heterogeneity of the natural sorbents is very important. It appears that, in addition to multiple sorption mechanisms acting simultaneously, one sees the influence of more than one solid surface type. Thus, the data require fits using several Langmuir isotherms, some for ion exchange interactions and possibly others for the non-ion exchange mechanism, but presumably each reflecting the involvement of different solid materials that make up the complex medium we simply call a soil, subsoil, or sediment.

Hemimicelles We conclude this discussion of amphiphilic sorption by discussing a special phenomenon called hemimicelle formation (Fuerstenau, 1956; Somasundaran et al., 1964; Chandar et al., 1983, 1987). This hemimicelle formation plays a critical role in amphiphile "sorption" to minerals when the organic ions are present at relatively high dissolved concentrations [about 0.001–0.01 of their *c*ritical *m*icelle *c*oncentrations (CMC), i.e., the level at which they self-associate in the bulk solution]. When the organic sorbate levels are low, the sorption mechanism is like the ion exchange mechanism we discussed above (Fig. 11.27, I). At some point in a titration of sorbents by micelle-forming compounds, presumably due to both electrostatic and hydrophobic effects, amphiphile concentrations build up in the near-particle region to a point where it seems likely that self-aggregation of the molecules occurs in that thin water layer (Fig. 11.27, IIa). This in turn would allow the rapid coagulation of the aggregated amphiphiles with the oppositely charged particle surface, smothering that subarea of the particle's surface charge with what have been called hemimicelles (Fig. 11.27, IIb). Electrophoretic mobility measurements clearly demonstrate the neutralization of the particle's charges in this steep portion of the isotherm, even going so far as to reverse the surface charge (e.g., Chander et al., 1987). The onset of this particle coating by hemimicelles occurs at different dissolved concentrations for various amphiphiles, but is near millimolar levels ($\gtrsim 100$ mg/L) for decyl-substituted amphiphiles and is near micromolar levels ($\gtrsim 100\ \mu$g/L) for octadecyl derivatives. In all cases, the bulk solution concentration is much less than the CMC. It appears that the elevated near-surface concentrations, derived from accumulation of these amphiphiles in the thin film of water near the particle surface by factors of 100 or more, is compensating just enough to achieve critical micelle concentrations in this near-surface water layer. Continued increase in amphiphile concentration results in the particle surface becoming increasingly coated by hemimicelles, apparently while the near-surface water maintains its concentration near that of the CMC (Fig. 11.27 II). Finally, the entire particle surface is covered with a bilayer of amphiphile molecules; the particle's surface charge is now that of the surfactant; and the addition of more amphiphile to the solution does not yield any higher sorbed loads (Fig. 11.27, III). This especially extensive degree of sorption may be the cause of macroscopic phenomena such as dispersion of coagulated colloids and particle flotation.

An Example: Calculation of K_d of a Charged Organic Chemical We conclude the discussion of sorption of charged organic sorbates by examining the situation for

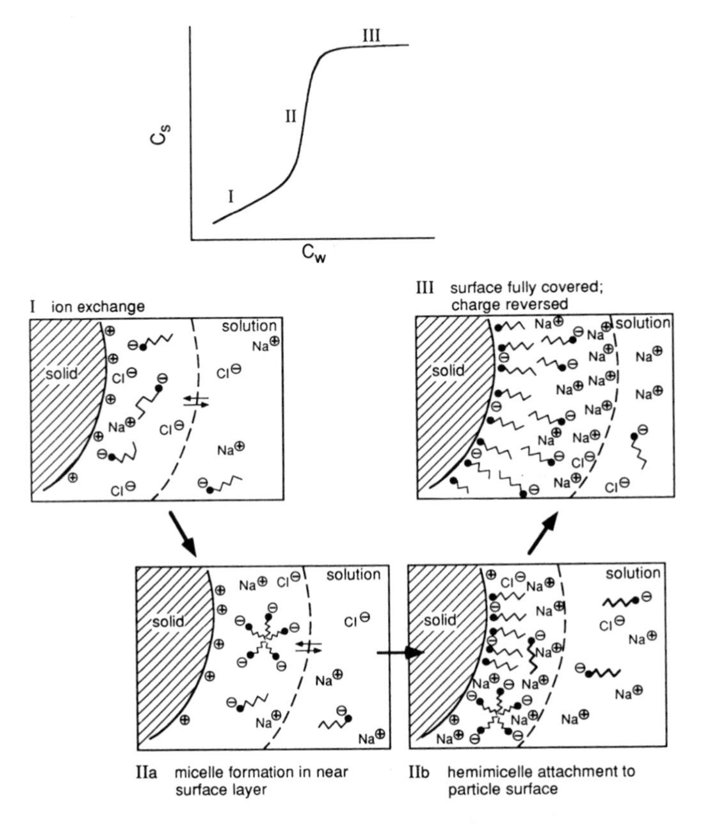

Figure 11.27 Relationship between sorbed and dissolved amphiphile concentration (upper isotherm plot). These different parts of the isotherm reflect changes in the solid surface as sorption proceeds, possibly explainable by the following: in portion (I) with low dissolved concentrations, sorption occurs via ion exchange and related mechanisms. At some point, sufficient near-surface concentration enhancement occurs that micelles form there (IIa) and rapid coagulation between oppositely charged micelles and the surface follows (IIb). When the surface becomes fully coated with such micelles, additional sorption is stopped (III). In portion II, the solid surface charge is converted from one sign to the other.

soils and sediments where both charged mineral surfaces and natural organic matter are present. The studies of Zachara et al. (1986) using quinoline as the sorbate are illustrative. This aromatic compound may occur in aqueous solution both as ionized

quinoline (log $K_{ow} = 2.0$ for neutral species, p$K_a = 4.94$ for the protonated species)

and neutral species; thus we must consider simultaneous sorption of both. Recalling

Eq. 11-9, we may write for quinoline:

$$K_d = \frac{[Q]_{om} \cdot f_{om} + [QH^+]_{ie} \cdot \sigma_{ie} \cdot A}{[Q] + [QH^+]} \qquad (11\text{-}103)$$

where

$[Q]_{om}$ is the concentration of quinoline associated with the natural organic matter,
$[Q]_{ie}$ is the concentration of ionized quinoline opposite positions of negative charge on the solid surface,
$[Q]$ is the concentration of the dissolved neutral quinoline species, and
$[QH^+]$ is the concentration of the dissolved protonated quinoline species.

Here we assume sufficient f_{om} so that Q_{min} will be unimportant. We also assume that the structure of this substance does not enable Q_{rxn} to form. To simplify Eq. 11-103 we relate all species to the neutral dissolved concentration of quinoline, $[Q]$. First, using the acid–base equilibrium we express $[QH^+]$ as

$$[QH^+] = [Q][H^+](10^{+4.94}) \qquad (mol \cdot L^{-1}) \qquad (11\text{-}104)$$

Thus, information on solution pH will be necessary to calculate the fraction of quinoline present as charged molecules, and to evaluate the particle surface charge density σ_{ie}. Next, we can relate the concentration of neutral quinoline sorbed to natural organic matter to the dissolved neutral quinoline concentration via Eq. 11-15:

$$[Q]_{om} = K_{om} \cdot [Q] \qquad (mol \cdot kg^{-1}\ om) \qquad (11\text{-}105)$$

On the whole–solid basis, this sorption to the natural organic matter contributes $f_{om} \cdot K_{om} \cdot [Q]$. Using a correlation expression such as that for aromatic hydrocarbons (Eq. 11-22a), we can estimate K_{om} using quinoline's octanol–water partition constant:

$$\log K_{om} = 1.01(\log K_{ow}) - 0.72 \qquad (11\text{-}22a)$$
$$= 1.01(2.0) - 0.72$$
$$= 1.3$$
$$K_{om} = 20$$

Thus, we have for a soil of interest: $[Q_{om}] \approx f_{om} \cdot 20 \cdot [Q]$.

Finally, we must characterize QH^+ sorbed by ion exchange in terms of $[Q]$. For the particular studies of Zachara et al. (1986), the aqueous solution was maintained at 10^{-2} M $CaCl_2$; consequently, we consider the ion exchange reaction in which QH^+ replaces Ca^{2+}:

$$2QH^+ + Ca{:}(surf)_2 \rightleftharpoons 2QH{:}surf + Ca^{2+} \qquad (11\text{-}106)$$

The ion exchange equilibrium expression for this reaction is

$$K_{ie}(\text{mol}\cdot\text{kg}^{-1})\cdot(\text{mol}\cdot\text{L}^{-1})^{-1} = \frac{[\text{QH:surf}]^2[\text{Ca}^{2+}]}{[\text{QH}^+]^2[\text{Ca:(surf)}_2]} \tag{11-107}$$

Note that the reaction stoichiometry gives us a K_{ie} expression somewhat different than that if Na^+ were the competing cation. To eliminate [Ca:(surf)$_2$], we use the fact that the particle surface charge density must equal the accumulated concentrations of counterion sorbates:

$$A\cdot\sigma_{ie}(\text{mol}\cdot\text{kg}^{-1}) = [\text{QH:surf}] + 2[\text{Ca:(surf)}_2] \tag{11-108}$$

Using this charge balance equation in Eq. 11-107, we obtain

$$K_{ie} = \frac{[\text{QH:surf}]^2[\text{Ca}^{2+}]}{[\text{QH}^+]^2(0.5)(A\cdot\sigma_{ie} - [\text{QH:surf}])} \tag{11-109}$$

which on rearrangement yields

$$[\text{QH:surf}] = \frac{(0.5)K_{ie}\cdot\sigma_{ie}\cdot A\cdot[\text{QH}^+]^2}{[\text{QH:surf}][\text{Ca}^{2+}] + (0.5)K_{ie}\cdot[\text{QH}^+]^2} \tag{11-110}$$

We see a result which is similar to Eq. 11-82; but owing to the double charge of calcium ions, this expression is somewhat more complex. At very low concentrations of QH^+ we have $[\text{QH:surf}]\cdot[\text{Ca}^{2+}] \gg (0.5)K_{ie}\cdot[\text{QH}^+]^2$; thus, under these conditions the ratio of bound-to-dissolved QH^+ species is constant:

$$\frac{[\text{QH:surf}]}{[\text{QH}^+]} = \left[\frac{K_{ie}\cdot\sigma_{ie}\cdot A}{2[\text{Ca}^{2+}]}\right]^{0.5} \tag{11-111}$$

On the other hand, at high concentrations of QH^+ we see the sorbed species concentration asymptotically approaches $\sigma_{ie}\cdot A$.

We can also see how sorption via this ion exchange mechanism will be a function of pH. Using Eq. 11-104 to replace $[QH^+]$ in Eq. 11-110, we arrive at

$$[\text{QH:surf}] = \frac{(0.5)K_{ie}\sigma_{ie}A[\text{Q}]^2[\text{H}^+]^2(10^{+4.94})^2}{[\text{QH:surf}][\text{Ca}^{2+}] + (0.5)K_{ie}[\text{Q}]^2[\text{H}^+]^2(10^{+4.94})^2} \tag{11-112}$$

Or at low concentrations of Q, and hence QH^+, we have

$$[\text{QH:surf}] = \left[\frac{K_{ie}\cdot\sigma_{ie}\cdot A}{2[\text{Ca}^{++}]}\right]^{0.5}[\text{Q}][\text{H}^+](10^{+4.94}) \tag{11-113}$$

whereas at high concentrations of [Q] we still approach

$$[\mathrm{QH{:}surf}] = \sigma_{\mathrm{ie}} \cdot A$$

We also note for K_{ie}:

$$K_{\mathrm{ie}} = \frac{[\mathrm{QH{:}surf}]^2[\mathrm{Ca}^{2+}]}{[\mathrm{QH}^+]^2[\mathrm{Ca{:}surf}_2]} \tag{11-107}$$

$$= \frac{K_{\mathrm{d}}^2 \text{ of } \mathrm{QH}^+}{K_{\mathrm{d}} \text{ of } \mathrm{Ca}^{2+}} \tag{11-114}$$

$$= \frac{(e^{-F\psi/RT} e^{-\Delta G_{\mathrm{hydrophob}}/RT})^2}{(e^{-2F\psi/RT})} \tag{11-115}$$

$$\approx (e^{-2\Delta G_{\mathrm{hydrophob}}/RT}) \tag{11-116}$$

This result implies that the exchange of two QH^+ for Ca^{2+} will favor the organic sorbate by an amount related to the *square* of the "desire" of the nonpolar portion of the molecule to escape the bulk water in favor of the near-particle-surface solution. As we saw earlier (Eqs. 11-88 to 11-96), this led us to relate K_{ie} to $[C_{\mathrm{w}}^{\mathrm{sat}}(\mathrm{l, L})]^{-0.2}$ of the comparable hydrocarbon making up the rest of the amphiphile's structure. Using naphthalene's subcooled liquid solubility ($10^{-3.06}$ M) to estimate quinoline's K_{ie}, we find that $[(10^{-3.06})^{-0.2}]^2 \approx 20$.

With these results, we can now rewrite the distribution coefficient expression 11-103 entirely in terms of Q for situations in which there are only low levels of this sorbate:

$$K_{\mathrm{d}} = \frac{f_{\mathrm{om}}K_{\mathrm{om}}[\mathrm{Q}] + (K_{\mathrm{ie}}\,\sigma_{\mathrm{ie}} \cdot A/2[\mathrm{Ca}^{2+}])^{1/2}([\mathrm{H}^+]/K_{\mathrm{a}})[\mathrm{Q}]}{[\mathrm{Q}] + ([\mathrm{H}^+]/K_{\mathrm{a}})[\mathrm{Q}]}$$
$$= \frac{f_{\mathrm{om}}K_{\mathrm{om}} + (K_{\mathrm{ie}}\,\sigma_{\mathrm{ie}} \cdot A/2[\mathrm{Ca}^{2+}])^{1/2}([\mathrm{H}^+]/K_{\mathrm{a}})}{1 + ([\mathrm{H}^+]/K_{\mathrm{a}})} \tag{11-117}$$

At high levels of $[\mathrm{QH}^+]$ (relative to the competing cations), we have

$$K_{\mathrm{d}} = \frac{f_{\mathrm{om}}K_{\mathrm{om}}[\mathrm{Q}] + \sigma_{\mathrm{ie}} \cdot A}{[\mathrm{Q}] + ([\mathrm{H}^+]/K_{\mathrm{a}})[\mathrm{Q}]} \tag{11-118}$$

These mixed-sorption-mechanism K_{d} expressions now allow estimation of the overall extent of sorption in terms of compound properties (K_{om}, K_{ie}, K_{a}), particle properties (f_{om}, σ_{ie}, A), and solution properties ($[\mathrm{Ca}^{2+}]$, pH). They also enable a side-by-side comparison of which species dominate in the particulate and dissolved phases.

Let us now use these results to estimate solid–solution distributions of quinoline.

Zachara et al. (1986) used a B horizon soil with $f_{om} \simeq 0.0048$, CEC $\simeq 8.4 \times 10^{-2}$ mol·kg^{-1}, and 0.01 M $CaCl_2$ background electrolyte to examine quinoline sorption. Assuming the CEC was due to a substantial portion of clay (28% in the particle size distribution), we expect that this parameter is not sensitive to solution pH. To estimate quinoline's K_d for pH 4.2 and again at pH 7.5 for the situation where quinoline is present in very low concentrations, we use Eq. 11-117 and obtain

$$K_d(\mathrm{pH}=4.2) = \frac{(0.0048)(20) + (20 \cdot 8.4 \times 10^{-2}/2 \times 10^{-2})^{1/2}(10^{-4.2}/10^{-4.94})}{1 + (10^{-4.2}/10^{-4.94})}$$

$$= \frac{0.096 + 48}{1 + 5.2} \tag{11-119}$$

$$= 7.8\,(\mathrm{mol \cdot kg^{-1}}) \cdot (\mathrm{mol \cdot L^{-1}})^{-1}$$

and

$$K_d(\mathrm{pH}=7.5) = \frac{(0.0048)(20) + (20 \cdot 8.4 \times 10^{-2}/2 \times 10^{-2})^{1/2}(10^{-7.5}/10^{-4.94})}{1 + (10^{-7.5}/10^{-4.94})}$$

$$= \frac{0.096 + 0.025}{1 + 0.0028} \tag{11-120}$$

$$= 0.12\,(\mathrm{mol \cdot kg^{-1}}) \cdot (\mathrm{mol \cdot L^{-1}})^{-1}$$

Zachara et al. (1986) measured $K_d(\mathrm{pH}=4.2) = 3.5$ and $K_d(\mathrm{pH}=7.5) = 0.75$, so it is obvious that our estimates of some inputs (e.g., σ_{ie} and K_{ie}) may not be accurate. Nonetheless we can get within a factor of a few, and even start to discern which sorption mechanisms are most important. For example, the K_d expressions show that at pH 4.2 most of the sorbed quinoline is accumulated through the ion exchange mechanism (i.e., the second term in the numerator is greatest), while at pH 7.5 similar amounts are sorbed to the natural organic matter and included as counterions in the film of water surrounding the particle surface.

Estimation of the Contribution of Surface Reactions

Organic Sorbate–Natural Organic Matter Reactions Until this point in the discussion, we have focused on cases where we could neglect chemical bond formation between the sorbate and materials in the solid phase. However, at least two kinds of *surface reactions* are known to be important for sorption of some chemicals (referred to as *chemisorption*). First, some organic sorbates can react with organic moieties contained within the natural organic matter of a particulate phase. Especially prominent in this regard are organic bases like substituted anilines (Hsu and Bartha, 1974, 1976). Due to their low pK_a's(~ 5), the aromatic amine functionality is mostly not protonated at natural water pH's. When compounds like 3,3′-dichlorobenzidine are mixed with sediment, they become irretrievable using organic solvents that should remove them from sorbed positions within natural organic matter or using salt solutions that should displace them from ion exchange sites (Appleton et al., 1980). Conditions that promote hydrolysis (see Chapter 12) do release much of these added aniline derivatives. Thus, it appears that reactions between the basic amine and carbonyl functionalities in the natural organic matter explain the strong sorption seen (Stevenson, 1976):

$$X\text{-}C_6H_3(R)\text{-}\ddot{N}H_2 + O{=}C\langle\ \text{natural organic matter} \longrightarrow X\text{-}C_6H_3(R)\text{-}N{=}C\langle\ \text{natural organic matter} + H_2O \tag{11-121}$$

Such reactions often proceed slowly over hours, days, and even years, so the extent of this sorption due to organic chemical:organic chemical reactions is difficult to predict. Furthermore, such bond-forming sorption is sometimes irreversible on the timescales of interest, and we might not wish to include these effects in a K_d expression reflecting sorption equilibrium.

Organic Sorbate–Inorganic Solid Surface Reactions A second type of reaction exhibited by some organic chemicals involves bonding with atoms (e.g., metals) contained on the surface of the solid. Examples of such reactions are shown in Table 11.6. In these cases a hydroxyl bound to a metal in the solid is displaced by the organic sorbate. Given this additional sorption mechanism, the distribution coefficient becomes more complicated:

$$K_{\substack{\text{d, ion exchange}\\ \text{and surface}\\ \text{reaction}}} = \frac{\begin{bmatrix}\text{organic counterion}\\ \text{near the surface}\end{bmatrix} + \begin{bmatrix}\text{organic ion bound}\\ \text{to the surface}\end{bmatrix}}{[\text{organic ion in solution}]} \tag{11-122}$$

Here we assume that we can neglect neutral species, though this may not always be true. We can separate Eq. 11-122 into parts:

$$K_{\substack{\text{d, ion exchange}\\ \text{and surface}\\ \text{reaction}}} = \frac{[\text{organic counterion}]}{[\text{organic ion in solution}]} + \frac{[\text{organic ion bound to surface}]}{[\text{organic ion in solution}]} \tag{11-123}$$

and use a previous result (e.g., Eq. 11-83) to write

$$K_{\substack{\text{d, ion exchange}\\ \text{and surface}\\ \text{reaction}}} = \frac{\sigma_{ie} \cdot A \cdot K_{ie}}{[\text{competing ion}] + K_{ie}\,[\text{organic ion in solution}]} + \frac{[\text{organic ion bound to surface}]}{[\text{organic ion in solution}]} \tag{11-124}$$

TABLE 11.6 Examples of Organic Sorbates Reacting with Mineral Surfaces[a]

Surface		Sorbate		Surface complex		Ref.[b]
$\equiv$Fe—OH	+	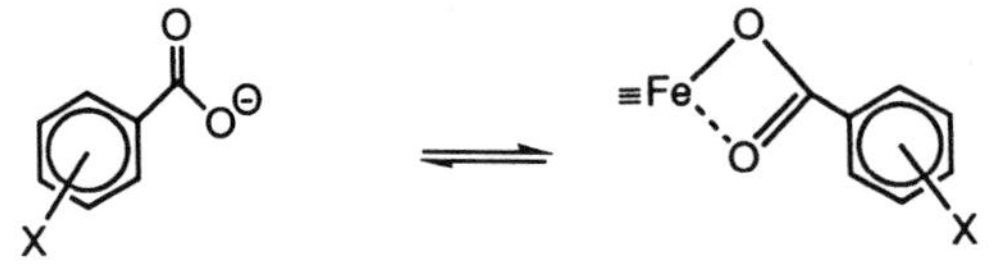 Substituted Benzoates	$\rightleftharpoons$		$+OH^-$	1
$\equiv$Fe—OH	+	 Salicylate	$\rightleftharpoons$		$+OH^-$	2, 3
$\equiv$Fe—OH	+	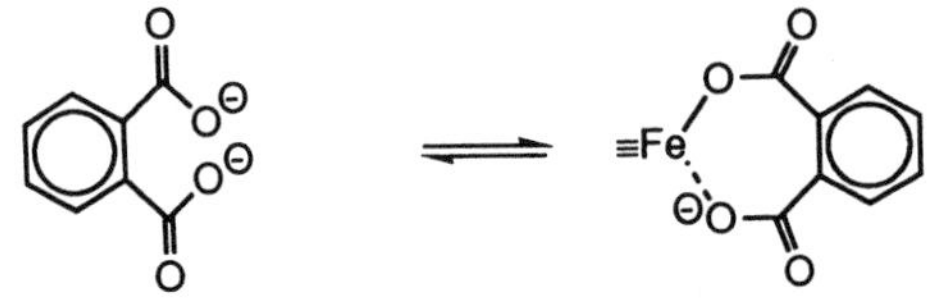 *o*-Phthalate	$\rightleftharpoons$		$+OH^-$	2, 4
$\equiv$Al—OH	+	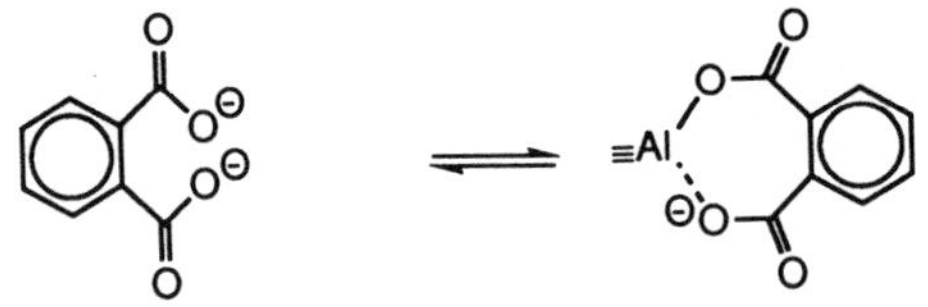 *o*-Phthalate	$\rightleftharpoons$		$+OH^-$	5
$\equiv$Al—OH	+	 Salicylate	$\rightleftharpoons$		$+OH^-$	5

[a]Only limited information is available regarding the bonding of species to water-wet surfaces; thus the bonding of the sorbates shown here is conjecture.

[b]1. Kung and McBride, 1989. 2. Balistrieri and Murray, 1987. 3. Yost et al., 1990. 4. Lövgren, 1991. 5. Stumm et al., 1980.

Now our task is to develop an expression to predict the last term. To do this, we begin by writing the reaction involved:

$$\text{R:surface} + \text{L—M}{\equiv}\text{surface} \rightleftharpoons \text{R—M}{\equiv}\text{surface} + \text{L:surface} \qquad (11\text{-}125)$$

where R—M≡surface and L—M≡surface are an organic compound and an inorganic ligand like —OH bonded to the solid as indicated by the hyphen. The ions

R:surface and L:surface are present at the concentrations in the immediate vicinity of the reaction site, that is, in the diffuse double layer (Region II in Fig. 11.18). Such a reaction reflects a free energy change that we will refer to as ΔG_{rxn} and a corresponding equilibrium expression:

$$K_{rxn} = \frac{[\text{R—M}{\equiv}\text{surface}][\text{L:surface}]}{[\text{R:surface}][\text{L—M}{\equiv}\text{surface}]} \tag{11-126}$$

If we can assume that there are a finite number of reactive sites on the solid, σ_{rxn} ($\text{mol}\cdot\text{m}^{-2}$), then we have

$$A\cdot\sigma_{rxn} = [\text{R—M}{\equiv}\text{surface}] + [\text{L—M}{\equiv}\text{surface}] \tag{11-127}$$

with A equal to the specific particle surface area (meter squares per kilogram). Therefore, we can rewrite Eq. 11-126:

$$K_{rxn} = \frac{[\text{R—M}{\equiv}\text{surface}][\text{L:surface}]}{[\text{R:surface}](A\cdot\sigma_{rxn} - [\text{R—M}{\equiv}\text{surface}])} \tag{11-128}$$

We also recall from Eqs. 11-86 and 11-87 that the concentrations of ions in the layer of water next to the particle surface can be related to the corresponding species in the bulk solution:

$$[\text{L:surface}] = [\text{L}^-]_{bulk}\cdot e^{-\Delta G_{elect}/RT} \tag{11-129}$$

and

$$[\text{R:surface}] = [\text{R}^-]_{bulk}\cdot e^{-\Delta G_{elect}/RT}\cdot e^{-\Delta G_{hydrophob}/RT} \tag{11-130}$$

Using these relations in Eq. 11-128, along with Eq. 11-88, we have

$$\begin{aligned} K_{rxn} &= \frac{[\text{R—M}{\equiv}\text{surface}][\text{L}^-]_{bulk}e^{-\Delta G_{elect}/RT}}{(A\cdot\sigma_{rxn} - [\text{R—M}{\equiv}\text{surface}])[\text{R}^-]_{bulk}e^{-\Delta G_{elect}/RT}e^{-\Delta G_{hydrophob}/RT}} \\ &= \frac{[\text{R—M}{\equiv}\text{surface}][\text{L}^-]_{bulk}}{(A\cdot\sigma_{rxn} - [\text{R—M}{\equiv}\text{surface}])[\text{R}^-]_{bulk}K_{ie}} \end{aligned} \tag{11-131}$$

Simplifying and rearranging, we then find:

$$[\text{R—M}{\equiv}\text{surface}] = \frac{\sigma_{rxn}\cdot A\cdot K_{ie}\cdot K_{rxn}\cdot[\text{R}^-]_{bulk}}{[\text{L}^-]_{bulk} + K_{ie}\cdot K_{rxn}\cdot[\text{R}^-]_{bulk}}$$

Thus another Langmuir isotherm is expected with the maximum bound concentrations equal to $\sigma_{rxn}\cdot A$ and the $K_{Langmuir}$ given by $K_{rxn}\cdot K_{ie}\cdot[\text{L}^-]_{bulk}^{-1}$. This result is very similar in form to that seen before for an ion exchange process (Eq. 11-83). Returning to our overall K_d expression (Eq. 11-124), we can now write

$$K_{\substack{\text{d, ion exchange}\\ \text{and surface}\\ \text{reaction}}} = \frac{\sigma_{\text{ie}} \cdot A \cdot K_{\text{ie}}}{[\text{competing ion}] + K_{\text{ie}}[\text{organic ion}]} + \frac{\sigma_{\text{rxn}} \cdot A \cdot K_{\text{ie}} \cdot K_{\text{rxn}}}{[\text{competing ligand}] + K_{\text{ie}} \cdot K_{\text{rxn}} \cdot [\text{organic ion}]} \quad (11\text{-}132)$$

As for nonreacting organic ions, we need information on the ion exchange tendency of the chemical of interest (K_{ie} or $\Delta G_{\text{hydrophob}}$); now we also need a means to assess K_{rxn}.

We can evaluate K_{rxn} recognizing that the tendency to form chemical linkages to solid surface atoms correlates with the likelihood of forming comparable complexes in solution (Stumm et al., 1980; Schindler and Stumm, 1987; Dzombak and Morel, 1990). That is, the free energy change associated with the exchange shown by Eq. 11-125 appears energetically similar to that for a process occurring between two dissolved components:

$$\text{M—L}^{+z} + \text{R}^- \rightleftharpoons \text{M—R}^{+z} + \text{L}^- \quad (11\text{-}133)$$

where z + 1 would be the charge of the free metal in aqueous solution. This entirely solution-phase exchange reaction is characterized by an equilibrium constant:

$$K_{\substack{\text{ligand exchange}\\ \text{in solution}}} = \frac{[\text{M—R}^{+z}][\text{L}^-]}{[\text{M—L}^{+z}][\text{R}^-]} \quad (11\text{-}134)$$

A substantial data base is available to quantify such solution equilibria (e.g., Martell and Smith, 1977; Morel, 1983). Let us examine a specific case. For example, we might be interested in the replacement of a bicarbonate ligand (HCO_3^-) by acetate ($CH_3CO_2^-$) on a calcite ($CaCO_3$) surface:

$$\underset{\substack{\text{calcite}\\ \text{surface}}}{\text{surface}{\equiv}\text{Ca—O}\overset{\overset{\text{O}}{\|}}{\text{C}}\text{OH}} + \text{CH}_3\text{CO}_2\text{:surface} \rightleftharpoons \text{surface}{\equiv}\text{Ca—O}\overset{\overset{\text{O}}{\|}}{\text{C}}\text{CH}_3 + \text{HCO}_3\text{:surface} \quad (11\text{-}135)$$

This surface reaction of acetate enables an adsorbed species to be formed which is different from acetate ions simply contained within the electric double layer. To assess this surface reaction, we examine the comparable solution case:

$$\text{Ca—O}\overset{\overset{\text{O}}{\|}}{\text{C}}\text{OH}^+ + \text{CH}_3\overset{\overset{\text{O}}{\|}}{\text{C}}\text{O}^- \rightleftharpoons \text{Ca—O}\overset{\overset{\text{O}}{\|}}{\text{C}}\text{CH}_3{}^+ + \text{HCO}_3{}^- \quad (11\text{-}136)$$

For the case of acetate exchange with bicarbonate in aqueous solution, we can find for aqueous solution (Morel, 1983):

$$\mathrm{CaO\overset{\overset{\displaystyle O}{\|}}{C}OH^{+} \rightleftharpoons Ca^{2+} + HCO_3^{-}} \qquad K = 10^{-1.26} \qquad (11\text{-}137)$$

$$\mathrm{Ca^{2+} + CH_3\overset{\overset{\displaystyle O}{\|}}{C}O^{-} \rightleftharpoons CaO\overset{\overset{\displaystyle O}{\|}}{C}OH^{+}} \qquad K = 10^{1.20} \qquad (11\text{-}138)$$

and combining:

$$\mathrm{CaO\overset{\overset{\displaystyle O}{\|}}{C}OH^{+} + CH_3\overset{\overset{\displaystyle O}{\|}}{C}O^{-} \rightleftharpoons CaO\overset{\overset{\displaystyle O}{\|}}{C}CH_3{}^{+} + HCO_3^{-}} \qquad K = 10^{-0.06} \qquad (11\text{-}139)$$

This result for the overall solution equilibrium implies that the free energy change of this particular ligand exchange in solution is near zero ($\Delta G_{rxn} = -RT \ln 10^{-0.06} = 0.3\,\mathrm{kJ \cdot mol^{-1}}$). This may not be too surprising in light of the chemical similarity of bicarbonate and carboxylate anions. Further, it seems very likely that other longer-chain carboxylic acids (i.e., fatty acids) will also exhibit the similar solution-phase reactivity with calcite.

Now we make use of the very important idea that such solution-phase equilibria can be used to estimate ΔG_{rxn} for comparable surface ligand exchanges. For the specific case of acetate binding to calcite, we would estimate

$$\Delta G_{rxn} \simeq \Delta G_{\substack{\text{ligand}\\ \text{exchange}\\ \text{in solution}}} = 0.3\,\mathrm{kJ \cdot mol^{-1}} \qquad (11\text{-}140)$$

Therefore, we estimate that K_{rxn} is 0.9 and use it in Eq. 11-132 for calculating K_d for the sorption of acetate on calcite.

The procedure for other charged organic chemicals is analogous; and by using the results in Eq. 11-132, we begin to build an overall estimate of charged organic chemical sorption to minerals. Generally, it seems that the tendencies of monodentate organic ligands (e.g., $RCOO^-$, ϕOH, $R\ddot{N}H_2$) to diplace inorganic ligands at mineral surfaces is not very great, and it may be reasonable to neglect such surface-bound species when considering the overall K_d of such simple organic sorbates. However,

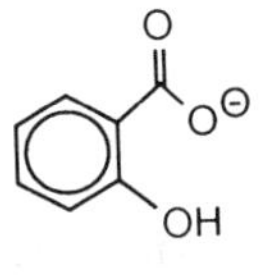

Salicylate

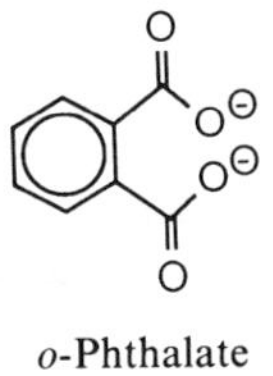

o-Phthalate

since even a small degree of adsorption can be important to the rate of heterogeneous transformations (e.g., Ulrich and Stone, 1989), in some cases we may need to deal with this. Other charged organic sorbates like salicylate or *o*-phthalate, which may form two bonds with the mineral, are much more likely to exhibit significant ΔG_{rxn} contributions to the overall ΔG_{ads} (Schindler and Stumm, 1987).

Summary Comments Regarding Sorption of Charged Organic Compounds

Before we conclude, we should note that the simple K_d formulations used here may not reflect all the possible mechanisms of charged organic chemical sorption. Frequently we assume sufficient f_{om} to neglect associations of the neutral organic species with mineral surfaces. Similarly, we presumed that charged organic molecules would not "dissolve" into natural organic matter, and Schellenberg et al. (1984) observed sorption of trichlorophenolates consistent with this picture. However, these workers also found that pentachlorophenolate exhibited K_d's far in excess of expectations. Further studies (Westall et al., 1985) indicated that this may be due to the sorptive uptake of the potassium phenolate complex or ion pair

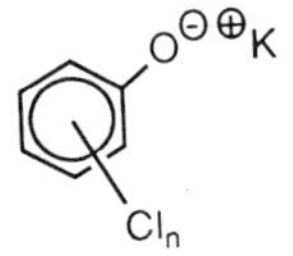

into the natural organic matter. If true, this would require the inclusion of more species in the initial K_d formulation and also thermodynamic expressions relating these species to the phenolate dissolved. In another case, Jung et al. (1987) interpreted oleic acid adsorption to iron oxides as including some role of a neutral oleic acid:oleate complex:

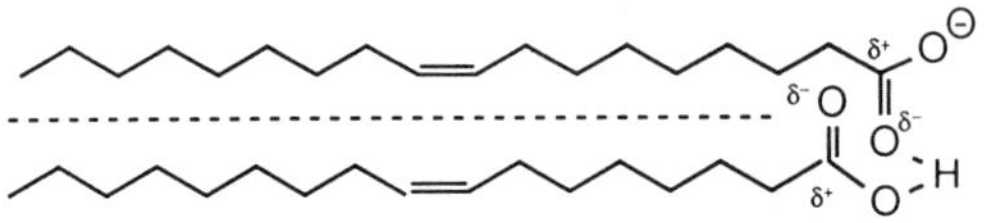

This is reminiscent of the hemimicelle cases we discussed, where now substantial aggregation in solution occurs for only pairs of molecules. Under solution conditions where these "new" species arise, obviously we must (1) adjust the starting K_d expressions to include them, and (2) utilize equilibrium information relating them to the nonassociated species.

11.7 SORPTION KINETICS

Thus far we have focused on situations in which the time was sufficient to allow solid–water exchange of chemicals to achieve equilibrium. However, sometimes we are concerned with cases where the solids do not remain in contact with the solutions of interest for long. Examples of such short-contact situations include storm-related sediment resuspensions or soil erosion events where, after calm is restored, the particles quickly settle to the sediment bed. In another case, zooplankton fecal pellets of approximately 100 μm diameter fall through a body of water at about 100 m/day, conceivably too fast to permit sorbates to establish sorption equilibrium with each depth interval of the surrounding water. Infiltrating groundwater may experience only brief contact with the soil grains by which these solutions pass. Even groundwater moving at natural slow flow rates may not always expose the aquifer solids long enough to permit complete sorption equilibrium (Roberts et al., 1986; Ball and Roberts, 1991). In all of these cases, certain molecules may simply not have enough time to exchange before the solution or the solid moves away from the other phase. Sometimes slow desorption also limits transformations like biodegradation (Rijnaarts et al., 1990).

Two types of processes could act as the bottleneck, inhibiting sorptive equilibrium. First, chemical reactions (*chemisorption*) between the sorbate and the point of association in the particles might limit the overall approach to sorptive equilibrium. An example is salicylate sorption to very small alumina particles (Kummert and Stumm, 1980). Most of the sorbed salicylate is thought to be chemically bound to the alumina surface:

$$\text{salicylate (O}^{\ominus}\text{, OH)} + HO-Al\equiv \rightleftharpoons \text{salicylate–Al surface complex} + H_2O \qquad (11\text{-}141)$$

During the short interval (< 2 h) after initial mixing of the salicylate with the alumina, dissolved salicylate concentration drops quickly and then decreases more gradually (Fig. 11.28*a*). Apparently, the rate of salicylate reaction with surface sites on the exterior of the particles is controlling the rate of overall sorption during the early portion of this timecourse. Another example of chemisorption might be slow condensation reactions of organic amines with carbonyl moieties of the solid phase natural organic matter. To quantify the kinetics of these particular chemisorption processes, one needs to deal with the rates of the individual reactions involved.

The second type of limitation occurs when the sorbate molecules do not physically have enough time to move to all the points of contact in and on the solids where they would become associated. We often refer to this physical limitation as a *mass transfer limited process*. For example, we can imagine charged sorbate molecules needing time to diffuse into a clayey floc before they can associate with oppositely charged surface sites in the interior (Fig. 11.28*b*). Similarly, we can easily see that some of the natural organic matter that absorbs nonpolar compounds may be located at somewhat inaccessible positions within silty aggregates (Fig. 11.28*c*). Even in the case of ligand exchange, as for salicylate ions, diffusion into porous alumina particles where there are more reactive sites can be slow (Fig. 11.28*a* for times greater than a few hours). Since sorptive equilibrium is only reached when each subpart of a solid has accumulated enough sorbate to itself be equilibrated with the solution on the outside, we can

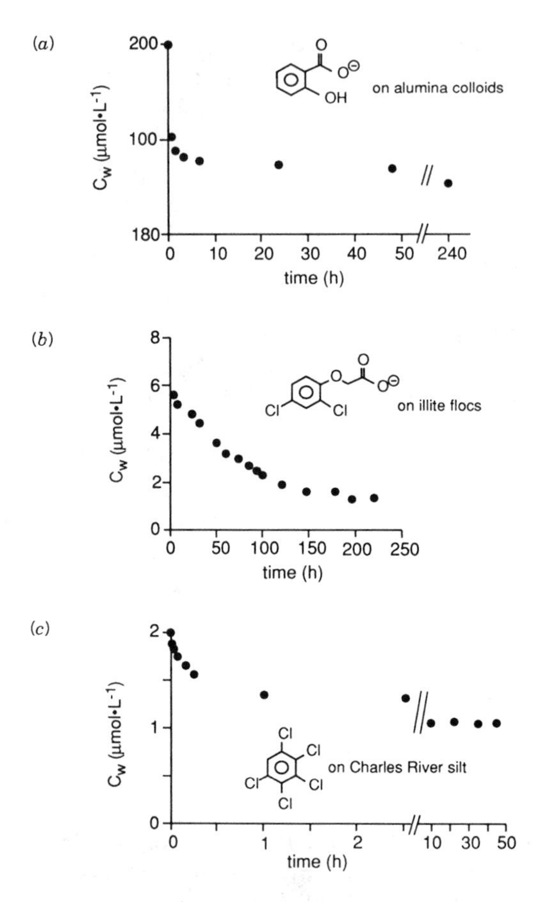

Figure 11.28 Some observed time courses of organic compounds associating with suspended solids. (*a*) Salicylate sorption to 20 nm γ-Al_2O_3 (data from Kummert and Stumm, 1980). (*b*) 2,4-D sorption to clay flocs (data from Haque et al., 1968). (*c*) Pentachlorobenzene sorption to silt-sized river sediment (data from Wu and Gschwend, 1986).

recognize that simply delivering molecules to all the internal solid "binding" sites must take a finite amount of time. Once the molecules have arrived at the particulate locations where they will be bound, then interactions may occur on the molecular scale. If the average time spent arriving at these local points of association is long compared to the time required to make these molecule-scale attachments, then we must focus on the rate of molecular penetration to the binding sites to describe the overall sorption kinetics. Many reports suggest that the molecule-scale sorbate–sorbent association is fast relative to diffusion within porous solids (e.g., Helfferich, 1962, or Adamson, 1982, regarding ion exchange; e.g., Brusseau and Rao, 1989 review of nonpolar organic chemical sorption). Therefore, in the following discussions, we focus on cases of mass transfer limitations to solid–water exchange and the resultant approach one may use to estimate overall sorption rates.

Sorption Kinetics for Mass Transfer Limited Mechanisms

We begin by noting that soil and sediment particles are often actually present as *aggregates* of individual solid phases (Fig. 11.29). Not only does this refer to the coalescence of many fine grains such as those seen in flocculated clay aggregates, but may also include parent mineral grains coated with other phases like iron oxides and natural organic matter. The resultant solids are then composed of subregions with differential capabilities to sorb organic compounds. A key exception to this extremely porous visualization involves sands or larger rocks which consist of individual grains (e.g., quartz). Even such "single solid" natural particles contain some micropores (e.g., Wood et al., 1990; Ball et al., 1990), and passing through these pores sometimes appears to be critical for accessing the quantitatively most important sorption sites.

For molecules in solution to distribute themselves between the dissolved and sorbed phases, several steps in series are required. First, there may be a mass transfer limitation due to diffusion across a poorly mixed water layer surrounding the outside of each particle (Fig. 11.29). The timescales of such external diffusion are probably on the order of seconds in most turbulent situations. Using Eq. 9-31, we can get an estimate of this timescale:

$$\tau \sim \frac{(\text{thickness of boundary layer surrounding aggregate})^2}{2D_w} \tag{11-142}$$

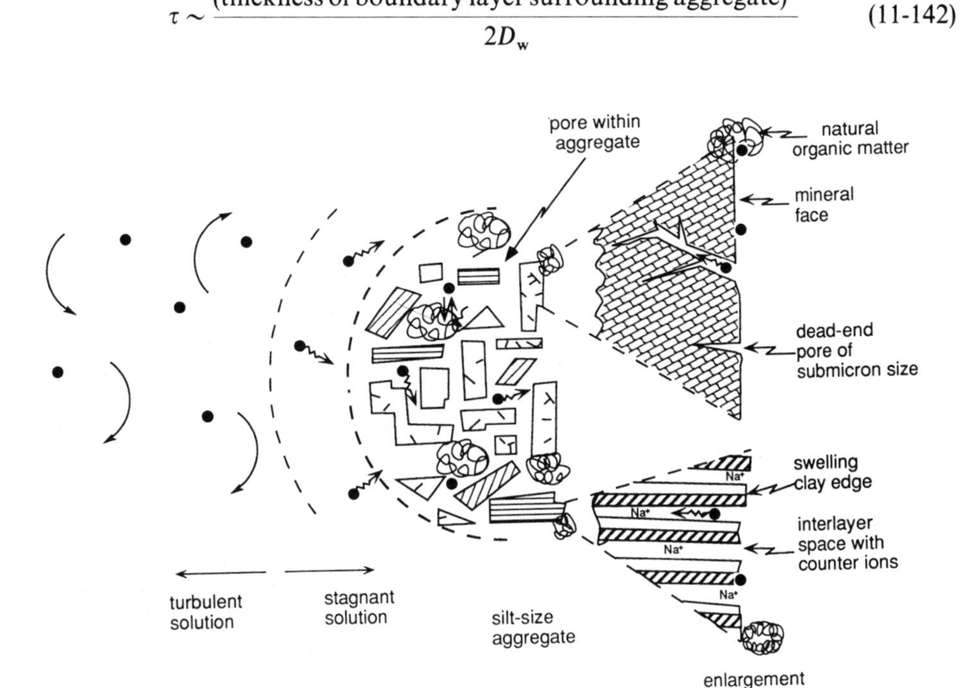

Figure 11.29 Conceptualization of the sequence of steps required to move sorbate molecules (pictured as black dots) between a turbulent solution and all the sorptive sites within a typical soil or sediment aggregate.

For relatively large particles ($>200\,\mu m$), like sands whose surfaces are large enough to impart significant frictional interaction with the fluids, such a boundary layer thickness is probably on the order of $200\,\mu m$ in thickness. For smaller particles ($\lesssim 200\,\mu m$), the diffusion limited zone shrinks in proportion to the particle size. Thus $200\,\mu m$ might typify the largest boundary thickness we encounter. Together with a reasonable value of D_w (see Fig. 9.7), we find that external diffusion requires on the order of

$$\tau \sim \frac{(2 \times 10^{-2}\,\text{cm})^2}{2(10^{-5}\,\text{cm}^2/\text{s})} \sim 20\,\text{s} \tag{11-143}$$

Obviously, if the solution outside the particles is mixed at all, there will not be much time spent by molecules reaching the aggregate exterior. In nonturbulent solutions like groundwater or sediment porewaters, the distance between grain surfaces is probably about the same as the grain size. Consequently, use of an equation analogous to Eq. 11-142 would show a timescales of diffusion in the intergranular solutions of only seconds to minutes at most. If the primary site of attachment is on the exterior surface of the particles, then the overall sorption process will be characterized by this external diffusion time plus that required to complete the local binding reaction. This is probably the appropriate image of what is controlling sorption of salicylate to alumina at early times, as we discussed above (Fig. 11.28*a*).

For many natural particles of interest (e.g., soils, suspended solids), the sorbates must continue their transport to their ultimate points of attachment by diffusing in the *immobile* fluids filling the interstices of the aggregates (Fig. 11.29; Rao et al., 1980, 1982; Wu and Gschwend, 1986, 1988; Weber et al., 1991). Generally, such pore spaces are large relative to the molecules themselves (i.e., these pores are probably about the same size as the parent particles of which the aggregate is made); thus diffusion in these spaces occurs somewhat like molecular diffusion in aqueous solution. An exception may be that some pore spaces are blocked by natural organic matter, and molecular transfer to points deeper in the aggregate interior or even deep inside the natural organic matter may require diffusion through these organic polymers (Brusseau and Rao, 1989; Brusseau et al., 1991; Brusseau and Rao, 1991). If the majority of solid sorbent is accessed after diffusion through "large pores", then the overall process may be effectively described by focusing on this step.

Other times, the bulk of the sorption occurs at quite inaccessible positions such as in *dead-end pores* or between the layers of aluminosilicate minerals (Ball and Roberts, 1991). These channels differ from those of the aggregate in that they exhibit openings that are comparable in size to the sorbate molecules of interest. This results in a *steric limitation* to diffusive transport. Also, these channels are probably not as extensively interconnected (hence the "dead-end" reference) as those between the grains making up aggregates. This also contributes to inhibiting exchange with the exterior. When the sorption sites of interest involve these kinds of solid positions, then we may consider the mass transfer to involve diffusion coefficients that are greatly reduced relative to the free solution values. Various reports of extremely slow desorption kinetics (decade timescales) such as those seen for residues of the soil fumigant, ethylene dibromide, appear to be best explained by release from such poorly accessible nanometer-sized pores (Steinberg et al., 1987).

Modeling Mass Transfer Limited Sorption

Since sorption kinetics experiments usually reveal an early period of extensive exchange, followed by a prolonged time of slowly proceeding uptake (or release) by particles (Leenheer and Ahlrichs, 1971; Connolly, 1980; Karickhoff, 1980; Wu and Gschwend, 1986; Brusseau and Rao, 1989; Ball and Roberts, 1991), the sorption process has been characterized with a two-box model (Karickhoff, 1980). In this conceptualization, one portion of the solid phase is taken to be rapidly equilibrated with the solution phase; the other part of the solid is described using a rate equation. One can easily extend this idea by imagining an infinite series of "boxes" in the particles, each in successively less direct contact with the exterior solution. That is the approach used here. Thus our problem entails diffusing sorbate molecules *radially* via the interconnecting porewater channels into soil or sediment aggregates. Such a picture may be well suited to cases in which most of the organic sorbate of concern will eventually be associated with relatively accessible solid phases like natural organic matter. Although real-world particles are not spherical, and the points of sorptive attachment may not be spread evenly throughout the natural particle aggregate, these characteristics are reasonable approximations.

To describe this transfer of sorbate molecules into or out of such aggregates, we first consider a single porous sphere of radius R (Fig. 11.30). The solute concentration

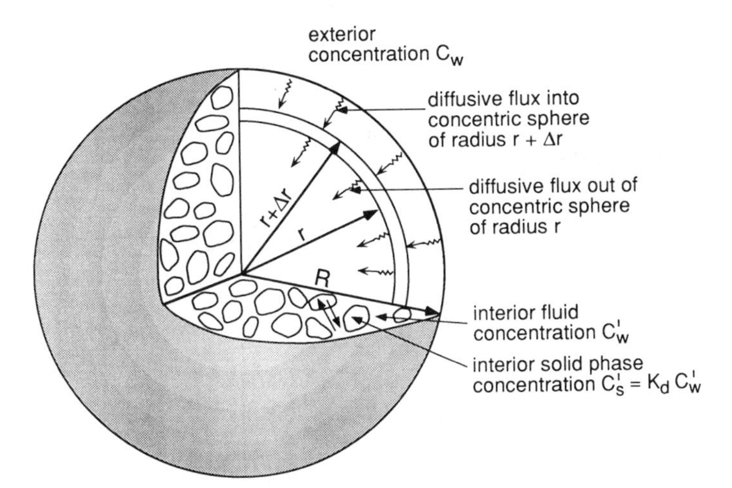

Figure 11.30 Diffusion-limited transfer into an idealized soil or sediment aggregate of radius R and interior concentrations $C'_w(r,t)$ in the water within the aggregate. Sorption is modeled using a mass balance equation for any concentric shell of thickness Δr (see Eq. 11-153 in text).

inside the aggregate pores, C'_w, is always at equilibrium with the adjacent sorbed phase, C'_s:

$$C'_s = K^*_d\, C'_w \tag{11-144}$$

where K^*_d is the *in situ* (microscopic) distribution coefficient. Ultimately, we are interested in the "macroscopic" distribution coefficient that relates the (macroscopic) sorbed concentration, C_s, to the dissolved concentration in the exterior water, C_w:

$$C_s(t) = K_d(t) \cdot C_w \tag{11-145}$$

where, owing to finite kinetics of macroscopic sorption, K_d is time dependent.

Note that C_s is determined as the total mass of the compound in the dried particle divided by the dry particle mass. Thus, C_s includes the dissolved fraction of the chemical in the porewater when the particles are dried. Due to the spherical symmetry, the evolving concentration distributions only depend on the radial distance r and on time t. The total concentration in the particle at distance r from the center, C_{tot}, on a particle volume basis is given by

$$C_{tot}(r) = \rho_s(1 - \phi)C'_s(r) + \phi C'_w(r) \qquad (\text{mol} \cdot \text{L}^{-1}) \tag{11-146}$$

where ρ_s is the density of (dry) solid (kg·L^{-1} solid) and ϕ is the porosity of particle aggregate (L water·L^{-1} total).

Since C'_s and C'_w are related by Eq. 11-144, we get

$$C_{tot}(r) = [K^*_d \cdot (1 - \phi) \cdot \rho_s + \phi] \cdot C'_w(r) \tag{11-147}$$

When K^*_d, ρ_s, and ϕ are constant everywhere within the particle aggregate, the mean total concentration in the particle is

$$\bar{C}_{tot} = [K^*_d \cdot (1 - \phi) \cdot \rho_s + \phi] \cdot \bar{C}'_w \tag{11-148}$$

where

$$\bar{C}'_w = \frac{\int_0^R 4\pi r^2 \cdot C'_w(r)dr}{4\pi R^3/3} = \frac{3}{R^3}\int_0^R r^2 C'_w(r)dr \tag{11-149}$$

is the volume-weighted mean of the radial concentration field $C'_w(r)$. The (macroscopic) sorbed concentration on a particle mass basis now can be calculated:

$$C_s = \frac{\bar{C}_{tot}}{\rho_{bulk}} = \frac{\bar{C}_{tot}}{(1-\phi)\rho_s} = \left[K_d^* + \frac{\phi}{(1-\phi)\rho_s}\right]\bar{C}'_w \tag{11-150}$$

where C_s is a function of time like $\bar{C}'_w$.

To make the last step to Eq. 11-145, we have to relate the time-dependent mean concentration $\bar{C}'_w$ to the external solute concentration C_w. If the sorbate molecules only diffuse in the pores, we recognize that the change in total sorbate concentration $\bar{C}_{tot}$ is controlled by diffusive uptake through the pore fluids driven by gradients in the internal sorbate concentration C'_w. Thus, we consider our single porous sphere of radius R to have sorbate molecules diffusing through successive concentric layers (Fig. 11.30). Diffusion into or out of the sphere begins at $t = 0$, and it occurs because of a difference in concentrations between the inside porewater and the outside bulk solution. If we consider a mass balance equation for any thin concentric shell within the spherical aggregate, bounded by spheres with radius r and $r + \Delta r$, we may write

$$\frac{\partial C_{tot}}{\partial t} = \frac{\text{Flux at inner sphere} - \text{Flux at outer sphere}}{\text{shell volume}} \tag{11-151}$$

Using Eq. 9-2, multiplied by the areas of the relevant shells, to quantify the fluxes in Eq. 11-151, and assuming that dissolved concentration changes along the radial directions are responsible for diffusive transport, we have

$$\begin{aligned}\frac{\partial C_{tot}}{\partial t} &= \frac{-\left(\phi\cdot D\cdot 4\pi r^2\cdot\frac{\partial}{\partial r}(C'_w)\right) + (\phi\cdot D\cdot 4\pi(r+\Delta r)^2)\cdot\frac{\partial}{\partial r}\left(C'_w + \Delta r\frac{\partial C'_w}{\partial r}\right)}{(4/3)\pi(r+\Delta r)^3 - (4/3)\pi r^3}\\ &= \frac{-4\pi D\phi(r)^2\cdot\left(\frac{\partial C'_w}{\partial r}\right) + 4\pi D\phi(r+\Delta r)^2\left(\frac{\partial C'_w}{\partial r} + \Delta r\frac{\partial^2 C'_w}{\partial r^2}\right)}{(4/3)\pi(r+\Delta r)^3 - (4/3)\pi r^3}\end{aligned} \tag{11-152}$$

where we have estimated the concentration gradient at the radius $r + \Delta r$ to be approximated by the linear expansion, $C'_w(r+\Delta r) \simeq C'_w(r) + \Delta r\frac{\partial C'_w}{\partial r}$. We have also reduced the fluxes in proportion to the aggregate porosity ϕ to account for the diminished cross-sectional area available for diffusion. Neglecting all terms of order $(\Delta r)^2$ or $(\Delta r)^3$, we can simplify to find

$$\frac{\partial C_{tot}}{\partial t} = \phi D\left(\frac{\partial^2 C'_w}{\partial r^2} + \frac{2}{r}\frac{\partial C'_w}{\partial r}\right) \tag{11-153}$$

To use this equation we need to consider what factors dictate the magnitude of D. First, of course, we know from our physical conceptualization that we are interested in molecules diffusing in aqueous solutions in the aggregate pores; hence D should be closely related to D_w, the molecular diffusivity in water. Unlike diffusion in water though, the sorbate molecules must navigate around the various fine grains making up the skeleton of our natural particle aggregate. This elongated path issue is treated by using a tortuosity factor f which reduces D to some value below D_w. If the size of the pores is small enough to be comparable to the diffusing molecules themselves, we also need to reduce D_w again using what is called a constrictivity factor (Satterfield et al., 1973; Ball and Roberts, 1991). It is also possible that diffusion occurs, not only in the water filling the pores, but also on the walls of the component particles in a two-dimensional process called surface diffusion. Here we neglect this process because of the discontinuous nature of the aggregates.

Returning to Eq. 11-153, C_{tot} can be replaced by C'_w using Eq. 11-147. Note again that ϕ, K_d^*, and ρ_s are assumed to be spatially constant. Thus,

$$\frac{\partial C'_w}{\partial t} = D^* \left[\frac{\partial^2 C'_w}{\partial r^2} + \frac{2}{r} \frac{\partial C'_w}{\partial r} \right] \tag{11-154}$$

with

$$D^* = \frac{\phi f D_w}{K_d^*(1 - \phi)\rho_s + \phi} \tag{11-155}$$

D^* is sometimes referred to as the effective diffusivity.

The time-dependent solution for C'_w which satisfies Eq. 11-154 is an infinite sum of exponential terms (Crank, 1975). If we consider the relative approach to equilibrium using the total mass of sorbate inside the solid phase at time t, $\bar{C}_{tot}(t)$, versus the final mass accumulated at steady-state, $\bar{C}_{tot,\infty}$, then the ratio of these sorbed phase loadings, $\bar{C}_{tot}(t)/\bar{C}_{tot,\infty}$, depends on a nondimensional parameter, $D^* \cdot t/R^2$:

$$\frac{\bar{C}_{tot}(t)}{\bar{C}_{tot,\infty}} = F\left(\frac{D^* t}{R^2}\right) \tag{11-156}$$

where F is a monotonically increasing function with $F(0) = 0$ and $F(1) = 1$. The larger is the sorbate's effective diffusivity, the faster will be the approach to equilibrium. Conversely, the larger the particle, the slower will be the overall sorptive equilibrium since diffusion pathlengths are longer.

We can picture the corresponding concentration profiles (Fig. 11.31) for the case where the exterior solution exhibits a constant concentration C_w and the molecules are diffusing into a spherical aggregate of interest. Upon initial mixing, the outermost portion of the sphere equilibrates with the solution so that this part of the sorbent exhibits sorbed concentration $K_d \cdot C_w$. Subsequently, more interior portions of the sphere accumulate the sorbate, and it is these integrated concentrations, weighted for

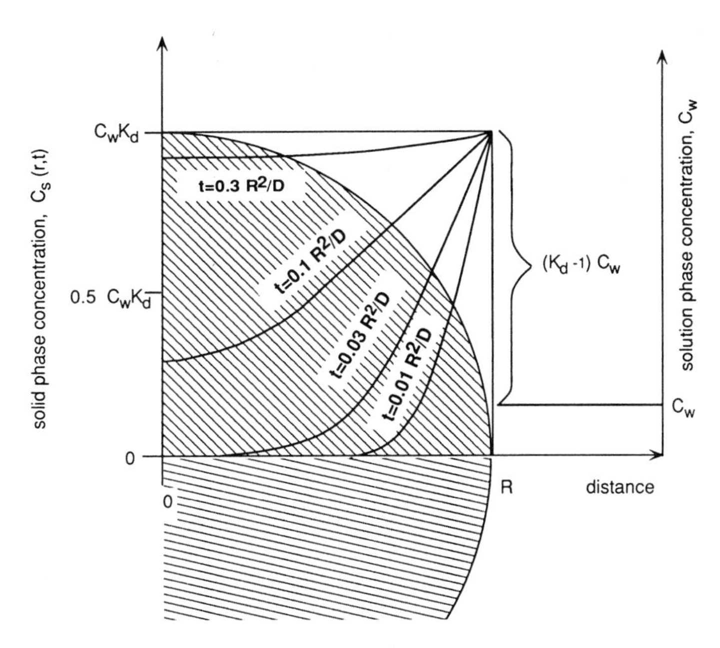

Figure 11.31 Concentration profiles resulting from the diffusion of a sorbate into a porous particle from a solution of constant concentration C_w. Successive profiles are characterized by times given as a function of R, the particle radius, and D^*, the effective diffusivity of the sorbate moving into the particle.

the concentric volumes they occupy, that correspond to the time-varying sorbed load. When $t \approx 0.03\, R^2/D^*$, the total sorbed load is about 50% of its equilibrium value, and when $t \approx 0.3\, R^2/D^*$, the process is 97% complete. Put another way, the solid phase is virtually "filled up" with sorbate (Fig. 11.31), and we might define an approximate *sorption rate constant for this case of constant exterior solution concentration:*

$$k_{\text{sorb}}(C_w = \text{constant}) \simeq \frac{\ln 2}{t_{50\%}} \simeq \frac{0.69\, D^*}{0.03\, R^2} \simeq 23\, D^*/R^2 \qquad [\text{T}^{-1}] \qquad (11\text{-}157)$$

Use of such an approximate first-order rate constant yields a result similar to that obtained from the complete solution (Crank, 1975), but it underpredicts sorptive exchange at short times and overpredicts sorption at long times (Fig. 11.32).

In the cases where C_w does not remain constant in the exterior solution (because a significant portion of the dissolved load is sorbed), the expression for k_{sorb} becomes a little more complicated (Crank, 1975; Wu and Gschwend, 1988). Figure 11.33 shows how the progress of sorptive exchange proceeds for several different values of $K_d \cdot r_{sw}$ (which characterizes the proportion of the total chemical in the system eventually sorbed). As $K_d \cdot r_{sw}$ increases from the infinite bath case (i.e., $C_w = \text{constant}$), we see that the times required to reach equilibrium decrease. For cases of sorptive uptake,

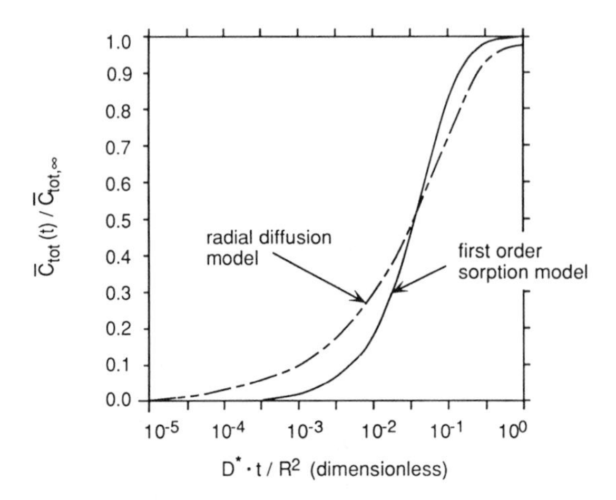

Figure 11.32 Comparison of the exact solution to the radial diffusion model of sorptive exchange from Crank (1975) with the first-order model approximation. $\bar{C}_{tot}/\bar{C}_{tot,\infty}$ is the ratio of the current average solid-phase sorbate concentration to that achieved at equilibrium; D^* is the effective diffusivity of a sorbate moving within the sorbent; R is the sorbent radius; and t is time.

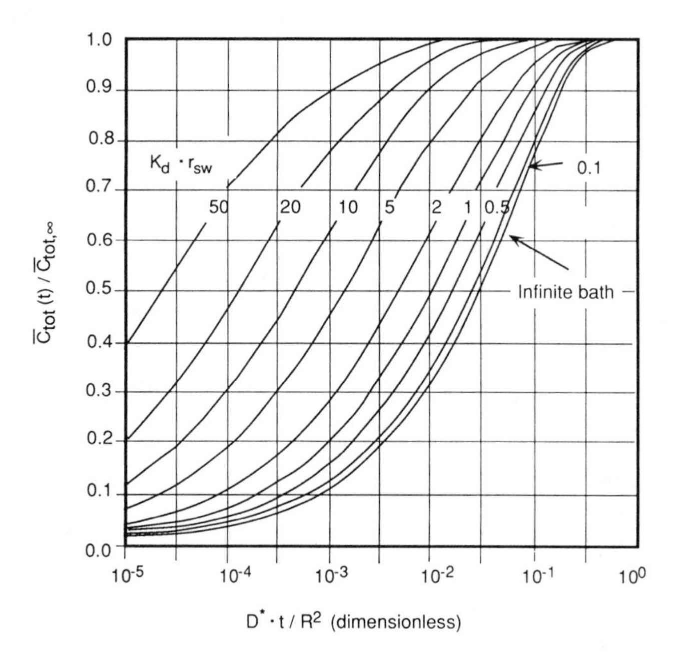

Figure 11.33 Timecourses for the diffusive uptake or release of sorbates by spherical particles suspended in closed systems. The numbers on the curves show the final ratio of the mass sorbed on solids to the mass dissolved in the solution.

this might be understood by recognizing that the gradients $\partial C'_w/\partial r$ will be steeper for longer if there are more solids or greater tendencies to sorb. For cases of desorption, this shortened time may be explained by realizing that each of the aggregates need only release a little sorbate to build up equilibrium levels in the solution.

Examining the curves for nonconstant C_w allows us to specify k_{sorb} when C_w is not a constant. For example, we may be interested in a case where $r_{sw} \cdot K_d$ is 10. To reach 50% sorption completion, we can use Figure 11.33 to see that the time required will be

$$t_{50\%}(K_d \cdot r_{sw} = 10) \simeq 4 \times 10^{-4}\, R^2/D^* \tag{11-158}$$

or that the approximate first-order rate constant should be

$$k_{sorb}(K_d \cdot r_{sw} = 10) \simeq \frac{D^* \ln 2}{4 \times 10^{-4}\, R^2} \simeq 2000\, D^*/R^2 \tag{11-159}$$

In general, the relation of k_{sorb} to $K_d \cdot r_w$, optimized to fit the midpoint of the sorptive exchange, is given (Wu and Gschwend, 1988) by

$$k_{sorb}(K_d \cdot r_{sw}) \simeq (11 \cdot K_d \cdot r_{sw} + 23)\, D^*/R^2 \tag{11-160}$$

We should reiterate that this approximate solution is well suited for describing the midregion of the solid–water exchange timecourse, but does poorly for early and late times. Use of Figure 11.33 can allow one to be more accurate at $t_{10\%}$, $t_{90\%}$, or whatever extent of approach to sorptive equilibrium is of interest.

If one believes that a retarded radial diffusion mass transfer is limiting overall sorption, as did the authors who produced the data shown in Figure 11.28*b* and *c*, we now can see how such data can be used to deduce the critical ratio R^2/D^* for situations of interest. For example, the uptake of 2,4-D by illite flocs, illustrated in Figure 11.28*b*, exhibits a $t_{50\%}$ of about 50 h. Since these particular data were obtained in experiments with $K_d \cdot r_{sw}$ of about 4, we may estimate

$$k_{sorb} \simeq \frac{\ln 2}{t_{50\%}} \simeq \frac{\ln 2}{50\,\text{h}} \sim 0.014\,\text{h}^{-1} \tag{11-161}$$

If we know the radius of the illite flocs, this result would allow us to estimate the effective diffusivity of 2,4,-D in those aggregates.

These results also enable us to understand how K_d will vary with time. First, since $\bar{C}'_w$ is proportional to $\bar{C}_{tot}$ and increases until it equals C_w, we can write

$$\bar{C}'_w = (\bar{C}_{tot}/\bar{C}_{tot,\infty})C_w \tag{11-162}$$

Inserting Eq. 11-162 into Eq. 11-150, we find

$$K_d(t) = \left(K_d^* + \frac{\phi}{(1-\phi)\rho_s} \right) (\bar{C}_{tot}/\bar{C}_{tot,\infty}) \tag{11-163}$$

$$= K_d^{\infty} \cdot (\bar{C}_{tot}/\bar{C}_{tot,\infty}) \tag{11-164}$$

with $K_d^{\infty} = K_d^* + \phi/(1-\phi)\rho_s$. Note that the second term on the right-hand side of our K_d^{∞} expression simply reflects the uptake due to pore water reaching concentrations like that of the exterior solution. As for the variation of $\bar{C}_{tot}$, the value of K_d approximately represents an equilibrium condition at (e.g., five half-lives)

$$t \simeq \frac{0.3 \cdot R^2}{D^*} = 0.3 \cdot R^2 \cdot \frac{K_d^*(1-\phi)\rho_s + \phi}{\phi f D_w} \tag{11-165}$$

An Example: Calculation of the Desorption Kinetics of an Organic Chemical Let us conclude this discussion with an example calculation to see how we might evaluate sorption kinetics in a particular case. For example, suppose we are concerned that some PCB-contaminated sediments will spill and settle through a water body during dredging operations. First, we note that the volume of water involved is probably very large compared to the mass of solids that we will spill; hence $K_d \cdot r_{sw}$ is a small number. In this event, we focus our attention on the timecourse in Figure 11.33 labeled "infinite bath" where $K_d \cdot r_{sw}$ is less than 0.01. Now if we are interested in finding the time necessary to desorb half of the PCB contaminants from the settling sediment grains, we read across the chart from $\bar{C}_{tot}/\bar{C}_{tot,\infty} = 0.5$ and see an intersection with the infinite bath timecourse at $D^*t/R^2 = 0.03$. At this juncture we need to specify some of the properties of the chemicals and sediments. Suppose we have a particular PCB congener with $\log K_{ow} = 7$ and $D_w = 7 \cdot 10^{-6}\,\text{cm}^2 \cdot \text{s}^{-1}$, as well as a particular silty sediment of 100-μm diameter and organic carbon content of 4%. Assuming these dredged sediments also exhibit $\phi \cdot f \approx 0.02$ (as reported by Wu and Gschwend, 1986 for a few samples), we now calculate the time to 50% release of the PCB of interest:

$$t_{50\%} \simeq 0.03\, R^2/D^* \tag{11-166}$$

$$\simeq \frac{(0.03)(R^2)(\rho_s(1-\phi) \cdot f_{oc} \cdot K_{oc} + \phi)}{\phi f \cdot D_w} \tag{11-167}$$

$$\simeq \frac{(0.03)(50 \cdot 10^{-6}\,\text{m}^2)^2(2 \times 10^3\,\text{kg/m}^3 \cdot 0.04 \cdot 3.1\, K_{ow}^{0.72} \times 10^{-3}\,\text{m}^3/\text{kg})}{(0.02 \cdot 7 \cdot 10^{-10}\,\text{m}^2/\text{s})}$$

$$\simeq 1.5 \cdot 10^5 \cdot \text{s}^{-1} \text{ or about 2 days}$$

where we assume a bulk density $(= \rho_s(1-\phi))$ of $2\,\text{g} \cdot \text{cm}^{-3}$. Note that $[\rho_s(1-\phi)K_d]$ is much greater than any reasonable value of ϕ, so the porosity in the numerator can be neglected in our calculations. Since the time of release appears similar to that time expected for such spilled dredge particles to fall back to the bottom, one should expect a substantial release of even hydrophobic PCB congeners, but not equilibration.

Other Issues of Sorption Kinetics

The situation in which mass transfer into and out of aggregates limits the rate of sorptive exchange may be common, but this conceptualization will not cover every case where mass transfers are controlling. As alluded to earlier, sometimes the slowest process controlling sorption of organic molecules of interest to us involves their movements through pores of similar size to themselves (e.g., Wood et al., 1990; Ball and Roberts, 1991). If this is the case, we must reduce D in mass balance equations like Eq. 11-153 even further using a constrictivity factor (Satterfield et al., 1973). It is also possible that diffusion in small continuous pores occurs, not only in the water filling the pores, but also on the walls themselves (i.e., surface diffusion). Release of organic contaminants from these positions within the solids may require months to years (Coates and Elzerman, 1986; Steinberg et al., 1987; Pavlostathis and Jaglal, 1991).

Another possibility is that molecular diffusion into macromolecular organic matter may be the slowest step in the overall process in some cases (Brusseau et al., 1991; Brusseau and Rao, 1991). Due to the relatively "viscous" nature of such organic matter, the diffusion coefficients of organic sorbates moving in this polymeric matter may be significantly lower than comparable free-solution values. If the diffusion pathlengths are long enough, transfer of hydrophobic compounds into all parts of the humus of soils and similar media on other particles may control the overall sorptive exchange rate.

Until now, we have considered only linear partitioning of sorbates at each microscopic position within an aggregate. Obviously, some sorption mechanisms require us to examine cases of nonlinear surface–solution exchange. Such a case was considered in the uptake of alkyl benzene sulfonates by porous granular carbon particles (Weber and Rumer, 1965). In this situation $C'_s(r)$ was related to $C'_w(r)$ using a Langmuir isotherm to yield a result comparable to what we reached with Eq. 11-147.

We conclude this discussion of sorption kinetics by noting that the concepts of mass transfer limitation we have discussed are general. Their application to many situations, with appropriate modifications, will undoubtedly prove effective. A recent example of this is provided by Rounds and Pankow (1990), who successfully characterized organic vapor molecule–atmospheric particle exchange kinetics with a radial diffusion model.

CHAPTER 12

CHEMICAL TRANSFORMATION REACTIONS

12.1 INTRODUCTION

So far, we have been concerned primarily with the transfer of organic chemicals between different environments (e.g., water and air, water and solid phases). Such processes leave the molecular structure of a compound unaltered. In this and the following two chapters, we turn our attention to processes by which a compound is converted to one or several products. Hence, we talk about processes (*reactions*) in which chemical bonds are broken and new bonds are formed. For convenience, we can divide the structural transformation processes that organic chemicals undergo in the environment into three major categories: chemical, photochemical, and biologically mediated transformation reactions. The former two types of reactions are commonly referred to as *abiotic* transformation processes. *Chemical* reactions encompass all reactions that occur in the dark and without the mediation of organisms. They are the topic of this chapter. In Chapter 13 we address reactions in which a compound undergoes transformation either as a consequence of direct absorption of light (direct *photolysis*), or by reaction with highly reactive species (e.g., free radicals or singlet oxygen) that are formed as a result of the incidence of sunlight on natural waters (*indirect* or *sensitized photolysis*). Finally, in Chapter 14 we learn how and under which conditions organisms, in particular, microorganisms, transform xenobiotic organic chemicals. Note that *microbial* transformation reactions are usually the only processes by which a xenobiotic organic compound may be *mineralized* (i.e., converted to CO_2, H_2O, etc.) in the environment, while abiotic reactions commonly yield other organic compounds. We also note that many types of chemical reactions are also performed by microorganisms; and therefore it may not always be clear

TABLE 12.1 Examples of Environmentally Relevant Chemical Reactions

Reactants	Products	Equation Number
Nucleophilic substitution		
$C_6H_5{-}CH_2{-}Cl + H_2O$ →	$C_6H_5{-}CH_2{-}OH + H^{\oplus} + Cl^{\ominus}$	(12-1)
Benzyl chloride	**Benzyl alcohol**	
$CH_3Br + H_2O$ →	$CH_3OH + H^{\oplus} + Br^{\ominus}$	(12-2)
Methyl bromide	Methanol	
$CH_3Br + SH^{\ominus}$ →	$CH_3SH + Br^{\ominus}$	(12-3)
Methyl bromide	Methyl mercaptan	
Elimination		
$Cl_2HC{-}CHCl_2 + OH^{\ominus}$ →	$ClHC{=}CCl_2 + Cl^{\ominus} + H_2O$	(12-4)
1,1,2,2-Tetrachloroethane	Trichloroethene	
Ester hydrolysis		
Dibutyl phthalate $+ 2\,OH^{\ominus}$ →	Phthalate $+ 2\,HO{-}C_4H_9$	(12-5)
Dibutyl phthalate	Phthalate; Butanol	
$(C_2H_5O)_2P(=S){-}O{-}C_6H_4{-}NO_2 + OH^{\ominus}$ →	$(C_2H_5O)_2P(=S){-}O^{\ominus} + HO{-}C_6H_4{-}NO_2$	(12-6)
Parathion	*O,O*-Diethyl-thiophosphoric acid; *p*-Nitrophenol	
Oxidation		
$2\,CH_3SH + 1/2\,O_2$ →	$H_3C{-}S{-}S{-}CH_3 + H_2O$	(12-7)
Methyl mercaptan	Dimethyl disulfide	
Reduction		
$C_6H_5{-}NO_2$ + "reduced species" $+ 6H^{\oplus}$ →	$C_6H_5{-}NH_2$ + "oxidized species" $+ 2H_2O$	(12-8)
Nitrobenzene	Aniline	

whether, in a given environmental system, a reaction occurs strictly abiotically, whether it is medicated by microorganisms, or whether both types of processes play a role.

Let us now focus our discussion on chemical reactions. We may classify these reactions according to whether there is a net electron transfer between the reactants (i.e., redox reactions) or not. We start out with the latter case and then proceed to discuss abiotic redox reactions. A few examples illustrating the type of reactions discussed in this chapter are given in Table 12.1. We focus our attention on reactions in aqueous solution (called homogeneous reactions because we assume the medium is the same throughout its volume). In so doing, we do not mean to imply that heterogeneous reactions (e.g., at interfaces of solids) are not important, but there is presently just too little data available to derive general rules for describing such processes. Before we start discussing the chemical transformation reactions, we need to review some of the general concepts used to describe and quantify the rates of chemical reactions.

12.2 KINETIC ASPECTS OF CHEMICAL TRANSFORMATION REACTIONS

In Chapter 3 we saw that, based on thermodynamic concepts, we can evaluate to what extent a reversible chemical reaction has proceeded when equilibrium is reached. We applied these concepts to proton transfer reactions in aqueous solution (Chapter 8), and assumed that, as compared to all other processes, proton transfer occurs at such a high rate that, from a practical point of view, we can always consider equilibrium to be established instantaneously. For the reactions discussed in this chapter, however, this assumption does not generally hold, since we are dealing with reactions that occur at much slower rates. Hence, in addition to energetic aspects, we now become interested in the mechanisms and kinetics of reactions (i.e., the pathway of molecular change and the rate).

Phenomenological Description of Chemical Kinetics

There are several questions that we need to address when dealing with transformation reactions of xenobiotic organic compounds in the environment:

1. Is there only one or are there several different reactions by which a given compound may be transformed under given environmental conditions, and what are the reaction products?
2. What are the kinetics of the different reactions, and what is the resultant overall rate by which the compound is "eliminated" from the system?
3. What is the influence of important environmental variables such as temperature, pH, redox condition, ionic strength, presence of certain solutes, or concentration and type of solids on the transformation behavior of a given compound?

To answer all these questions, we generally need to know something about the mechanism(s) of the reaction(s) by which a compound is transformed. A *reaction mechanism* is defined as a set of *elementary molecular changes* describing the sequence in which chemical bonds are broken and new bonds are formed to convert the compound to the observed product(s). Hence, when just writing down the stoichiometry of an overall reaction, as we have done in Table 12.1, we do not say anything about the number of reaction steps and possible reaction intermediates occurring during the conversion of educts [i.e., the compound of interest plus reactant(s)] to products.

From experimental data or by analogy to the reactivity of compounds of related structure, we can often derive an empirical rate law for the transformation of a given compound. The *rate law* is a mathematical function, specifically a differential equation, describing phenomenologically the turnover rate of the compound of interest as a function of the concentrations (more precisely, the activities) of the various species participating in the reaction. Hence, the rate law describes the overall reaction on a *macroscopic* level. *Note that when dealing with molecular transformation reactions in aqueous solution, we commonly use the infinite dilution state as the reference state for the solutes, and, if not stated otherwise, assume activity coefficients to be 1.* Thus organic chemical activities are given by their concentrations, in contrast to our system used earlier in the book regarding phase transfers in which pure liquids are taken to be the reference state and activity coefficients were typically much greater than 1.

Let us now have a look at some typical rate laws encountered with chemical reactions in aqueous solution. We first perform a simple experiment in which we observe how the concentration of benzyl chloride (Table 12.1, Eq. 12-1) changes as a function of time in aqueous buffer solutions of pH 3, 6, and 9 at 25°C. When plotting the concentration of benzyl chloride (denoted as A) as a function of time, we find that, independent of pH, we get an *exponential* decrease in concentration as schematically depicted in Figure 12.1*a*. Hence, we find that, at any time, the turnover rate of benzyl chloride is solely proportional to its actual concentration, which can be expressed mathematically by a so-called *first-order rate* law:

$$\frac{d[\mathrm{A}]}{dt} = -k[\mathrm{A}] \tag{12-9}$$

where k is referred to as the *first-order rate constant*, and has the dimension $[\mathrm{T}^{-1}]$. Integration of Eq. 12-9 from $[\mathrm{A}] = [\mathrm{A}]_0$ at $t = 0$ to $[\mathrm{A}] = [\mathrm{A}]_t$ at $t = t$ yields the mathematical description of our observed curve in Figure 12.1*a*:

$$[\mathrm{A}]_t = [\mathrm{A}]_0 \cdot e^{-kt} \tag{12-10}$$

From Eq. 12-10 we can see that, if a reaction obeys a first-order rate law, a plot of the natural logarithm of $[\mathrm{A}]_t/[\mathrm{A}]_0$ versus time should yield a straight line with the slope $-k$. Such plots, as given for our example in Figure 12.1*b*, are useful to evaluate whether a reaction shows first-order kinetics, and to determine the rate constant k using a linear regression analysis. We note that in the case of first-order kinetics, the *half-life* $t_{1/2}$ of the compound (i.e., the time in which its concentration

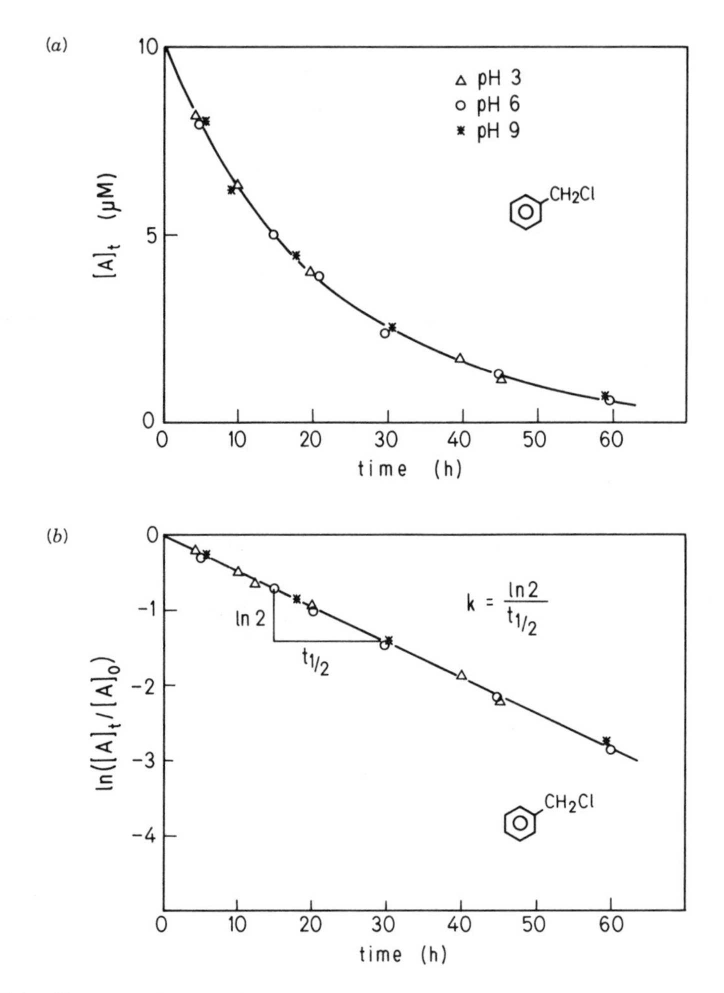

Figure 12.1 Decrease in benzyl chloride concentration (plotted directly in upper graph and logarithmically in lower graph) as a function of time in aqueous solutions at three different pH's.

is lowered by a factor of 2 with respect to the reaction considered), is *independent of concentration* and is deduced from Eq. 12-10 to be

$$t_{1/2} = \frac{\ln 2}{k} = \frac{0.693}{k} \tag{12-11}$$

When looking at the stoichiometry of the transformation of benzyl chloride to benzyl alcohol as depicted by Eq. 12-1 (Table 12.1), we notice that a water molecule is involved in the reaction, but that we have not included this reactant in our rate law (Eq. 12-9). If, however, the water molecule is involved in the slowest step of this reaction, that is, in the *rate-determining* step, we describe our reaction by a *second-order rate* law:

$$\frac{d[\mathrm{A}]}{dt} = -k'[\mathrm{A}][\mathrm{B}] \tag{12-12}$$

where we have denoted H_2O as B, and k' is now referred to as *second-order rate constant*, and has the dimension ($M^{-1}T^{-1}$). Note that the *total order of a rate law* is deduced from the number of chemical species involved in the rate-determining step(s) and, therefore, amounts to the sum of the exponents with which the concentrations of the various reactants appear in the rate law.

Back to our example, we immediately realize, of course, that water is present in a large excess. Its concentration (~ 55 M) is not altered significantly during the course of the reaction, that is, $[\mathrm{B}]_t = [\mathrm{B}]_0 \sim 55$ M. Hence, by setting $k = k'[\mathrm{B}]_0$, we again obtain a first-order rate law (Eq. 12-9). In this and all other cases in which we simplify the rate law by assuming the concentrations of certain species to be constant, we use the prefix *pseudo-* to indicate that, on a molecular level, more species take part in the reaction than are incorporated in the rate law. In our benzyl chloride case, without knowing more about the reaction mechanism, we cannot say whether our rate law is *pseudo-first order* or truly first order. In the former case, we refer to k as the *pseudo-first order rate constant.*

Since most reactions with which we are concerned are not truly first order, it is convenient for modeling purposes to make assumptions that allow us to reduce the order of the reaction law, ideally to pseudo-first order. For example, when considering reactions such as Eqs. 12-2 to 12-6 in Table 12.1, by assuming that the concentrations of OH^- (i.e., pH) or SH^- are constant in a given water, the reactions could be expressed by pseudo-first order rate laws. The rate law for the oxidation of methyl mercaptan to dimethyl disulfide (Eq. 12-7) could be simplified to pseudo-second order by setting the concentration of the oxidant (e.g., O_2) constant ($k'[O_2] = k$):

$$\frac{d[\mathrm{A}]}{dt} = -k[\mathrm{A}]^2 \tag{12-13}$$

In this particular case, where only the concentration of one species is incorporated into the rate law, the integrated rate law is still relatively simple and is given by

$$\int_{\mathrm{A}_0}^{\mathrm{A}_t} \frac{d[\mathrm{A}]}{[\mathrm{A}]^2} = \int_0^t -k\,dt$$

$$-\frac{1}{[\mathrm{A}]}\Bigg|_{\mathrm{A}_0}^{\mathrm{A}_t} = -kt$$

$$-\frac{1}{[\mathrm{A}]_t} + \frac{1}{[\mathrm{A}]_0} = -kt$$

$$\frac{1}{[\mathrm{A}]_t} = kt + \frac{1}{[\mathrm{A}]_0}$$

or

$$[A]_t = [A]_0(1 + k[A]_0 t)^{-1} \quad (12\text{-}14)$$

We note that in this case the half-life of a compound is dependent on the initial concentration at the beginning of the time period considered. Here it is given by

$$t_{1/2} = \frac{1}{k[A]_0} \quad (12\text{-}15)$$

For all non-first-order reactions, the half-life of A is inversely related to the concentrations of the substances with which it reacts. Thus for rate expression 12-12, we find

$$t_{1/2} \text{ for loss of A} = \frac{\ln 2}{k'[B]} \quad (12\text{-}16)$$

There are, of course, some cases where still higher-order rate laws have to be applied to describe adequately the reaction kinetics of a xenobiotic compound. The mathematics involved in solving these more complex situations may be found in textbooks on chemical kinetics (e.g., Frost and Pearson, 1961; Laidler, 1965). For our purposes, first- and second-order expressions are in most cases sufficient.

Arrhenius Equation and Transition State Theory

If we want to understand and describe the influence of environmental factors, especially temperature, on thermochemical reaction rates, and if we want to see how transformation rates vary as a function of the chemical structure of a compound, we need to take a closer look at these reactions on a molecular (microscopic) level. As we have already stated, a chemical reaction proceeds in several sequential elementary steps. In the reactions considered here, often one step in the reaction sequence occurs at a much slower rate than all the others. This step acts as a "bottleneck" in the overall process and determines the overall rate of the reaction. Consequently, this slow step is generally the one in which we are most interested.

Let us begin by taking a look at the effect of temperature on the rate of a chemical reaction. Experimentally, we commonly find that the reaction rate constant varies as an exponential function of temperature, which can be mathematically expressed by the so-called *Arrhenius equation*:

$$k = A \cdot e^{-E_a/RT} \quad (12\text{-}17a)$$

where A is called the *pre-exponential factor* or *frequency factor*, and E_a is referred to as the *activation energy*. The units of A correspond to the units of the observed rate constant, and A and E_a can be derived for a given reaction by linear regression of a plot of $\ln k$ versus $1/T(K)$:

$$\ln k = \ln A - E_a/RT \quad (12\text{-}17b)$$

TABLE 12.2 Temperature Dependence of Reaction Rates as a Function of the Activation Energy E_a[a]

Activation Energy E_a ($kJ \cdot mol^{-1}$)	Rate Relative to $T = 25°C$ Deduced from $k(T_1)/k(T_2) = e^{E_a(1/T_2 - 1/T_1)/R}$ (%)						Average Increase (Decrease) Factor per 10°C Increase (Decrease) in Temperature
	5°C	10°C	15°C	20°C	25°C	30°C	
40	31	42	57	76	100	130	1.8
50	23	34	50	71	100	139	2.0
60	18	28	43	66	100	149	2.3
70	13	22	37	62	100	159	2.7
80	10	18	33	58	100	170	3.1
90	7.4	15	28	54	100	181	3.5
100	5.5	12	25	50	100	193	4.0
110	4.1	9.5	21	47	100	207	4.7
120	3.1	7.7	19	44	100	221	5.4
130	2.3	6.2	16	41	100	236	6.2

[a] Relative rates are given as percentage of the rate at 25°C.

Note that when deriving and using Eq. 12-17 to calculate rate constants at different temperatures, one assumes that A and E_a are temperature independent, which is a reasonable first approximation if the temperature range considered is not too large, and if we are dealing with only one reaction that causes the compound to disappear. If E_a is known for a reaction, then the effect of temperature on the rate constant, and thus on the reaction rate, can be calculated. Table 12.2 shows that, depending on the magnitude of E_a, an increase (decrease) of 10°C may accelerate (slow down) a reaction by a factor of between 2 and 6.

For a qualitative interpretation of the Arrhenius equation, we consider a simple elementary biomolecular reaction in aqueous solution:

$$B + C \rightarrow D + E \tag{12-18}$$

An example of such a reaction is Eq. 12-3 in Table 12.1. Let us now deduce the factors that control the rate of conversion of B and C to D and E by imagining the transformation process is portrayed well by what is known as a collision rate model. (Strictly speaking, this collision rate model applies to gas phase reactions; here we use it to describe interactions in solution where we are not specifying the roles played by the solvent molecules.) First, in order to be able to react, the molecules B and C have to encounter each other and collide. Hence the rate of reaction depends on the frequency of encounters of B and C, which is proportional to their concentrations and to how fast B and C move around (diffuse, see Chapter 9) in the aqueous solution. Second, the rate is proportional to the probability that B and C meet with the "right orientation" to be able to react, which we may refer to as the "orientation probability." Third, only a fraction of collisions have a sufficient amount of energy to break the relevant bonds in B and C (i.e., greater than E_a) that make the reactants change to products D and E. The fraction of species exhibiting an energy greater than E_a is given by $e^{-E_a/RT}$, which corresponds exactly to the exponential term in the Arrhenius equation. Consequently, the collision frequency and the orientation probability factors are included in the pre-exponential factor A and the concentration dependence in the rate law:

$$\text{rate} = -A \cdot e^{-E_a/RT} \cdot [\text{B}] \cdot [\text{C}] \tag{12-19}$$

This simple collision rate model has allowed us to rationalize qualitatively the Arrhenius equation. However for a more quantitative approach to chemical kinetics, in particular for understanding and applying linear free-energy relationships to estimate transformation rates of organic chemicals from structure, we need to acquaint ourselves with a different, somewhat more sophisticated theoretical framework for treating chemical kinetics, the so-called *activated complex* or *transition state* theory. We can view this theory as a thermodynamic approach to chemical kinetics. To explain, we consider the highest-energy state (i.e., the transition state) that the reactants of an elementary reaction go through on their way to form products, and we try to come up with a proposal for the "structure" of this high-energy species that is commonly referred to as an activated complex. We discuss specific examples of activated complexes in the following sections. Since the transition state represents the point of bond changing highest in energy, and, therefore, the point most difficult to reach, the rate of formation of the activated complex determines the overall rate of the elementary reaction (and of the overall reaction if we consider the rate-limiting step). The quantitative treatment of this approach hinges on the postulate that *the reactants establish an equilibrium with the activated complex*. This assumption is somewhat unusual in that the activated complex has only a transitory existence since it lies at an energy maximum (rather than a minimum). The second assumption, which is derived from statistical mechanics, is that *all activated complexes proceed on to products with a fixed first-order rate constant which is given by* $\mathbf{k}T/h$, where $\mathbf{k}$ and h are the Boltzmann (1.38×10^{-23} J·K^{-1}) and Planck (6.63×10^{-34} J·s) constants, respectively (for details see Atkins, 1978).

Based on these two postulates, the rate law of a reaction is then simply given as the product of the universal rate constant times the concentration of the activated complex. By denoting the activated complex of our bimolecular reaction Eq. 12-18 as $BC^{\ddagger}$ (we use the double dagger superscript to represent the activated complex and other features of the transition state), we get

$$\text{rate} = \left(\frac{\mathbf{k}T}{h}\right)[\text{BC}^{\ddagger}] \tag{12-20}$$

By assuming activity coefficients of 1 for all species involved, $[BC^{\ddagger}]$ can be expressed in terms of the reactant concentrations and the equilibrium constant $K^{\ddagger}$:

$$K^{\ddagger} = \frac{[\text{BC}^{\ddagger}]}{[\text{B}][\text{C}]} \tag{12-21}$$

and

$$\text{rate} = \left(\frac{kT}{h}\right) K^{\ddagger}[\text{B}][\text{C}] \tag{12-22}$$

From Chapter 3 we also recall that $K = e^{-\Delta G^{\circ}/RT}$. Hence, we get

$$\text{rate} = \left(\frac{\mathbf{k}T}{h}\right) \cdot e^{-\Delta G^{\ddagger}/RT}[\text{B}][\text{C}] \tag{12-23}$$

where $\Delta G^{\ddagger}$ is referred to as the standard free energy of activation.
Since $\Delta G^{\ddagger} = \Delta H^{\ddagger} - T\Delta S^{\ddagger}$, we can also write

$$\text{rate} = \left(\frac{\mathbf{k}T}{h}\right) e^{\Delta S^{\ddagger}/R} \cdot e^{-\Delta H^{\ddagger}/RT}[\text{B}][\text{C}] \tag{12-24}$$

where $\Delta S^{\ddagger}$ and $\Delta \text{H}^{\ddagger}$ are the entropy and enthalpy of activation, respectively. The rate constant k is then given by

$$\text{rate constant} = \left(\frac{\mathbf{k}T}{h}\right) e^{\Delta S^{\ddagger}/R} \cdot e^{-\Delta H^{\ddagger}/RT} \tag{12-25}$$

which resembles very closely the Arrhenius rate law (Eq. 12-16). Since E_a represents the *potential* energy of activation, the difference between $\Delta H^{\ddagger}$ and E_a is simply given by the *kinetic* energy of activation, which is usually small as compared to E_a. For our bimolecular reaction, this difference is given by RT (e.g., Atkins, 1978), and we can rewrite Eq. 12-25 as

$$\begin{aligned}\text{rate constant} &= \frac{\mathbf{k}T}{h} e^{\Delta S^{\ddagger}/R} \cdot e^{-(E_a - RT)/RT} \\ &= \frac{\mathbf{k}T}{h} e^{\Delta S^{\ddagger}/R} \cdot e^{1} \cdot e^{-E_a/RT}\end{aligned} \tag{12-26}$$

Thus, based on transition state theory, the preexponential factor A in the Arrhenius equation is interpreted to encompass the universal rate constant and the entropy of activation, the latter factor corresponding to what we have called "orientation probability" in our simple collision rate model. From Eq. 12-26 we also recognize that A is linearly dependent on temperature, but it is easy to see how that is usually masked by the overwhelming exponential temperature dependence of the activation energy term. Finally, we note that A values may range over several orders of magnitude, depending on the reaction considered. Unimolecular dissociation or elimination reactions usually exhibit large preexponential factors (A between 10^{12} and $10^{16}\,\text{s}^{-1}$) because the entropy of activation is only slightly negative or even positive. Bimolecular reactions, on the other hand, commonly exhibit A values between about 10^7 and $10^{12}\ M^{-1}\ s^{-1}$ (Harris and Wamser, 1976; Mabey and Mill, 1978).

Linear Free-Energy Relationships

When considering the kinetics of a chemical reaction, the transition state theory provides us with a very handy framework to evaluate the impact of structural moieties on the relative rates at which structurally related compounds are transformed. In Chapter 8 we applied linear free-energy relationship (LFERs) to relate equilibrium constants, in that case, acidity constants of aromatic acids. By applying the Hammett equation (Eq. 8-20), we were able to express quantitatively the impact of substituents on the relative stability, that is, the relative energy status of the educts (e.g., neutral species) and products (e.g., anionic species) of a chemical reaction (i.e., proton dissociation reaction), and thus on the equilibrium constant:

$$\log\left(\frac{K_a}{K_{aH}}\right) = \rho \sum_i \sigma_i \tag{8-20}$$

where K_a and K_{aH} represented the acidity constants of the compound of interest and the reference compound, respectively. We also recall that σ_i are *electronic* substituent constants (Hammett σ constants), and that ρ, the susceptibility factor, was an expression of how sensitive the dissociation reaction of the acid functionality considered (e.g., phenolic group) was to substitution as compared with the reference system, in this case, substituted benzoic acids.

In a similar fashion, we can now try to quantify the influence of substituents on the reaction rate. However, now we proceed by linearly correlating $\Delta G^{\ddagger}$ values (and hence rate constants), instead of ΔG^0 values. Thus, for a given type of reaction, we need to define the "structure" of the activated complex and to select a set of substituent constants that gives us a quantitative measure of the change in $\Delta G^{\ddagger}$ upon substitution. In other words, we have to try to quantify how much a substituent enhances or decreases the free energy of activation relative to $\Delta G^{\ddagger}$ of the unsubstituted compound. Although this may sound straightforward, quantification of *electronic* [inductive (polar), resonance] and *steric* effects of substituents on the free energy of transition states is often quite difficult because of the complicated nature of the activated complex, particularly when dealing with more complex reactions in aqueous solution. Nevertheless, for many of the reactions of interest to us, LFERs are useful tools for relating kinetic data, as illustrated by the following two examples. We discuss a few more examples of the application of LFERs for relating reaction rate constants in the following sections. For a more comprehensive treatment of this topic we refer, however, to the literature (e.g., Wells, 1963; Shorter, 1973; Hansch and Leo, 1979; Williams, 1984).

Hammett Relationship

In our first example we evaluate the influence of meta and para ring substituents on the base-catalyzed hydrolysis of substituted benzoic acid ethyl esters:

$$\text{(R, X)-}C_6H_3\text{-C(=O)-O-}CH_2\text{-}CH_3 + HO^{\ominus} \xrightarrow{k} \text{(R, X)-}C_6H_3\text{-C(=O)-}O^{\ominus} + HO\text{-}CH_2\text{-}CH_3 \quad (12\text{-}27)$$

In this particular case, the activated complex is thought to be represented by (see Section 12.3)

$$\left[\text{(R, X)-}C_6H_3\text{-C}(\overset{\ominus}{O})(OH)\text{-O-}CH_2\text{-}CH_3 \right]^{\ddagger}$$

I

and hence is believed to exhibit a negative charge that is not significantly delocalized into the ring system. Thus, intuitively, we expect that through an *inductive* effect, electron-withdrawing ring substituents ($-X$) will stabilize the negatively charged activated complex relative to the uncharged ground state; that is, they will decrease $\Delta G^{\ddagger}$ as compared to the unsubstituted compound and, therefore, increase k relative to k_H. Conversely, an electron-donating substituent exerts the opposite effect. As we have seen in Chapter 8, the inductive effect of aromatic substituents in meta or para positions may be quantitatively expressed by the σ_{meta} and σ_{para} substituent constants (see Table 8.4). It comes, therefore, as no surprise that, in this case, where we only deal with inductive effects (i.e., no resonance, no steric effects), the rate constants of the base-catalyzed hydrolysis of meta and para monosubstituted benzoic acid ethyl esters can, as shown by Figure 12.2, be related successfully by the Hammett equation:

$$\log\left(\frac{k}{k_H}\right) = \rho\sigma_{m,p} \quad (12\text{-}28)$$

From the data shown in Figure 12.2, a ρ value of 2.55 (Wells, 1963) is obtained at 25°C from a linear regression analysis, indicating a substantial influence of the substituents on the reaction rate. For example, a nitro group in the meta or para position increases the rate of hydrolysis by a factor of 100. It should be pointed out that the data shown in Figure 12.2 were not obtained in pure water, but in a mixture of ethanol and water (85:15). Therefore, the absolute rate constants as well as the ρ value are not directly applicable to natural waters. Nevertheless, this example fulfills its major purpose, that is, to illustrate the applicability of the Hammett equation and substituent constants to relate quantitatively kinetic data.

Impact of Solution Composition on Reaction Rates

At this point we need to make a few comments on the impact of solution composition on reaction rates and LFERs. Unfortunately, because many organic compounds are only sparingly soluble in water, a large portion of the kinetic data available in the

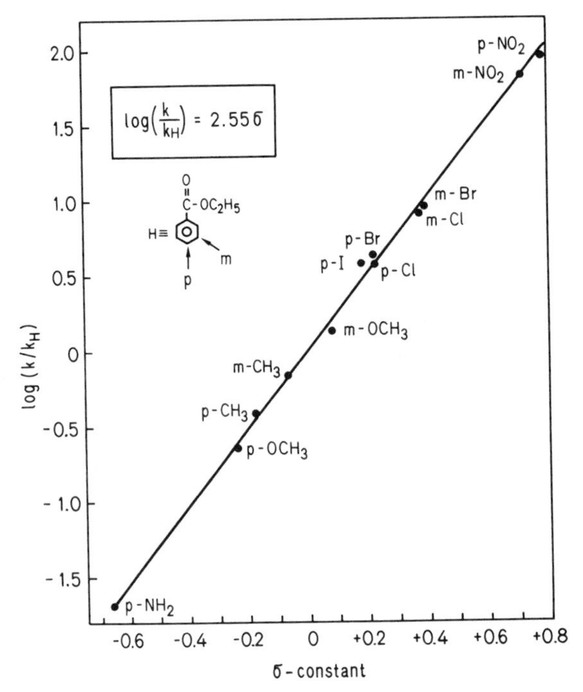

Figure 12.2 Effects of substituents on the base-catalyzed hydrolysis of benzoic acid ethyl esters in ethanol: water (85:15). Relative reaction rates are correlated with Hammett σ constants (data from Tinsley, 1979).

literature on reactions that are of interest to us have been determined in organic solvents or in organic solvent–water mixtures instead of pure water. Since the intermolecular interactions of reactants and of activated complexes (particularly if they are charged) with solvent molecules (called solvation) may involve significant energies, making or breaking these interactions can have an important effect on $\Delta G^{\ddagger}$. Thus, significantly different rate data are obtained for different solvent systems, in particular for reactions involving ionic species, when we compare polar solvents like water with nonpolar ones. As illustrated by the reaction rate constants given in Table 12.3 for the base-catalyzed hydrolysis of benzoic acid ethyl ester in various organic solvent–water systems, differences of more than one order of magnitude may be found. Similarly, different ρ values are obtained when deriving LFERs from different solvent systems. For the example discussed above, a ρ value of 2.38 is found at 25°C in acetone/water (3:2) as compared to 2.55 in ethanol/water (85:15). Furthermore, we should note that ρ is a function of temperature since, as implied by Eq. 12-26, the effect of temperature on the rate constant is different for compounds exhibiting different $\Delta H^{\ddagger}$ (or E_a) values, the relative size of which is determined by the type of substituent(s) present. In the case of the reaction in ethanol/water (85:15), ρ decreases

TABLE 12.3 Reaction Rate Constants for the Base-Catalyzed Hydrolysis of Benzoic Acid Ethyl Ester in Various Organic Solvent–Water Mixtures[a]

Mixture	k_B ($M^{-1} s^{-1}$)
Water	3.0×10^{-2}
60% Ethanol/40% water	1.2×10^{-3}
60% Acetone/40% water	2.8×10^{-3}
40% Dioxane/60% water	7.0×10^{-3}
70% Dioxane/30% water	3.3×10^{-3}

[a] Data from Mabey and Mill (1978).

from 2.55 at 25°C to 2.13 at 50°C (Wells, 1963). In summary, since solvation plays an important role in determining $\Delta G^{\ddagger}$ of a reaction, we should recognize that we have to be very cautious when applying kinetic data obtained in organic solvents and/or organic solvent–water systems to purely aqueous solution, particularly for reactions involving polar reactants. In every case, the feasibility of extrapolating rate constants from nonaqueous to aqueous solutions has to be checked carefully (for further discussion of solvent effects see Shorter, 1973).

In aqueous solutions, the influence of dissolved organic and inorganic species (e.g., buffer solutions used in laboratory experiments, the major ions and dissolved organic matter present in natural waters) on transformation rates of organic chemicals is in many cases only of secondary importance (Macalady et al., 1989). As long as these species are neither reactants nor catalysts, that is, as long as they do not participate in a rate-determining step, we neglect their impact on reaction rates.

Taft Relationship

We now examine a second example illustrating the application of LFERs to kinetic data. We shall see how an approach in which both electronic and steric effects of substituents on reaction rate constants can be considered. This approach, developed by Taft (1956), has been applied mostly to aliphatic compounds, although some applications involving reactions in aromatic systems (e.g., the hydrolysis of ortho-substituted benzoic acid esters) have been reported (for details see Shorter, 1973). In discussing this example we take the opportunity to briefly sketch the way substituent constants are derived. This allows us to have a closer look at the assumptions on which such LFERs are built, and thus increases our ability to evaluate more critically the applicability of a given LFER to a given problem.

The major assumption in the approach taken by Taft (1956) and Pavelich and Taft (1957) is that the influence of a given group on $\Delta G^{\ddagger}$ of a specific reaction can be separated into independent contributions from polar (inductive) and steric effects: $\Delta G^{\ddagger} = \Delta G^{\ddagger}_{\text{ref}} + \Delta G^{\ddagger}_{i,\text{electronic}} + \Delta G^{\ddagger}_{i,\text{steric}}$, where $\Delta G^{\ddagger}_{\text{ref}}$ is the free energy of activation for the reference compound. Hence, in formal analogy to the Hammett equation, but by

TABLE 12.4 Examples of Taft Polar and Steric Substituent Constant for Aliphatic Systems[a]

Substituent	σ^*	E_s	Substituent	σ^*	E_s
—H	0.49	1.24	$—CH_2C_6H_5$	0.22	−0.38
$—CH_3$	0.00	0.00	$—CH_2CH_2C_6H_5$	0.08	−0.38
$—C_2H_5$	−0.10	−0.07	$—CH_2F$	1.10	−0.24
—*n*-C_3H_7	−0.12	−0.36	$—CHF_2$	2.05	−0.67
—*i*-C_3H_7	−0.19	−0.47	$—CH_2Cl$	1.05	−0.24
—*n*-C_4H_9	−0.13	−0.39	$—CHCl_2$	1.94	−1.54
—*i*-C_4H_9	−0.13	−0.93	$—CCl_3$	2.65	−2.06
—*s*-C_4H_9	−0.21	−1.13	$—CH_2CH_2Cl$	0.39	−0.90
—*t*-C_4H_9	−0.30	−1.54	$—CH_2Br$	1.00	−0.27
—cyclo-C_6H_{11}	−0.15	−0.79	$—CHBr_2$		−1.86
$—CH_2$-cyclo-C_6H_{11}	−0.06	−0.98	$—CBr_3$		−2.43
$—CH{=}CH_2$	0.36	−1.63	$—CH_2OCH_3$	0.52	−0.19
$—C_6H_5$ (phenyl)	0.60	−2.55	$—CH_2OC_6H_5$	0.85	−0.33

[a] Data taken from a more comprehensive data set given by Williams (1984).

adding a steric term, it is postulated that

$$\log\left(\frac{k}{k_{\text{ref}}}\right) = \rho^*\sigma^* + \delta E_s \tag{12-29}$$

where σ^* and E_s are referred to as polar and steric substituent constants, respectively. Equation 12-29 is commonly called the Taft equation. The reference compound (subscript ref) is usually chosen as the one with a methyl substituent at the atom at which substitution is considered [i.e., $\sigma^*(CH_3) \equiv E_s(CH_3) \equiv 0$]. By analogy to the Hammett equation, ρ^* and δ are measures of how sensitive the reaction of interest is to substitution as compared to the reference reaction series from which the substituent constants have been derived. Note that we now have two variables (ρ^*, δ) for fitting experimental data (which requires more data points and a multiple linear correlation analysis), and that we could also apply Eq. 12-29 to relate equilibrium constants (e.g., acidity constants of aliphatic acids).

Let us now take a brief look at how the Taft substituent constants σ^* and E_s were originally derived (some examples are given in Table 12.4). Since, in contrast to the Hammett equation, the Taft equation contains two constants for each substituent, for the derivation of these constants one needs two different reactions that have to fulfill the following requirements: (1) in one of the reactions, both electronic and steric effects have to be important; (2) in the second reaction, only one of these effects should predominate. In addition, this latter effect has to be of similar magnitude in both reactions. Taft found that these criteria were reasonably met when choosing the acid- and base-catalyzed hydrolyses of a carboxylic acid ester function (II) as the model reaction. The activated complexes of the rate-determining steps of these two reactions are postulated as structures III and IV, respectively (for reaction mechanisms see Section 12.3):

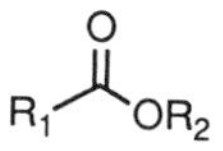

II 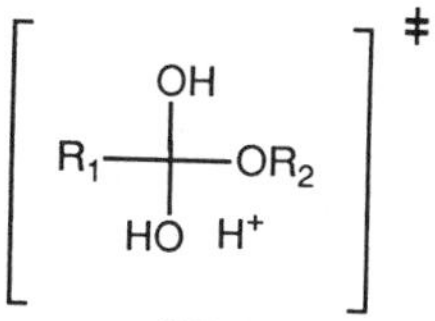 III 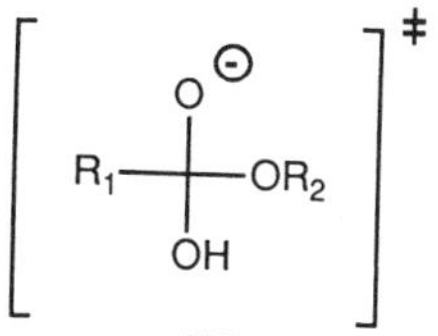 IV

By varying R_1, but keeping R_2, solvent, and temperature constant, Taft proposed that the steric effect of R_1 as compared to methyl may be derived directly from the reaction rate constant k_A of the acid-catalyzed hydrolysis reaction:

$$E_s = \log\left(\frac{k_A}{k_{A,ref}}\right) \tag{12-30}$$

This implies that for the acid-catalyzed hydrolysis (in contrast to the base-catalyzed reaction), the inductive effect of R (as compared to methyl) on $\Delta G^{\ddagger}$ can be neglected. We will rationalize this assumption later when we talk about the mechanism of this type of reaction (Section 12.3). Experimentally, the assumption is supported by the very small Hammett ρ-values ($\rho \ll 0.5$) found for the acid-catalyzed hydrolysis of meta- and para-substituted benzoic acid esters (Shorter, 1973) as compared to the base-catalyzed reaction for which ρ values of about 2.5 are obtained (see example discussed above). It should be added, however, that in cases in which conjugation of R_1 with $COOR_2$ is important, E_s as determined by Eq. 12-30 may contain an electronic resonance contribution (i.e., stabilization of the ground state).

For determining the inductive (polar) substituent constant σ^*, one then needs to simply subtract the steric effect (Eq. 12-30) from the total (polar plus steric) effect observed in the base-catalyzed reaction. The σ^* value is thus defined as

$$\sigma^* = \log\left(\frac{k_B}{k_{B,ref}} - \log\frac{k_A}{k_{A,ref}}\right) \tag{12-31}$$

Note that in some textbooks, $\sigma^{*\prime} = \sigma^*/2.48$ values are used in order to get values of similar magnitude as the Hammett σ values (for more details see Shorter, 1973). As already pointed out, σ^* expresses exclusively inductive effects only if the steric effect of the substituent is similar in both the acid- and base-catalyzed hydrolysis of R_1COOR_2. This appears reasonable since from a geometrical point of view, both reactions start out with the same (trigonal) initial state (II), and the (tetrahedral) transition states (III and IV) differ by only two protons that are very small in size.

From Table 12.4 it can be seen that, as we would have expected from our experience with the Hammett σ constants, substituents containing highly electronegative atoms (e.g., $—CCl_3$, $—CHCl_2$, $—CHF_2$) have large positive σ^* values; that is, such substituents significantly increase the rates of reactions with a positive ρ^* value (e.g., hydrolysis reactions of carboxylic acid derivatives) provided that there is not a large steric effect (see discussion below). Similarly, electron-donating substituents (alkyl groups) have negative but generally small σ^* values. From the σ^* values of the multiply substituted groups (e.g., $—CH_2Cl$ vs. $—CHCl_2$ vs. $—CCl_3$), we can also see that the polar effects are approximately additive.

But exactly what does the steric constant E_s measure? Looking at the transition states of both the acid- and the base-catalyzed hydrolysis (III, IV), we can see that the activated complexes are more "crowded" then the initial state, since a water molecule or hydroxide ion has moved in. Compared to the ground state, one can, therefore, expect an increase in repulsions between the different groups or atoms (a potential energy effect), as well as an increase in mutual interference of these groups or atoms with each other's motions (a kinetic energy effect). Both effects lead to an increase in the free energy of activation, which is reflected in the negative E_s values found for all substituents expect H (see Table 12.4). Table 12.4 also shows that, as we would intuitively expect, more "bulky" substituents have more negative E_s values (e.g., compare *n*-butyl and *t*-butyl). Hence, as we show in the following section, bulky substituents that are close to the reaction center and that are not strongly electron withdrawing significantly decrease the reaction rate. Such steric effects are commonly referred to as "*steric hindrance*" of a reaction.

As can be seen from Table 12.4, substituents that are conjugated with $COOR_2$ [e.g., $CH_3—CH{=}CH—$, phenyl ($C_6H_5—$)] have rather negative E_s values. In these cases, the E_s values are governed to a large extent by a resonance effect that is important only in the ground state but not in the transition state. Breaking up such conjugation leads to an increase in $\Delta G^{\ddagger}$ as compared to a nonconjugated substituent of the same size.

In summary, we note that LFERs such as the Hammett and the Taft equations are very useful tools for evaluating the effect of structural moieties on the rate at which a given type of reaction occurs at a given functional group, and for predicting reaction rates from structurally related compounds. However, when applying LFERs to kinetic (as well as to equilibrium) data, one has to be very careful to use a set of substituent constants that *properly reflects the electronic and steric effects* of the substituents on the reaction considered. We have seen that substituent constants are derived from a particular reaction, and it is logical that these constants can only be applied to other reactions if the different substituents exhibit the same relative effects on the free energy of activation, or, when considering reaction equilibria, on the standard free energy change of these reactions. In the case of electronic substituents, the major concern is to evaluate whether resonance effects play a role. As discussed in Chapter 8, when applying the Hammett equation to relate pK_a values of phenols, we needed to use a different set of σ substituent constants (σ^- constants) to account for resonance. Nevertheless, it is usually possible to treat electronic effects quite reasonably. The quantification of steric effects is commonly more difficult. The steric effect of a substituent depends on the geometry of the molecule in the ground state and in the activated complex. In our example of the carboxylic acid ester hydrolysis, the geometry at the carbon atom at which the reaction takes place changes from trigonal in the ground state to tetrahedral in the transition state. Hence, as demonstrated by various applications (see e.g., Shorter, 1973), E_s values are applied with satisfactory success to reactions involving such a geometric change. This includes the hydrolysis of all carboxylic acid derivatives such as esters, amides, and carbamates. In contrast, the E_s values deduced from ester hydrolysis seem to be less applicable to second-order nucleophilic substitution of alkyl halides which involve a change in geometry from tetrahedral to trigonal bipyramidal.

After this brief review of some important aspects of chemical kinetics, we are now prepared to have a closer look at some specific chemical reactions of organic compounds in the environment.

12.3 NONREDUCTIVE CHEMICAL REACTIONS INVOLVING NUCLEOPHILIC SPECIES

General Remarks

In Chapter 2 we noted that covalent bonds between two atoms of different electronegativity (e.g., carbon and halogens, carbon and oxygen) are polar; that is, one of the atoms carries a partial positive charge (usually carbon), whereas the other one exhibits a partial negative charge (e.g., halogen, oxygen). In organic molecules, such a polar bond may now become the site of a chemical reaction in that either a *nucleophilic* species (nucleus-liking and, hence, positive-charge-liking species) is attracted by the electron-deficient atom of the bond, or an *electrophilic* species (electron-liking and, hence, negative-charge-liking species) is attracted by the electron-rich atom. In the environment, the majority of the chemical species that may chemically react with organic compounds are inorganic *nucleophiles* (see examples given in Table 12.5). Because of the large abundance of such nucleophiles in the

TABLE 12.5 Examples of Important Environmental Nucleophiles and Their Nucleophilicities Relative to Water Based on Reaction with Methyl Bromide[a]

Nucleophile	n
ClO_4^-	<0
H_2O	0.0
NO_3^-	1.0
F^-	2.0
SO_4^{2-}	2.5
CH_3COO^-	2.7
Cl^-	3.0
HCO_3^-	3.8
HPO_4^{2-}	3.8
Br^-	3.9
OH^-	4.2
I^-	5.0
CN^-	5.1
HS^-	5.1

[a]See Eq. 12-34. From Hine, 1962.

environment, reactive *electrophiles* are very short-lived, and, therefore, reactions of organic compounds with such species occur usually only in light-induced or biologically mediated processes (see Chapters 13 and 14).

As can be derived from Table 12.5, nucleophilic species possess a partial or full negative charge and/or have nonbonded valence electrons. As a consequence of an encounter with an organic molecule exhibiting a polar bond, the electron-rich atom of the nucleophile may form a bond with the electron-deficient atom in the organic molecule, thus causing a modification of the organic compound. Since a new bond is formed by this process, another bond has to be broken at the atom at which the reaction occurs, which usually (but not always) means that a group (or atom) is split off from the organic compound. Such a group (or atom) is commonly referred to as a *leaving group*.

Because of its great abundance, water plays a pivotal role among the nucleophiles present in the environment. A reaction in which a water molecule (or hydroxide ion) substitutes for another atom or group of atoms present in an organic molecule is commonly called a *hydrolysis* reaction. We have already encountered this very important chemical process in natural waters (see examples given in Table 12.1). We note that in a hydrolysis reaction, the compound is transformed into more polar products that have quite different properties, and, therefore, a different environmental behavior than the starting chemical. We also note that the products of hydrolysis are usually of somewhat less environmental concern as compared to the parent compound. This is, however, not necessarily true for the products of reactions involving nucleophiles other than water or hydroxide ion. Finally, hydrolysis reactions usually exhibit rather large negative ΔG values. Consequently, as a practical matter we assume that these and many other reactions involving nucleophiles proceed in only one direction; that is, we consider them to be irreversible.

Nucleophilic Displacement of Halogens at Saturated Carbon Atoms

With our first example of chemical reactions, we want to get acquainted with a very important type of reaction in organic chemistry, that is, with *nucleophilic substitution* at a *saturated* carbon atom. Since halogens are very common constituents of man-made organic chemicals, we consider their displacement by environmentally relevant nucleophiles; that is, we look at cases in which a halogen plays the role of the leaving group.

To describe aliphatic nucleophilic substitution reactions, it is useful to consider two different reaction mechanisms representing two extreme cases. In the first case shown in Figure 12.3, we picture the reaction to occur because a nucleophile (e.g., Y^-) "attacks" the carbon atom from the side opposite to the leaving group, X^- (e.g., halogen). In the transition state, which is postulated to exhibit a trigonal bipyramidal geometry, the nucleophile is then thought to be partly bound to the carbon atom, and the leaving group is postulated to be partly dissociated. Hence, in such a simple picture, we consider the nucleophile to push the leaving group out of the molecule. We note that two electrons of the new bond formed are provided by the nucleophile, while the leaving group takes two electrons of the bond that is broken.

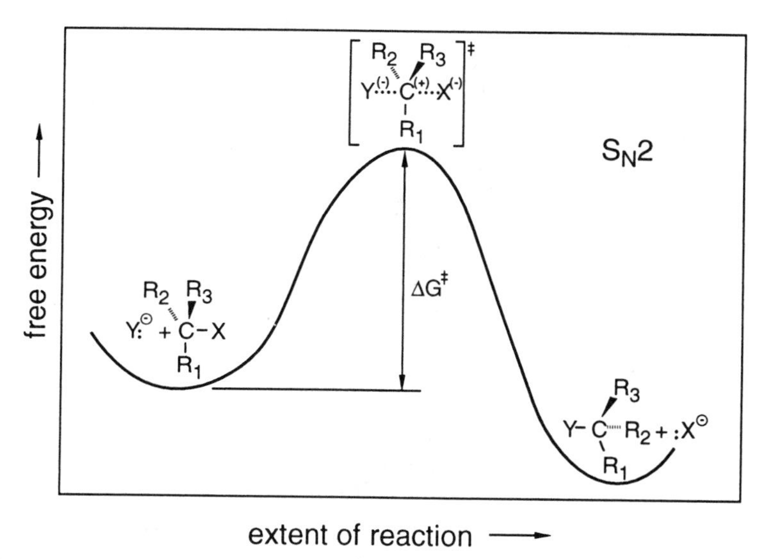

Figure 12.3 Two-dimensional portrayal of relative free energies exhibited by the educts, activated complex, and products of an S_N2 reaction.

S_N2 *Mechanism* In this first case, the free energy of activation $\Delta G^\ddagger$, and thus the rate of the reaction, depends strongly on both the capability of the nucleophile to initiate a substitution reaction and the willingness of the organic molecule to undergo that reaction. The former factor may be expressed by the relative nucleophilicity of the nucleophile, an entity that can be quantified (see discussion below). The latter contribution to $\Delta G^\ddagger$, however, is more difficult to quantify since it incorporates various electronic and steric factors that are strongly determined by the structure of the organic molecule. As we can imagine, $\Delta G^\ddagger$, depends upon the facility with which the nucleophile can get to the site of reaction (i.e., how much steric hindrance there is), upon the charge distribution at the reaction center, and upon how easily the leaving group will split from the molecule.

If a reaction occurs by this first mechanism, it is commonly termed an S_N2 reaction (i.e., substitution, nucleophilic, bimolecular). It represents an example of a simple elementary bimolecular reaction, as we discussed in the previous section, and it therefore follows a second-order kinetic rate law:

$$\text{rate} = -k[R_1R_2R_3CX][Y^-] \tag{12-32}$$

where k is a second-order rate constant (e.g., $M^{-1}\,s^{-1}$).

S_N1 *Mechanism* A second mechanism, differing substantially from the first, is one in which we postulate that the substitution reaction occurs in two steps. As illustrated by Figure 12.4, in the first (rate-determining) step, the leaving group is completely dissociated from the organic compound, and because it takes both electrons with it,

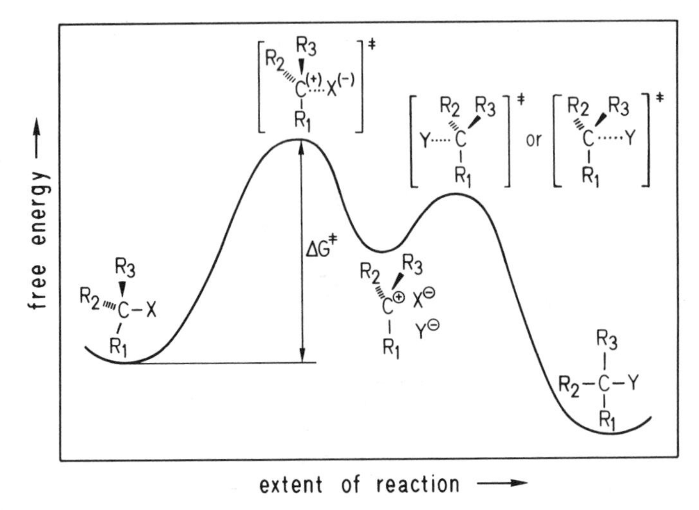

Figure 12.4 Two-dimensional portrayal of the relative free energies exhibited by educt, activated complexes, an intermediate, and product of an S_N1 reaction.

a (planar) carbocation is formed as an *intermediate.* In the second, faster step, this reactive carbocation then combines with a nucleophile to form a product. In this case, the reaction rate depends solely on how easy the leaving group dissociates from the molecule. Since the "structure" of the activated complex can be assumed to resemble the structure of the intermediate (see Fig. 12.4), an important factor determining $\Delta G^{\ddagger}$ is the stability of the carbocation formed. Hence this mechanism is favored in cases where the carbocation is stabilized, for example, by resonance.

If a reaction occurs exclusively by this second mechanism, the observed rate law is first-order:

$$\text{rate} = -k[R_1R_2R_3CX] \tag{12-33}$$

where k is now a first-order rate constant (e.g., s^{-1}). The reaction is then said to occur by an S_N1 (i.e., substitution, nucleophilic, unimolecular) *mechanism.* Note that in aqueous solution, the S_N1 mechanism will, in general strongly favor the formation of the hydrolysis product (i.e., substitution of —X by —OH) because the nucleophiles are not involved in the rate-limiting step, and water molecules are present in such overwhelming abundance that one of them will have the greatest likelihood of available nucleophiles to collide with the reactive carbocation.

Relative Nucleophilicity of Inorganic Nucleophiles Let us first look at some reactions that occur predominantly by an *S_N2 mechanism.* This will allow us to evaluate the relative nucleophilicities of some important environmental nucleophiles. We consider nucleophilic substitution in aqueous solution of methyl halides (CH_3X, X = F, Cl, Br, I) which are important gases in the marine and freshwater environments (Zafiriou, 1975; Pearson, 1982a). In Figure 12.5, the second-order rate constants at

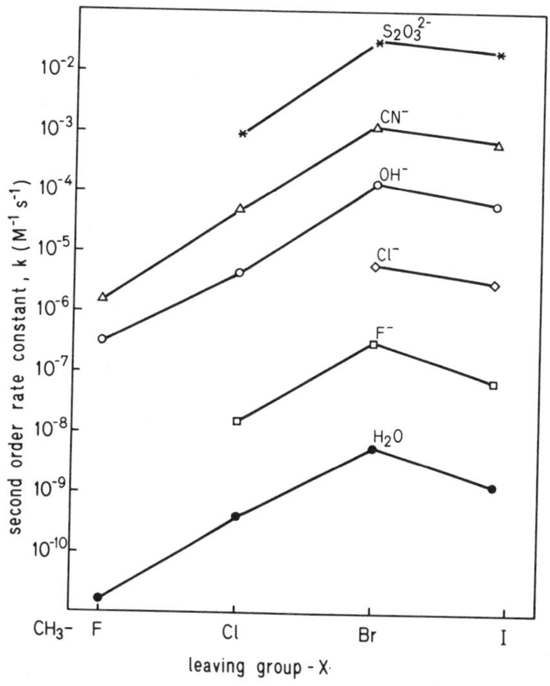

Figure 12.5 Rate constants for reactions of methyl halides with various nucleophiles (data from Hughes, 1971; Mabey and Mill, 1978).

25°C for the reactions of the methyl halides with various nucleophiles are plotted for each X. From these data, we may derive two important, quite generally applicable, conclusions. First, we recognize that a given methyl halide shows the same relative reactivity toward the various nucleophiles as the other methyl halides. These findings have been quantified in a linear free-energy relationship by Swain and Scott (1953):

$$\log\left(\frac{k}{k_{H_2O}}\right) = s \cdot n \tag{12-34}$$

where k is the *second-order* rate constant for a nucleophilic displacement by a nucleophile of interest, k_{H_2O} in the *second-order* rate constant for nucleophilic attack by water (the standard nucleophile), n is a measure of the attacking aptitude or nucleophilicity of the nucleophile of interest, and s reflects the sensitivity of the organic molecule to nucleophilic attack. The n values of some important environmental nucleophiles are given in Table 12.5. The major factors influencing the nucleophilicity of a chemical species (and thus the magnitude of its n value) are the ability of the bonding atom to donate its electrons to form the transition state and the ease with which the nucleophile can leave the solution to get to the reaction center. Hence, nucleophilicity increases with decreasing solvation energy of the nucleophile. Since the valence electrons of larger atoms (e.g., S, I) are more polarizable (they are further away from the nucleus), and since larger species have, in general, lower solvation energies, they are better nucleophiles as compared to smaller species. Thus, we can qualitatively understand why, for example, nucleophilicity increases from F^- to Cl^- to Br^- to I^- (Table 12.5), and why SH^- is a stronger nucleophile than OH^-.

As a standard for the sensitivity values, s is set equal to 1.0 for S_N2 reactions of methyl bromide. Hence, a compound for which the reaction rate of nucleophilic substitution is more dependent than methyl bromide on the nucleophilicity of the attacking group will have an s value greater than 1, and one that is less dependent will have a smaller value. From the few s values available, we can conclude that for many of the halogenated compounds of interest to us, we may, as a rough approximation, assume s to be close to 1. By doing so, we can now estimate at what concentration a given nucleophile must be present in a natural water in order to compete with H_2O in an S_N2 reaction with an alkyl halide. For some nucleophiles, Table 12.6 gives the calculated concentrations, $[\text{nucl}]_{50\%}$, at which the two reactions are equally important; that is, $k[\text{nucl}]_{50\%} = k_{H_2O}[H_2O]$:

$$[\text{nucl}]_{50\%} = 55.5 \cdot 10^{-n}\,\text{M} \qquad (12\text{-}35)$$

Although the values given in Table 12.6 represent only order-of-magnitude estimates, they allow some important conclusions. First, in uncontaminated freshwaters (where bicarbonate typically occurs at about 10^{-3} M, chloride and sulfate occur at about 10^{-4} M, and hydroxide is micromolar or less, Stumm and Morgan, 1981), the concentrations of nucleophiles are usually too small to compete with water in S_N2 reactions involving aliphatic halides. Hence the major reaction will be the displacement of the halide by water molecules. In salty or contaminated waters, however, nucleophilic substitution reactions other than hydrolysis may occur. Zafiriou (1975) for example, has demonstrated that in seawater ($[Cl^-] \approx 0.5$ M), a major sink for methyl iodide is transformation to methyl chloride:

$$CH_3I + Cl^- \rightarrow CH_3Cl + I^- \qquad (12\text{-}36)$$

TABLE 12.6 Estimated Concentration of Nucleophile Required to Compete with Water in an S_N2 Reaction with Alkyl Halides[a]

Nucleophile	$[\text{nucl}]_{50\%}$ (M)	Nucleophile	$[\text{nucl}]_{50\%}$ (M)
NO_3^-	≈ 6	Br^-	$\approx 7 \times 10^{-3}$
F^-	$\approx 6 \times 10^{-1}$	OH^-	$\approx 4 \times 10^{-3}$
SO_4^{2-}	$\approx 2 \times 10^{-1}$	I^-	$\approx 6 \times 10^{-4}$
Cl^-	$\approx 6 \times 10^{-2}$	SH^-	$\approx 4 \times 10^{-4}$
HCO_3^-	$\approx 9 \times 10^{-3}$	CN^-	$\approx 4 \times 10^{-4}$
HPO_4^{2-}	$\approx 9 \times 10^{-3}$		

[a]Here $[\text{nucl}]_{50\%}$ is the concentration at which the two reactions are equally important. We use Eq. 12-35 with the n values given in Table 12.5

The half-life with respect to thermochemical transformation of CH_3I in seawater at 20°C was determined to be 20 days, as compared with about 200 days in freshwater (reaction with H_2O yielding CH_3OH). In a case of a groundwater contamination with several alkyl bromides, Schwarzenbach et al. (1985) reported the formation of dialkyl sulfides under sulfate-reducing conditions in an aquifer. They postulated that in an initial reaction, primary alkyl bromides reacted with HS^- by an S_N2 mechanism to yield the corresponding mercaptans (thiols):

$$R{-}CH_2{-}Br + HS^- \rightarrow R{-}CH_2{-}SH + Br^- \qquad (12\text{-}37)$$

These mercaptans then reacted further to yield rather hazardous products. We return to this case later.

A final conclusion that we may draw from Table 12.6 is that the reaction of aliphatic halides with OH^- should be unimportant at pH values below about 10. Since the hydrolysis of a carbon–halogen bond is commonly not catalyzed by acids, one can assume that in most cases, the hydrolysis rate of aliphatic halides will be *independent of pH* at typical ambient conditions. Hence, regardless of whether hydrolysis occurs by an S_N1 or S_N2 mechanism (or a mixture of both, see below), the reaction may be described by a first-order rate law (Eq. 12-33). The first-order rate constant is then commonly denoted as k_N $[=k\cdot(H_2O)]$ to express neutral hydrolysis.

Leaving Groups We now return to Figure 12.5 to learn something about the various halogens as leaving groups. It is tempting to assume that the weaker a nucleophile, the better leaving group it should be. Hence we would expect the reactivities of the methyl halides to decrease in the order $CH_3F > CH_3Cl > CH_3Br > CH_3I$. However, what is experimentally found (Fig. 12.5) is almost the opposite, namely, the reaction rate decreases in the order $CH_3Br \sim CH_3I > CH_3Cl > CH_3F$. The major reason for these findings is the increasing strength of the C–X bond (that has to be broken) when going from C–F to C–I (Table 2.2). This bond-strength factor proves to be dominant in determining the much slower reaction rates of C–Cl and, in particular, C–F bonds as compared to C–Br and C–I.

Let us now look at some examples to illustrate what we have discussed so far, and to get a feeling of how structural moieties influence the mechanisms and the rates of nucleophilic substitution reactions of halogenated hydrocarbons in the environment. In Table 12.7 the (neutral) hydrolysis half-lives are given for various monohalogenated compounds at 25°C. Also indicated are the postulated reaction mechanisms with which these compounds undergo nucleophilic substitution reactions. As anticipated, we can see that for a given type of compound, the carbon–bromine and carbon–iodine bonds hydrolyze fastest, about 1–2 orders of magnitude faster than the carbon–chlorine bond. Furthermore, we note that for the compounds of interest to us, hydrolysis of carbon–fluorine bonds is likely to be too slow to be of great environmental significance.

When comparing the hydrolysis half-lives of the alkyl halides in Table 12.7, we notice that the reaction rates increase dramatically when going from primary to secondary to tertiary carbon–halogen bonds. In this series, increasing the stabilization

TABLE 12.7 Postulated Reaction Mechanisms and Hydrolysis Half-Lives at 25°C of Some Monohalogenated Hydrocarbons at Neutral pH[a]

Compound	$t_{1/2}$(hydrolysis)				Dominant Mechanisms in Nucleophilic Substitution Reactions
	X = F	Cl	Br	I	
$R-CH_2-X$	≈ 30 yr[b]	340 d[b]	20–40 d[c]	50–110 d[d]	S_N2
$(H_3C)_2CH-X$		38 d	2 d	3 d	$S_N2 \ldots S_N1$
$(CH_3)_3C-X$	50 d	23 s			S_N1
$H_2C{=}CH-CH_2-X$		69 d	0.5 d	2 d	$(S_N2) \ldots S_N1$
$C_6H_5-CH_2-X$		15 h	0.4 h		S_N1

[a]Data taken from Mabey and Mill, 1978.
[b]R=H.
[c]R=H, C_1 to C_5-*n*-alkyl.
[d]R=H, CH_3.

of the carbocation by the electron-donating methyl groups decreases the activation energy needed to form this intermediate, thereby shifting the reaction to an increasingly S_N1-like mechanism. Similarly, faster hydrolysis rates and increasing S_N1 character can be expected if stabilization is possible by resonance with a double bond or an aromatic ring. As indicated by the denotation $S_N2 \cdots S_N1$ in Table 12.7, it is in some cases not possible (nor feasible) to assign a strict S_N2 or S_N1 character to a given nucleophilic substitution reaction. We recall that we refer to an S_N2 mechanism if the nucleophile plays the most important role it can play in the nucleophilic substitution reaction. In the other extreme, in the S_N1 case, the nucleophile is not relevant at all for determining the reaction rate. It is now easy to imagine that depending on the nucleophile and on various steric (e.g., steric hindrance) and electronic (e.g., stabilization by conjugation) factors, the relative importance of the nucleophile may well lay somewhere in between these two extremes. We may, therefore, simply look at such cases as exhibiting properties intermediate between S_N1 and S_N2 mechanisms.

With respect to possible product formation, we have seen that other nucleophiles may compete with water only if present in appreciable concentrations (see Table 12.6), and if the reaction occurs by an S_N2-like mechanism. An interesting example illustrating the above mentioned intermediate situation is the already mentioned case study

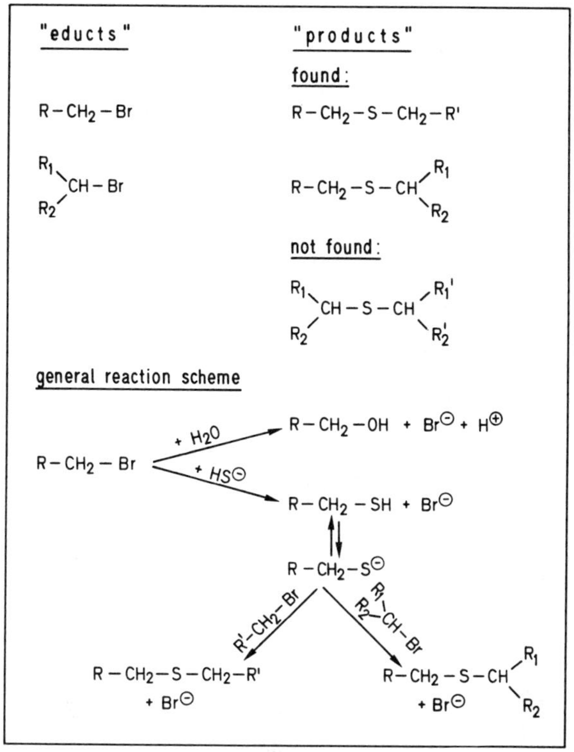

"educts"

$R-CH_2-Br$

R_1R_2CH-Br

"products"

found:

$R-CH_2-S-CH_2-R'$

$R-CH_2-S-CHR_1R_2$

not found:

$R_1R_2CH-S-CHR_1'R_2'$

general reaction scheme

$R-CH_2-Br \xrightarrow{+H_2O} R-CH_2-OH + Br^{\ominus} + H^{\oplus}$

$R-CH_2-Br \xrightarrow{+HS^{\ominus}} R-CH_2-SH + Br^{\ominus}$

$R-CH_2-SH \rightleftharpoons R-CH_2-S^{\ominus}$

$R-CH_2-S^{\ominus} \xrightarrow{R'-CH_2-Br} R-CH_2-S-CH_2-R' + Br^{\ominus}$

$R-CH_2-S^{\ominus} \xrightarrow{R_1R_2CH-Br} R-CH_2-S-CHR_1R_2 + Br^{\ominus}$

Figure 12.6 Alkyl bromides leaked into groundwater and thioethers found several years later; the reaction scheme shown can account for the products seen (for details, see Schwarzenbach et al., 1985).

of a groundwater contamination by primary and secondary alkyl bromides. In this case, a series of short-chain alkyl bromides (Fig. 12.6) were introduced continuously into the ground by wastewater also containing high concentrations of sulfate (SO_4^{2-}). Due to the activity of sulfate-reducing bacteria (see Section 12.4), hydrogen sulfide (H_2S/HS^-) was formed which, in turn, reacted with the alkyl bromides to yield alkyl mercaptans or thiols (Fig. 12.6). The mercaptans (RSH/RS^-), which are even better nucleophiles than H_2S/HS^-, then reacted further with other alkyl bromide molecules resulting in the formation of a whole series of dialkyl sulfides and other hazardous products (for more details see Schwarzenbach et al., 1985). Of interest to us here is the fact that all possible dialkyl sulfides exhibiting at least one primary alkyl group were found, but that no compounds with two secondary alkyl groups could be detected. These results suggest that the secondary alkyl bromides were reacting chiefly via an S_N1 mechanism, thereby yielding secondary alcohols, and it was not until the primary alkyl mercaptanates, which are particularly strong nucleophiles, appeared that the secondary bromides also became involved in a more S_N2-like reaction.

Table 12.8 Kinetic Data on Nucleophilic Substitution and Nonreductive Elimination (Dehydrohalogenation) Reactions of Some Polyhalogenated Hydrocarbons in Aqueous Solution[a]

Compound	Major Product(s)	$k_N(25°C)$ (s^{-1})	$k_B(25°C)$ ($M^{-1}s^{-1}$)	$t_{1/2}$ at pH 7 and 25°C	log A (s^{-1} or $M^{-1}s^{-1}$)	E_a ($kJ\ mole^{-1}$)
CH_2Cl_2[b]	(CH_2O)	3×10^{-11}	2×10^{-8}	≈700 yr		
$CHCl_3$[b]	(HCOOH)	7×10^{-13}	7×10^{-5}	≈3500 yr		
$CHBr_3$[b]	(HCOOH)		3×10^{-4}	≈700 yr		
$BrCH_2$–CH_2Br[c,d]	$HOCH_2CH_2OH$ (>75%) and $CH_2{=}CHBr$	6×10^{-9}		≈4 yr	10.5	105
Cl_2CH–$CHCl_2$[c]	$ClCH{=}CCl_2$		2×10^{0}	40 d		
CH_3–CCl_3[c,d]	CH_3–COOH (≈80%) and $CH_2{=}CCl_2$ (≈20%)	2×10^{-8}		≈400 d	13	118
$BrCH_2$–CHBr–CH_2Cl[e]	$CH_2{=}CBr$–CH_2OH (>95%)	$\approx 10^{-10}$	6×10^{-3}	≈40 yr	14[f]	93[f]

[a]Note that most data is extrapolated from experimental data obtained at elevated temperatures.
[b]Data from Mabey and Mill (1978).
[c]Data from Haag and Mill (1988).
[d]Kinetic data from disappearance of parent compound.
[e]Data from Burlinson et al. (1982).
[f]Determined at pH 6.8, where base-catalyzed reaction is still dominant.

Polyhalogenated Hydrocarbons—Elimination Mechanisms So far, we have considered only monohalogenated compounds. As we have seen in Chapter 2 there are, however, a variety of polyhalogenated alkanes that are of great environmental concern. Table 12.8 summarizes some of the kinetic data available on the reactivity of such compounds in aqueous solution. Some important conclusions can be drawn from Table 12.8. Firstly, we notice that polyhalogenated methanes hydrolyze extremely slowly under environmental conditions. This result is mostly due to steric hindrance and to back-bonding by the relatively electron-rich bulky halogens (Hughes, 1971). Hence, nucleophilic substitution reactions of such compounds are of minor environmental significance. However, as we will see later, the polyhalogenated methanes as well as other polyhalogenated compounds may, under certain environmental conditions, react by another reaction pathway, namely, reductive dehalogenation (see Section 12.4).

From the reaction products of the polyhalogenated ethanes and propanes shown in Table 12.8 we notice that halogenated compounds may react in aqueous solution by yet another type of reaction, so-called *β-elimination*. In this reaction, in addition to the leaving group —X, a proton is lost from an adjacent carbon atom (hence the prefix *β*-) and a double bond is formed:

$$\underset{\text{H}\quad\text{X}}{-\overset{|}{\underset{|}{C}}-\overset{|}{\underset{|}{C}}-} \xrightarrow[\text{-HX}]{\beta\text{-elimination}} \;\rangle C{=}C\langle \qquad (12\text{-}38)$$

If X = halogen, this type of reaction in which HX is eliminated from a molecule is referred to as *dehydrohalogenation*. Thus, when assessing the fate of halogenated compounds is natural waters, this process has to be considered in addition to nucleophilic substitution. The question then is which structural features and environmental conditions determine whether only one or both of these two competing types of reactions will be important.

From the few studies conducted to date, it is not possible to derive quantitative structure–reactivity relationships for predicting the relative importance of elimination versus nucleophilic substitution reactions of halogenated compounds in water. Generally, *β*-elimination is only important in molecules in which nucleophilic substitution is sterically hindered and/or in which relatively acidic protons are present at carbon atoms adjacent to the carbon carrying the leaving group. These criteria are optimally met in 1,1,2,2-tetrachloroethane where the four electron-withdrawing chlorine atoms render the hydrogens more acidic and, simultaneously, these relatively large halide substituents hinder nucleophilic attack. In water 1,1,2,2-tetrachloroethane is converted more or less quantitatively to trichloroethylene (Haag and Mill, 1988) by a so-called *E2* (elimination, bimolecular) *elimination*; that is, the elimination takes place in a "concerted reaction" in which there is only one transition state; and the reaction, therefore, follows a second-order kinetic rate law (Eq. 12-32 with, e.g., $Y^- = OH^-$):

$$\text{Cl}_2\text{HC–CHCl}_2 + \text{HO}^{\ominus} \longrightarrow [\ldots]^{\ddagger} \longrightarrow \text{Cl}_2\text{C=CHCl} + H_2O + Cl^{\ominus} \quad (12\text{-}39)$$

We can picture the E2 reaction in a very similar way as we have done with the S_N2 reaction. Here, however, the nucleophile (usually OH^-) plays the role of a base that induces the breaking of a C—H bond at a β-carbon by leaving both electrons to the carbon atom. The resulting activated complex then contains a carbon which is partially (or, in the extreme case, fully) negatively charged. Hence, any group that stabilizes the negative charge at this carbon atom by induction or resonance will enhance the reaction rate. Note that this is equivalent to our earlier statement that the reaction occurs faster the more acidic the proton(s) is at the β-carbon(s). The electrons of this breaking (or broken) C—H bond now play the role of a nucleophile by attacking the leaving group from the backside (as the electrons of the nucleophile in an S_N2 reaction), thus causing the breaking of the C–X bond and the formation of a double bond. The steric requirements for optimal E2 elimination are, therefore, an antiplanar configuration of the atoms involved in the reaction as depicted in Eq. 12-39. Consequently, in ring systems, elimination might in some cases be hindered owing to such steric factors (the inability of the β-protons to be antiplanar to leaving groups as in β-hexachlorocyclohexane, see structure in Chapter 2).

The role of the leaving group in elimination reactions can, in general, be looked at in a very similar way as in S_N reactions. As illustrated by the relative amounts of elimination products formed by the base-catalyzed reactions of the pesticide, 1,2-dibromo-3-chloropropane (DBCP, Eq. 12-40), bromide is a better leaving group as compared to chloride (Burlinson et al., 1982):

(DBCP) —(-HBr)→ 95% (BCP) —(H_2O)→ (BAA)

(DBCP) —(-HCl)→ 5% (DBP) —(H_2O)→ (BAA) (12-40)

We note that in this case the elimination products, that is, 2-bromo-3-chloropropene (BCP) and 2,3-dibromopropene (DBP), are allylic halides that hydrolyze in relatively fast steps most likely via S_N1 reactions (see the example given in Table 12.7) to form 2-bromoallyl alcohol (BAA).

As indicated in Table 12.8, 1,2-dibromoethane ($BrCH_2$–CH_2Br) and 1,1,1-trichloroethane (CH_3–CCl_3) are examples in which both hydrolysis and elimination are important. If in such cases the reactions occur by S_N2 and E2 mechanisms, respectively, the ratio of the hydrolysis versus elimination products should vary with varying pH and temperature, since, in general, the two competing reactions will exhibit different pH and temperature dependencies. On the other hand, if the reaction mechanisms were more S_N1 and E1-like, a much less pronounced effect of temperature or pH on product formation would be expected, since the rate-determining step in aqueous solution may be considered to be identical for both reactions:

$$\underset{\mathrm{H}\quad\ \mathrm{X}}{-\overset{|}{\underset{|}{\mathrm{C}}}-\overset{|}{\underset{|}{\mathrm{C}}}-} \xrightarrow[-\mathrm{X}^{\ominus}]{\text{slow}} \underset{\text{intermediate}}{\underset{\mathrm{H}}{-\overset{|}{\underset{|}{\mathrm{C}}}-\overset{\oplus}{\mathrm{C}}\!\lt}} \quad \begin{cases} \xrightarrow[\text{fast } S_N1]{H_2O} & \underset{\mathrm{H}\quad\ \mathrm{OH}}{-\overset{|}{\underset{|}{\mathrm{C}}}-\overset{|}{\underset{|}{\mathrm{C}}}-} \\ \xrightarrow[-H^{\oplus}]{\text{fast } E_1} & \gt\mathrm{C}=\mathrm{C}\lt \end{cases} \tag{12-41}$$

We note that in Eq. 12-41 we have introduced the E1 (elimination, unimolecular) reaction, which commonly competes with the S_N1 reaction provided that an adjacent carbon atom carries one or several hydrogen atoms that may dissociate. We also note that similar to what we have stated earlier for nucleophilic substitution reactions, elimination reactions may occur by mechanisms representing borderline cases between E2 and E1.

From the experimental data available for the reactivities of 1,2-dibromoethane (EDB) and 1,1,1-trichloroethane in water (TrCE), it is not possible to draw sound conclusions on the mechanisms and on the pH and temperature dependence of product formation of the reactions of these compounds in water. It is, however, interesting to note that the overall reaction rate of TrCE was found to be pH independent below pH 11, and that temperature had no significant influence on the product formation in the temperature range between 25 and 80°C (Haag and Mill, 1988). These findings indicate that this compound undergoes S_N1 and E1 type reactions in aqueous solution. It should also be pointed out that the primary hydrolysis products of both EDB (i.e., $BrCH_2$–CH_2OH) and TrCE (i.e., CH_3–CCl_2OH) subsequently hydrolyze again in relatively fast reactions to yield the final products, ethylene glycol and acetic acid, respectively (see Table 12.8).

With these examples we conclude our discussion on nucleophilic substitution and β-elimination reactions involving saturated carbon–halogen bonds in environmental chemicals. Before we go on discussing another group of reactions, we need, however, to make some final remarks on S_N and E reactions of halogenated compounds. First, we note that the activation energies of the reactions in which a halogen is removed from a saturated carbon in an organic molecule by an S_N or E mechanism are between 80 and 120 $kJ \cdot mole^{-1}$. Hence, these reactions are quite sensitive to temperature; that is, a difference in 10°C means a difference in reaction rate of a factor of 3–5 (see Table 12.2). Secondly, we have seen that a compound may react by several competing reactions. In these cases, the general rate law will be a composite of the rate laws of the individual reactions:

$$\text{rate} = -\left\{k_N + k_{EN} + (k_B + k_{EB})[OH^-] + \sum_i k_i[\text{nucl.}i]\right\} C_w \tag{12-42}$$

where C_w is the concentration of the dissolved halogenated compound in water, k_N and k_{EN} are the (pseudo) first-order, and k_B and k_{EB} are the second-order rate constants for the neutral and base-catalyzed hydrolysis and elimination reactions, respectively, and k_i is the second-order rate constant of the S_N2 reaction with any other particular nucleophile i. Note that k_i may be estimated using the Swain–Scott relationship, Eq. 12-34. We recall that by assuming constant pH and constant nucleophile concentration(s), Eq. 12-42 can be reduced to a pseudo-first-order rate law with a pseudo-order rate constant k_{obs} that is given by

$$k_{obs} = k_N + k_{EN} + (k_B + k_{EB})[OH^-] + \sum_i k_i[\text{nucl.}i] \tag{12-43}$$

We should point out, however, that depending on the relative importance of the various reactions, k_{obs} may not be a simple function of pH and temperature, and that product formation may strongly depend on these two variables. Furthermore, we note that many environmentally important organic compounds exhibit halogen atoms bound to a carbon–carbon double bond, be it an olefinic (e.g., chlorinated ethylenes) or an aromatic (e.g., chlorinated benzene, PCB) system. Under environmental conditions, these carbon–halogen bonds undergo S_N or E reactions at extremely slow rates, and we may, therefore, consider these reactions to be unimportant. Finally, it should be pointed out that, so far, no catalysis of the hydrolysis of alkyl halides by solid surfaces has been observed (El-Amamy and Mill, 1984; Haag and Mill, 1988).

Hydrolytic Reactions of Acid Derivatives

In this section we consider a second important type of reaction in which a nucleophile attacks a carbon atom (or P or S) doubly bound to a heteroatom and singly bound to at least one other heteroatom. The major difference from the cases discussed in the previous section is that we are now considering nucleophilic reactions involving an *unsaturated* atom (e.g., C, P, S) exhibiting multiple bonds to other more electronegative atoms. Also, structural parts connected by singly bound heteroatoms may serve as leaving groups. Since in the environment such functional groups react predominantly with the nucleophiles H_2O and OH^-, we confine ourselves to hydrolytic reactions. As an illustration we may consider the reaction of an *acid derivative* with

OH^-, a reaction that, in many cases, occurs by the general reaction mechanism:

$$\begin{array}{c}X\\ \|\\ -Z-L\\ \uparrow\\ HO^{\ominus}\end{array} \longrightarrow \begin{array}{c}X^{\ominus}\\ |\\ -Z-L\\ |\\ HO\end{array} \longrightarrow \begin{array}{c}X\\ \|\\ -Z-OH\\ \\ +L^{\ominus}\end{array} \rightleftharpoons \begin{array}{c}X\\ \|\\ -Z-O^{\ominus}\\ \\ +HL\end{array} \qquad (12\text{-}44)$$

where Z is commonly C, P, or S, and X is O, S, or NR. The most common leaving groups, L^-, include RO^-, $R_1R_2N^-$, RS^-, and Cl^-. Figure 12.7 gives some examples of acid derivative functions frequently encountered in environmental chemicals. We note that if hydrolysis of such a functionality occurs by a mechanism similar to Eq. 12-44, the reaction products include the acid (under basic conditions usually present as the conjugate base) and the leaving group, which, in most cases of interest to us, is an alcohol, thiol, or an amine.

In the following discussion we look closely at the hydrolytic reactions of only a few of the functional groups shown in Figure 12.7. Our selection is based on two criteria, the relative importance of the functional groups and the availability of kinetic data for aqueous solutions. These data allow us to acquire a fairly good general idea of how structural features determine the hydrolysis behavior of such compounds in the environment.

Carboxylic Acid Esters Ester functions are among the most common acid derivatives present in natural as well as man-made chemicals (e.g., lipids, plasticizers, pesticides). An ester bond is defined as

$$\begin{array}{c}X\\ \|\\ -Z-O-R\end{array}$$

ester bond

where R is a carbon-centered substituent. Hence, hydrolysis of an ester bond yields the corresponding acid and the alcohol. If O is replaced by S, the functional group is referred to as thioester. Such thioesters are quite common in phosphoric acid and thiophosphoric acid derivatives (Fig. 12.7) that are used as pesticides. We first consider, however, the hydrolysis of a more familiar groups of esters that we encountered earlier in this chapter, the *carboxylic acid esters*:

$$R_1-C(=O)-O-R_2$$

We use this type of functionality to discuss some general mechanistic and structural aspects of hydrolysis that are valid not only for esters but also for other carboxylic and carbonic acid derivatives.

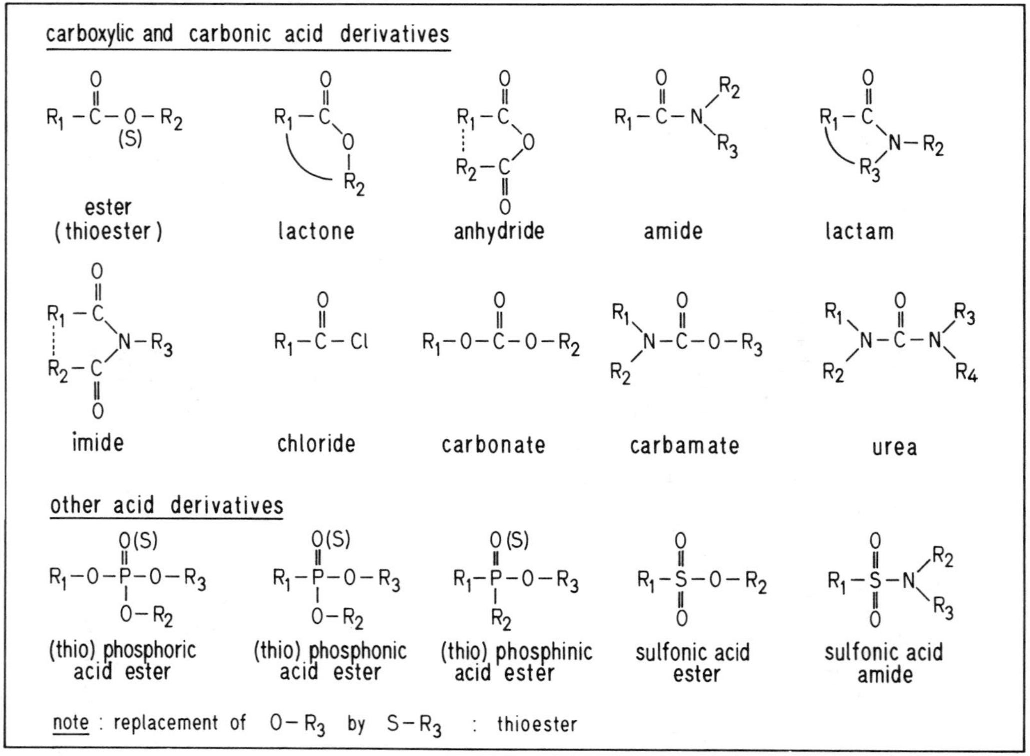

carboxylic and carbonic acid derivatives

ester (thioester) | lactone | anhydride | amide | lactam

imide | chloride | carbonate | carbamate | urea

other acid derivatives

(thio) phosphoric acid ester | (thio) phosphonic acid ester | (thio) phosphinic acid ester | sulfonic acid ester | sulfonic acid amide

note : replacement of $O-R_3$ by $S-R_3$: thioester

Figure 12.7 Examples of acid derivatives.

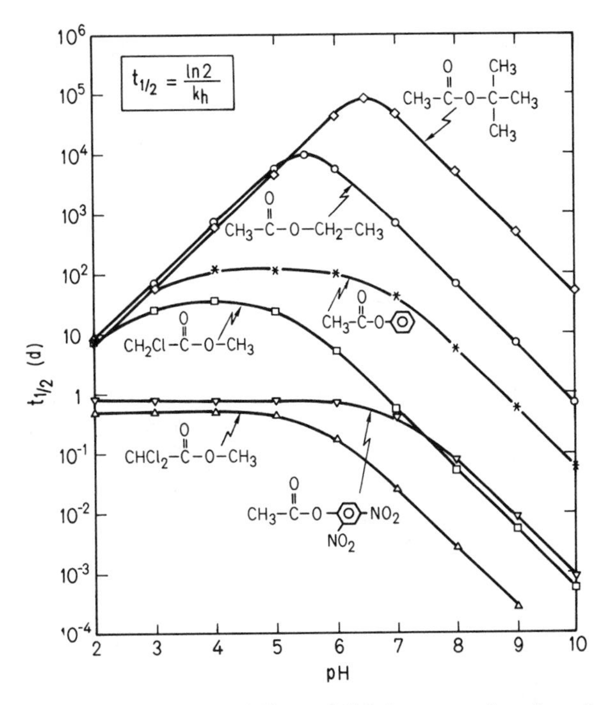

Figure 12.8 Variation of hydrolysis half-life at 25°C for several carboxylic acid esters as a function of solution pH caused by changing contributions of the acid-catalyzed, neutral, and base-catalyzed mechanisms.

Figure 12.8 shows the hydrolysis half-lives

$$t_{1/2(\text{hydrolysis})} = \frac{\ln 2}{k_h} \tag{12-45}$$

for some simple carboxylic acid esters as a function of pH, where k_h is the pseudo first-order hydrolysis rate constant at a given pH. From Figure 12.8 we can see that, in general, the hydrolysis rate of carboxylic acid esters is pH dependent over the ambient pH range. Recognizing that the curve sections that decrease with a slope of -1 as a function of pH reflect reactions mediated by OH^-, we notice that for all compounds, reaction with OH^- ("base catalysis") is important even at pH values below pH 7, and that acid catalysis (curve portions with slope of $+1$) is relevant only at relatively low pH's and only for compounds showing rather slow hydrolysis kinetics. By taking into account the acid-catalyzed (k_A, e.g., $M^{-1}\,s^{-1}$), neutral (k_{H_2O}, e.g., $M^{-1}\,s^{-1}$), and base-catalyzed (k_B, e.g., $M^{-1}\,s^{-1}$) reactions, we can express the observed (pseudo-first-order) hydrolysis rate constant, k_h (e.g., s^{-1}), as

$$k_h = k_A[H^+] + k_{H_2O}[H_2O] + k_B[OH^-] \tag{12-46}$$

and since $[H_2O]$ generally remains constant, we can simplify to

$$k_h = k_A[H^+] + k_N + k_B[OH^-] \quad (12\text{-}47)$$

where

$$k_N = k_{H_2O} \cdot [H_2O] \quad (12\text{-}48)$$

If k_A, k_N, and k_B are known for a given compound, we can calculate the pH values at which two reactions are equally important. As is schematically shown in Figure 12.9*a* and *b*, these pH values are given by the intersections, I, of the lines representing the contributions of each reaction to the overall reaction rate as a function of pH. Note that Figure 12.9 is drawn on a logarithmic scale. Hence, for example, I_{AB} is the pH at which $k_A[H^+] = k_B[OH^-]$. If we set $pH = -\log[H^+]$ and $[OH^-] = k_w/[H^+]$, then we obtain $I_{AB} = 0.5 \log(k_A/k_B K_w)$. If the neutral reaction (pH independent reaction with H_2O) is dominant over a wider pH range (case shown in Fig. 12.9*a*),

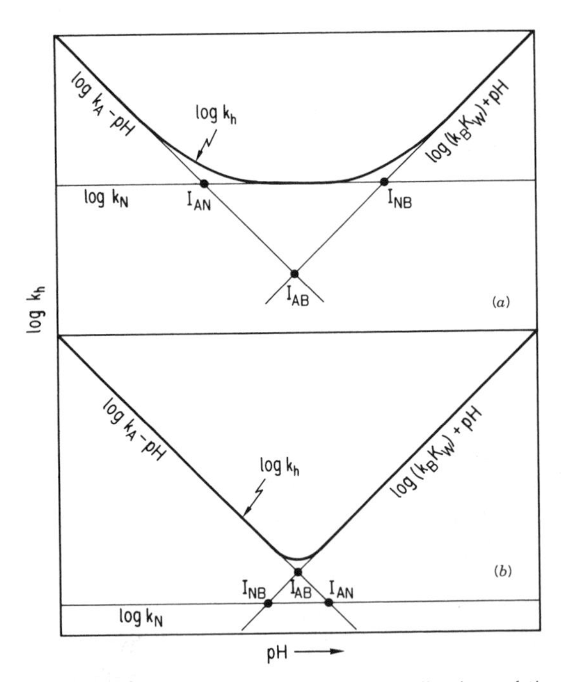

Figure 12.9 Schematic representation of the relative contribution of the acid-catalyzed, the neutral, and the base-catalyzed reactions to the overall hydrolysis rate as a function of solution pH: (*a*) neutral reaction rate is significant over some pH range, (*b*) the contributions of the neutral reaction can always be neglected.

then I_{AB} is only of theoretical value since both acid- and base-catalyzed reactions are unimportant at this pH. Similarly, if the neutral reaction is never important (case shown in Fig. 12.9*b*), then both I_{AN} and I_{NB} do not have much practical meaning. Before we discuss hydrolysis reactions in more detail, we should stress again that the neutral, acid-catalyzed, and base-catalyzed reactions are three very different reactions exhibiting different reaction mechanisms and, hence, different kinetic parameters. Thus, for example, because of different activation energies, the relative importance of each reaction (as is expressed by the I values) depends on temperature. Furthermore, as we have already seen during our discussion of the Taft LFER (Section 12.2), substituents usually have quite different effects on the neutral, acid-catalyzed, and base-catalyzed hydrolyses. Consequently, one has to be very careful to apply LFERs only to the kinetics of each individual reaction and not to the overall reaction (unless, of course, the overall reaction reflects only one dominant mechanism). Let us now look at each reaction more closely.

In Table 12.9, the hydrolysis rate constants and I values at 25°C are given for some carboxylic acid esters including the compounds shown in Figure 12.8. Note that the activation energies of ester hydrolysis reactions (data not shown) span quite a wide range between about 40 and 80 $kJ \cdot mole^{-1}$ (Kirby, 1972; Mabey and Mill, 1978). Hence, depending on structure and reaction mechanisms, reaction rates will change by a factor of between 2 and 3 for a 10-degree change in temperature. The data in Table 12.9 illustrate some general findings about the influence of structural moieties on the rate of the different hydrolytic reactions. First, we see that between the various compounds, relatively small differences are observed in the magnitude of k_A, which is in contrast to the large differences found for the k_N and k_B values, respectively. We have already noticed this fact when deriving the steric parameters E_s in Section 12.2, where we postulated that the acid-catalyzed hydrolysis of carboxylic acid esters is quite insensitive to electronic substituent effects. There we considered effects of substituents in the acid part of the ester functionality. From Table 12.9 we also see that structural differences in the leaving group (i.e., the alcohol) seem not to have a big influence on k_A, suggesting that dissociation of the leaving group is not rate determining. Let us try to rationalize these findings by looking at the reaction mechanisms of *acid-catalyzed hydrolysis*. We consider the mechanism believed to reflect the situation for most carboxylic acid esters; that is, the one in which the reaction proceeds through a tetrahedral intermediate. Figure 12.10 shows the postulated reaction scheme for this reaction. We recall that each of the elementary reaction steps is, in principle, reversible, and that the overall reaction rate of any chemical reaction is determined by the rate(s) of the slowest step(s).

Acid-Catalyzed Hydrolysis From Figure 12.10 we see that the species that undergoes the rate-determining step in the acid-catalyzed ester hydrolysis is the protonated ester and not the neutral compound. When the molecule is in this form, the general depletion of electrons near the central carbon promotes the approach of an electron-rich oxygen of a water molecule. Hence, the hydrolysis rate depends on how many of the compound molecules are protonated, that is, on how strong a base

TABLE 12.9 Rate Constants k_A, k_N, and k_B, Half-Lives at pH 7, and I Values for Hydrolysis of Some Carboxylic Acid Esters at 25°C[a]

Compound $R_1-C(=O)-O-R_2$								
R_1	R_2	k_A ($M^{-1}s^{-1}$)	k_N (s^{-1})	k_B ($M^{-1}s^{-1}$)	$t_{1/2}$ (pH 7)	I_{AN}[b,e]	I_{AB}[c,e]	I_{NB}[d,e]
CH_3-	$-CH_2CH_3$	1.1×10^{-4}	1.5×10^{-10}	1.1×10^{-1}	2 yr	(5.9)	5.5	(5.1)
CH_3-	$-C(CH_3)_3$	1.3×10^{-4}		1.5×10^{-3}	140 yr		6.5	
CH_3-	$-CH=CH_2$	1.4×10^{-4}	1.1×10^{-7}	1.0×10^{1}	7 d	3.1	(4.6)	6.0
CH_3-	$-C_6H_5$	7.8×10^{-5}	6.6×10^{-8}	1.4×10^{0}	38 d	3.1	(4.8)	6.7
CH_3-	$-C_6H_3(NO_2)_2$ (2,4-)		1.1×10^{-5}	9.4×10^{1}	10 h			7.1
CH_2Cl-	$-CH_3$	8.5×10^{-5}	2.1×10^{-7}	1.4×10^{2}	14 h	2.6	(3.9)	5.2
$CHCl_2-$	$-CH_3$	2.3×10^{-4}	1.5×10^{-5}	2.8×10^{3}	40 min	1.2	(3.5)	5.7
$CHCl_2-$	$-C_6H_5$		1.8×10^{-3}	1.3×10^{4}	4 min			7.1

[a] Data from Mabey and Mill (1978).
[b] $I_{AN} = \log(k_A/k_N)$.
[c] $I_{AB} = 1/2 \log(k_A/k_B K_W)$.
[d] $I_{NB} = \log(k_N/k_B K_w)$.
[e] Parentheses indicate that one or both of the processes is too slow to contribute significantly to the overall rate.

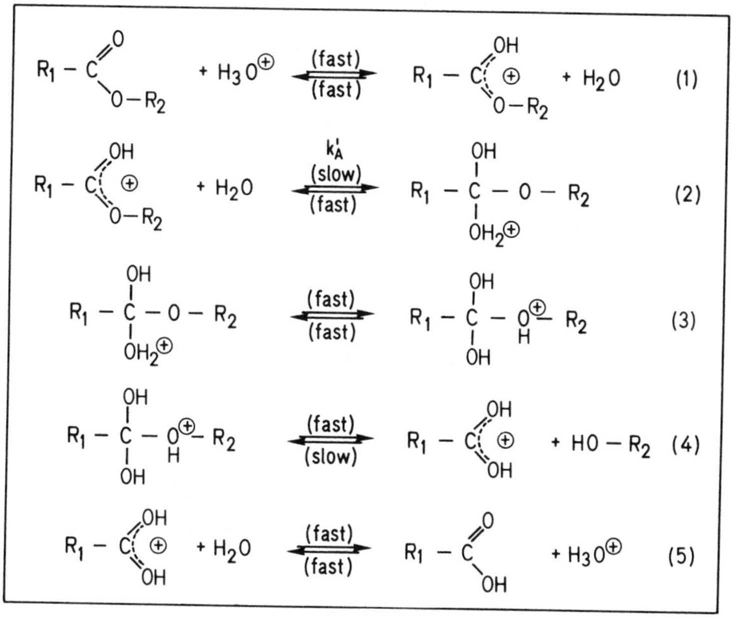

$$R_1-C(=O)-O-R_2 + H_3O^{\oplus} \underset{\text{(fast)}}{\overset{\text{(fast)}}{\rightleftharpoons}} R_1-C^{\oplus}(OH)-O-R_2 + H_2O \quad (1)$$

$$R_1-C^{\oplus}(OH)-O-R_2 + H_2O \underset{\text{(fast)}}{\overset{k'_A \ \text{(slow)}}{\rightleftharpoons}} R_1-C(OH)(OH_2^{\oplus})-O-R_2 \quad (2)$$

$$R_1-C(OH)(OH_2^{\oplus})-O-R_2 \underset{\text{(fast)}}{\overset{\text{(fast)}}{\rightleftharpoons}} R_1-C(OH)(OH)-O^{\oplus}(H)-R_2 \quad (3)$$

$$R_1-C(OH)(OH)-O^{\oplus}(H)-R_2 \underset{\text{(slow)}}{\overset{\text{(fast)}}{\rightleftharpoons}} R_1-C^{\oplus}(OH)-OH + HO-R_2 \quad (4)$$

$$R_1-C^{\oplus}(OH)-OH + H_2O \underset{\text{(fast)}}{\overset{\text{(fast)}}{\rightleftharpoons}} R_1-C(=O)-OH + H_3O^{\oplus} \quad (5)$$

Figure 12.10 Reaction scheme for the acid-catalyzed hydrolysis of carboxylic acid esters.

the ester function is. If we define an acidity constant K'_a for the protonated species $\left[R_1-C^{\oplus}(O)OR_2\right]$:

$$K'_a = \frac{\left[R_1-C(=O)OR_2\right] \cdot [H^+]}{\left[R_1-C^{\oplus}(O)OR_2\right]} \tag{12-49}$$

we can express the concentration of the protonated ester molecule as

$$\left[R_1-C^{\oplus}(O)OR_2\right] = \frac{1}{K'_a}\left[R_1-C(=O)OR_2\right] \cdot [H^+] \tag{12-50}$$

As indicated in Figure 12.10, the slowest, and therefore rate-determining, reaction step is then the nucleophilic attack of a water molecule at the carbonyl carbon of the protonated species, which is much more susceptible to nucleophilic attack than the neutral ester. Since the dissociation of the (protonated) leaving group ($HO-R_2$) is fast (forward reaction in reaction 4 in Fig. 12.10), the disappearance rate of the ester by acid catalyzed hydrolysis is given by

$$-\frac{d\left[R_1{-}C({=}O)OR_2\right]}{dt} = k'_A \left[R_1{-}C({\cdots}O)({\cdots}OR_2)^{\oplus}\right] \cdot [H_2O] \tag{12-51}$$

or, when substituting Eq. 12-50 into Eq. 12-51:

$$-\frac{d\left[R_1{-}C({=}O)OR_2\right]}{dt} = \frac{k'_A}{K'_a}\left[R_1{-}C({=}O)OR_2\right] \cdot [H^+] \cdot [H_2O]$$

$$= k_A \left[R_1{-}C({=}O)OR_2\right] \cdot [H^+] \tag{12-52}$$

Hence the second-order rate constant k_A is given by a combination of other constants:

$$k_A = \frac{k'_A \cdot [H_2O]}{K'_a} \tag{12-53}$$

Now we are in a better position to understand, at least qualitatively, why acid-catalyzed ester hydrolysis is relatively insensitive to electronic substituent effects. When considering the influence of an electron-withdrawing substituent on k_A, we can easily see that this substituent has two effects that compensate each other. On the one hand, the substituent will decrease the $\Delta G^{\ddagger}$ of the rate-limiting step (i.e., increase k'_A, see Reaction 2 in Fig. 12.10), while on the other hand, it will render the ester group more acidic, that is, increase the K'_a of the protonated ester. As a result, any electron-withdrawing substituents make the neutral and base-catalyzed reactions more effective than the associated acid-catalyzed mechanism at near-neutral pH conditions (see discussion below). Said another way, acid-catalyzed hydrolysis will only be important for esters exhibiting neither electron-withdrawing substituents nor good leaving groups (i.e., also not electron-withdrawing in nature), as is the case, for example, for alkyl esters of aliphatic carboxylic acids.

Before we turn to discussing neutral and base-catalyzed hydrolysis of ester functions, we need to reflect on what structural features determine how good a leaving group a given alcohol moiety is. We have postulated that under acidic conditions, the alcohol dissociates as a neutral molecule, and that the dissociation step is not rate determining. However, in some cases under neutral, and always under basic, conditions, the alcohol moiety leaves as an anionic species (i.e., RO^-). In these cases, the rate of dissociation of the alcohol moiety may influence the overall reaction rate. As a rule of thumb, we can relate the ease with which the RO^- group dissociates with the ease with which the corresponding alcohol dissociates in aqueous solution, expressed by its pK_a value. We recall from Chapter 8 that the pK_a of an acid is a measure of the relative stability of its nondissociated versus dissociated form. Hence, we can expect

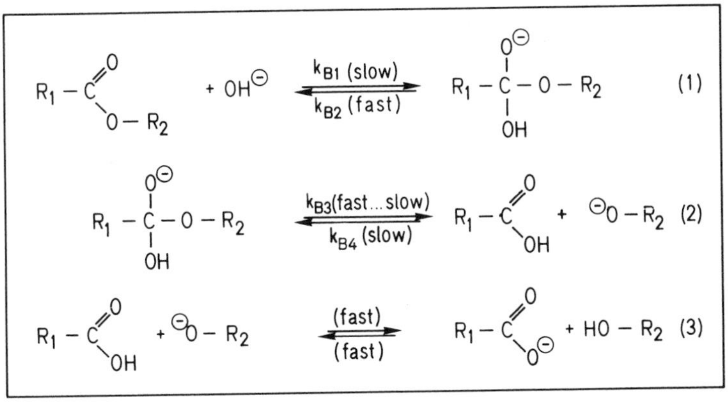

$$R_1-C(=O)-O-R_2 + OH^{\ominus} \underset{k_{B2}\,(\text{fast})}{\overset{k_{B1}\,(\text{slow})}{\rightleftharpoons}} R_1-C(O^{\ominus})(OH)-O-R_2 \qquad (1)$$

$$R_1-C(O^{\ominus})(OH)-O-R_2 \underset{k_{B4}\,(\text{slow})}{\overset{k_{B3}(\text{fast...slow})}{\rightleftharpoons}} R_1-C(=O)-OH + {}^{\ominus}O-R_2 \qquad (2)$$

$$R_1-C(=O)-OH + {}^{\ominus}O-R_2 \underset{(\text{fast})}{\overset{(\text{fast})}{\rightleftharpoons}} R_1-C(=O)-O^{\ominus} + HO-R_2 \qquad (3)$$

Figure 12.11 Reaction scheme for the base-catalyzed hydrolysis of carboxylic acid esters.

alcohols with low pK_a values to be better leaving groups than those with high pK_a's. Note that we use here again a thermodynamic argument to describe a kinetic phenomena.

Base-Catalyzed Hydrolysis Let us now look at the reaction of a carboxylic ester with OH^-, that is, the *base-catalyzed hydrolysis.* The reaction scheme for the most common reaction mechanism is given in Figure 12.11. As indicated in Reaction Step 2, in contrast to the acid-catalyzed reaction (Fig. 12.10), the breakdown of the tetrahedral intermediate, I, may be kinetically important. Thus we write for the overall reaction rate:

$$-\frac{d\left[R_1-C(=O)OR_2\right]}{dt} = k_{B3} \cdot [I] \qquad (12\text{-}54)$$

If the chain of events "backs up" at the tetrahedral intermediate (I), then this species quickly reaches an unchanging or steady-state concentration and we may write:

$$\frac{d[I]}{dt} = 0 = +k_{B1}\left[R_1-C(=O)OR_2\right] \cdot [HO^{\ominus}] - k_{B2} \cdot [I] - k_{B3} \cdot [I] + k_{B4}\left[R_1-C(=O)OH\right] \cdot [RO_2^{\ominus}] \qquad (12\text{-}55)$$

Recognizing that R_1COOH is usually very quickly removed by deprotonation, we may neglect the fourth term in Eq. 12-55. Thus we solve for the abundance of the intermediate I at steady state:

$$[I] = \frac{k_{B1}\left[R_1-C(=O)OR_2\right] \cdot [HO^{\ominus}]}{(k_{B2} + k_{B3})} \qquad (12\text{-}56)$$

and substituting in the overall rate expression:

$$-\frac{d\left[R_1-C(=O)-OR_2\right]}{dt} = \frac{k_{B1} \cdot k_{B3}}{k_{B2} + k_{B3}} \left[R_1-C(=O)-OR_2\right] \cdot [HO^{\ominus}] \tag{12-57}$$

we derive a rate law in terms of starting compounds and

$$k_B = \frac{k_{B1}\,k_{B3}}{k_{B2} + k_{B3}} \tag{12-58}$$

For good leaving groups ($k_{B3} \gg k_{B2}$), k_B is equal to k_{B1}, meaning that solely the formation of the tetrahedral intermediate is rate determining. This is usually the case for esters exhibiting an *aromatic* alcohol moiety (e.g., for phenyl esters, where pK_a of the phenol is < 10). In the hydrolysis of alkyl esters ($pK_a > 15$), however, k_{B3} may be even smaller than k_{B2}, reflecting the loss of an alkoxy versus hydroxide ion from the tetrahedral intermediate. In these cases, k_B is not be equal to k_{B1}, but is be given by Eq. 12-58.

To illustrate some of these points we compare the Hammett relationships (Eq. 12-28) derived for the k_B values of a series of meta- and para-substituted benzoic acid ethyl esters (V) and a series of meta- and para-substituted acetic acid phenyl esters (VI). Note that in both of these sets of hydrolyses, the use of a Hammett relationship to describe the effect of substituents is appropriate since we may reasonably assume insignificant steric effects of moieties in meta and para positions.

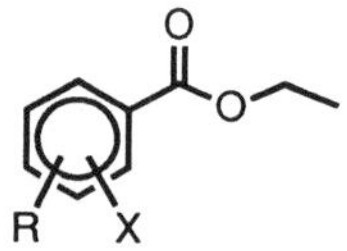

O O R X R X H3C O O

V 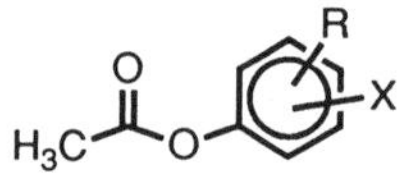 VI

What is of interest to us here are the susceptibility factors ρ of the two relationships since, as we recall, ρ reflects the sensitivity of the reaction rate toward electronic effects of the substituents. In the case of the benzoic acid esters (V), where we leave the alcohol moiety ($-O-R_2$) invariant, ρ primarily expresses structural effects on the k_{B1} terms. Depending on the solvent–water systems investigated, ρ values on the order of 2.1–2.6 have been found for this reaction (see Section 12.2). For the acetic acid phenyl esters (VI), the situation is somewhat different, since the substituents exhibit effects on k_{B1} as well as on k_{B3}. These effects are paralleling each other. For example, an electron-withdrawing substituent increases k_{B1} by an inductive effect and, at the same time, it renders the alcohol moiety a better leaving group (it decreases the pK_a of the alcohol). If the dissociation of the leaving group (i.e., the phenolate species) is rate determining, we expect a ρ value similar or even greater than that found for the benzoic acid esters (although there is an oxygen between the phenyl

TABLE 12.10 Second-Order Rate Constants k_B for the Base-Catalyzed Hydrolysis at 30°C of Some Dialkyl Phthalates Together with σ^* and E_s Values for the Corresponding Alkyl Groups

Compound (phthalate, $C_6H_4(COOR)_2$) R	k_B^a ($M^{-1}s^{-1}$)	σ^*(R)	E_s(R)
$-CH_3$	6.9×10^{-2}	0.00	0.00
$-CH_2-CH_3$	2.5×10^{-2}	−0.10	−0.07
$-CH_2-CH_2-CH_2-CH_3$	1.0×10^{-2}	−0.13	−0.39
$-CH_2-CH(CH_3)_2$	1.4×10^{-3}	−0.13	−0.93
$-CH_2-CH(CH_2-CH_3)(CH_2-CH_2-CH_2-CH_3)$	1.1×10^{-4}	−0.21	−1.13

[a]Data from Wolfe et al. (1980a,b).

group and the carbonyl carbon which renders the electronic effects of substituents on k_{B1} smaller as compared to the benzoic acid ethyl esters). The ρ value derived from k_B values of some substituted phenyl acetic acid esters (Bruice et al., 1962) is, however, very small, that is, on the order of 1.1. This result implies that the effect of electron-withdrawing substituents in enhancing the combined k_B expression (Eq. 12-58) is significantly less than the impact of the same substituents on k_{B1} rates for benzoic acid esters, suggesting that for the phenyl esters, the rate of dissociation of the leaving group is not significant in determining the overall reaction rate.

An example illustrating the case in which dissociation of the leaving group is rate determining is the base-catalyzed hydrolysis of dialkyl phthalates (Eq. 12-59). Table 12.10 gives the k_B values for the hydrolysis of five dialkyl phthalates to the corresponding monoalkyl esters:

$$C_6H_4(COOR)_2 + HO^- \xrightarrow{k_B} C_6H_4(COO^-)(COOR) + HOR \quad (12\text{-}59)$$

Also included in Table 12.10 are the Taft polar (σ^*) and steric (E_s) substituent constants for the various alkyl groups (R). Although one would prefer to derive a Taft correlation (three unknowns: ρ^*, δ, and k_{ref}) on more than five k_B values, the magnitude of the susceptibility factors ρ^* (= 4.59) and δ (= 1.52) reported by Wolfe et al. (1980a, b) are instructive:

$$\log \frac{k_B}{k_B\,(R{=}CH_3)} = 4.59\,\sigma^* + 1.52\,E_s \quad (12\text{-}60)$$

In Section 12.2 we have noticed that for the reference reaction, that is, the basic hydrolysis of a series of esters with the general structure (VII)

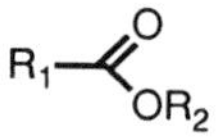

$$R_1-C(=O)-OR_2$$

VII

where the alcohol moiety (—O—R_2) is invariant, ρ^* and δ were set to 1.00. As we might have expected, and is reflected in the larger δ value found for the dialkyl phthalates (1.52 versus 1.00), the steric effect of R when present in two adjacent alcohol moieties is larger than in the case where only one R is present, even though this group was directly connected to the carbonyl carbon (VII) at which the reaction occurs. A much larger ρ^* value (4.59) is obtained for the electronic effect which, in this case, is an electron-donating (rate-decreasing) effect. These findings strongly suggest that the rate of dissociation of the alkoxy group (expressed by k_{B3}) is rate determining, since the effect of R on k_{B1} of the dialkyl phthalates would be expected to be smaller as compared to the reference reaction, where R is directly bound to be reaction center. These examples nicely demonstrate that LFERs are not only useful tools for relating and predicting rate constants, but that they may also help evaluating which reaction steps are rate determining. We may also realize at this point that when applying LFERs, we have to be careful when trying to predict rate constants for compounds for which k_{B3} is not rate determining from an equation derived from compounds for which k_{B3} is important (and vice versa).

Since dialkyl phthalates are very important environmental chemicals (see Chapter 2), it is worth spending some time looking at the rates at which such compounds undergo hydrolysis under typical environmental conditions. Because there are no rate constants available for the neutral and acid-catalyzed hydrolysis of dialkyl phthalates, we need to make some assumptions. From the hydrolysis rate data reported for benzoic acid alkyl esters we may conclude that neutral hydrolysis of dialkyl phthalates is unimportant (Case b in Fig. 12.9), and that k_B will be some three to four orders of magnitude larger than k_A. Hence, the I_{AB} value will be well below pH 6. Therefore, at pH 7, we may assume that only the base-catalyzed reaction is important. For dimethyl phthalate, for example, we then obtain a hydrolysis half-life at pH 7 and 30°C of about 3 years. Assuming a rate decrease of about a factor of 2 per 10°C decrease in temperature (Wolfe et al., 1980a,b), one obtains a half-life of 6 years at 20°C and 12 years at 10°C, respectively, for this compound. Since among the dialkyl phthalates, the dimethyl ester hydrolysis fastest (alkyl groups exhibit negative σ^* and E_s values, see Table 12.4), we can conclude that chemical hydrolysis of these compounds will, in general, be rather slow as compared to other processes. We also note that the monoalkyl phthalates (the hydrolysis products of the dialkyl phthalates) hydrolyze more slowly (about a factor of 10; Wolfe et al., 1980a,b) than the corresponding diesters.

We finish our discussion on the major hydrolysis mechanisms of carboxylic acid esters by looking at the neutral (pH independent) reaction at the carbonyl carbon. From the reaction scheme given in Figure 12.12, we see that, very similar to what

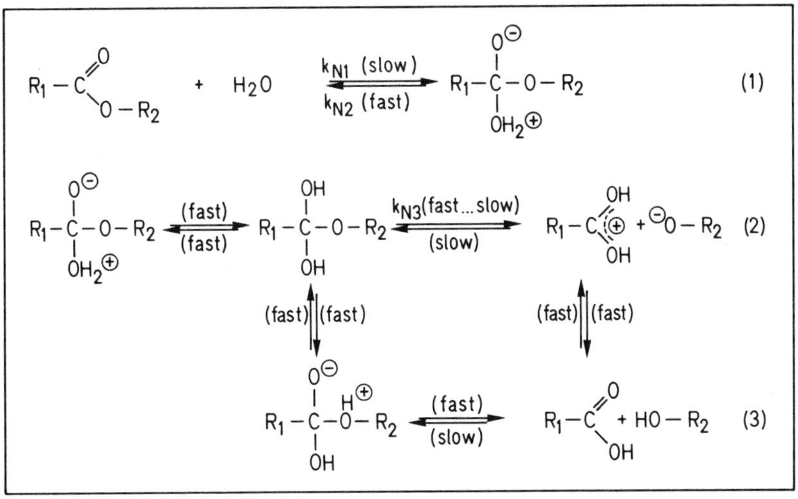

$$R_1-C(=O)-O-R_2 + H_2O \underset{k_{N2}\ (fast)}{\overset{k_{N1}\ (slow)}{\rightleftharpoons}} R_1-C(O^{\ominus})(OH_2^{\oplus})-O-R_2 \quad (1)$$

$$R_1-C(O^{\ominus})(OH_2^{\oplus})-O-R_2 \underset{(fast)}{\overset{(fast)}{\rightleftharpoons}} R_1-C(OH)(OH)-O-R_2 \underset{(slow)}{\overset{k_{N3}(fast\ldots slow)}{\rightleftharpoons}} R_1-C^{\oplus}(OH)(OH) + {}^{\ominus}O-R_2 \quad (2)$$

$$R_1-C(O^{\ominus})(OH)-O(H^{\oplus})-R_2 \underset{(slow)}{\overset{(fast)}{\rightleftharpoons}} R_1-C(=O)-OH + HO-R_2 \quad (3)$$

Figure 12.12 Reaction scheme for the neutral hydrolysis of carboxylic acid esters.

we have postulated for the base-catalyzed reaction, the dissociation of the leaving group (expressed by k_{N3}) may be rate determining. In the neutral case, however, the situation is somewhat more complicated since, particularly, for poor leaving groups (i.e., alkoxy groups), the alcohol moiety may leave as a neutral species and not as an anion (reaction path 3 in Figure 12.12). This might have to be taken into account when deriving or applying LFERs to k_N values.

Let us now compare the relative importance of the neutral versus base-catalyzed hydrolysis of carboxylic acid esters. Inspection of Table 12.9 and Figure 12.8 reveals that the relative importance of these two processes (i.e., the magnitude of the I_{NB} value) depends on both the goodness of the leaving group and on substitution in the acid part of the molecule. From the examples given in Table 12.11 we can see that structural changes, particularly with respect to the leaving group, but also to a certain extent with respect to substitution in the acid part, have a greater impact on k_N as compared to k_B. We can intuitively rationalize these findings by imagining that structural differences influencing the $\Delta G^{\ddagger}$ of the reaction will be more strongly felt by the weak nucleophile H_2O as compared to the much stronger (more pushy) nucleophile OH^-. Consequently, carboxylic acid esters exhibiting good leaving groups and/or electron-withdrawing substituents in the acid part of the molecule will have relatively high I_{NB} values (I_{NB} values of up to 7–7.5). In these cases, neutral hydrolysis has to be considered at ambient pH values, and, as we recall from our earlier discussion, the acid-catalyzed reaction can be neglected (see examples given in Fig. 12.8 and Table 12.9). On the other hand, rate-decreasing substituents (i.e., alkyl groups) will decrease I_{NB} and increase I_{AN}, leading to a situation as depicted in Figure 12.9*b*.

So far we have confined our discussion to the most common case of ester hydrolysis, that is, the case in which the reaction takes place at the carbonyl carbon. In some cases, however, an ester may also react in water by an S_N- or E-type mechanism (see

TABLE 12.11 Comparison of k_N and k_B Values of Some Carboxylic Acid Esters at 25°C. Influence of Leaving Group and Polar Substituents on k_N and k_B[a]

Compound	pK_a of ROH	Relative Value k_N	Relative Value k_B	k_B/k_N (M^{-1})
$CH_3-C(=O)-O-CH_2CH_3$	≈16	1	1	7.3×10^8
$CH_3-C(=O)-O-C_6H_5$	9.98	440	13	2.1×10^7
$CH_3-C(=O)-O-C_6H_3(NO_2)_2$ (2,4-dinitrophenyl)	3.96	73300	850	8.5×10^6
$ClCH_2-C(=O)-OCH_3$	≈15	1	1	6.6×10^8
$Cl_2CH-C(=O)-OCH_3$	≈15	71	20	1.9×10^8
$Cl_2CH-C(=O)-O-C_6H_5$	9.98	8570	93	6.3×10^6

[a] Data derived from Table 12.9.

Section 12.3) with the acid moiety (i.e., $^-OOC-R_1$) being the leaving group. The S_N-type reactions occur primarily with esters exhibiting a tertiary alcohol group. The products of this reaction are the same as the products of the common hydrolysis reaction. In the case of elimination, however, products are different since the ester is converted to the corresponding acid and olefin:

$$R_1C(=O)O-CR_2R_3-CHR_4R_5 + {}^-OH \xrightarrow{\text{elimination}} R_1COO^- + R_2R_3C{=}CR_4R_5 \qquad (12\text{-}61)$$

Elimination according to Eq. 12-61 will be important for compounds exhibiting acidic protons vicinal to the alcoholic carbon forming the ester bond.

Finally, if the α-carbon of the acid moiety (i.e., the carbon bound to the carbonyl carbon) is substituted by an electron-withdrawing group that renders the α-hydrogens more acidic, the ester may hydrolyze by an elimination mechanism involving a ketene inter-mediate:

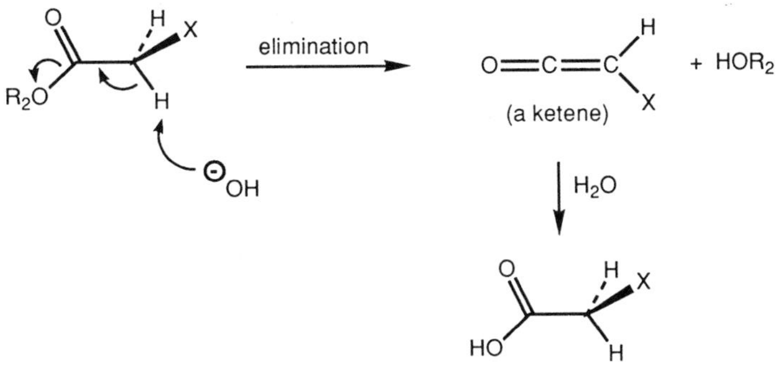

elimination

$O=C=C(H)(X)$ (a ketene) + HOR_2

H_2O

(12-62)

where the second step of reaction 12-62 is an addition of H_2O to the carbon–carbon double bond (for more details see, e.g., March, 1985). We encounter analogous mechanisms when discussing the hydrolysis of carbamates.

We should also note at this point that certain transition metals, as well as certain oxide surfaces, may have a catalytic effect on the hydrolysis rates of carboxylic acid esters (and , therefore, possibly also on the hydrolysis rate of other acid derivatives). In the case of heterogeneous hydrolysis, the observed acceleration of hydrolysis has been attributed to either reaction with surface hydroxyl groups (Hoffman, 1990) or to reaction with hydroxide ions postulated to be present at higher concentrations in the diffuse layer near the surface as compared to the bulk solution (Stone, 1989). From the few data available to date, it is, however, difficult to judge how important such catalytic effects are in a given natural environment.

With the background that we have acquired from our rather extensive discussion of the hydrolysis of carboxylic acid esters in homogeneous solution, it should not be too difficult to evaluate the hydrolysis behavior of other carboxylic or carbonic acid derivatives including carboxylic acid amides and carbamates.

Carboxylic Acid Amides Amide functions are very important linkages in natural compounds (e.g., in proteins), and some simple amides are used in industry. Furthermore, quite a few herbicides contain amide groups (Kearny and Kaufman, 1976). Generally an amide bond is defined as

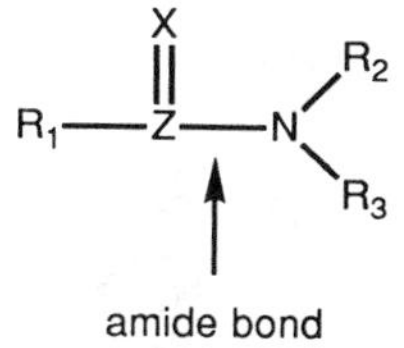

$R_1-Z(=X)-NR_2R_3$

amide bond

where R_2 and R_3 are hydrogens or carbon-centered substituents.

The hydrolysis of carboxylic acid amides (i.e., $Z=C$ and $X=O$) can be treated in a very similar way as the hydrolysis of carboxylic acid esters, that is, a similar structure–reactivity pattern is found (Talbot, 1972). Compared to ester functions, however, amide functions are in general much less reactive since the $-NR_2R_3$ group is less electronegative than the $-OR_2$ group, and, even more importantly, it is a much poorer leaving group [the pK_a's of amines ($R_1R_2NH \rightleftharpoons R_1R_2N^-$) are much larger than those of alcohols ($ROH \rightleftharpoons RO^-$)]. As a consequence of these factors, and because amide groups are quite basic, neutral hydrolysis is usually unimportant relative to the acid- or base-catalyzed reaction (case *b* in Fig. 12.9). Furthermore, because the amide group

$$R_1{-}C(=O)NR_2R_3 \rightleftharpoons R_1{-}C(=\overset{+}{O}H)NR_2R_3$$

is more basic than the ester group

$$R_1{-}C(=O)OR_2 \rightleftharpoons R_1{-}C(=\overset{+}{O}H)OR_2$$

the I_{AB} values of amides are commonly higher than those of ester functions. In many cases, k_A and k_B are of similar magnitude, unless the amide function is substituted with electron withdrawing groups or atoms (see examples given in Table 12.12). Note that because the hydrolysis half-lives of most of the compounds shown in Table 12.12 are very large under ambient conditions, many of the values given are only order-of-magnitude estimates that have been extrapolated from measurements conducted at elevated temperatures and extreme pH values. Finally, we note that activation energies are typically between 80 and 90 kJ mol^{-1} for the acid-catalyzed hydrolysis of amide functions, and between 50 and 80 kJ mol^{-1} for the base-catalyzed reaction (Mabey and Mill, 1978).

Carbamates The next group of compounds that we want to look at more closely are derivatives of carbamic acid ($HO{-}CO{-}NH_2$), that is, the carbamates. Carbamates are widely used as herbicides and insecticides (Kearny and Kaufman, 1976). The carbamate function exhibits both an ester and amide-type bonds:

TABLE 12.12 Rate Constants k_N and k_B, Half-Lives at pH 7, and I_{NB} Values for Hydrolysis of Some Simple Amides at 25°C[a]

Compound $R_1{-}C(=O)NR_2R_3$						
R_1	R_2	R_3	k_A ($M^{-1}s^{-1}$)	k_B ($M^{-1}s^{-1}$)	$t_{1/2}$ (pH 7)	I_{AB}
CH_3-	$-H$	$-H$	8.4×10^{-6}	4.7×10^{-5}	4000 yr	6.6
i-C_3H_9-	$-H$	$-H$	4.6×10^{-6}	2.4×10^{-5}	7700 yr	6.6
	$-H$	$-H$	2.3×10^{-5}	1.7×10^{-5}	5500 yr	7.1
CH_2Cl-	$-H$	$-H$	1.1×10^{-5}	1.5×10^{-1}	1.5 yr	4.9
CH_3-	$-CH_3$	$-H$	3.2×10^{-7}	5.5×10^{-6}	40,000 yr	6.4
CH_3-	$-CH_3$	$-CH_3$	5.2×10^{-7}	1.1×10^{-5}	20,000 yr	6.3

[a]Data from Mabey and Mill (1978).

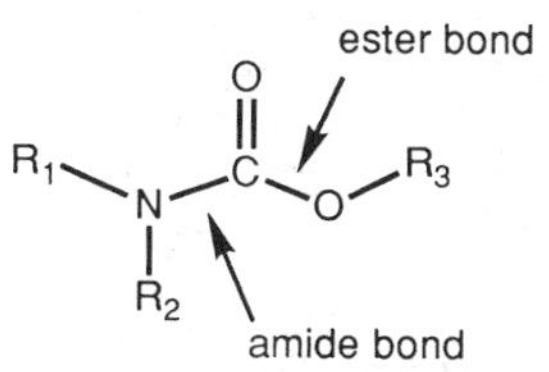

ester bond

amide bond

where R_1 and R_2 are hydrogens or carbon-centered substituents, and R_3 is a carbon-centered substituent. Hence, a carbamate has two potential leaving groups: an alcohol and an amine moiety. Since, in most cases, the alcohol moiety will be the better leaving group, the initial hydrolysis reaction occurs commonly by cleavage of the ester bond. Initial breaking of the amide bond may, for example, occur if R_3 is an alkyl group, and R_1 and R_2 are aromatic rings that are substituted with electron-withdrawing substituents (and thus stabilize the $^-NR_1R_2$ anion in aqueous solution). However, regardless of the reaction mechanism, the hydrolysis of carbamates eventually yields the alcohol (R_3OH), the amine (R_1R_2NH), and CO_2 (see below).

Since base catalysis plays an important role in the hydrolytic breakdown of carbamates, this process has been investigated quite extensively. Very little data is, however, available on the neutral reaction which, in some cases, may be significant at ambient pH values. The acid-catalyzed reaction, on the other hand, can, in general, be neglected (presumably because so many electron-withdrawing atoms surround the central carbon that protonation of the carboxyl oxygen insignificantly enhances its susceptibility to nucleophilic attack).

When considering the base-catalyzed hydrolysis of carbamate functions, the critical question is whether one of the groups bound to the nitrogen (R_1, R_2) is a hydrogen atom or not. This becomes obvious when we compare the k_B values of compounds VIII and IX (Williams, 1972):

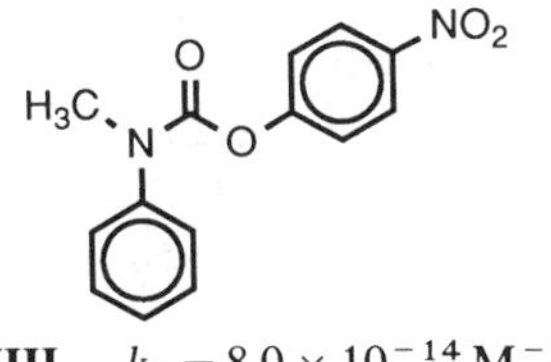

VIII $k_B = 8.0 \times 10^{-14}\ M^{-1}\ s^{-1}$ 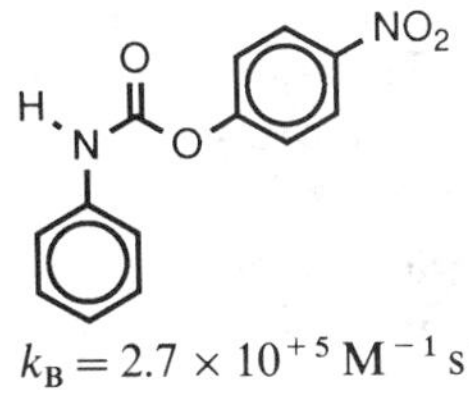 **IX** $k_B = 2.7 \times 10^{+5}\ M^{-1}\ s^{-1}$

First, we realize that although the *p*-nitrophenol group is a good leaving group, the base-catalyzed hydrolysis of 4-nitrophenyl *N*-methyl-*N*-phenyl carbamate (VIII) is very slow. In this case, by analogy to what we have postulated for most ester and amide functions, the rate-determining step is the formation of a tetrahedral intermediate (see reaction step 1 in Fig. 12.13). We note that the hydrolysis of the ester bond is generally followed by a fast decarboxylation reaction yielding the corresponding amine (reaction step 4 in Fig. 12.13). An analogous reaction occurs if the amide is the leaving group, yielding an unstable carbonic acid monoester that also hydrolyzes rapidly. As is illustrated by the k_B values of some other *N*,*N*-disubstituted carbamates

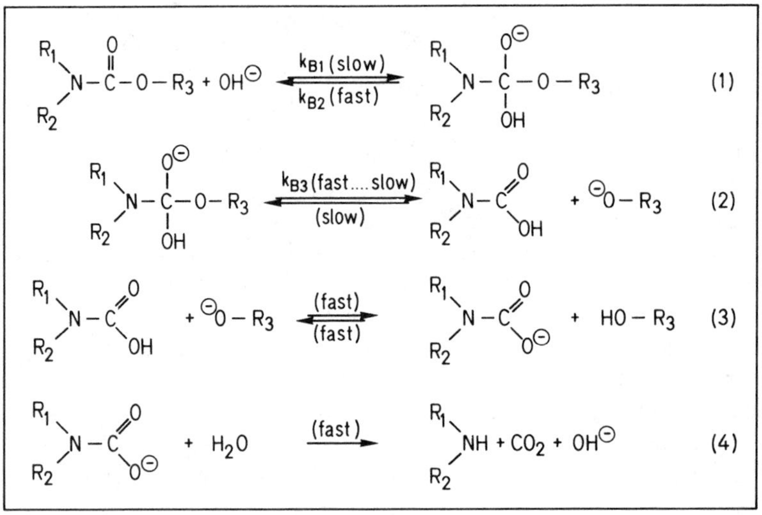

$$R_1R_2N{-}C(=O){-}O{-}R_3 + OH^{\ominus} \underset{k_{B2}\,(\text{fast})}{\overset{k_{B1}\,(\text{slow})}{\rightleftharpoons}} R_1R_2N{-}C(O^{\ominus})(OH){-}O{-}R_3 \quad (1)$$

$$R_1R_2N{-}C(O^{\ominus})(OH){-}O{-}R_3 \underset{(\text{slow})}{\overset{k_{B3}\,(\text{fast....slow})}{\rightleftharpoons}} R_1R_2N{-}C(=O){-}OH + {}^{\ominus}O{-}R_3 \quad (2)$$

$$R_1R_2N{-}C(=O){-}OH + {}^{\ominus}O{-}R_3 \underset{(\text{fast})}{\overset{(\text{fast})}{\rightleftharpoons}} R_1R_2N{-}C(=O){-}O^{\ominus} + HO{-}R_3 \quad (3)$$

$$R_1R_2N{-}C(=O){-}O^{\ominus} + H_2O \xrightarrow{(\text{fast})} R_1R_2NH + CO_2 + OH^{\ominus} \quad (4)$$

Figure 12.13 Reaction scheme for the base-catalyzed hydrolysis of carbamates when the mechanism involves a tetrahedral intermediate.

(see Table 12.13), we can conclude that such compounds are generally quite resistant to base-catalyzed hydrolysis. From the data in Table 12.13 we can also see that the base-catalyzed hydrolysis of *N,N*-disubstituted carbamates is somewhat insensitive to the nature of the alcohol moiety, indicating that dissociation of the leaving group is not rate determining. These findings are better illustrated by Figure 12.14, where the logarithms of the k_B values are plotted for a series of carbamates versus the pK_a of the leaving group alcohol. The acidity constants of these alcohols serve as a measure of how good a leaving group it is. Furthermore, since the pK_a of the alcohol is determined by the nature of the carbon skeleton and by the substituents present in the molecule, it also expresses the electronic (and to a certain extent steric) effects of the alcohol moiety on the $\Delta G^{\ddagger}$ of the formation of the tetrahedral intermediate. Consequently, as is illustrated by Figure 12.14, quite good correlations are obtained between log k_B and pK_A of the alcohol, even when combining aliphatic and aromatic ester groups.

From Figure 12.14 we can see that the slope, ρ, of the LFER,

$$\log k_B = -\rho \cdot pK_a + C \qquad (12\text{-}63)$$

is very small ($\rho = 0.25$) for the few *N*-methyl-*N*-phenyl carbamates for which k_B values are available. A similarly small ρ value ($\rho = 0.17$) is found for a series of *N,N*-dimethyl carbamates (Wolfe et al., 1978). In contrast, a much larger ρ value ($\rho = 1.15$) is obtained for the *N*-phenyl carbamates (Fig. 12.14), as well as for a series of *N*-methyl carbamates ($\rho = 0.91$, Wolfe et al., 1978). Obviously, in these cases, the reaction occurs much faster and the dissociation of the leaving group is rate determining. The reaction

TABLE 12.13 Rate Constant k_B (and k_N), Half-Lives at pH 7, and I_{NB} Values for Hydrolysis of Some Simple Carbamates at 25°C[a,b]

Compound: $R_1R_2N{-}C(=O){-}O{-}R_3$

R_1	R_2	R_3	k_N (s^{-1})	k_B ($M^{-1}s^{-1}$)	$t_{1/2}$[c] (pH 7)	I_{NB}
CH_3-	CH_3-	$-CH_2CH_3$	NA	4.5×10^{-6}	50,000 yr	
CH_3-	C_6H_5-	$-CH_2CH_3$	NA	4.0×10^{-6}	55,000 yr	
CH_3-	CH_3-	$-C_6H_4-NO_2$	NA	4.0×10^{-4}	550 yr	
CH_3-	C_6H_5-	$-C_6H_4-NO_2$	NA	8.0×10^{-4}	275 yr	
H–	CH_3-	$-CH_2CH_3$	NA	5.5×10^{-6}	40,000 yr	
H–	C_6H_5-	$-CH_2CH_3$	NA	3.2×10^{-5}	7,000 yr	
H–	CH_3-	$-C_6H_4-NO_2$	NA	6.0×10^{2}	3 h	
H–	C_6H_5-	$-C_6H_4-NO_2$	NA	2.7×10^{5}	25 s	
H–	CH_3-	$-C_6H_4-CH_3$	6.0×10^{-8}	5.6×10^{-1}	70 d[d]	7.01
H–	CH_3-	$-C_{10}H_7$ (naphthyl)	9.0×10^{-7}	5.0×10^{1}	33 h[d]	6.25

[a]Data from Dittert and Higuchi (1963), Williams (1972 and 1973), Vontor et al. (1972), and El-Amamy and Mill (1984).

[b]NA = not available.

[c]Half-life for base-catalyzed reaction, actual half-life may be shorter.

[d]Half-life for neutral and base-catalyzed reaction.

must therefore, proceed by a different mechanism. The generally accepted mechanism (e.g., Bender and Homer, 1965; Williams; 1972) involves deprotonation of the amide function (similarly to esters with acidic protons at the carbon adjacent to the ester group, see Eq. 12-62), with subsequent elimination of the alkoxy group (Fig. 12.15). The resulting isocyanate ($R_1N{=}C{=}O$) is then converted to the amine and CO_2 by addition of water and subsequent decarboxylation.

Comparison of the k_B values of *N,N*-disubstituted versus *N*-monosubstituted carbamates (Table 12.13, Fig. 12.14) shows that if the alcohol moiety is a good leaving group (i.e., an aromatic ring-carrying electron-withdrawing substituents), differences in half-lives for the base-catalyzed reaction of up to 10 orders of magnitude may be

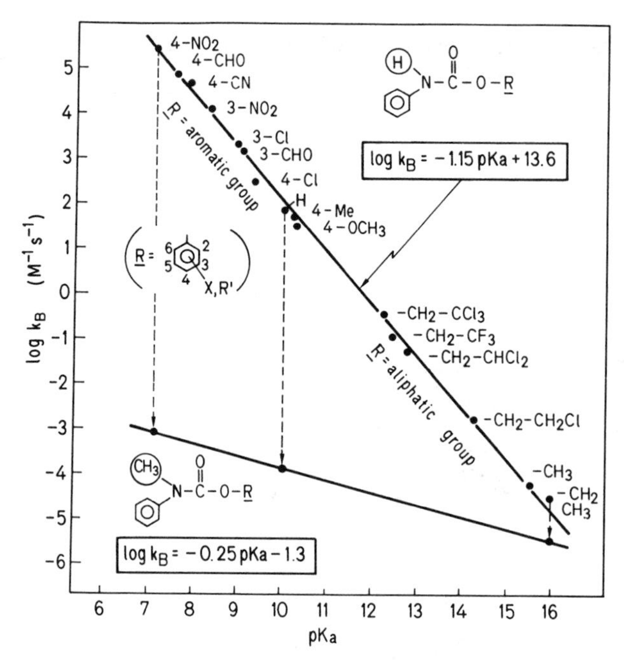

Figure 12.14 Correlation of base-catalyzed hydrolysis rates (log k_B) of carbamates as a function of the pK_a of the alcohol moiety for a series of *N*-phenyl and *N*-methyl-*N*-phenyl carbamates. Data from references given in Table 12.13.

found. Only if the leaving group is very poor (e.g., $R_3 = $ alkyl) can the reaction occurring via a tetrahedral intermediate (Fig. 12.13) compete with the reaction involving an elimination step (Fig. 12.15).

From Table 12.13 we also see that the *N*-methyl carbamates have significantly smaller k_B values as compared to the corresponding *N*-phenyl carbamates. We can rationalize these findings by the anion-stabilizing effect of the phenyl group, an effect also reflected by the N–H proton being more acidic than the *N*-methyl compound. Thus, the *N*-phenyl carbamates have a greater fraction in the deprotonated species available to undergo elimination.

As is indicated in Table 12.13, half-lives of the base-catalyzed hydrolysis of particularly the *N*,*N*-disubstituted carbamates (but also of the monosubstituted carbamates exhibiting bad leaving groups) are very large at ambient pH values. Although there are virtually no rate data available on the neutral reaction of such slowly hydrolyzing compounds, it can be assumed that I_{NB} might be fairly large ($I_{NB} > 8$), but in general, k_N will be too small to be of environmental significance. For the more reactive *N*-monoalkyl carbamates, however, the neutral reaction might have to be considered. From the few compounds for which k_N values have been reported (Table 12.13), and by using our chemical intuition, we may conclude that the relative importance to the neutral reaction increases (i.e., the I_{NB} value decreases) with increasing

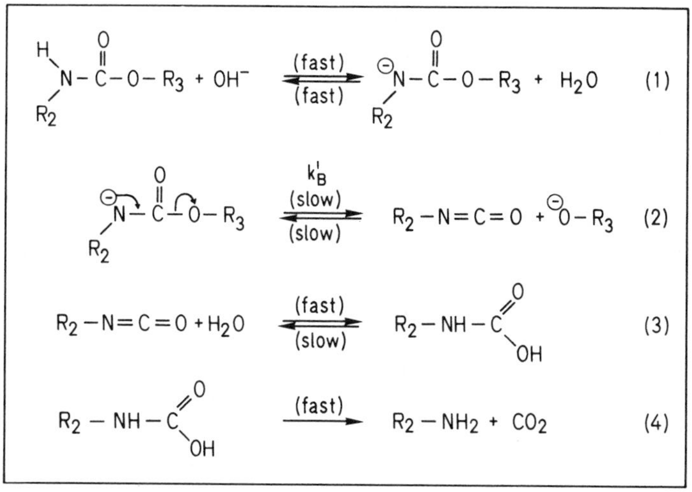

$$\begin{array}{lcll}
\text{H(R}_2\text{)N}-\overset{\displaystyle O}{\overset{\|}{C}}-O-R_3 + OH^- & \underset{\text{(fast)}}{\overset{\text{(fast)}}{\rightleftharpoons}} & {}^{\ominus}\text{N(R}_2\text{)}-\overset{\displaystyle O}{\overset{\|}{C}}-O-R_3 + H_2O & (1) \\
{}^{\ominus}\text{N(R}_2\text{)}-\overset{\displaystyle O}{\overset{\|}{C}}-O-R_3 & \underset{\text{(slow)}}{\overset{k'_B \ \text{(slow)}}{\rightleftharpoons}} & R_2-N=C=O + {}^{\ominus}O-R_3 & (2) \\
R_2-N=C=O + H_2O & \underset{\text{(slow)}}{\overset{\text{(fast)}}{\rightleftharpoons}} & R_2-NH-C(=O)OH & (3) \\
R_2-NH-C(=O)OH & \xrightarrow{\text{(fast)}} & R_2-NH_2 + CO_2 & (4)
\end{array}$$

Figure 12.15 Reaction scheme for the base-catalyzed hydrolysis of carbamates when the mechanism involves an elimination step.

reactivity. From our discussion of neutral ester hydrolysis, we recall that the relative importance of the dissociation of the alcohol moiety decreases with decreasing pK_a of the corresponding alcohol. Hence, since in the elimination step the dissociation of $^-O—R_3$ is always rate limiting, the k_N/k_B ratio will decrease with increasing goodness of the leaving group (i.e., decreasing pK_a). Furthermore, any group or substituent that increases the acidity of the N–H proton will have a much greater impact on k_B than on k_N (only electronic effect). From Table 12.13 we see that for moderately reactive carbamates, I_{NB} values around 7 are found. Consequently, for these and for less reactive compounds, neutral hydrolysis has to be considered at ambient pH values. Finally, we note that activation energies for the base-catalyzed hydrolysis of carbamates span a wide range of between 50 and 100 kJ mol^{-1} (Christenson, 1964).

Phosphoric and Thiophosphoric Acid Esters In the last part of our discussion on the hydrolysis of acid derivatives in homogeneous aqueous solution, we want to look at a few examples of compounds that are not carboxylic or carbonic acid derivatives. We use these examples to gain some insights into the reactivity of compounds exhibiting a pentavalent phosphorus atom. Because of their significant biological activity (cholinesterase inhibition), esters and thioesters of phosphoric acid and thiophosphoric acid (see Fig. 12.7) are widely applied as insecticides (Khan, 1980). Furthermore, some trialkyl and triaryl phosphates are used in very large quantities in fire-resistant hydraulic fluids and as fire-retardant plasticizers. Consequently, such compounds are of great environmental significance and concern. Interestingly, despite the widespread use of phosphate and thiophosphate esters and thioesters, there is surprisingly little data available in the open literature on the reactivity of such compounds in aqueous solution. In addition, rate constants reported for a given compound often differ by more than an order of magnitude between different authors. Nevertheless, from the data available, it is possible to draw some important general conclusions concerning the hydrolytic decomposition of this group of compounds. Furthermore, the following examples shall give us an additional opportunity to deepen our knowledge of organic reactions involving nucleophilic species.

Table 12.14 gives kinetic constants for the hydrolytic reaction of some phosphoric and thiophosphoric acid triester. We note that in the following discussion we are concerned primarily with acid triesters although the hydrolysis products of these compounds, that is, the di- and monoesters are also of environmental concern, inasmuch as they usually react at slower rates as compared to the triesters (Mabey and Mill, 1978; Wolfe, 1980).

When trying to understand the reactivity of phosphate and thiophosphate esters, it is important to realize that such compounds react with a nucleophile by nucleophilic displacement (S_N2) at both the phosphorus atom (with an alcohol moiety being the leaving group) as well as at the carbon bound to the oxygen of an alcohol moiety (with the diester being the leaving group):

Nu: → P(=O(S))–O–R or P(=O(S))–O–R ← :Nu (12-64)

Note that the reaction at the phosphorus atom occurs by an S_N2 (no intermediate formed) rather than by an addition mechanism such as we encountered with carboxylic acid derivatives (Kirby and Warren, 1967). As we learned in Section 12.3, for attack at a saturated carbon atom, OH^- is a better nucleophile than H_2O by about a factor of 10^4 (Table 12.5). Toward phosphorus, however, the relative nucleophilicity increases dramatically. For triphenyl phosphate, for example, OH^- is about 10^8 times stronger than H_2O as a nucleophile (Barnard et al., 1961). Note that in the case of triphenyl phosphate, no substitution may occur at the carbon bound to the oxygen of the alcohol moiety, and, therefore, neutral hydrolysis is much less important as compared to the other cases (see Table 12.14). Consequently, the *base-catalyzed reaction* generally occurs at the phosphorus atom leading to the dissociation of the alcohol moiety that is the best leaving group (P–O cleavage) as is illustrated by the reaction of parathion with OH^-:

$HO^{\ominus}$ + $(CH_3CH_2O)_2P(=S)–O–C_6H_4–NO_2$ → $^{\ominus}O–P(=S)(OCH_2CH_3)_2$ + $HO–C_6H_4–NO_2$ (12-65)

On the other hand, depending on the alcohol moieties present [i.e., goodness of leaving group(s), presence of an aliphatic alcohol moiety], the *neutral reaction* may proceed by nucleophilic substitution at a carbon atom (C–O cleavage), as is the case for trialkyl phosphates such as trimethyl- and triethyl phosphate:

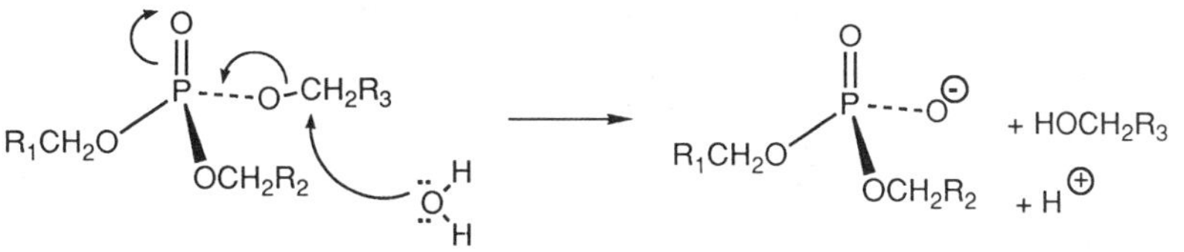

$$R_1CH_2O-P(=O)(OCH_2R_2)\cdots O-CH_2R_3 + H_2O \longrightarrow R_1CH_2O-P(=O)(OCH_2R_2)\cdots O^{\ominus} + HOCH_2R_3 + H^{\oplus} \qquad (12\text{-}66)$$

Note that if the reaction occurs by mechanism Eq. 12-66, in analogy to what we have encountered with S_N2 reactions of primary alkyl halides, methyl esters will react faster than the corresponding ethyl- or other primary alkyl esters. Of course, if a good leaving group is present, the neutral reaction may proceed by both reaction mechanisms, that is, C–O as well as P–O cleavage. For example, Weber (1976) found that at 70°C and pH 5.9, parathion reacted 90% by C–O cleavage, while at lower temperatures, a higher proportion of the neutral reaction occurred by P–O cleavage. This observation can be explained by the higher activation energy of the reaction involving C–O cleavage as compared to P–O cleavage. This simple example already shows us that when dealing with phosphoric acid and thiophosphoric acid derivatives, we have to be aware that under different conditions, different hydrolysis mechanisms may be involved.

In most cases, hydrolysis of phosphoric and thiophosphoric acids is quite insensitive to acid catalysis unless there is a base function present in one of the alcohol moieties, which, when protonated, enhances the reactivity. Examples are the two insecticides diazoxon and diazinon (Table 12.14), where protonation of one of the nitrogens of the pyrimidine ring renders the alcohol moiety a much better leaving group.

Comparison of the relative reactivities of parathion and paraoxon, or diazinon and diazoxon (Table 12.14), shows that the thiophosphoric acid esters hydrolyze somewhat more slowly than the corresponding phosphoric acid esters. As one would expect, the effect is much more pronounced for substitution at the phosphorus atom (i.e., P–O cleavage) than for the reaction involving C–O cleavage. Hence, differences will generally be larger in the k_B values as compared to the k_N values, unless the neutral reaction occurs also predominantly by P–O cleavage.

There are quite a number of phosphoric and thiophosphoric acid derivatives exhibiting one thioester and two (often identical) ester groups:

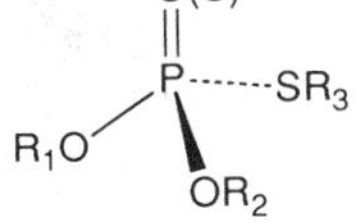

$$R_1O-P(=O(S))(OR_2)\cdots SR_3$$

where often $R_1 = R_2$ = methyl or ethyl. In these cases, the situation is now even more complicated since, depending on R_1, R_2, and R_3, the compound may react by P–O, P–S, C–O, and C–S cleavage, giving rise to a variety of possible products. If R_1 and R_2 are methyl or ethyl, the base-catalyzed reaction generally occurs by P–S cleavage with $^-S{-}R_3$ being the leaving group. The neutral reaction, however, may proceed by either or all P–S, C–O, and C–S cleavage. The C–S cleavage may preferably occur

TABLE 12.14 Rate Constants k_A, k_N, and k_B, Half-Lives at pH 7, and I Values for Hydrolysis of Some Phosphoric and Thiophosphoric Acid Triesters at 25°C[a]

Compound Name	Structural Formula	k_A[b] ($M^{-1}s^{-1}$)	k_N (s^{-1})	k_B ($M^{-1}s^{-1}$)	$t_{1/2}$ (pH 7)	I_{AN}	I_{NB}
Trimethylphosphate	$(CH_3O)_2P(=O)\text{-}OCH_3$	NI	1.8×10^{-8}	1.6×10^{-4}	1.2 yr		10.0
Triethylphosphate	$(CH_3CH_2O)_2P(=O)\text{-}OCH_2CH_3$	NI	$\approx 4 \times 10^{-9}$	8.2×10^{-6}	$\approx$5.5 yr		10.7
Triphenylphosphate	$(C_6H_5\text{-}O)_2P(=O)\text{-}O\text{-}C_6H_5$	NI	$< 3 \times 10^{-9}$	2.5×10^{-1}	320 d		< 6
Paraoxon	$(CH_3CH_2O)_2P(=O)\text{-}O\text{-}C_6H_4\text{-}NO_2$	NI	7.3×10^{-8}	3.9×10^{-1}	72 d		7.3
Parathion	$(CH_3CH_2O)_2P(=S)\text{-}O\text{-}C_6H_4\text{-}NO_2$	NI	8.3×10^{-8}	5.7×10^{-2}	89 d		8.2

Methylparathion	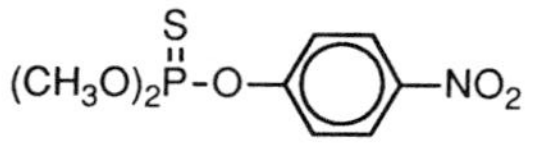 $(CH_3O)_2P(=S)$-O-C_6H_4-NO_2	NI	1.2×10^{-7}	1.1×10^{-2}	67 d		9.0
Thiometon[c]	$(CH_3O)_2P(=S)$-$SCH_2CH_2SCH_2CH_3$	NI	1.1×10^{-7}	6.4×10^{-3}	73 d		9.4
Disulfoton[c]	$(CH_3CH_2O)_2P(=S)$-$SCH_2CH_2SCH_2CH_3$	NI	1.4×10^{-7}	2.0×10^{-3}	57 d		10.0
Diazoxon[c]	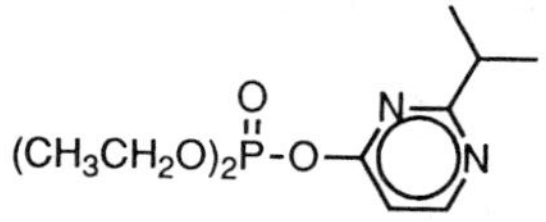 $(CH_3CH_2O)_2P(=O)$-O-(isopropylpyrimidinyl)	6.5×10^{-1}	2.8×10^{-7}	7.6×10^{-2}	23 d	6.4	8.6
Diazinon[c]	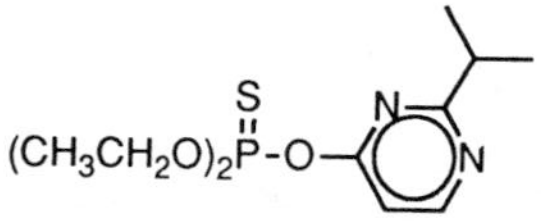 $(CH_3CH_2O)_2P(=S)$-O-(isopropylpyrimidinyl)	2.1×10^{-2}	4.3×10^{-8}	5.3×10^{-3}	178 d	5.7	8.9

[a] Data from Faust and Gomaa (1972), Mabey and Mill (1978), and Wanner et al. (1989).
[b] NI = not important.
[c] At 20°C.

if the R_3 moiety contains a nucleophilic group, which, by internal nucleophilic attack (S_Ni) may favor this reaction pathway, and thus accelerate the overall disappearance rate of the compound (Eq. 12-67). Examples are the systox-type compounds that contain a nucleophilic sulfide group:

$$(RO)_2P(=O(S))\text{---}S\text{---}CH_2CH_2\ddot{S}CH_2CH_3 \xrightarrow{S_Ni} \underset{+\ (RO)_2P(=O(S))\text{---}S^{\ominus}}{H_2C\text{---}CH_2\ \text{(cyclic)}\ \overset{\oplus}{S}\text{---}CH_2CH_3} \xrightarrow{H_2O} HOCH_2CH_2SCH_2CH_3 \quad (12\text{-}67)$$

$(R = CH_3\text{-},\ H_3C\text{-}CH_2\text{-})$

As can be seen from Figure 12.16, the rate of neutral hydrolysis of, for example, demeton S is faster than that of the corresponding sulfoxide and sulfone, respectively, although in the latter two cases, the $^-S\text{–}R_3$ moiety should be a better leaving group when considering P–S cleavage. However, both the —SO— and —SO_2— group are much weaker nucleophiles than —S—, and will, therefore, not favor C—S cleavage by an S_Ni mechanism. On the other hand, the k_B values of both the sulfoxide

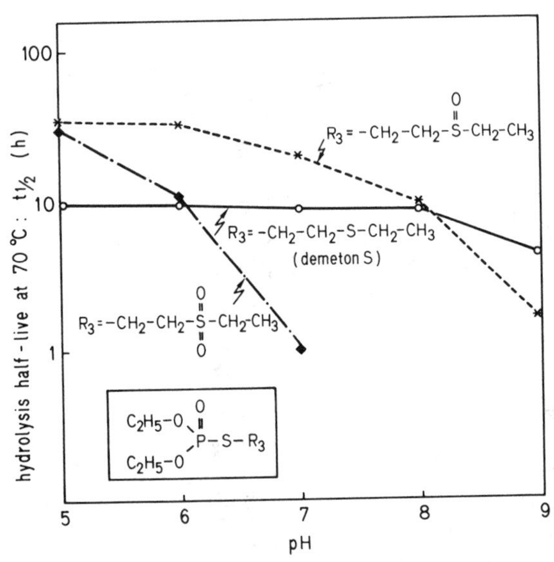

Figure 12.16 Variation of hydrolysis half-life of three thiophosphoric acid esters with solution pH indicating the relative insensitivity exhibited by demeton S due to the importance of an S_Ni mechanism for that compound. (Data from Muhlmann and Schrader, 1957).

and, particularly, the sulfone are much larger than that of demeton S. This may result from two factors, the abovementioned differences in $^{-}S—R_3$ as leaving groups, and, perhaps more importantly, the effect of the —SO— and $—SO_2—$ groups on the acidity of the protons at the adjacent carbon atoms. This factor may allow yet another reaction mechanism to become important, that is, β-elimination, similar to what we discussed earlier for carboxylic acid esters (Eq. 12-61):

$$(RO)_2P(=O(S))\text{-}S\text{-}CH_2\text{-}CH_2\text{-}SO_2\text{-}CH_2\text{-}CH_3 + {}^{\ominus}OH \xrightarrow{\beta\text{-elimination}} (RO)_2P(=O(S))\text{-}S^{\ominus} + H_2C{=}CH\text{-}SO_2\text{-}CH_2\text{-}CH_3 \qquad (12\text{-}68)$$

Note that (as is unfortunately true for most of the studies reported in the literature) the studies reflected by the data shown in Figure 12-16 did not include analysis for any transformation products. Therefore, our conclusions regarding reaction mechanisms are only speculative. Nevertheless, from the discussion above we may conclude that when dealing with the hydrolysis of phosphoric and thiophosphoric acid derivatives, as well as with other phosphorus-containing hydrolyzable functionalities (see Fig. 12.7), one has to be aware that various reaction mechanisms may apply. Consequently, depending on the environmental conditions prevailing, product distribution, at least with respect to intermediates formed, may vary considerably.

With these remarks we conclude our discussion of hydrolysis reactions of acid derivatives. Of course, as is indicated by Figure 12.7, there are a variety of other acid derivatives for which hydrolysis may be important in a given environmental system. The general knowledge that we have acquired in this section should, however, put us in a better position to evaluate the importance of hydrolytic reactions of such functionalities. Some quantitative rate data for a few representative compounds of acid derivatives other than the ones discussed here may be found in the literature (e.g., Mabey and Mill, 1978).

12.4 OXIDATION AND REDUCTION REACTIONS

Overview

So far we have confined ourselves to transformation reactions in which no net electron transfer from (i.e., oxidation) or to (i.e., reduction) the organic compound of interest occurred. The major pathways by which organic chemicals are transformed in the environment include, however, oxidative and reductive steps, especially when we consider photochemical and biologically mediated transformation processes. Some of these reactions may, however, also occur abiotically in the dark. We should note that some of the reactions we discuss may be catalyzed by biological molecules (e.g., iron porphyrins, quinoid compounds) that are no longer associated with living organisms (e.g., released after cell lysis). This has led to a certain confusion with respect to the use of the term "abiotic" for such reactions. For the following discussion, we adopt the definition of Macalady et al. (1986), who suggest that a reaction is abiotic if it does not directly involve the participation of metabolically active organisms. This does, of course, not imply that "abiotic" redox reactions are not heavily influenced by biological (particularly microbial) activity since the redox conditions, and thus the availability of suitable reactants for electron transfer reactions, are determined largely by biological processes.

At this point, we should first ask ourselves how we can recognize whether an organic compound has been oxidized or reduced during a reaction. The easiest way to do that is to check whether there has been a net change in the oxidation state(s) of the atom(s) (commonly C, N, or S, see Chapter 2) involved in the reaction. For example, if a chlorine atom in an organic molecule is substituted by a hydrogen atom, as is observed in the transformation of DDT to DDD (e.g., Macalady et al., 1986):

$$\text{DDT } [\text{(+III) } CCl_3\text{; (-I)}] + H^+ + 2e^- \longrightarrow \text{DDD } [\text{(+I) } CHCl_2\text{; (-I)}] + Cl^{\ominus} \qquad (12\text{-}69)$$

the oxidation state of the carbon atom at which the reaction occurs changes from +III to +I, whereas the oxidation states of all other atoms remain the same. Hence, conversion of DDT to DDD requires a total of two electrons to be transferred from an electron donor to DDT. This type of reaction is termed a *reductive* dechlorination. Note that the species that donates the electrons is oxidized during this process. Thus, in any electron transfer reaction, one of the reactants is oxidized while the other one is reduced, and it is only logical that we term such reactions *redox* reactions. Since our focus is on the organic pollutant, we speak of an oxidation reaction if the pollutant is oxidized, and of a reduction reaction if the pollutant is reduced.

Let us now compare the reaction discussed above (Eq. 12-69) with another reaction that we discussed in Section 12.3, dehydrochlorination. Here, as is illustrated by the transformation of DDT to DDE,

$$\text{DDT } [\text{(+III) } CCl_3\text{; (-I)}] + HO^{\ominus} \longrightarrow \text{DDE } [\text{(+II) } =CCl_2\text{; (0)}] + H_2O + Cl^{\ominus} \qquad (12\text{-}70)$$

the change in oxidation state of one of the carbon atoms involved in the reaction is compensated by the change in oxidation state of the adjacent carbon atom. Hence, dehydrochlorination requires no net electron transfer from or to the compound, and we shall, therefore, consider this reaction not to be a redox reaction. [We should note that it has been suggested that the transformation of DDT to DDE (Eq. 12-70) might also occur by a free radical mechanism involving a reduction and subsequent oxidation step (Quirke et al., 1979).] Another elimination reaction, the *dihalo-elimination*, is, however, again a redox reaction. If we consider, for example, the transformation of hexachloroethane (HCE) to tetrachloroethylene (perchloroethylene, PCE), a reaction that has been observed to occur in groundwater systems (e.g., Criddle et al., 1986),

$$\underset{\text{HCE}}{Cl_3\overset{(+III)}{C}-\overset{(+III)}{C}Cl_3} + 2e^- \longrightarrow \underset{\text{PCE}}{Cl_2\overset{(+II)}{C}=\overset{(+II)}{C}Cl_2} + 2Cl^{\ominus} \qquad (12\text{-}71)$$

we realize that during this reaction, the oxidation states of both carbon atoms are altered by −I. Hence, as in our first example (Eq. 12-69), the reduction of HCE to PCE requires two electrons to be transformed from an electron donor to HCE.

Let us now take a brief look at some important redox reactions of organic pollutants that may occur abiotically in the environment. We first note that only a few functional groups are oxidized or reduced abiotically. This contrasts biologically mediated redox processes by which, as we will see in Chapter 14, organic pollutants may, for example, be completely mineralized to CO_2, H_2O and so on. Table 12.15 gives some examples of functional groups that may be involved in chemical redox reactions. We discuss some of these reactions in detail later. In Table 12.15 only overall reactions are indicated, the species that act as a sink or source of the electrons (i.e., the oxidants or reductants, respectively) are not specified. Hence, Table 12.15 gives no information about the actual reaction mechanism that may consist of several reaction steps. Furthermore, we should point out that in the environment, it is in many cases not known which species act as electron donors or acceptors in an observed redox reaction of a given organic pollutant and, therefore, it is often not possible to assess exact reaction pathways and to derive kinetic data that can be generalized. In contrast to the reactions discussed in Section 12.3, we are, therefore, in a much more difficult position with respect to quantification of reaction rates. Consequently, with our present knowledge of redox reactions of organic pollutants in the environment, we frequently have to content ourselves with a rather qualitative description of such processes which may include an assessment of the environmental (redox) conditions that must prevail to allow a reaction to occur spontaneously, and an assessment of the relative reactivities of a series of related compounds in a given system.

Energetic Considerations of Redox Reactions

When discussing reactions such as hydrolysis, we have (correctly) assumed that under typical environmental conditions, the free energy change, ΔG, of the reaction considered is negative, that is, that the reaction occurs spontaneously in one direction. We, therefore, did not bother evaluating the thermodynamics of such reactions. When looking at redox reactions of organic pollutants, however, the situtation is

TABLE 12.15 Examples of Some Simple Redox Reactions That May Occur Chemically in the Environment[a]

Oxidized Species	Reduction ⇌ Oxidation	Reduced Species	Equation Number
	Change in oxidation state of carbon atom(s)		
$R{-}COOH + 2H^{\oplus} + 2e^-$	←	$R{-}CHO + H_2O$	(12-72)
$O{=}\langle\text{ring}\rangle{=}O + 2H^{\oplus} + 2e^-$ (1,4-benzoquinone)	⇄	$HO{-}\langle\text{ring}\rangle{-}OH$	(12-73)
$-\overset{\vert}{\underset{\vert}{C}}-X$ (X=Cl,Br,I) $+ H^{\oplus} + 2e^-$	→	$-\overset{\vert}{\underset{\vert}{C}}-H + X^{\ominus}$	(12-74)
$-\overset{\vert}{\underset{X}{C}}-\overset{\vert}{\underset{X}{C}}-$ (X=Cl,Br,I) $+ 2e^-$	→	$>C{=}C< + 2X^{\ominus}$	(12-75)
$2{-}\overset{\vert}{\underset{\vert}{C}}-X$ (X=Cl,Br,I) $+ 2e^-$	→	$-\overset{\vert}{\underset{\vert}{C}}-\overset{\vert}{\underset{\vert}{C}}- + 2X^{\ominus}$	(12-76)
	Change in oxidation state of nitrogen atom(s)		
$R,X{-}C_6H_3{-}NO_2 + 6H^{\oplus} + 6e^-$	⇄ (dotted reverse)	$R,X{-}C_6H_3{-}NH_2 + 2H_2O$	(12-77)
$R,X{-}C_6H_3{-}N{=}N{-}C_6H_3{-}R,X + 2H^{\oplus} + 2e^-$	⇄	$R,X{-}C_6H_3{-}NH{-}NH{-}C_6H_3{-}R,X$	(12-78)
$R,X{-}C_6H_3{-}NH{-}NH{-}C_6H_3{-}R,X + 2H^{\oplus} + 2e^-$	⇄	$2\ R,X{-}C_6H_3{-}NH_2$	(12-79)
	Change in oxidation state of sulfur atom(s)		
$R{-}S{-}S{-}R + 2H^{\oplus} + 2e^-$	⇄	$2R{-}SH$	(12-80)
$R{-}\overset{O}{\overset{\Vert}{S}}{-}R + 2H^{\oplus} + 2e^-$	⇄ (dotted reverse)	$R{-}S{-}R + H_2O$	(12-81)

[a] Note that some reactions are reversible (indicated by ⇌), whereas others are irreversible under environmental conditions. The dotted arrow indicates that, in principle, a reaction is possible, but no clear evidence exists showing that the reaction proceeds abiotically in the dark.

quite different. Here, depending on the redox conditions prevailing in a given (micro)environment, a spontaneous electron transfer from or to an organic pollutant may be thermodynamically possible or not. Or, in other words, depending on the redox conditions (which are predominantly determined by microbially mediated processes, see below), electron acceptors (oxidants) or donors (reductants) that may react abiotically in a thermodynamically favorable reaction with a given organic chemical may or may not be present in sufficient abundance. Of course, as we have seen when discussing hydrolysis reactions, for kinetic reasons, a reaction may still not occur at a significant rate. Nevertheless, thermodynamic considerations are very helpful as a first step in evaluating the redox conditions under which a given organic pollutant might undergo an oxidation or reduction reaction. Furthermore, since most of the redox reactions in the environment are biologically mediated, the evaluation of how much energy an organism may derive from a given reaction may provide very useful hints on the sequence in which important biological redox reactions occur in the environment, or, in turn, what kind of organisms may be expected under given conditions (see Thauer et al., 1977; Hanselmann, 1986). Hence, the following remarks on some thermodynamic aspects of redox reactions also form an important base for our discussions of biological transformation processes in Chapter 14.

Half Reactions and (Standard) Reduction Potentials We start out by looking at a simple reversible redox reaction for which we are able to measure directly the free energy change, ΔG, with a galvanic cell. This example helps us to introduce the concept of using (standard) reduction potentials for evaluating the energetics (i.e., the free energies) of redox processes. Let us consider the reversible interconversion of 1,4-benzoquinone (BQ) and hydroquinone (HQ) (reaction 12-73 in Table 12.15). We perform this reaction at the surface of an inert electrode (e.g., platinum, graphite) that is immersed in an aqueous solution buffered at pH 7 (i.e., $[H^+] = 10^{-7}$ M) containing a certain concentration of BQ and HQ, respectively (see Fig. 12.17). The electron transfer occurs through a wire that connects the electrode with another inert electrode (e.g., platinum) that is immersed in an aqueous solution maintained at pH 0 (i.e., $[H^+] = 1$ M) and bubbled with molecular hydrogen (partial pressure $p_{H_2} =$ 1 atm). The latter electrode is commonly referred to as the standard (SHE) or normal (NHE) hydrogen electrode. At this electrode, hydrogen is oxidized or H^+ is reduced depending upon the direction of electron flow:

$$2\,H^+ + 2\,e^- \rightleftharpoons H_2(g) \qquad (12\text{-}82)$$

Note that by convention we always write a *half reaction* such as 12-82 as a reduction, that is, the oxidized species appears on the left side of the equation. Also, we omit the denotion (aq) for the dissolved species. In our experimental setup the reaction occurring at the other electrode is

$$\text{O=C}_6\text{H}_4\text{=O} + 2\,H^+ + 2\,e^- \longrightarrow \text{HO–C}_6\text{H}_4\text{–OH} \qquad (12\text{-}73)$$

BQ HQ

The platinum electrodes serve to transport electrons from H_2 on one side to BQ in the other. Hence, the reaction we are actually considering is given by

$$BQ + H_2(g)(p_{H_2} = 1\text{ atm}) + 2H^+(10^{-7}\text{ M}) \rightleftharpoons HQ + 2H^+(1\text{ M}) \qquad (12\text{-}83)$$

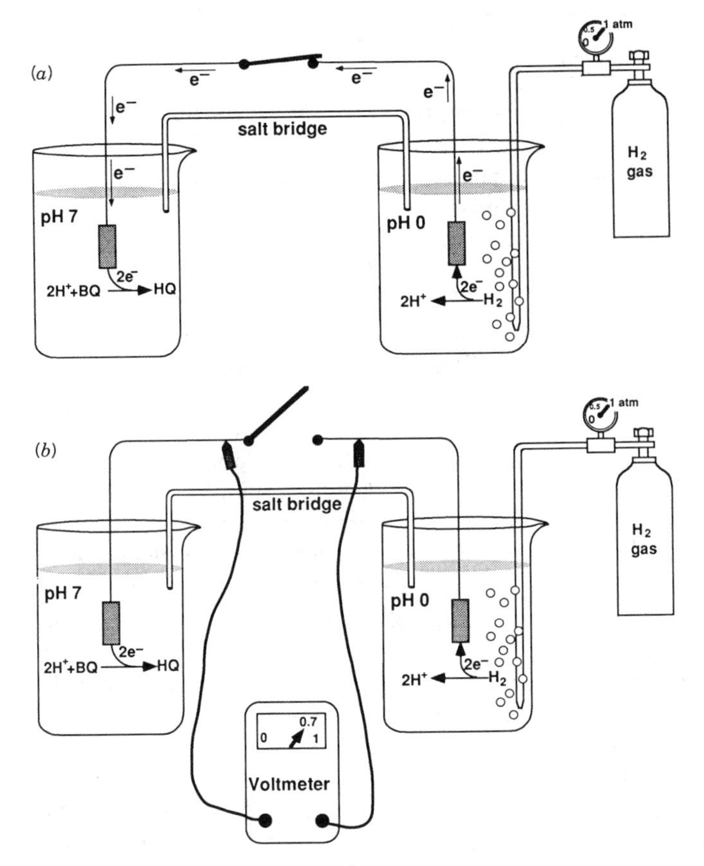

Figure 12.17 Illustration of electrochemical processes occurring in solutions with benzoquinone-hydroquinone and proton–hydrogen couples: (*a*) processes cycling electrons through connected system, and (*b*) voltage measured between separated half-reactions.

The free energy change ΔG of reaction 12-83 may be represented by (see Chapter 3):

$$\Delta G = \Delta G^0 + RT \ln \frac{[\mathrm{HQ}][\mathrm{H}^+(1\,\mathrm{M})]^2}{[\mathrm{BQ}][\mathrm{H}^+(10^{-7}\,\mathrm{M})]^2\, p_{\mathrm{H}_2}} \tag{12-84}$$

since $[\mathrm{H}^+(1\,\mathrm{M})] = 1$ and $p_{\mathrm{H}_2} = 1$, we obtain

$$\Delta G = \Delta G^0 + RT \ln \frac{[\mathrm{HQ}]}{[\mathrm{BQ}]\cdot 10^{-14}} \tag{12-85}$$

or for any other proton activity (pH) in the second cell,

$$\Delta G = \Delta G^0 + RT \ln \frac{[HQ]}{[BQ][H^+]^2} \tag{12-86}$$

With a potentiometer we can now measure the potential difference ΔE between the two electrodes, a measurement in which we compensate for the free energy change, ΔG, that is driving the reaction. Consequently, during our potentiometric measurement, no current is flowing between the two electrodes, that is, no net electron transfer occurs. If we now assume electrochemical equilibrium at the electrode surface (which implies that we have a fast reversible half reaction at each electrode), then the applied potential difference, ΔE, is directly related to the ΔG of the reaction (e.g., Lewis and Randall, 1961, Chapter 24) by

$$\Delta G = -nF\Delta E \tag{12-87}$$

where n is the number of electrons transferred ($n = 2$ in our example) and F is the Faraday constant (= electric charge of 1 mole of electrons = 96490 Coulomb (C)·mol^{-1}]. Note that we assign a positive sign to ΔE if the reaction as it is written in Eq. 12-83 proceeds spontaneously from left to right, that is, if the oxidized species (i.e., BQ) is spontaneously reduced by $H_2(g)$.

Since in experiments such as the one we have just discussed it is only possible to determine potential differences between two electrodes (and not the absolute potential of each half cell), it is now useful to choose a reference system to which all measured potential differences may be related. In accord with the IUPAC 1953 Stockholm convention, the standard hydrogen electrode (SHE) is commonly selected as the reference electrode to which we can arbitrarily assign a zero value of electrical potential. This is equivalent to assigning arbitrarily a standard free energy change, ΔG^0, of zero to the half reaction:

$$H^+ + e^- \rightleftharpoons \tfrac{1}{2}H_2(g) \qquad K = 1,\ \Delta G^0 = 0 \tag{12-88}$$

Note that this is also equivalent to setting the standard free energy of formation, ΔG_f^0, of the proton (as well as of the electron) in aqueous solution equal to zero.

By using this convention, we may now assign the measured ΔE value completely to the reaction occurring at the other electrode, in our example, to the half reaction Eq. 12-73. Instead of ΔE, we then use the term E_H, the subscript H indicating that the potential is given relative to the SHE. Hence we can rewrite Eq. 12-87 as

$$\Delta G = -nFE_H \tag{12-89}$$

Substitution of Eq. 12-89 into Eq. 12-86 and conversion to decadic logarithms yields for our half reaction 12-73:

$$E_H = E_H^0 + \frac{2.303\,RT}{nF} \log \frac{[BQ][H^+]^2}{[HQ]} \tag{12-90}$$

where $E_H^0 = -\Delta G^0/nF$. This type of equation is commonly referred to as the *Nernst equation* of an electrode reaction. Here E_H^0 is called the *standard redox* or, more appropriately, *reduction potential* of the half reaction of interest; that is, the potential we would measure against the SHE if all species involved in the (reversible) reaction were in their standard states of unit activity (note that we use the "infinite dilution state" as the reference state). In our example, E_H would, of course, also be equal to E_H^0 if [BQ] = [HQ] and $[H^+] = 1$ M. The E_H^0 value of reaction 12-73 is -0.70 V at 25°C. Hence, at standard conditions, the value for $\Delta G^0 (= -nFE_H^0)$ of the half reaction Eq. 12-73 (actually of the reaction of BQ with gaseous hydrogen under standard conditions) is -135 kJ mol^{-1} corresponding to an equilibrium constant $K = 10^{+23.7}$. This value indicates that, at pH 0 we would thermodynamically be able to reduce BQ more or less completely to HQ using molecular hydrogen at 1 atm pressure.

Since we are dealing with redox reactions occurring in natural waters we should, of course, be more interested in redox potential values (or ΔG values) that are more representative for typical natural conditions. We can calculate such values easily by assigning a typical concentration (or, more precisely, activity) value to the major water constituents that are involved in a given redox reaction. For example, we can define a E_H^0(W) value (the W indicating conditions typical for natural waters) by setting the pH equal to 7, the concentration of chloride to 10^{-3} M (if we consider a dechlorination reaction, see Table 12.15), of bromide to 10^{-5} M, and so on, but by leaving oxidant and reductant at unit activity. As we can see from Eq. 12-90, in our example, only the hydrogen ion activity is relevant. At 25°C the term $2.303\,RT/F$ has a value of 0.059 V. Hence, the E_H^0(W) value for the half reaction Eq. 12-73 is $E_H^0(W) = E_H^0 - 14 \times 0.059/2 = 0.28$ V, corresponding to a ΔG^0(W) value of -54.0 kJ mol^{-1}, and an equilibrium constant $K(W) = 10^{+9.5}$. In the following, we primarily use E_H^0(W) values for evaluating the energetics of redox reactions under natural conditions.

In our example, we considered a *reversible* redox reaction with an overall transfer of two electrons. Since, in most *abiotic* multi-electron redox processes, particularly if organic compounds are involved, the actual electron transfer occurs by a sequence of one electron transfer step (Eberson, 1987), there are intermediates formed which are often very reactive, that is, they are not stable under environmental conditions. In our example, BQ is first reduced to the corresponding semiquinone (SQ) which is then reduced to HQ:

$$\text{BQ} + H^+ + e^- \rightleftharpoons \text{SQ} + H^+ + e^- \rightleftharpoons \text{HQ} \qquad (12\text{-}91)$$

BQ SQ HQ

Each of these subsequent one-electron steps has, of course, its own E_H^0(W) value (Neta, 1981), which we denote E_H^1(W) and E_H^2(W):

$$\begin{aligned} &\text{BQ} + H^+ + e^- \leftrightharpoons \text{SQ}; \qquad E_H^1(W) = +0.10 \text{ V} \\ &\text{SQ} + H^+ + e^- \leftrightharpoons \text{HQ}; \qquad E_H^2(W) = +0.46 \text{ V} \end{aligned} \qquad (12\text{-}92)$$

[Note that in the literature one often finds the notation E_{m7}^1 and E_{m7}^2 for E_H^1(W) and

$E_H^2(W)$, respectively.] From these values we see that the free energy change is much less negative [smaller $E_H^0(W)$ value] for the transfer of the first electron to BQ as compared to the transfer of the second electron to SQ. Conversely, there is more energy required to oxidize HQ to SQ as compared to the oxidation of SQ to BQ. In general, we can assume that the formation of an organic radical is much less favorable from an energetic point of view, as compared to the formation of an organic species exhibiting an even number of electrons. From this we may conclude that the first step of a two-electron transfer between an organic chemical and an electron donor or acceptor is frequently the rate-limiting step. Thus, when we are interested in relating thermodynamic and kinetic data (e.g., through LFERs), we need to consider primarily the E_H values of this rate-limiting step, that is, the E_H value of the first one-electron transfer. We should be aware that if this first step is endergonic (i.e., positive ΔG^0 value), the overall reaction may still be exergonic (i.e., negative ΔG value), and the whole reaction may nevertheless proceed spontaneously (Eberson, 1987). Therefore, for our evaluation, whether or not a given redox reaction is possible under given conditions, we need to consider the E_H values of the overall reaction.

Table 12.16 summarizes standard reduction potentials of some environmentally important redox couples. It should be pointed out that many of the half reactions that we consider do not occur reversibly at an electrode surface, so that we would

TABLE 12.16 (Standard) Reduction Potentials at 25°C of Some Redox Couples That Are Important in Natural Redox Processes[a]

Half-Reaction: Oxidized Species = Reduced Species	E_H^0 (V)	$E_H^0(W)$ (V)	$\Delta G_H^0(W)/n^c$ (kJ mol^{-1})
(1) $O_2(g) + 4H^+ + 4e^- = 2H_2O$	+1.22	+0.81	−78.3
(2) $2NO_3^- + 12H^+ + 10e^- = N_2(g) + 6H_2O$	+1.24	+0.74	−71.4
(3) $MnO_2(s) + HCO_3^-(10^{-3}\,M) + 3H^+ + 2e^- = MnCO_3(s) + 2H_2O$		+0.52[b]	−50.2[b]
(4) $NO_3^- + 2H^+ + 2e^- = NO_2^- + H_2O$	+0.83	+0.42	−40.5
(5) $NO_3^- + 10H^+ + 8e^- = NH_4^+ + 3H_2O$	+0.88	+0.36	−34.7
(6) $FeOOH(s) + HCO_3^-(10^{-3}\,M) + 2H^+ + e^- = FeCO_3(s) + 2H_2O$		−0.05[b]	+4.6[b]
(7) Pyruvate $+ 2H^+ + 2e^- =$ lactate		−0.19	+18.3
(8) $SO_4^{-2} + 9H^+ + 8e^- = HS^- + 4H_2O$	+0.25	−0.22	+21.3
(9) $S(s) + 2H^+ + 2e^- = H_2S(g)$	+0.17	−0.24	+23.5
(10) $CO_2(g) + 8H^+ + 8e^- = CH_4(g) + 2H_2O$	+0.17	−0.25	+23.5
(11) $2H^+ + 2e^- = H_2(g)$	0.00	−0.41	+39.6
(12) $6CO_2(g) + 24H^+ + 24e^- = C_6H_{12}O_6$(glucose) $+ 6H_2O$	−0.01	−0.43	+41.0

[a] Note that most of the electron transfer reactions involving these redox couples are biologically mediated. The reactions are ordered in decreasing $E_H^0(W)$ values. Data from Thauer et al. (1977) and Stumm and Morgan (1981).

[b] Note that these values correspond to $[HCO_3^-] = 10^{-3}$ M.

[c] n = number of electrons transferred.

not be able to measure the corresponding E_H values using a galvanic cell. Nevertheless, it is very convenient to express the free energy change of a half reaction by assigning the appropriate reduction potentials, that is, $E_H = \Delta G/nF$. One possibility is to calculate such reduction potentials from thermodynamic data, for example, from (estimated) standard free energies of formation (ΔG_f^0 or $\mu^{0\prime}$) of the various species involved in the half reaction. To illustrate, we consider the half reaction

$$2NO_3^- + 12H^+ + 10e^- \rightleftharpoons N_2(g) + 6H_2O \tag{12-93}$$

which is catalyzed by microorganisms and is commonly referred to as denitrification. From compilations of thermodynamic data (e.g., Thauer et al., 1977; Stumm and Morgan, 1981) we can extract the ΔG_f^0 value for NO_3^- (aq) (-111.3 kJ mol^{-1}) and for H_2O (-237.2 kJ mol^{-1}). Furthermore, $\Delta G_f^0 = 0$ for N_2(g) (see Chapter 3), and we have arbitrarily set the standard free energy of formation of H^+ (and of e^-) equal to zero. Hence as already pointed out, the ΔG^0 value (ΔG_H^0) that we calculate for a half reaction is actually the ΔG^0 of the reaction of the oxidized species (i.e., NO_3^-) with H_2. For the reaction Eq. 12-93, we obtain for $\Delta G_H^0 = 6(-237.2) - 2(-111.3) = -1200.6$ kJ mol^{-1}. Since $E_H^0 = -\Delta G_H^0/nF$ ($n = 10$), the calculated standard reduction potential for the half reaction is $+1.24$ V. To calculate the reaction potential at pH 7, we need to use the Nernst equation for the reaction, which at 25°C is given by

$$E_H = E_H^0 + \frac{0.059}{10} \log \frac{[NO_3^-]^2 [H^+]^{12}}{P_{N_2} \cdot [H_2O]^6} \tag{12-94}$$

With all species except H^+ (10^{-7} M) at standard conditions, we obtain

$$E_H^0(W) = E_H^0 + \frac{0.059}{10} \log (10^{-7})^{12}$$

$$= 1.24 - 0.50 = +0.74 \text{ V} \tag{12-95}$$

We come back to this kind of calculation of reduction potentials when discussing E_H^0(W) values of organic pollutants. Let us now look at some of the important redox processes that determine the redox conditions in the environment.

Processes Determining the Redox Conditions in the Environment From the data in Table 12.16 we may get a general idea about the maximum free energy that microorganisms may gain from catalyzing redox reactions. We recall that on our earth, the maintenance of life resulting directly or indirectly from a steady input of solar energy is the main cause for nonequilibrium redox conditions. In the process of photosynthesis, organic compounds exhibiting reduced states of carbon, nitrogen, and sulfur are synthesized, and at the same time oxidized species including O_2 (oxic photosynthesis) or oxidized sulfur species (anoxic photosynthesis) are produced. Using glucose as a model organic compound, we can express oxic photosynthesis by combining Eqs. (12) and (1) in Table 12.16. Note that we have to take the reversed form of Eq. (1). Since we are looking at the overall process, it is convenient to write the reaction with a stoichiometry corresponding to the transfer of one electron:

$$\tfrac{1}{4}CO_2(g) + \tfrac{1}{4}H_2O \rightleftharpoons \tfrac{1}{24}C_6H_{12}O_6 + \tfrac{1}{4}O_2(g) \qquad (12\text{-}96)$$

The standard free energy change per electron transferred, $\Delta G^0(W)/n$, of reaction Eq. 12-96 can now be simply derived from Table 12.16 by adding the $\Delta G^0_H(W)$ value of reaction (12) ($= +41.0\,kJ\,mol^{-1}$) and reversed reaction (1) ($= +78.3\,kJ\,mol^{-1}$): $\Delta G^0(W)/n = +119.3\,kJ\,mol^{-1}$. Thus, on a "per electron basis," under standard conditions (pH 7), we have to invest $119.3\,kJ\,mol^{-1}$ to (photo)synthesize glucose from CO_2 and H_2O. In our standard redox potential picture using $E^0_H(W)$ values, this is equivalent to promoting one mole of electrons from a potential of $+0.81$ to -0.43 V (see Table 12.16).

The chemical energy stored in reduced chemical species (including organic pollutants) can now be utilized by organisms that are capable of catalyzing energy yielding redox reactions. For example, from Table 12.16 we can deduce that in the oxidation of glucose [reversed reaction (12)], oxygen is the most favorable oxidant (i.e., electron acceptor) from an energetic point of view, at least, if O_2 is reduced all the way to H_2O (which is commonly the case in biologically mediated processes). The $\Delta G^0(W)/n$ value for the reaction of glucose with O_2 (reversed reaction Eq. 12-96) is, of course, $-119.3\,kJ \cdot mol^{-1}$. The next "best" electron acceptors would be NO_3^- (if converted to N_2), then $MnO_2(s)$, and so on going down the list in Table 12.16.

Interestingly, the chemical reaction sequence given in Table 12.16 (that is based on standard free energy considerations) is, in essence, paralleled by a spatial and/or temporal succession of different microorganisms in the environment. In other words, in a given (micro)environment, those organisms will be dominant that are capable of utilizing the "best" electron acceptor(s) available, that is, the electron acceptor(s) exhibiting the most positive reduction potential. These microorganisms then in turn determine the redox conditions in that (micro)environment. This is illustrated in Figure 12.18 where we look at the dynamics of some redox species along the flow path of a contaminant plume in the ground. For simplicity, we assume a situation where we have a constant input of reduced (e.g., organic compounds, NH_4^+) and oxidized species (e.g., O_2, NO_3^-, SO_4^{2-}). As is shown in Figure 12.18, natural or synthetic organic compounds (the major electron donors) are degraded over the whole length of the plume. As long as there is molecular oxygen present, *aerobic respiration* takes place, which includes the oxidation of organic compounds and NH_4^+ (to NO_3^-) and the consumption of O_2. We should point out that in aerobic respiration, oxygen not only plays the role of a terminal electron acceptor, but it is also a cosubstrate in many important biologically catalyzed reactions. This is the reason why we usually make such a sharp distinction between aerobic (or oxic) and anaerobic (suboxic, anoxic) conditions (see also Chapter 14).

Once the oxygen is consumed, *denitrification* (Eq. 12-93) is observed until no nitrate is present any more. In the region where denitrification occurs, one often observes the reductive dissolution of oxidized manganese phases [e.g., $MnO_2(s)$, $MnOOH(s)$], which may or may not be biologically catalyzed. Under those conditions iron is still

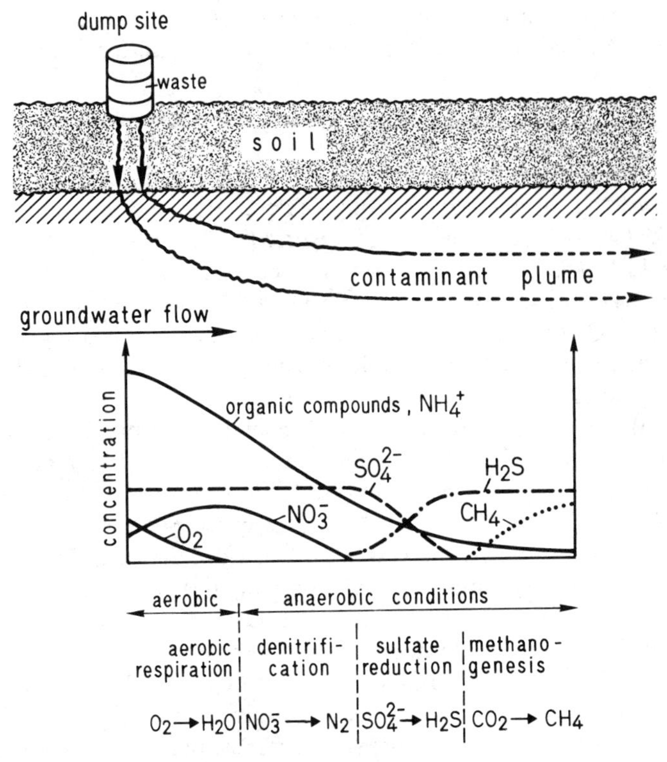

Figure 12.18 Variation in concentration of important redox species along the flowpath of a contaminant plume in groundwater. This sequence results in several zones of characteristic microbial metabolism and corresponding redox conditions (Adapted from Bouwer et al., 1984).

present in oxidized forms [e.g., FeOOH(s)]. Then, a marked decrease in redox potential occurs when only electron acceptors are left in significant abundance that exhibit low reduction potentials (see Table 12.16). This redox sequence has led to a somewhat different terminology in that one speaks of the *oxic* (aerobic), *suboxic* (denitrification, manganese reduction), and *anoxic* conditions (low redox potential). In our discussions, we will mostly use the terms aerobic and anaerobic and express the redox condition by referring to the corresponding major microbially mediated redox process (see Fig. 12.18). Processes involving electron acceptors exhibiting a low redox potential include, in sequence: *iron reduction*, *sulfate respiration* (or sulfate reduction), and *fermentation* including *methanogenesis*.

The temporal and/or spatial succession of redox processes as is illustrated in Figure 12.18 for the groundwater environment is also observed in other environments in which access to oxygen and other electron acceptors is limited, for example, in sediments of lakes, rivers, and the oceans (e.g., Hanselmann, 1986; Drever, 1988). For a more detailed discussion of the biogeochemical processes that determine the redox conditions in natural systems, we refer to corresponding textbooks (e.g., Stumm and Morgan, 1981; Morel, 1983; Drever, 1988).

Environmental Redox Conditions and Reduction Potentials of Half-Reactions Involving Organic Pollutants Let us now come back to our question of how to assess whether or not a given organic pollutant may, in principle, undergo a redox reaction in a given environmental system. For such an assessment we obviously need to know the standard reduction potential of the half reaction involving the compound of interest and its oxidized or reduced transformation product. Furthermore, since we often do not know the oxidant or reductant, we need to assign an E_H value to the environmental system we are considering. Unfortunately, unlike the situation with proton transfer reactions, where we may use pH as a master variable, it is usually not possible to assign an unequivocal E_H value to a given natural water (Stumm and Morgan, 1981). As we have already stated, many environmentally significant redox processes are slow and we, therefore, cannot assume equilibrium between all redox couples present. That also means that measurements of redox potentials of natural waters using an inert electrode and a reference electrode are often difficult to interpret, inasmuch as many important redox pairs do not show reversible electrochemical behavior at the electrode surface. This is particularly true for more oxidizing environments (aerobic conditions, denitrifying conditions) since the electrode does not respond to redox couples involving oxygen or inorganic nitrogen species. Under more reducing conditions, E_H measurements may be of some value, since there are often certain redox couples present to which the electrode does respond. Such couples include manganese species (Mn^{III}, Mn^{IV}/Mn^{II}), iron species (Fe^{III}/Fe^{II}), and reduced sulfur species as well as certain organic compounds (e.g., quinones, hydroquinones) and hydrogen. When measuring redox potentials in the field as well as in the laboratory, for practical reasons, the SHE is seldom used as a reference electrode. The most common reference electrodes are the saturated calomel electrode (SCE, $E_H^0 = +0.24$ V at 25°C) and the silver–silver chloride electrode ($E_H^0 = +0.22$ V at 25°C). The measured potentials are, however, easily converted to the hydrogen scale by adding the appropriate E_H^0 value of the reference electrode (e.g., +0.24 V in the case of SCE) to the measured value.

Owing to the difficulties in assigning a meaningful E_H value to a given natural system, it is helpful to use the $E_H^0(W)$ values of the most important biogeochemical redox processes (Table 12.16) as a framework for evaluating under which general redox conditions a given organic pollutant might undergo a certain redox reaction. To illustrate, we again consider the (reversible) reduction of benzoquinone to hydroquinone (Eq. 12-73). The $E_H^0(W)$ value of this redox couple is +0.28 V. Inspection of Table 12.16 shows that the reduction of BQ to HQ is only feasible at redox conditions typical for environments in which iron reduction, sulfate reduction, or fermentation occurs. On the other hand, since we are considering a reversible reaction, under aerobic, denitrifying, and manganese reducing conditions, HQ may serve as an electron donor and, hence, become oxidized to BQ.

At this point, we might wonder how reasonable it is to use $E_H^0(W)$ values as given in Table 12.16 for characterization of the redox conditions in a natural system, since the species that define these conditions will, of course, usually not be present at standard concentrations (i.e., at 1 M). To evaluate this problem, let us compare the E_H value of a 10^{-4} M aqueous hydrogen sulfide solution at pH 8 with the $E_H^0(W)$ value of reaction (9) in Table 12.16, which is defined for gaseous H_2S at 1 atm. We assume that any oxidation of hydrogen sulfide in our solution will yield elementary sulfur to which, by convention, we assign an activity of 1. However, in contrast to reaction (9) in Table 12.16, we do not consider hydrogen sulfide in the gaseous form but dissolved in water, which means that we also have to take into account that H_2S dissociates in aqueous solution (the pK_a of H_2S is 7.0 at 25°C). The Nernst equation for our system is then given by

$$E_H = E_H^0[H_2S(aq)] + 0.03 \log \frac{[H^+]^2}{[H_2S_{aq}]} \tag{12-97}$$

with $E_H^0[H_2S(aq)] = +0.14$ V (CRC, 1985/86). The fraction of the total hydrogen sulfide present as H_2S_{aq} is given by (see Chapter 8)

$$[H_2S(aq)] = \frac{1}{1 + K_a/[H^+]}[H_2S]_{tot} \tag{12-98}$$

Substitution of $E_H^0[H_2S(aq)]$ and Eq. 12-98 into Eq. 12-97 yields the Nernst equation for calculating the E_H value at 25°C at a given pH for a total hydrogen sulfide concentration $[H_2S]_{tot}$:

$$E_H = +0.14 \text{ V} + 0.03 \text{ V} \log \frac{[H^+]\{[H^+] + K_a\}}{[H_2S]_{tot}} \tag{12-99}$$

In our example, we set $[H^+] = 10^{-8}$, $K_a = 10^{-7}$, and $[H_2S]_{tot} = 10^{-4}$, and we obtain an E_H value of -0.19 V which is not that much different from the $E_H^0(W)$ value of reaction (9) in Table 12.16.

Table 12.17 contains $E_H^0(W)$ values for some organic half-reactions involving organic pollutants. Again, we consider only the free-energy change of the overall reaction (which is not necessarily reversible under ambient conditions), and, therefore, we do not necessarily consider the rate-limiting step(s). The $E_H^0(W)$ values for the rate-limiting step of a redox reaction, which usually involves the formation of a radical, are quite difficult to obtain, particularly for reactions in aqueous solution. Furthermore, most of the $E_H^0(W)$ values given in Table 12.17 are calculated from (estimated) ΔG_f^0 values of the educt(s) and product(s) of the reaction. Hence, they should be considered as rough estimates. Note that in the literature (e.g., Dean, 1979; Wagman et al., 1982), ΔG_f^0 values of organic compounds are usually given for the gaseous species and/or for the pure organic phase (i.e., liquid, solid). Since we are interested in aqueous-phase free energy of formation values, we need to add the free energy contribution for transferring the compound from the gaseous or pure liquid (solid) phase to the aqueous phase. To illustrate, we discuss two examples.

TABLE 12.17 Examples of Standard Reduction Potentials at 25°C and pH 7 of Some Organic Redox Couples

Half-Reaction				$E_H^0(W)^a$
Oxidized Species		Reduced Species		(V)
CCl_3-CCl_3	$+ 2e^- =$	$CCl_2{=}CCl_2$	$+ 2Cl^-$	+1.13
CCl_4	$+ H^+ + 2e^- =$	$CHCl_3$	$+ Cl^-$	+0.67
$CHCl_3$	$+ H^+ + 2e^- =$	CH_2Cl_2	$+ Cl^-$	+0.56
CBr_4	$+ H^+ + 2e^- =$	$CHBr_3$	$+ Br^-$	+0.83
$CHBr_3$	$+ H^+ + 2e^- =$	CH_2Br_2	$+ Br^-$	+0.61
$C_6H_5-NO_2$	$+ 6H^+ + 6e^- =$	$C_6H_5-NH_2$	$+ 2H_2O$	+0.42
$CH_3-S(=O)-CH_3$	$+ 2H^+ + 2e^- =$	CH_3-S-CH_3	$+ H_2O$	+0.16
$CH_3-S(=O)_2-CH_3$	$+ 2H^+ + 2e^- =$	$CH_3-S(=O)-CH_3$	$+ H_2O$	−0.24
R—S—S—R (cystine)	$+ 2H^+ + 2e^- =$	2 R—SH (cysteine)		−0.39

[a]Estimated from thermodynamic data (Dean, 1979) and E_H data from Clark (1960); for $[H^+] = 10^{-7}$ M, $[Cl^-] = 10^{-3}$ M, and $[Br^-] = 10^{-5}$ M.

First, we consider the reductive dechlorination of hexachloroethane to tetrachloroethylene, a reaction that we have already encountered earlier:

$$C_2Cl_6 + 2e^- \rightleftharpoons C_2Cl_4 + 2Cl^- \quad (12\text{-}71)$$

The ΔG_H^0 value for this reaction is given by

$$\Delta G_H^0 = \Delta G_f^0\,[C_2Cl_4(aq)] + 2\Delta G_f^0\,[Cl^-(aq)] - \Delta G_f^0\,[C_2Cl_6(aq)] \quad (12\text{-}100)$$

(Note that we assume reaction reversibility to make this calculation, although one never sees any backreaction in practice.) In the literature (Dean, 1979), we find $\Delta G_f^0\,[C_2Cl_4(g)] = +20.5\ \text{kJ mol}^{-1}$, $\Delta G_f^0\,[C_2Cl_6(g)] = -54.9\ \text{kJ mol}^{-1}$, and $\Delta G_f^0\,[Cl^-(aq)] = -131.3\ \text{kJ mol}^{-1}$. From Chapter 6, we recall that the free-energy change for transferring a molecule from the gaseous phase to the aqueous phase, $\Delta G_{g \to a}^0$ $[= \Delta G_f^0(aq) - \Delta G_f^0(g)]$, is related to the Henry's Law constant K_H by $\Delta G_{g \to a}^0 = + RT \ln K_H$ (the positive sign is used because we consider the transfer from air to water and not from water to air). Hence, we may write

$$\Delta G_f^0(\text{aq}) = \Delta G_f^0(\text{g}) + RT \ln K_H \tag{12-101}$$

The Henry's Law constants at 25°C for C_2Cl_6 and C_2Cl_4 are 3.9 atm L mol^{-1} and 27.5 atm L mol^{-1}, respectively (Munz and Roberts, 1987). With $R = 8.314$ J K^{-1} mol^{-1}, we obtain $\Delta G_f^0\,[C_2Cl_4(\text{aq})] = 28.7$ kJ mol^{-1} and $\Delta G_f^0\,[C_2Cl_6(\text{aq})] = -51.5$ kJ mol^{-1}. Substitution of these values into Eq. 12–100 yields a ΔG_H^0 value for the half-reaction Eq. 12-71 of -182.4 kJ mol^{-1}. Hence, we obtain an E_H^0 value for reaction Eq. 12-71 of $+0.95$ V (see Eq. 12-89). As mentioned earlier, we are most interested in E_H values typical for natural conditions, that is, in $E_H^0(\text{W})$ values where we choose a typical concentration (activity) of the major water constituents that take part in the reaction. In our example, we need to choose a common freshwater chloride concentration, say 10^{-3} M. Using the Nernst equation for reaction 12-71 at 25°C:

$$E_H = E_H^0 + \frac{0.059}{n} \log \frac{[C_2Cl_6]}{[C_2Cl_4][Cl^-]^2} \tag{12-102}$$

We then obtain an $E_H^0(\text{W})$ value of $E_H^0(\text{W}) = 0.95\text{ V} + (0.059/2)(-2)(-3)\text{ V} = +1.13$ V. This value tells us that, from a thermodynamic point of view, the reduction of hexachloroethane to tetrachloroethylene may occur under any environmental conditions including aerobic conditions. Indeed, the aerobic transformation of hexachloroethane to tetrachloroethylene has been observed in the field (Criddle et al., 1986). In this case, it was, however, not clear whether the reaction occurred abiotically or was biologically mediated (or both).

In our second example we consider the reduction of nitrobenzene (NB) to aniline (AN):

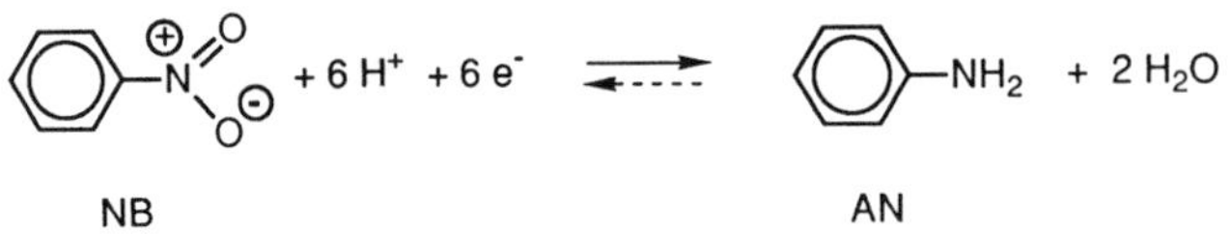

$$C_6H_5\text{-}NO_2 + 6\,H^+ + 6\,e^- \rightleftharpoons C_6H_5\text{-}NH_2 + 2\,H_2O \tag{12-103}$$

NB AN

In this case, ΔG_f^0 values of the pure liquid compounds at 25°C are available for all species involved in the reaction (Dean, 1979). Since for the solvent H_2O the reference state is the pure liquid, we may directly use $\Delta G_f^0\,[H_2O(\text{l})]$. For NB and AN, however, we need to calculate ΔG_f^0 (aq), that is, the standard free energy of formation in aqueous solution at a concentration of 1 M. From Chapter 5, we recall that the free energy for transforming a solute from its pure liquid to water is given by

$$\Delta G_s = RT \ln x_w + RT \ln \gamma_w \tag{5-5}$$

where $RT \ln x_w$ is the contribution of the entropy of ideal mixing, and $RT \ln \gamma_w$ is the partial molar excess free energy. Thus, we obtain

$$\Delta G_f^0\,(\text{aq}) = \Delta G_f^0\,(\text{l}) + RT \ln x_w\,(1\text{M}) + RT \ln \gamma_w \tag{12-104}$$

and for the ΔG_H^0 of reaction 12-103,

$$\begin{aligned}\Delta G_H^0 = {} & \Delta G_f^0 [AN(l)] + 2\Delta G_f^0 [H_2O(l)] - \Delta G_f^0 [NB(l)] \\ & + RT \ln x_w ([AN] = 1M) - RT \ln x_w ([NB] = 1M) \\ & + RT \ln \gamma_w (AN) - RT \ln \gamma_w (NB)\end{aligned} \qquad (12\text{-}105)$$

To approximate ΔG_H^0, we may set the contributions of the entropies of ideal mixing of NB and AN equal, and assume that the activity coefficients of both compounds in water are independent of concentration (which is not necessarily true, particularly, for compounds such as AN, see Chapter 5). Then, $\gamma_w = \gamma_w^{sat}$, and since $\gamma_w^{sat} \cong (C_w^{sat} V_w)^{-1}$, (Eqs. 5-8 and 5-9 in Chapter 5), we obtain

$$\begin{aligned}\Delta G_H^0 = {} & \Delta G_f^0 [AN(l)] + 2\Delta_f G^0 [H_2O(l)] - \Delta G_f^0 [NB(l)] \\ & + RT \ln [C_w^{sat}(NB)/C_w^{sat}(AN)]\end{aligned} \qquad (12\text{-}106)$$

Substituting the values for ΔG_f^0 $[AN(l)] = +149.1 \text{ kJ mol}^{-1}$, $\Delta G_f^0 [H_2O(l)] = -237.2 \text{ kJ mole}^{-1}$, $\Delta G_f^0 [NB(l)] = +146.2 \text{ kJ mol}^{-1}$, C_w^{sat} (NB) $= 0.017$ M, and C_w^{sat} (AN) $= 0.39$ M, a ΔG_H^0 value of $-479.2 \text{ kJ mol}^{-1}$ is obtained for the half reaction Eq. 12-103. This corresponds to an E_H^0 value of $+0.83$ V, and an $E_H^0(W)$ value of $0.83 - (0.059) \cdot 42/6 = +0.42$ V.

Instead of using $E_H^0(W)$ values calculated from thermodynamic data for evaluating the energetics of a given redox reaction, it is also quite common to try to relate polarographic half-wave potentials to reduction potentials (e.g., Henglein, 1976). Polarographic data may be very useful to get a rough idea about the reduction potential of a given half reaction; however, one has to be aware of the difficulties that may arise in the interpretation of polarographic half-wave potentials. Particularly in aqueous solution, polarographic measurements of many organic pollutants involve irreversible steps in water at pH 7 (e.g., Zuman, 1967). Furthermore, it may not be always clear how many electrons are transferred at the electrode. In other words, even within a series of related compounds, the products of the reaction at the electrode may be different. For example, the reduction of substituted azobenzenes may either yield the corresponding hydrazobenzene (2-electron transfer) or the corresponding anilines (4-electron transfer):

R1, X1 –N=N– R2, X2 → ($2H^+ + 2e^-$) R1, X1 –NH–HN– R2, X2

R1, X1 –N=N– R2, X2 → ($4H^+ + 4e^-$) R1, X1 –NH_2 + H_2N– R2, X2 (12-107)

Electron-withdrawing substituents (e.g., —COOH, —SO_3^-) seem to favor the reduction to hydrazobenzene, while electron-donating substituents [e.g., —OH, —NH_2, —$N(CH_3)_2$] seem to lead to the formation of the anilines (Stradins and Glezer, 1979). Consequently, we note that polarographic data for estimating reduction or oxidation potentials may be useful but should be interpreted and used with caution.

Kinetics of Redox Reactions

After considering the reduction potentials of "overall" half reactions, we are in a position to decide whether or not a given compound may, in principle, undergo oxidation or reduction in a given environment. We now have to tackle the more difficult part, that is, describe the kinetics of such reactions. As pointed out earlier, in most studies in which redox reactions of organic pollutants with natural materials have been investigated, the natural oxidants or reductants are not known. Furthermore, since there is very little kinetic data available on such "natural" reactions, in our following discussion, we will often have to rely on the results of laboratory studies conducted with well-defined systems. Before we look at some of these studies, we need to make some general remarks on the kinetics of redox reactions.

Factors Determining the Rate of Redox Reactions From our earlier discussion, we recall that oxidative or reductive transformation of an organic compound commonly requires two electrons to be transferred to yield a stable product. In the majority of abiotic redox reactions, the two electrons are transferred in sequential steps (Eberson, 1987). With the transfer of the first electron, a radical species is formed which is, in general, much more reactive than the parent compound. Hence, very often, the rate-determining step will be the transfer of the first electron from or to the pollutant. To illustrate, we consider the reductive dechlorination of trichloromethane ($CHCl_3$) and tetrachloromethane (CCl_4) by reduced hematin, a reduced iron porphyrin hydroxide [Fe^{II}(Porph)]. These reactions have been studied by Klecka and Gonsior (1984) in buffered aqueous solutions. If, for simplicity, we assume that the major reaction products are dichloromethane (CH_2Cl_2) from the reduction of $CHCl_3$, and $CHCl_3$ from the reduction of CCl_4, we may express the two reactions as

$$CHCl_3 + 2Fe^{II}(Porph) + H^+ \rightleftharpoons CH_2Cl_2 + 2Fe^{III}(Porph) + Cl^- \quad (12\text{-}108)$$

and

$$CCl_4 + 2Fe^{II}(Porph) + H^+ \rightleftharpoons CHCl_3 + 2Fe^{III}(Porph) + Cl^- \quad (12\text{-}109)$$

For calculating the $\Delta G^0(W)$ values for reactions Eq. 12-108 and 12-109, we need to know the reduction potentials of hematin and of the half-reactions of the reduction of $CHCl_3$ to CH_2Cl_2 and CCl_4 to $CHCl_3$, respectively. The latter two reduction potentials are given in Table 12.17. The reduction potential of hematin in aqueous solution in not well known; it may, however, be assumed to be in the order of $+0.1$ V. Using this number, we obtain $\Delta E_H^0(W)$ values of $+0.46$ V and $+0.57$ V, and, since $\Delta G_H^0(W) = -nF\Delta E_H(W)$ (see Section 12.4), $\Delta G_H^0(W)$ values of $-89\ \text{kJ mol}^{-1}$ and $-110\ \text{kJ mol}^{-1}$ may be estimated for reactions Eq. 12-108 and 12-109, respectively.

Hence, both reactions are exergonic with both exhibiting large negative $\Delta G^0(W)$ values. However, the rates of transformation of $CHCl_3$ and CCl_4 are very different. In an aqueous 0.1 M Na_2S/0.1 M K_2HPO_4 solution of pH 7 containing 8 μM hematin and 4 μM $CHCl_3$ or CCl_4, respectively, no significant loss of $CHCl_3$ was observed within 42 days, while over 95% of the CCl_4 reacted in less than 8 hours (Klecka and Gonsior, 1984). This result is not too surprising when we compare the large difference in the willingness of the two compounds to accept the first electron, which is reflected in the large difference in the respective one-electron reduction potentials, determined by polarographic methods: -1.44 V for $CHCl_3$ and -0.54 V for CCl_4 (Stackelberg and Stracke, 1949). Note that these reduction potentials were measured in an organic solvent/water mixture (dioxane/water).

In a simple way, we may express a one-electron transfer reaction (e.g., the transfer of the first electron from a reductant R to a pollutant P) schematically as (Eberson, 1987):

$$\underset{\text{educts}}{P + R} \rightleftharpoons \underset{\substack{\text{precursor}\\\text{complex}}}{(PR)} \rightarrow \underset{\substack{\text{transition}\\\text{state}}}{[PR \leftrightarrow P^{\overline{\cdot}} R^{\dot{+}}]^{\ddagger}} \rightarrow \underset{\substack{\text{successor}\\\text{complex}}}{(P^{\overline{\cdot}} R^{\dot{+}})} \rightleftharpoons \underset{\text{products}}{P^{\overline{\cdot}} + R^{\dot{+}}} \tag{12-110}$$

If there is a strong electronic coupling between P and R in the transition state, one commonly speaks of an *inner-sphere* mechanism; conversely, if the interaction is weak, we refer to the reaction as occurring by an *outer-sphere* mechanism.

From Eq. 12-110 we see that we may divide a one-electron transfer into various steps (maybe somewhat artificially). First, a precursor complex has to be formed, that is, the reactants have to meet and interact. Hence, electronic as well as steric effects determine the rate and extent at which this precursor complex formation occurs. Furthermore, in many cases, redox reactions occur at surfaces, and therefore, the sorption behavior of the compound may also be important for determining the rate of transformation. In the next step, the actual electron transfer between P and R occurs. The activation energy required to allow this electron transfer to happen depends primarily on the "willingness" of the two reactants to lose and gain, respectively, an electron. A measure of the "willingness" of a given chemical species to lose or gain an electron is given by the reduction potential of the respective *one-electron* half reaction. Hence, it can be expected that the reaction rate will depend on the magnitude of the one-electron reduction and oxidation potential, respectively, of the two reactants. Finally, in the last steps of reaction sequence Eq. 12-110, a successor complex is postulated which decays into the products. These last steps may be important, particularly in heterogeneous systems where the creation of new reactive surface sites may be rate determining.

From the discussion above, we may conclude that for relating oxidation or reduction rate data of organic pollutants, we need to be interested in the reduction potentials for the transfer of the first electron from or to the compounds of interest [and not in overall $E_H^0(w)$ values]. Unfortunately, for aqueous solution, such data are very scarce. The best investigated groups of compounds include phenolic compounds, anilines, and a series of nitroaromatic compounds, the latter because of their importance as radiosensitizers in radiotherapy (Wardman, 1977). One-electron oxidation or reduction potentials are commonly determined either by pulse radiolysis techniques or by electrochemical methods including polarographic techniques. For a description of these methods and for data compilations, we refer to the literature (e.g., Meisel and Czapski, 1975; Meisel and Neta, 1975; Neta, 1981; Suatoni et al., 1961; Bard and Lund, 1978–1984 Wardman, 1989). We will present some one-electron reduction potentials during our discussion of redox reactions.

Oxidation Reactions When we think about the oxidation of organic pollutants in the environment, we immediately wonder about the importance of molecular oxygen in such reactions. From our daily experience, we know that most organic compounds (fortunately) do not react spontaneously at significant rates with molecular oxygen, although the overall reactions would, in general, be exergonic. Hence, the reason for the inertness of organic pollutants with respect to molecular oxygen (i.e., not activated by, for example, photolytic or biological processes, see Chapters 13 and 14) must be a kinetic one. Indeed, the standard reduction potential for transferring one electron to molecular oxygen yielding superoxide [$pK_a(O_2H^{\bullet}) = 4.88$; Ilan et al., 1987]:

$$O_2(1M) + e^- \rightleftharpoons O_2^{\bullet -}(1M) \qquad E_H^0(W) = -0.16\ V \tag{12-111}$$

shows that in aqueous solutions at pH 7, molecular oxygen in only a very weak oxidant (note that we are again using a thermodynamic entity for a kinetic argument). Consequently, only compounds that are very easily oxidized will react with molecular oxygen at significant rates. Examples of such compounds include mercaptans (R-SH, reverse reaction Eq. 12-80), and anilines ($ArNH_2$) that are substituted with electron-donating groups (e.g., reversed reaction, Eq. 12-79).

Besides molecular oxygen, iron(III) and manganese (III/IV) oxides are the abundant oxidizing agents that may undergo abiotic reactions with organic pollutants in natural systems. Stone (1987) and Ulrich and Stone (1989) investigated the oxidation of a series of substituted phenols (ArOH) by well-characterizd manganese oxide particles. In this process, as in any reaction involving particle surfaces, adsorption–desorption phenomena have to be considered in addition to the actual electron transfer. Hence, the rate of transformation is dependent on the tendency of the compound to bind to the surface as well as on its tendency to lose an electron. Using Mn(III) as a model oxidant, Stone (1987) proposed the following reaction scheme for oxidation of phenolic compounds by manganese oxides:

(1) Precursor complex formation:

$$\text{surface} \equiv \text{Mn(III)} - \text{OH} + \text{ArOH} \rightleftharpoons \text{surface} \equiv \text{Mn(III)} - \text{OAr} + H_2O \tag{12-112}$$

where the symbol "≡" denotes bonds between surface metal centers and the oxide lattice.

(2) Electron transfer:

$$\text{surface} \equiv \text{Mn(III)} - \text{OAr} \rightleftharpoons \text{surface} \equiv \text{Mn(II)}, {}^{\bullet}\text{OAr} \quad (12\text{-}113)$$

(3) Release of phenoxy radical:

$$(\text{surface} \equiv \text{Mn(II)}, {}^{\bullet}\text{OAr}) + H_2O \rightleftharpoons \text{Mn(II)} - OH_2 + {}^{\bullet}\text{OAr} \quad (12\text{-}114)$$

(4) Release of reduced Mn(II):

$$\text{surface} \equiv \text{Mn(II)} - OH_2 \rightleftharpoons Mn^{2+}(aq) + \text{free underlying site} \quad (12\text{-}115)$$

Note that in the last step (Eq. 12-115), a new surface site with an oxidized manganese species is uncovered. Hence, this last step is important for oxidation of additional phenol molecules, and could therefore, also play a role in determining the overall rate of transformation. The phenoxy radical ($ArO^{\bullet}$) released in reaction Eq. 12-114 may undergo further oxidation to yield a quinone, or may get involved in coupling reactions leading to the formation of dimers and polymeric products (Stone, 1987). All these reactions may be considered to be fast.

Since the speciation of both the phenolic species and the surface sites is pH-dependent, generally, a relatively complicated relationship between transformation rate and pH was obtained. However, the rate of phenol oxidation by manganese oxide decreases with increasing pH at near neutral pH by about an order of magnitude per pH unit. For a discussion of the mathematical framework proposed to describe the kinetics of this reaction, we refer to the papers by Stone (1987) and Ulrich and Stone (1989). Here, we confine ourselves to a few highlights of their results.

In their study, Stone (1987) and Ulrich and Stone (1989) measured the production of Mn^{2+} rather than the disappearance of the phenolic compounds for quantification of reaction rates. This renders an interpretation of differences found in reaction rates between different compounds a little bit more difficult, since the reaction stoichiometry, that is the number of Mn^{2+} produced per oxidized phenol, may vary between compounds. However, for the compounds discussed, these difference should not exceed a factor of 2. Hence, we will assume a constant reaction stoichiometry for the following discussion. Table 12-18 summarizes the dissolution rates of a specific manganese oxide (for characterization see Stone, 1987) in the presence of various different substituted phenols. Note that these rates represent initial dissolution rates at a given manganese oxide and phenol concentration. Consequently, these rates cannot be easily extrapolated to other conditions. Nevertheless, from the data in Table 12-18 we may draw some interesting conclusions with respect to the effect of substituents on the rate of oxidation of phenolic compounds by manganese oxides, and we may try to get some idea about the importance of this process in the environment.

Considering that the dissolution rates given in Table 12-18 were measured during the initial stage of the experiments, we may assume that these rates were primarily determined by the tendency of the compound to form a precursor complex, and by

TABLE 12.18 Properties of Substituted Phenols and Rates of Reductive Dissolution of Manganese Oxide in the Presence of These Phenols[a]

Compound name[b]	Structure	pK_a	$E_{1/2,ox}$[c] (V)	Dissolution Rate[d] (mol Mn^{2+} L^{-1} min^{-1})	Dissolution Rate Relative to 4-Methylphenol
4-Methylphenol	 OH	10.26	0.78	2.1×10^{-6}	1.0
4-Ethylphenol	 OH	10.21	0.81	2.0×10^{-6}	0.95
3-Methylphenol	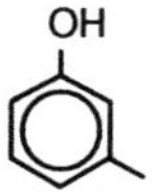 OH	10.01	0.85	4.5×10^{-7}	0.21
2-Chlorophenol	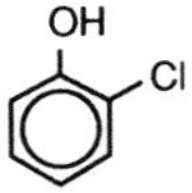 OH Cl	8.29	0.87	2.5×10^{-7}	0.12
Phenol	 OH	9.98	0.87	2.2×10^{-7}	0.10
4-Chlorophenol		9.42	0.89	3.9×10^{-7}	0.19

4-Hydroxybenzoic acid	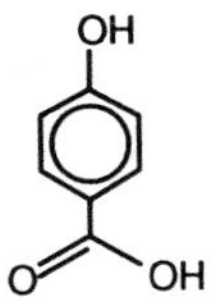 9.46 (4.58)	0.96	1.9×10^{-8}	0.009
3-Chlorophenol	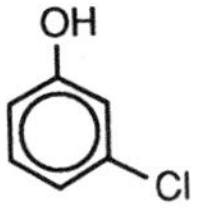 9.13	0.98	3.8×10^{-8}	0.018
4-Hydroxyacetophenone	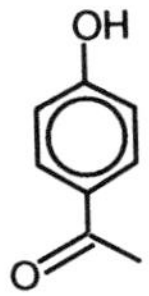 8.05	1.03	1.4×10^{-9}	0.0007
2-Hydroxybenzoic acid	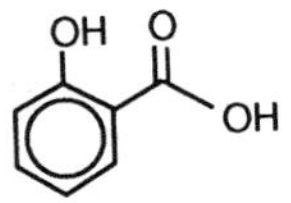 13.74 (2.97)	1.09	$<1 \times 10^{-9}$	<0.0005
4-Nitrophenol	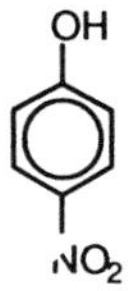 7.15	1.17	1.2×10^{-8}	0.006

[a] Data from Stone (1987) and references cited therein.
[b] Compounds ordered according to increasing $E_{1/2,ox}$ values.
[c] Determined in 1.0-M acetate buffer/50% isopropanol, apparent pH 5.6, values relative to SHE (Suatoni et al., 1961).
[d] Expressed as appearance rate of Mn^{2+}, measured in pH 4.4 (10^{-3} M acetate buffer) solutions containing 5×10^{-2} M NaCl, 10^{-4} M substituted phenol, and 4.8×10^{-5} M total manganese added as oxide particles.

its willingness to lose an electron. As discussed earlier, a quantitative measure for the willingness of a compound to accept or lose an electron is given by the reduction or oxidation potential of the corresponding half reaction. Note that we have decided to write all half-reactions as reduction reactions, the oxidation potential of a phenol is equal to the reduction potential of the half-reaction:

$$\mathrm{ArO}^{\bullet} + \mathrm{e}^- + \mathrm{H}^+ \rightleftharpoons \mathrm{ArOH} \qquad E^1_{\mathrm{H}}(\mathrm{ArO}^{\bullet}) \tag{12-116}$$

That is, $E^1_{\mathrm{H,ox}}(\mathrm{ArOH}) = E^1_{\mathrm{H}}(\mathrm{ArO}^{\bullet})$. Note that we have introduced the superscript 1 (instead of 0) to indicate the standard reduction potential of a one-electron transfer. For this reaction, E^1_{H} is a positive number. Note that the more positive $E^1_{\mathrm{H,ox}}$, the more difficult it is to oxidize the compound. For the phenolic compounds in Table 12.18, polarographic half-wave potentials determined at pH 5.6, $E_{1/2,\mathrm{ox}}$ are available (Suatoni et al., 1961). Since the pK_a values of most phenols considered are well above 7, these values should reasonably parallel the oxidation potentials of the compounds at pH 4.4, the pH at which the experiments summarized in Table 12.18 were conducted. The exceptions are 2-hydroxybenzoic acid (p$K_{a,1} = 2.97$) and 4-hydroxybenzoic acid (p$K_{a_1} = 4.58$).

When comparing the $E_{1/2,\mathrm{ox}}$ values of the substituted phenols to that of unsubstituted phenol, we see that electron donating groups such as methyl or ethyl decrease the oxidation potential and increase the reaction rate. This is due to the stabilizing effect of the substituent on the phenoxy radical. Conversely, electron-withdrawing substituents increase the oxidation potential and decrease the reaction rate. In the case of the chlorophenols, we also note a substantial difference in substituent effects between the meta and para (or ortho) positions; this is mostly due to resonance effects (see Chapters 2 and 8). For example, a chloro substituent has an inductive electron-withdrawing effect, and an electron-donating resonance effect. The latter effect is possible in para and ortho but not in the meta position:

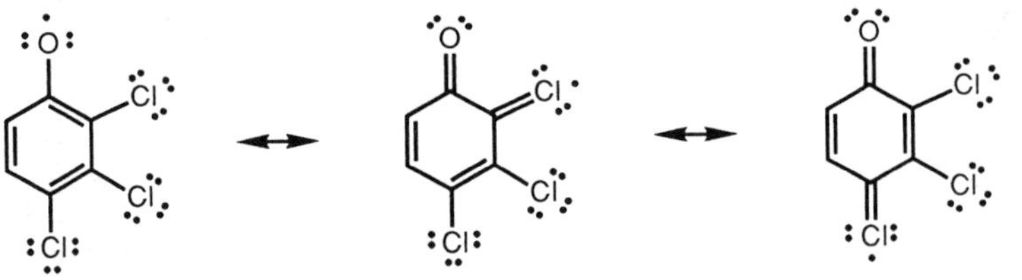

(12-117)

As can be seen from Table 12.18, the two effects almost compensate for each other in the case of 2- and 4-chlorophenol, respectively, while 3-chlorophenol exhibits a much more positive oxidation potential as compared to unsubstituted phenol. We present an example of a more quantitative treatment of the correlation of reaction rates with one-electron reduction potentials when discussing the reduction of nitroaromatic compounds.

Let us conclude this example by trying to assess whether or not oxidation of phenolic compounds by manganese oxides is an important process in the environment. This task is somewhat difficult because the reaction rates given in Table 12.18 represent initial rates determined with freshly precipitated manganese oxide, and measured at a given phenol and oxide concentration. Nevertheless, if, in a natural system, we assume a constant concentration of manganese oxide surface sites equivalent to those present at the beginning of the experiments described above ($\sim 5 \times 10^{-5}$ M total manganese added as oxide particles exhibiting a specific surface area of about $25\,m^2\,g^{-1}$), we may define a specific surface reaction rate for oxidation of a phenol by dividing the dissolution rate by the initial concentration of the phenolic species. For the most reactive compound investigated by Stone (1987), that is, 4-methylphenol, a dissolution rate, r_{diss}, of about 10^{-8} moles Mn^{+2} per liter per minute was determined at pH 7. At pH 8, the rate was about a factor of 10 smaller. If we assume a reaction stoichiometry of 1, we obtain a specific reaction rate, $r_{diss}/C_{initial}$, of $10^{-4}\,min^{-1}$ for the oxidation of 4-methylphenol, which corresponds to a half-life of about 5 days. Hence, if the initial rate of transformation does not drop significantly with time (e.g., because of a decrease of available sites), for a reactive phenol such as 4-methylphenol, oxidation by manganese oxides could very well be important in the environment, in as much as the manganese concentrations used in this study are quite representative for certain natural systems.

We conclude by summarizing that chemical oxidations of organic pollutants are only important for easily oxidizable compounds. Included in this group are the so-called antioxidants, particularly the polyalkylated phenols and anilines. It is the relative ease with which these compounds are oxidized that permits them to protect other organic products from prolonged exposures to environmental oxidants. We also recall that the major natural oxidants to be considered include molecular oxygen and manganese oxides.

Reduction Reactions As we have just noted, most organic compounds are quite inert with respect to chemical oxidation. This is not surprising since most compounds, whether they are of natural or synthetic origin, are designed to "survive" in an aerobic world. When introduced into an anaerobic environment, however, quite a few classes of compounds may undergo reduction.

We start out with a discussion of some studies in which the reduction of a series of organic compounds has been investigated in anaerobic systems containing natural materials. These examples serve to address some of the problems arising when trying to determine rates of chemical reductions of organic pollutants in natural systems. One major difficulty encountered when working with natural materials is to distinguish between abiotic and biologically mediated reactions. In cases in which the reactants and their abundances are well known (e.g., hydrolysis reactions), it is possible to derive rate constants from well-controlled sterile systems, and by comparison with data from natural systems, determine whether or not biological mediation is important. Unfortunately, for most reduction processes, we neither know the identities nor the abundance of the reductants responsible for transformation of a given organic pollutant. Common methods used to distinguish between biological and abiotic processes include autoclaving the system, or addition of poisons such as formaldehyde or sodium azide. In general, however, such studies yield quite contrasting results. For example, the rate of reductive deiodation of 1,2-diiodoethane ($CH_2I{-}CH_2I$) in an anaerobic sediment–water slurry decreased by a factor of 10 after autoclaving, while in the same system, autoclaving had a stimulating effect on the reductive dechlorination of hexachloroethane (Jafvert and Wolfe, 1987). For both compounds, significant inhibition was observed upon addition of formaldehyde and sodium azide. In another study, Weber and Wolfe (1987) found that poisoning had little effect on the rate of reduction of azobenzene in an anaerobic sediment–water system, while autoclaving decreased the rate of transformation by more than an order of magnitude. From the results of these and other studies (e.g., Zoro et al., 1974; Wahid et al., 1980; Wolfe et al., 1986), we may conclude that chemical reduction (possibly mediated by thermolabile constituents released from microorganisms) is an important abiotic process for a variety of reducible organic pollutants. There is also strong evidence that solid surfaces play a pivotal role in these processes.

But let us look at a few examples first, and then hypothesize which naturally occurring reductants may be important for reduction of organic pollutants. Table 12.19 gives the pseudo-frist-order rate constants determined by Jafvert and Wolfe (1987) for the disappearance of some polyhalogenated ethanes in an anaerobic sediment–water mixture. To avoid saturation of reactive sites and/or fast consumption of the natural reductants, the experiments were conducted at low compound concentrations and by using a relatively high sediment-to-water ratio (i.e., 0.075 g sediment per gram of water). For all compounds listed in Table 12.19, the most important reaction mechanism is vicinal dehalogenation (cf. Eq. 12.75 in Table 12.15). Again, to discuss the relative rates at which these reactions occur, we need to consider the factors that influence precursor complex formation as well as the actual electron transfer step(s). From Table 12.19 we see that in this system an appreciable fraction of all compounds (particularly the hydrophobic compound hexachloroethane) was present in sorbed form. We should note, however, that strong sorption of a hydrophobic compound does not necessarily mean enhanced reactivity. In fact, from their study of the reduction of a series of substituted azobenzenes (reaction Eq. 12-104) in a very similar sediment–water system as that used to study reductive dehalogenation, Weber and Wolfe (1987) hypothesized that sorption to nonreactive sites (e.g., into organic material) may significantly decrease the overall reduction rate of hydrophobic compounds. For now we assume that differences in reactivity found between the polyhalogenated ethanes listed in Table 12.19 are not only due to differences in precursor complex formation, but to a great extent to differences in the "willingness" of the compound to accept an electron. Since for the compounds in Table 12.19, there are no data available on one-electron reduction potentials in aqueous solution, we confine our discussion to a qualitative evaluation of the factors that determine the free-energy of activation for adding an electron to a polyhalogenated ethane.

Let us imagine what happens when an electron is added to a carbon–halogen (C–X) bond. Obviously, this electron causes a partial dissociation of the C–X bond that is subsequently cleaved to yield a carbon radical and a halide anion. The halide takes the extra electron because it is very stable in this negatively charged form. Hence, the factors that will determine the free-energy of the activated complex include the tendency of the C–X bond to accept an electron (which, in our case, is largely determined by the electronegativity of the halogen), the ease with which the C–X

TABLE 12.19 Rates of Disappearance of Some Halogenated Ethanes in an Anaeorbic Sediment–Water Slurry[a]

Compound Name	Structure	k_{obs} (s^{-1})	$t_{1/2}$ (h)	Sediment–Water Distribution coefficient K_d ($L\ kg^{-1}$)	Percent Sorbed	Initial Concentration ($mol\ L^{-1}$ of suspension)
1,2-Dichloroethane	$CH_2Cl\text{–}CH_2Cl$	$\ll 2 \times 10^{-7}$	$\gg 950$	1.3	9	1.0×10^{-5}
1,2-Dibromoethane	$CH_2Br\text{–}CH_2Br$	3.5×10^{-6}	55	2.0	13	3.0×10^{-7}
1,2-Diiodoethane	$CH_2I\text{–}CH_2I$	4.8×10^{-4}	0.4	3.5	21	2.0×10^{-7}
1,1,2,2-Tetrachloroethane	$CHCl_2\text{–}CHCl_2$	1.2×10^{-6}	160	3.4	20	3.5×10^{-7}
Hexachloroethane	$CCl_3\text{–}CCl_3$	3.2×10^{-4}	0.6	29	69	1.0×10^{-7}

[a]Data taken from Jafvert and Wolfe, 1987. Sediment to Water ratio = 0.075 (s:w), pH 6.5, apparent E_H value = − 0.140 V.

bond is cleaved (i.e., the strength of the C–X bond), and the stability of the carbon radical formed. From the relative reactivities of dichloroethane (DCA), dibromoethane (DBA), and diiodoethane (DIA), we may conclude that the latter two factors are dominating, since the electronegativities of the halogen decrease from chlorine to bromine to iodine. However, as we have already noticed when discussing halogens as leaving groups in S_N2 reactions (Section 12.3), the bond strengths also decrease significantly in the same sequence. Finally, we may also speculate that stabilization of the radical by bridging of the second halogen (March, 1985, p. 612) increase from chlorine to bromine to iodine:

$$\overset{\bullet}{\text{C}}\text{—}\overset{\ddot{\text{X}}}{\overset{|}{\text{C}}} \longleftrightarrow \text{C}\overset{\dot{\text{X}}}{\triangle}\text{C} \qquad (12\text{-}118)$$

The increasing reactivity of the chlorinated ethanes with increasing number of chlorines (Table 12.19) may be rationalized by the decreasing C–Cl bond dissociation energy with increasing chlorination (Goldfinger and Martens, 1961), and probably by the availability of an increasing number of halogens to stabilize the radical.

Let us now turn to the question of which water or solid constituents might cause the reduction of organic pollutants in anaerobic environments. Transformation rates of a specific chemical have been determined in different anaerobic sediment (soil)–water systems in only a few studies. Examples of such studies include the reductive dechlorination of hexachloroethane (Jafvert and Wolfe, 1987), the reduction of the nitro group of methyl parathion (structure of methyl parathion is given in Table 12.14) to amino methyl parathion (Wolfe et al., 1986), and the reduction of azobenzene to aniline (Weber and Wolfe, 1987) in anaerobic sediment–water systems. In these studies, no evident relationship between bulk properties of the systems (e.g., apparent redox potential, pH, total particulate organic carbon, total iron, etc.) and transformation rates could be established. This is actually not too surprising, since one can imagine that there may be a great variety of different reduced species present that may react with a given reducible organic pollutant. Such species may or may not be indicated by a bulk parameter, particularly by a measurement of the apparent redox potential.

The most abundant natural reductants in anaerobic soils and sediments include reduced inorganic forms of iron and sulfur, such as iron (II) sulfides, iron(II) carbonates, and hydrogen sulfide. Although some of these reductants have been found to react with reducible organic pollutants including, for example, nitrobenzenes (Schwarzenbach et al., 1990), the reaction rates are, in general, much too slow to account for the extremely rapid transformation rates often observed in natural systems. For example, for the reduction of parathion (Wahid et al., 1980) or methyl parathion (Wolfe et al., 1986) in anaerobic soils and sediments, half-lives of seconds to minutes have been determined. Consequently, there must be more reactive reductants available. Such reactive species may not be present in large abundances, but may play the role of electron transfer mediators; that is, after electron transfer to the pollutant, they may be rapidly reduced again by the bulk of reductants present. Hence, in abiotic systems, similarly to what is a well-known phenomenon in biological systems, such species may play the role of "electron transfer mediators":

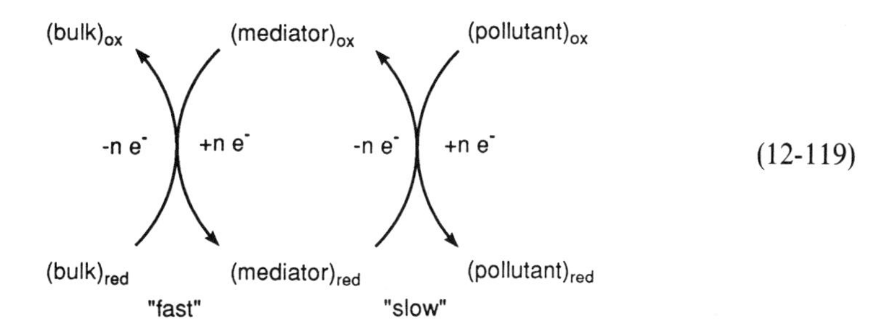

(12-119)

In biological systems, species acting as "mediators" include quinoid-type compounds, a variety of transition metal complexes, NAD/NADH, NADP/NADPH, flavins and so on (see Chapter 14). Since quinoid-type compounds, as well as transition metal complexes, in particular, iron and cobalt complexes (e.g., iron porphyrins, corrinoids) are also very likely to exist as constituents of natural organic materials (Thurman, 1985; Buffle and Altmann, 1987), such species have attracted a lot of interest as possible mediators of chemical reductions of organic pollutants in the environment. In a variety of laboratory studies, it has been demonstrated, for example, that transition metal complexes may efficiently mediate reductive dehalogenation reactions. We have already seen in one of our earlier examples that reduced hematin, an iron(II) porphyrin, is capable of rapidly transforming CCl_4 in an aqueous hydrogen sulfide buffer solution (Eq. 12-109). Other examples involving iron(II) porphyrins are the reductive dechlorination of DDT (Zoro et al., 1974), of mirex, a polychlorinated polycyclic aliphatic pesticide (Holmstead, 1976), and of a series of polyhalogenated C_1- and C_2-hydrocarbons (Wade and Castro, 1973a, b; Castro et al., 1988). In addition, the reduction of the insecticide lindane (γ-1,2,3,4,5,6-hexachlorocyclohexane, see Chapter 2) and of some chlorinated C_1-hydrocarbons by other transition metal complexes including cobalt complexes have been reported (Marks et al., 1989; Krone et al., 1989a, b). Finally, it has been shown that both iron(II) porphyrins and hydroquinones may reduce nitroaromatic compounds (Ong and Castro, 1977; Tratnyek and Macalady, 1989; Schwarzenbach et al., 1990). This latter case, the quinone and iron porphyrin mediated reduction of nitroaromatic compounds, provides closer look at the factors that determine the rates of such mediated reduction reactions.

In our example, we consider the reduction of a series of substitued nitrobenzenes in two homogeneous aqueous model systems at 25°C: (1) in 5 mM aqueous hydrogen sufide solution containing small concentrations of 8-hydroxy-1,4,-naphthoquinone ("juglone", JUG), and (2) in 5 mM aqueous cysteine solution containing small concentrations of meso-tetra-(*N*-methyl-pyridyl) iron porphyrin ("FeP"). Hence, the reduced sulfur species represent the bulk electron donors, and the quinone or the iron porphyrin play the role of the mediators. A description of the two-model system is given in Table 12.20. For details of this study, we refer to Schwarzenbach et al. (1990).

From the reduction potentials, indicated in Table 12.20, it can be concluded that,

TABLE 12.20 Model Systems for the Quinone- and Iron Porphyrin-Mediated Reduction of Nitroaromatic Compounds[a]

Redox Buffer	E_H (pH 7) of Redox buffer (V)	Mediator: Oxidized Form (Abbrev.)		Mediator: Reduced Form (Abbrev.)	E_H^0 (V)	$pK_a^{(ox)}$ (ox. form)	$pK_{a,1}^{(red)}$ (red. form)	$pK_{a,2}^{(red)}$
Hydrogen Sulfide (5 mM)	−0.19[b]	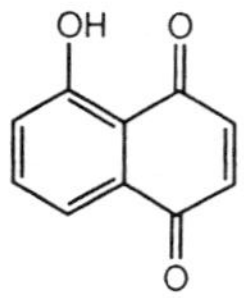 "JUG"	$+ 2 e^- + 2 H^+ =$	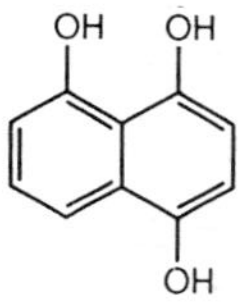 "H_2JUG"	0.43[c]	8.00	6.60	10.60
Cysteine (5 mM)	< −0.38[c]	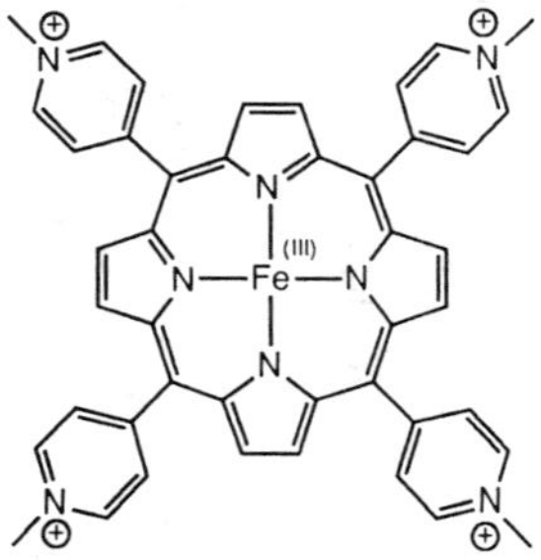 ("$Fe^{(III)}$P")	$+ e^- =$	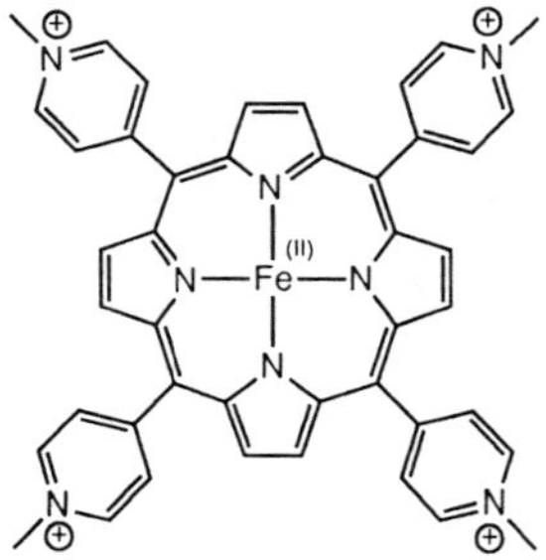 ("$Fe^{(II)}$P")	0.17[e]	5.21	(> 10)	

[a]The E_H and pK_a values for juglone are from Clark (1960) and for the iron porphyrin from Schoder (1975).
[b]For pH dependence of E_H see Eq. 12-99.
[c]The Nernst equation for the juglone redox couple is given in Eq. 12-120.
[d]E_H^0 values for $[\text{cysteine}]^2/[\text{cystine}] = 1$.
[e]$E_H = E_H^0 + 0.059 \log\{[\text{Fe(III)P}]/[\text{Fe(II)P}]\} + 0.059 \log\{[H^+]/([H^+] + K_a^{(ox)})\}$.

at equilibrium, both juglone and the iron porphyrin are always completely reduced within the pH range considered. To illustrate, we calculate the fraction of reduced juglone present in 5 mM hydrogen sulfide at pH 7. Since in our model system hydrogen sulfide is present in large excess as compared to juglone, it determines the E_H of the system, that is, it acts as a redox buffer. In Section 12.4 we have already calculated the E_H values of a hydrogen sulfide solution as a function of pH (Eq. 12-99). By setting $[H_2S]_{tot} = 5 \times 10^{-3}$ and $[H^+] = 10^{-7}$, an E_H values of -0.20 V is obtained. The Nernst equation for the juglone redox couple at 25°C is given by (see Clark, 1960):

$$E_H = E_H^0 + 0.03 \log \frac{[JUG]}{[H_2JUG]} + 0.03 \log \frac{[H^+]^3 + K_{a,1}^{(red)}[H^+]^2 + K_{a,1}^{(red)} K_{a,2}^{(red)}[H^+]}{[H^+] + K_a^{(ox)}} \quad (12\text{-}120)$$

where [JUG] and $[H_2JUG]$ are the concentrations of the oxidized and the reduced species, respectively (see Table 12.20), and the E_H^0 value as well as the acidity constants of JUG and H_2JUG are given in Table 12.20. From Eq. 12-120 we may now calculate the $[JUG]/[H_2JUG]$ ratio at equilibrium in 5 mM H_2S by setting E_H in Eq.12-120 equal to -0.20 V. The result of this calculation is $\log([JUG]/[H_2JUG]) = -7.5$, indicating that virtually all juglone is present in reduced form (i.e., as hydroquinone).

As is illustrated by Figure 12.19 for 4-chloronitrobenzene, in both model systems the rate of disappearance of a given substituted nitrobenzene ($ArNO_2$) could always be described by a first-order rate law (Eq. 12-121), even when the concentration of the quinone or the iron porphyrin was small ($< 10\,\mu$M) as compared to the initial

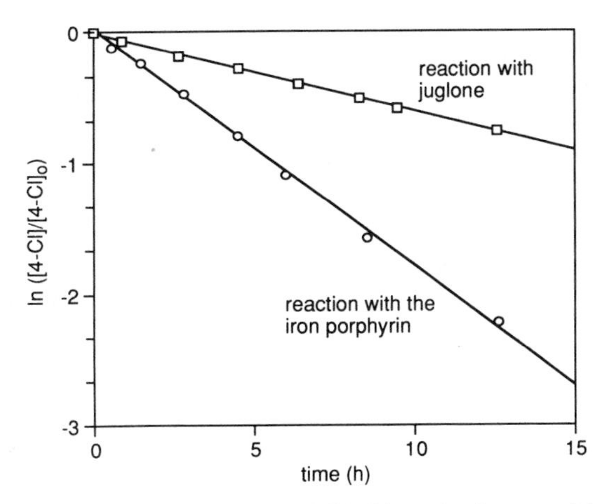

Figure 12.19 Plot of ln ([4-chloronitrobenzene]/[4-chloronitrobenzene]$_0$) versus time showing the rate of disappearance of 4-chloronitrobenzene (4-Cl) at 25°C in the two model systems. The conditions were: juglone (50 μM, pH 7.08 ± 0.02); iron porphyrin (20 μM, pH 7.03 ± 0.02); initial concentration of 4-Cl: 100 μM.

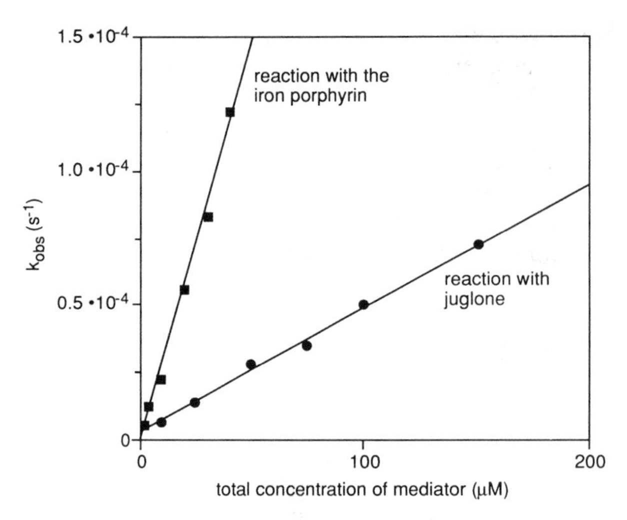

Figure 12.20 Reduction of 4-chloronitrobenzene (4-Cl) in the two model systems. Observed pseudo-first-order rate constant at 25°C versus total electron mediator concentration. The conditions were juglone (pH 7.18 ± 0.02); iron porphyrin (pH 7.03 ± 0.02); initial concentration of 4-Cl: 100 μM.

concentration of the nitroaromatic compound (100 μM):

$$\text{rate} = -\frac{d[\text{ArNO}_2]}{dt} = k_{\text{obs}}[\text{ArNO}_2] \qquad (12\text{-}121]$$

where k_{obs} is the observed pseudo-first-order rate constant under the given conditions. These findings, together with the fact that k_{obs} was linearly related to the juglone or iron porphyrin concentration (Fig. 12.20), indicate that both the quinone and the iron porphyrin acted as effective mediators in the two model systems. From the intercepts of the regression lines in Figure 12.20, it can also be seen that in the absence of a mediator, the reactions of 4-chloro nitrobenzene with hydrogen sulfide and cysteine, respectively, are very slow. An interesting difference observed between the two-model systems is that the reduction of the substituted benzenes led to different (initial) products. The reduction of an aromatic nitro compound to the corresponding aniline (I → IV in Eq. 12-122) is commonly assumed to occur in three steps with the nitroso (II) and hydroxylamine (III) compounds being the intermediates (see March, 1985):

$$\underset{\text{I}}{\text{ArNO}_2} \xrightarrow[-\text{H}_2\text{O}]{+2e^- +2H^+} \underset{\text{II}}{\text{ArNO}} \xrightarrow{+2e^- +2H^+} \underset{\text{III}}{\text{ArNHOH}} \xrightarrow[-\text{H}_2\text{O}]{+2e^- +2H^+} \underset{\text{IV}}{\text{ArNH}_2} \qquad (12\text{-}122)$$

In the system with the iron porphyrin, the aniline was formed at the same rate as the nitro compound disappeared. In contrast, the reaction with juglone led to accumulation of an initial product that was the slowly converted to aniline. This product was postulated to be the hydroxylamine (Schwarzenbach et al., 1990).

The evaluation of the pH dependence of k_{obs} showed that in the case of juglone (and of other hydroquinones), only the reaction with the hydroquinone monophenolate ($HJUG^-$) and the biphenolate (JUG^{2-}) species are important. The nondissociated hydroquinone was found to exhibit a very low reactivity toward nitroaromatic compounds. Hence, for the reaction with juglone, k_{obs} may be expressed by

$$k_{obs} = k_{HJUG^{2-}}[HJUG^-] + k_{JUG^{2-}}[JUG^{2-}] \tag{12-123}$$

where k_{HJUG^-} and $k_{JUG^{2-}}$ are the second-order rate constants for the reaction of the nitrobenzene with $HJUG^-$ and JUG^{2-}, respectively. It was found that JUG^{2-} was about 500 times more reactive than $HJUG^-$, that is, $k_{JUG^{2-}} \approx 500\, k_{HJUG^-}$. However, since at ambient pH values the concentration of JUG^{2-} is very small [$pK_{a,2}^{(red)} = 10.60$] as compared to the concentration of $HJUG^-$ [$pK_{a,1}^{(red)} = 6.60$], reaction with JUG^{2-} is only important at pH values greater than about 7.

In the case of the iron porphyrin, k_{obs} may be expressed by

$$k_{obs} = k_{Fe(II)P}[Fe(II)P] \tag{12-124}$$

where $k_{Fe(II)P}$ is the second-order rate constant and [Fe(II)P] is the total concentration of the reduced iron porphyrin. Note, however, that in contrast to k_{HJUG^-} and $k_{JUG^{2-}}$, $k_{Fe(II)P}$ is pH-dependent since the reduction potential of the iron porphyrin decreases with increasing pH (see footnote in Table 12.20), which is due to stabilization of the iron(III) porphyrin as a consequence of deprotonation of an axial water ligand.

Let us now look at the effect of substituents on the rate of reduction of the nitro group. In Table 12.21, the k_{HJUG^-} and $k_{Fe(II)P}$ (at pH 7) values of a series of monosubstituted nitrobenzenes are given. Also included are the one-electron reduction potentials, $E_H^{1'}$ ($ArNO_2$), of the nitro compounds,

$$ArNO_2 + e^- = ArNO_2^{\cdot -} \qquad E_H^{1'}(ArNO_2) \tag{12-125}$$

where the superscript (′) has been introduced to denote ambient pH conditions (i.e., pH 6–8). We first consider the reaction series with juglone. As can be seen from the data in Table 12.21, an electron-withdrawing substituent (i.e., Cl, $COCH_3$) leads to a less negative value of $E_H^{1'}(ArNO_2)$, which is paralleled by an increase in reaction rate (i.e., an increase in k_{HJUG^-}). A rather small effect on $E_H^{1'}(ArNO_2)$ and on k_{HJUG^-} is observed for methyl substituents in the meta and para position. Remarkable, however, is the drastic rate-diminishing effect of an ortho substituent as compared to the para substituted compound, which is also reflected in the one-electron reduction potential. The most plausible explanation for these findings is that reasonance of the nitro group with the aromatic ring is hindered by an ortho substituent due to steric interactions, leading to a decrease in the stabilization of the $ArNO_2^{\cdot -}$ species.

Based on the pH dependence of the reaction rates of the substituted nitrobenzenes

TABLE 12.21 One-Electron Reduction Potentials and Second-Order Rate Constants for Reaction with $HJUG^-$ and Fe(II) P for a Series of Monosubstituted Nitrobenzenes at 25°C

			Second-Order Rate Constants	
Compound	Abbreviation	$E_H^{1'}$ [a]	k_{HJUG^-} [b] $(M^{-1}s^{-1})$	$k_{Fe(II)P}$ [c] $(M^{-1}s^{-1})$
Nitrobenzene	H	−0.485	7.9×10^{-2}	9.6×10^{-1}
2-Methylnitrobenzene	2-Me	−0.590	3.0×10^{-3}	2.5×10^{-1}
3-Methylnitrobenzene	3-Me	−0.475	1.2×10^{-1}	1.1
4-Methylnitrobenzene	4-Me	−0.500	5.1×10^{-2}	6.7×10^{-1}
2-Chloronitrobenzene	2-Cl	(−0.485)[d]	1.5×10^{-1}	8.8
3-Chloronitrobenzene	3-Cl	−0.405	1.3	5.3
4-Chloronitrobenzene	4-Cl	−0.450	5.0×10^{-1}	2.7
2-Acetylnitrobenzene	2-$COCH_3$	(−0.470)[d]	3.5×10^{-1}	23.8
3-Acetylnitrobenzene	3-$COCH_3$	(−0.405)[d]	2.5	6.3
4-Acetylnitrobenzenes	4-$COCH_3$	−0.360	3.0×10^{1}	19.3

[a] Values from Meisel and Neta (1975), Neta and Meisel (1976), Wardman (1977), and Kemula and Krygowski (1979).
[b] Determined at pH 6.79 ± 0.02 in 5 mM H_2S.
[c] Determined at pH 7.01 ± 0.02 in 5 mM cysteine.
[d] Predicted from the reactivity of the compound with 2-hydroxy-1,4-naphthoquinone (see Schwarzenbach et al., 1990).

with juglone, it may be postulated that for the reaction with $HJUG^-$ (and similarly with JUG^{2-}), the transfer of the first electron from $HJUG^-$ (or JUG^{2-}) to the nitroaromatic compound is the rate-determining step (Schwarzenbach et al., 1990):

$$ArNO_2 + HJUG^- \rightarrow ArNO_2^{\cdot -} + HJUG^{\cdot} \quad (12\text{-}126)$$

The standard free-energy change, $\Delta G^{0'}$, of this reaction may be expressed by the difference of the one-electron reduction potentials of the two half-reactions:

$$\begin{array}{ll} ArNO_2 + e^- = ArNO_2^{\cdot -}; & E_H^{1'}(ArNO_2) \\ HJUG^{\cdot} + e^- = HJUG^-; & E_H^{1'}(HJUG^{\cdot}) \\ \hline ArNO_2 + HJUG^- = ArNO_2^{\cdot -} + HJUG^{\cdot}; & \Delta G^{0'} = -F[E_H^{1'}(ArNO_2) - E_H^{1'}(HJUG^{\cdot})] \end{array} \quad (12\text{-}127)$$

where $F = 96.5 \ kJ \cdot V^{-1} \cdot mol^{-1}$.

For a quantitative evaluation of the relationship between the one-electron reaction potentials and the second-order rate constants for reaction with $HJUG^-$ of the nitrobenzenes listed in Table 12.21, it is suitable to check whether a linear free-energy relationship exists between the free-energy of activation of reaction Eq. 12-126, $\Delta G^{\ddagger}$,

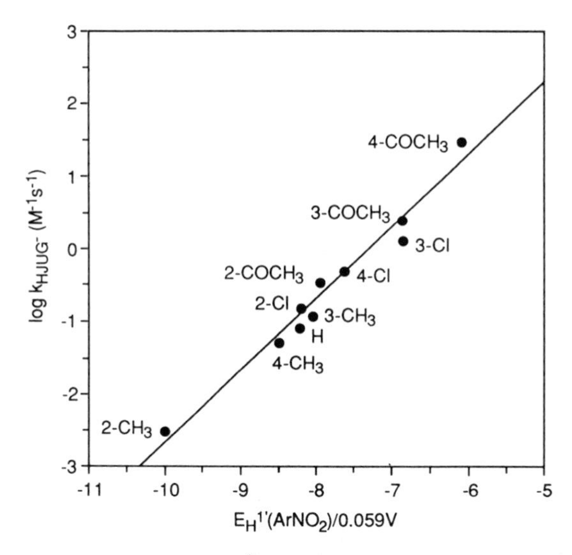

Figure 12.21 Plot of log k_{HJUG^-} versus $E_H^{1'}$ $(ArNO_2)/0.059$ V for the substituted nitrobenzenes listed in Table 12.21. The linear regression analysis yields: log $k_{HJUG^-} = 1.0\, E_H^{1'}\, (ArNO_2)/0.059\,V + 7.21$ ($R^2 = 0.99$).

and $\Delta G^{0\prime}$ as defined in Eq. 12-127:

$$\Delta G^{\ddagger} = a\Delta G^{0\prime} + b' \tag{12-128}$$

Since $\log k_{HJUG^-} = -\Delta G^{\ddagger}/RT +$ constant (see Section 12.2) and since $E_H^{1'}(HJUG^{\bullet})$ is a constant for the reaction series considered, we may rewrite Eq. 12-128 as

$$\log k_{HJUG^-} = a\frac{F}{RT}\cdot E_H^{1'}(ArNO_2) + b \tag{12-129}$$

At 25°C, $F/RT = 1/0.059$ V, and thus

$$\log k_{HJUG^-} = a\frac{E_H^{1'}(ArNO_2)}{0.059\text{ V}} + b \tag{12-130}$$

Figure 12.21 shows that a good linear correlation is found between k_{HJUG^-} and $E_H^{1'}(ArNO_2)$ for the series of nitrobenzenes considered. The linear regression analysis yields a slope a of close to 1, indicating that electron transfer between $HJUG^-$ and $ArNO_2$ occurs by an outer-sphere mechanism, and that precursor complex formation is not too important in determining the relative overall reaction rates. A very similar result has been obtained for the reaction of the nitrobenzenes with another hydroquinone monophenolate (see Schwarzenbach et al., 1990).

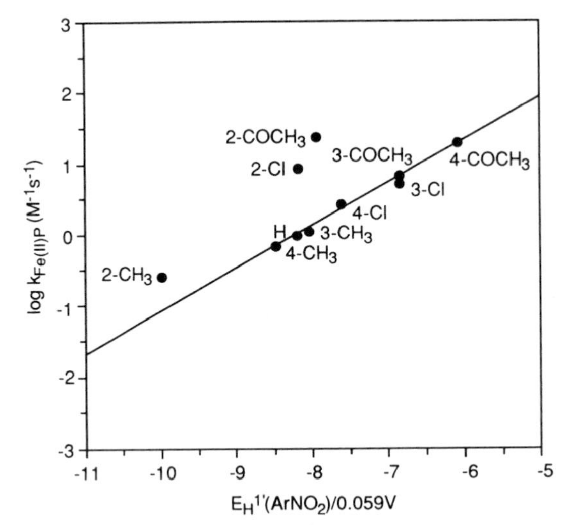

Figure 12.22 Plot of log $k_{Fe(II)P}$ versus $E_H^{1'}$ ($ArNO_2$)/0.059 V for the substituted nitrobenzenes listed in Table 12.21. The linear regression analysis yields for the meta- and para-substituted compounds: $\log k_{Fe(II)P} = 0.60\ E_H^{1'}$ ($ArNO_2$)/0.059 V + 4.95 ($R^2 = 0.99$).

Inspection of the second-order rate constants $k_{Fe(II)P}$ for the reaction of the nitrobenzenes with the iron porphyrin (Table 12.21) reveals that qualitatively similar substituent effects are observed for the meta- and para-substituted compounds. However, as Figure 12.22 shows, for these compounds the linear regression yields a slope of only 0.6. Furthermore, as compared to the meta- and para-substituted compounds, the nitrobenzenes exhibiting an ortho-substituent are reduced much faster than would be expected for the LFER shown in Figure 12.22. These findings suggest that, in contrast to the reaction with the hydroquinone monophenolates, precursor complex formation is a much more important factor in determining the overall reaction rate.

From these examples, we have seen that electron-transferring species such as iron porphyrins and quinoid-type compounds may be quite effective mediators for the abiotic reduction of nitroaromatic compounds in homogeneous aqueous solution. It is likely that such types of electron mediator molecules play an important role in reductive transformations of organic pollutants in natural systems. When considering the large structural variety of possible electron carriers in the environment, each of which exhibits a different (pH-dependent) reactivity, it may be concluded that prediction of reduction rates of organic pollutants from measurements of bulk parameters in a given natural system could be rather difficult. Furthermore, as is demonstrated by the different relative reactivities of the nitrobenzenes in the two-model systems discussed above, even predictions of relative rates may not always be feasible. On the other hand, LFERs such as illustrated by Eq. 12-130 may be helpful tools for evaluating the relative importance of the various factors that determine the rate of reduction of a series of related organic pollutants in a given environment.

This concludes our discussion of chemical reactions of organic pollutants in the environment. Many of the topics that we have addressed in this chapter will be important for our discussions in the following two chapters. We now proceed to look at the impact of light on the transformations of organic compounds in natural waters.

CHAPTER 13

PHOTOCHEMICAL TRANSFORMATION REACTIONS

13.1 INTRODUCTION

In Chapter 12, we discussed chemical reactions in which the reacting molecules were in their so-called electronic *ground state*. We termed these reactions "chemical reactions", mainly because temperature had a significant effect on reaction rates. We might recall that heat has primarily an impact on the translational, rotational, and vibrational energy of an organic molecule, but that a (small) change in temperature leaves the electronic ground state of the molecule essentially unchanged. In Chapter 2 we classified in a very simplified manner the electronic ground state of a molecule by assigning the valence electrons of the various atoms to three categories, that is, part of a σ-bond, a π-bond (or a delocalized π-bond system), or a nonbonding electron localized on an atom, usually a heteroatom. In the ground state, one commonly assigns the electrons that are engaged in chemical bonds to *bonding orbitals* (i.e., σ- or π-orbitals), whereas the electrons localized on an atom are said to occupy so-called *nonbonding orbitals* (i.e., n-orbitals). If a molecule is now exposed to ultraviolet (uv) or visible (vis) light (of interest to us is the wavelength range of solar radiation that may promote phototransformations of organic pollutants at the earth surface, i.e., 290–600 nm, see Section 13.3), electrons may get promoted from bonding or nonbonding orbitals to so-called *antibonding orbitals* (i.e., σ^*- or π^*-orbitals). The molecule is then said to be in an *excited state*, that is, it has become a much more reactive species as compared to the reactivity it exhibits in the electronic ground state. As we will see in Section 13.2, an excited species may undergo a variety of processes.

In this chapter we are concerned with two different types of photochemical processes that may lead to the transformation of organic pollutants in the aquatic environment.

First, we consider the case in which a given pollutant absorbs light and, as a consequence of that light absorption, undergoes transformation. This process is commonly referred to as *direct photolysis.* In our discussions we present the concepts used to quantify this process, rather than trying to evaluate reaction pathways and transformation products. The second type of photochemical process that we address is called *indirect photolysis.* As has been illustrated by many examples, the photolytic half-life of a compound may be much shorter in natural water as compared to distilled water. In fact, a compound may not undergo photochemical transformation at all in distilled water, but may react at a significant rate in water containing other constituents. The reasons for this behavior are either an energy transfer from another excited species (sensitized photolysis) or (chemical) reactions of the (nonexcited) compound with very reactive, short-lived species (e.g., hydroxyl radicals, peroxy-radicals, singlet oxygen) that are formed in the presence of light owing to reactions of, for example, excited humic or fulvic materials or nitrate. Again, our focus is on quantification of these processes rather than on product analysis. Before we can begin discussing direct and indirect photolysis of organic pollutants, we need, however, to review a few basic principles of photochemistry.

13.2 SOME BASIC PRINCIPLES OF PHOTOCHEMISTRY

Light Absorption by Chemical Species—Molar Extinction Coefficients

To picture the process of light absorption by a chemical species, we recall that we may look upon light as having both wave- and particle-like properties (see, e.g., Turro, 1978; Finlayson-Pitts and Pitts, 1986). As a wave, we consider light to be a combination of oscillating electric and magnetic fields perpendicular to each other and to the direction of propagation of the wave. The distance between two consecutive maxima is the wavelength λ, which is inversely proportional to the frequency ν, commonly expressed by the number of complete cycles passing a fixed point in 1 sec:

$$\lambda = \frac{c}{\nu} \tag{13-1}$$

where c is the speed of light in a vacuum, $3.0 \times 10^8\ \mathrm{m\,s^{-1}}$.

A more particle-oriented consideration of light shows that light is quantized and is emitted, transmitted, and absorbed in discrete units, so-called *photons* or *quanta.* The energy E of a photon or quantum (the unit of light on a molecular level) is given by

$$E = h\nu = h\frac{c}{\lambda} \tag{13-2}$$

where h is the Planck constant, $6.63 \times 10^{-34}\ \mathrm{J \cdot s}$. Note that the energy of a photon is dependent on its wavelength. On a molar basis the unit of light is commonly called an *einstein,* although the IUPAC has recently decided to discard this term. Hence, 1 einstein is nothing else than the equivalent of 6.02×10^{23} (= 1 mol) photons or quanta. The energy of light of wavelength λ (nm) is

$$E = 6.02 \times 10^{23} \cdot h \frac{c}{\lambda} = \frac{1.196 \times 10^5}{\lambda} \text{ kJ einstein}^{-1} \qquad (13\text{-}3)$$

It is instructive to compare the light energies at different wavelengths with bond energies typically encountered in organic molecules. Table 13.1 shows that the energy of uv and visible light is of the same order of magnitude as that of covalent bonds. Thus, in principle, such bonds could be cleaved as a consequence of light absorption. Whether or not reactions take place depends on the probability with which a given compound absorbs light of a given wavelength, and on the probability that the excited species undergoes a particular reaction.

When a photon passes close to a molecule, there is an interaction between the electromagnetic field associated with the molecule and that associated with the radiation. If, and only if, the radiation is absorbed by the molecule as a result of this interaction, can the radiation be effective in producing photochemical changes (Grotthus–Draper law, see, e.g., Finlayson-Pitts and Pitts, 1986). Therefore, the first thing we need to be concerned about is the probability witn which a given compound absorbs uv and visible light. This information is contained in the compounds *uv/vis absorption spectrum*, which is often readily available or can be easily measured with a spectrophotometer.

Let us consider the absorption of light by a solution of a given chemical in particle-free water contained in a transparent vessel, for example, in a quartz cuvette. The quantitative description of light absorption by such a system is based on two empirical

TABLE 13.1 Typical Energies for Some Single Bonds and the Approximate Wavelengths of Light Corresponding to This Energy[a]

Bond	Bond Energy E^b (kJ mol^{-1})	Wavelength λ (nm)
O–H	465	257
H–H	436	274
C–H	415	288
N–H	390	307
C–O	360	332
C–C	348	344
C–Cl	339	353
Cl–Cl	243	492
Br–Br	193	620
O–O	146	820

[a]Compare Eq. 13.3.
[b]Values from Table 2.2.

laws. The first law, Lambert's law, states that the *fraction* of radiation absorbed by the system is independent on the intensity of that radiation. Note that this law is not valid when very high intensities of radiation are employed (e.g., when using lasers). The second law, Beer's law, states that the *amount* of radiation absorbed by the system is proportional to the number of molecules absorbing the radiation. Beer's law is valid as long as long as there are no significant interactions (e.g., associations) between the molecules. From these two laws, the well-known *Beer–Lambert law* is obtained that relates the light intensity $I(\lambda)$ emerging from the solution to the incident light intensity $I_0(\lambda)$ (e.g., in einstein cm^{-2}):

$$I(\lambda) = I_0(\lambda) \times 10^{-[\alpha(\lambda)+\varepsilon(\lambda)C]l} \quad \text{or} \quad \text{absorbance } A \equiv \log\frac{I_0(\lambda)}{I(\lambda)} = [\alpha(\lambda)+\varepsilon(\lambda)C]l \quad (13\text{-}4)$$

In these equations, C is the concentration of the compound of interest in moles per liter (M), l is the path length of the light in the solution commonly expressed in centimeters, $\alpha(\lambda)$ is the decadic absorption or attenuation coefficient of the medium (i.e., of the water that may or may not contain other light-absorbing species), and $\varepsilon(\lambda)$ is the *decadic molar extinction coefficient* of the compound at wavelength λ. Hence, the units of $\alpha(\lambda)$ and $\varepsilon(\lambda)$ are in this case per centimeter and liter per mole per centimeter, respectively. Here $\varepsilon(\lambda)$ is a measure of the probability that the compound absorbs light at a particular wavelength.

Using a spectrophotometer and an appropriate solvent and reference solution (i.e., the same liquid phase as the one containing the compound so that absorption effects other than by the chemical cancel), the absorbance A of a solution of the compound

$$A = \varepsilon(\lambda) \cdot C \cdot l \quad (13\text{-}5)$$

can be measured as a function of wavelength in a cuvette exhibiting a specific width (e.g., 1, 5, or 10 cm). In the spectrophotometer, it can be assumed that the path length of the light within the cell is more or less identical with the cell width (provided that there is no light scattering occurring within the cell, for example, due to the presence of particles). In this case, $\alpha(\lambda)$ in Eq. 13-4 is commonly referred to as *beam attenuation coefficient*.

Figure 13.1*a* gives an example of an electronic absorption spectrum, that is, a *uv/vis spectrum* of an organic compound. From this spectrum, ε may be calculated for each wavelength and plotted as a function of wavelength as shown in Figure 13.1*b*. Other examples are given in Figures 13.2–13.5 and discussed in the next section. Note that in these figures, $\varepsilon(\lambda)$ is expressed on a logarithmic scale since it may range over several orders of magnitude.

As we can see from the spectra shown in Figures 13.2–13.5, organic compounds may absorb light over a wide wavelength range exhibiting one or several absorption maxima. Each absorption maximum can be assigned to a specific electron transition, for example, a $\pi \rightarrow \pi^*$ or $n \rightarrow \pi^*$ transition. Note that particularly in the excited state, members of a population of molecules are distributed among various vibrational and rotational states, and that, therefore, broad absorption bands (resulting from

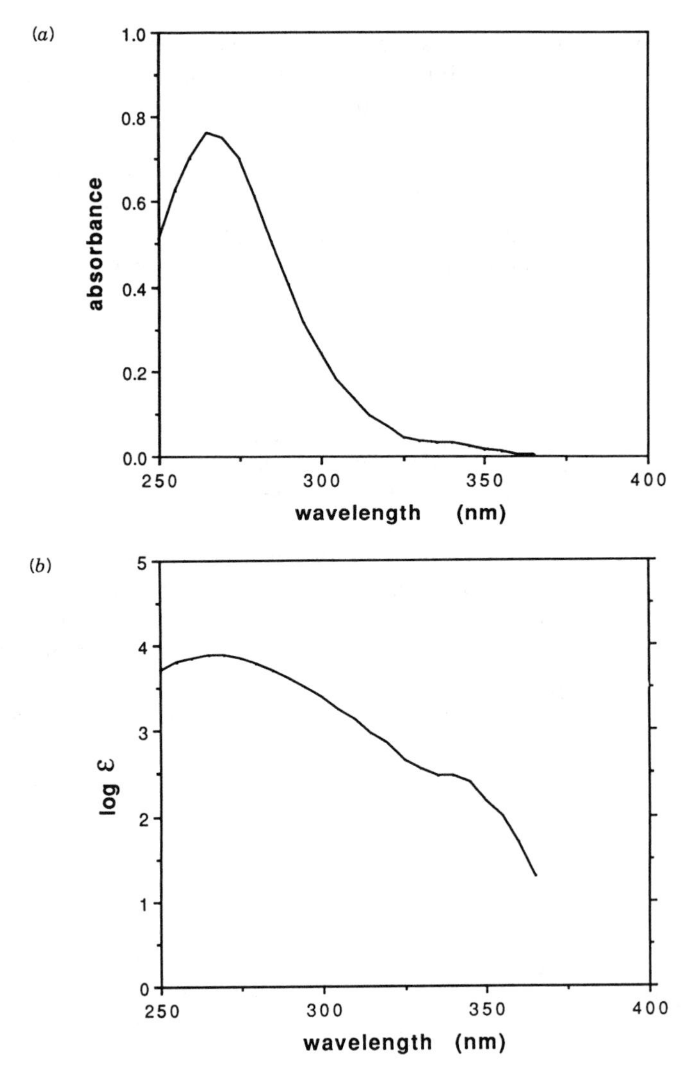

Figure 13.1 Electronic absorption spectrum of nitrobenzene in aqueous solution. (*a*) Absorbance as a function of wavelength for a 0.1 mM solution. (*b*) Log ε as a function of wavelength.

numerous unresolved narrow bands) rather than sharp absorption lines are observed in the uv/vis spectrum.

Often in the literature, only the wavelengths (λ_{max}) and corresponding ε values (ε_{max}) of the absorption maxima are reported. These values can be used for a preliminary assessment of whether or not a compound might absorb ambient light. For quantification of photochemical processes, however, the whole spectrum must be known.

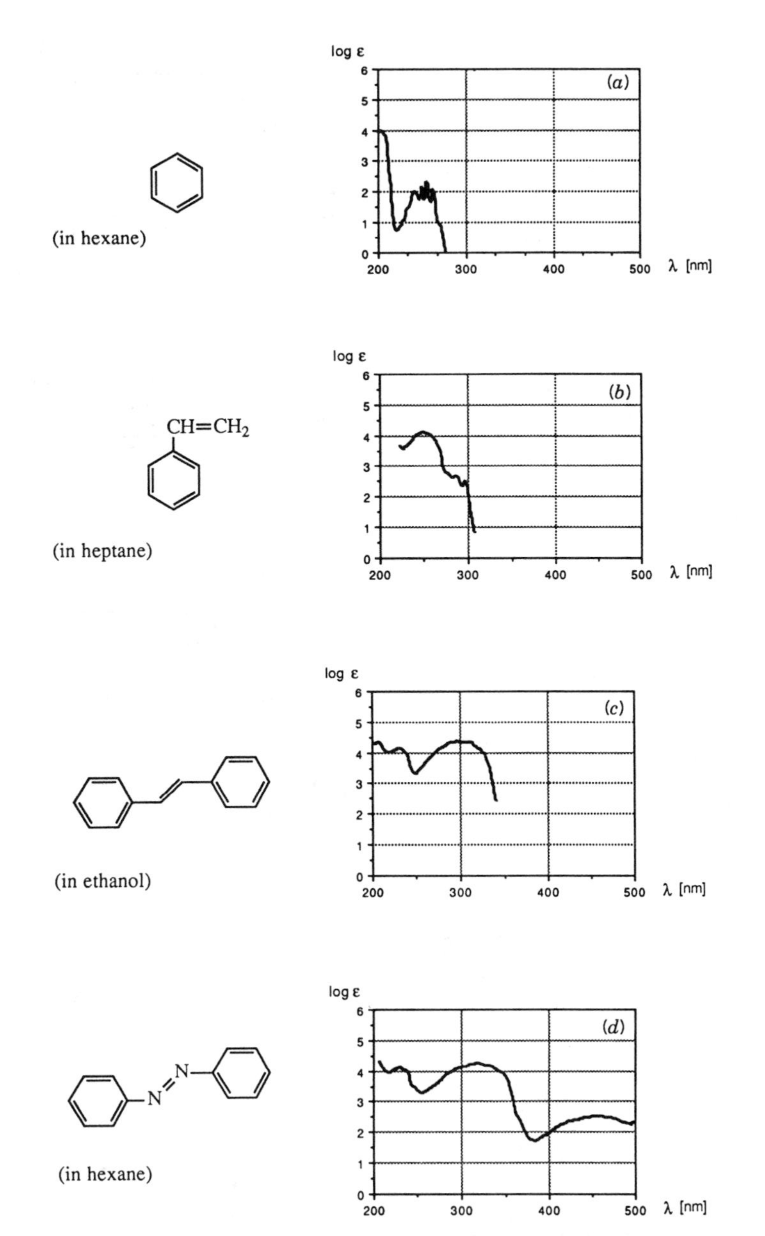

Figure 13.2 Electronic absorption spectra of (*a*) benzene, (*b*) styrene, (*c*) *trans*-stilbene, and (*d*) azobenzene (data from Pretsch et al., 1983).

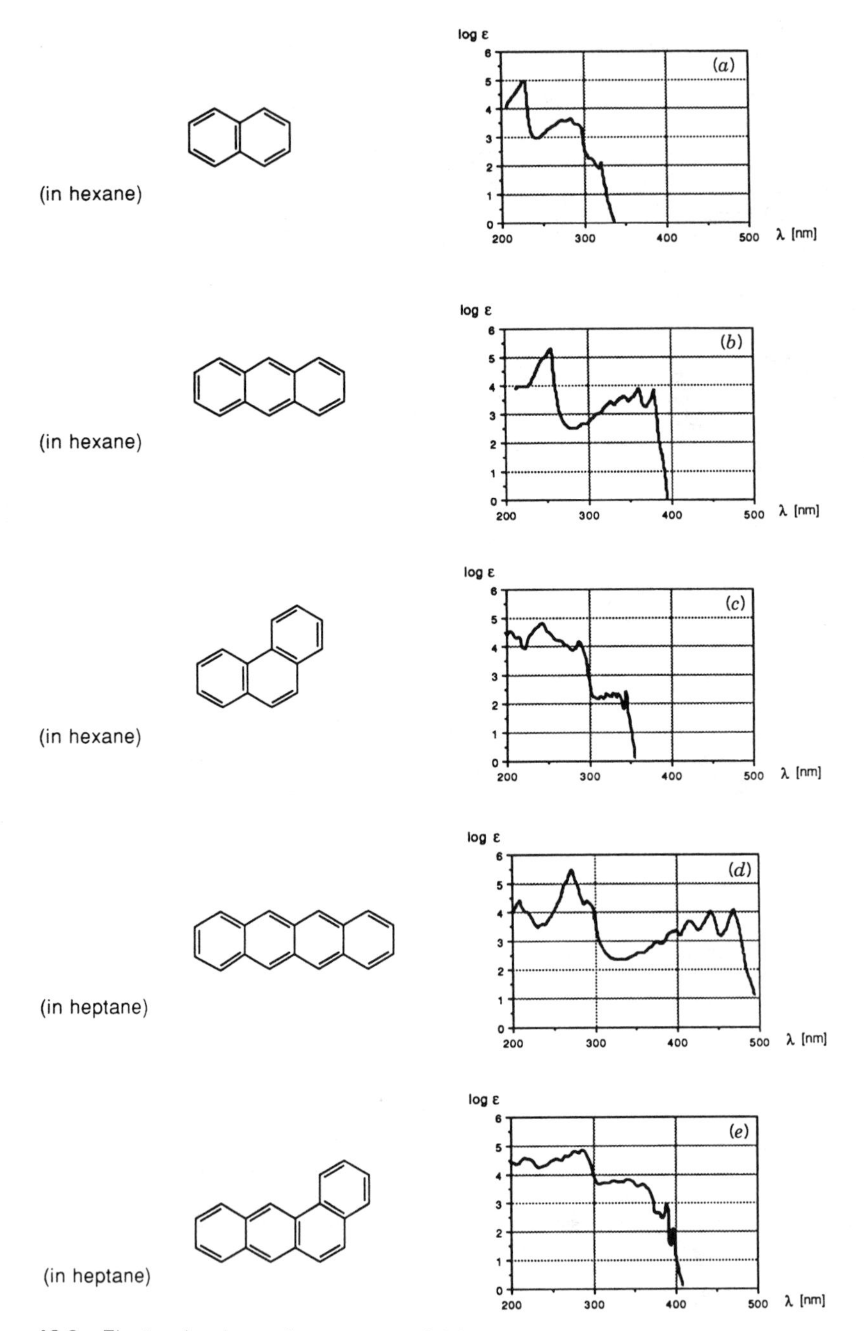

Figure 13.3 Electronic absorption spectra of (*a*) naphthalene, (*b*) anthracene, (*c*) phenanthrene, (*d*) naphthacene and (*e*) benz(a)anthracene (data from Pretsch et al., 1983).

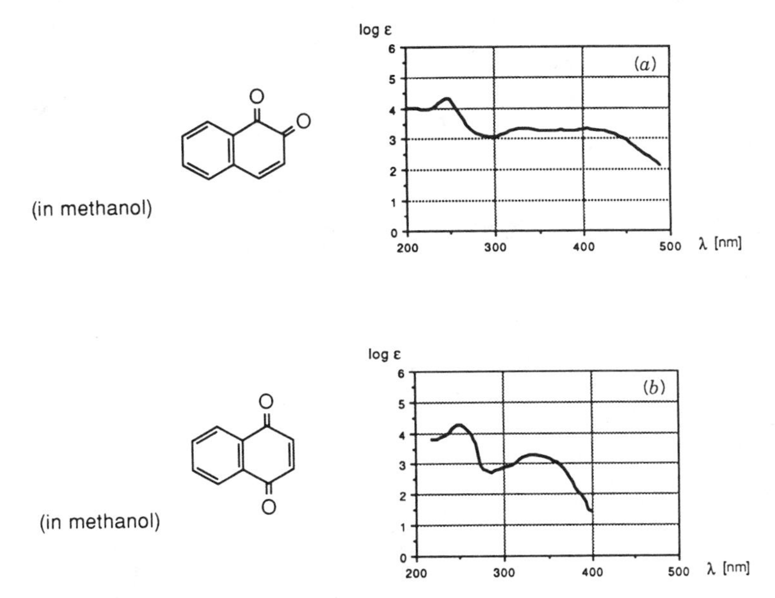

Figure 13.4 Electronic absorption spectra of (*a*) 1,2-naphthoquinone and (*b*) 1,4-naphthoquinone (data from Pretsch et al., 1983).

There is in fact a large body of uv/vis spectra of organic compounds available in the literature, although most of these data were obtained in organic solvents (e.g., Pretsch et al., 1983). It should be noted that absorption spectra are susceptible to solvent effects, especially if the solute undergoes hydrogen bonding with the solvent molecules. Nevertheless, particularly if a spectrum was recorded in a polar organic solvent (e.g., methanol, acetonitrile) or an organic solvent–water mixture, it can be used at least for a crude assessment of the rate of light absorption of a given chemical in aqueous solution when exposed to sunlight. This is, of course, only feasible if the speciation of the chemical is the same in the organic solvent and in water, which may not be the case if the compound exhibits acid or base functionalities.

Chemical Structure and Light Absorption

As indicated by the examples given in Figures 13.1–13.5, light absorption of organic compounds in the wavelength range of interest to us (i.e., 290–600 nm) is in most cases associated with the presence of a delocalized π-electron system. Hence, aromatic rings and conjugated double bonds in particular may form a so-called *chromophore*, a structural moiety that exhibits a characteristic uv/vis absorption spectrum. In such systems, the most probable electron transitions are promotions of π electrons from bonding to antibonding π orbitals. Such transitions are commonly referred to as $\pi \rightarrow \pi^*$ transitions, and they give, in general, rise to the most intense absorption bands in the spectrum. If a π-electron system contains atoms with nonbonding electrons (i.e., hetero-

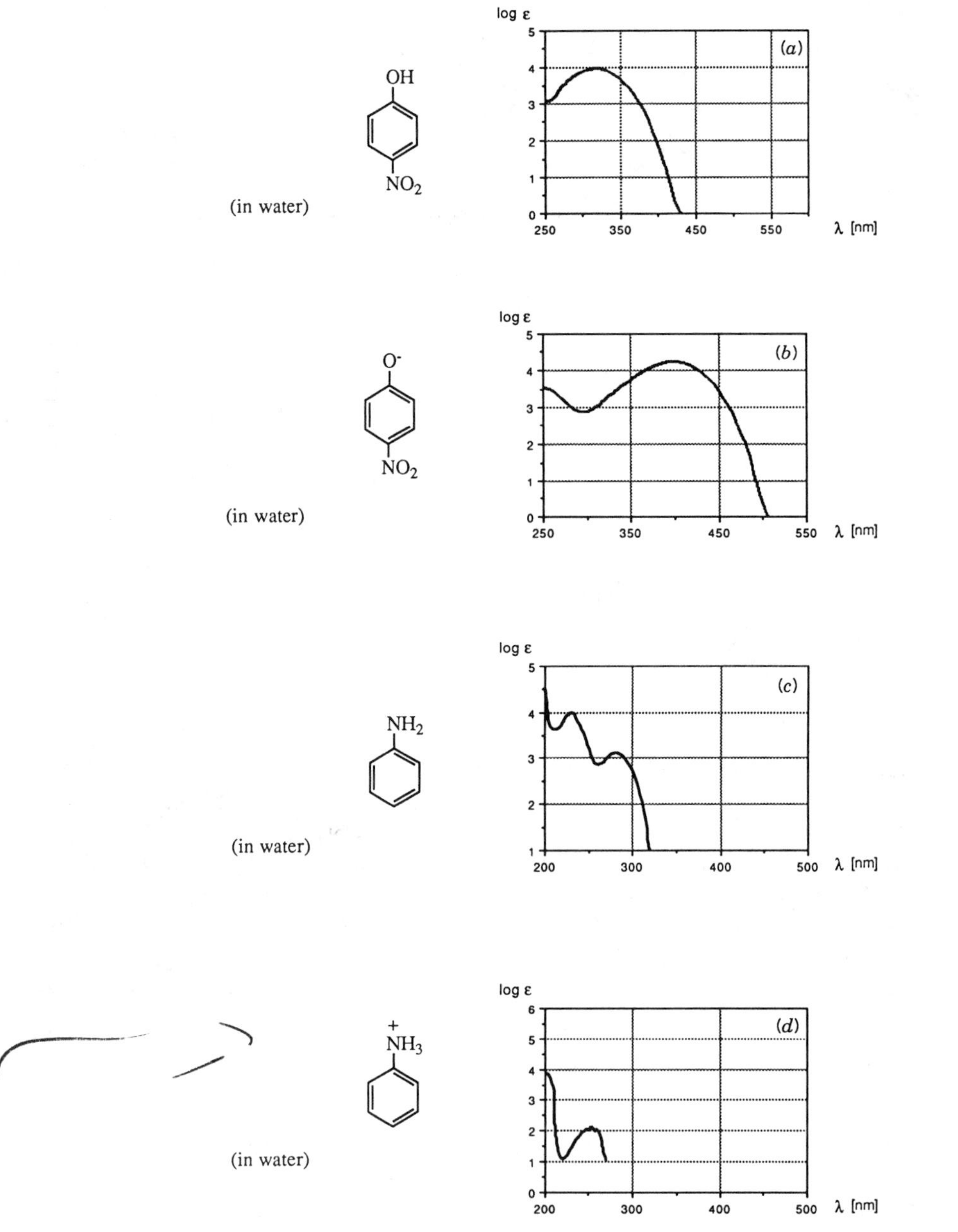

Figure 13.5 Electronic absorption spectra of (*a*) 4-nitrophenol, (*b*) 4-nitrophenolate, (*c*) aniline, and (*d*) anilinium ion (data from Pretsch et al., 1983).

atoms), so-called $n \rightarrow \pi^*$ transitions may also be observed. These transitions commonly occur at longer wavelengths (i.e., lower energy) as compared to the $\pi \rightarrow \pi^*$ transitions, and they usually exhibit a significantly smaller molar extinction coefficient. For example, in the spectrum of nitrobenzene (Fig. 13.1), the absorption band with a maximum at 267 nm ($\varepsilon_{max} \cong 7500\ M^{-1}\ cm^{-1}$) may be assigned to a $\pi \rightarrow \pi^*$ transition, while the much less intense band at 340 nm ($\varepsilon_{max} \cong 150\ M^{-1}\ cm^{-1}$) is due to an $n \rightarrow \pi^*$ transition.

Let us now consider some aspects concerning the relationship between chemical structure and light absorption of organic compounds. We will confine ourselves to a few general remarks. For a more detailed discussion of this topic, we refer to the literature (Williams and Fleming, 1980).

We start out by looking at light absorption by chromophores that consist of a series of conjugated double bonds. Such chromophores are not very frequently encountered in xenobiotic organic compounds, but they play an important role in natural materials, for example, in pigments present in photosynthetic cells (e.g., carotenoid pigments, porphyrins). In straight-chain polyenes, each additional conjugated double bond shifts the absorption maximum of the lowest energy (i.e., highest wavelength) $\pi \rightarrow \pi^*$ transition by about 30 nm to higher wavelengths (a so-called bathochromic shift). This is a general phenomenon, and it may be stated that, in general, the more conjugation in a molecule, the more the absorption is displaced toward higher wavelengths. This may also be seen when comparing the absorption spectra of benzene (Fig. 13.2*a*), styrene (Fig. 13.2*b*), and stilbene (Fig. 13.2*c*), or the spectra of a series of polycyclic aromatic hydrocarbons exhibiting different numbers of rings (Fig. 13.3). Note that in the case of polycyclic aromatic compounds, not only the number of rings but also the way in which the rings are "fused" together determines the absorption spectrum. As Figure 13.3 shows, large differences exist, for example, between the spectra of anthracene (Fig. 13.3*b*) and phenanthrene (Fig. 13.3*c*), or between 2,3-benzanthracene (Fig. 13.3*d*) and 1,2-benzanthracene (napthacene) (Fig. 13.3*e*).

Comparison of the absorption spectra of stilbene (Fig. 13.2*c*) and (di)azobenzene (Fig. 13.2*d*), shows another interesting feature. The replacement of the two double-bonded carbon atoms by two nitrogen atoms leads to additional, intensive $n \rightarrow \pi^*$ transitions, which, in this case, lay in the visible wavelength range (i.e., above 400 nm, see Table 13.2). Substituted azobenzenes are widely used dyes, and it has recently been recognized that large amounts of such compounds enter the environment (Weber and Wolfe, 1987).

Another important group of chromophores that absorb light over a wide wavelength range that includes visible light are the quinoid type chromophores (Fig. 13.4). Quinoid-type chromophores are important constituents of naturally occurring organic material (e.g., humic and fulvic acids), and they are partly responsible for the yellow color of natural waters containing high concentrations of dissolved organic matter. In Chapter 12 we discussed the role of quinoid-type compounds in abiotic reduction processes of organic pollutants. From a photochemical point of view, quinoid compounds are interesting because they may act as sensitizers for indirect photolytic processes.

We conclude our short discussion of relationships between chemical structure and

TABLE 13.2 Correlation between Wavelength of Absorbed Radiation by a Given Object (e.g., an Aqueous Solution) and Observed Color of the Object When Exposed to White Light[a]

Absorbed Light		
Wavelength (nm)	Corresponding Color	Observed Color
400	Violet	Yellow-green
425	Indigo Blue	Yellow
450	Blue	Orange
490	Blue-green	Red
510	Green	Purple
530	Yellow-green	Violet
550	Yellow	Indigo blue
590	Orange	Blue
640	Red	Blue-green
730	Purple	Green

[a]From Pretsch et al., 1983.

light absorbance by considering some cases in which an acid or base function forms part of a chromophore. Important examples of compounds exhibiting such chromophores are phenols and anilines. As is evident from the spectra shown in Figure 13.5, deprotonation of a phenolic group results in a substantial bathochromic shift (shift to longer wavelengths), which is due mostly to delocalization of the negative charge (see Chapter 8). Consequently, depending on the pK_a of a given phenol, light absorption by the phenolic species may vary significantly over the ambient pH range. In the case of aromatic amines (Fig. 13.5*c*,*d*), protonation of the amino group results in a so-called hypsochromic shift (shift to shorter wavelengths), because, as a consequence of protonation, the chromophore is altered in that the nitrogen atoms no longer possess nonbonded electrons that may delocalize into the aromatic system. Since protonation of aromatic amines occurs only at relatively low pH values (pH $< \sim 5$, see Chapter 8), this effect is only important in acidic waters (e.g., in an acidic rain droplet).

In conclusion, from an environmental photochemistry point of view, the most important chromophores present in organic compounds consist of conjugated π-electron systems that may or may not contain heteroatoms. In addition, there are certain cases in which xenobiotic organic compounds that themselves do not absorb light above 300 nm undergo *charge transfer transitions* when complexed to a transition metal. A prominent example is the iron(III)–EDTA complex that absorbs light above 300 nm. As a consequence, EDTA, a widely used complexing agent that absorbs no light above 250 nm and that is very resistant to microbial and chemical degradation, may undergo direct photolytic transformation in surface waters, provided that enough iron(III) is available (Frank and Rau, 1990). Finally, compounds that have two or more noninteracting chromophores exhibit an absorption spectrum corresponding to the superposition of the spectra of the individual chromophores.

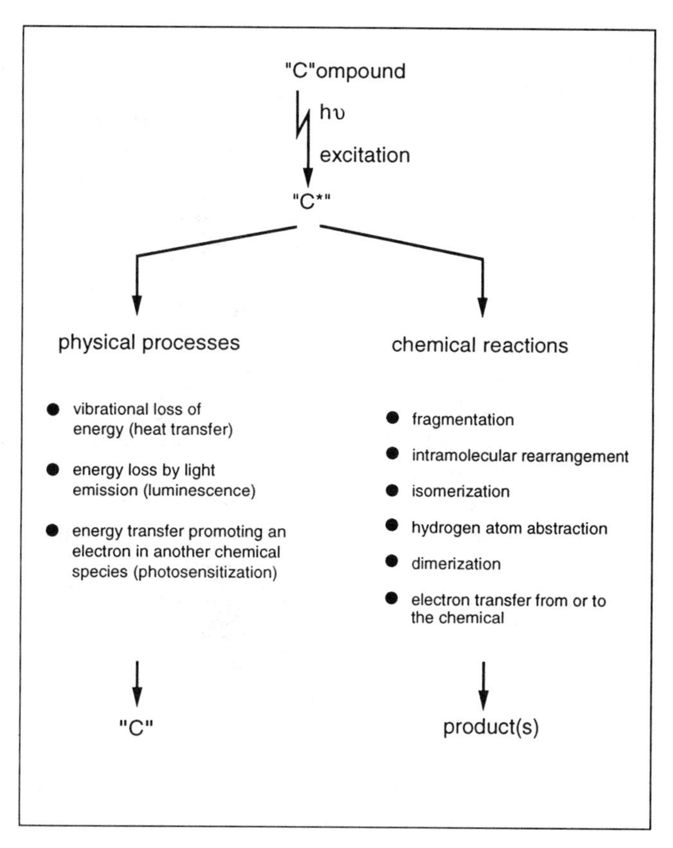

Figure 13.6 Physical processes and chemical reactions of a photochemically excited organic species.

The Fate of Excited Chemical Species—Quantum Yields

When a chemical species has been promoted to an excited state, it does not remain there for long. There are various physical or chemical processes that the excited species may undergo. Figure 13.6 summarizes the most important reaction pathways. As is indicated, there are several *physical processes* by which an excited species may return to the ground state; that is, it is not structurally altered by these processes. For example, a species in the first excited state (the state it is commonly promoted to as a consequence of light absorption) may convert to a high vibrational level of the ground state, and then "cascade" down through the vibrational levels of the ground state by giving off its energy in small increments of heat to the environment. This process is referred to as *internal conversion.* Alternatively, an excited molecule may directly, or after undergoing some change to another excited state (by so-called *intersystem crossing*), drop to some low vibrational level of the ground state all at once by giving off the energy in the form of light. These luminescent processes are called fluorescence and phosphorescence, respectively. Finally, an excited species may transfer its excess energy to another molecule in the environment in a process called *photosensitization.* The excited species thus drops to its ground state while the other molecule becomes excited. Compounds that, after light absorption, efficiently transfer their energy to other chemical species are referred to as *photosensitizers.* The chemical species that efficiently accept the electronic energy are called *acceptors or quenchers.* We will come back to photosensitized processes later when discussing indirect photolysis of organic pollutants (Section 13.4). A more detailed treatment of the various physical processes of excited species is given by Roof (1982) and by March (1985). An extensive discussion may be found in an appropriate text book (e.g., Calvert and Pitts, 1966; Turro, 1978).

In addition to the physical processes mentioned above, there are a variety of *chemical reactions* that an excited species may undergo (Fig. 13.6). These reactions are of interest when considering direct photolysis of organic pollutants, because only chemical reactions lead to a transformation and thus to a "removal" of the compound from a given system. Note that the chemical processes indicated in Figure 13.6 represent primary steps in the photolytic transformation of a given compound, and that the products of these primary steps may further react by either photochemical, chemical, or biological processes. Consequently, it can be very difficult to identify and quantify all photochemical transformation products, particularly in natural waters or in soils where a variety of possible reactants are present. Some examples of photochemical transformations are given in Figure 13.7. For more examples and detailed discussions we refer to the literature (e.g., March, 1985; Harris and Hayes, 1982; Mill and Mabey, 1985). It should be noted that the pathway(s) and the rate(s) of photochemical transformations of excited species in solution commonly depend strongly on the solvent, and in many cases also on the solution composition (e.g., pH, oxygen concentration, ionic strength; Mill and Mabey, 1985). Thus, it is advisable to use data from experiments conducted in solutions with water as the major ($>90\%$) or even sole solvent and with a solution composition representative of the natural system considered in assessing the photochemical transformation of a given compound in the environment. In this context it is necessary to point out that certain organic cosolvents (e.g., acetone) are good sensitizers and may, therefore, strongly influence the photolytic half-life of the compound.

Finally, compared to the chemical reactions discussed in the previous chapter, photochemical transformations of organic compounds usually exhibit a much weaker temperature dependence. Reactions of excited species in aqueous solutions have activation energies of between 10 and 30 kJ mol^{-1} (Mill and Mabey, 1985). Hence, a 10° increase (decrease) in temperature accelerates (slows down) a reaction only by a factor of between 1.15 and 1.5 (see Chapter 12).

As we have seen, an excited organic molecule may undergo several physical and chemical processes. The relative importance of the various processes depends, of course, on the structure of the compound and on its environment (e.g., type of solvent, presence of solutes). For each individual process i, we may, for a given environment, define a *quantum yield* $\Phi_i(\lambda)$ which denotes the fraction of the excited molecules of a

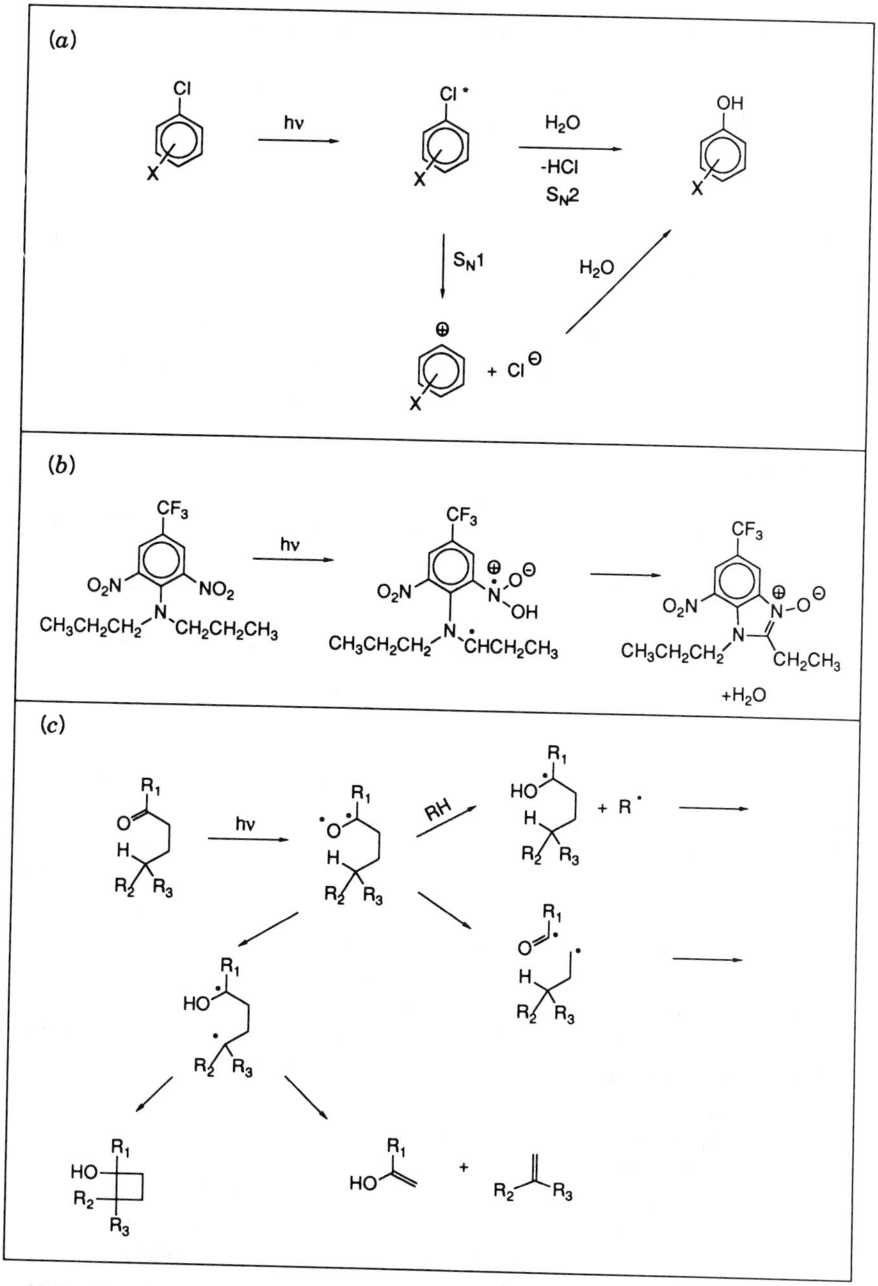

Figure 13.7 Examples of direct photochemical reaction pathways: (*a*) substituted chlorobenzenes, (*b*) trifluralin, and (*c*) a ketone (from Mill and Mabey, 1985).

given compound that react by that particular physical or chemical pathway:

$$\Phi_i(\lambda) = \frac{\text{number of molecules reacting by pathway } i}{\text{total number of molecules excited by absorption of radiation of wavelength } \lambda} \tag{13-6}$$

Since the absorption of light by an organic molecule is, in general, a one-quantum process, we may also write Eq. 13-6 as

$$\Phi_i(\lambda) = \frac{\text{number of molecules reacting by pathway } i}{\text{total number of photons (of wavelength } \lambda\text{) absorbed by the system owing to the presence of the compound}} \tag{13-7}$$

From an environmental chemist's point of view, it is often not necessary to determine all the individual quantum yields for each reaction pathway (which is, in general, a very difficult and time-consuming task), but to derive a "lumped" quantum yield which encompasses all reactions that alter the structure of the component. This lumped parameter is commonly referred to as *reaction quantum yield* and is denoted as $\Phi_r(\lambda)$:

$$\Phi_r(\lambda) = \frac{\text{total number (i.e., moles) of molecules transformed}}{\text{total number (i.e., moles) of photons (of wavelength } \lambda\text{) absorbed by the system due to the presence of the compound}} \tag{13-8}$$

Unfortunately, there are no simple rules to predict reaction quantum yields from chemical structure, and, therefore, $\Phi_r(\lambda)$ values have to be determined experimentally. We will address such experimental approaches in the next section, and confine ourselves here to a few general remarks. First, we should note that, in principle, reaction quantum yields may exceed unity in cases in which the absorption of a photon by a given compound causes a chain reaction to occur that consumes additional compound molecules. Such cases are, however, very unlikely to happen with organic pollutants in natural waters, mainly because of the rather low pollutant concentrations, and because of the presence of other water constituents that may inhibit chain reactions (Roof, 1982). Consequently, in the discussions that follow we always assume maximum reaction quantum yields of one.

A second aspect that needs to be addressed is the wavelength dependence of Φ_r. Although vapor phase reaction quantum yields differ considerably between different wavelengths, they are in many cases approximately wavelength-independent (at least over the wavelength range of a given absorption band, corresponding to one mode of excitation) for reactions of organic pollutants in aqueous solutions (Zepp, 1982). Hence, reaction quantum yields determined at a given wavelength (preferably at a wavelength at or near the maximum specific light absorption rate of the compound, see below) may be used for a crude assessment of the transformation rate of a given compound. Note, however, that if a compound absorbs light over a broad wavelength range exhibiting several maxima of light absorption (e.g., azo dyes), quantum yields may have to be determined for various wavelengths (e.g., Haag and Mill, 1987).

13.3 DIRECT PHOTOLYSIS OF ORGANIC COMPOUNDS IN NATURAL WATER

Light and Light Attenuation in Natural Water Bodies

When dealing with the exposure of a natural water body to sunlight, unlike the situation encountered in a spectrophotometer, one cannot consider the radiation to enter the water perpendicular to the surface as a collimated beam. Sunlight at the surface of the earth consists of direct and scattered light (the latter is commonly referred to as sky radiation) entering a water body at various angles. The solar spectrum at a given point at the surface of the earth depends on many factors including the geographic location (latitude, altitude), season, time of day, weather conditions, air pollution above the region considered, and so on (Finlayson-Pitts and Pitts, 1986). In our discussion we address some of the approaches taken for either calculating or measuring light intensities at the surface or in the water column of natural waters. For a more detailed treatment of this topic we refer, however, to the literature (e.g., Smith and Tyler, 1976, Zepp and Cline, 1977; Zepp, 1980; Baker and Smith, 1982; Leifer, 1988).

Let us consider a well-mixed water body of volume V (cubic centimeter) and (horizontal) surface area A (square centimeter) that is exposed to sunlight. Recall that we speak of a well-mixed water body when mixing is fast compared to all other processes. This means that the system is homogeneous with respect to all water constituents and properties, including optical properties such as the light attenuation coefficient of the medium. This is true in many cases such as in shallow water bodies or when we are interested in only the surface layer of a given natural water. If we denote the incident light intensity (i.e. the total light intensity at the water surface) as $W(\lambda)$ (e.g., in einstein per square centimeter per second), we can express the light intensity at $z_{mix} = V/A$ (the mean depth of the mixed water body in centimeter) by applying the Lambert–Beer law (see Section 13.2):

$$W(z_{mix}, \lambda) = W(\lambda) \cdot 10^{-\alpha_D(\lambda) z_{mix}} \tag{13-9}$$

where $\alpha_D(\lambda)$ (in per centimeter) is commonly referred to as the apparent or *diffuse attenuation coefficient*. The diffuse attenuation coefficient [which often is also denoted as $K_T(\lambda)$] can be determined *in situ* by measuring the light intensities at the surface and at the depth z_{mix} (e.g., Baker and Smith, 1982; Winterle et al., 1987).

$$\alpha_D(\lambda) = \frac{1}{z_{mix}} \log \frac{W(\lambda)}{W(z_{mix}, \lambda)} \tag{13-10}$$

Hence, $\alpha_D(\lambda)$ is a measure of how much radiation is absorbed by the mixed water layer *over a vertical distance* z_{mix}. As schematically depicted in Figure 13.8, light enters the water at various angles and is then refracted at the air–water interface. Less than 10% of the incident light is usually lost due to backscatter and reflection (Zafiriou, 1985). Within the water column, it is (1) scattered (e.g., by nonabsorbing particles), (2) absorbed by (organic) particles, and, (3) absorbed by dissolved species, especially dissolved organic matter. From Figure 13.8 it can easily be deduced that the average

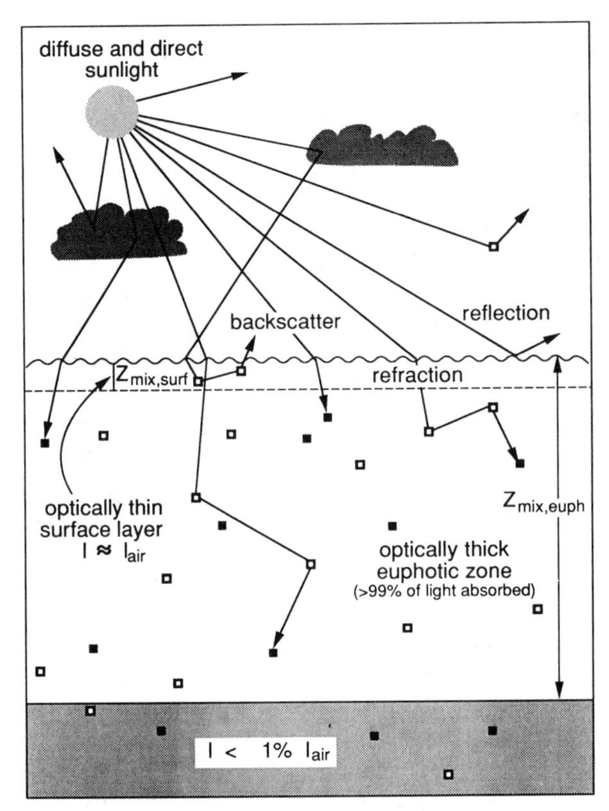

Figure 13.8 Fate of photons in a natural water body (adapted from Zafiriou, 1983). Open squares indicate reflective particles; filled squares are absorptive particles or molecules.

pathlength of light at any wavelength λ will be larger than z_{mix}. For a given well-mixed water body, we can define a distribution function, $D(\lambda)$, as the ratio of the average light pathlength $l(\lambda)$ and z_{mix}:

$$D(\lambda) = \frac{l(\lambda)}{z_{\text{mix}}} \tag{13-11}$$

Recall that the beam attenuation coefficient $\alpha(\lambda)$ (Eq. 13.4) of a given solution is a measure of the attenuation of a collimated beam entering the solution perpendicular to the surface (e.g., to the surface of a cuvette in a spectrophotometer). In such a case, the pathlength of the light, $l(\lambda)$, is approximately equal to the width of the cuvette (which we could also denote as z_{mix}), provided that no significant scattering occurs and thus $D(\lambda)$ is equal to 1. Scattering is mainly due to particles present in the water. For our discussion and calculations, we initially assume a situation in which particles play a minor role, such as encountered in nonturbid waters. Later we address the impact of particles on light attenuation in a natural water body, as well as the effect of particles on photolytic transformation rates of organic pollutants.

For a nonturbid water, we may determine $\alpha(\lambda)$ in a spectrophotometer, and use this entity to estimate the diffuse attenuation coefficient $\alpha_D(\lambda)$ by multiplication with $D(\lambda)$:

$$\alpha_D(\lambda) = D(\lambda)\alpha(\lambda) \tag{13-12}$$

Hence, $\alpha(\lambda)$ represents the attenuation coefficient of the water per unit pathlength. Application of Eq. 13-12 requires, however, that we can obtain a good estimate for $D(\lambda)$, which is not easily done, especially when dealing with deeper water columns and for very turbid waters. For shallow depths (e.g., the top 50 cm of a nonturbid natural water body), $D(\lambda)$ is primarily determined by the ratio of direct and sky radiation, and by the angle of refraction of direct radiation. For these cases, $D(\lambda)$ may be calculated by computer programs (see Zepp, 1980 and references cited therein). Estimated values of $D(\lambda)$ for near-surface uv and blue light (450 nm) range between 1.05 and 1.30, depending on the solar zenith angle (Zepp, 1980). In very turbid waters where light scattering by particles is significant, $D(\lambda)$ values of up to 2.0 have been determined (Miller and Zepp, 1979a).

Let us now return to our well-mixed water body and ask how much light *of a given wavelength* λ is absorbed by the water column per unit surface area and time. We may calculate this rate of light absorption by simply calculating the difference between the incident light intensity and the light intensity at depth z_{mix} (Eq. 13-9):

$$\begin{aligned}\text{Rate of light absorption by}\\ \text{the water body } \textit{per unit surface area} &= W(\lambda) - W(z_{mix}, \lambda) \\ &= W(\lambda)\left[1 - 10^{-\alpha_D(\lambda) z_{mix}}\right]\end{aligned} \tag{13-13}$$

The average rate of light absorption *per unit volume* is then obtained by multiplying Eq. 13-13 by the total irradiated surface area (yielding the total number of photons absorbed per unit time by the whole water body), and dividing by the total volume:

$$\begin{aligned}\text{Rate of light absorption by}\\ \text{the water body } \textit{per unit volume} &= W(\lambda)\left[1 - 10^{-\alpha_D(\lambda) z_{mix}}\right]\frac{A}{V} \\ &= \frac{W(\lambda)\left[1 - 10^{-\alpha_D(\lambda) z_{mix}}\right]}{z_{mix}}\end{aligned} \tag{13-14}$$

Specific Rate of Light Absorption by an Organic Pollutant

If we now add a pollutant exhibiting a molar extinction coefficient $\varepsilon(\lambda)$ to a given water body, the attenuation coefficient $\alpha(\lambda)$ [not $\alpha_D(\lambda)$!] is altered to $\alpha(\lambda) + \varepsilon(\lambda)C$, where C is the concentration of the pollutant in moles per liter. In most cases, however, the pollutant concentration in a natural water will be low, and light absorption by the pollutant will be small as compared to the light absorption by all other chromophores present. Consequently, the rate of sunlight absorption by the water body (Eq. 13-14) will essentially be unchanged. The (small) fraction of light, $F_c(\lambda)$ absorbed by the pollutant is given by

$$F_c(\lambda) = \frac{\varepsilon(\lambda)C}{\alpha(\lambda) + \varepsilon(\lambda)C} \tag{13-15}$$

or, since $\varepsilon(\lambda)C \ll \alpha(\lambda)$,

$$F_c(\lambda) \cong \frac{\varepsilon(\lambda)}{\alpha(\lambda)} C \tag{13-16}$$

Multiplication of the rate of light absorption by the water body per unit volume (Eq. 13-14) with $F_c(\lambda)$ (Eq. 13-16) then yields the entity that we are most interested in, that is, the rate of light absorption by the compound per unit volume denoted as $I_a(\lambda)$:

$$\begin{aligned} I_a(\lambda) &= \frac{W(\lambda) \cdot \varepsilon(\lambda) \cdot [1 - 10^{-\alpha_D(\lambda) z_{mix}}]}{z_{mix} \cdot \alpha(\lambda)} C \\ &= k_a(\lambda) \cdot C \end{aligned} \tag{13-17}$$

where $k_a(\lambda)$ is commonly referred to as the *specific rate of light absorption* (not a first-order rate constant!) of a given compound in a given system. Hence, $k_a(\lambda)$ expresses the amount (moles) of photons of wavelength λ that are absorbed per unit time per mole of compound present in the system considered (which in our case is a well-mixed water body of mean depth z_{mix}). Using the units introduced above, that is, $W(\lambda)$ in einsteins per square centimeter per second, $\varepsilon(\lambda)$ in liter per mole per centimeter, z_{mix} in centimeter, and $\alpha(\lambda)$ and $\alpha_D(\lambda)$ in per centimeter, $k_a(\lambda)$ has the units

$$\begin{aligned} [k_a(\lambda)] &= \frac{\text{einstein cm}^{-2}\,\text{s}^{-1}\,\text{cm}^{-1}\,(\text{mol compound})^{-1}\,\text{L}}{\text{cm} \cdot \text{cm}^{-1}} \\ &= \text{einstein cm}^{-3}\,\text{L}\,(\text{mol compound})^{-1}\,\text{s}^{-1} \\ &= 10^3\,\text{einstein}\,(\text{mol compound})^{-1}\,\text{s}^{-1} \end{aligned} \tag{13-18}$$

Thus, we have to express $W(\lambda)$ in *millieinstein* per square centimeter per second to obtain the (desired) units of einstein (per mol compound per second for $k_a(\lambda)$. The unit of $I_a(\lambda)$ is then einstein per liter per second.

Near-Surface Specific Rate of Light Absorption of an Organic Pollutant

There are two extreme cases for which Eq. 13-17 can be simplified. The first case applies to the situation where very little light (e.g., less than 5%) is absorbed by the system, that is, the situation in which $\alpha_D(\lambda) z_{mix} < 0.02$. This is true for a very shallow mixed water body (i.e., the top few centimeters of a natural water body, laboratory tubes), or a water body exhibiting a very low $\alpha_D(\lambda)$ value (e.g., distilled water, open ocean water). If $\alpha_D(\lambda)z_{mix} < 0.02$, we can make the following approximation (with $\ln 10 = 2.3$):

$$1 - 10^{-\alpha_D(\lambda)z_{mix}} \cong 2.3\alpha_D(\lambda)z_{mix} \tag{13-19}$$

and

$$k_a^0(\lambda) = \frac{2.3\,W(\lambda)\alpha_D^0(\lambda)\cdot\varepsilon(\lambda)}{\alpha(\lambda)} \tag{13-20}$$

where $k_a^0(\lambda)$ is the *near-surface* (superscript "0") specific rate of light absorption at wavelength λ.

Since $\alpha_D^0(\lambda) = D^0(\lambda)\cdot\alpha(\lambda)$ (Eq. 13-12), we obtain

$$k_a^0(\lambda) = 2.3\cdot W(\lambda)D^0(\lambda)\varepsilon(\lambda) \tag{13-21}$$

or $k_a^0(\lambda) = 2.3\cdot Z(\lambda)\varepsilon(\lambda)$ where $Z(\lambda) = W(\lambda)\,D^0(\lambda)$.

As pointed out earlier, for shallow depths $D^0(\lambda)$ can be approximated by computer calculations. Also, with the same computer programs (SOLAR, or GCSOLAR, which is an updated version of SOLAR; see Zepp and Cline, 1977; Leifer, 1988), $W(\lambda)$ values may be estimated for a given geographic location, season, and time of the day. The programs also allow one to take into account the effects of overcast skies. Tables 13.3 and 13.4 give calculated $W(\lambda)$ as well as $Z(\lambda)$ values for midday (noon) at sea level at latitude 40°N for a midseason clear summer and winter day, respectively. In addition, the tables contain 24-hour averaged $Z(\lambda)$ values that take into account diurnal fluctuations in sunlight intensity. This data set is important for comparison of photolysis rates with the rates of other processes that determine the fate of a given compound in a water body. Note again that the $Z(\lambda)$ values given in Tables 13.3 and 13.4 are only applicable to shallow waters (i.e., $z_{mix} \leqslant \sim 50$ cm).

To allow simple "back of the envelope" calculations, all $W(\lambda)$ and $Z(\lambda)$ values given in Tables 13.3 and 13.4 are *integrated values* over a specified wavelength range. The indicated wavelength number represents the center of a given wavelength range. For example, W(noon, 310 nm) is the total number of photons (expressed in millieinsteins) per square centimeter of surface and per second at midday integrated over the wavelength range between 308.75 and 311.25 nm. The midday average light intensity within this wavelength range is then given by W(noon, 310 nm)/$\Delta\lambda$ with $\Delta\lambda = 2.5$ nm.

Example Illustrating the Calculation of the Near-Surface Specific Rate of Light Absorption of an Organic Pollutant

Let us use a practical example to get acquainted with some aspects of solar irradiance in natural waters, and to illustrate how to use light data as presented in Tables 13.3 and 13.4 for estimating specific light absorption rates of organic pollutants. We first want to calculate the near-surface specific light absorption rate k_a^0 of *para*-nitro-acetophenone (PNAP) at 40°N latitude (sea level) at noon on a clear midsummer day. Note that for this geographic location, the result of our calculation will represent the maximum k_a^0

TABLE 13.3 $W(\lambda)$ and $Z(\lambda)$ Values for a *Midsummer* Day at 40°N Latitude (Sea Level) under Clear Skies[a]

λ (Center) (nm)	λ Range ($\Delta\lambda$) (nm)	$W(\text{noon}, \lambda)$[b,d] (millieinstein $cm^{-2}\,s^{-1}$)	$Z(\text{noon}, \lambda)$[b,d] (millieinstein $cm^{-2}\,s^{-1}$)	$Z(24\,h, \lambda)$[c,d] (millienstein $cm^{-2}\,d^{-1}$)
297.5	2.5	1.08(−9)	1.19(−9)	2.68(−5)
300.0	2.5	3.64(−9)	3.99(−9)	1.17(−4)
302.5	2.5	1.10(−8)	1.21(−8)	3.60(−4)
305.0	2.5	2.71(−8)	3.01(−8)	8.47(−4)
307.5	2.5	4.55(−8)	5.06(−8)	1.62(−3)
310.0	2.5	7.38(−8)	8.23(−8)	2.68(−3)
312.5	2.5	1.07(−7)	1.19(−7)	3.94(−3)
315.0	2.5	1.43(−7)	1.60(−7)	5.30(−3)
317.5	2.5	1.71(−7)	1.91(−7)	6.73(−3)
320.0	2.5	2.01(−7)	2.24(−7)	8.12(−3)
323.1	3.75	3.75(−7)	4.18(−7)	1.45(−2)
330.0	10	1.27(−6)	1.41(−6)	5.03(−2)
340.0	10	1.45(−6)	1.60(−6)	6.34(−2)
350.0	10	1.56(−6)	1.71(−6)	7.03(−2)
360.0	10	1.66(−6)	1.83(−6)	7.77(−2)
370.0	10	1.86(−6)	2.03(−6)	8.29(−2)
380.0	10	2.06(−6)	2.24(−6)	8.86(−2)
390.0	10	2.46(−6)	2.68(−6)	8.38(−2)
400.0	10	3.52(−6)	3.84(−6)	1.20(−1)
420.0	30	1.40(−5)	1.51(−5)	4.77(−1)
450.0	30	1.77(−5)	1.90(−5)	6.04(−1)
480.0	30	1.91(−5)	2.04(−5)	6.52(−1)
510.0	30	1.99(−5)	2.12(−5)	6.82(−1)
540.0	30	2.10(−5)	2.22(−5)	7.09(−1)
570.0	30	2.13(−5)	2.25(−5)	7.14(−1)
600.0	30	2.13(−5)	2.24(−5)	7.19(−1)
640.0	50	3.54(−5)	3.72(−5)	1.22

[a] $Z(\lambda) = W(\lambda) \cdot D^0(\lambda)$. Midsummer refers to a solar declination of +20° (late July).
[b] Values derived from data of Zepp and Cline (1977).
[c] Values derived from Leifer, 1988; note: $Z(24\,h, \lambda) = L(\lambda)/2.303$.
[d] Numbers in parentheses are powers of 10.

value for PNAP to be expected in a natural water exposed to sunlight. Figure 13.9 is a graphical representation of Eq. 13-21 derived from the $Z(\lambda)$ [$= W(\lambda) D^0(\lambda)$] and $\varepsilon(\lambda)$ values for PNAP given in Table 13.5. As mentioned above, all $W(\lambda)$ and $Z(\lambda)$ values are integrated values over a given wavelength range $\Delta\lambda$, and $\varepsilon(\lambda)$ is the average molar extinction coefficient of the compound within this range. Hence, since

$$k_a^0(\lambda) = 2.303 Z(\lambda)\, \varepsilon(\lambda) \qquad (13\text{-}22)$$

TABLE 13.4 $W(\lambda)$ and $Z(\lambda)$ Values for a *Midwinter* Day at 40°N Latitude (Sea Level) under Clear Skies[a]

λ(Center) (nm)	λ Range ($\Delta\lambda$) (nm)	W(noon, λ)[b,d] (millieinstein $cm^{-2} s^{-1}$)	Z(noon, λ)[b,d] (millieinstein $cm^{-2} s^{-1}$)	Z(24 h, λ)[c,d] (millienstein $cm^{-2} d^{-1}$)
297.5	2.5	0.00(0)	0.00(0)	0.00(0)
300.0	2.5	1.00(−10)	1.22(−10)	2.22(−6)
302.5	2.5	4.98(−10)	6.11(−10)	1.31(−5)
305.0	2.5	2.31(−9)	2.83(−9)	5.17(−5)
307.5	2.5	6.12(−9)	7.46(−9)	1.47(−4)
310.0	2.5	1.16(−8)	1.42(−8)	3.26(−4)
312.5	2.5	2.41(−8)	2.94(−8)	6.04(−4)
315.0	2.5	3.69(−8)	4.50(−8)	9.64(−4)
317.5	2.5	4.92(−8)	6.02(−8)	1.39(−3)
320.0	2.5	6.78(−8)	8.28(−8)	1.84(−3)
323.1	3.75	1.23(−7)	1.51(−7)	3.58(−3)
330.0	10	4.63(−7)	5.68(−7)	1.37(−2)
340.0	10	5.66(−7)	6.97(−7)	1.87(−2)
350.0	10	6.03(−7)	7.46(−7)	2.16(−2)
360.0	10	6.36(−7)	7.95(−7)	2.47(−2)
370.0	10	6.94(−7)	8.63(−7)	2.70(−2)
380.0	10	7.48(−7)	9.34(−7)	2.94(−2)
390.0	10	1.07(−6)	1.33(−6)	2.75(−2)
400.0	10	1.55(−6)	1.93(−6)	3.95(−2)
420.0	30	6.19(−6)	7.75(−6)	1.58(−1)
450.0	30	7.92(−6)	1.00(−5)	2.03(−1)
480.0	30	8.59(−6)	1.09(−5)	2.21(−1)
510.0	30	9.02(−6)	1.15(−5)	2.31(−1)
540.0	30	9.37(−6)	1.20(−5)	2.40(−1)
570.0	30	9.47(−6)	1.21(−5)	2.40(−1)
600.0	30	9.57(−6)	1.23(−5)	2.44(−1)
640.0	50	1.65(−5)	2.12(−5)	4.25(−1)

[a] $Z(\lambda) = W(\lambda) \cdot D^0(\lambda)$. Midwinter refers to a solar declination of −20° (late January).
[b] Values derived from data of Zepp and Cline (1977).
[c] Values derived from Leifer, 1988; note: $Z(24\,h, \lambda) = L(\lambda)/2.303$.
[d] Numbers in parentheses are powers of 10.

the $k_a^0(\lambda)$ values calculated in Table 13.5 are also integrated values over the indicated $\Delta\lambda$ range. The curves drawn in Figure 13.9 have been constructed by using the average values within a given $\Delta\lambda$ range, that is, by using $Z(\lambda)/\Delta\lambda$ and $k_a(\lambda)/\Delta\lambda$ values, respectively.

As can be seen from Figure 13.9 and from Tables 13.3 and 13.4, the solar irradiance at the surface of the earth shows a sharp decrease in the uv-B region (uv-B: 280–320 nm) with virtually no intensity below 290 nm. Hence, only compounds absorbing light above 290 nm undergo direct photolysis. Owing to the sharp decrease in light intensity in the uv-B region, compounds absorbing light primarily in the uv-B and lower uv-A

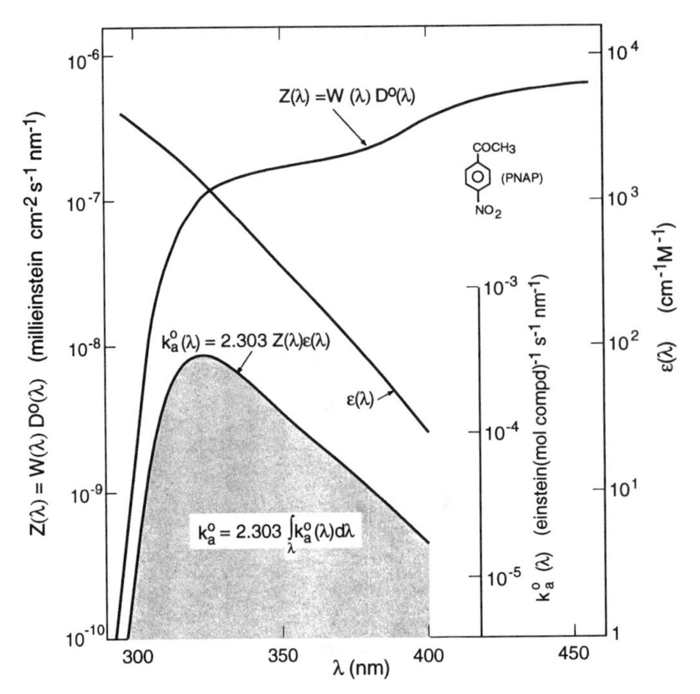

Figure 13.9 Graphical representation of the calculation of the near-surface specific light absorption rate, k_a^0, for *para*-nitroacetophenone (PNAP) for a clear-sky midday, midsummer at 40°N latitude. The shaded area corresponds to the total rate. Note that the *y* axes are on logarithmic scales.

regions (uv-A: 320–400 nm) show a maximum in $k_a^0(\lambda)$ even if they do not exhibit a maximum in $\varepsilon(\lambda)$ in that wavelength region. For example, as Figure 13.9 and Table 13.5 show, PNAP absorbs sunlight primarily in the wavelength range between 305 and 370 nm, with a maximum $k_a^0(\lambda)$ value between 320 and 330 nm. Thus, when evaluating the direct photolytic transformation of PNAP within the water column of a natural water body, we may need to know $\alpha_D(\lambda)$ values only for a relatively narrow wavelength range (see discussion below).

Integration of Eq. 13-22 over the wavelength range over which the chemical absorbs light (i.e., 295–420 nm for PNAP) yields the near-surface specific rate of light absorption by the compound (see hatched area in Fig. 13.9):

$$k_a^0 = \int_\lambda k_a^0(\lambda)\, d\lambda = 2.303 \int_\lambda Z(\lambda)\, \varepsilon(\lambda)\, d\lambda \qquad (13\text{-}23)$$

When using integrated $Z(\lambda)$ values as given in Tables 13.3–13.5, the integral in Eq. 13.23 is approximated by a sum

$$k_a^0 \simeq \Sigma\, k_a^0(\lambda)\, d\lambda = 2.303\, \Sigma\, Z(\lambda)\cdot\varepsilon(\lambda) \qquad (13\text{-}24)$$

TABLE 13.5 Calculation of the Near-Surface Total Specific Light Absorption Rate k_a^0 of *p*-Nitroacetophenone (PNAP) at 40°N Latitude at Noon on a Clear Midsummer Day

	Solar Irradiance		PNAP	
λ (Center) (nm)	λ Range ($\Delta\lambda$) (nm)	$Z(\text{noon}, \lambda)^a$ (millieinstein $cm^{-2} s^{-1}$)	$\varepsilon(\lambda)^b$ ($cm^{-1} M^{-1}$)	$k_a^0(\lambda) = 2.303 Z(\lambda)\varepsilon(\lambda)$ [einstein (mol PNAP)$^{-1}$ s^{-1}] $10^3\, k_a^0(\lambda)$
297.5	2.5	1.19(−9)	3790	0.01
300.0	2.5	3.99(−9)	3380	0.03
302.5	2.5	1.21(−8)	3070	0.09
305.0	2.5	3.01(−8)	2810	0.20
307.5	2.5	5.06(−8)	2590	0.30
310.0	2.5	8.23(−8)	2380	0.45
312.5	2.5	1.19(−7)	2180	0.60
315.0	2.5	1.60(−7)	1980	0.73
317.5	2.5	1.91(−7)	1790	0.79
320.0	2.5	2.24(−7)	1610	0.83
323.1	3.75	4.18(−7)	1380	1.33
330.0	10	1.41(−6)	959	3.12
340.0	10	1.60(−6)	561	2.06
350.0	10	1.71(−6)	357	1.42
360.0	10	1.83(−6)	230	0.97
370.0	10	2.03(−6)	140	0.66
380.0	10	2.24(−6)	81	0.41
390.0	10	2.68(−6)	45	0.28
400.0	10	3.84(−6)	23	0.22
420.0	30	1.51(−5)	0	0
450.0	30	1.90(−5)	0	0
				$k_a^0 = \sum k_a^0(\lambda) = 14.5 \cdot 10^{-3}$ einstein (mol PNAP)$^{-1}$ s^{-1}

[a] Values from Table 13.3; numbers in parentheses are powers of 10.
[b] Values are taken from Leifer, 1988.

For PNAP, the calculation of k_a^0(noon) for a midsummer day at 40°N latitude is performed in Table 13.5. The result $k_a^0(\text{noon}) = 14.5 \times 10^{-3}$ einstein (mol compound)$^{-1}$ s^{-1} indicates that, near the surface of a natural water body, a total of 14.5 millieinstein are absorbed per second per mole of compound present in dilute solution. The corresponding calculated near-surface specific rate of light absorption averaged over one day is $k_a^0(24\text{h}) = 532$ einstein (mol compound)$^{-1}$ d^{-1} (calculation not shown), which is about 40% of the midday value extrapolated to 24h [i.e., $k_a^0(\text{noon}) = 1250$ einstein (mol compound)$^{-1}$ d^{-1}].

Figure 13.10 shows the variation in the k_a^0(24h) values of PNAP as a function of season and decadic latitude in the northern hemisphere. As can be seen, little differences are found in the summer between different latitudes. During the other seasons, however, a significant decrease in k_a^0 is observed with increasing latitude. For example, at 30°N, k_a^0(24h) of PNAP is only twice as large in the summer as compared to the

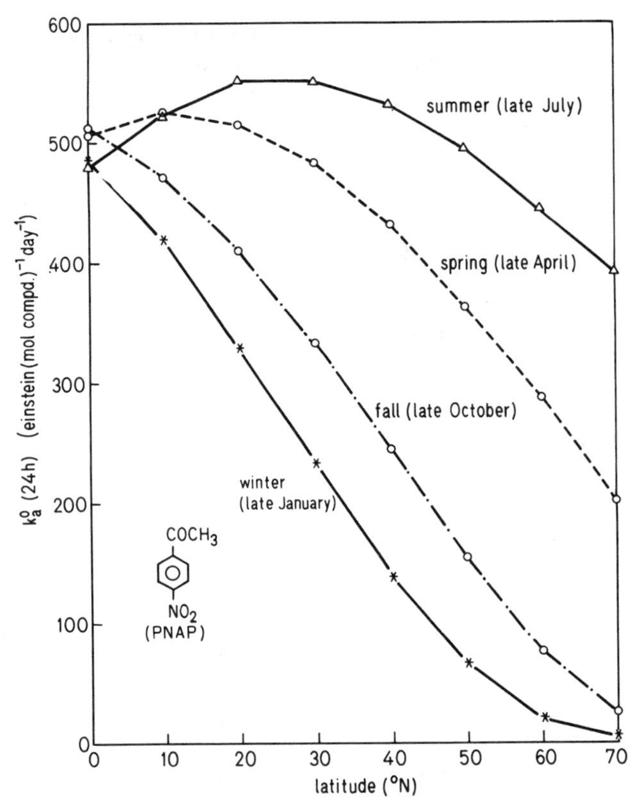

Figure 13.10 Calculated 24-h averaged near-surface specific light absorption rates, k_a^0 (24 h), for PNAP as a function of season and latitude in the northern hemisphere (data from Leifer, 1988).

winter, while at 60°N, the difference is more than a factor of 20. It should be pointed out that temporal and geographical variations in light intensity are most pronounced in the uv-B and low uv-A regions. Consequently, for compounds such as PNAP that absorb light mostly in that wavelength range, the rate of light absorption is very sensitive to diurnal, seasonal, and geographic changes (Zepp and Cline, 1977).

Specific Rate of Light Absorption of an Organic Pollutant in a Well-Mixed Water Body

So far we have considered the rate of light absorption by a pollutant present at low concentration near the surface of a water body, that is, within a zone in which only very little light is absorbed [i.e., $\alpha_D(\lambda) z_{mix} \leqslant 0.02$]. The opposite extreme is the situation in which nearly all of the light is absorbed in the mixed water body considered [i.e.

$\alpha_D(\lambda) z_{mix} \geqslant 2$]. In this case, $1-10^{-\alpha_D(\lambda) z_{mix}} \cong 1$, and Eq. 13-17 simplifies to

$$I_a^t(\lambda) = \frac{W(\lambda)\varepsilon(\lambda)}{z_{mix}\alpha(\lambda)} \cdot C \tag{13-25}$$

and

$$k_a^t(\lambda) = \frac{W(\lambda)\varepsilon(\lambda)}{z_{mix}\alpha(\lambda)} \tag{13-26}$$

where the superscript t denotes *t*otal light absorption rate. Note that Eqs. 13-25 and 13-26 are valid only if $\varepsilon(\lambda)C \ll \alpha(\lambda)$, (i.e., for a dilute solution of the pollutant), and that $k_a^t(\lambda)$ is dependent on $\alpha(\lambda)$ and not on $\alpha_D(\lambda)$ [the pathlength of the light has no effect on $k_a(\lambda)$, since all light is absorbed within the mixed water body considered].

In most natural waters, light and particularly uv-light (which is of most importance for photolytic transformations of organic pollutants), is absorbed primarily by organic constituents, especially by dissolved organic matter. In Figure 13.11, α values (i.e., beam attenuation coefficients) determined per milligram per liter of dissolved organic carbon (DOC) for filtered water samples derived from various Swiss lakes and rivers are plotted as a function of wavelength. These examples illustrate that natural organic materials interact with solar light primarily between 300 and 600 nm, that $\alpha(\lambda)$ generally decreases with increasing wavelength (particularly in the uv-B and uv-A region), but

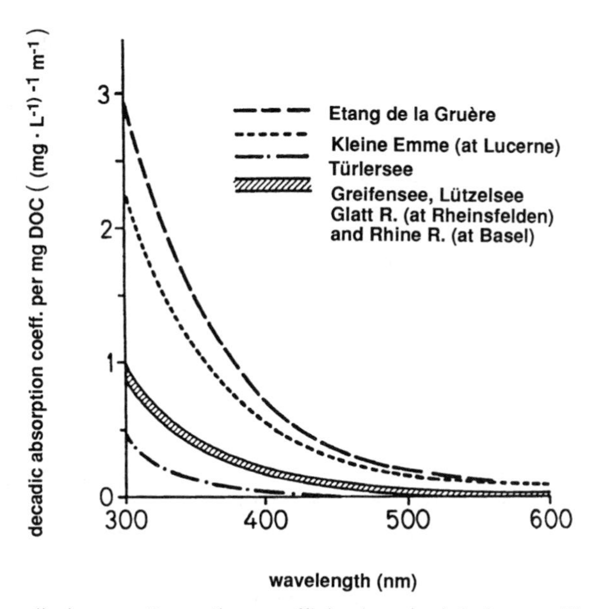

Figure 13.11 Decadic beam attenuation coefficients calculated per milligram of DOC as a function of wavelength for a series of water samples from various Swiss lakes and rivers (data from Haag and Hoigné, 1986).

that significant differences exist between the specific light attenuation coefficients ($\alpha(\lambda)/[\mathrm{DOC}]$) of waters from different origins. These latter findings are not too surprising since, for example, humic and fulvic acids derived from different sources have been postulated to exhibit different types and relative abundances of chromophores (Zafiriou et al., 1984).

We may calculate the k_a^t value for a well-mixed water body in which virtually all light is absorbed by

$$k_a^t = \sum k_a^t(\lambda) = \frac{1}{z_{mix}} \sum \frac{W(\lambda)\varepsilon(\lambda)}{\alpha(\lambda)} \tag{13-27}$$

In Table 13.6, $k_a^t(24\,\mathrm{h})$ is calculated for PNAP for the well-mixed epilimnion of a small eutrophic lake (our example is Greifensee in Switzerland; $z_{mix} = 5\,\mathrm{m}$, DOC $= 4\,\mathrm{mg\,C\,L^{-1}}$, $\alpha(\lambda)$ values are given in Table 13.6) on a clear midsummer day at 47.5°N latitude. The result is $k_a^t(24\,\mathrm{h}) = 22.5$ einstein (mol PNAP)$^{-1}$ day^{-1}. Before we can be sure that our assumption that mixing (typical vertical mixing rates in the epilimnion of Swiss lakes are, for example, between 1 and 10 day^{-1}; Imboden and Schwarzenbach, 1985) is fast as compared to the photolytic transformation of PNAP in the epilimnion of this lake is correct, we must discuss quantum yields.

Light-Screening Factors

Before we turn to discussing reaction quantum yields, we want to introduce an approximation for calculating $k_a(\lambda)$ values of a given compound in a well-mixed water body in which neither very little light [situation described by the near-surface value $k_a^0(\lambda)$] nor all light [described by $k_a^t(\lambda)$] is absorbed. For this purpose, we introduce a *light-screening factor*, $S(\lambda)$, which is defined as the ratio of the $k_a(\lambda)$ value for a mixed water body of depth z_{mix} (Eq. 13-17) and the near-surface specific rate of light absorption $k_a^0(\lambda)$ (Eq. 13-21); therefore,

$$\begin{aligned} S(\lambda) &= \frac{W(\lambda)\varepsilon(\lambda)[1 - 10^{-\alpha_D(\lambda) z_{mix}}]}{z_{mix}\,\alpha(\lambda) \cdot 2.303 \cdot W(\lambda)\, D^0(\lambda)\, \varepsilon(\lambda)} \\ &= \frac{[1 - 10^{-\alpha_D(\lambda) z_{mix}}]}{2.303\, z_{mix}\, \alpha_D^0(\lambda)} \end{aligned} \tag{13-28}$$

Note that $S(\lambda)$ is a function of both $\alpha_D^0(\lambda)$ (the diffuse attenuation coefficient near the surface) and of $\alpha_D(\lambda)$ (the average diffuse attenuation coefficient for the whole water column of depth z_{mix}). Using $S(\lambda)$, we may express $k_a(\lambda)$ as

$$k_a(\lambda) = k_a^0(\lambda) \cdot S(\lambda) \tag{13-29}$$

and approximate k_a by

$$k_a = \Sigma k_a^0(\lambda)\, S(\lambda) \tag{13-30}$$

For crude estimates of $k_a(\lambda)$, we may make two assumptions that simplify the calculation of $S(\lambda)$. First, we may set $\alpha_D(\lambda) \cong \alpha_D^0(\lambda)$ if the water body considered is not too deep (e.g., only a few meters). Second, for nonturbid waters, we may assume an average $D(\lambda)$ value of $D \cong 1.2$ (see Zepp and Cline, 1977). Then, Eq. 13-28 simplifies to

$$S(\lambda) = \frac{[1 - 10^{-(1.2)\alpha(\lambda)z_{mix}}]}{(2.303)(1.2)z_{mix}\,\alpha(\lambda)} \qquad (13\text{-}31)$$

where $\alpha(\lambda)$ is the beam attenuation coefficient that we may easily determine with a spectrophotometer. Finally, if the compound of interest absorbs light only over a relatively narrow wavelength range (e.g., PNAP, Fig. 13.11), one may use a single (average) $\alpha(\lambda)$ [or $\alpha_D(\lambda)$] value for this wavelength range to calculate the effect of light attenuation on the specific rate of light absorption of the compound. For example, if we choose the α value at the wavelength λ_m of the *maximum specific rate of light absorption* [not the maximum $\varepsilon(\lambda)$!] of the compound (i.e., 323 nm in the case of PNAP, see Fig. 13.11), k_a may be approximated by

$$\begin{aligned} k_a &\simeq S(\lambda_m)\Sigma k_a^0(\lambda) \\ &\simeq S(\lambda_m)k_a^0 \end{aligned} \qquad (13\text{-}32)$$

For practical applications, Eq. 13-32 is extremely useful, since in many cases, experimental data is available only for the near-surface specific rate of light absorption or, even more frequently, for the total near-surface direct photolytic transformation rate of a given pollutant.

Direct Photolysis Rates

In Section 13.2 we defined a reaction quantum yield, $\Phi_r(\lambda)$, describing the total number of compound molecules (e.g., moles compound) transformed by a chemical reaction per total number of photons (e.g., einsteins) absorbed by a given system resulting from the presence of the compound (Eq. 13-8). In Eq. 13-17, we denoted the rate of light absorption of wavelength λ by the pollutant per unit volume (e.g., einstein per liter per second) as $I_a(\lambda)$. It is now easy to see that the product of these two entities describes the number of compound molecules transformed per unit volume per time [e.g., (mol compound) per liter per second], which is nothing but the concentration change per unit time in a given system, or the rate of transformation of the pollutant:

Rate of direct *p*hotolysis

$$\begin{aligned} \text{(subscript, "p") at wavelength } \lambda &= -\left(\frac{dC}{dt}\right)_{p,\lambda} \\ &= \Phi_r(\lambda)I_a(\lambda) \\ &= \Phi_r(\lambda)k_a(\lambda)C \\ &= k_p(\lambda)C \end{aligned} \qquad (13\text{-}33)$$

TABLE 13.6 Calculation of the 24-h Averaged Total Specific Light Absorption Rate k_a^t of *p*-Nitroacetophenone (PNAP) at ~47.5°N Latitude on a Clear Midsummer Day in the Well-Mixed Epilimnion of Greifensee in Switzerland

λ(Center) (nm)	λ Range ($\Delta\lambda$) (nm)	Intensity of Solar Radiation at the Surface $W(24h, \lambda)^a$ (millieinstein cm^{-2} d^{-1})	Average Beam Attenuation Coefficient $\alpha(\lambda)^b$ (cm^{-1})	PNAP[c]	
				$\varepsilon(\lambda)$ (cm^{-1} M^{-1})	$k_a^t(\lambda) = W(24\,h, \lambda) \cdot \varepsilon(\lambda)/z_{mix} \cdot \alpha(\lambda)$ [einstein (mol PNAP)$^{-1}$ d^{-1}]
297.5	2.5	1.12(−5)	0.0430	3790	0.00
300.0	2.5	5.92(−5)	0.0415	3380	0.01
302.5	2.5	2.10(−4)	0.0395	3070	0.03
305.0	2.5	5.49(−4)	0.0375	2810	0.08
307.5	2.5	1.14(−3)	0.0355	2590	0.17
310.0	2.5	1.99(−3)	0.0335	2380	0.28
312.5	2.5	3.06(−3)	0.0320	2180	0.42
315.0	2.5	4.26(−3)	0.0305	1980	0.55
317.5	2.5	5.52(−3)	0.0290	1790	0.68
320.0	2.5	6.75(−3)	0.0275	1610	0.79
323.1	3.75	1.23(−2)	0.0260	1380	1.31
330.0	10	4.34(−2)	0.0220	959	3.86
340.0	10	5.53(−2)	0.0185	561	3.35

350.0	10	6.20(−2)	0.0150	357	2.95
360.0	10	6.87(−2)	0.0125	230	2.53
370.0	10	7.34(−2)	0.0100	140	2.05
380.0	10	7.86(−2)	0.0083	81	1.53
390.0	10	7.38(−2)	0.0069	45	0.96
400.0	10	1.06(−1)	0.0055	23	0.89
420.0	30	4.23(−1)	0.0042	0	0.00
450.0	30	5.37(−1)	0.0028	0	0.00
480.0	30	5.80(−1)	0.0019	0	0.00
Σ495.5 – 600.0	114.5	1.85(0)	0.0010	0	0.00
		$\sum W(\lambda) = 3.99(0)$ millieinstein $cm^{-2}\,d^{-1}$			$k_a^t = \sum k_a^t(\lambda) = 22.5$ einstein (mol PNAP)$^{-1}\,d^{-1}$

[a] $W(\lambda)$ values estimated from Leifer (1988) for 50°N latitude; values in parentheses indicate powers of 10.
[b] Average value for the indicated λ-range; see Figure 13.9.
[c] Well-mixed epilimnion with mean depth $z_{mix} = 500$ cm; $\varepsilon(\lambda)$ values taken from Leifer (1988).

where $k_p(\lambda) = \Phi_r(\lambda)k_a(\lambda)$ is the direct photolysis first-order rate constant at wavelength λ, and has the unit of time^{-1}, (e.g., per second or per day). We recall that we may express direct photolysis as a first-order process only for a dilute solution of the chemical. The (total) rate of direct photolytic transformation of a given pollutant in a well-mixed water body is then given by

$$\text{(total) rate} = -\left(\frac{dC}{dt}\right)_p = [\textstyle\sum k_p(\lambda)]C \tag{13-34}$$

If we assume that the quantum yield is independent of wavelength, we simply multiply the total specific light absorption rate of the compound in a well-mixed water body by Φ_r to obtain k_p:

$$k_p = \Phi_r \cdot k_a \tag{13-35}$$

Similarly, the near-surface first-order rate constant k_p^0 for direct photolysis is given by

$$k_p^0 = \Phi_r \cdot k_a^0 \tag{13-36}$$

Finally, using Eq. 13-32, k_p and k_p^0 may be related by

$$k_p = k_p^0 \cdot S(\lambda_m) \tag{13-37}$$

As indicated by Eq. 13-35, to estimate the rate of direct photolysis of a pollutant in a given system, one needs to know the k_a value as well as the reaction quantum yield for the compound considered. As we have discussed extensively above, k_a values may be estimated with the help of computer programs. However, there are presently no rules for predicting Φ_r values from chemical structure. Thus, quantum yields have to be determined experimentally. We shall briefly outline the most widely used procedures, but refer to the literature for a more detailed discussion (Zepp, 1978; Roof, 1982; Zepp, 1982; Mill and Mabey, 1985; Leifer, 1988).

Determination of Quantum Yields—Chemical Actinometry

In the most common procedures used to determine reaction quantum yields, an (oxygenated) *dilute* solution of the compound (preferably in distilled water or distilled water containing a low amount of a polar organic solvent) is irradiated by *constant intensity monochromatic radiation* in a photochemical apparatus (e.g., optical bench, merry-go-round reactor). In the laboratory, various light sources are available to investigate photolytic processes and to determine quantum yields. The most common lamps include low-, medium-, and high-pressure mercury lamps, xenon lamps, and lasers. These lamps are used in connection with various filter systems to obtain the desired monochromatic or polychromatic light (Calvert and Pitts, 1966; Zepp, 1982; Mill and Mabey, 1985). For determination of quantum yields in the uv-B and uv-A region, two-filter systems (a short description is given by Mill and Mabey, 1985) are widely used in connection with medium- and high-pressure mercury lamps to isolate the 313- and 366-nm bands. Because many important environmental pollutants absorb light primarily in the uv-region, a large number of quantum yields reported in the literature have been determined at 313 and/or 366 nm.

Distilled rather than natural water is often used as the solvent for determination of quantum yields for two major reasons. First, the total absorbance of the solution at the wavelength of irradiation should not exceed 0.02. Second, and more important, the presence of natural water constituents (e.g., humic material, nitrate) could enhance the total photolytic transformation rate by indirect photolytic processes as described in Section 13.4. Zepp and Baughman (1978) have argued that for many chemicals Φ_r obtained in distilled water is nearly the same as that observed in natural waters (at least in uncontaminated freshwaters), because concentrations of natural water constituents that could undergo reactions with or quench photolysis of excited pollutants are generally very low. Furthermore, the effects of molecular oxygen, which may act as a quencher, can also be studied in distilled water.

From measurements of the concentration C of the compound as a function of exposure time, the first-order photolysis rate constant, $k_p(\lambda)$, is then determined by calculating the slope of a plot of $\ln(C/C_0)$ versus time (see Section 12.2). Since the absorbance of the solution is less than 0.02, $k_p(\lambda)$ is given as [Zepp, 1978; see also analogy to Eq. 13-21 with $D(\lambda) = (A/V)\cdot l(\lambda)$]:

$$k_p(\lambda) = 2.3\cdot W(\lambda)\frac{A}{V}\cdot\varepsilon(\lambda)\cdot l(\lambda)\cdot\Phi_r(\lambda) \qquad (13\text{-}38)$$

where $W(\lambda)(A/V)$ is the incident light intensity per unit volume of the cell (e.g., a quartz vessel with total surface A and volume V), and $l(\lambda)$ is the cell pathlength that can be determined experimentally for the selected λ value (Zepp, 1978). Hence, $\Phi_r(\lambda)$ can be calculated by

$$\Phi_r(\lambda) = \frac{k_p(\lambda)}{2.3\cdot W(\lambda)(A/V)\cdot\varepsilon(\lambda)\cdot l(\lambda)} \qquad (13\text{-}39)$$

provided that the light intensity term $W(\lambda)(A/V)$ is known. This light intensity term may be determined by exposing a *chemical actinometer* to the light source in the same way and at the same time that the compound of interest is exposed. A chemical actinometer is a solution of a photoreactive reference compound (subscript R) that reacts with a well-defined reaction quantum yield, $\Phi_{r,R}(\lambda)$, preferably with an approximately similar half-life as the compound for which $\Phi_r(\lambda)$ is to be determined. There are two types of chemical actinometers: (1) concentrated solutions of some chemicals that absorb virtually all of the incident light, and (2) chemical actinometers that only weakly absorb the monochromatic radiation (i.e., for which the absorbance is less than 0.02). In the first case, the reaction proceeds by zero-order kinetics and the reaction rate is given by [see Eq. 13-25 with $W(\lambda)/z_{mix} = W(\lambda)\cdot A/V$, and

$\alpha(\lambda) = \varepsilon_R(\lambda)C_R]$:

$$\text{rate}_R = W(\lambda)\frac{A}{V}\cdot\Phi_{r,R}(\lambda) \tag{13-40}$$

and

$$W(\lambda)\frac{A}{V} = \frac{\text{rate}_R}{\Phi_{r,R}(\lambda)} \tag{13-41}$$

which may be substituted into Eq. 13-39 to calculate $\Phi_r(\lambda)$. Classical actinometers that are used in this way include the potassium ferrioxalate actinometer that can be employed both in the uv and visible spectral region, the Reinecke's salt actinometer (visible region), and the *ortho*-nitrobenzaldehyde actinometer (uv region). For a description of these actinometers we refer to the literature (e.g., Leifer, 1988, pp. 148–151).

For the dilute solution actinometer (i.e., absorbance < 0.02), Eq. 13-38 applies also for the description of $k_{p,R}(\lambda)$. From this equation, $W(\lambda)(A/V)\cdot l(\lambda)$ may be determined by

$$W(\lambda)\frac{A}{V}\cdot l(\lambda) = \frac{k_{p,R}(\lambda)}{2.3\,\varepsilon_R(\lambda)\cdot\Phi_{r,R}(\lambda)} \tag{13-42}$$

Substitution of Eq. 13-42 into Eq. 13-39 then yields the reaction quantum yield of the compound of interest at wavelength λ:

$$\Phi_r(\lambda) = \frac{k_p(\lambda)\cdot\varepsilon_R(\lambda)}{k_{p,R}(\lambda)\cdot\varepsilon(\lambda)}\cdot\Phi_{r,R}(\lambda) \tag{13-43}$$

To obtain an "environmental" quantum yield or quantum efficiency, Dulin and Mill (1982) suggested exposing dilute solutions of both the pollutant and the chemical actinometer to sunlight. The quantum efficiency in this way represents an averaged value over the wavelength range over which the compound absorbs sunlight. It can be estimated from the measured first-order rate constants k_p and $k_{p,R}$, and from the ratio of the specific sunlight absorption rates of pollutant and reference compound *calculated* (Eq. 13-24) for the time and locations of the experiments:

$$\Phi_r = \frac{k_p k_{a,R}^{\text{calc}}}{k_{p,R} k_a^{\text{calc}}}\Phi_{r,R} \tag{13-44}$$

Eq.13-44 assumes that the ratio of the total light absorbed by the pollutant and the chemical actinometer is constant over changes in seasons, latitudes, and sky conditions. The validity of this approach depends, of course, also on the reliability of the simulated solar spectral irradiances. Since variations in sunlight intensities as a consequence of weather, diurnal, and/or seasonal changes are most pronounced in the uv-B and low uv-A region, the largest errors arising with this approach can be expected for chemicals that have a maximum specific light absorption rate in this wavelength region (i.e., 290–350 nm). Nevertheless this "outdoor" approach to determine Φ_r of a given compound may be very useful, particularly in cases in which the quantum yield is wavelength-dependent.

In principle, any organic compound [with known $\Phi_r(\lambda)$ or Φ_r(sunlight)] that absorbs light in the appropriate wavelength range, and that exhibits a photolytic half-life similar to that of the compound of interest could be used as a dilute solution actinometer (Zepp, 1978 and 1982). In practice, however, such compounds are often difficult to find. Dulin and Mill (1982) discussed the criteria that need to be fulfilled by a good chemical actinometer, particularly when used for sunlight experiments. They described a binary chemical actinometer approach that is applicable primarily to measure radiation intensities in the uv-region. The major advantage of this type of actinometer is that the quantum yield and thus the half-life of the actinometer chemical is adjustable, thus ensuring that both actinometer and pollutant are exposed to the same levels of light. This is particularly important in cases where the compounds are exposed to sunlight over a longer period of time (e.g., several hours to days).

The basic principle of a binary actinometer lays in a bimolecular photoreaction of a photosensitive species (the reference compound R) with a photoinsensitive reactant R_i:

$$R + R_i \xrightarrow{h\nu} \text{defined products} \tag{13-45}$$

The rate of the reaction, that is, the rate of conversion of R (which is measured) is then given by

$$\text{rate}(\lambda) = -\left(\frac{d[R]}{dt}\right)_\lambda = k_{a,R}(\lambda)\cdot\Phi_{r,R}(\lambda)[R]$$
$$\text{with } \Phi_{r,R}(\lambda) = \Phi^0_{r,R}(\lambda) + k_i[R_i] \tag{13-46}$$

$\Phi^0_{r,R}(\lambda)$ is the extrapolated quantum yield in the absence of R_i, and k_i is a measure for the yield of the reaction of the excited R with R_i, and has the units of (mol R converted) einstein^{-1}M^{-1}. For practical purposes, R_i should be present in excess concentration (i.e., $[R_i] \gg [R]$). Thus, if over a reasonable concentration range of R_i, $\Phi^0_{r,R}(\lambda) \ll k_i[R_i]$, $\Phi_{r,R}(\lambda)$ (and thus the photolytic half-live of R at a given light intensity) can be varied linearly with $[R_i]$. Two useful examples of such binary actinometers are *p*-nitroanisole(PNA)/pyridine and *p*-nitroacetophenone(PNAP)/pyridine. In their excited states, both PNA and PNAP undergo a nucleophilic displacement reaction with pyridine:

$$\text{R-C}_6\text{H}_4\text{-NO}_2 + \text{C}_5\text{H}_5\text{N} \xrightarrow{h\nu} \text{R-C}_6\text{H}_4\text{-N}^{\oplus}\text{C}_5\text{H}_5 + NO_2^- \tag{13-47}$$

R = -OCH$_3$ (PNA)
R = -COCH$_3$ (PNAP)

The reaction follows the kinetics described by Eq. 13-46. Both PNA and PNAP absorb light in the uv region (< 400 nm) and show a constant $\Phi_{r,R}(\lambda)$ over this wavelength range.

The PNA/pyridine actinometer is useful for very fast reactions. In sunlight, it can be adjusted to half-lives between a few minutes and about 12 h. The upper time limit is determined by the (very high) $k_{a,0}$ value of PNA [~ 5000 einstein (mol PNA)$^{-1}$d^{-1} for a midsummer day at 40°N latitude], and the (rather small) $\Phi^0_{r,R}$ value of 3×10^{-4} (Dulin and Mill, 1982; Leifer, 1988). In comparison, as we discussed earlier, for the same geographic latitude and time, PNAP exhibits a $k_{a,0}$ value that is about 10 times smaller [i.e., 532 einstein (mol PNAP)$^{-1}$ day^{-1}], and it has an even smaller quantum yield as compared to PNA ($\Phi^0_{r,R} < 10^{-5}$, Dulin and Mill, 1982). Hence, although PNAP absorbs sunlight at an appreciable rate, its (direct) photolytic half-life would be very large in a natural water. In the presence of pyridine in a test vessel, however, the photolytic half-life can be adjusted to range between a few hours and several months (for details see Dulin and Mill, 1982, or Leifer, 1988).

At this point we should note that when extrapolating photolysis rates determined in test vessels to natural water bodies, the geometry of the test vessel must be taken into account. For example, sunlight photolysis rates measured in cylindrical test tubes have been found to be greater by a factor of 1.5–2.2 as compared to the rates determined

TABLE 13.7 Direct Photolysis Reaction Quantum Yields of Some Selected Organic Pollutants in Aqueous Solution

Compound	Wavelength[a] (nm)	Solvent Other Than Water[b] (pH)	Reaction Quantum Yield (Φr)[c]	Ref[d]
Naphthalene	313		1.5×10^{-2}	1
Phenanthrene	313		1.0×10^{-2}	1
Anthracene	313		3.0×10^{-3}	1
1,2-Benzanthracene	313, 366, sun	1% AN	3.2×10^{-3}	2
Benzo(a)pyrene	313, 366, sun	1–20% AN	8.9×10^{-4}	2
3,4-Dichloroaniline	313, polychr.	pH 7–10	4.4×10^{-2}	3
3,5-Dichloroaniline	313, sun	pH 4–10	5.2×10^{-2}	3
Pentachlorophenolate	314, polychr.	pH 8–10	1.3×10^{-2}	3
4-Nitrophenol	313, polychr.	pH 2–4	1.1×10^{-4}	3
4-Nitrophenolate	365, polychr.	pH 9–10	8.1×10^{-6}	3
Nitrobenzene	313		2.9×10^{-5}	4
4-Nitrotoluene	366		5.2×10^{-3}	4
2,4-Dinitrotoluene	313		2.0×10^{-3}	4
2,4,6-Trinitrotoluene	313, 366, sun		2.1×10^{-3}	5

[a] Wavelength or wavelength range at which Φ_r has been determined, sun = sunlight, polychr = polychromatic artificial light (> 290 nm).

[b] AN = acetonitrile.

[c] Moles of compound converted per moles of photons absorbed.

[d] (1) Zepp and Schlotzhauer (1979); (2) Mill et al. (1981); (3) Lemaire et al. (1985); (4) Simmons and Zepp (pers. comm.). (5) Mabey et al. (1983)

in flat dishes (Dulin and Mill, 1982; Haag and Hoigné, 1986). The major reason for these findings are lens effects of the curved glass and the fact that the tubes are exposed to light from all sides.

In Table 13.7 the reaction quantum yields are given for some selected organic pollutants. As can be seen, reaction quantum yields vary over many orders of magnitude, with some compounds exhibiting very small Φ_r values. However, since the reaction rate is dependent on both k_a and Φ_r (Eq. 13-33), a low reaction quantum yield does not necessarily mean that direct photolysis is not important for that compound. For example, the near-surface direct photolytic half-life of 4-nitrophenolate ($\Phi_r = 8.1 \times 10^{-6}$) at 40°N latitude is estimated to be on the order of only a few hours, which is similar to the half-life of the neutral 4-nitrophenol, which exhibits a Φ_r more than 10 times larger (Lemaire et al., 1985). The reason for the similar half-lives is the much higher rate of light absorption of 4-nitrophenolate as compared to the neutral species, 4-nitrophenol (compare uv/vis spectra in Fig. 13.5). As a second example, comparison of the near-surface photolytic half-lives (summer, 40°N latitude) of the two isomers phenanthrene ($t_{1/2} \sim 60$ days) and anthracene ($t_{1/2} \sim 5$ days, Zepp and Schlotzhauer, 1979) shows that the smaller Φ_r value of anthracene (as compared to phenanthrene) is by far outweighed by its much higher k_a value (compare uv/vis spectra in Figs. 13.3*b* and *c*). These two examples illustrate again that both k_a and Φ_r are important factors in determining the rate of direct photolysis in natural waters.

13.4 INDIRECT (SENSITIZED) PHOTOLYSIS OF ORGANIC COMPOUNDS

Overview

So far, we have dealt with reactions of organic pollutants occurring as a consequence of direct light absorption by the pollutant. In a natural water, there are, however, other light-induced processes that may lead to the transformation of a given xenobiotic compound. Such processes are initiated through light absorption by other chemicals present in the system, and they are, therefore, commonly referred to as *indirect* or *sensitized* photolysis (note that the term "sensitized" is sometimes used only to refer to an indirect photoreaction involving energy transfer, see Fig. 13.6).

The most important light absorbers that may induce indirect photolytic transformations of organic pollutants are the chromophores present in dissolved organic material. A little calculation illustrates that each of these chromophores is excited numerous times during one day. Let us look at the well-mixed epilimnion of Greifensee (small eutrophic lake in Switzerland at 47.5°N, latitude), our model lake that we use throughout the remainder of this chapter. From the $\alpha(\lambda)$ values given in Table 13.6 we can deduce that virtually all light between 290 and 600 nm is absorbed within the epilimnion ($z_{mix} = 5$ m) of this lake. As is indicated by Table 13.6, on a clear summer day, the total number of photons absorbed in the wavelength range that is important for indirect photolysis (290–600 nm) is about 4 millieinstein $cm^{-2}d^{-1}$. This corresponds to an average absorption rate of 8 millieinstein $L^{-1}d^{-1}$. Hence, per liter of epilimnion water, 8×10^{-3} moles of photons are absorbed per day. The question is now how many chromophores are present that absorb these photons. We may make an upper estimate of this number by assuming that each chromophore contains at least 10 carbon atoms. With a dissolved organic carbon content (DOC) of $4\,mg\,C\,L^{-1}$ we obtain a maximum chromophore concentration of about 30 μM. This means that, on the average, each chromophore present in the epilimnion of Greifensee would be excited 270 times per day or more than 10 times per hour. As one can easily imagine, reactions of these excited chromophores may lead to alterations within the dissolved organic matter, and, as we will see, to transformations of organic pollutants. In our discussion, we confine ourselves to presenting only a few examples of such reactions. For a more detailed discussion, and a literature survey on this quite complex topic, we refer to some excellent recent reviews (Zepp, 1988; Cooper et al., 1989; Hoigné et al., 1989).

As we discussed in Section 13.2, an excited chemical species may undergo various physical or chemical processes (Fig. 13.6). Figure 13.12 depicts in a simplified manner the most important physical (i.e., sensitized C path, $^1O_2$ path) and chemical (i.e., formation of radical species, e_{aq}^-) pathways that may lead to the transformation of an organic chemical (C) as a consequence of the excitation of an unknown chromophore (UC) in a natural water (or soil). By far the most important acceptor (quencher) of excited UCs is molecular oxygen in its ground-state (triplet oxygen, $^3O_2$). Since promotion of $^3O_2$ to its first excited state (singlet oxygen, $^1O_2$) requires only

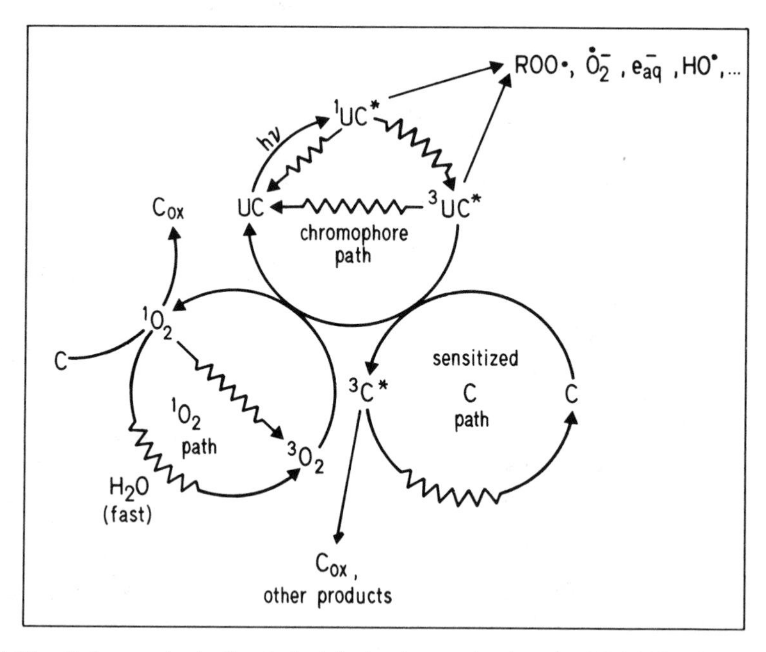

Figure 13.12 Pathways for indirect photolysis of organic chemical (C). UC refers to unknown chromophores. Wavy arrows symbolize radiationless transition (adapted from Zafiriou et al., 1984).

94 kJ mole^{-1}, almost all chromophores absorbing light in the uv as well as in the visible wave-length range may (after intersystem crossing to a triplet state) transfer their absorbed light energy to an oxygen molecule. The resulting $^1O_2$ may then react by chemical reactions with certain organic pollutants.

In contrast to $^3O_2$, the energy required to excite organic pollutants is much higher ($> \sim 250$ kJ mol^{-1}, Zepp et al., 1985), and it is estimated that only about one third of the $^3UC^*$ (the longer-lived triplet states of the excited DOM chromophores that are formed by intersystem crossing from the singlet states, $^1UC^*$) have sufficient energies to perform this process (Zepp et al., 1985). Furthermore, since organic pollutants are in competition with $^3O_2$ for the available absorbed light energy, energy transfer from $^3UC^*$ to a given organic pollutant is most important in waters of low oxygen concentration, and is most probable with pollutants exhibiting low excited triplet state energies (Mill and Mabey, 1985). The latter requirement is met in compounds with delocalized π-electron systems, such as those encountered in polyenes, polycyclic aromatic compounds, and nitroaromatic compounds (Zepp et al., 1985; Mill and Mabey, 1985). Among the very few clear-cut, known examples of indirect photolysis of organic compounds by energy transfer from $^3UC^*$ are the photoisomerization of 1,3-pentadiene (Zepp et al., 1981) and the photolysis of 2,4,6-trinitrotoluene (TNT, a widely used explosive) (Mabey et al., 1983). For 2,4,6-trinitrotoluene, up to a factor of 100 faster photolysis was observed in river water as compared to distilled water when exposed to sunlight (Mabey et al., 1983).

In addition to energy transfer to either molecular oxygen or to organic compounds (including organic pollutants) chemical reactions of excited UCs ($^1UC^*$, $^3UC^*$) may lead to other reactive species that may react with organic pollutants (Fig. 13.12). Such processes include the formation of reactive DOM species (DOM*), the reaction with $^3O_2$ to form DOM-derived peroxy radicals (ROO·), the transfer of an electron to $^3O_2$ to yield superoxide anions ($O_2^{\cdot -}$), the formation of "solvated electrons" (e_{aq}^-), the formation of hydrogen peroxide (H_2O_2), and the formation of hydroxyl radicals (HO·). In the case of hydroxyl radicals, however, DOM also reacts with hydroxyl radicals and is probably a more important sink than source. One major path for production of hydroxyl radicals is thought to be the photolysis of nitrate (Zafiriou, 1974; Zepp et al., 1987b; see also discussion below).

From the rather scarce data available, one may conclude that e_{aq}^-(Zepp et al., 1987a), as well as $O_2^{\cdot -}$ and H_2O_2 (Cooper et al., 1989) are not too significant with respect to tansformations of organic pollutants. We will, therefore, confine our discussion to the "photoreactants" $^1O_2$, HO·, and ROO·, which are all electrophiles that may act as oxidants. Our task will be to quantify these oxidants in a given natural water exposed to sunlight, and to derive rate constants for the (chemical) reactions of organic compounds with these photooxidants.

General Kinetic Approach

For describing the kinetics of indirect photolysis of organic pollutants involving the *photooxidants* $^1O_2$, HO·, and ROO·, we adopt the approach suggested by Hoigné et al. (1989) and Mill (1989). The rate of formation of a given photooxidant (Ox) by radiation of wavelength λ may be described by

$$r_{f,ox}(\lambda) = \left(\frac{d[Ox]}{dt}\right)_{\lambda} = k_{a,A}(\lambda)\Phi_{r,A}(\lambda)[A] \tag{13-48}$$

where $k_{a,A}(\lambda)$ [einstein (mol A)$^{-1}$s^{-1}] is the specific light absorption rate of the bulk of the chemical(s) exhibiting the chromophore(s) (i.e., in the case of DOM the UCs) for the production of Ox, $[A]$ is the (bulk) concentration of the responsible chemicals (e.g., [DOM] or [DOC] in the case of dissolved organic matter; $[NO_3^-]$), and $\Phi_{r,A}(\lambda)$ is the overall quantum efficiency for the production of Ox. Hence, $\Phi_{r,A}(\lambda)$ is a lumped parameter taking into account the reactions of the excited chromophores with other chemical species including $^3O_2$. Consequently, $\Phi_{r,A}(\lambda)$ is only a constant if all relevant parameters (e.g., the concentration of $^3O_2$) are kept constant in the system considered. The total rate of production of Ox is then given by integration of Eq. 13-48 over the wavelength range that is significant for the formation of Ox (i.e., range over which A absorbs light of sufficient energy for production of Ox):

$$\begin{aligned} r_{f,ox} = \frac{d[Ox]}{dt} &= \int_{\lambda} k_{a,A}(\lambda)\Phi_{r,A}(\lambda)[A]d\lambda \\ &\cong \left(\sum k_{a,A}(\lambda)\Phi_{r,A}(\lambda)\right)[A] \end{aligned} \tag{13-49}$$

This wavelength range may be very narrow as, for example, for the production of HO from NO_3^- ($\lambda = 290$–340 nm), or it may be rather broad as, for example, for the prodution of $^1O_2$ from DOM ($\lambda = 290$–~ 600 nm).

Since photooxidants are quite reactive species, they are also "consumed" by various processes including physical quenching by the water itself (in the case of $^1O_2$), or by chemical reactions with various water constituents (e.g., with DOM, HCO_3^-/CO_3^{2-}). If we assign to each of these Ox-consuming processes 'i' a pseudo-first order rate constant, $k_{ox,i}$ (hence, we also keep the concentrations of all consuming species constant), we may describe the rate of consumption, $r_{c,ox}$, of Ox by

$$r_{c,ox} = -\frac{d[Ox]}{dt} = \sum_i (k_{ox,i})[Ox] \tag{13-50}$$

Note that the Ox-consuming processes are chemical processes and that, therefore, $r_{c,ox}$ is light-independent.

Let us now consider a shallow water body that is exposed to a light source, say, to noon sunlight. If we keep everything in the system constant, we will reach a steady state in which $r^0_{f,ox} = r^0_{c,ox}$, that is, the photooxidant will be present at a steady-state concentration $[Ox]^0_{ss}$ of

$$[Ox]^0_{ss} = \frac{\sum k_{a,A}(\lambda)\Phi_{r,A}(\lambda)[A]}{\sum_i k_{ox,i}} \tag{13-51}$$

where we have introduced the superscript '0' to indicate near-surface light conditions. Now we have an easy way of describing the indirect photolysis of a pollutant by a pseudo-first-order rate law, provided that the compound considered does not significantly affect $[Ox]_{ss}^0$ (i.e., $k_{ox} \ll \sum k_{ox,i}$), and that we are able to measure or estimate $[Ox]_{ss}^0$. The near-surface rate of loss of the pollutant is then given by

$$\text{near-surface rate of loss of pollutant} = -\left(\frac{dC}{dt}\right)_{ox}^{0}$$

$$= k'_{ox}[Ox]_{ss}^0 C = k_{ox}^0 C \qquad (13\text{-}52)$$

where k'_{ox} and k_{ox}^0 are the second-order and near-surface pseudo-first-order reaction rate constants, respectively, for reaction of the pollutant with Ox.

Unfortunately, it is in most cases not possible to quantify a given photooxidant by a direct measurement. By analogy to chemical actinometry, however, one may use a probe compound (subscript p) with known $k'_{ox,p}$ to determine $[Ox]_{ss}^0$ in a given natural water. This involves adding the chemical at a known concentration to the water, illuminating, and measuring the compound's disappearance. Since the probe compound disappearance kinetics also obeys Eq. 13-52, $[Ox]_{ss}^0$ can then be calculated from the slope of a correlation of ln $[P]$ versus time:

$$\ln\frac{P}{P_0} = -k'_{ox,p}[Ox]_{ss}^0 t \qquad (13\text{-}53)$$

A requirement for such a measurement is, of course, that the probe compound does not react by any other pathway. We discuss some of these probe systems later when discussing some specific photoreactants. At this point we recall, however, that if $[Ox]_{ss}^0$ values are determined in a cuvette or other photochemical vessel, the geometry of the vessel has to be taken into account when extrapolating the values to natural water bodies (see Section 13.3). The average steady-state photooxidant concentrations for longer periods of time (e.g., one to several days) may be roughly estimated from the measured value by multiplication with the ratio of the (computed or measured) integrated average light intensities (integrated over the wavelength range of maximum production of Ox) prevailing during the two time periods:

$$[Ox]_{ss}^0\begin{pmatrix}\text{estimated}\\ \text{period}\end{pmatrix} = [Ox]_{ss}^0\begin{pmatrix}\text{experimental}\\ \text{period}\end{pmatrix}\frac{\sum Z(\text{est.period}, \lambda)}{\sum Z(\text{exp.period}, \lambda)} \qquad (13\text{-}54)$$

We also recall that when considering near-surface light conditions, we have to apply $Z(\lambda)$ [and not $W(\lambda)$ values]. For example, for a summer day at 40°N latitude, we may use the $Z(\lambda)$ values given in Table 13.3 to estimate the 24 h average Ox steady-state concentration from the concentration measured at noon by

$$[Ox]_{ss}^0(24\,\text{h}) = [Ox]_{ss}^0(\text{noon})\frac{\sum Z(24\,\text{h}, \lambda)}{86400\sum Z(\text{noon}, \lambda)} \qquad (13\text{-}55)$$

Note that a conversion factor of 86400 has to be introduced to make the two sets of $Z(\lambda)$ values compatible with respect to their conventional units (s^{-1} and d^{-1}, respectively).

In principle, by analogy to the direct photolytic processes, measurements of near-surface steady-state concentrations of photooxidants may be used to estimate average Ox concentrations in a well-mixed water body by applying an (average) light-screening factor (see Eqs. 13-28–13-32) to the near-surface rate of Ox production [and thus to $[Ox]^0_{ss}$; see Eq. 13-51]:

$$[Ox]_{ss} = [Ox]^0_{ss} S(\lambda_{max}) \tag{13-56}$$

However, to apply Eq. 13-56, an appropriate λ_{max} has to be selected; that is, one has to know in which wavelength region maximum Ox production takes place. As illustrated by the following examples, this is not always easy.

Reactions with Singlet Oxygen ($^1O_2$)

As indicated in Figure 13.12, $^1O_2$ is formed primarily by energy transfer from $^3UC^*$ to $^3O_2$. The most important consumption mechanism for $^1O_2$ is physical quenching by water. At DOM concentrations typical for most surface waters (DOC < 20 mg C L^{-1}), quenching of $^1O_2$ by DOM can be neglected (Haag and Hoigné, 1986). Hence, the steady-state concentration of $^1O_2$ in a natural water is directly proportional to the DOM concentration. Note, however, that different types of aquatic DOM may exhibit quite different overall quantum yields for $^1O_2$ production. This is illustrated by Figure 13.13 in which $[^1O_2]^0_{ss}$ values for various lake and river waters are plotted against the DOC concentration. As can be seen from this plot, in waters exhibiting DOC values between 3 and 4 mg L^{-1}, maximum $^1O_2$ steady-state concentrations in the order of 7 to 11×10^{-14} M are detected on a summer day at 47.5°N latitude. Since the variation in $[^1O_2]^0_{ss}$ is broader than the DOC concentration range, and since $^1O_2$ consumption rates (via quenching by water) is the same for all waters, this variation in $[^1O_2]^0_{ss}$ must be due to differences in the light absorbance by the UCs present and/or in the quantum yields for production of $^1O_2$.

The $[^1O_2]^0_{ss}$ values shown in Figure 13.13 were determined by using furfuryl alcohol (FFA, I) as a probe compound (Haag et al., 1984a). Another frequently used "trapping agent" for $^1O_2$ determination is 2,5-dimethylfuran (2,5-DMF, II) (Zepp et al., 1977).

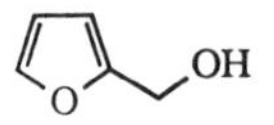

FFA

I

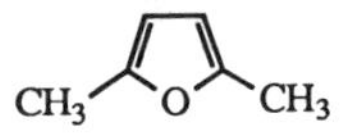

DMF

II

With these compounds (as with other dienes), $^1O_2$ undergoes a so-called *Diels–Alder-reaction* (March, 1985), forming an endoperoxide intermediate that reacts further to yield various products. For example, the reaction of FFA with $^1O_2$ is (Haag et al.,

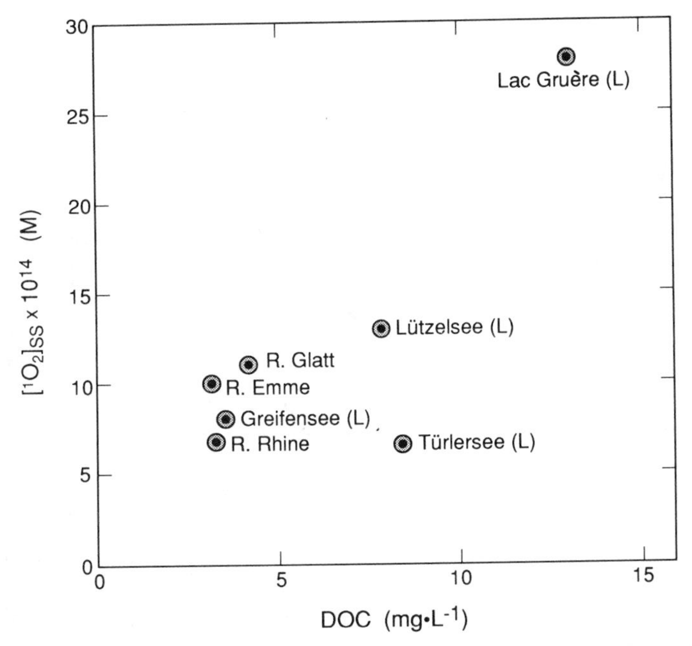

Figure 13.13 Observed $[^1O_2]^0_{ss}$ in water samples from some Swiss rivers (R) and lakes (L) as a function of the dissolved organic carbon (DOC) concentrations of these waters. The results apply for noontime light intensity on a clear summer day at 47.5°N (data from Haag and Hoigné, 1986).

1984a):

$$+\ {}^1O_2 \longrightarrow \qquad (13\text{-}57)$$

85%   <10%   <10%

Besides its properties as a reactant in addition reactions, $^1O_2$ is also a significantly better electron acceptor (oxidant) as compared to $^3O_2$ $\{[^1O_2(1M) + e^- = O_2^{\cdot -}(1M);\ E^0_H(W) = +0.79V]$ (Eberson, 1987) as compared to -0.16 V for $^3O_2$, see Section 12.4$\}$. However, because of its low steady-state concentration in natural waters, it is an important photooxidant only for a few very reactive types of organic compounds. Such compounds include those exhibiting structural moieties that may undergo Diels–Alder reactions, those containing electron-rich double bonds (i.e., double bonds that are substituted with electron-donating groups), or compounds exhibiting functional groups

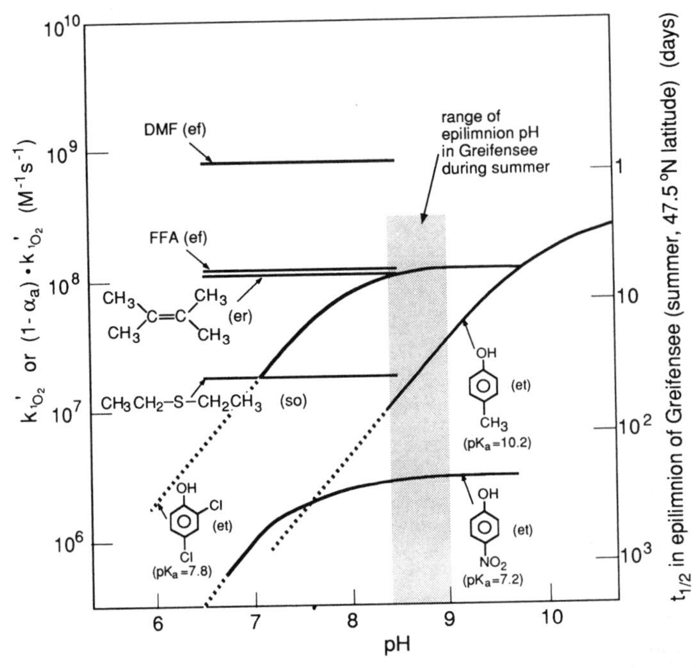

Figure 13.14 Second-order rate constants [multiplied by $(1-\alpha_a)$ for phenols] for reactions of several compounds with $^1O_2$ (left scale) as a function of pH. The abbreviations in parentheses indicate the reaction type: ef = endoperoxide formation (Eq. 13-57); er = ene reaction; et = electron transfer; so = oxidation of sulfide or sulfoxide. The scale on the right indicates the half-lives of the compounds in the epilimnion of Greifensee on a clear summer day (data from Scully and Hoigné, 1987).

that are easily oxidized including reduced sulfur groups (e.g., sulfides, mercaptans), anilines, and phenols (see also Section 12.4). Figure 13.14 summarizes some of the kinetic data available for reactions of organic compounds with $^1O_2$ in water. As can be seen, for phenolic compounds the transformation rate is pH-dependent, since (as we noted in Section 12.4) the phenolate species (A^-) is much more reactive toward oxidation as compared to the neutral phenol (HA) [i.e., $k'_{^1O_2}(HA) \ll k'_{^1O_2}(A^-)$].

Since $^1O_2$ behaves as an *electrophile*, one can assume that electron-donating substituents on an organic compound will, in general, increase its reactivity, while electron-withdrawing substituents will have the opposite effect. In the case of phenolic compounds, the effect(s) of the substituent(s) on the pK_a (see Chapter 8), and thus on the abundance of the reactive phenolate species present at a given pH, may be more important than the effect(s) of the substituent(s) on $k'_{^1O_2}$. In this case, the overall transformation rate is dominated by the rate of transformation of the anionic species (A^-):

$$\text{rate} = -\left(\frac{dC_t}{dt}\right)_{^1O_2} = (1-\alpha_a)k'_{^1O_2}[^1O_2]_{ss}C_t \qquad (13\text{-}58)$$

where C_t is the total phenol concentration ($[A^-]+[HA]$) and $(1-\alpha_a)=[1+10^{(pK_a-pH)}]^{-1}$ is the fraction in the dissociated (anionic) form (see Eq. 8-16). This effect is illustrated in Figure 13.14, where for the phenolic compounds, $(1-\alpha_a)\cdot k'_{^1O_2}$ is plotted as a function of pH.

The right-hand scale in Figure 13.14 gives the calculated half-lives for indirect photolysis involving $^1O_2$ for the various compounds in the well-mixed epilimnion of Greifensee ($z_{mix}=5$ m) on a clear summer day. The half-lives are based on a measured $[^1O_2]^0_{ss}$ (noon) value of 8×10^{-14} M, corresponding to a $[^1O_2]^0_{ss}$ (24 h) value of about 3×10^{-14} M (Eq. 13-55). When assuming that virtually all light is absorbed within the epilimnion {i.e., $S(\lambda_m)\simeq[2.3(1.2)\alpha(\lambda_m)z_{mix}]^{-1}$}, and taking the α value at 410 nm [$\alpha(410\,\text{nm})\simeq0.005\,\text{cm}^{-1}$; see Table 13.6], an average $^1O_2$ concentration $[^1O_2]_{ss}$ (24 h) $=4\times10^{-15}$ M is calculated (Eq. 13-56). The choice of 410 nm is based on the findings by Haag et al. (1984b) that some humic and fulvic materials exhibit a maximum in $^1O_2$ production around this wavelength.

With the $[^1O_2]^0_{ss}$ (24h) value calculated above, the half-life of a compound with respect to photooxidation by $^1O_2$ in the epilimnion of Greifensee is then given by

$$t_{1/2}=\frac{\ln 2}{(1-\alpha_a)\,k'_{^1O_2}\cdot(4\times10^{-15})}\,\text{sec}=\frac{2\times10^{9}}{(1-\alpha_a)k'_{^1O_2}}\,\text{days} \tag{13-59}$$

From the half-lives indicated in Figure 13.14 it can be seen that for most pollutants, the assumption of a well-mixed epilimnion (typical mixing rates 1–10 d^{-1}; see Chapter 15) with respect to indirect photolysis with $^1O_2$ is a reasonable assumption. Furthermore, for compounds exhibiting $k'_{^1O_2}$ values [or $(1-\alpha_a)k'_{^1O_2}$ values for phenolic compounds] greater than $10^7\,M^{-1}\,s^{-1}$, during the summer, photooxidation by $^1O_2$ is equal to or more important than depletion of the concentration by dilution with inflowing water [$t_{1/2}$(dilution) in the epilimnion of Greifensee on the order of 70 days]. We should recall, however, that only a few compound classes exhibit such large $k'_{^1O_2}$ values, and that, therefore, $^1O_2$ must be considered to be a rather selective photo-oxidant.

Reactions with Hydroxyl Radical (HO·)

Because of its high reactivity, direct observation of hydroxyl radicals is very difficult in natural waters. Most of the evidence for the existence of HO· derives from product analysis studies, and from studies of relative photolytic reactivities of a series of compounds. Because HO· reacts with many organic compounds at nearly diffusion-controlled rates (H atom abstraction or hydroxylation, i.e., addition of OH), a great variety of organic substrates may be used as probe molecules. One frequently applied compound is *n*-butyl chloride (Haag and Hoigné 1985), which does not undergo direct photolysis and which is inert against other photoreactants except e_{aq}^- (which is usually present at too low concentrations to be important, Zepp et al., 1987a).

There are several possible mechanisms by which hydroxyl radicals may be formed in surface waters including the photolysis of nitrate, nitrite, H_2O_2, from reactions of excited humic materials and from the reaction of H_2O_2 with Fe(II) (Fenton's reaction,

Cooper et al., 1989). In freshwaters, particularly in waters exhibiting high NO_3^- concentration, the major source for HO· production appears to be the photolysis of nitrate (Zepp et al., 1987b):

$$NO_3^- \xrightarrow{h\nu} NO_3^{-*} \begin{cases} \nearrow NO_2^- + O \\ \searrow NO_2 + \dot{O}^- \xrightarrow{H_2O} HO^{\bullet} + OH^- \end{cases} \qquad (13\text{-}60)$$

Note that the major fate of atomic oxygen (O), which is in a nonexcited state, is likely to be the reaction with $^3O_2$ to form ozone, which is then rapidly consumed by, for example, reaction with NO_2^- (Hoigné et al. 1985) or by decomposition to HO· (Staehelin and Hoigné, 1985).

If the major pathway of HO· production is given by Eq. 13-60, the rate of HO· formation, $k_{f,HO\cdot}$ can be estimated from the specific rate of sunlight absorption of NO_3^- (that may be calculated as we calculated the k_a value of PNAP in Table 13.5) and from the reaction quantum yield (see Eq. 13-49) determined at a specific wavelength. The NO_3^- absorbs light in a rather narrow wavelength range (290–340 nm), with a maximum specific light absorption rate near 320 nm. For example, the maximum near-surface specific rate of light absorption of NO_3^- (i.e., noon, midsummer) at 47.5°N is 1.8×10^{-5} einsteins (mole $NO_3^-)^{-1}\,s^{-1}$. The quantum yield for HO· production measured at 313 nm is 0.017 (Zepp et al. 1987b). Thus, the near-surface rate of HO· production in our model lake (Greifensee) is given by

$$r^0_{f,HO\cdot}(\text{noon}) = (3 \times 10^{-7})[NO_3^-] \qquad (M\,s^{-1}) \qquad (13\text{-}61)$$

The major sink for HO· in natural water is the dissolved organic material (DOM). In waters exhibiting very low DOM concentrations (i.e., [DOC] < 1 mg C L^{-1}) and a high carbonate content (i.e., $[HCO_3^{2-}] > 1$ mM), bicarbonate and carbonate may also become important HO· scavengers (Zafiriou, 1974; Hoigné and Bader, 1978). The second-order rate constants for reaction of HO· with DOM, HCO_3^-, and CO_3^{2-} are $\sim 2.5 \times 10^4$ L (mg C)$^{-1}\,s^{-1}$, $1.5 \times 10^7\,M^{-1}\,s^{-1}$, and $4.2 \times 10^8\,M^{-1}\,s^{-1}$ (Larson and Zepp, 1988). Using Eq. 13-61, we may now calculate the maximum near-surface steady-state HO· concentration in Greifensee (pH = 8.4):

$$[HO\cdot]^0_{ss}(\text{noon}) = \frac{(3 \times 10^{-7})[NO_3^-]}{(1.5 \times 10^7)[HCO_3^-] + (4.2 \times 10^8)[CO_3^{2-}] + (2.5 \times 10^4)[DOC]} \qquad (13\text{-}62)$$

With $NO_3^- = 0.1$ mM, $[HCO_3^-] = 1.2$ mM, $[CO_3^{2-}] = 0.014$ mM, and [DOC] = 4 mg C L^{-1}, we obtain a $[HO\cdot]^0_{ss}$ (noon) value of 2.5×10^{-16} M, corresponding to a 24h averaged $[HO\cdot]^0_{ss}(24\text{h}) \cong 1 \times 10^{-16}$ M. Finally, for calculating the average HO· steady-state concentration in the epilimnion of Greifensee, we need to apply a screening factor $S(\lambda_{max})$(Eq. 13-56). In the case of nitrate photolysis we choose $\lambda_{max} = 320$ nm.

The α value of the water in the epilimnion of Greifensee at this wavelength is α (320 nm) = 0.0275 cm^{-1} (see Table 13.6). This yields an S(320 nm) value of 0.026. Hence, the average steady-state concentration of HO· is the epilimnion of Greifensee on a midsummer day is $[HO\cdot]_{ss}$ (24 h) $\cong 3 \times 10^{-18}$ M. Therefore, despite the very high reactivity of HO· toward many organic pollutants, in Greifensee, indirect photolysis involving HO· will not be a dominant process, as can be easily seen from the following calculation.

For many organic pollutants, the second-order rate constant for reaction with HO· is on the order of 6×10^9 $M^{-1} s^{-1}$ or somewhat smaller (for a compilation of $k'_{HO\cdot}$ values see Farhataziz and Ross, 1977). Using this $k'_{HO\cdot}$ value and the above average HO· steady-state concentration, one obtains a pseudo-first-order rate constant of 1.8×10^{-8} s^{-1}, corresponding to a half-life of 446 days (as compared to 70 days for dilution, see the discussion of $^1O_2$). We should note, however, that in shallow, nitrate-rich waters with small DOM and carbonate concentrations, much higher (factor 100 and more) HO· steady-state concentrations may be obtained, and that, therefore, in such waters, indirect photolysis of organic pollutants involving HO· may be important.

Reactions with Organic Peroxy Radicals (ROO·)

We conclude our discussion on indirect photolysis by one example in which the photoreactant is not well defined, and, therefore, prediction of absolute reaction rates is difficult. In a study concerning the photolytic transformation of alkylated phenols in aqueous solution containing fulvic acid, Faust and Hoigné (1987) found a relationship between the transformation rate and fulvic acid concentration. Since the observed rates of disappearance of the phenols could not be attributed to direct photolysis or to reaction with singlet oxygen, hydroxy radicals or excited DOM species, they postulated that the transient oxidants responsible for the transformation of these phenols could be peroxy radicals (ROO·) that are formed by the reaction of excited DOM chromophores with $^3O_2$. These ROO· species seem not to be scavenged significantly by DOM (at least for DOC concentrations < 5 mg C L^{-1}), and they react predominantly with easily oxidizable compounds including alkyl phenols, aromatic amines, thiophenols, and imines. The most likely reaction mechanism is abstraction of a hydrogen atom H· since peroxy radicals act as electrophilic species. For example, in the initial step, reaction of 2,4,6-trimethyl-phenol (2,4,6-TMP) with ROO· would yield the corresponding phenoxy radical and ROOH:

$$(CH_3)_3C_6H_2\text{–}OH + ROO^\bullet \longrightarrow (CH_3)_3C_6H_2\text{–}O^\bullet + ROOH \qquad (13\text{-}63)$$

Considering the wide variety of DOM chromophores present in a natural water, it is easy to understand that the peroxy radicals formed will cover a wide range of reactivities. Thus, very much akin to the situation that we encountered when discussing chemical reduction reactions involving quinoid-type compounds or transition metal complexes (see Section 12.4), the reactive environmental species are unknown. Since absolute reaction rates are dependent on the type of DOM-derived peroxy radicals present in a given natural water, the only way to arrive at a prediction of transformation rates of organic pollutants involving photolytically produced peroxy radicals in a natural water is to assess first the relative reactivity of a series of compounds in a given model system. Then, by using an appropriate probe compound for which the transformation rate is measured in the system of interest, rough predictions of the reaction rates of related compounds should be possible.

Table 13.8 gives the relative reactivities of various alkylphenols (normalized to phenol) determined in the laboratory in a fulvic acid solution (4.1 mg C L^{-1}) using a mercury lamp as the light source (for details see Faust and Hoigné, 1987). Also given in Table 13.8 are the estimated 24-h averaged half-lives of the compounds in the well-mixed epilimnion of Greifensee for a clear day in June. These half-lives are based on measurements by Faust and Hoigné (1987) using 2,4,6-trimethylphenol as probe compound. Furthermore, for calculating the depth-averaged half-lives, a screening factor $S(366 \text{ nm}) \cong 0.066$ has been applied.

From the data in Table 13.8, it can be seen that with increasing number of alkyl substituents, the reactivity of a given phenol increases, which is consistent with our expectation that electron-donating substituents facilitate oxidation by electrophilic peroxy radicals. An interesting observation can be made when comparing the relative reactivities of the various 4-alkylphenols. As can be seen, the reactivity decreases with increasing number of carbon atoms, that is, with increasing hydrophobicity of the compound. An explanation offered by Faust and Hoigné (1987) for these findings is that because of hydrophobic interactions of the alkyl group with parts of the DOM, the probability of encounters of the phenol with the more polar peroxy radicals decreases. Nevertheless, as the half-lives estimated for the various phenols in the epilimnion of Greifensee indicate, reaction with peroxy radical species may be an

TABLE 13.8 Relative Reactivity of a Series of Alkylphenols (Normalized to Phenol) and Estimated Half-Lives for Reaction with ROO· in the Epilimnion of Greifensee for Clear Summer Day Conditions[a]

Compound	Reactivity Relative to Phenol	Estimated Half-life in the Epilimnion of Greifensee[b] (days)
Phenol	1	200
4-Nonylphenol	2.6	78
4-Isopropylphenol	5.3	38
2-Methylphenol	7.3	27
4-Ethylphenol	8.9	22
4-Methylphenol	12.4	16
2,6-Dimethylphenol	23.8	8
2,4,6-Trimethylphenol	40.6	5

[a] Derived from data reported by Faust and Hoigne', 1987.

[b] 24 h-averaged light intensity on a clear day in June at 47°N latitude.

important transformation pathway, at least for compounds that are as reactive as the di- and trimethyl phenols. Such compounds are typically used as antioxidants, and are known to find their way into natural waters (e.g., Lopez-Avila and Hites, 1981).

13.5 EFFECTS OF PARTICLES ON PHOTOLYTIC TRANSFORMATIONS OF ORGANIC COMPOUNDS IN NATURAL WATERS

We conclude this chapter by addressing briefly the effects of particles on photolytic transformations of organic pollutants in natural waters. As we have already mentioned, particles may contribute to the light attenuation in a water body, both by light absorption and light scattering. Depending on which effect is predominant, the rate of direct photolysis of a *dissolved* species in a given system may be decreased or enhanced. In most cases, a decrease in direct photolysis rates of dissolved organic pollutants is observed with increasing particle concentrations, indicating that light absorption is the more important factor (Miller and Zepp, 1979b). In some clay suspensions, however, Miller and Zepp (1979b) observed an increase in the photolysis rate of a ketone which was attributed to increases in the mean light path length caused by scattering.

A more complicated issue is the photolytic transformation of sorbed compounds. Predictions of direct photolysis rates of organic pollutants sorbed to a solid surface are impeded by the fact that a compound may be shielded from the light. In addition, the uv/vis absorption spectrum of a given compound may be significantly different in the sorbed state as compared to the dissolved state (e.g., Parlar, 1980). Similar effects may be observed when a (hydrophobic) pollutant is associated with dissolved or colloidal humic and fulvic material. Furthermore, owing to the different molecular environment, sorbed species may exhibit very different reaction quantum yields and photoreaction distributions. For example, kinetic and product studies of photoreactions of some highly hydrophobic nonionic organic chemicals indicated that the compounds were in a microenvironment that was less polar than water and that was a considerably better hydrogen donor (Miller and Zepp, 1979b). This is consistent with our picture that hydrophobic compounds preferably sorb in organic components of natural solids (see Chapter 11).

In addition to direct heterogeneous photoreactions of organic pollutants, indirect photoprocesses involving solid surfaces may also be important. Such processes may involve reactive intermediates produced by surface photolysis, particularly at materials possessing semiconductor properties, or reactions with electronically excited surface species (quite similar to excited humic or fulvic acid species in heterogeneous solution). Although the role of semiconductors appears to be not too significant for transformations of organic pollutants in natural waters, indirect photolysis of sorbed compounds may be important (Zepp and Wolfe, 1987). For example, the rate of photolytic transformation of a series of alkylated anilines was found to be significantly accelerated by algae (Zepp and Wolfe, 1987). The rate constants for these photoreactions were virtually unaffected by heat killing of the cells or by addition of a metabolic inhibitor. The accelerating effect increased with increasing hydrophobicity of the compound, indicating that sorption of the compound was rate determining.

In summary, solids may have various effects on the photolytic transformation of organic compounds in the aquatic environment, both by influencing the light attenuation in a water body and by affecting the reactivity of sorbed species. Because of the complexity of these processes and the scarcity of studies dealing with heterogeneous phototransformations of organic compounds in natural waters, quantification of such processes is so far not possible.

CHAPTER 14

BIOLOGICAL TRANSFORMATION REACTIONS

14.1 INTRODUCTION

Another set of transformations that remove organic compounds from the environment is that group of reactions mediated by organisms. As for chemical and photochemical reactions, these biochemical processes change the structure of the organic chemical of interest, thereby removing that particular compound from an environmental system of interest. The resulting one or more products exhibit their own properties, reactivities, fates, and effects. We should note that when we speak of biologically mediated transformations of organic compounds, we are not necessarily implying that they are fully mineralized. *Mineralization* involves the complete degradation of an organic chemical to stable inorganic forms of C, H, N, P, and so on; consequently, mineralization generally entails several successive biological transformations to complete. As a result, experimental observations relying solely on the appearance of CO_2 (e.g., ^{14}C-labelled CO_2) to quantify the loss rate of a specific chemical only place a lower limit on the initial transformation rate of this chemical [$dCO_2/dt \leqslant -d(\text{organic compound})/dt$].

Biochemical transformations of organic compounds are especially important because many reactions, although thermodynamically feasible, occur extremely slowly because of kinetic limitations. Organisms enable such reactions to proceed via two important approaches. The first approach involves the use of special proteins, called *enzymes*, that serve as catalysts. These reusable "biological tools" facilitate educt interactions and thereby lower the free energy of activation that determines the transformation rate (Fig. 14.1). Enzymes can lower the activation energy of reactions by several tens of kilojoules per mole, thereby speeding the transformations by factors of 10^9 or more. Some of this reaction rate enhancement arises because enzymes complex with the

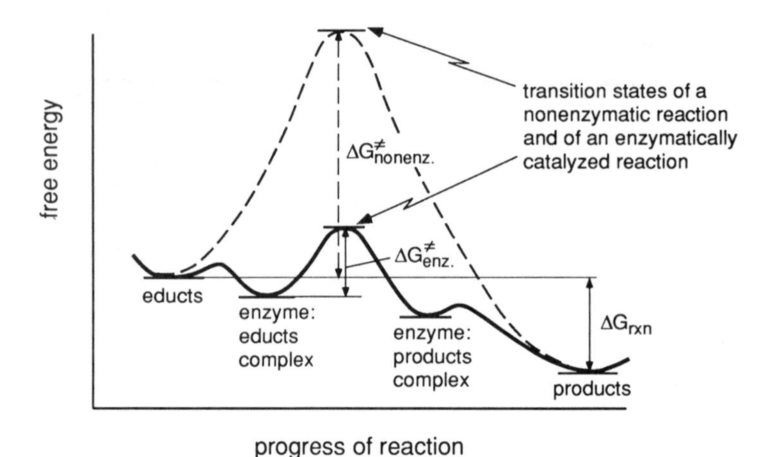

Figure 14.1 Schematic representation of the change in activation energy barriers for an enzymatically mediated reaction as compared to the analogous noncatalyzed chemical reaction.

educts, and by holding the reacting compounds in an advantageous orientation with respect to one another, they reduce entropic limitations to educt interaction. Secondly, organisms may invest energy to convert educts of interest into more reactive species. For example, molecular oxygen (O_2) is converted to a more reactive species by a biochemical reducing agent before it is used to oxidize hydrocarbons (see Section 14.3). This scheme is similar to one previously discussed in photochemical transformations where, by absorption of light, activated species are formed that are much more reactive (Chapter 13).

In this chapter we focus on microorganisms as the most ubiquitous agents of biochemical processes that determine the fate of organic chemicals in the environment. We first describe a few general principles pertaining to the ecology and enzymatic adaptability of microorganisms, especially insofar as these characteristics influence what we can expect these organisms to do. Next, we focus on three major strategies microorganisms use to initiate xenobiotic compound transformations: oxidations, reductions, and hydrolyses. In this section, we look closer at the mechanisms of these biochemical transformation reactions to develop a better understanding of what structural features of organic chemicals are susceptible to microbial attack. These relatively detailed descriptions also provide some basis for mathematical formulations with which rates of such processes may be described.

Finally, we discuss how we might approach predicting the overall rate of any particular biodegradation. Unfortunately, several very different steps in the overall process may end up controlling the overall rate (Fig. 14.2); by picturing the whole sequence involved here, we may recognize the relevance of various discussions as they appear in this chapter. For example, processes can be limiting that control the rate of

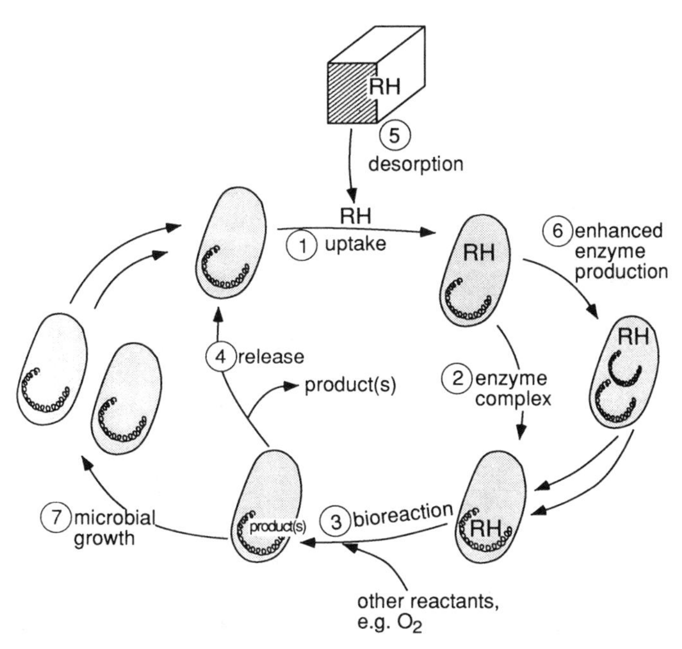

Figure 14.2 Sequence of events in the overall process of biotransformations: (1) bacterial cell containing enzymes takes up organic chemical RH, (2) RH binds to suitable enzyme, (3) enzyme: RH complex reacts producing the transformation product(s) of RH, and (4) the product(s) is(are) released from the enzyme. Several additional processes may influence the overall rate such as: (5) desorption of RH from solids making it available to the microorganisms, (6) production of new or additional enzyme capacity (e.g., due to turning on genes {induction}, due to removing materials which prevent enzyme operation {activation}, or due to acquisition of new genetic capabilities via mutation or plasmid transfer), and (7) growth of the total microbial population carrying out the biotransformation of RH.

delivery of substrate molecules (RH or O_2 as examples in Fig. 14.2) to microorganisms. This effect is often cited as the reason why lower solubility aromatic hydrocarbons are oxidized more slowly in the environment than their more soluble congeners. Mass transport has also been recognized as important for special situations in which microorganisms exist in such abundance that they form associations where some cells are buried beneath many others as in biofilms (Rittman et al., 1980; Wanner and Gujer, 1986). In these cases, it is frequently the rate of molecular diffusion of the chemical itself or of necessary cosubstrates like molecular oxygen across intervening media (e.g., through polymeric microbial secretions) that dictates the overall rate of xenobiotic chemical biotransformation. Obviously, one must know the average distribution of cells and other media, the tendencies of chemicals to partition between phases (e.g., solid–water partitioning), and the abilities of these molecules to move in each phase (e.g., by diffusion) to describe mathematically these transport processes. Probably, this is most critical whenever we are dealing with situations where most of a chemical of concern is sorbed, such as encountered in sediments, soils, and aquifers for very hydrophobic compounds.

Returning to Figure 14.2, we imagine other potential limitations to biological transformations of organic compounds. First, the rate of organic compound uptake by the isolated cells may be rate limiting. Since the chemicals of interest to us are often foreign to the microorganisms, in general the cells may not have systems associated with their exterior membranes to pick up the substance *actively* and rapidly carry it to the interior of the cell. However, for nonpolar substances, the lipid-rich cell membranes, designed to separate charged metabolites within the cell from the ionic aqueous medium on the outside, allow hydrophobic chemicals to dissolve in these cellular boundaries (recall our discussions of octanol–water partitioning, Chapter 7). Thus we see what is called *passive* uptake as nonpolar species diffuse through the membranes to the interior of microbial cells. Conceivably, this uptake rate may limit the overall biotransformations of interest. Castro et al. (1985) interpreted the relative rates found for reduction of a series of organohalides by cultures of *Pseudomonas putida* as reflecting permeability limitations. In such cases, the implicit assumption is that once the chemical arrives at the cellular site of enzymatic processing, it is rapidly transformed, so the overall chemical depletion rate is governed by the rate of substrate permeation into the cells. If permeation is the bottleneck, organic compound structural features affecting this uptake rate will be most important. Note that it is also possible that intracellular transport could further limit biotransformation rates, if the degrading enzymes were localized in subcellular regions or organelles (e.g., mitochondria in eucaryotic microorganisms like fungi).

Once the chemical and enzymes coexist, however, other processes still remain which may govern the overall biodegradation (Fig. 14.2). First, the presence of the chemical may cause the organism to produce more enzyme units suited to degrading this substance. Such an increase in useful enzymes may be due to: (1) *induction* causing genes to be turned on for enzyme manufacture, (2) *derepression* causing existing enzymes to be enabled for their catalytic roles, or even (3) mutation causing new genetic codes to arise which luckily alter a resultant enzyme to operate more effectively. Obviously, if greater quantities or more effective enzymes result, the overall internal processing of the chemical will speed up. On the other hand, the requisite enzymatic apparatus may always be in place (called *constitutive*), for example, in response to the presence of like metabolites. In this case, the rate now may depend on the specific interactions of the compound with the enzyme: both in so far as they associate with one another and then as they become involved in bond breaking and making. These interactions are best quantified using *Michaelis–Menten and related types of enzyme kinetics*, which we discuss further below.

Lastly, if the metabolism of the chemical results in substantial energy yield and/or cell-building materials, then the microorganism may increase in cell numbers in response. Consequently, the additional microorganisms will cause the system degradation rate of the substance to increase. Overall, the rate of biodegradation will be dictated by the rate of microbial population increase. In these cases, *microbial population dynamics* need to be discussed which can be done by an approach developed by Monod (1949). This too will be expanded upon below.

In sum, we can imagine that biotransformations can be limited by: (1) delivery of the chemical to the organisms' metabolic apparatus capable of transforming the chemical, (2) the enzyme's ability to mediate the initial transformation of the chemical, or (3) the growth of a population of microorganisms in response to the presence of a new substrate. Depending on what limits the rate of biotransformation, different mathematical frameworks are required to describe the kinetics of the process both with respect to the nature of the equations and the parameters they require.

This chapter does not attempt to review the immense literature on biodegradation rates, nor does it deal with the primary metabolic pathways that are the focus of general biochemistry texts, (e.g., Lehninger, 1970; Mahler and Cordes, 1971). We also do not describe much about microbial ecology, since excellent texts are available (Stanier et al., 1976; Gottschalk, 1986). Rather, by treating the issue from an organic chemist's point of view we believe useful generalities result, deducible by considering the structure of the chemical of interest. Thus, although biotransformation rates are probably the least well understood inputs to chemical fate modeling, we may use these generalities to better anticipate, or at least understand after the fact, biologically mediated transformations in environmental systems.

14.2 SOME IMPORTANT CONCEPTS ABOUT MICROORGANISMS

Issues of Microbial Ecology and Interaction

Although plants and animals can transform xenobiotic organic chemicals, from a quantitative point of view, microorganisms are the most important organisms in determining the degradation of organic compounds in the environment. Microorganisms are comprised of many forms including bacteria, protozoans, fungi, and microalgae (e.g., Stanier et al., 1976), and they are present virtually everywhere in nature, even under extreme conditions of temperature, pressure, pH, salinity, oxygen, nutrients, and even low water content.

However, the types of microorganisms present are dependent on the conditions of their environment. Since these organisms are competing with one another, both dramatic environment-to-environment differences such as oxic versus anoxic conditions, and even subtle ones like high versus low trace metal activities, affect the particular mixture of species which preside at any one place and time. Fortunately with respect to biotransformations of organic compounds, subtle effects may not be too important since many organisms exhibit similar biochemical pathways and capabilities. However, important environmental factors, such as whether oxygen is present or not, have an extremely important influence on the combination of microorganisms living at a site (recall Section 12.4). Such variations in the species present are known to cause major differences in metabolic capabilities. As an example, we cite the biodegradation of hydrocarbons in oxic settings; without oxygen present, these compounds are much more persistent (Fig. 14.3). A complementary example would be the much more efficient transformation of some chlorinated solvents under highly reducing conditions as compared to oxic situations where many of these compounds are very persistent (Bouwer and McCarty, 1983b; Wilson and Wilson, 1985; Fogel et al., 1986). Thus we

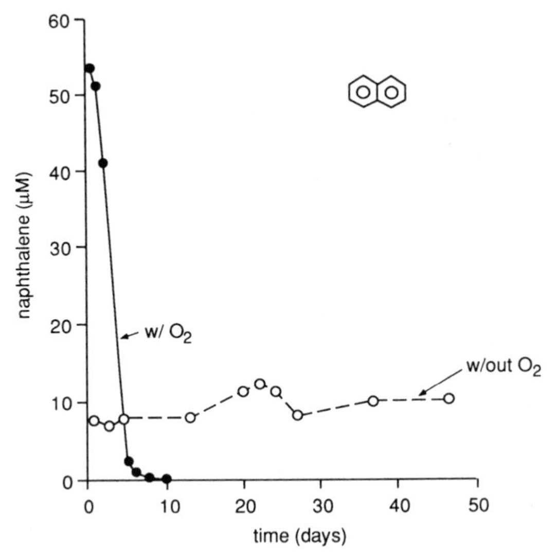

Figure 14.3 Variation in timecourses of naphthalene degradation by microorganisms in laboratory soil–water incubations with molecular oxygen present (●) or no molecular oxygen present (○) (data from Mihelcic and Luthy, 1988).

should not be surprised if environmental factors like temperature, pH, ionic strength, and oxygen concentration not only influence the rates of biologically mediated transformations, but sometimes dictate whether these processes operate at all within the time frame considered, since these factors affect the composition, growth rate, and enzymatic processes of the microbial community (Stanier et al., 1976).

Another important aspect to consider when dealing with biochemical transformations is how microorganisms cooperate with one another. It has long been recognized that cooperating species, termed *consortia*, may be required to execute a particular sequence of transformations (Gray and Thornton, 1928). For example, 3-chlorobenzoate is degraded by a consortium of bacteria (Dolfing and Tiedje, 1986). The first bacterium removes the chlorine from the aromatic ring:

$$\underset{\text{3-chlorobenzoate}}{ClC_6H_4COO^{\ominus}} \xrightarrow{2\,e^- \;+\; H^{\oplus}} \underset{\text{benzoate}}{C_6H_5COO^{\ominus}} + Cl^{\ominus} \tag{14-1}$$

The degradation of the benzoate is continued by another bacterial species:

$$\underset{\text{benzoate}}{C_6H_5COO^{\ominus}} + 5\,O_2 \longrightarrow \longrightarrow \longrightarrow \underset{\text{acetate}}{H_3C\text{-}COO^{\ominus}} + 5\,CO_2 + H_2 \tag{14-2}$$

The H_2 produced by this second bacterium is subsequently used by the first microorganism. Finally, methanogenic microorganisms utilize the acetate releasing methane:

$$H_3C{-}C(=O){-}O^{\ominus} + H^{\oplus} \longrightarrow \longrightarrow \longrightarrow CH_4 + CO_2 \qquad (14\text{-}3)$$

acetate methane

This example illustrates how cooperation of different bacteria, each exhibiting specific enzymatic capabilities, may enable the mineralization of a compound that would not have been performed by a single organism. Another case has been reported in which seven different microorganisms were needed to process the herbicide Dalapon (2,2-dichloro-proprionic acid; Senior et al., 1976). As a result, if certain species are absent or inactive, overall biochemical pathways can be shut down.

Another recognized means by which microorganisms interact involves the exchange between species of genetic material called *plasmids* (Gottschalk, 1986; Boyle, 1989; Chaudhry and Chapalamadugu, 1991). These relatively short lengths of DNA have been found to code for a variety of degradative enzymes; and, unlike the "normal" DNA which is passed from mother to daughter during cell multiplication, plasmid DNA moves "horizontally" between nonprogeny organisms. Thus, such exchange allows additional metabolic tools or combinations of tools to be acquired and utilized in a particular microorganism species. If the resulting metabolic capability proves beneficial, for example, by providing a new energy or carbon source or by eliminating a critical toxicant, the newly enhanced microorganism will maintain the genetic code and the derived enzymatic facility. However, should the extra tools prove not very worthwhile (i.e., not helping confer competitive advantage), then the newly acquired approaches will probably be lost. Through an amazing (but poorly understood) balance of metabolic and genetic cooperation, the microbial communities in the environment seem to work to maximize their ability to live there. Since the introduction of organic chemicals to these sites could present new nutritional opportunities or toxic threats, it is obvious that microorganisms and their community structure may change in response to unfamiliar compounds in their milieu.

Issues of Enzymology

In light of these insights, some comments are needed on principles controlling the types of enzymatic tools which microorganisms maintain. First, since organisms evolved to deal optimally with particular chemicals like amino acids, sugars, and fatty acids which are very common and necessary for growth and metabolism, we should expect most of their enzymatic apparatus is suited to performing the metabolic job of transforming these compounds. The corollary of this point is that recently invented chemicals, which have structures that differ from those usually processed by organisms and therefore are called *xeno*biotic compounds (*xeno*: from the Greek word for stranger), are often not met with a ready and abundant arsenal of suitable enzymatic tools when micro-organisms first encounter them. We too have experienced this phenomena: our own catabolic capabilities are specialized for handling L-amino acids; and thus synthetic amino acids made of the same atoms linked together in the same way but in the mirror image or D-form can be used as food additives to stimulate our taste buds but not add to our caloric intake! The expectation then is that structurally unusual chemicals will be somewhat refractory in the environment with respect to microbial transformations.

R, COOH, H, NH_2 — D-amino acid | HOOC, R, H_2N, H — L-amino acid

A second key principle is that many enzymes exhibit imperfect substrate specificity. By this we mean that enzymes, although "designed" for binding and inducing reactions of particular chemicals, may also have some ability to bind and induce reactions in structurally similar compounds. This is a well-known principle to biochemists who use chemicals called *competitive inhibitors* to block the active sites of enzymes (Walsh, 1979); these compounds are sufficiently like the enzyme's appropriate substrate to bind, but may be somewhat or even completely unreactive. A very interesting example of such enzyme inhibition appears to occur during the microbial dehalogenation of 3-chlorobenzoate (Suflita et al., 1983). In this case, 3,5-dichlorobenzoate is initially transformed to 3-chlorobenzoate:

$$\text{3,5-dichlorobenzoate} \xrightarrow{+H^{\oplus} + 2e^-} Cl^{\ominus} + \text{3-chlorobenzoate} \xrightarrow{+H^{\oplus} + 2e^-} Cl^{\ominus} + \text{benzoate} \qquad (14\text{-}4)$$

Subsequent degradation of the 3-chlorobenzoate does not proceed until most of the 3,5-dichlorobenzoate is transformed. A plausible explanation for this finding is that the dichloro aromatic substrate competes with the monochloro compound for the same enzyme active site. As a result, 3,5-dichlorobenzoate acts as a competitive inhibitor of the biochemical removal of 3-chlorobenzoate. Once the dichloro compound is fully degraded, the 3-chlorobenzoate transformation to benzoate proceeds. This result shows that the rates of a particular chemical's biotransformation may be a function of unanticipated factors such as the presence of competing substrates.

The phenomenon of imperfect enzyme specifity also suggests that chemicals not usually included in the mixture of regularly metabolized substrates, but exhibiting some structural part very like those substrates, may undergo some biotransformation, albeit possibly at a reduced overall effectiveness. This would occur due to the action of enzymes present for processing more common chemicals. For example, 5-phenyl-pentanoic acid may become involved with the enzymatic apparatus designed to handle fatty acids (Fig. 14.4, Lehninger, 1970). Thus imperfect enzyme specificity suggests that

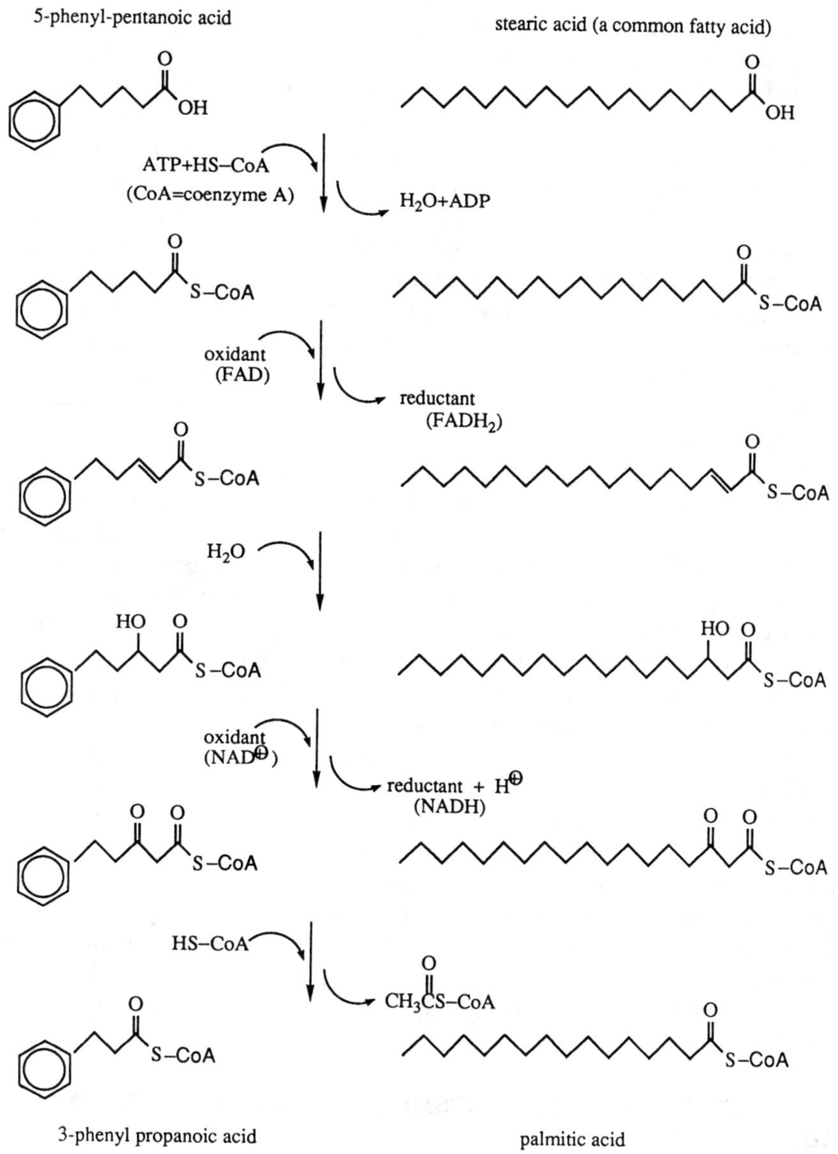

5-phenyl-pentanoic acid

stearic acid (a common fatty acid)

ATP+HS–CoA
(CoA=coenzyme A)

H_2O+ADP

S–CoA

oxidant
(FAD)

reductant
($FADH_2$)

H_2O

HO O

oxidant
($NAD^{\oplus}$)

reductant + $H^{\oplus}$
(NADH)

HS–CoA

CH_3CS–CoA

3-phenyl propanoic acid

palmitic acid

Figure 14.4 Parallel transformation pathways of a substituted benzene, 5-phenylpentanoic acid, and a fatty acid, stearic acid (Lehninger, 1970).

a certain level of biodegradation may occur for some xenobiotic chemicals with structural similarities to natural products.

Now we note a third generality regarding the metabolic aptitude of organisms. Organisms always seem to have some relatively nonspecific enzymes available just for the purpose of attacking unexpected or unwanted compounds. This may be analogous to our carrying a Swiss Army knife readily available to pry, hammer, pick at, slice, uncork, punch, and tweeze as the occasion presents itself. Organisms have long been bombarded with chemicals made by other species, and consequently, they have always needed to eliminate some of this chemical noise. It seems the strategy for many bacteria often entails an oxidative initial step, converting the insulting chemical signal into something more polar. The resultant product may now fit into the common metabolic pathways; or since it is more water soluble, it may be returned to the environment. The lack of substrate specificity designed into this enzyme capability, concomitantly results in its not being especially abundant in most organisms. From the organism's point of view, it would simply be energetically too expensive to maintain a high concentration of this enzyme (like carrying *ten* Swiss Army knives instead of one!). Nonetheless, this principle implies that, at least at some low rate, most organic chemicals should be slowly degradable if such enzymes can get at them together with suitable cosubstrates (e.g., O_2).

Finally, we should recognize that not all of the enzymes which organisms are genetically capable of producing are always present and at constant cellular concentrations (i.e., they are not *constitutive* or essential). In response to a new stimulus, such as the introduction of an organic compound, organisms can turn on the production of appropriate enzymes. Those enzymes are referred to as *inducible*. This process of gearing up for a particular metabolic activity may make the description of the rate of a pollutant's biotransformation somewhat complicated because a time of no apparent activity, or a *lag period*, would be seen. There are other possible reasons for lag periods including: (1) already available enzymes are "repressed" (made to be ineffective) and some time must pass or some condition must change before they begin to act, (2) some time may be required for a few bacteria to multiply to significant numbers (see Section 14.4), (3) an interval may be necessary for mutations in genetic codes to enable development of enzymes able to perform new or more efficient transformations, (4) plasmid transfers may be required to allow existing microorganisms to develop or combine suitable enzymatic tools, or (5) it may simply be that particular species of microorganisms must "immigrate" to the environmental region of interest. One or more of these causes of delay may make it difficult to predict how fast a chemical in a particular environmental setting will undergo even the initial step of biodegradation.

14.3 STATEGIES OF MICROORGANISMS TO INITIATE METABOLISM OF XENOBIOTIC COMPOUNDS

Although organisms are capable of executing a wide array of chemical transformations, there are three types of reactions that microorganisms frequently use to *initiate the breakdown* of xenobiotic compounds. These are: (1) oxidation using an electrophilic form of oxygen, (2) reduction by a nucleophilic form of hydrogen or by direct electron delivery, and (3) hydrolysis via an enzymatically mediated nucleophilic attack. In the first two approaches, organisms invest biochemical energy to form very reactive species that can interact with the organic compounds via mechanisms not operating in abiotic dark environments. The third approach entails catalysis of hydrolytic pathways.

The goal of these initial reactions is to transform the xenobiotic compound into a product(s) that is structurally more similar to chemicals with which microorgnisms are used to metabolizing. As a result, after one or just a few initial transformations, it is common for the resulting chemical product(s) to be included in the more common metabolic pathways and be fully degraded. A good example of this is the bacterial degradation of substituted benzenes (Table 14.1). Initially these aromatic hydrocarbons are oxidized to catechol (*ortho*-dihydroxybenzene) or its derivatives by a dioxygenase and a dehydrogenase:

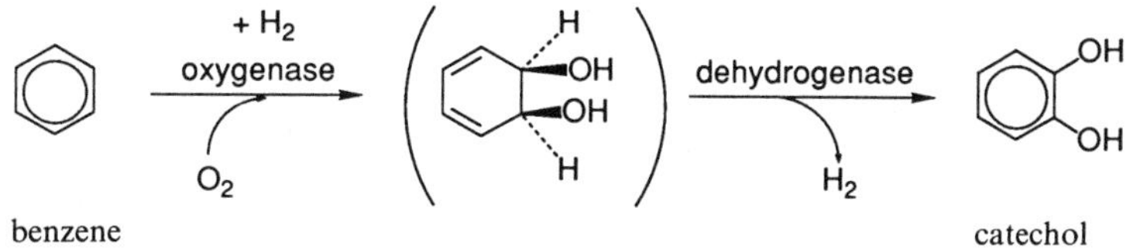

benzene —($+ H_2$, O_2; oxygenase)→ [cis-dihydrodiol] —(dehydrogenase; H_2)→ catechol (14-5)

Catechol and its derivatives are also produced in the metabolism of numerous natural aromatic compounds, like salicylate or vanillate:

salicylate vanillate

As a result, in many microoganisms, pathways are available for processing dihydroxybenzenes that are formed. Thus after the initial oxidation of the aromatic ring, continued breakdown proceeds via enzymatic pathways which open the ring between the two hydroxyl substituents (ortho cleavage) or adjacent to them (meta cleavage), and then convert the resulting chain compounds to small, potentially useful, metabolites (Table 14.1).

Some exceptions exist to this tendency for ready incorporation of the initial transformation products of xenobiotic compounds into a common pathway. First, occasionally a product is formed which is unreactive in subsequent steps. Such partially degraded compounds have been referred to as *dead-end metabolites* (Knackmuss, 1981). An example of this is the 5-chloro-2-hydroxymuconic acid semialdehyde produced by the meta cleavage of 4-chlorocatechol by a pseudomonad:

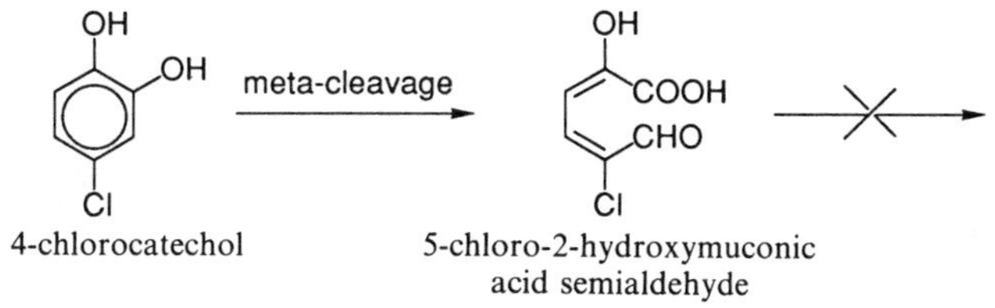

4-chlorocatechol —(meta-cleavage)→ 5-chloro-2-hydroxymuconic acid semialdehyde —✕→ (14-16)

Apparently, the presence of the chloro substituent blocks the next reaction, which

TABLE 14.1 Aerobic Degradation of Substituted Benzenes Proceeding through Catechol or Catechol-like Intermediates

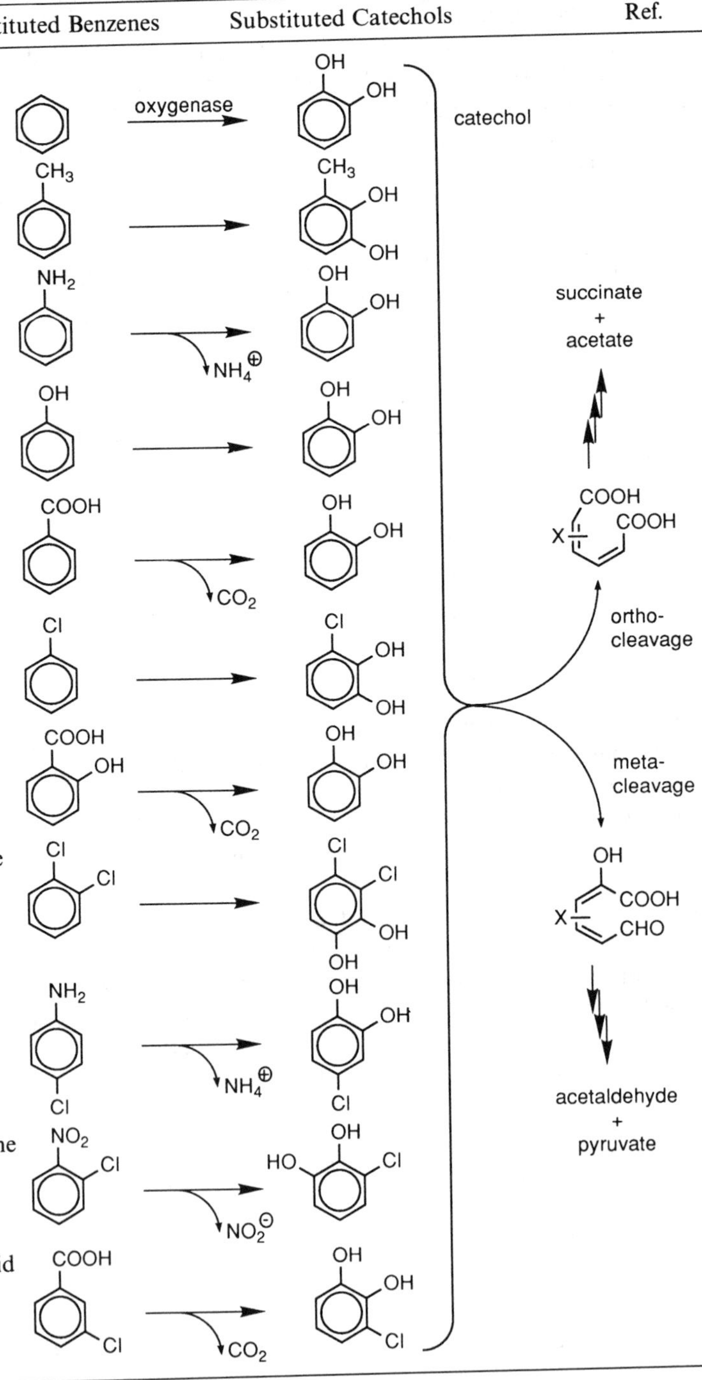

Equation	Substituted Benzenes		Substituted Catechols	Ref.
Benzene (14-5)		oxygenase →	OH, OH	1
Toluene (14-6)	CH_3	→	CH_3, OH, OH	2
Aniline (14-7)	NH_2	→ $NH_4^{\oplus}$	OH, OH	3
Phenol (14-8)	OH	→	OH, OH	4
Benzoic acid (14-9)	COOH	→ CO_2	OH, OH	5
Chlorobenzene (14-10)	Cl	→	Cl, OH, OH	6
Salicylic acid (14-11)	COOH, OH	→ CO_2	OH, OH	7
1,2-Dichloro benzene (14-12)	Cl, Cl	→	Cl, Cl, OH, OH	8
4-Chloroaniline (14-13)	NH_2, Cl	→ $NH_4^{\oplus}$	OH, OH, Cl	9
2-Chloronitrobenzene (14-14)	NO_2, Cl	→ $NO_2^{\ominus}$	OH, HO, Cl	10
3-Chlorobenzoic acid (14-15)	COOH, Cl	→ CO_2	OH, OH, Cl	11

catechol → ortho-cleavage → X, COOH, COOH → succinate + acetate

catechol → meta-cleavage → OH, X, COOH, CHO → acetaldehyde + pyruvate

References: 1. Gibson et al., 1968. 2. Gibson et al., 1970. 3. Bachofer et al., 1975. 4. Stanier and Orston, 1973. 5. Reiner and Hegeman, 1971; Reiner, 1971. 6. Reineke and Knackmuss, 1984. 7. Stanier and Orston, 1973. 8. Haigler et al., 1988. 9. Zeyer et al., 1985. 10. Zeyer and Kocher, 1988. 11. Reineke and Knackmuss, 1988.

normally operates on 2-hydroxymuconic acid semialdehyde to produce 2-oxo-pent-4-enoic acid:

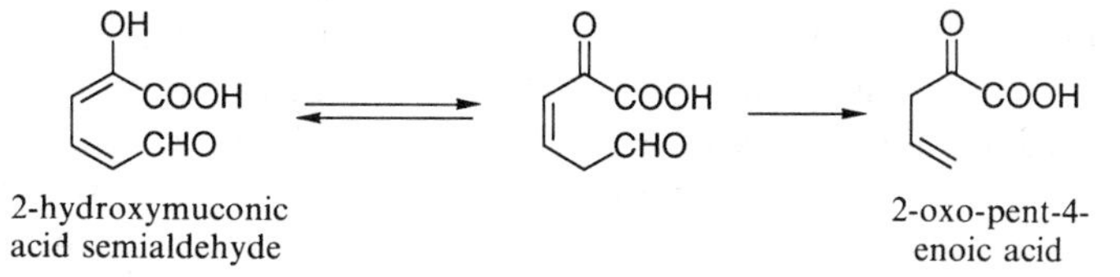

2-hydroxymuconic acid semialdehyde ⇌ → 2-oxo-pent-4-enoic acid (14-17)

If a meta-cleavage pathway is the only one available to a particular microorganism, then the 5-chloro-2-hydroxymuconic acid semialdehyde will accumulate unless another "initial" biotransformation is performed on it.

Another exception to the tendency for initial biotransformation products to be readily directed into subsequent steps in common metabolic pathways involves the production of so-called *suicide metabolites* (Knackmuss, 1981). These chemicals result when the biological transformation yields a product which subsequently attacks the enzymes producing or degrading it. If this attack debilitates one of these enzymes, the successful operation of the relevant metabolic pathway is stopped. An example of this type of problem is the production of acylhalides from 3-halocatechols:

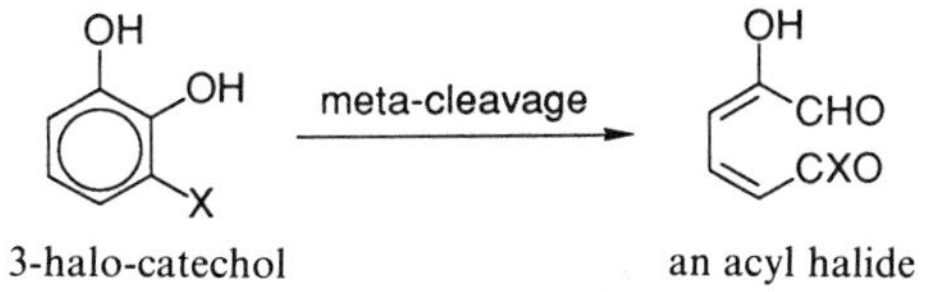

3-halo-catechol —meta-cleavage→ an acyl halide (14-18)

Such acylhalides react rapidly with nucleophiles, and consequently these compounds may bind to nucleophilic moieties (e.g., —SH) of the enzyme near the site of their initial production:

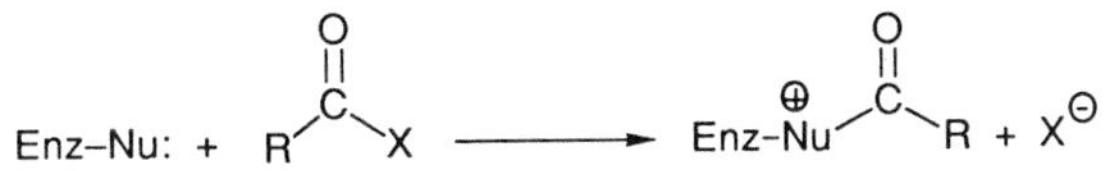

$$\text{Enz–Nu:} + \text{R–C(=O)–X} \longrightarrow \text{Enz–Nu}^{\oplus}\text{–C(=O)–R} + \text{X}^{\ominus} \qquad (14\text{-}19)$$

The resulting change in the enzyme's structure may be sufficient to prevent its continued operation, such as is seen for the *meta*-dioxygenase which formed an acylhalide from 3-halo-catechol. Thus, if biodegradation of the original compound (e.g., 3-halo-catechol) is to proceed, another metabolic approach must be utilized.

Despite these potential problems, many xenobiotic compounds are successfully biotransformed, especially after an initial enzymatic attack is made. Thus our major focus in the discussions below is on the first transformation step(s). We will pay particular attention to some of the details of these enzymatic reactions in order to anticipate where in a molecule they will occur and to provide a basis for understanding how mathematical expressions describing the associated rates should be derived.

Oxidations Involving O_2

We begin by considering microbially mediated oxidations of xenobiotic compounds in the presence of molecular oxygen. After considering a variety of empirical obervations (Table 14.2), we recognize a pattern regarding where the initial point of attack in a molecule is located: the moiety with the most readily available electrons. Often these are π electron systems of aromatic rings and double bonds. In other structures, these are the nonbonded electrons of sulfur or nitrogen. In the absence of these structural elements, oxidation can involve the σ electrons of carbon–hydrogen bonds (Ullrich, 1972; White and Coon, 1980). To explain these biochemical transformations, we recognize that organisms could achieve them by using enzymes carrying an electrophilic form of oxygen that reacts with electron-rich centers in the organic molecule:

dioxygenase: δ- O=O δ+ + C=C ⟶ dioxygenase + O–C / O–C (dioxetane) —reduction→→ HO–C–C–OH, a diol (14-26)

monooxygenase(Fe^{III}):[Ö]$^{\oplus}$ + C=C ⟶ monooxygenase(Fe^{III}) + C–C with bridging O, an epoxide (14-27)

monooxygenase(Fe^{III}):[Ö]$^{\oplus}$ + R_1–S̈–R_2 ⟶ monooxygenase(Fe^{III}) + R_1–S(=O)–R_2, a sulfoxide (14-28)

monooxygenase(Fe^{III}):[Ö]$^{\oplus}$ + R_1–C(H)(R_2)–R_3 ⟶ monooxygenase(Fe^{III}) + R_1–C(OH)(R_2)–R_3 (14-29)

As indicated in these schematic reactions, if both atoms of oxygen from O_2 are transferred, the enzymes are called *dioxygenases*; *monooxygenases*, in contrast, only deliver a single oxygen atom. Additionally, the electrophilic form of oxygen could seek out a C–H bond for hydride ($H:^-$) or hydrogen atom ($H\cdot$) removal and subsequent rapid return of OH^- or $\cdot OH$ (Guengerich and Macdonald, 1984). Whether one oxygen atom (monooxygenases) or two oxygen atoms (dioxygenases) are transferred to the organic chemical, the key that permits these reactions to occur, while they commonly do not proceed abiotically in the environment, is the attraction of electrons of the

TABLE 14.2 Some Microbially Mediated Oxidations

Substrate	Product	Equation	Reference
π electron attack			
 toluene	cis-1,2-dihydroxy-3-methyl-cyclohexa-3,5-diene	(14-20)	Gibson et al., 1970; Yeh et al., 1977
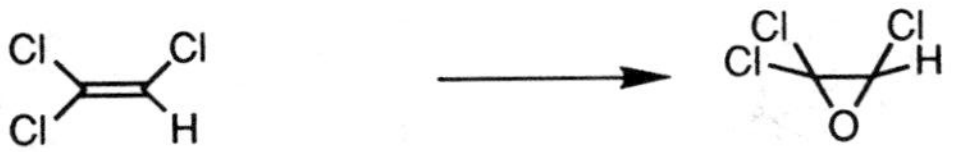 trichloroethylene	trichloroethylene epoxide	(14-21)	Little et al., 1988
n electron attack			
 aldicarb		(14-22)	Jones, 1976

TABLE 14.2 *(Continued)*

Substrate	Product	Equation	Reference
σ electron attack			
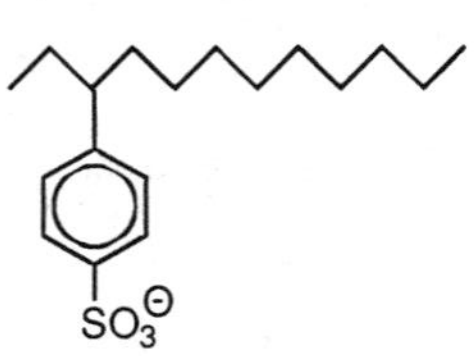 $SO_3^{\ominus}$ 3-dodecyl-benzene sulfonate →	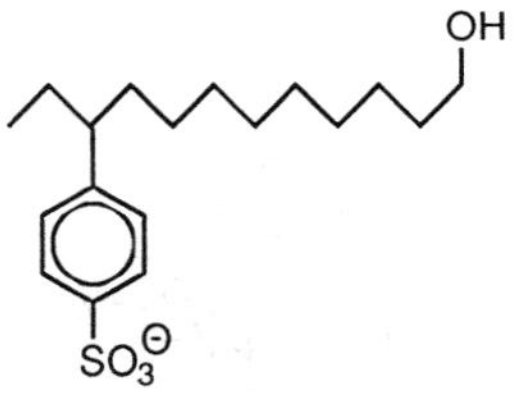 $SO_3^{\ominus}$ OH	(14-23)	Swisher, 1987
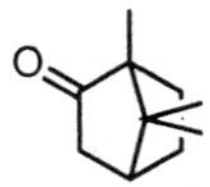 camphor →	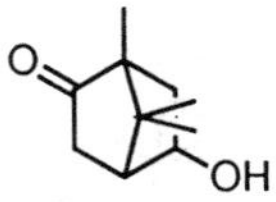 OH	(14-24)	Hedegaard and Gunsalus, 1965; Conrad et al., 1965
$R{-}CH_3$ → $R{-}CH_2OH$ → (alkane → alcohol)	$R{-}CHO$ → $R{-}COOH$ (aldehyde → acid)	(14–25)	Ratledge, 1984

organic chemical to the biologically produced oxidant. Thus it is clear that organisms can accomplish this by using an enzymatically prepared form of electrophilic oxygen.

Recognition of these approaches of microorganisms for executing oxygenation reactions enables us to make important predictions regarding such biotransformations. For example, if we consider attack of phenol (Eq. 14-9) by the electrophilic oxygen of a monooxygenase, we are not surprised to see the addition ortho (but not meta) to the hydroxyl moiety:

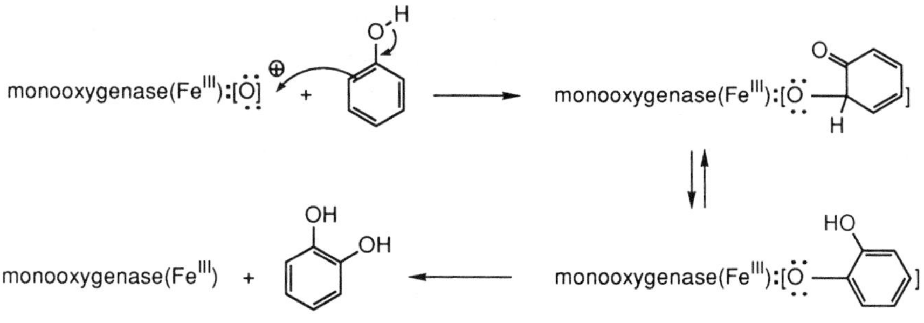

monooxygenase(Fe^{III}):[O]$^{\oplus}$ + phenol → monooxygenase(Fe^{III}):[O–(cyclohexadienone, H)]

⇅ (14-30)

monooxygenase(Fe^{III}) + catechol ← monooxygenase(Fe^{III}):[O–(hydroxyphenyl)]

For alkenes, the presence of π electrons tends to cause reaction to occur at the double bond rather than at saturated positions of the molecule:

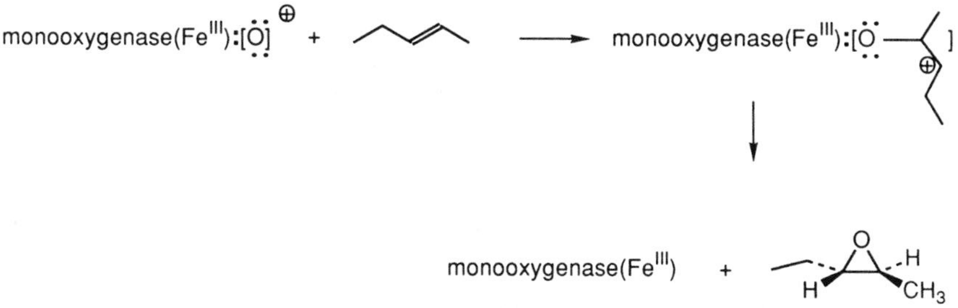

monooxygenase(Fe^{III}):[O]$^{\oplus}$ + (alkene) → monooxygenase(Fe^{III}):[O–(carbocation)]

↓ (14-31)

monooxygenase(Fe^{III}) + (epoxide, H, H, CH_3)

Furthermore, one might expect that electron-withdrawing substituents on benzoate make the rate of attack on that ring slower, and this has been seen where steric factors do not dominate (Knackmuss, 1981):

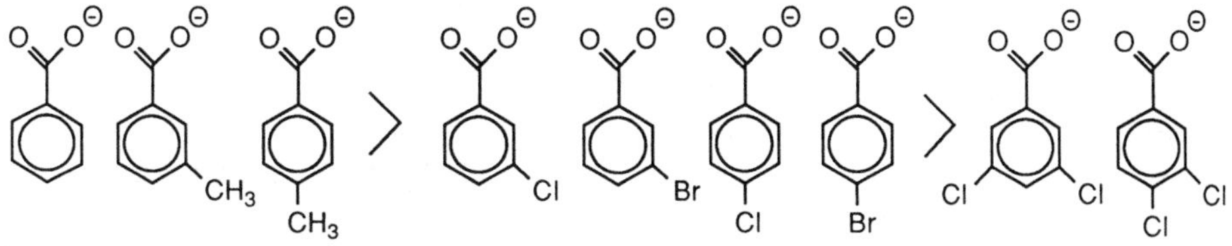

benzoate, 3-CH_3-benzoate, 4-CH_3-benzoate > 3-Cl-benzoate, 3-Br-benzoate, 4-Cl-benzoate, 4-Br-benzoate > 3,5-Cl-benzoate, 3,4-Cl-benzoate

This trend is consistent with a situation in which the rate-limiting step involves an electrophilic attack on the π electrons of the ring. This result illustrates how variations in chemical structure may affect the rates of biotransformations. We develop this further below.

An Example of a Biological Oxygenation: Cytochrome P450 Monooxygenase

Organisms have developed a very interesting approach for preparing and using electrophilic oxygen. We illustrate this methodology with the case of cytochrome P450 monooxygenases (Ullrich, 1972; White and Coon, 1980; Guengerich and MacDonald, 1984), a widespread and well-studied biooxidation system. The active site of these enzymes is an iron porphyrin (the combination is called a heme) carried within a protein environment:

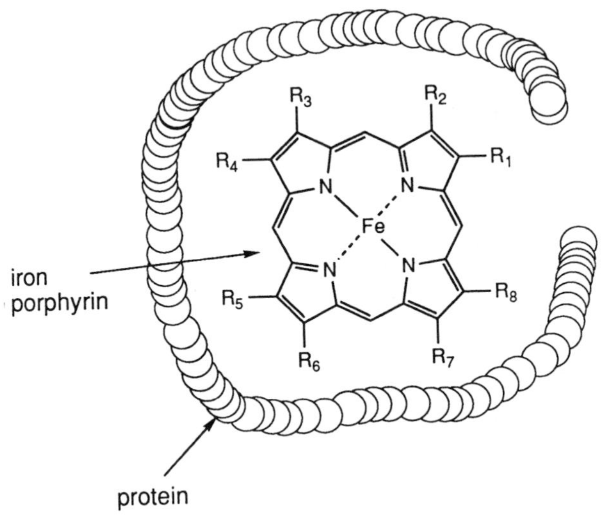

The macromolecular region in which the oxidation reactions occur is nonpolar; as a result, organic substrates are transferred from aqueous solution into the enzymes's active site primarily due to their hydrophobicity. This important feature seems very appropriate: nonpolar molecules will readily associate with this enzyme, and these are precisely the ones that are most difficult for organisms to avoid accumulating from a surrounding aquatic environment.

After substrate:enzyme complex formation, the enzyme begins to be prepared for reaction (Fig. 14.5). The iron atom is converted from Fe^{III} to Fe^{II} by an enzyme called a reductase which supplies one electron (ultimately from a reduced nicotinamide adenine dinucleotide, NAD(P)H, generated by another energy-yielding metabolism). Then molecular oxygen is bound by the iron. Next a second electron is transferred to the iron–oxygen complex (again ultimately from NAD(P)H). The resulting anionic oxygen quickly protonates, and we now have a biologically produced analog to hydrogen peroxide (HO–OH). The details of the subsequent reaction steps are only poorly understood, but it is clear that some form of electrophilic oxygen (illustrated by O^+ in Fig. 14.5) is produced, and this highly reactive species can now attract electrons from a neighboring source. Not coincidentally, since the enzyme has bound

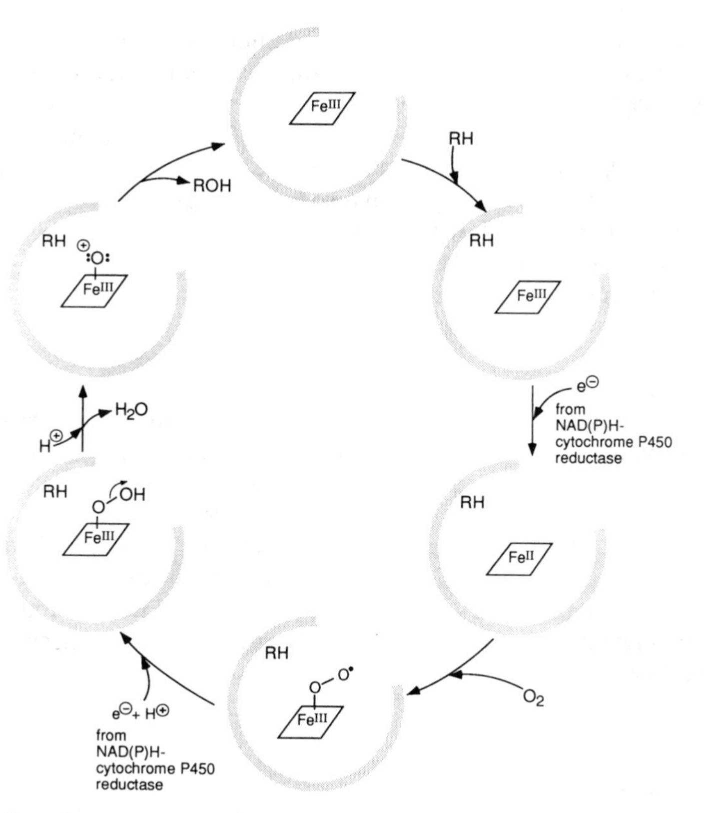

Figure 14.5 Reaction sequence for oxidation of an organic substrate, RH, by a cytochrome-P450 monooxygenase (after Ullrich, 1972). The exact oxidation states of the Fe and the O in the final reactive enzyme (shown as Fe(III) and O^+) are not known (e.g., could be Fe(IV) and O·).

a hydrophobic organic chemical earlier in the sequence, the electrophilic oxygen sees this chemical as its first opportunity to acquire electrons. The ultimate result is the addition of an oxygen atom to the substrate. Often the product is itself reactive (e.g., epoxides); and it may subsequently add H_2O to form a still more polar product that will tend to escape the nonpolar protein to the surrounding aqueous cytoplasm if it is not suitable for continued metabolic processing.

In sum, enzymes like cytochrome P450 monooxygenases are not very selective since association with the active site chiefly involves van der Waals attractions. Since biological energy (in the form of NAD(P)H) has to be invested to make the overall oxidation work, most organisms do not maintain a great stock of this type of enzymatic capability. An exception would be some very specialized microorganisms that survive on their ability to execute difficult oxidations (e.g., the monooxygenation of methane by bacteria called methanotrophs, or the oxidation of lignin by white rot fungi who use a peroxidase). These organisms must exhibit quite high activities of such enzymes to allow them to survive. Interestingly, methane–monooxygenase and lignin peroxidase have been found to be quite capable of oxidizing organic pollutants too (e.g., Wilson and Wilson, 1985; Fogel et al., 1986; Bumpus et al., 1985; Hammel et al., 1986). Though most other organisms do not maintain high levels of nonspecific monooxygenases, it seems such metabolic tools are ubiquitous in aerobic organisms. Thus, we can anticipate that in the presence of molecular oxygen, biooxidation of xenobiotic organic compounds will occur, albeit sometimes slowly relative to biological processing of more typical metabolites. The products of such oxidation reactions may be more effectively returned to the aquatic medium or catabolized in subsequent enzymatic steps.

Reductions

The second prominent set of biologically mediated reactions that are used for the initial transformations of xenobiotic compounds are reductions. As we discussed in Chapter 12, this type of reaction entails transferring electrons to the organic substrate of interest. Fundamentally, these biotransformations are similar to the reductions described in Chapter 12. Hence, microbially mediated reductive transformations (Table 14.3) involve the same structural moieties that we found to be susceptible to abiotic reductions. The common characteristic for the structures at the point of reduction is that electron-withdrawing heteroatoms are causing the formation of oxidized central atoms (recall Table 2.4):

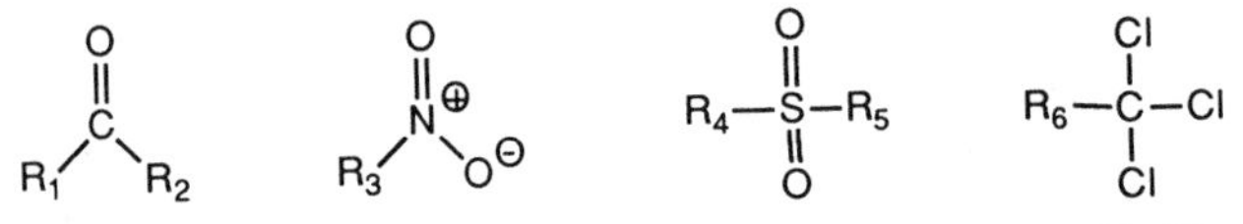

$R_1C(=O)R_2$ C(+II) $R_3N^{\oplus}(=O)O^{\ominus}$ N(+III) $R_4S(=O)_2R_5$ S(+II) R_6CCl_3 C(+III)

Biological Reduction by Alcohol Dehydrogenases Observation of bioreductions of atoms doubly bound to oxygen are consistent with a mechanism for attack that would involve a nucleophilic form of hydrogen. Indeed such bioreductants occur in organisms; the reactive portion of one such reductant, the nicotinamide ring of NADH or NADPH [henceforth referred to both as NAD(P)H], is shown below:

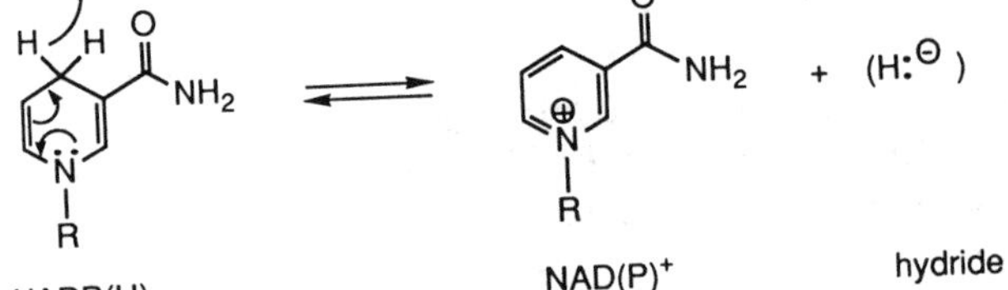

$$\text{NADP(H)} \rightleftharpoons \text{NAD(P)}^{+} + (\text{H:}^{\ominus}) \text{ hydride} \qquad (14\text{-}38)$$

This molecule has the marvelous property of holding a nucleophilic form of hydrogen on the ring at the position opposite to the nitrogen atom. This hydrogen is easily split off as hydride since, using the nonbonded nitrogen electrons, a stabilized aromatic π-electron system can be established. The hydride may react with the electron-deficient

TABLE 14.3 Some Microbially Mediated Reductions

Substrate	Product(s)	Equation	Reference
carbonyl group			
$H_3C\text{-}C(=O)\text{-}H$ acetaldehyde	→ $CH_3\text{-}CH_2\text{-}OH$ ethanol	(14-32)	Gottschalk, 1986
nitro group			
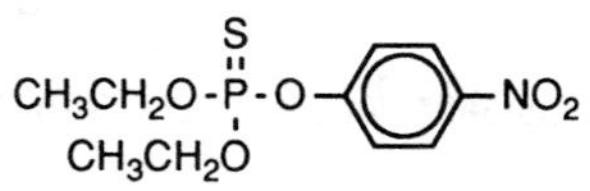 $(CH_3CH_2O)_2P(=S)\text{-}O\text{-}C_6H_4\text{-}NO_2$ parathion	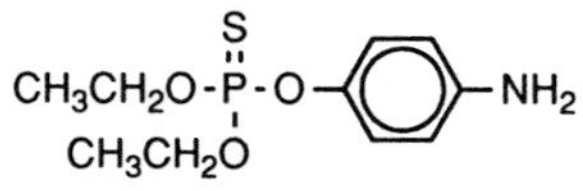 → $(CH_3CH_2O)_2P(=S)\text{-}O\text{-}C_6H_4\text{-}NH_2$ amino-parathion	(14-33)	
sulfoxide and sulfone groups			
$H_3C\text{-}S(=O)_2\text{-}C(CH_3)_2\text{-}CH{:}N\text{-}O\text{-}C(=O)\text{-}NH\text{-}CH_3$ aldicarb sulfone	→ $H_3C\text{-}S(=O)\text{-}C(CH_3)_2\text{-}CH{:}N\text{-}O\text{-}C(=O)\text{-}NH\text{-}CH_3$ aldicarb sulfoxide		
	→ $H_3C\text{-}S\text{-}C(CH_3)_2\text{-}CH{:}N\text{-}O\text{-}C(=O)\text{-}NH\text{-}CH_3$ aldicarb	(14-34)	

TABLE 14.3 *(Continued)*

Substrate		Product(s)	Equation	Reference
$CH_3-S(=O)-CH_3$ dimethyl sulfoxide (DMSO)	→	CH_3-S-CH_3 dimethyl sulfide	(14–35)	Weiner et al., 1988 Zinder and Brock, 1978
halides (reductive dehalogenation)				
CCl_4 carbon tetrachloride	→	$CHCl_3$ trichloromethane (chloroform)	(14-36)	Castro et al., 1985
$BrCH_2CH_2Br$ ethylene dibromide	→	$CH_2=CH_2$ ethene	(14-37)	Castro et al., 1968

part of an organic substrate being reduced. NAD(P)H can later be "reloaded" using energies derived from other metabolic processes.

To illustrate the role of NAD(P)H in reductive reactions, we consider a class of enzymes called alcohol dehydrogenases (Fersht, 1985). This group of enzymes is ubiquitous, and they have been studied extensively in a variety of organisms (e.g., *Bacillus* bacteria, yeasts). These enzymes catalyze the conversion of alcohols to aldehydes or ketones and vice versa (e.g., Eq. 14-32 in Table 14.3) by facilitating the interactions of the alcohol–ketone of interest with $NAD(P)^+$/NAD(P)H. In the following, we examine this interaction in some detail to gain insights into the factors governing the overall rates.

Combining information from studies using alcohol dehydrogenases isolated from a variety of organisms, the general modes of operation allowing NAD(P)H to reduce carbonyls can be pictured as shown in Figure 14.6. First, the enzyme binds the reductant (i.e., NAD(P)H) at a position advantageous for interaction with the compound to be reduced. Next, the carbonyl oxygen associates with an acidic moiety; in the figure this involves the carbonyl associating as a ligand with a zinc in the

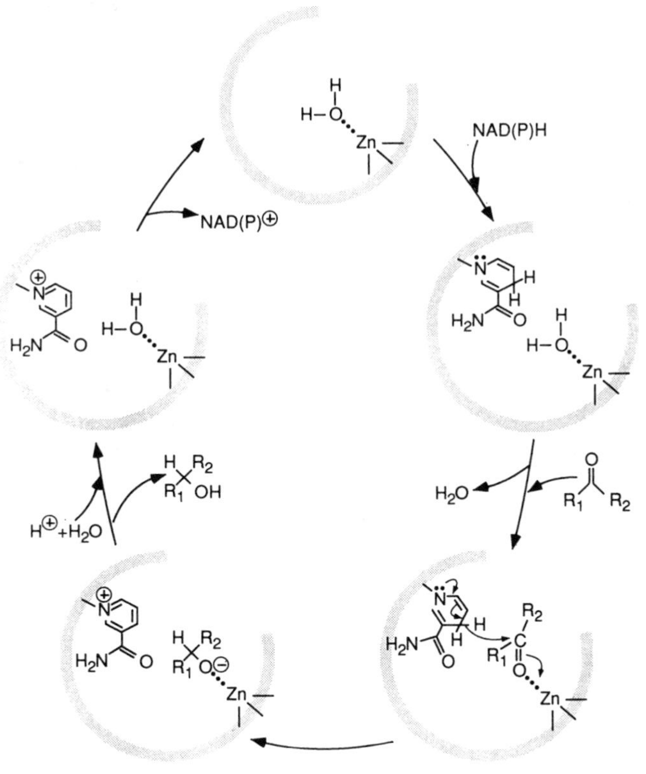

Figure 14.6 Reaction sequence for reduction of a carbonyl compound by a dehydrogenase (Fersht, 1985).

enzyme (displacing a water molecule that previously occupied the position). The complexation of the oxygen of the organic substrate with the zinc serves to draw more electron density away from the carbonyl carbon, as well as to position the organic substrate suitably for interaction with the previously bound NAD(P)H. Now the nucleophilic form of hydrogen, the hydride carried by NAD(P)H, can attack the carbonyl carbon reducing this functional group to an alcoholic moiety in a two-electron transfer step. Acquisition of a proton (ultimately from the aqueous medium, but initially from acidic groups near the reaction site) completes the structural changes and results in the release of the alcohol product from the enzyme. Finally, the oxidized $NAD(P)^+$ must be discharged from the enzyme; a new reductant must be bound before the reaction can proceed again.

Several of the abovementioned steps could limit the overall rate of this transformation: binding of NAD(P)H, binding of the carbonyl compound, bond making and breaking in the reduction reaction, transfer of the proton to the alcoholate moiety, release of the alcohol, and release of the $NAD(P)^+$. For the reduction of aromatic aldehydes with horse liver alcohol dehydrogenase, it is the rate of releasing the product alcohol that controls the overall rate of the process; but for acetaldehyde reduction by this same enzyme, it is the rate of hydride attack that is rate limiting. In an alcohol dehydrogenase obtained from a yeast, the hydride transfer step is also the slowest. These differences not only demonstrate some variations within this class of enzymes, but also suggest that particular enzymes have evolved to exhibit some specificity with respect to the chemicals they process. Thus we may anticipate carbonyl reduction rates in the environment to differ from compound to compound and site to site as different organisms are involved. Overall, the transformation proceeds quickly both because organisms invest energy to form reactive reductants (NAD(P)H) and because the organization and association provided by the active enzyme site lowers the activation energy for the reaction to proceed (recall Fig. 14.1). We should note that this enzymatic transformation process also operates in reverse (note the name of the enzyme is dehydrogenase referring to the oxidation reaction).

Biological Reduction With Cytochromes Although alcohol dehydrogenase reactions are illustrative of reductions of chemical moieties susceptible to hydride attack, some reductions occur on organic structures that are not well suited to such attack even though very oxidized carbons are involved. For example, in the case of bromotrichloromethane the bulky chloride moieties sterically block a nucleophile's approach, which might push out the bromide:

NAD(P)–H ✗→ CCl_3Br (Cl, Cl, Cl, C—Br)

C(+IV)

(14-39)

Yet, microbially mediated reductions of this and other polyhalogenated compounds have long been seen to occur (Hill and McCarty, 1967; Parr and Smith, 1976; Bouwer and McCarty, 1983a,b; Vogel et al., 1987; Egli et al., 1987). Various metallo-porphyrins

TABLE 14.4 Half Reactions and Standard Reduction Potentials of Some Bacterial Cytochromes and of Several Halogenated Methanes at 25°C Showing the Thermodynamic Capability of Cytochromes to Reduce These Chlorine- and Bromine-Containing Compounds

Oxidized Species		Reduced Species	E_H° (V)	$E_H^\circ(W)$[a] (V)	Ref.[b]
$CBrCl_3 + H^+ + 2e^-$	=	$CHCl_3 + Br^-$	+0.83	+0.77	1
$CCl_4 + H^+ + 2e^-$	=	$CHCl_3 + Cl^-$	+0.79	+0.67	1
$CHCl_3 + H^+ + 2e^-$	=	$CH_2Cl_2 + Cl^-$	+0.68	+0.56	1
$CH_2Cl_2 + H^+ + 2e^-$	=	$CH_3Cl + Cl^-$	+0.61	+0.49	1
$CH_3Cl + H^+ + 2e^-$	=	$CH_4 + Cl^-$	+0.59	+0.47	1
Cytochrome a(FeIII) + e^-	=	cytochrome a(FeII)		+0.38	2
Cytochrome b(FeIII) + e^-	=	cytochrome b(FeII)		+0.25	2
Cytochrome c(FeIII) + e^-	=	cytochrome c(FeII)		+0.07	2

[a]pH 7; $Cl^- = 10^{-3}$ M; $Br^- = 10^{-5}$ M.
[b]References: 1. Vogel et al., 1987; 2. Gottschalk, 1986.

which contain reduced Fe, Co, or Ni have been found to be chemically capable of reducing such polyhalogenated chemicals (e.g., Wood et al., 1968; Zoro et al., 1974; Krone et al., 1989a, b).

Castro and co-workers (Wade and Castro, 1973ab; Castro and Bartnicki, 1975; Bartnicki et al., 1978; and Castro et al., 1985) have extensively documented the reactivities of one such set of bioreductants, the heme proteins. In their Fe(II) state, the hemes of several very common cell constituents like cytochromes have sufficiently low reduction potentials to reduce many halogenated compounds (Table 14.4). These hemes have also been shown capable of reducing other functional groups such as nitro substituents (Ong and Castro, 1977). Based on studies of hemes outside their protein carriers, it seems that halogenated compounds begin the reaction sequence by complexing with the reduced iron atom in an axial position (Fig. 14.7). Lysis of the carbon–halogen bond occurs after an inner sphere (i.e., direct) single electron transfer from the iron to the halogen. For isolated hemes in solution, this is the rate-limiting step. The result is a very stable halide product (X^-) and a highly reactive carbon radical ($R\cdot$). Since free radicals are formed, dimers can occur as products:

$$CCl_4 \xrightarrow{+e^-} \cdot CCl_3 + Cl^- \tag{14-40}$$

$$2 \cdot CCl_3 \longrightarrow Cl_3C{-}CCl_3 \tag{14-41}$$

When $R\cdot$ escapes and collides with a good $H\cdot$ donor (e.g., unsaturated lipids), a hydrogen is abstracted, yielding RH and a damaged cellular component (Hanzlik, 1981):

$$R_1^{\cdot} + R_2CH(H)CH{=}CHR_3 \longrightarrow R_1{-}H + \left(R_2\dot{C}HCH{=}CHR_3 \longleftrightarrow R_2CH{=}CH\dot{C}HR_3\right) \tag{14-42}$$

allylic hydrogen   damage

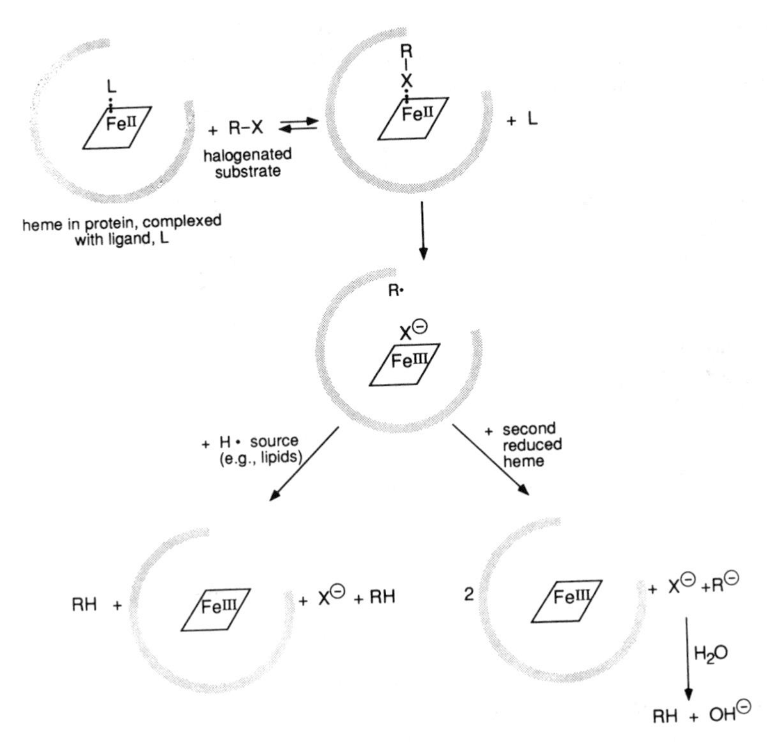

Figure 14.7 Schematized reaction sequence showing reduction of an organic halide by a heme-protein such as cytochrome P450 (after Hanzlik, 1981; Castro et al., 1985).

Finally, should a second reduced heme provide a second electron to R·, a carbanion would be formed which would immediately be protonated on entering any surrounding aqueous cytoplasm or solution:

$$R\cdot \xrightarrow{\text{Fe(II)heme} \rightarrow \text{Fe(III)heme}} R{:}^{\ominus} \xrightarrow{H_2O \rightarrow OH^{\ominus}} RH \tag{14-43}$$

Alternatively, if a suitable leaving group is present at a vicinal position, elimination may result:

$$Cl_3C{-}CCl_3 \xrightarrow{\text{Fe(II)heme} \rightarrow \text{Fe(III)heme}} Cl^{\ominus} + \cdot CCl_2{-}CCl_3 \tag{14-44}$$

$$\cdot CCl_2{-}CCl_3 \xrightarrow{\text{Fe(II)heme} \rightarrow \text{Fe(III)heme}} \overset{\ominus}{:}CCl_2{-}CCl_3 \longrightarrow Cl_2C{=}CCl_2 + Cl^{\ominus} \tag{14-45}$$

Interestingly, several factors may limit the overall rate of such reductive reactions.

First, as we noted above for hemes and alkyl halides in solution, the electron transfer may be limiting. In considering hemes within their carrier proteins, there is apparently some extra limitation with respect to access of the halocarbon to the reduced iron. Mass transfer limitation is even more important in intact bacterial cells, since the overall rate of some specific reductive dehalogenations appears to be independent of the heme content of the cells (Castro et al., 1985).

In summary, organisms contain biological reductants such as NAD(P)H and reduced metallo-proteins, capable of hydride transfer and single electron donation, respectively. Although not necessarily intended to interact with xenobiotic compounds, when such substances do confront these bioreductants *in vivo*, reductive reactions can occur.

Hydrolysis

Another important metabolic approach used by microorganisms to initiate transformations of xenobiotic compounds involves hydrolysis of bonds. By examining some of the transformations executed by microorganisms (Table 14.5), we see that these xenobiotic compounds react at the same structural positions as in reaction with hydroxide ion or water (see Chapter 12):

Alkyl halides

$$H_2O + H{-}CH_2{-}X \longrightarrow H^{\oplus} + X^{\ominus} + HO{-}CH_2{-}H \qquad \left(\begin{array}{c}\text{see Eq.}\\ \text{14-46}\end{array}\right)$$

Acid derivatives

$$H_2O + R{-}Z({=}Y){-}Y{-}R \longrightarrow R{-}Z({=}Y){-}OH + HY{-}R \qquad \left(\begin{array}{c}\text{see Eq.}\\ \text{14-49}\end{array}\right)$$

where X is a halogen, Y is O, NR, or S, and Z is C, P, or S.

As for abiotic hydrolyses, these biochemical transformations replace halogens by hydroxyl groups or break up esters (or other acid derivatives) to yield acids and alcohols or amines. Such transformations facilitate expulsion of the xenobiotic compound from the organism back into its aquatic environment and/or enable the continued catabolic breakdown of the substance. Hydrolyses are found to proceed faster when mediated by microbes (e.g., Munnecke, 1976; Wolfe et al., 1980c). Since the cytoplasmic pH of organisms is not particularly different from the range of pH values seen in natural waters, these organisms must utilize special methods to facilitate these reactions. What we will see is that organisms use strong nucleophiles to make the initial attack, or they bind the hydrolyzable moiety with an electron-withdrawing substituent to enhance the rate of attack in a manner analogous to what we have seen in acid-catalyzed hydrolyses (Section 12.3). Also, the enzymes involved hold the reacting species in the right positions with respect to one another, thereby enhancing the rate of their

TABLE 14.5 Some Microbially Mediated Hydrolysis Reactions

Substrate	Product(s)	Equation	Reference
alkyl halides			
Cl + H_2O	⟶ OH + $Cl^{\ominus}$ + $H^{\oplus}$	(14-46)	Scholtz et al., 1987a, b
Cl Cl + H_2O	⟶ Cl OH + $Cl^{\ominus}$ + $H^{\oplus}$	(14-47)	Keuning et al., 1985
CH_2Cl_2 + H_2O	⟶ H_2CO + $2Cl^{\ominus}$ + $2H^{\oplus}$	(14-48)	Kohler-Staub and Leisinger, 1985
esters			
permethrin (insecticide)	⟶ (acid, OH) + (alcohol, OH)	(14-49)	Maloney et al., 1988
amides			
propanil (herbicide)	⟶ (Cl, Cl, NH_2) + HO (O)	(14-50)	Chisaka and Kearney, 1970 Bartha, 1971
carbamates			
carbofuran (insecticide)	⟶ CH_3NH_2 + CO_2 + (HO, O)	(14-51)	Chaudry and Ali, 1988 Ramanand et al., 1988

ureas

$(CH_3)_2N{-}C(=O){-}NH{-}C_6H_3Cl_2$ → $(CH_3)_2NH + CO_2 + H_2N{-}C_6H_3Cl_2$ (14-52) Engelhardt et al., 1973

linuron (herbicide)

(thio)phosphates

$(CH_3O)_2P(=S){-}S{-}CH(COOCH_2CH_3)CH_2COOCH_2CH_3$ → $(CH_3O)_2P(=S){-}OH + HS{-}CH(COOCH_2CH_3)CH_2COOCH_2CH_3$ (14-53) Rosenberg and Alexander, 1979

malathion (insecticide)

$(CH_3CH_2O)_2P(=O){-}O{-}C_6H_4{-}NO_2$ → $(CH_3CH_2O)_2P(=O){-}OH + HO{-}C_6H_4{-}NO_2$ (14-54) Munnecke, 1976

paraoxon (insecticide)

interaction by reducing any unfavorable entropy change of reaction (i.e., making a negative ΔS_{rxn} more positive). The combination of such influences greatly hastens the biological process over nonenzymatic mechanisms (Chapter 12). Xenobiotic compound–enzyme complexes are subsequently separated by attack on them by water. The result of the overall sequence is the same as if the reaction occurred by direct attack of water.

Biological Hydrolysis of Alkyl Halides with Glutathione A first hydrolysis approach involves the use of the tripeptide, glutathione:

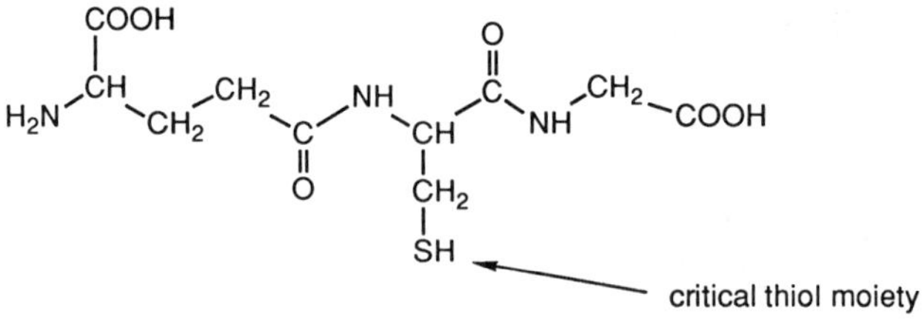

$$H_2N-CH(COOH)-CH_2-CH_2-C(=O)-NH-CH(CH_2SH)-C(=O)-NH-CH_2-COOH$$

critical thiol moiety

which we will refer to as GSH. This tripeptide (γ–glutamic acid–cysteine–glycine) contains a thiol moiety which is an excellent nucleophile. As a result, some compounds like alkyl halides are susceptible to nucleophilic attack by glutathione. For example, this appears to be the case for dichloromethane (Stucki et al., 1981; Kohler-Staub and Leisinger, 1985):

$$GS{-}H + H_2CCl_2 \longrightarrow GS{-}CH_2Cl + Cl^{\ominus} + H^{\oplus} \qquad (14\text{-}57)$$

dichloromethane

This interaction seems to be mediated by an enzyme called a glutathione transferase, which facilitates the encounter of GSH and the compound it is attacking (Mannervik, 1985). Formation of the glutathione adduct permits the attack of water on the previously chlorinated carbon:

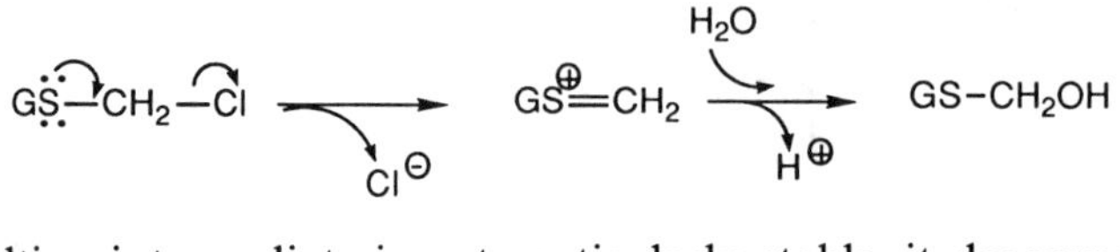

$$GS{-}CH_2{-}Cl \xrightarrow{-Cl^{\ominus}} GS^{\oplus}{=}CH_2 \xrightarrow[-H^{\oplus}]{H_2O} GS{-}CH_2OH \qquad (14\text{-}58)$$

Since the resulting intermediate is not particularly stable, it decomposes, releasing formaldehyde and glutathione in a reaction much like the dehydration of gem-diols (see Section 10.5, case 3):

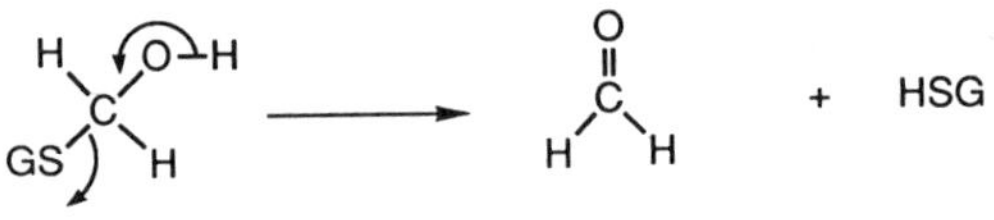

$$GS{-}CH_2{-}O{-}H \longrightarrow H_2C{=}O + HSG \qquad (14\text{-}59)$$

The overall result is equivalent to hydrolysis of both of the original carbon–chlorine bonds:

$$CH_2Cl_2 + H_2O \longrightarrow CH_2O + 2H^{\oplus} + 2Cl^{\ominus} \tag{14-60}$$

In sum, the excellent bionucleophile, glutathione, is used to get the reaction started, and subsequent hydrolysis steps cause it to go to completion. The process works well enough that one bacterial species, a *Hyphomicrobium* isolate, can even manage to grow on methylene chloride as its sole source of carbon (Stucki et al., 1981). This microorganism is also capable of degrading other dihalomethanes, CH_2BrCl, CH_2Br_2, and CH_2I_2 (Kohler-Staub and Leisinger, 1985). Monohalogenated compounds act as inhibitors, possibly because the glutathione adduct formed is difficult to hydrolyze:

GSH + Cl–alkyl ⟶ GS–alkyl + $Cl^{\ominus}$ + $H^{\oplus}$ (14-61)

GS–alkyl + H_2O $\xrightarrow{\text{very slow}}$ (14-62)

Interestingly, a similar enzymatic approach to hydrolyzing monohaloalkanes, which appears to rely on cysteine(s) contained within the protein structure, has also been observed (Keuning et al., 1985; Scholtz et al., 1987a, b). To operate, these enzymes first form the cysteine adduct, similar to the glutathione-based approach:

$$\text{Enz–cys–SH} + \text{R–X} \longrightarrow \text{Enz–cys–S–R} + H^{\oplus} + X^{\ominus} \tag{14-63}$$

The enzyme adduct then detaches the alkyl group as an alcohol product.

Enzymatic Hydrolysis Reactions of Esters We now shift our focus to xenobiotic compounds containing carboxylic acid esters or other acid derivatives in their structures (e.g., several shown in Table 14.5). To understand how enzymatic steps can be used to transform these substances, it is instructive to consider the biological apparatus used by organisms to hydrolyze naturally occurring analogs (e.g., fatty acid esters or amides in peptides). It is very likely that the same chemical procedures, and possibly even some of the same enzymes themselves, are involved in the hydrolysis of xenobiotic substrates.

We begin by considering the enzymes using the hydroxyl group of serine (Bruice et al., 1962) or the thiol group of cysteine to initiate the hydrolytic attack (Fig. 14.8):

OH–CH_2–CH(NH)–C(=O)   SH–CH_2–CH(NH)–C(=O)

serine in a peptide chain   cysteine in a peptide chain

As for glutathione, these functional groups play the role of nucleophiles and attack electron-deficient central carbons or other central atoms (e.g., P, S) in esters or ester

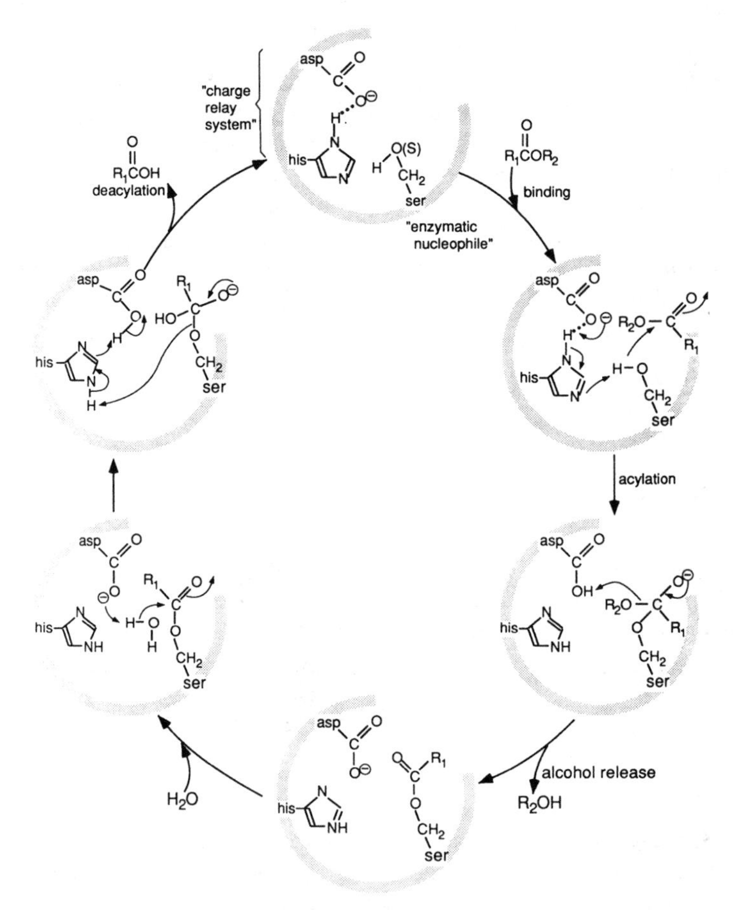

Figure 14.8 Schematized reaction sequence showing the hydrolysis of an ester by a serine-hydrolase (Fersht, 1985).

analogs. Thus, the general process proceeds by the following steps (Fig. 14.8). First, the substrate associates with the enzyme in a position suited for nucleophilic attack. To improve the ability of the nucleophile to form a bond with the electron-deficient atom, other amino acids (e.g., histidine and aspartic acid) may assist in this and subsequent steps by proton transfers (this group of amino acids is often referred to as the "charge relay system"). This may be likened to converting the serine or the cysteine to the more nucleophilic RO^- or RS^-, respectively. The result of the attack is the formation of a tetrahedral intermediate much as we saw previously in the chemical hydrolysis process (Section 12.3). Decomposition of this intermediate leads to the release of an alcohol or an amine from the original compound. Continued processing involves attack of water on the enzyme adduct (again with the help of the charge relay system) and release of the acid compound. Experience with various carboxylic acid esters and amides shows that either initial attack (acylation) or release of the acid compound (deacylation) can be the rate-limiting step, depending on the kind of compound hydrolyzed. For amides, the initial attack and release of an amine is often the slow step (Fersht, 1985). This is the case because these nitrogen compounds are very strong bases and poor leaving groups. In contrast, esters often have the deacylation steps as the bottleneck to reaction completion (Fersht, 1985). Since ionizable groups of various amino acids play such critical roles in these hydrolysis reactions, such enzymes exhibit marked sensitivity to the medium pH. Generally, these enzymes operate best at near-neutral pH's, since more acidic conditions protonate the histidine and thereby negate its involvement.

The reaction of a series of phenylamides and phenylureas with a hydrolyzing enzyme purified from a *Bacillus* sp. (Englehardt et al., 1973) is an example of this type of hydrolysis:

linuron → 3,4-dichloroaniline (NH_2) + $HO{-}C(=O){-}N(CH_3)OCH_3$ —fast→ $(CO_2 + CH_3NHOCH_3)$ (14-64)

Since, in this case, the enzyme activity decreased when reagents were added that complex with thiols, it was concluded that cysteine was probably the key enzymatic nucleophile. The production of this enzyme could be induced by exposure of the bacteria to various phenylamide- or phenylurea-containing herbicides and fungicides, and the enzyme was capable of hydrolyzing a variety of related compounds, albeit at somewhat different rates (Table 14.6).

The serine- or cysteine-based approach contrasts the approaches used in metal-containing hydrolases (Fig. 14.9). In these enzymes, a metal, for example, a zinc atom, is included in the active site of the enzyme (Fersht, 1985). The process begins with the association of the carbonyl oxygen with the fourth ligand position on this metal. Now as we found for the acid-catalyzed hydrolyses mechanism discussed in Chapter 12, the carbonyl carbon is made even more electron deficient. The result is a very susceptible position for attack by other nucleophilic moieties, such as the carboxylate of a nearby glutamate. An anhydride-type tetrahedral intermediate is formed, resulting in the release of the alcohol portion of the original ester. The new enzyme–acid complex is more suited to attack by water than the original compound. Thus attack by water ultimately leads to the severance of the acid's covalent linkage to the glutamate. Ligand exchange at the zinc permits the release of the acid product. In order for this enzyme to work well, the attacking glutamic acid must be ionized ($pK_a \sim 4$).

We contrast these two enzymatic approaches to ester hydrolysis, because it is in knowing the nature of the mechanisms that we can anticipate how the rates of transformations of related xenobiotic compounds will vary. This should help us understand and/or develop predictive relationships between the rates of biohydrolyses and comparable chemical hydrolyses as long as these processes are rate-limited by similar mechanisms.

TABLE 14.6 Variation in Hydrolysis Capability of a Crude Extract from *Bacillus sphaericus* Acting on Several Amides and Ureas[a]

Substrate	Hydrolysis Rates Relative to That Seen for Propanil
linuron	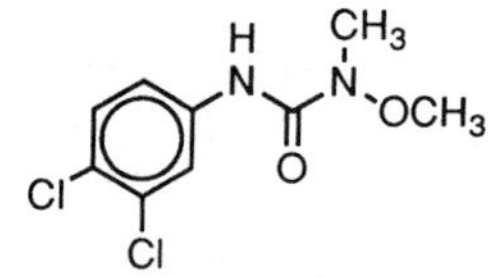 65
maloran	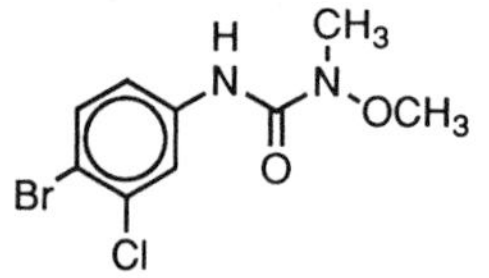 7.5
monalide	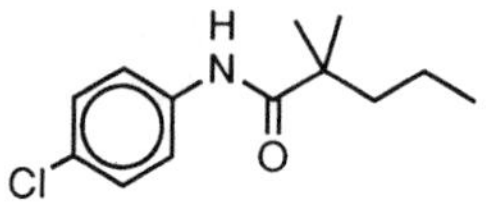 15
propanil	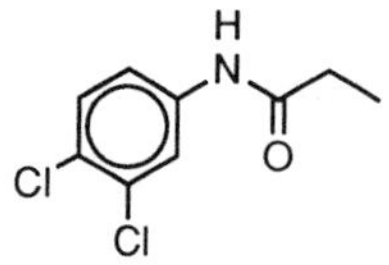 1
2-chlorobenzanilide	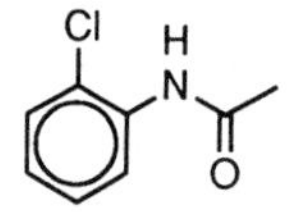 15
2,5-dimethylfuran-3-carboxanilide	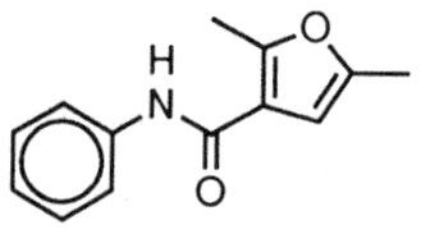 25
propham	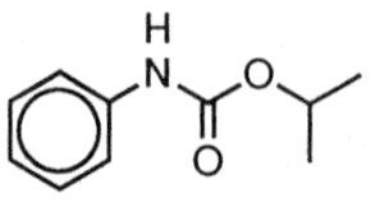 4

[a]From Engelhardt et al., 1973.

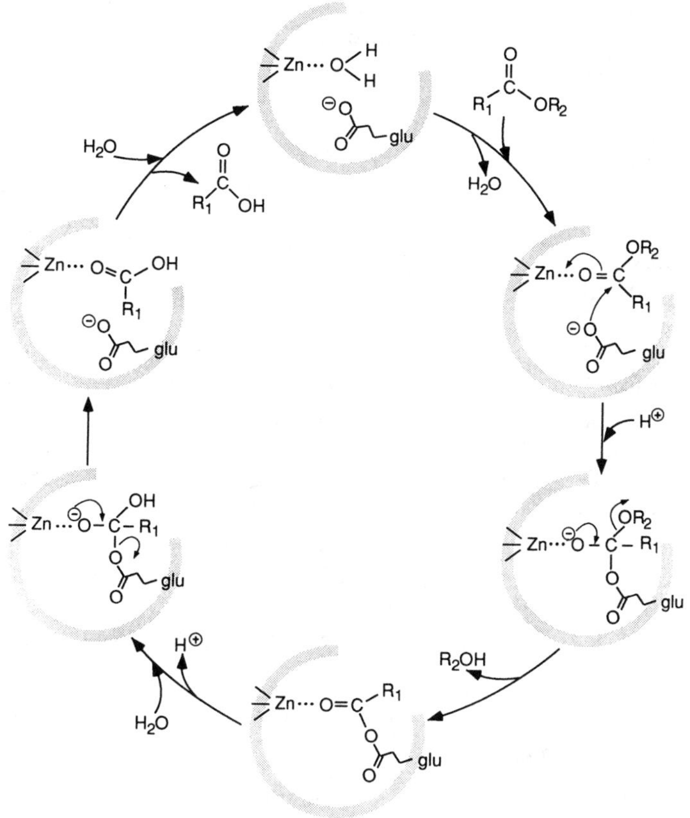

Figure 14.9 Reaction sequence for hydrolysis of an ester by a zinc-containing hydrolase (Ferscht, 1985).

Enzymatic hydrolyses do not seem to be very selective within a related group of xenobiotic compounds. This observation has been noted for enzymes induced by various kinds of substrates including: halides (Keuning et al., 1985; Scholtz et al., 1987a, b), phenyl amides and ureas (Englehardt et al., 1973), and phosphoric and thiophosphoric esters (Munnecke, 1976; Rosenberg and Alexander, 1979). This may indicate that the enzymatic apparatuses are newly evolved and suited to responding to unusual chemicals. Also, hydrolysis enzymes suited for attack of xenobiotic compounds appear to be constitutive (i.e., always present in the cells). Thus Wanner et al. (1989) were not surprised to see immediate biodegradation (i.e., no lag period) of some hydrolyzable insecticides:

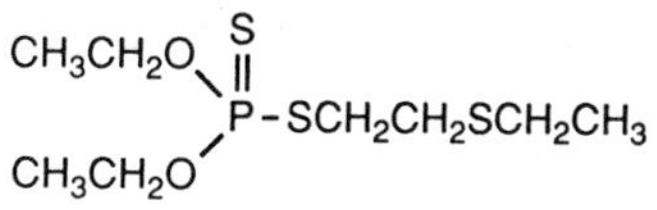

$(CH_3CH_2O)_2P(=S)-SCH_2CH_2SCH_2CH_3$

disulfoton

$(CH_3O)_2P(=S)-SCH_2CH_2SCH_2CH_3$

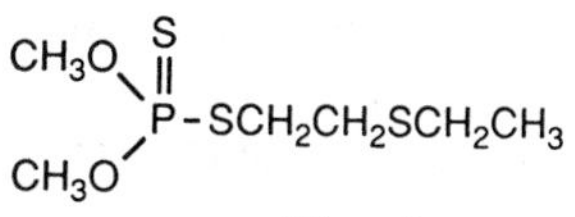

thiometon

upon accidental input of these compounds into the Rhine River. Finally, hydrolases may be important outside of microorganism cells, since some hydrolases play a role in releasing pieces of biopolymers.

14.4 RATES OF BIOTRANSFORMATIONS

Now that we have seen some of the major types of biologically mediated initial transformations of xenobiotic compounds, we begin to consider the factors that determine the rates of such processes.

To distinguish the Michaelis–Menten type from the Monod type cases, we note that the latter necessarily depend on the idea that the substrate of concern is acting as the limiting factor for growth of the population. Obviously, this requirement means: (1) the chemical must be sufficiently abundant relative to the biomass of the microbial community to permit significant energy and/or biomass building blocks to be acquired, and (2) the chemical must be sufficiently easy to metabolize. If either the compound is present at only trace levels or it is recalcitrant, then although biodegradation may occur, it will not support significant changes in cell numbers. In this case, it is the detailed kinetics of chemical–bioreagent interactions about which we care. Interestingly, for both cases very similar forms of mathematical model are used; thus model fits of data do not allow us to discern which picture is actually more appropriate for any case of interest.

Passive Uptake Limited Kinetics

As we noted earlier, many xenobiotic compounds probably are taken up by microorganisms from the environment by passive means (Nikaido, 1979). For nonpolar compounds, the rate of this uptake involves the rate of *diffusive* delivery across membranes separating the outside from the inside of the cells (Konings et al., 1981; see Fig. 14.10). Thus we can express the flux across a unit area of microorganism membrane using Fick's Law (see Chapter 9):

$$\text{Flux}\ (\text{mol}\cdot\text{m}^{-2}\cdot\text{s}^{-1}) = \frac{D_m([RH]_{out} - [RH]_{in})}{\Delta z} \qquad (14\text{-}65)$$

where

D_m is the molecular diffusivity of nonpolar compounds in the membrane ($m^2\cdot s^{-1}$),
$[RH]_{out}$ is the compound concentration in the outermost portion of the membrane ($mol\cdot m^{-3}$),
$[RH]_{in}$ is the compound concentration in the innermost part of the membrane ($mol\cdot m^{-3}$), and
Δz is the membrane thickness (m).

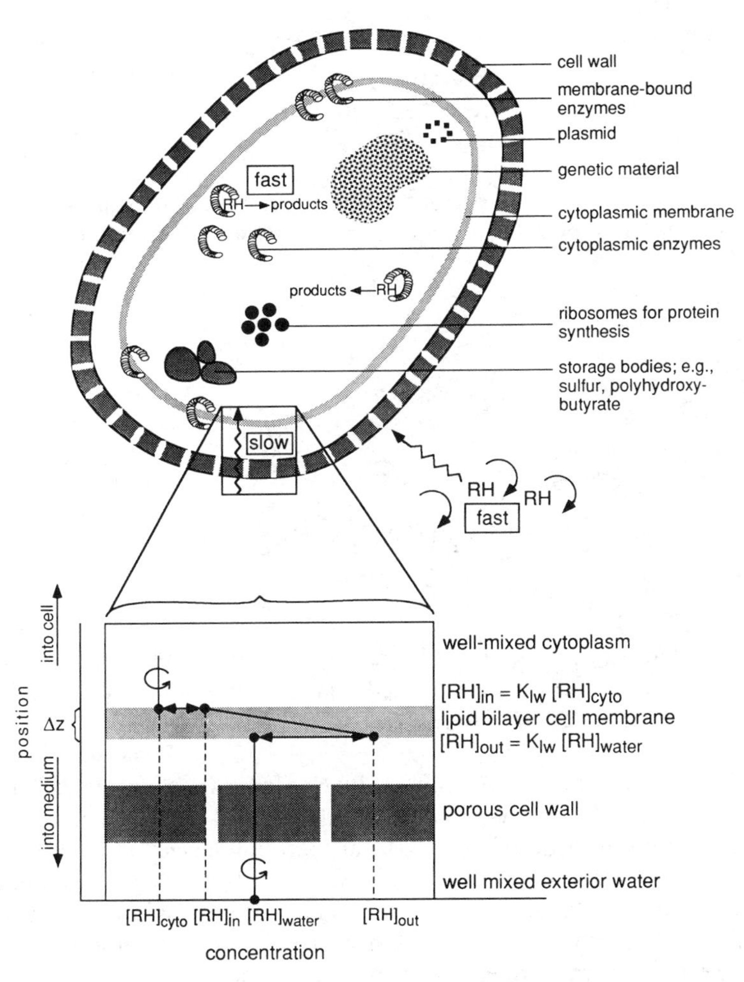

Figure 14.10 Image of the rate-determining step of biotransformation of a nonpolar organic compound RH whose permeation into cells limits its overall transformation. Upper: picture of the components of a gram-positive cell (after Schlegel, 1988); lower: expansion of portion of cell surface limiting transport from outside to inside. See text for definitions of symbols.

Here we define a positive flux as one that involves a net transport into the cells. If the *trans*-membrane transport is limiting overall degradation, then one can presume the compound concentration inside the cell, and by extension $[RH]_{\text{in}}$ will be maintained quite low ($[RH]_{\text{in}} \approx 0$). Further, the compound concentration in the outer part of the membrane can be related to the exterior water concentration, $[RH]_{\text{water}}$, by a lipid–water partition coefficient K_{lw} (not unlike a K_{ow}):

$$[RH]_{\text{out}} = K_{\text{lw}} \cdot [RH]_{\text{water}} \tag{14-66}$$

Thus the flux expression becomes

$$\text{Flux} \simeq \frac{D_m \cdot K_{lw}}{\Delta z} \cdot [RH]_{water} \tag{14-67}$$

Finally overall uptake from a particular environmental volume will be related to the number of cells involved and the areas of their surfaces allowing such passive chemical entrance:

$$\frac{d[RH]_{water}}{dt} = -\text{Flux} \cdot A \cdot [B] \qquad (\text{mol} \cdot \text{m}^{-3} \cdot \text{s}^{-1}) \tag{14-68}$$

$$= -\frac{D_m \cdot K_{lw} \cdot A \cdot [B] \cdot [RH]_{water}}{\Delta z} \tag{14-69}$$

$$= -k_{up} \cdot [RH]_{water} \tag{14-70}$$

where

$[B]$ is the "concentration" of degrading microorganisms ($\text{cells} \cdot \text{m}^{-3}$),
A is their individual surface area ($\text{m}^2 \cdot \text{cell}^{-1}$) participating in uptake, and
k_{up} (s^{-1}) is the combination of terms, $D_m \cdot K_{lw} \cdot A \cdot B \cdot \Delta z^{-1}$.

Nikaido (1979) notes that a portion of microbial surfaces are constructed from hydrophilic components (e.g., lipopolysaccharides and proteins) and these areas can inhibit nonpolar compound penetration; thus the total microbial surface area reflects an upper limit of A. Experiments with various single compounds have revealed values of k_{up} ranging between $0.005 - 0.009\ \text{min}^{-1}$ for microbial species with low area fractions and $0.1 - 1.8\ \text{min}^{-1}$ for microbial species with high areas permitting nonpolar transport. Thus uptake rates may vary widely, probably due to this area term.

Several interesting conclusions can be drawn from this simple passive uptake picture. First, for a given microorganism we can expect that the relative uptake rates for a series of nonpolar compounds will follow:

$$\frac{\text{uptake rate compound \#1}}{\text{uptake rate compound \#2}} = \frac{D_1 \cdot K_1}{D_2 \cdot K_2} \tag{14-71}$$

where subscripts on D and K refer to the compound number. This is indeed the type of relation seen (Collander, 1949). Next, such a bottleneck to overall biodegradation (i.e., causing the chemical concentration inside the cell to remain negligibly small) causes overall chemical losses to be directly proportional to the compound concentration in the surrounding water for all concentrations, $[RH]_{water}$. This contrasts other situations where enzyme kinetics or microbial population dynamics control biotransformations; in these cases saturating concentrations can be found where increases in

$[RH]_{\text{water}}$ do not cause increased rates. Since passive permeation into cells appears to occur on timescales of hours or less, *nonpolar compounds* that are biologically recalcitrant (i.e., lasting for years and decades in the environment) are probably not so because of their inability to enter microorganisms.

We should also note that microorganisms appear to have "water-filled pores" through their exterior membranes which permit the passive entrance of small hydrophilic substances (Nikaido, 1979). In studies of enteric bacteria, which appear to have smaller pores than other gram-negative species, passive glucose uptake exhibited transmembrane uptake timescales of less than a millisecond. Thus, as for nonpolar compound transport into microorganisms, the rate of passive uptake of *small, hydrophilic molecules* (<500 molecular mass units) via membrane pores of bacteria is not likely to cause them to avoid biodegradation for prolonged times. Such pores appear to be partially selective insofar as larger hydrated species diffuse in more slowly and *negatively charged chemicals* experience electrostatic repulsions that limit their ability to enter. This may explain the situation for highly charged organic compounds like ethylene diamine tetraacetic acid (EDTA) which are thought to be recalcitrant because of very slow uptake:

$$({}^{\ominus}OOC-CH_2)_2N-CH_2-CH_2-N(CH_2-COO^{\ominus})_2$$

Michaelis–Menten Type Enzyme Kinetics

We now turn to the situation in which degradation of the compound of interest is not limited by chemical uptake and does not result in an increase in the microorganism numbers or enzyme concentration. This lack of growth may be due to low concentrations of the substrate which are insufficient to support microbial multiplication, or it may be that the substrate is not easy to metabolize. Returning to Figure 14.2, this corresponds to the situation where the limit to biotransformation involves the rates of interactions between chemicals to be degraded and the enzymes and coenzymes performing the initial structural changes. Such situations may arise because organisms have alternative labile and abundant substrates to use (e.g., acetate), and hence the metabolism of the xenobiotic compounds are incidental processes occurring as secondary metabolism. Alternatively, a population of organisms may be living on a particular substrate, and they also transform compounds with similar structures at the same time. An example of this situation is illustrated in Figure 14.11 where a culture of bacteria grew on quinoline, while also degrading some benzo(f)quinoline if it was present (Smith et al., 1978):

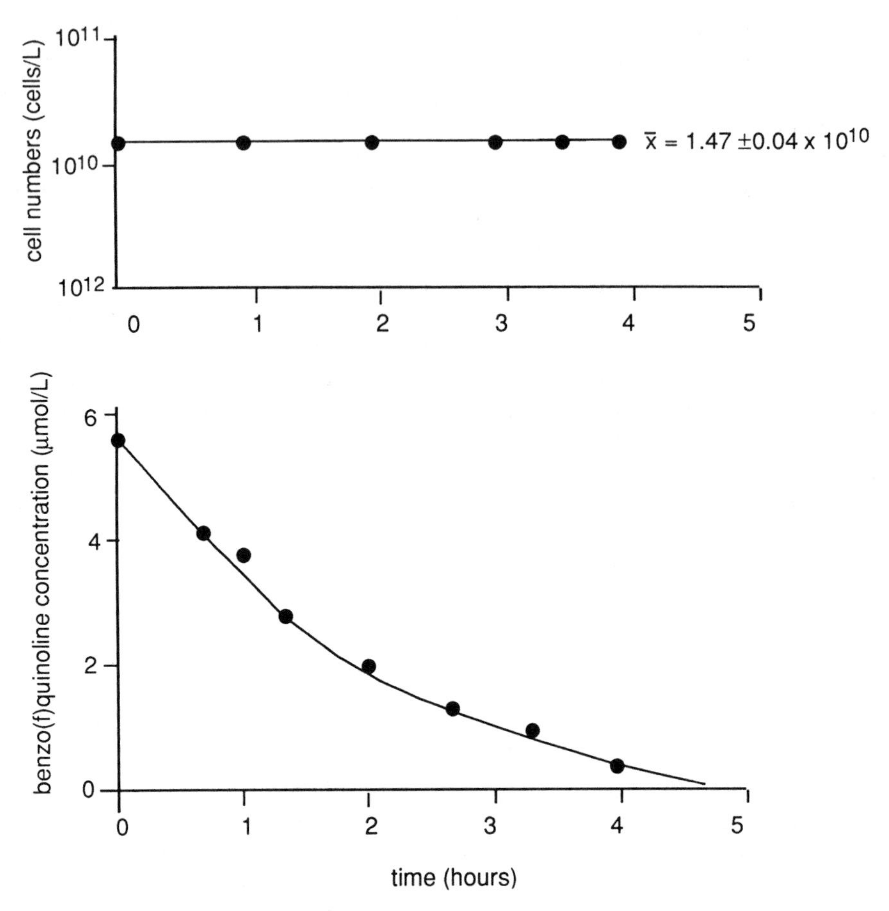

Figure 14.11 Timecourses for cell numbers and benzo(f)quinoline concentrations in a batch culture experiment (Smith et al., 1978).

If quinoline was removed, the cultures usually died since they could not survive metabolizing benzo(f)quinoline alone. This type of degradation has been called *cometabolism* (Alexander, 1981). The organisms have an enzyme or series of enzymes capable of processing quinoline so as to maintain and multiply the population. Possibly the same initial enzymatic reaction takes place with benzo(f)quinoline, but something about benzo(f)quinoline degradation makes it unsuited to support microbial growth. Thus a batch culture of microorganisms at a particular cell number growing on quinoline does not change in population size upon addition of benzo(f)quinoline (Fig. 14.11). Further, additions of benzo(f)quinoline to the quinoline-degrading cultures result in virtually immediate benzo(f)quinoline losses (i.e., there is no lag period) since cell numbers and enzyme abundances are not changing (Fig. 14.11). When Smith et al. (1978) observed benzo(f)quinoline removal in cultures whose cell numbers remained constant, the benzo(f)quinoline concentration declined immediately and exponentially in time (Fig. 14.11). If the abundance of the bacteria was varied, the rate of benzo(f)-quinoline removal was found to change accordingly. Consequently, the removal of benzo(f)quinoline could be described by a second-order rate law:

$$\frac{d[\text{benzo(f)quinoline}]}{dt} = -k_{\text{bio}} \cdot [\text{cells}] \cdot [\text{benzo(f)quinoline}] \qquad (14\text{-}72)$$

where k_{bio} is the observed removal rate constant (e.g., $\text{L} \cdot \text{cell}^{-1} \cdot \text{h}^{-1}$), and [cells] is the concentration of bacterial cells degrading benzo(f)quinoline ($\text{cells} \cdot \text{L}^{-1}$). The experimental data of Smith et al. (1978) yielded a second-order rate constant, k_{bio}, of $3.6 \times 10^{-11}\ \text{L} \cdot \text{cell}^{-1} \cdot \text{h}^{-1}$. In this case, a pseudo-first-order rate constant, k_{obs}, for benzoquinoline could be derived since the bacterial numbers did not change:

$$k_{\text{obs}} = k_{\text{bio}} \cdot [\text{cells}] = 0.5\ \text{h}^{-1} \qquad (14\text{-}73)$$

In the following discussion, we examine the types of biochemical factors that determine the rate constant k_{bio}. Although one is still unable to make *a priori* estimates of such rate constants, it is possible to develop some insights into the factors governing their magnitude. To illustrate, we develop some detailed kinetic expressions for one case of enzyme-mediated transformations. Examination of these results will help us to see how structural features of xenobiotic compounds may affect rates. Further, this example may serve to show how we could use probe compounds to ascertain biotransformation rates by a particular microbial population in a specific environmental setting, and to utilize these rate data for predicting rates for structurally related chemicals that are transformed by the same mechanism and are limited at the same biodegradation step.

An Example of Enzyme Kinetics: Hydrolases We begin by considering organic compounds whose degradation begins with a hydrolysis reaction (e.g., Fig. 14.8). The steps of such a reaction sequence may be written:

Step 1:

Enzyme:substrate association

$$\text{Enz} - \ddot{\text{N}}\text{u} + \text{R} - \text{X} \underset{k_{-1}}{\overset{k_1}{\rightleftharpoons}} (\text{Enz} - \ddot{\text{N}}\text{u}:\text{R} - \text{X}) \qquad (14\text{-}74)$$

where the colon indicates the noncovalent enzyme-substrate association.

Step 2:

First nucleophilic addition reaction, release of leaving group

$$(\text{Enz} - \ddot{\text{N}}\text{u}:\text{R} - \text{X}) \underset{k_{-2}}{\overset{k_2}{\rightleftharpoons}} (\text{Enz} - \text{Nu}^+ - \text{R}) + \text{X}^- \qquad (14\text{-}75)$$

Step 3:

Second nucleophilic addition reaction, release of remainder of substrate

$$(\text{Enz}-\text{Nu}^{+}-\text{R})+\text{H}_2\text{O} \underset{k_{-3}}{\overset{k_3}{\rightleftharpoons}} \text{Enz}-\ddot{\text{N}}\text{u}+\text{R}-\text{OH}+\text{H}^{+} \quad (14\text{-}76)$$

The general kinetic description of such a reaction scheme (that is referred to as a ping pong bi bi system) can be seen in specialized texts such as that of Segel (1975, pp. 606–625). Under special conditions, simplified kinetic expressions can be derived (Table 14.7). For example, it is often justified to treat the enzyme:substrate binding step as fast relative to a subsequent nucleophilic addition step. Therefore, enzyme–substrate complex formation can be modeled as an equilibrium process:

$$\frac{d[\text{Enz}-\ddot{\text{N}}\text{u}:\text{R}-\text{X}]}{dt} \simeq 0 \quad (14\text{-}87)$$

$$\simeq -k_1[\text{Enz}-\ddot{\text{N}}\text{u}][\text{R}-\text{X}] + k_{-1}[\text{Enz}-\ddot{\text{N}}\text{u}:\text{R}-\text{X}] \quad (14\text{-}88)$$

and we may define an equilibrium constant:

$$K_1 = \frac{k_1}{k_{-1}} \quad (14\text{-}89)$$

$$= \frac{[\text{Enz}-\ddot{\text{N}}\text{u}:\text{R}-\text{X}]}{[\text{Enz}-\ddot{\text{N}}\text{u}][\text{R}-\text{X}]} \quad (14\text{-}78)$$

Thus the possibilities reduce to those cases in which the initial nucleophilic reaction is the slowest step in the overall process, or the overall sequence bottleneck is at Step 3, the second nucleophilic addition (Fig. 14.12). When Step 2 is slowest and we neglect its back reaction, the kinetic expression simplifies to (see Table 14.7 left-hand side for derivation):

$$\frac{d[\text{R}-\text{X}]}{dt} = -\frac{k_2[E_\text{T}][\text{R}-\text{X}]}{K_1^{-1}+[\text{R}-\text{X}]} \quad (14\text{-}81)$$

Alternatively, if Step 3 is slowest and we can neglect its back reaction, we may establish an additional equilibrium expression (see Table 14.7 right-hand side for derivation):

$$K_2 = \frac{k_2}{k_{-2}} \quad (14\text{-}90)$$

$$= \frac{[\text{Enz}-\text{Nu}^{+}-\text{R}][\text{X}^{-}]}{[\text{Enz}-\ddot{\text{N}}\text{u}:\text{R}-\text{X}]} \quad (14\text{-}83)$$

TABLE 14.7 Derivation of Enzymatic Hydrolysis Kinetic Expressions Neglecting Back Reactions and Product Inhibition When Either the First Nucleophilic Reaction (Acylation or Alkylation) or the Second Nucleophilic Reaction (Deacylation and Dealkylation) Is the Slowest Step in the Overall Process

2nd Step Slowest	Equation	3rd Step Slowest	Equation
(i) $\frac{d[\mathrm{R-X}]}{dt} = -k_2[\mathrm{Enz} - \ddot{\mathrm{N}}\mathrm{u{:}R-X}]$	(14-77)	$\frac{d[\mathrm{R-X}]}{dt} = -k_3[\mathrm{H_2O}][\mathrm{Enz-Nu^+ - R}]$	(14-82)
(ii) Assume the complexation step is describable by an equilibrium constant:		Assume the complexation step and the first nucleophilic reaction are describable by equilibrium constants:	
$K_1 = \frac{[\mathrm{Enz} - \ddot{\mathrm{N}}\mathrm{u{:}R-X}]}{[\mathrm{Enz} - \ddot{\mathrm{N}}\mathrm{u}][\mathrm{R-X}]}$	(14-78)	$K_1 = \frac{[\mathrm{Enz} - \ddot{\mathrm{N}}\mathrm{u{:}R-X}]}{[\mathrm{Enz} - \ddot{\mathrm{N}}\mathrm{u}][\mathrm{R-X}]}$	(14-78)
		$K_2 = \frac{[\mathrm{Enz-Nu^+ - R}][\mathrm{X^-}]}{[\mathrm{Enz} - \ddot{\mathrm{N}}\mathrm{u{:}R-X}]}$	(14-83)
(iii) $E_T = [\mathrm{Enz} - \ddot{\mathrm{N}}\mathrm{u}] + [\mathrm{Enz} - \ddot{\mathrm{N}}\mathrm{u{:}R-X}]$ so substitute from (ii) and rearrange	(14-79)	$E_T = [\mathrm{Enz} - \ddot{\mathrm{N}}\mathrm{u}] + [\mathrm{Enz} - \ddot{\mathrm{N}}\mathrm{u{:}R-X}] + [\mathrm{Enz-Nu^+ - R}]$ so substitute from (ii) and rearrange	(14-84)
$[\mathrm{Enz} - \ddot{\mathrm{N}}\mathrm{u{:}R-X}] = \left[\frac{E_T[\mathrm{R-X}]}{K_1^{-1} + [\mathrm{R-X}]}\right]$	(14-80)	$[\mathrm{Enz-Nu^+ - R}] = [E_T]\frac{[\mathrm{R-X}]}{[\mathrm{X^-}]K_1^{-1}K_2^{-1} + [X^-][\mathrm{R-X}]K_2^{-1} + [\mathrm{R-X}]}$	(14-85)
(iv) Substitute (iii) into (i)		substitute (iii) into (i)	
$\frac{d[\mathrm{R-X}]}{dt} = -\frac{k_2[E_T][\mathrm{R-X}]}{K_1^{-1} + [\mathrm{R-X}]}$	(14-81)	$\frac{d[\mathrm{R-X}]}{dt} = -\frac{k_3[\mathrm{H_2O}][E_T][\mathrm{R-X}]}{[\mathrm{X^-}]K_1^{-1}K_2^{-2} + [\mathrm{X^-}][\mathrm{R-X}]K_2^{-1} + [\mathrm{R-X}]}$	(14-86)

(a)

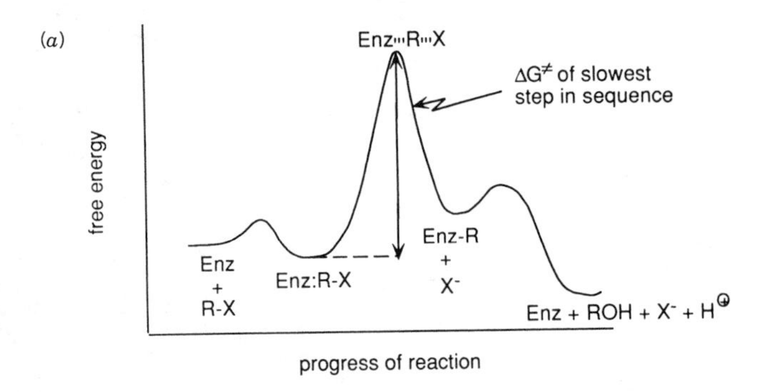

(b)

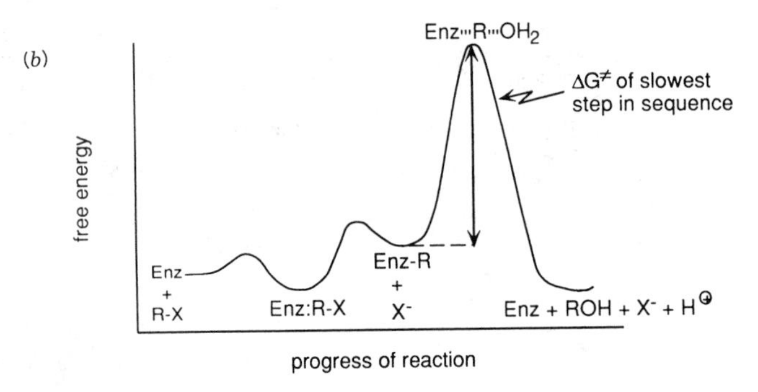

Figure 14.12 Schematic energy profiles for hydrolyses reactions carried out by enzymes when (*a*) the first nucleophilic reaction is rate-limiting and (*b*) the second nucleophilic step is the slowest process in the sequence.

and deduce that

$$\frac{d[\mathrm{R-X}]}{dt} = -\frac{k_3'[E_\mathrm{T}][\mathrm{R-X}]}{K_1^{-1}K_2^{-2}[\mathrm{X}^-] + K_2^{-1}[\mathrm{X}^-][\mathrm{R-X}] + [\mathrm{R-X}]} \qquad (14\text{-}86)$$

where

$[E_\mathrm{T}]$ is total enzyme concentration (e.g., mol enzyme $\cdot \mathrm{L}^{-1}$),

k_2 and k_3' ($= k_3 \cdot [\mathrm{H_2O}]$) are the rate constants associated with the slowest steps in the enzyme-catalyzed hydrolysis schemes (s^{-1}, s^{-1}, respectively),

K_1 is the equilibrium constant quantifying enzyme:substrate association, and

K_2 is the equilibrium constant quantifying the ratio of species established in step 2 whenever step 3 proves to be the slowest in the sequence.

We note that both of these rate expressions are hyperbolic as a function of substrate concentration [R – X]. That is, at low concentrations of [R – X] the rate linearly

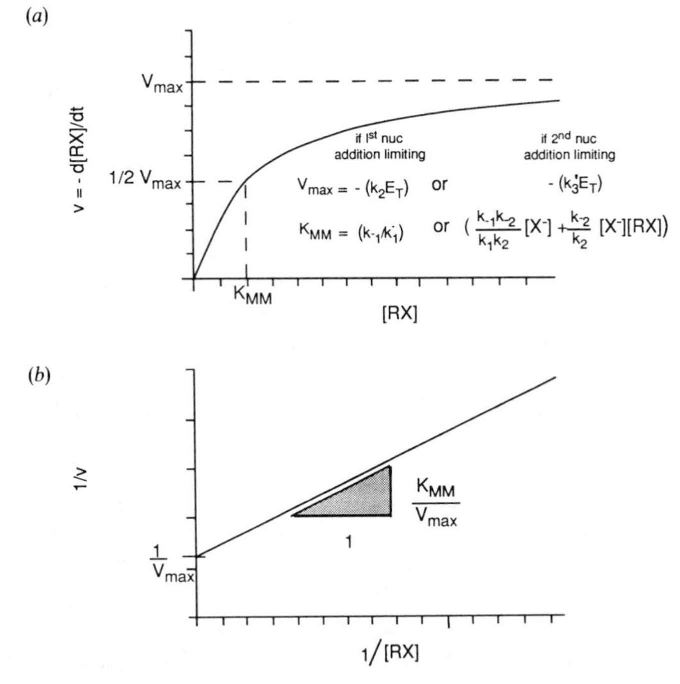

Figure 14.13 Relationships of biodegradation rate, $v = -[R - X]/dt$, to substrate concentration $[R - X]$ when Michaelis–Menton type enzyme kinetics is appropriate: (*a*) when plotted as hyperbolic relationship (Eq. 14-91 in text), or (*b*) when plotted as inverse equation, $1/v = (K_{MM}/V_{max})/[R - X] + (1/V_{max})$. Factors governing V_{max} and K_{MM} differ depending on rate-limiting step in the enzymatic sequence.

increases with $[R - X]$; but at high concentrations, the rate becomes independent of $[R - X]$.

These kinetic expressions now enable us to understand what the chemical or physical steps are embroiled in such hydrolyses within the familiar Michaelis–Menten type formulation typically used to describe enzyme kinetics (Fig. 14.13):

$$v = \frac{V_{max} \cdot [R - X]}{K_{MM} + [R - X]} \tag{14-91}$$

where

v is the rate of substrate removal ($= -d[R - X]/dt$),
V_{max} is the fastest possible removal rate ($= -k_2[E_T]$ or $-k_3'[E_T]$), and
K_{MM} is the concentration of substrate at which removal is half the fastest possible

$$K_{MM} = K_1^{-1} \quad \text{or} \quad ([X^-]/K_1K_2 + [X^-][R - X]/K_2).$$

First we see that V_{max} (the rate when $[R - X]$ is large relative to K_{MM}) is different for our two cases of hydrolysis limitation. This parameter always reflects the product of the rate constant of the slowest step and the concentration of total hydrolyzing enzyme present:

$$V_{max}\,(\text{1st nucleophilic attack slowest}) = -k_2[E_T] \tag{14-92}$$

or

$$V_{max}(\text{2nd nucleophilic attack slowest}) = -k_3'[E_T] \tag{14-93}$$

Not surprisingly, this maximal rate is always directly related to the total amount of hydrolyzing enzyme present. If such enzymes occur at a constant level in microorganisms, this result implies that V_{max} may also be directly related to cell numbers in a sample.

We also note the importance of the rate constants k_2 and k_3' for these two rate-limiting situations. For conditions where hydrolyzable substrates are always present in excess of their respective K_{MM}'s, we should be able to understand the relative rates of hydrolysis by using our knowledge of how variations in chemical structure are known to influence the ease of nucleophilic attacks. An interesting case involves a set of carboxylic acid esters with the same acid moiety (Fig. 14.14; Zerner et al., 1964); these compounds turn out to be limited at the second nucleophilic attack (deacylation) step. Since the *same Enz–Nu–acid complex* is cleaved for this set of esters in this second rate-limiting step, we expect all of these esters to exhibit nearly the same overall rate. This is consistent with the rate data that vary by about $\pm 15\%$ (Fig. 14.14). Conversely, when a family of esters is hydrolyzed with the first nucleophilic attack (i.e., enzyme acylation) as the slowest step in the process (Epand and Wilson, 1963), the rates are seen to vary more ($\pm 40\%$) even between related compounds (Fig. 14.14). In this case, as the alcohol moiety becomes larger (more steric hindrance, see E_s values in Table 12.4) and more electron donating (see σ^* values in Table 12.4), the rates of enzyme attack generally decrease (with the exception of the isobutyl compound). When an electron-withdrawing substituent (methyl pyridine) is present, the rate of enzyme acylation increases so much that the step is no longer rate-limiting. Obviously, if different steps are rate-limiting within a compound family (e.g., Eq. 14-101 versus Eqs. 14-97 through 14-100), we also see a different rate.

The study of Scholtz et al. (1987a, b) is another instructive example of an enzymatic hydrolysis. These investigators found that in cell-free extracts (i.e., cells broken open for experimentation) from an *Arthrobacter* sp., alkyl halides hydrolyzed at similar rates as in pure cultures of the intact bacteria. Hence transport limitations can be neglected in this case. Further, these investigators proposed an enzymatic sulfhydryl group to be involved in the reaction leading ultimately to the corresponding alcohols. The V_{max} values derived from their cell-free extract data for a series of alkyl halides are summarized in Table 14.8. If the first nucleophilic reaction step (enzyme alkylation) of the halidohydrolase was the slow part, we would expect that alkyl chlorides would be degraded slower than alkyl bromides and alkyl iodides (recall Fig. 12.5). Further, we might expect the longer carbon chain compounds to react somewhat more slowly due

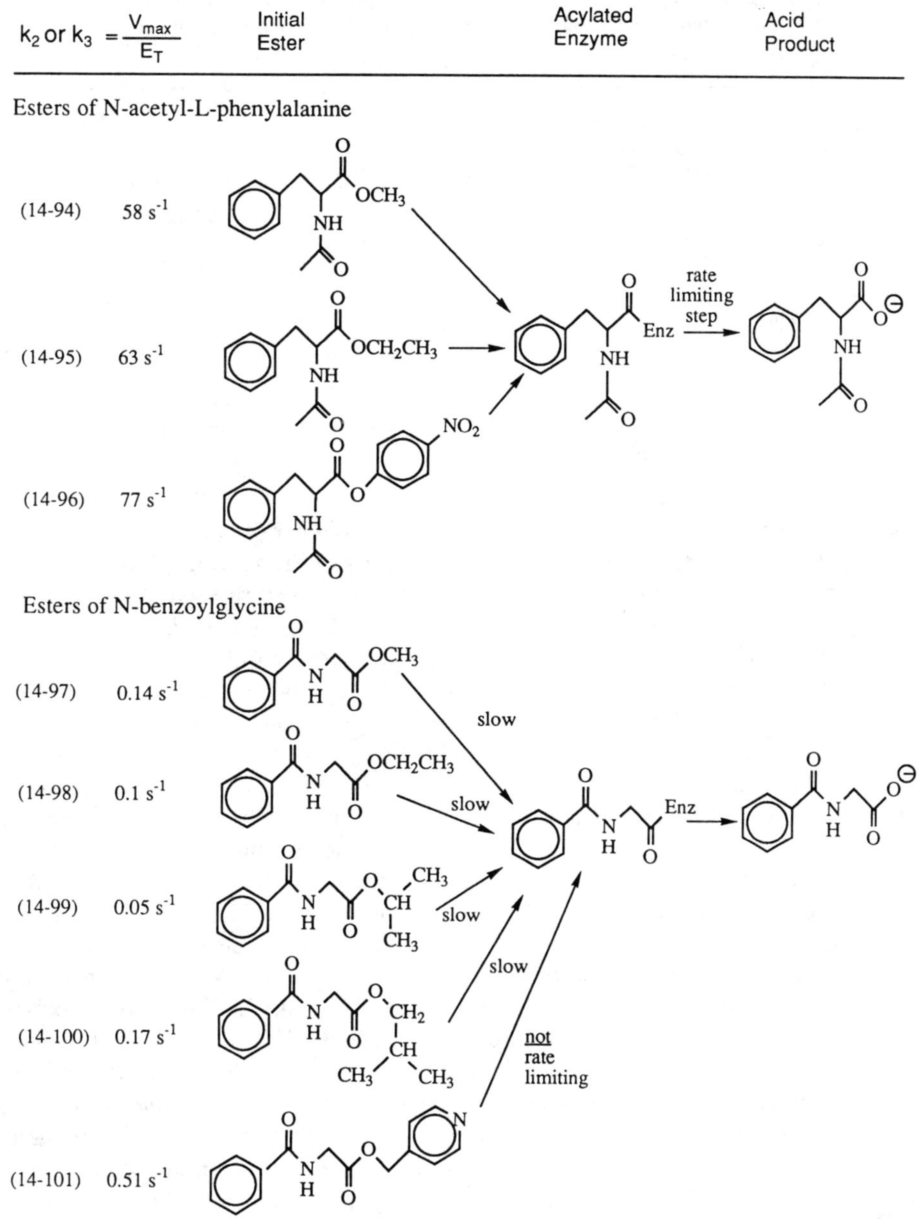

Figure 14.14 Rates of hydrolysis of two families of esters by a hydrolase, chymotrypsin. The esters of *N*-acetyl-L-phenylalanine exhibit very similar rates because the process in each case is limited by the same enzyme deacylation reaction (Zerner et al., 1964). The esters of *N*-benzoyl glycine exhibit rates varying by more than a factor of 3 because their hydrolyses are mostly limited by the initial enzyme acylation step (Epand and Wilson, 1963).

TABLE 14.8 Observed Maximal Rates (nmol·min^{-1}·mg^{-1} protein) of Hydrolysis in Cell-Free Extracts of *Arthrobacter sp.* Grown with 1-Chlorobutane for Primary Halide Compounds Present at 5 mM (in great excess of their K_{MM}).[a]

$$\text{R-X} + \text{Enz-SH} \xrightarrow[\text{H}^{\oplus},\ \text{X}^{\ominus}]{\text{enzyme alkylation}} \text{R-S-Enz} \xrightarrow[\text{enzyme dealkylation}]{\text{H}_2\text{O}} \text{R-OH} + \text{Enz-SH}$$

	Methyl-	Ethyl-	*n*-Propyl	*n*-Butyl-
—Cl	—	—	—	40
—Br	—	70	70	40
—I	50	60	80	50

[a] Data from Scholtz et al., 1987a, approximately corrected for fraction dissolved in the experimental media.

to steric hindrance of the initial nucleophilic attack. Generally the data in Table 14.8 indicate all of the halides shown reacted at about the same rate, which does not support either of these expectations. Thus, as in the case of the esters discussed above, the initial nucleophilic attack is apparently not the rate-limiting step. Dealkylation (i.e., release of the alkyl group bound covalently to the enzyme after the initial attack) may indeed be slow; the inability to break up such adducts is frequently blamed for the toxic effects of alkylating agents. Thus we focus on the second nucleophilic attack as determining the rate of these alkyl halide hydrolyses. In this case, the leaving group (i.e., the enzyme itself) has become identical for all compounds involved so we would not expect any rate variations because of this factor. Further, if owing to the action of the enzyme, the steric arrangement of the atoms involved in the second nucleophilic substitution reaction has already been optimized, then there might not be too much difference in steric accessibility between the various homologues. Thus the data shown in Table 14.8 support the conclusion that the third step (i.e., release of the enzymatic sulfhydryl group) is the rate-determining step in this reaction. We should note, however, that if another step in the dehalogenation of these alkyl halides is the same, that such a step could be rate-limiting. In either case, these data suggest it is possible to make reasonable estimates of V_{max} values for other alkyl halides for these conditions. In sum, if one is correct in the assumption that *the enzyme reaction mechanism is the same for all compounds in a family of interest*, then it may be reasonable to predict rates for new compounds in that family based on data from related data on other compounds.

We return to the hyperbolic rate laws (Eqs. 14-81 and 14-86) and consider the rates of biotransformation when [R – X] is very small. For hydrolyses in which the first nucleophilic attack is the slow step, we obtain the result:

$$\frac{d[\mathrm{R}-\mathrm{X}]}{dt} = -[k_2 K_1][E_\mathrm{T}][\mathrm{R}-\mathrm{X}] \qquad (14\text{-}102)$$

Alternatively, if the dealkylation is slowest, we have

$$\frac{d[\mathrm{R-X}]}{dt} = -\left[\frac{k_3[\mathrm{H_2O}]K_1K_2}{[\mathrm{X^-}]}\right][E_{\mathrm{T}}][\mathrm{R-X}] \tag{14-103}$$

Both of these results are very reminiscent of the mathematical form we used earlier to describe the biodegradation of benzo(f)quinoline (Eq. 14-72) and expressions used by others to model the hydrolysis of other pollutants (e.g., Paris et al., 1981, for the butoxyethylester of 2,4-dichlorophenoxy acetic acid; Wanner et al., 1989 for disulfoton and thiometon at low concentrations):

$$k_{\mathrm{bio}}\cdot[\mathrm{cells}] = [k_2K_1]\cdot[E_{\mathrm{T}}] \tag{14-104}$$

or

$$= \left[\frac{k_3[\mathrm{H_2O}]K_1K_2}{[\mathrm{X^-}]}\right][E_{\mathrm{T}}] \tag{14-105}$$

depending on which part of the transformation mechanism is rate-limiting. That is, a product of reaction constant information and a measure of biological abundance is used in both cases. Consequently, we now recognize the factors aggregated to yield a k_{bio}. [Note that the fit of chemical biodegradation data to such expressions does not mean that we have correctly identified the limiting process; benzo(f)quinoline is undoubtedly oxidized, not hydrolyzed, in its initial biotransformation. Such an oxidation could be described using kinetic derivations similar to those developed for hydrolysis in Table 14.7.]

As we saw for V_{max}, this result facilitates our understanding of how fast a set of related compounds should be hydrolyzed by a particular enzymatic system. First compounds with structures that encourage their binding to the hydrolase (i.e., increase K_1) will also be processed quicker. Interestingly, some hydrolases such as chymotrypsin (Berezin et al., 1970) are known to bind their substrates in large measure owing to hydrophobic interactions. Put another way, compound binding in such instances should correlate with measures of compound hydrophobicity (e.g., aqueous activity coefficient or K_{ow}). Additionally, if we know something about the rate-limiting nucleophilic substitution step, we can understand how structural variations within a compound class should influence either k_2 or k_3. Thus working with insights on susceptibility to nucleophilic attacks (e.g., chemical hydrolysis rates by hydroxyl anion) and hydrophobicity, one might find predictive relationships to describe relative biological hydrolysis rates of related sets of compounds.

Examination of Eqs. 14-81 and 14-86 also enables us to understand why Wanner et al. (1989) observed a fractional order (i.e., rate dependent on an exponent between 0 and 1) in the removal rate of disulfoton and thiometon when those compounds occurred at high concentrations:

$$\frac{d[\mathrm{R-X}]}{dt} \simeq -k_{\mathrm{obs}}[\mathrm{R-X}]^{0.5} \tag{14-106}$$

In this case $[R-X]$ was present at a level above that where first-order kinetics in compound concentration apply, but below that where the zero-order limit was appropriate.

Calibration of Hydrolyses Using Probe Compounds The preceeding brief discussions of enzyme kinetics suggest how we might approach the prediction of the *relative magnitudes* of biological hydrolyses under particular conditions. It is presently difficult to determine the total enzyme concentration $[E_T]$ (Eqs. 14-81 and 14-86) or the number of active microorganisms [cells] (Eq. 14-72) that metabolize a given compound in the environment, so approaches to obtain *a priori* estimates of microbially mediated hydrolysis rates are not forthcoming. However, as we saw for photochemically formed oxidants whose concentrations were difficult to determine directly (Section 13.4), we can now imagine using probe compounds to calibrate a given natural system of interest for its microbiological potential to hydrolyze the type of compounds considered. For example, to evaluate the hydrolysis of a set of structurally related compounds known to be present at concentrations which are small relative to their respective K_{MM} values and which we believe are limited in their overall transformation by the first nucleophilic attack rate, we could investigate one (or a few) probe chemical of the compound class added to the natural water of interest. Knowing its (their) removal rate(s), we could then approximate those of the related substances based on

$$\frac{\dfrac{d[\text{unknown}]/dt}{[\text{unknown}]}}{\dfrac{d[\text{probe}]/dt}{[\text{probe}]}} = \frac{k_{\text{obs,unknown}}}{k_{\text{obs,probe}}} \tag{14-107}$$

$$= \frac{k_{2,\text{unknown}} K_{1,\text{unknown}}}{k_{2,\text{probe}} K_{2,\text{probe}}} \tag{14-108}$$

If we assume that the rates of hydrolase nucleophilic attack vary from compound to compound in proportion to the rates of hydroxide ion reaction (e.g., if serine is the relevant nucleophile), then we have

$$\frac{k_{2,\text{unknown}}}{k_{2,\text{probe}}} \simeq \text{coeff}_1 \cdot \frac{k_{\text{OH,unknown}}}{k_{\text{OH,probe}}} \tag{14-109}$$

where the coefficient (coeff_1) reflects the differential nucleophilicities involved (recall Eq. 12-34). Further, if enzyme: compound complex formation is fast and driven largely by hydrophobic influences (such as in the case of chymotrypsin, Berezin et al., 1970), so that a relation of the form $\log(K_1) = \text{coeff}_2 \cdot \log K_{ow} + \text{coeff}_3$ holds (recall Eq. 7-12), then we may estimate:

$$K_{1,\text{unknown}} \simeq \left[\frac{K_{\text{ow,unknown}}}{K_{\text{ow,probe}}}\right]^{\text{coeff}_2} \cdot K_{1,\text{probe}} \tag{14-110}$$

where this second coefficient ($coeff_2$) characterizes the differential hydrophobicities of the enzyme-active site and octanol. Using these relations in Eq. 14-108, we expect

$$k_{obs,unknown} \simeq -\,coeff_1 \cdot \frac{k_{OH,unknown}}{k_{OH,probe}} \cdot \left[\frac{K_{ow,unknown}}{K_{ow,probe}}\right]^{coeff_2} \cdot k_{obs,probe} \qquad (14\text{-}111)$$

A similar expression can be derived for other cases, such as when compound concentrations are much greater than K_{MM} values. In this case, if acylation is rate-limiting, we find

$$k_{obs,unknown} = \frac{k_{2,unknown}}{k_{2,probe}} \cdot k_{obs,probe} \qquad (14\text{-}112)$$

$$\simeq coeff_1 \cdot \frac{k_{OH,unknown}}{k_{OH,probe}} \cdot k_{obs,probe} \qquad (14\text{-}113)$$

In contrast, if enzyme deacylation is the slowest step, then we may have

$$k_{obs,unknown} = \frac{k_{3,unknown}}{k_{3,probe}} \cdot k_{obs,probe} \qquad (14\text{-}114)$$

$$\simeq coeff_4 \cdot \frac{k_{OH,unknown}}{k_{OH,probe}} \cdot k_{obs,probe} \qquad (14\text{-}115)$$

where $coeff_4$ reflects the relation of the variability of nucleophilic attack from compound to compound in the deacylation step. A study described by Wolfe et al. (1980c), in which rates of microbial hydrolysis of a family of phthalate esters were compared to chemical reactions with hydroxide ion, suggests that correlations of biologically mediated hydrolysis rates may indeed correlate with chemical hydrolysis rates.

Thus calibration with a probe compound and knowledge of the enzyme chemistry involved in the substrate binding and reaction may allow us to predict biologically mediated hydrolysis reaction rates in environmental settings of interest. Note again the need for several assumptions to hold: (1) unknown and probe both at levels above or below their respective K_{MM}'s; (2) neither compound limited by its rate of transport to the enzymes, and (3) enzymatic processing is controlled by the same rate-limiting step.

Summary Comments on Michaelis–Menten Type Enzyme Kinetics Before ending our discussion of biodegradation rates that can be described using Michaelis–Menten type enzyme kinetics, we need to emphasize several points of caution. In order to have confidence in a technique like that illustrated above, to evaluate the relative reactivities of a series of related compounds, we must know several facts about the system. First, we must be sure that transport limitations are not controlling; that is, desorption from nearby solids or transport into the cell must not be the slowest step in the process

TABLE 14.9 Apparent Michaelis Menten Parameters Reported for Microbial Degradation of Various Substrates[a]

Substrate	K_{MM}	V_{max} ($mol \cdot kg^{-1} protein \cdot s^{-1}$)	Reference
	A. Natural populations		
Toluene in seawater	18 nM	6×10^{-10} $\left(1.5 \frac{pmol}{L \cdot h}\right)$	Reichardt et al., 1981
Biphenyl in seawater	1.5 nM	4×10^{-8} $\left(100 \frac{pmol}{L \cdot h}\right)$	
m-Cresol in estuarine seawater	6–17 nM	$5\text{–}4000 \times 10^{-9}$ $\left(4\text{–}1300 \frac{pmol}{L \cdot h}\right)$	Bartholomew and Pfaender, 1983
Chlorobenzene in estuarine seawater	9–46 nM	$2\text{–}4 \times 10^{-8}$ $\left(15\text{–}130 \frac{pmol}{L \cdot h}\right)$	
Trichlorobenzene in estuarine seawater	25–38 nM	$1\text{–}2 \times 10^{-8}$ $\left(13\text{–}43 \frac{pmol}{L \cdot h}\right)$	
Nitrilotriacetic acid in estuarine seawater	290–580 nM	$4\text{–}400 \times 10^{-7}$ $\left(300\text{–}2600 \frac{pmol}{L \cdot h}\right)$	

	B. Intact microorganisms		
3-Chloro-benzoate in enrichment of methanogenic consortium	67 μM	1.1×10^{-4} $\left(24 \frac{\mu mol}{L \cdot h}\right)$	Suflita et al., 1983
3,5-Dichloro-benzoate in enrichment of methanogenic consortium	47 μM	3.6×10^{-5} $\left(7.7 \frac{\mu mol}{L \cdot h}\right)$	
4-Amino-3,5-dichlorobenzoate in enrichment of methanogenic consortium	30–60 μM	$1.1\text{–}1.4 \times 10^{-5}$ $\left(2.5\text{–}3.1 \frac{\mu mol}{L \cdot h}\right)$	
	C. Cell-free extracts or isolated enzymes		
4-Chlorobenzoic acid → 4-hydroxybenzoic acid	30 μM	2×10^{-3}	Marks et al., 1989
Fluoroacetate → hydroxyacetate (glycolate)	2.4 mM	—	Goldman, 1965
Chloroacetate →	20 mM	—	
1-Chlorohexane → 1-hexanol	40 μM	4.7×10^{-4}	Scholtz et al., 1987a
1-Chloropentane → 1-pentanol	50 μM	6.3×10^{-4}	Scholtz et al., 1987b
1-Chlorobutane → 1-butanol	60 μM	7.0×10^{-4}	
1-Chloropropane → 1-propanol	120 μM	1.7×10^{-4}	
1-Bromopropane → 1-propanol	20 μM	1.1×10^{-3}	
1-Iodopropane → 1-propanol	80 μM	1.3×10^{-3}	
CH_2Cl_2	30 μM	1.7×10^{-2}	Kohler-Staub and Leisinger, 1985
CH_2BrCl	15 μM	1.5×10^{-2}	
CH_2Br_2	13 μM	1.6×10^{-2}	
CH_2I_2	5 μM	4.2×10^{-3}	
Linuron → (3,4-Dichlorophenyl)-1-methoxy-1-methylurea	2 μM	2.5×10^{-3}	Englehardt et al., 1973
o-Nitrophenol → catechol	8 μM	8×10^{-2}	Zeyer and Kocher, 1988

[a] For natural microbial populations, cell counts are converted to protein assuming 1×10^{-16} kg of protein per cell.

(steps 1 or 5 in Fig. 14.2). Also we must assume the enzymes acting on the substrates of interest are essentially unchanging in their abundance (i.e., E_T is not increasing as indicated by steps 6 or 7 of Fig. 14.2). We must understand the enzymatic mechanism involved, particularly the rate-limiting step and the forces governing chemical binding to the enzyme active site. Finally, we must know whether substrates are present at high or low levels relative to their K_{MM} values. Although the available data are very limited, some organic compounds exhibit K_{MM} values in the nanomolar and micromolar ranges (Table 14.9). Therefore, for these chemicals we may expect millimolar and higher levels of substrates to reflect saturation enzyme kinetics (i.e., zero-order with respect to the organic substrate concentration). Conversely, nanomolar and lower concentrations may be degraded with rate constants given by the ratio V_{max}/K_{MM} and exhibit first-order kinetics with respect to substrate concentration. The V_{max} data, reported per volume of culture or natural water (Table 14.9), are of course highly dependent on the numbers of active microorganisms present in any particular water sample investigated. Researchers often try to normalize such V_{max} results to the cell counts or mass of protein present to enable comparisons. Recognizing that not much of the protein in a medium is directly involved in compound degradation, we can easily understand why this still leaves such a wide range of V_{max} results. Thus, while K_{MM} data may be applicable from case to case, one should not expect the same to hold for V_{max} information. Unfortunately, although a great deal of biodegradation data has been collected in real world samples, one must be cautious in presuming to know whether enzymatic limitations explain the reported rates.

Monod Kinetics We conclude this chapter by considering the case where a microorganism population increases because the chemical of interest to us supports that microorganism's growth (step 7 in Fig. 14.2). An experiment reported by Smith and colleagues (1978) is illustrative (Fig. 14.15). In this case, microorganisms that were present in water from a pond and that could grow on *para*-hydroxytoluene (*p*-cresol) were enriched. When a low abundance of these microorganisms, 10^7 cells/L, were exposed to a 40–50 μmolar solution of the *p*-cresol, it initially *appeared* that the chemical was not being metabolized; that is, a lag period was seen. In this instance, since the microorganisms were chosen to be able to degrade this phenolic compound, it can be assumed that the lag phase was not due to the need to induce the necessary enzyme(s). Rather, as indicated in the upper portion of Figure 14.15, initially the cell numbers were too low to have any discernible impact on the relatively high *p*-cresol concentrations. Obviously, the cells multiplied very quickly in the period from 2 to 16 hours, and when they finally reached abundances greater than about 10^9 cells per liter, enough degraders were present to cause significant substrate depletion. Thus, to describe the timecourse of this chemical concentration variation in this case, the microbial population dynamics in the system has to be quantified. This can be done using the approach suggested by Monod (1949).

We begin by considering the relationship of cell numbers in time for a growing population like our *p*-cresol degraders. We find that for a certain time, the cell numbers increase exponentially, and this period of so-called exponential growth can be described

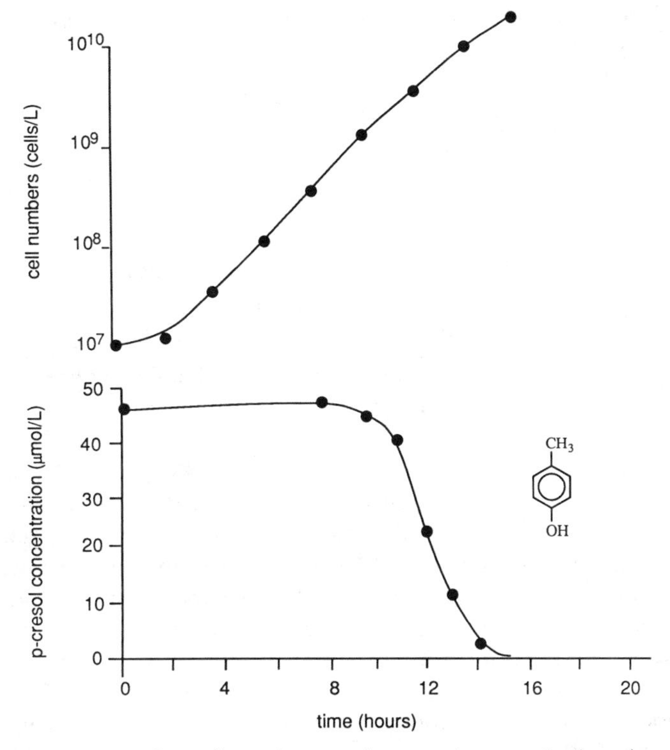

Figure 14.15 Timecourses for cell numbers and *p*-cresol concentrations in a batch culture experiment (Smith et al., 1978).

by

$$\frac{d[B]}{dt} = \mu \cdot [B] \quad \text{or} \quad [B] = [B]_0 e^{\mu t} \tag{14-116}$$

where $[B]$ is the cell abundance (cells/L), and μ is the specific growth rate (h^{-1}). Hence, $\ln [B]$ will change in direct proportion to t:

$$\ln B(t_2) = \ln B(t_1) + \mu \cdot (t_2 - t_1) \tag{14-117}$$

Put another way, during exponential growth the population will double in number for every time interval, $t = (\ln 2)/\mu$.

Monod (1949) recognized that the growth rate of a microorganism community was related to the abundance of a critical substance sustaining its growth. More "food" means faster growth, at least up to a certain point when the *maximal growth rate*, μ_{max}, is achieved (or some other factor becomes limiting). He mathematically related this growth response to the concentration of the substance limiting growth with the expression

$$\mu = \frac{(\mu_{max})[RH]}{K_M + [RH]} \qquad (14\text{-}118)$$

where

μ_{max} is the fastest possible growth rate (h^{-1}) corresponding to the situation when the limiting chemical (e.g., *p*-cresol) is present in excess,

$[RH]$ is the concentration (mol/L) of the growth-limiting chemical, and

K_M is Monod constant (mol/L) equivalent to the chemical concentration at which population growth is half maximal.

This formulation yields a hyperbolic relation of μ on $[RH]$ (Fig. 14.16). That is, when there is a surplus level of the "food" chemical, other factors limit the rate of population increase. For example, the supply of another nutrient such as nitrogen, may control growth. Consequently, we see a plateau in the relationship of growth rate versus concentration of substrate of interest to us. In the experiment of Smith et al. (1978), *p*-cresol concentration did not change in the first 10 hours; this chemical was obviously present in excess and the population growth was at its μ_{max}.

Now, rather than emphasize what is happening to the numbers of microorganisms, we wish to focus on how the limiting chemical is changing in concentration. To do this we need to relate growth to changes in compound concentration. This is done by recognizing that processing a certain amount of chemical mass enables a proportional enhancement in microbial biomass:

$$\frac{d[B]/dt}{d[RH]/dt} = Y\left[\frac{\text{cells grown}}{\text{moles of substrate used}}\right] \qquad (14\text{-}119)$$

This proportionality is called the *yield* of the particular biological process and it is commonly denoted as *Y*. Using this yield information, we can now relate the production rate of new cells, $\mu(h^{-1})\cdot B(\text{cells}\cdot L^{-1})$, to the disappearance rate of the chemical of concern:

$$\mu\cdot[B] = Y\,(\text{cells}\cdot\text{mol}^{-1})\cdot\frac{-d[RH]}{dt}(\text{mol}\cdot L^{-1}\cdot h^{-1}) \qquad (14\text{-}120)$$

And, upon rearranging:

$$\frac{d[RH]}{dt} = -\frac{\mu\cdot[B]}{Y} \qquad (14\text{-}121)$$

Substituting Monod's mathematical description of microbial growth (Eq. 14-118) into

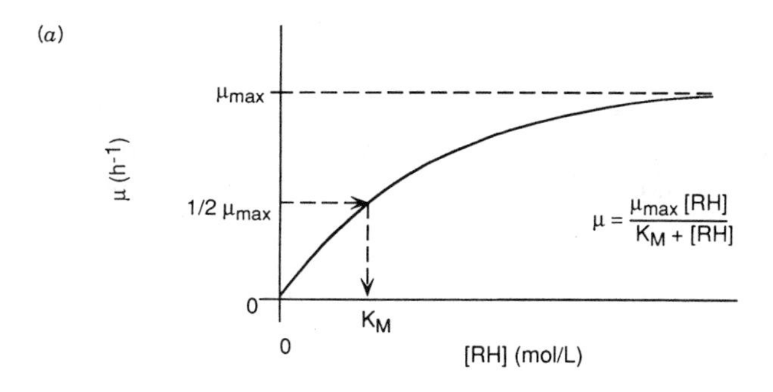

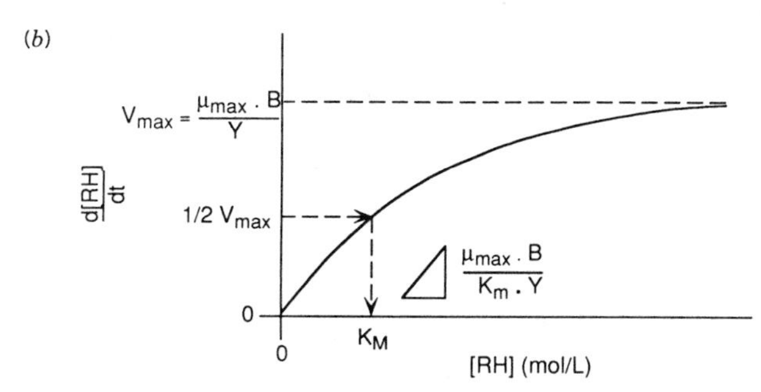

Figure 14.16 Schematic relationships of: (*a*) microbial population specific growth rate μ versus substrate concentration after Monod (1949), and (*b*) consequent substrate disappearance rate, $d[RH]/dt$, versus substrate concentration.

Eq. 14-121, we have

$$\frac{d[RH]}{dt} = -\frac{\mu_{\max}\cdot[B]\cdot Y^{-1}\cdot[RH]}{K_{\mathrm{M}}+[RH]} \tag{14-122}$$

This variation of chemical removal rate $d[RH]/dt$, with its concentration, $[RH]$, has the same hyperbolic form as we found for Michaelis–Menten type enzyme kinetics; consequently, we note that one cannot deduce whether enzyme chemistry or microbial growth determines the rate of chemical degradation by examining the influence of $[RH]$ on $d[RH]/dt$.

As illustrated in Figure 14.16, the relationship of microbial growth and chemical removal (Eq. 14-122) implies that when the chemical is present at low levels ($[RH] \ll K_{\mathrm{M}}$), its instantaneous rate of degradation (i.e., at a particular microbial abundance $[B]$) is linearly proportional to its concentration and the "concentration" of the microorganisms "reacting" with it. The proportionality is equal to a combination of factors describing microbial community growth:

$$\frac{d[RH]}{dt} = -\left[\frac{\mu_{\max}}{K_{\mathrm{M}}\cdot Y}\right][B][RH] \tag{14-123}$$

$$= -k_{bio}[B][RH] \tag{14-124}$$

where k_{bio} is equal to $\mu_{max} \cdot K_M^{-1} \cdot Y^{-1}$ when $[RH] \ll K_m$. In contrast, when the chemical is present in large amounts relative to the microbial community needs ($[RH] \gg K_M$), then its rate of removal becomes independent of its concentration:

$$\frac{d[RH]}{dt} = -\left[\frac{\mu_{max}}{Y}\right][B] \tag{14-125}$$

Consequently, k_{bio} is equal to μ_{max}/Y for $[RH] \gg K_M$. In both of these limiting cases, and obviously for the transitional conditions between, we need information on the factors quantifying microbial growth, μ_{max}, K_M, and Y, to be able to predict transformation rates of the compound that limits population growth.

From the example shown in Figure 14.15, we can see how some of these biological parameters are deduced. First, we recognize that in the early part of the experiment, when *p*-cresol levels do not change (< 10 hr), the substrate is not limiting growth, so the changing cell numbers reflect μ_{max}. From the upper portion of the figure, we see that cell numbers increase from about 10^7 cells/L at 2 hours to about 10^9 cells/L at 10 hours; using Eq. 14-117 we estimate

$$\mu_{max} = \ln\left[\frac{[B(t_2)]}{[B(t_1)]}\right] \Big/ [t_2 - t_1] \tag{14-126}$$

$$\simeq [\ln[100]]/8\,\text{h}$$

$$\simeq 0.6\,\text{h}^{-1}$$

Examining the results between about 10 and 14 hours, we can estimate the yield factor:

$$Y = \frac{[B(14\,\text{h})] - [B(10\,\text{h})]}{[p\text{-cresol}(10\,\text{h})] - [p\text{-cresol}(14\,\text{h})]} \tag{14-127}$$

$$= \frac{9.4 \times 10^9\,\text{cells/L} - 1.3 \times 10^9\,\text{cells/L}}{4.4 \times 10^{-6}\,\text{mol/L} - 0.3 \times 10^{-6}\,\text{mol/L}}$$

$$\simeq 2 \times 10^{14}\,\text{cells/mol}$$

Since bacterial cells typically weigh on the order of 0.1 to 1 pg dry weight/cell and are about half carbon, this yield appears reasonable (i.e., produce 20–200 g of cells from about 100 g of *p*-cresol).

Finally, to deduce K_M we need to examine growth rates at relatively low levels of *p*-cresol, as well as the relatively high concentration conditions exhibited in Figure 14.15. Smith et al. (1978) have performed such incubations. Inversion of Eq. (14-118) yields

$$\frac{1}{\mu} = \left[\frac{K_M}{\mu_{max}}\right]\left[\frac{1}{[RH]}\right] + \left[\frac{1}{\mu_{max}}\right] \tag{14-128}$$

Hence a fit of $[RH]^{-1}$ versus μ^{-1} yields from the intercept $\mu_{max} = 0.69\ h^{-1}$, and from the slope divided by the intercept $K_M = 6.4 \times 10^{-6}$ mol/L.

Using these microbial population dynamics, we are now in a position to estimate biodegradation rates for compounds supporting growth like *p*-cresol. In the situation depicted in Figure 14-15, we have for the early part of the experiment,

$$k_{bio} = \frac{\mu_{max}}{Y} \qquad \text{since } [RH] \gg K_m \tag{14-129}$$

$$\simeq \frac{0.6\ h^{-1}}{(2 \times 10^{14}\ \text{cells/mol})}$$

$$\simeq 3 \times 10^{-15}\ \text{mol} \cdot \text{cell}^{-1} \cdot h^{-1}$$

Thus, early in the timecourse when $[RH] \gg K_M$, the rate of *p*-cresol removal was continuously changing as microorganism growth occurred, ranging from about 3×10^{-8} $mol \cdot L^{-1} \cdot h^{-1}$ at 2 h [when $[B(2\ h)] \sim 10^7$ cells/L] to about $3 \times 10^{-6}\ mol \cdot L^{-1} \cdot h^{-1}$ at 10 h [when $[B(10\ h)] \sim 10^9$ cells/L]. Subsequently, the rate of *p*-cresol removal started to become a function of the concentration of this substrate; so near the end of the incubation (say 14 hours) we have

$$\frac{d[p\text{-cresol}]}{dt} = -\frac{(\mu_{max})[B(14\ h)][p\text{-cresol}]}{(K_M + [p\text{-cresol}])(Y)} \tag{14-130}$$

$$\simeq -\frac{(0.6\ h^{-1})(10^{10}\ \text{cell/L})(3 \times 10^{-6}\ \text{mol/L})}{(6 \times 10^{-6}\ \text{mol/L} + 3 \times 10^{-6}\ \text{mol/L})(2 \times 10^{14}\ \text{cells/mol})}$$

$$= 1 \times 10^{-5}\ \text{mol} \cdot L^{-1} \cdot h^{-1}$$

Various xenobiotic compounds have been studied for their ability to be the sole support of growth for certain microorganisms (Table 14.10). In these cases, specialized bacteria are obtained by forcing the species to survive when only the organic chemical of interest is provided as an energy and/or carbon source. Sometimes this "enrichment" procedure simply isolates a preexisting subpopulation of bacteria from the mixture of organisms present in a natural sample; however other times a mutation must occur which permits survival on the chemical provided (e.g., Brunner et al., 1980). From the data shown in Table 14.10, a few cautious generalizations may be suggested. First, maximum cell growth rates for acclimated cultures appear to correspond to doubling times of one to several hours. This is not too different from cells grown on "normal" substrates like glucose. Next, K_M values reported for the few xenobiotic compounds

TABLE 14.10 Some Monod Biodegradation Parameters Obtained from Enrichment Cultures Grown on the Substrate Indicated

Substrate	Source of Microorganisms	μ_{max} (h^{-1})	K_M (μM)	Y (cells/mol)	Reference
Malathion	Bacterial enrichments from river water	0.37	2.2	4×10^{10}	Paris et al., 1975
p-Cresol	Bacterial enrichments from pond water	0.69	6.4	2×10^{14}	Smith et al., 1978
Quinoline	Enrichments from pond	0.74	1.2	2×10^{14}	Smith et al., 1978
Methyl parathion	Enrichments from creek	0.61	10	2×10^{14}	
Methylene chloride	Enrichments of *Pseudomonas sp.* mutants	0.11	—	$\approx \frac{10^{13}\text{–}10^{14}}{\text{mol}}$	Brunner et al., 1980
Glycerol	Pure cultures				
	Aerobacter	1.2	120		Jannasch, 1967
	Achromobacter	0.55	11		
Glucose	Pure cultures				
	Vibrio, Aerobacter, Achromobacter, Escherichia coli	0.40–0.65	17–46		Jannasch, 1968

studied in this manner are between μM and mM levels. Finally, the cell yields appear to fall in the range such that 10–100% of the xenobiotic compound mass is translated into biomass. An exception to this is seen for malathion, which was found to be hydrolyzed to the monoacid and ethanol. The ethanol was used as a growth substrate, but the acid product was accumulated without being used further. Obviously, the gross yield in this case would not be large.

Given the typical values of μ_{max}, K_M, and Y, under conditions where excess xenobiotic chemical is added to an environment containing some cells capable of living on it, we might expect k_{bio} to be about

$$k_{bio} \simeq \frac{0.1 \text{ to } 1\ \text{h}^{-1}}{(0.1\text{–}1\ \text{g cells/1 g compound})} \qquad (\text{assuming } [RH] \gg K_M) \qquad (14\text{-}131)$$

$$\simeq \frac{0.1 \text{ to } 1\ \text{h}^{-1}}{(10^{11}\text{–}10^{13}\ \text{cells}/10^{-2}\ \text{mol})}$$

$$\simeq 10^{-16}\text{–}10^{-13}\ \text{mol/cell}\cdot\text{h}$$

The factor controlling the time it would take for total compound degradation would be the abundance of the subpopulation of degraders. This, then, brings us to the major weakness in trying to quantify degradation limited by microbial growth: how does one know what subset(s) of microorganisms are involved and what their abundance is for any particular environmental situation of interest?

Let us conclude with an example calculation showing how we might use the Monod kinetics approach to evaluate the time required for biodegradation to remove a xenobiotic compound. Imagine the case where the compound (e.g., *p*-cresol) is added in a sharp input (e.g., a spill) and dispersed to a particular concentration, $[RH]_0$. Some subpopulation of the microorganisms may be capable of living on this chemical, and so they metabolize it and rapidly increase in number. As we saw in the experiment done by Smith et al. (1978), cell numbers finally increase to a point where their rate of increase causes a dramatic decrease in pollutant concentration. We estimate that this cell abundance, $[B_{crit}]$, is sufficient to consume all of the chemical during the next doubling interval:

$$[B_{crit}](\text{cells}\cdot\text{L}^{-1}) \simeq [RH]_0(\text{mol}\cdot\text{L}^{-1})\cdot\text{Y}(\text{cells}\cdot\text{mol}^{-1}) \qquad (14\text{-}132)$$

For the case illustrated in Figure 14.15, $[B_{crit}]$ would be about 3×10^9 cells$\cdot$L^{-1}. Since the population numbers are increasing exponentially, we deduce that the time period t_{crit} from the spill until there is a sharp drop in pollutant concentration can be estimated by calculating how long it will take for the microorganisms to reach $[B_{crit}]$:

$$t_{crit} \simeq [\ln([B_{crit}]/[B_0])]/\mu_{max} \qquad (14\text{-}133)$$

$$\simeq [\ln([RH]_0\cdot Y/[B_0])]/\mu_{max} \qquad (14\text{-}134)$$

To check the effectiveness of this expression, we apply it to the data of Smith et al. (1978) for *p*-cresol:

$$t_{crit} \simeq \left[\ln\left(\frac{46 \times 10^{-6}\,\text{mol}\cdot\text{L}^{-1}\cdot 2 \times 10^{14}\,\text{cells}\cdot\text{mol}^{-1}}{1 \times 10^{7}\,\text{cells}\cdot\text{L}^{-1}} \right) \right] \Big/ 0.7\,\text{h}^{-1} \qquad (14\text{-}135)$$

$$\simeq 10\,\text{h}$$

Recalling Figure 14.15, we see that indeed after approximately 10 hours the *p*-cresol levels dramatically drop! Thus, we have a nice approach to estimate this critical time interval using growth rate inputs (such as Y and μ_{max}) and information on initial degrader numbers $[B_0]$. Of course, all of this presumes that other factors do not limit the microbial population growth in this catastrophic incident-type case. Such factors would include difficulties of mass transport of pollutant molecules (e.g., in an oil slick or tar balls) to microorganisms in a water column. Also other critical nutrients such as nitrogen or phosphorus species must be present in sufficient quantities to permit unchecked microbial growth.

14.5 CLOSING REMARKS

We have now seen that microorganisms can initiate structural changes on organic chemicals using special tools like enzymatically formed electrophilic oxygen, nucleophilic hydrogens, or other nucleophilic enzyme moieties. The rates of microbial transformations may be limited by a variety of mechanisms, and often the mathematical models which fit such loss rates (e.g., $d[RH]/dt = -k_{obs}[RH]$) provide very little help in distinguishing between limitations. Although quantitative treatment of biodegradation remains elusive, there is a good case to be made for believing we can handle Michaelis–Menten type enzyme chemistry or Monod microbial growth dynamics if these are identified as critical. Situations requiring genetic changes in microorganisms to enable biotransformations appear much more difficult to predict at present.

CHAPTER 15

MODELLING CONCEPTS

15.1 INTRODUCTION

Models are popular, especially among environmental scientists. The technical literature is full of papers in which measurements are compared to "tailor-made" models ending with statements like "... the model can adequately describe the measurements." Often, the reader may then be tempted to conclude: that is nice, but are real data not better than any model?

Although the present use of the word "model" is relatively new, the concept behind it dates back to the origin of modern sciences. The success of physics is linked to the perception that understanding is only possible at the price of simplification. The fundamentals of classical mechanics, Newton's laws, could only be established by ignoring disturbing phenomena such as friction or heat. All the many rules which in physics and chemistry are commonly called "laws" are, in fact, models of some specific phenomena. We have discussed many of them throughout this book: Fickian diffusion, sorption, air–water exchange, and the theory of reaction kinetics, to name just a few; and each of these is a model for physicochemical processes relevant to the behavior of chemicals in environmental systems.

There is only a gradual difference between these "process models" and the models employed in environmental sciences. Of course, the latter are less general and less fundamental, but there is the same general philosophy behind them, which can be defined in the following way:

> A model is an imitation of reality which stresses those aspects that are assumed to be important and omits all properties considered to be nonessential.

Indeed, the contribution of this chapter is not really the modeling as such but the idea of integrating several processes into a larger structure under the general constraint of mass balance applied to an adequately chosen subunit of the environment. Let us, for instance, discuss the fate of phenanthrene in a lake.

phenanthrene

From the physicochemical properties of this compound we expect that the following transfer and transformation processes have to be considered: (1) exchange between the lake surface and the overlying air, (2) sorption–desorption between the aqueous solution and solids (sediments and suspended particles), (3) direct and indirect photolysis, and (4) biodegradation. Most of these processes can be quantified by the various process models developed earlier in this book. For instance, air–water exchange can be described by the stagnant-boundary model of Chapter 10, the solute–particle interaction with the sorption model of Chapter 11, and photolysis with the concepts of Chapter 13.

It may turn out that some of these processes, compared to others, do not significantly influence the fate of phenanthrene in the lake and can thus be omitted. How can the relevance of a process be quantified? How, for instance, shall we compare the effect of air–water exchange (a mass flux at the water surface) to that of biodegradation (a process probably occurring in the "bulk volume" of the water body) or photolysis, which strongly depends on light intensity and thus is spatially variable in the water body? How shall we deal with inputs via rivers, sewers, and wet and dry deposition onto the lake surface?

As explained in greater detail in the next section, the method for putting these processes together utilizes the principle of mass balance applied to the system as a whole or to some parts of it ("control volumes"). Control volumes are connected by internal transport processes such as diffusion or advection (see Chapter 9). The system as a whole is linked to the environment by external inputs and outputs. In fact, the combination of transport and transformation processes and the necessity for defining compartments and subcompartments are the essential characteristics associated with modeling of environmental systems.

> An environmental system is defined as a subunit of the environment separated by a boundary from the rest of the world. The description of the system is comprised of the relations within the system as well as those characterizing the action of the outside world on the system.

Note the hierarchy between system and environment hidden in the definition given above: the environmental "forces," which are considered to be given by some outside "power," drive the system without being driven by the system, in turn (Fig. 15.1). From a mathematical viewpoint this means that the external functions influence the

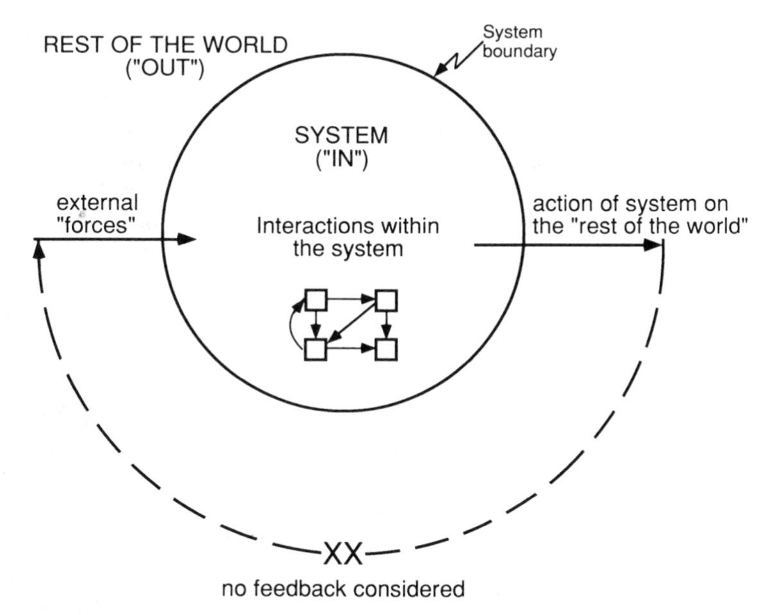

Figure 15.1 An environmental system is a subunit of the world defined by the system boundary separating the system ("in") from the rest of the world ("out"). The model describes the system as it responds to external influences ("forces") but neglects the influence of the system on the outside world.

dynamics of the system variables without themselves being controlled from inside the system.

Choosing a model first requires choosing a system, that is, a boundary between "in" and "out." Let us continue our example and search for an adequate model to describe the concentration of phenanthrene in a lake. A first and obvious candidate for the boundary of the system is the interface between the water body of the lake and its surroundings (lake sediments, atmosphere, inlets). Thus, this model needs information on driving forces from inputs of phenanthrene by rivers, through the atmosphere, and from the sediments of the lake. Obviously, the quality of the model result cannot be better than the quality of these input data. To assess the sedimentary input, information regarding the phenanthrene content of the sediments is needed. We may decide to include the sediments in the system and to shift the boundary of the model away from the sediment–water interface into the sediments. In a further step we may conclude that predicting the phenanthrene level in the lake is not possible without including the processes within the drainage area of the lake, thus moving the boundary of the system again. Eventually, we may even include in our system the political institutions which decide on the emission standards for phenanthrene and are thus the most important driving force for the future phenanthrene level in the lake!

The second step for the construction of a model involves choosing the complexity of the "internal" system description. Remember, a model is like a caricature of a real system. Depending on the intention of the cartoonist, his portrayal of a certain person may look very different. This is also true for models. There is no unique (or best) lake model. If we want to describe the fate of phenanthrene in lakes over decades, the most convenient model would certainly be very different from a model suitable for describing the daily concentration variations of this compound. Choosing the model structure is the cartoonist's task. There are no simple recipes for this; but there is at least one point to be mentioned, namely, that there should be a close relationship between the model structure and the model's purpose! In other words, before we can select a model we should know its purpose.

Let us elaborate on this from the environmental scientist's or engineer's point of view. The ultimate goal of modeling is to understand the relevant processes and to be able to make predictions regarding the impact of human activities on the environment. Within the framework of this book this often means predicting "exposures" in space and time, that is, concentration levels of chemicals in various environmental systems as a function of time.

There is an important difference between the classical and environmental sciences which is relevant to the way we apply models. Classical sciences primarily utilize controlled experiments. The laboratory is a synthetic world in which the external influences can be chosen in such a way that the outcome of an experiment can be reduced to the one cause which stands at the center of our current interest. Environmental sciences are based on laboratory knowledge as well. For instance, we can determine the relevant physicochemical properties of a compound, such as its aqueous solubility, its vapor pressure, or its molecular diffusion coefficient in water. However, there are other necessary pieces of information which do not allow us to retreat entirely into the laboratory. For instance, even with the best physicochemical knowledge, it would not have been possible to predict adequately the rapid distribution of DDT throughout the global environment. There was only one global DDT experiment, complex and unplanned in its set-up and not easy to reconstruct. If we want to learn something from this experiment (at least *a posteriori*), we have to combine the field observations with "model experiments," that is, with scenarios which we can invoke as alternative explanations for the one existing real-time experiment. Thus, to a certain degree modeling serves as a substitute for the controlled experiment which cannot be conducted in natural systems. In addition, models provide us with an understanding of the essential phenomena, the key parameters, and the most sensitive information needed for a given situation. This, in turn, can serve as guidance for the monitoring of environmental systems.

Let us summarize these considerations on the role of modeling in the following ways:

1. Mathematical models play a central role in the development of sciences. They allow us to combine knowledge acquired for different systems and/or different situations with the ultimate aim of constructing general theories. Models are also essential to making predictions, which then are verified by measurements.

2. Every classic theory began as a model, and it is only our long positive experience with some of these models which leads us to believe that they are more true and more fundamental than others and should thus bear the name "law". Doubtless there

are different levels of models. The "law-models" are nearly exclusively found in physics and chemistry. In biology some fundamental laws (in genetics and evolution theory) are making their way up Mount Olympus. The situation is less clear in ecology and environmental sciences, in spite of the richness of our understanding gained during the last few decades. And yet, the same scientific process guides the physicist and the ecologist, and thus their models play the same roles in their lives.

3. In environmental sciences models serve to sort out alternative explanations for the observations made in nature which cannot be controlled in the same way as we can control the conditions for a laboratory experiment. Thus, model calculations serve as substitutes for unrepeatable experiments such as the input of DDT into the global environment or the input of freons into the atmosphere.

4. Models that are developed only to reproduce some existing field data are of limited scientific use. Models should always be constructed to transfer the knowledge acquired in one system to another situation. They should be used to obtain a deeper understanding of the processes responsible for the measured data. A model should be applied to design new experiments or observation programs that are critical for the testing of hypotheses.

5. Real observations are always better than outputs produced by some model. But model results can help us examine various scenarios (e.g., what if DDT was still in use?) and likely future outcomes (e.g., what will happen to groundwater near a hazardous waste disposal site?) regarding the behavior of environmental systems for which real data would never be collected as frequently and as ubiquitously as we may need them.

15.2 MASS BALANCE—THE CORNERSTONE OF BOX MODELS

As mentioned in the preceding section, our goal in modeling will be to analyze all the relevant processes simultaneously. To this end, the concept of *mass balance* serves as the means to link everything together. To avoid lengthy theoretical discourses, the explanations in this and the following sections are based on one specific aquatic system, a lake. All the important phenomena affecting anthropogenic compounds occur in lakes, and lakes are still simple enough to allow the construction of manageable, but realistic cases. While proceeding from the simple to the more complex lake models, information on the functioning of lakes will be introduced where it is required for the understanding of the examples.

To use the idea of mass balance, the system is first divided into one or several "control volumes" (CV) which are connected with each other and with the rest of the world by mass fluxes. Next, for each CV and each chemical, a mass balance equation is written:

$$\begin{pmatrix}\text{Change of mass}\\ \text{in CV with time}\end{pmatrix} = \begin{pmatrix}\text{sum of}\\ \text{all inputs}\end{pmatrix} + \begin{pmatrix}\text{sum of all}\\ \text{internal sources}\end{pmatrix} - \begin{pmatrix}\text{sum of all}\\ \text{outputs}\end{pmatrix} - \begin{pmatrix}\text{sum of all}\\ \text{internal sinks}\end{pmatrix}$$

In mathematical terms:

$$\frac{dM_i}{dt} = I_i + P_i - O_i - R_i \tag{15-1}$$

where M_i is the mass of the chemical i in the CV, and I_i, P_i, O_i, R_i are the sums of the rates of all inputs, internal production processes (sources), outputs, and internal removal mechanisms (sinks) of the chemical i within the CV. If the lake system is chosen to have only one CV, all inputs and outputs are fluxes across the system boundary. In the language introduced in Figure 15.1, the inputs are then called "external forces". If the system is subdivided into several CVs, some of the input and output terms are external forces and others are fluxes connecting different CVs and thus belong to the internal structure of the system.

Because of its rigorous physical background (conservation of mass), mass balance is a "law concept". Our experience tells us that mass balance is "absolutely" true except for the relativistic equivalence of mass and energy, which is not relevant here. If such models go wrong, it is not because of the physical or mathematical concept, but because of an inadequate record of all terms of Eq. 15-1. In fact, the record may either be wrong or incomplete; that is, we may not be aware of all the processes (transport or transformation) which have to be considered on the right-hand side of Eq. 15-1.

Let us look at the example of perchloroethylene (PER, also called tetrachloroethylene or tetrachloroethene)

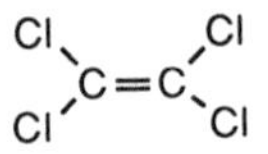

Cl Cl
C=C
Cl Cl

tetrachloroethene

in Greifensee, a small lake in Switzerland (Table 15.1) As a first step we treat this lake as a single completely mixed system. Thus, there is only one control volume, the whole lake. Such a scheme is called a *one-box model.* Perchloroethylene is assumed to enter the lake through inlets and to leave it through a single outlet (Fig. 15.2). Since the system does not have any internal structure, the location and number of inlets and outlets are not relevant. The water regime can be described solely by the total rate of water flowing through the lake. Furthermore, the exchange of PER between the atmosphere and the lake causes a flux G which can have either sign. It is defined as positive if the flux is directed from the lake to the atmosphere. In order to keep a certain generality, we imagine a reaction R, which may act to degrade PER within the lake (e.g., photolysis, biodegradation) and a term S describing the net removal to the sediments, though for the case of PER these terms are negligible.

The mass balance equation now takes the form

$$\frac{dM}{dt} = I - O - G - R - S \qquad [\mathrm{M \cdot T^{-1}}] \tag{15-2}$$

TABLE 15.1 Characteristic Data of Greifensee (Switzerland)[a]

Volume:	Total	V	$150 \times 10^6\,m^3$
	Epilimnion	V_E	$50 \times 10^6\,m^3$
	Hypolimnion	V_H	$100 \times 10^6\,m^3$
Area:	Surface	A_0	$8.6 \times 10^6\,m^2$
	At thermocline	A_{th}	$7.5 \times 10^6\,m^2$
Mean depth:	Total lake	$h = V/A_0$	17.4 m
	Epilimnion	$h_E = V_E/A_0$	5.8 m
	Hypolimnion	$h_H = V_H/A_{th}$	13.3 m
Throughflow of water		Q	$0.34 \times 10^6\,m^3 \cdot d^{-1}$
Mean water residence time		$\tau_w = V/Q$	440 d
Flushing velcoity		$q = Q/A_0$	$0.04\,m \cdot d^{-1}$
Turbulent diffusivity in thermocline		E_{th}	$0.2\,m^2 \cdot d^{-1}$
Thermocline thickness		h_{th}	4 m
Turbulent exchange velocity between epilimnion and hypolimnion		$v_{th} = E_{th}/h_{th}$	$0.05\,m \cdot d^{-1}$
Typical wind speed at 10 m above lake surface		u_{10}	$1\,m.s^{-1}$

[a]Figures for characteristic lake data (volume, surface area, etc.) and input data are rounded off to facilitate the quantitative considerations.

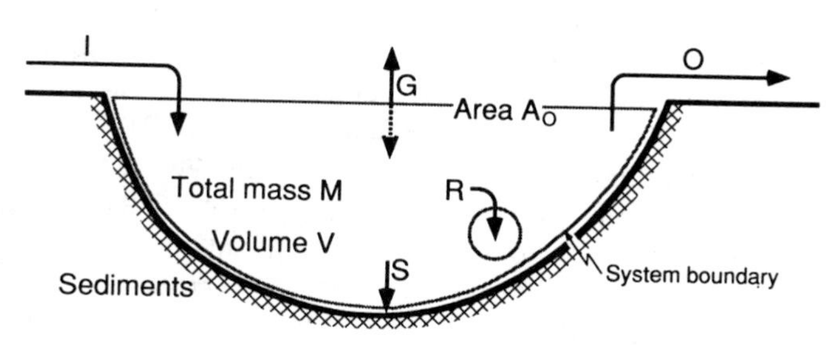

Figure 15.2 One-box lake model for tetrachloroethylene (PER) in Greifensee. The mass balance is given by Eq. 15-2. Definitions are: I, O = input and output of PER through rivers; G = air–water exchange (defined as positive if directed from the water into the air); R = in situ reaction; S = sedimentation. For the case of PER, R and S are negligible.

There are six terms involved in this expression. Obviously, some of the terms are quantified more easily than others. For instance, the river input and output terms I and O can be measured by monitoring water through-flow and concentration at the inlets and outlets (Table 15.2). The change of PER in the lake, dM/dt, follows from measurements within the water body of the lake. The best situation for determining M is met when the lake is completely mixed during the cold season (see below).

It is more difficult to quantify internal transformation processes (R) or boundary fluxes (G and S). As discussed below, from the physicochemical properties of PER it appears that in situ reactions and removal to the sediments are not significant for the mass balance of PER in lakes. For the only remaining flux, air–water exchange G, we

TABLE 15.2 Mass Balance for Perchloroethylene (PER) in Griefensee (Switzerland)

Mass of PER in lake	M	83 mol
Mean lake concentration	C	$0.55 \cdot 10^{-6}$ mol·m^{-3}
Mean concentration in air	C_a	$< 10^{-8}$ mol·m^{-3}
River input	I	0.90 mol·d^{-1}
Loss through outlet	O	0.18 mol·d^{-1}
In situ reaction	R	0
Removal to sediments	S	0
Net loss to atmosphere	G	
(1) Calculated assuming steady state[a]		0.72 mol·d^{-1}
(2) Calculated from Eq. 15-3[b]		1.20 mol·d^{-1}
Net change in lake (case 2)	dM/dt	-0.48 mol·d^{-1}
Mean residence time of PER in lake	$\tau = M/I$	90 d
Mean concentration in riverine input	$C_{in} = I/Q$	$2.6 \cdot 10^{-6}$ mol·m^{-3}

[a]Corresponds to $v_{tot} = 0.155$ m·d^{-1}.
[b]Calculated from wind speed $u_{10} = 1$ m·s^{-1}. Note $v_{tot} \sim v_w = 3 \cdot 10^{-4}$ cm·s^{-1} $= 0.26$ m·d^{-1}.

can assume a steady state as a first approximation. Therefore dM/dt is equal to 0. Since I and O were derived from direct measurements (Table 15.2), we can calculate $G = I - O = 0.72$ mol·d^{-1}.

Thus, the mass balance model provides us with an estimated flux of PER from the lake into the atmosphere, a value which is not easily available from direct measurements. The accuracy of this result depends on how well we know not only the other terms appearing in Eq. 15-2, that is input and outflow of PER, but also a possible change of the PER content in the lake, dM/dt. To simplify the discussion, let us neglect possible errors in I and O and assume an accuracy of the PER-measurement in the lake of 5‰, that is about $\pm 3 \cdot 10^{-8}$ mol·m^{-3}, equivalent to an error in M of ± 4.5 moles. Assume that PER in Greifensee was observed during 100 days and no change in M was found. Then the uncertainty in M corresponds to a total flux error of ± 4.5 moles per 100 days, that is ± 0.045 mol·d^{-1} or about 6% of the calculated value for G. Although other errors, for example, a nonhomogeneous distribution of PER in the lake, may add to the uncertainty in M, this example nevertheless shows that a mass balance, if extended over a long time period, can yield accurate values for unknown fluxes.

15.3 DYNAMIC BOX MODELS

The principle of mass balance derived in the last section enables us to compare all inputs to and all outputs from a control volume and to find a possible imbalance, $dM/dt \neq 0$, responsible for a temporal change of the mass within the CV. This concept is however not sufficient to predict, for instance, the behavior of a system under the influence of a changing input, $I(t)$. To give an example, let us address the question of how the PER concentration in Greifensee would respond to a sudden doubling of I at time t_0. A model that is able to describe the temporal change of a system variable [e.g., PER concentration, $C(t)$] under the influence of an arbitrary "external force", $I(t)$, is called "dynamic."

To turn a mass balance model into a *dynamic model* we need theories relating the internal processes (whether they refer to transport or transformation) to the state of the system expressed by concentrations C or inventories M. Such elements needed to build dynamic models are called *process models.* For instance, the process model for outflow, O, just consists of the assumption that the concentration is constant within the lake. Then, the concentration of the outlet must be equal to C which leads to

$$O = Q \cdot C \qquad [\mathrm{M \cdot T^{-1}}] \tag{15-3}$$

where Q is the outflow rate $[\mathrm{L^3 \cdot T^{-1}}]$.

Many of these process models can be developed from the theories derived in the proceeding chapters of this book. In the case of PER we are especially interested in getting an air–water exchange model for G. According to Chapter 10, Eq. 10-10, G can be expressed as

$$G = A_0 \cdot v_{tot} \cdot \left(C_w - \frac{C_a}{K'_H} \right) \qquad [\mathrm{M \cdot T^{-1}}] \tag{15-4}$$

where

A_0 is the surface area of the lake ($\mathrm{L^2}$; e.g., $\mathrm{m^2}$),
v_{tot} is the total air–water transfer velocity ($\mathrm{L \cdot T^{-1}}$; e.g., $\mathrm{m \cdot s^{-1}}$),
C_w is the concentration of PER at the water surface ($\mathrm{M \cdot L^{-3}}$; e.g., $\mathrm{mol \cdot m^{-3}}$),
C_a is the concentration of PER in the atmosphere ($\mathrm{M \cdot L^{-3}}$; e.g., $\mathrm{mol \cdot m^{-3}}$), and
K'_H is the nondimensional Henry's Law Constant.

The mass transfer velocity, v_{tot}, can be estimated using a process model which either leads to Eq. (10-12) for the stagnant two-film model or to Eq. 10-22 for the surface renewal model. In both cases v_{tot} depends on the wind velocity; thus this model requires a new kind of data, which may not be available.

We can nevertheless try to estimate the size of G. First, the PER concentration in air, C_a, is of the order of $10^{-8}\ \mathrm{mol \cdot m^{-3}}$ or less. Using the nondimensional Henry's Law constant for this compound listed in Table 6.3, $K'_H = 0.727$, we can show that C_a/K'_H can be neglected in Eq. 15-4 compared to the concentration found in the lake of about $0.5 \cdot 10^{-6}\ \mathrm{mol \cdot m^{-3}}$. Therefore G simplifies to the product $A_0 \cdot v_{tot} \cdot C_w$. Second, for a mean wind velocity $u_{10} = 1\ \mathrm{m \cdot s^{-1}}$, the exchange velocities for water vapor in air, $v_a(H_2O)$, and oxygen in water, $v_w(O_2)$, are 0.5 and $4.4 \cdot 10^{-4}\ \mathrm{cm \cdot s^{-1}}$, respectively (see Eqs. 10-28 and 10-32). These velocities can be adjusted to PER using the corresponding approximations, Eqs. 10-29 and 10-33. Yet, because the relatively large K'_H of PER, the liquid phase transfer velocity v_w is much smaller than the gas phase velocity v_a/K'_H. Thus we get

$$v_{tot} \sim v_w(\mathrm{PER}) \sim 4.4 \cdot 10^{-4}\ \mathrm{cm \cdot s^{-1}} \left[\frac{D_w(\mathrm{PER})}{D_w(O_2)} \right]^{0.57} = 3.0 \cdot 10^{-4}\ \mathrm{cm \cdot s^{-1}}$$

where D_w(PER) and $D_w(O_2)$ are the molecular diffusion coefficients in water of PER and dissolved oxygen, respectively. Finally, the assumption that the lake is completely mixed with respect to PER allows us to replace the concentration at the water surface, C_w, by the mean concentration C measured to be $0.55 \cdot 10^{-6}\ \text{mol} \cdot \text{m}^{-3}$. The net loss of PER to the atmosphere, G, thus becomes $1.2\ \text{mol} \cdot \text{d}^{-1}$.

This value is significantly larger than $0.72\ \text{mol} \cdot \text{d}^{-1}$, the value calculated from the steady-state mass balance model. The uncertainty regarding the size of v_{tot} represents one possible explanation for the discrepancy between the two values. Another possible source of error is that there may be a systematic difference between the mean and surface PER concentration which occurs if the lake is not always completely mixed with respect to PER. We come back to this possibility in the next section, when two-box models are discussed.

Let us now insert the process models, Eqs. 15-3 and 15-4, into the mass balance Eq. 15-2 (with $R = S = 0$) and rearrange the terms so that the expressions contain the concentration C appear in one group:

$$\frac{dM}{dt} = (I + A_0 v_{tot} C_a / K'_H) - (Q + A_0 v_{tot}) \cdot C \tag{15-5}$$

Unfortunately, we are still left with two sorts of system variables, M and C. Assuming a constant lake volume V, the two quantities are related by $M = V \cdot C$. Thus we can derive a dynamic equation for either the total mass in the lake

$$\frac{dM}{dt} = \left(I + A_0 v_{tot} \frac{C_a}{K'_H}\right) - \left(\frac{Q}{V} + \frac{A_0}{V} v_{tot}\right) \cdot M \tag{15-6a}$$

or for the mean concentration,

$$\frac{dC}{dt} = \left(\frac{I}{V} + \frac{A_0}{V} v_{tot} \frac{C_a}{K'_H}\right) - \left(\frac{Q}{V} + \frac{A_0}{V} v_{tot}\right) \cdot C \tag{15-6b}$$

Though these equations were derived for a particular situation (PER in Greifensee), they are, in fact, typical for box models. It is worthwhile to spend some time with this very special kind of equation called a "first-order linear inhomogeneous differential equation" (FOLIDE).

To learn to handle a FOLIDE wherever it appears, we should first strip Eq. 15-6 of all its unnecessary components originating from the special problem for which it was derived and keep the parts essential for its mathematical structure. In fact, the FOLIDE boils down to the form

$$\frac{dy}{dt} = J - k \cdot y \tag{15-7}$$

where y is the dynamic (state) variable to be modeled; in the preceding example y

stood either for the total mass M or for the concentration C. Here J is the so-called inhomogeneous term; for the case of box models it described the external "driving force," a quantity that is not affected by the state of the system (i.e., it is not influenced by M or C) but is controlled "by the rest of the world". Note that J can vary with time. The dimensions of J correspond to the dimensions of y multiplied by inverse time. Hence, the J's appearing in Eqs. 15-6a and b cannot be identical, since M and C have different dimensions (mass and mass per volume, respectively). In fact, we have

$$J_M = I + A_0 v_{tot} \frac{C_a}{K'_H} \tag{15-8a}$$

and

$$J_C = \frac{I}{V} + \frac{A_0}{V} v_{tot} \frac{C_a}{K'_H} \tag{15-8b}$$

where the indices refer to the variable to which the J term belongs. Thus, J_M is total input to the box (mass per unit time $[MT^{-1}]$), and J_C is total input per unit box volume and time $[ML^{-3}T^{-1}]$, that is, $J_C = J_M/V$. Two kinds of fluxes contribute to the total input: the input linked to the inflow of water I (rivers, man-made inlets) and the input from the atmosphere. In fact, we have separated the net flux to the atmosphere G (Eq. 15-3), into a flux to the atmosphere that depends on the box variable C, and into a flux from the atmosphere that depends on the atmospheric concentration C_a. The latter term appears in the general input term J. The temporal variation of J may originate from the variation of I, C_a, v_{tot}, or temperature (since K'_H depends on T). Since we have assumed a constant lake volume, the morphometric parameters A_0 and V cannot contribute to the variation of J.

Now, let us discuss the second term of Eq. 15-7 ($-k \cdot y$). This is the so-called homogeneous part of the differential equation since it links the temporal change of y, dy/dt, to the variable y itself. The parameter k, called the (first-order) rate constant (though the term "constant" may be somewhat misleading since k can also be time-dependent!), is identical for both Eqs. 15-6a and b:

$$k = \frac{Q}{V} + \frac{A_0}{V} v_{tot} \qquad [T^{-1}] \tag{15-9}$$

Since the dimensions of dy/dt and $(k \cdot y)$ have to be the same, the dimension of k is always $[T^{-1}]$ whatever the dimensions of the variable y may be.

For the case of linear box models (i.e., for model equations in which the state variable y only appears in the form y, but not as y^2 or y^{-1} or even as a more complicated expression), the rate constants always consist of one or more additive terms, each describing one particular (first-order) removal mechanism. In our example, k consists of two such terms, the first (Q/V) describes the removal through the outlet (flushing rate), and the second ($A_0 \cdot v_{tot}/V$) quantifies the removal to the atmosphere. Since both parts of k must also have the dimension $[T^{-1}]$ and thus can be interpreted as "subrates," we define a rate constant for flushing:

$$k_w = \frac{Q}{V} = \frac{1}{\tau_w} \qquad [T^{-1}] \tag{15-10}$$

(τ_w: mean residence time of water in the system), and a rate constant for loss to the atmosphere:

$$k_g = \frac{A_0}{V} v_{tot} = \frac{v_{tot}}{h} \qquad [T^{-1}] \tag{15-11}$$

($h = V/A_0$: mean depth of lake). Together these rate constants add to yield the overall system loss rate constant:

$$k = k_w + k_g \tag{15-12}$$

In the introduction to this chapter we discussed the problem of how to compare different (transport and transformation) processes with respect to their influence on the state of the system. Note that as a byproduct of our search for the solution of Eq. 15-6 we have found an answer to this problem. Since k_w and k_g have the same dimension, their relative size directly expresses the relative importance of the two removal pathways. As we will see, there are situations where many more subrates k_i simultaneously contribute. Thus, comparing their sizes will serve as a measure of their relative influence on the dynamic behavior of C or M.

In the case of PER in Greifensee, the ratio between the flux to the atmosphere and the flux through the outlet, respectively, is

$$\frac{k_g}{k_w} = \frac{v_{tot}}{Q/A_0} = \frac{v_{tot}}{q} \tag{15-13}$$

where $q = Q/A_0$, the water input per unit lake area, has the dimensions of a velocity ("flushing velocity"). For Greifensee, $q = 0.04\,\mathrm{m \cdot d^{-1}}$ (see Table 15.1) is significantly smaller than the gas exchange velocity v_{tot} (between 0.15 and $0.26\,\mathrm{m \cdot d^{-1}}$, see Table 15.2). In fact, only in lakes with very high flushing velocities can the removal through the outlet of a volatile compound compete with air–water exchange of chemicals with reasonably high Henry's Law constants.

We are still looking for the solution of the FOLIDE, Eq. 15-7. Depending on the time variability of J and k, three cases can be distinguished:

1. Both J and k are constant with time
2. J varies with time, but k is constant
3. Both J and k are time dependent

The solutions to all three cases are given in Table 15.3. In the following, we discuss the first two cases in some detail.

TABLE 15.3 Solutions to a First-Order Linear Inhomogeneous Differential Equation (FOLIDE)

$$\frac{dy}{dt} = J - k \cdot y \qquad k > 0$$

y^0 : Initial value at $t = 0$

a. Constant coefficient ($J, k = \text{const.}$)

(a) $$y(t) = y^\infty + (y^0 - y^\infty)e^{-kt} = y^0 e^{-kt} + y^\infty(1 - e^{-kt})$$

with $y^\infty = \frac{J}{k}$: steady state

b. Variable input $J(t)$, $k = $ const.

(b) $$y(t) = y^0 e^{-kt} + \int_0^t e^{-k(t-t')} J(t')dt'$$

c. Variable coefficients $J(t)$ and $k(t)$

(c) $$y(t) = y^0 e^{-\Phi(t)} + e^{-\Phi(t)} \int_0^t e^{\Phi(t')} J(t')dt'$$

where

$$\Phi(t) = \int_0^t k(t')dt'$$

One-Box Model with Constant Parameters

The solution of Eq. 15-7 for constant J and k has the form (remember, y stands for either M or C):

$$y(t) = y^\infty + (y^0 - y^\infty)e^{-kt} \quad \text{with} \quad k > 0 \tag{15-14}$$

where y^0 is the value at $t = 0$ (initial value), and

$$y^\infty = \frac{J}{k} \tag{15-15}$$

is the value attained as $t \to \infty$, that is when y has reached its steady state defined by $dy/dt = 0$.

The shape of $y(t)$ is shown in Figure 15-3 for three different cases. Note that the time dependence always enters through the factor e^{-kt}. For positive k and increasing time t, e^{-kt} tends to zero though, from a purely mathematical point of view, it reaches zero only for $t = \infty$. In practice the exponential expression becomes very small once the argument $(k \cdot t)$ has surpassed a certain value. For instance, e^{-3} is about 0.05, thus the time-dependent term of Eq. 15-14 drops to less than 5% of its initial

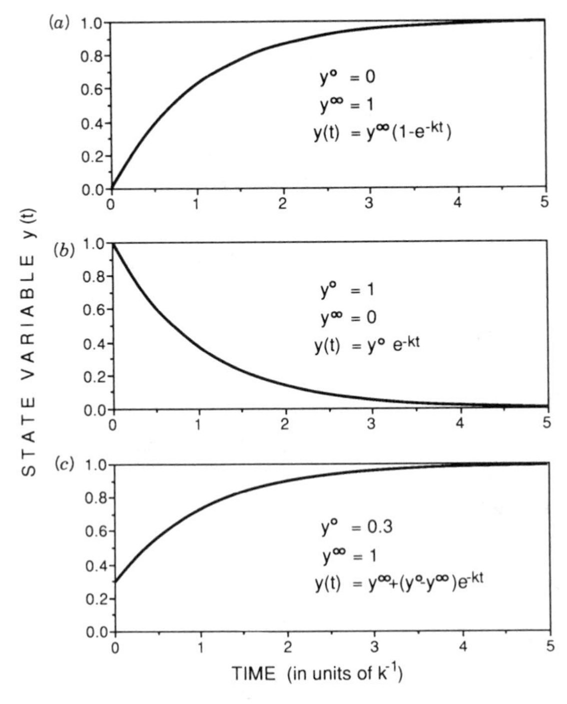

Figure 15.3 Solution to first-order linear inhomogeneous differential equation (FOLIDE), Eq. 15-7; y^0 = initial value, $y^\infty = J/k$ steady-state value. (*a*) Growth curve for $y^0 = 0$. (*b*) Linear decay of *y* without input ($J = y^\infty = 0$). (*c*) General case drawn for $y^0 < y^\infty$ showing that the initial difference $(y^0 - y^\infty)$ drops as e^{-kt}. Time to steady state $t_{ss} = 3/k$ defined as time when 95% of the initial difference has disappeared.

value once $(k \cdot t)$ becomes larger than 3. This corresponds to the condition

$$t > t_{ss} = \frac{3}{k} \tag{15-16}$$

where "the time to steady state", t_{ss}, is a measure for the response velocity of the system (remember that the real time to steady state would be infinite!). As shown in Figure 15.3, the initial difference between y^0 and the steady state, $(y^0 - y^\infty)$, decreases as e^{-kt}. The greater the rate k, the faster y moves toward its steady state.

For the case of PER in Greifensee, the steady-state concentration C^∞ and the corresponding time to steady state t_{ss} are given by:

$$C^\infty = \frac{J_C}{k_w + k_g} = \frac{I + A_0 v_{tot}(C_a/K'_H)}{Q + A_0 \cdot v_{tot}} \tag{15-17}$$

$$t_{ss} = \frac{3}{k_w + k_g} = \frac{3}{(Q/V) + v_{tot}(A_0/V)} \tag{15-18}$$

TABLE 15.4 Dynamic One-Box Model for PER in Greifensee (Switzerland)

Dynamic equations

Use Eq. 15-6b and the definitions 15-8b, 15-10, 15-11

$$\frac{dC}{dt} = J_C - (k_w + k_g)C$$

If the atmospheric PER concentration C_a is negligible, J_C can be expressed by the mean input concentration $C_{in} = I/Q$:

$$J_C = I/V = C_{in} \cdot k_w$$

Specific values

Gas transfer velocity	v_{tot}	$0.155\,m \cdot d^{-1}$
Flushing rate	$k_w = Q/V$	$2.3 \cdot 10^{-3}\,d^{-1}$
Gas exchange rate	$k_g = v_{tot}/h$	$8.9 \cdot 10^{-3}\,d^{-1}$
Atmospheric PER concentration	$C_a \sim 0$	
Mean input concentration	$C_{in} = I/Q$	$2.6 \cdot 10^{-6}\,mol \cdot m^{-3}$
Steady-state concentration	$C^\infty = \frac{k_w}{k_w + k_g} C_{in}$	$0.54 \cdot 10^{-6}\,mol \cdot m^{-3}$
Time to steady state	$t_{ss} = \frac{3}{k_w + k_g}$	270 d

(See values from Tables 15.1 and 15.2)

Values are given in Table 15.4 for the case $C_a \sim 0$. Since k_g is much larger than k_w, the flushing rate k_w has only little influence on the response time t_{ss}. In fact, it is always primarily the fastest removal reaction which determines t_{ss}.

Let us come back to the question raised at the beginning of this section. What would happen to the PER concentration in Greifensee if the external input I were suddenly doubled? Provided that the fluxes and concentrations listed in Table 15.2 correspond to a steady-state situation and thus the rates listed in Table 15.4 are correct, we can use Eq. 15-17 to deduce that a doubling of I means a steady state of twice the old value (that is $C^\infty = 1.1 \cdot 10^{-6}\,mol \cdot m^{-3}$). The time necessary to reach the new concentration within 5%, t_{ss}, would be 270 days or about 9 months (Table 15.4).

One-Box Model with Time-Dependent Input

Stepwise changes of the input function $I(t)$, as assumed in the preceding example, do not commonly occur in natural systems. More realistically, the input changes continuously with time. For instance, we may ask the question how would the PER concentration in Greifensee change under the influence of an exponentially increasing riverine input:

$$I(t) = I_0 e^{\alpha t} \tag{15-19}$$

where α is the input growth rate (dimension T^{-1}). Since we assumed that the input of PER through the atmosphere is negligible, Eq. 15-19 is equivalent to an exponential growth of the total input function $J(t) = J_0 e^{\alpha t}$.

The solution to Eq. 15-7 for a time-dependent inhomogeneous term $J(t)$ is given in Table 15.3 (Eq. b). Before applying this formula we should try to qualitatively understand it. The first term simply describes the exponential "decay" of the initial value y^0. A similar term appears in the solution for constant J (case a). The second term, an integral over the product of J and some exponential function, is more interesting. The integration variable t' runs from $t' = 0$, the beginning of the computation, to $t' = t$, the time in which we are interested. In the integral, the input J is summed up over this time interval. Yet, because of the term $e^{-k(t-t')}$, not all inputs have the same weight in this sum. At the upper limit of the interval ($t = t'$), we have $e^{-k \cdot 0} = 1$; thus the weight of the most recent input is greatest. The more we move backwards in time (by decreasing t' from t to 0), the larger $(t - t')$ becomes, and thus the smaller the weight $e^{-k(t-t')}$. In other words, inputs that are back in time have only little or no influence on the actual value $y(t)$. The speed at which the input is "forgotten" is again determined by the overall rate constant k. This sounds pretty logical: as for case (a) the inverse rate k^{-1} has the meaning of memory or response time of the dynamic system.

Now, let us formally solve Eq. 15-7 with $J(t)$ from Eq. (15-19) using Table 15.3 and assuming $y^0 = 0$. The integration yields

$$y(t) = J_0 \int_0^t e^{-k(t-t')} e^{\alpha t'} dt' = J_0 \frac{e^{\alpha t} - e^{-kt}}{k + \alpha} \tag{15-20}$$

If the input grows exponentially for long enough (more precisely, for a time longer than t_{ss} of Eq. 15-16), the term e^{-kt} can be neglected. Thus, using Eq. 15-19 we get

$$y(t) = \frac{J(t)}{k + \alpha}, \tag{15-21}$$

an expression which looks similar to the steady-state value of Eq. 15-15 except for the extra term α in the denominator.

We can now see how different combinations of α and k affect the temporal variation of $y(t)$. First, we consider the case $\alpha \ll k$, that is, a situation where the timescale of change of the external force $J(t)$ is much smaller than the timescale of overall system response k. From Eq. 15-21:

$$y(t) \sim \frac{J(t)}{k} = y^{\infty}(t) \qquad \text{for } \alpha \ll k \tag{15-22}$$

This result means that the system state at any time t is equal to the steady state $y^{\infty}(t)$, which corresponds to the input $J(t)$ at that time. Thus, the "external perturbation" of the system is so slow and smooth that the system variable $y(t)$ is able to adjust to the changing J in such a way that y remains in quasi steady state. Second, if α is not much smaller than k, α in Eq. 15-21 cannot be disregarded. The actual state $y(t)$ lags behind the input change; the system is not at steady state with respect to the input.

As an example we consider two different hypothetical scenarios for the PER input into Greifensee. First assume that PER input is increasing at an annual rate of 30%, and second assume the same loading curve but a total loss rate k which is reduced by a factor of 10. As a starting point we use $C_0 = 0.54 \cdot 10^{-6}\,\text{mol}\cdot\text{m}^{-3}$ and $I_0 = 0.9\,\text{mol}\cdot\text{d}^{-1}$. From Table 15.4 we get $k = k_w + k_g = 11.2 \cdot 10^{-3}\,\text{d}^{-1} = 4.1\,\text{yr}^{-1}$ for our first scenario and thus $1.12 \cdot 10^{-3}\,\text{d}^{-1}$ for our second case.

For case 1, $\alpha = 0.3\,\text{yr}^{-1}$; thus α is distinctly smaller than k, and we expect C to increase nearly as $e^{\alpha t}$, that is, in proportion to the input $J(t)$. As shown in Figure 15.4*a*, the actual concentration (curve 1) closely follows the respective steady-state value. In contrast, if $k = 0.41\,\text{yr}^{-1}$, the actual concentration (Fig. 15.4*a*, curve 2) lags far behind the steady state, since now α and k are of the same order. Note that the input $J(t)$ was also reduced by a factor of 10 in order to keep the steady-state concentration given by Eq. 15-15.

The response of $y(t)$ to a time-dependent input function $J(t)$ includes still another facet to be discussed. The preceding example dealt with a monotonously increasing $J(t)$. How does $y(t)$ respond if $J(t)$ is oscillating about some mean value $\bar{J}$? For instance,

$$J(t) = \bar{J}(1 + a \cdot \sin \omega t) \tag{15-23}$$

where $T = 2\pi/\omega$ is the period of the oscillation. Integration of Eq. (b) (Table 15.3) with this particular input function yields

$$y(t) = y^0 e^{-kt} + \frac{\bar{J}}{k}(1 - e^{-kt}) + \frac{\bar{J} \cdot a}{\sqrt{k^2 + \omega^2}} \sin(\omega t - \psi) \tag{15-24}$$

where $\psi = \tan^{-1}(\omega/k)$ ($\tan^{-1} \equiv$ arctan: inverse function of tan).

The first two terms are identical to the solution for constant input $\bar{J}$ [Eq. (b), Table 15.3]. For t much greater than t_{ss} they tend toward $\overline{y^\infty} = \bar{J}/k$. The third term of Eq. 15-24 expresses the influence of the oscillating external forcing. Again, the discussion of two extreme cases help us to understand this influence.

If the variation of the external forcing is much slower than the rate of the system response ($\omega \ll k$), then Eq. 15-24 eventually simplifies to

$$y(t) = \frac{\bar{J}}{k}(1 + a \sin \omega t) = \frac{J(t)}{k} = y^\infty(t) \qquad (\text{for } \omega \ll k) \tag{15-25}$$

(Note: $\psi \to 0$ for $(\omega/k) \to 0$).

This result indicates that for slowly varying external forces the system is always at its actual steady state. Again, the criterion of whether an oscillation is slow or not is determined by the overall response rate k. We could say that depending on the size of k every system has its "personal criterion" of slowness.

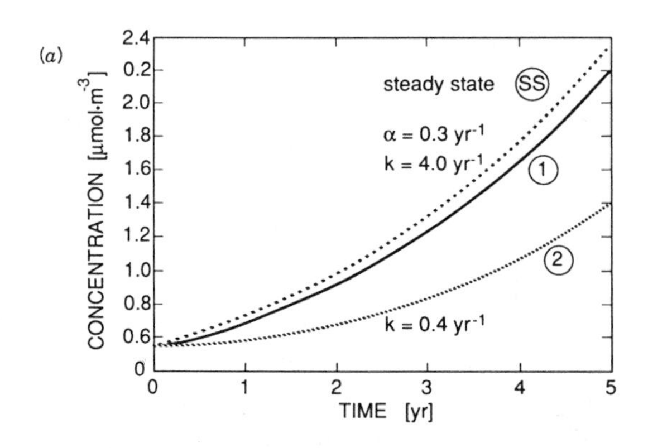

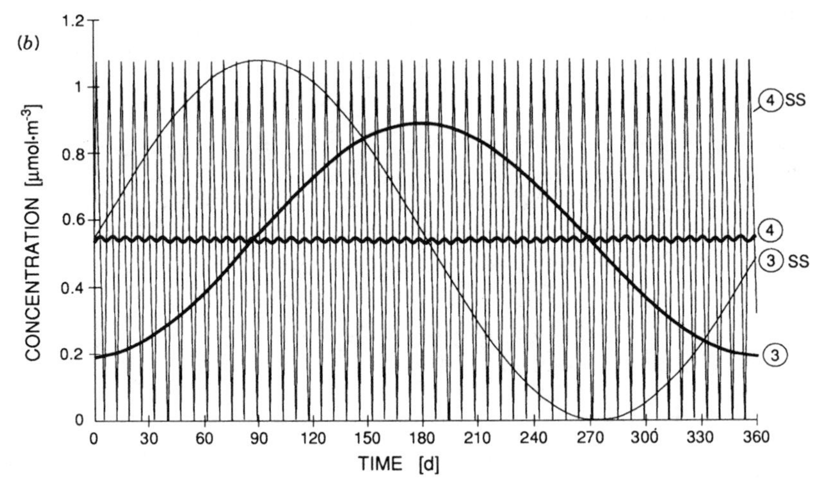

Figure 15.4 (*a*) Change of PER concentration in Greifensee under the influence of an exponentially increasing riverine input $J(t) = J_0 e^{\alpha t}$ with $\alpha = 0.3\,\text{yr}^{-1}$ (curve 1) compared to the corresponding steady-state concentration (curve SS). Curve 2 shows the response of the lake to the same input growth if total loss rate k and input J were both 10 times smaller. (*b*) As before, but with oscillatory input variation, $J(t) = \bar{J}\,(1 + \sin \omega t)$. Curve 3 and 3 SS (steady state) for period $T = 1$ yr($\omega = 0.0172\,\text{d}^{-1}$), curve 4 and 4 SS for $T = 1$ week ($\omega = 0.90\ \text{d}^{-1}$). In the case of weekly variation in inputs, the slow response of the lake dampens the resulting change in PER concentration almost entirely; in contrast, input changes occurring over times of about one year are similar enough to the lake responsiveness to result in large swings in lake concentrations.

In contrast, for ω much greater than k (use $t \gg t_{ss}$, as before):

$$y(t) = \frac{\bar{J}}{k} + a\frac{\bar{J}}{\omega}\sin(\omega t - \psi) \sim \frac{\bar{J}}{k} \qquad (\text{for } \omega \gg k) \tag{15-26}$$

The variation of the external forcing has become so fast that the system only feels the average force $\bar{J}$. By the way, when moving from $\omega \ll k$ (Eq. 15-25) to $\omega \gg k$ (Eq. 15-26), the phase lag ψ increases from 0 to $\pi/2$ [note $\tan^{-1}(\omega/k) = \pi/2$ for $(\omega/k) \to \infty$]. This means that the system variable $y(t)$ gets more and more out of phase with the external force until the lag becomes one quarter of a full period. However, at this point the oscillatory reaction of the system has become negligible.

The result of these considerations are of practical value and not just a mathematical game. As an example, we address the problem of temporal fluctuations of the PER input to Greifensee and their possible influence on PER concentration in the lake. Given the way PER is used and introduced into rivers and sewers, and how temperature and other meteorological factors may influence exchange to the atmosphere and removal during sewage treatment, we expect the riverine input I to show at least diurnal, weekly, and seasonal variations. From these considerations we can determine that only the seasonal input variations are able to produce significant concentration fluctuations in the lake (Fig. 15.4*b*).

In Figure 15.5 the complex behavior of the Greifensee-PER system is summarized using real data from the lake and input values that were measured and also estimated from the vertical distribution of PER in Greifensee. Since the lake is stratified during the summer, loss by gas exchange is overestimated by the one-box model. This explains why the decrease of the PER concentration in the lake is overestimated by the model.

We have spent quite some time discussing a type of equation which unfolded from the analysis of linear one-box models. Fortunately, this effert is not needed every time a new box model is constructed. In fact, the behavior of linear models mostly follows from the kind of arguments which we have developed so far (although non-linearities may lead to some surprisingly new phenomena as discussed at the end of this chapter!) We use the PER model of Greifensee to demonstrate how the box model concept can be extended to two and more dimensions.

The one-box approach is based on the assumption that the chemical concentration in the lake is always homogeneous. A short excursion into physical limnology may help us decide whether this assumption is realistic enough to still account for the essential properties of the PER–lake system. As discussed in Section 9.4 and Table 9.6, owing to the vertical density stratification of the water column, horizontal mixing by turbulent diffusion is commonly much faster than vertical mixing. Typical horizontal eddy diffusivities E_x in Greifensee are about $5 \cdot 10^3\,\text{cm}^2 \cdot \text{s}^{-1}$ (in accordance with Okubo's diffusion diagram (Fig. 9.10) for a length scale L of 1 km). From Eq. 9-31 (D replaced by E_x and with $L = 1$ km) we calculate the time for horizontal mixing to be about 10^6 s (12 d). Therefore the PER entering the lake through some local sources (sewage pipes, rivers) is horizontally mixed at the respective input depth within time intervals which are short compared to the mean residence time of PER in the lake (90 d). However, this may not always be true for the vertical distribution of the chemical. In most lakes the intensity of vertical mixing follows a characteristic annual pattern. From spring to autumn the water column is split into two regions, the well-mixed surface layer (called an epilimnion) whose depth typically reaches between 5 and 20 m below the surface, and a region underneath of weaker vertical (and horizontal) mixing (hypolimnion). The two water bodies are separated by a layer with a steep vertical temperature gradient called a thermocline (Fig. 15.6) in which

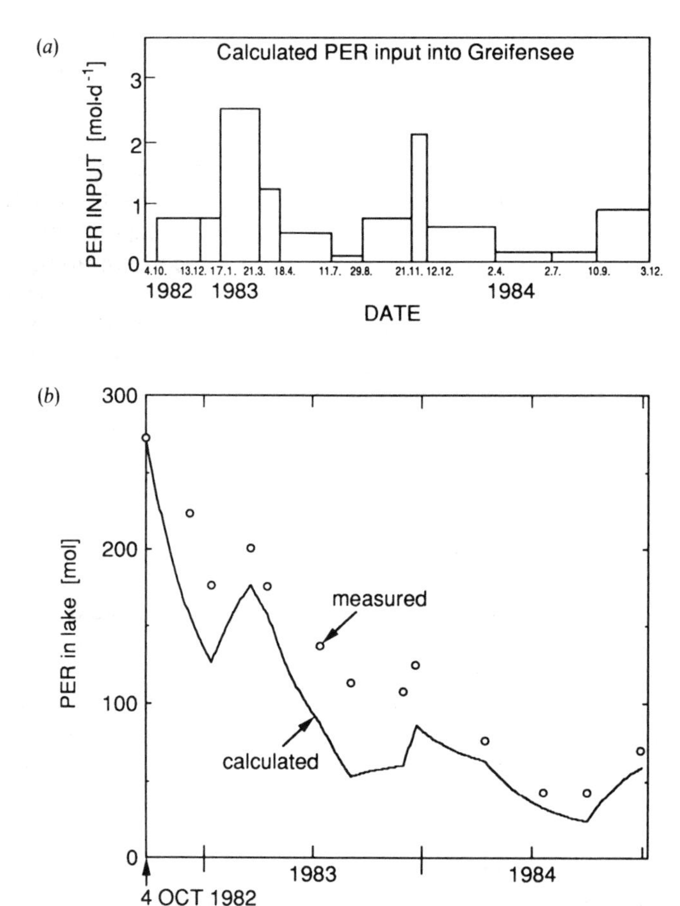

Figure 15.5 (*a*) Perchloroethylene (PER) input into Greifensee between October 1982 and December 1984 estimated from the one-dimensional lake model (see Fig. 15.6). (*b*) Comparison of measured mean mass in the lake with values calculated with the one-box model (Eq. b of Table 15.3) using the input given above. Note that the one-box model overestimates the loss of PER by air–water exchange and flushing during the summer, when the lake is stratified. This may explain the systematic difference between measurements and model calculation (see also Fig. 15.6).

vertical mixing is strongly suppressed according to the theory of turbulent diffusion discussed in Section 9.4 (see Eq. 9-50). The time necessary to replace hypolimnetic water by surface water, $\tau_{ex,H}$, can be estimated using the concept of diffusion across a layer as developed in Chapter 10 for air–water exchange (stagnant film model), except the the "piston velocity" (Eq. 10-11) is now computed from the thermocline thickness h_{th} and the turbulent eddy diffusivity across the thermocline E_{th}:

$$v_{th} = E_{th}/h_{th}$$

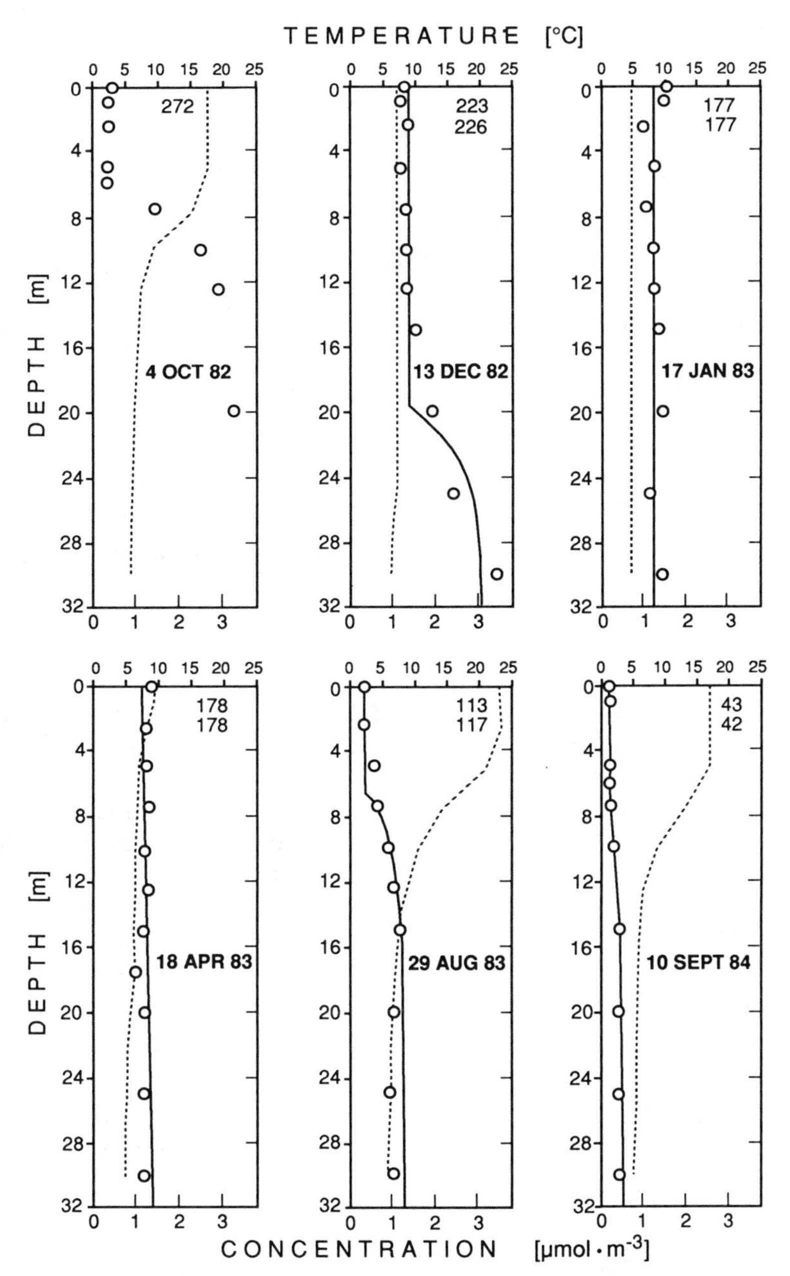

Figure 15.6 Vertical profiles of measured (circles) and modeled PER concentrations in Greifensee (Switzerland). The calculations were executed with the one-dimensional lake model CHEMSEE described in Section 15.5. The numbers in the upper right corner of each plot are total PER inventory in the lake in moles: measured (upper), calculated (lower) number. The break visible in most of the computed profiles indicates the depth to which the lake is fully mixed. This depth was determined from the temperature profiles (broken lines).

The exchange time is the quotient of the mean hypolimnion depth h_H and the piston velocity v_{th} (Schwarzenbach and Imboden, 1983):

$$\tau_{ex,H} = \frac{h_H}{v_{th}} = \frac{h_H}{E_{th}/h_{th}} \tag{15-27}$$

Since in Greifensee E_{th} is typically about $2 \cdot 10^{-2}\,\text{cm}^2\,\text{s}^{-1}$ or $0.2\,\text{m}^2 \cdot \text{d}^{-1}$, $h_{th} \sim 4\,\text{m}$, and $h_H \sim 12\,\text{m}$, we calculate $\tau_{ex,H} \sim 240\,\text{d}$. This time has to be compared with the time needed to flush the epilimnion by gas exchange, $\tau_{g,E} = h_E/v_{tot} \sim 37\,\text{d}$ (see Table 15.4).

From the relative size of the two characteristic times we conclude that during the stratification period, loss of PER from the epilimnion to the atmosphere is fast compared to the time needed for the hypolimnion to adjust its concentration to the decreasing epilimnetic PER content.

In fact, measured PER profiles (Fig. 15.6) show a decrease of the PER concentration at the surface relative to the deep water during the summer. Thus, for the sake of realism, sometimes we should sacrifice part of the simplicity of the one-box model by considering more than one CV in our system. We do this by looking at the so-called two-box model next.

Two-Box Model

The two-box model is a direct extension of the one-box model (Fig. 15.7). The mass balance for both boxes contains the terms used in Eq. 15-2, except for the absence of gas exchange in the hypolimnion ($G_H = 0$) and for the additional two internal transport terms, T_{EH} and T_{HE}. (Note that the two indices refer to the source and the recipient box, respectively. This notation will be used for the multibox models, as well.) As before, sedimentation and in situ reaction are omitted for the case of PER. The transport terms, T_{EH} and T_{HE}, are related to the exchange of water masses between the two volumes by turbulence. Therefore, the fluxes of all dissolved and suspended components should be proportional to their respective concentrations in the source box. As indicated in Fig. 15.7, the input of PER can occur in the hypolimnion, such

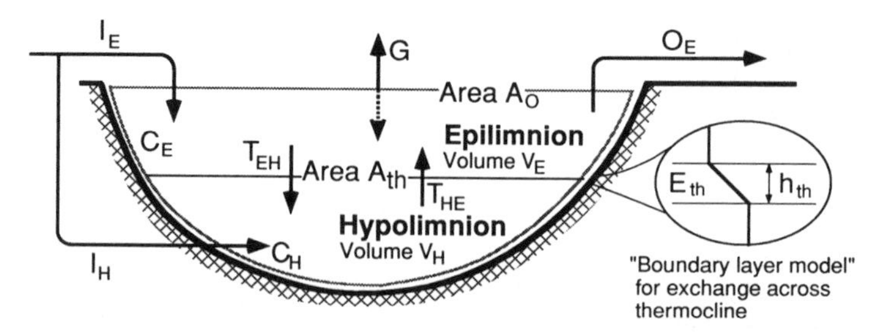

Figure 15.7 Two-box model for (seasonally or permanently) stratified lakes. Symbols as in Figure 15.2, expect for the indices E and H, referring to epilimnion (upper layer) and hypolimnion (lower layer), respectively. T_{EH} and T_{HE} represent vertical fluxes between the two boxes.

as when outfalls release water at depth or when rivers entering the lake plunge into the deep waters.

The mass balance equations for the two water volumes are

$$\frac{dM_E}{dt} = I_E - O_E - G - T_{EH} + T_{HE} \tag{15-28a}$$

$$\frac{dM_H}{dt} = I_H + T_{EH} - T_{HE} \tag{15-28b}$$

As before, to derive a dynamic model the mass fluxes have to be expressed in terms of concentrations. For gas exchange G and outflow O we can use the process models derived before (Eqs. 15-3 and 15-4). The water exchange fluxes are calculated as products of the "piston flow velocity" across the thermocline, of thermocline area, and of concentration in the box from which the flow originates:

$$T_{EH} = v_{th} \cdot A_{th} \cdot C_E \quad \text{and} \quad T_{HE} = v_{th} \cdot A_{th} \cdot C_H \tag{15-29}$$

If the water volumes V_E and V_H are constant, substituting these expressions into Eq. 15-28 and dividing by the respective box volumes yields two equations for the mean concentrations:

$$\frac{dC_E}{dt} = \frac{I_E}{V_E} + k_{g,E} \cdot \frac{C_a}{K'_H} - (k_{w,E} + k_{g,E} + k_{ex,E})C_E + k_{ex,E}C_H \tag{15-30a}$$

$$\frac{dC_H}{dt} = \frac{I_H}{V_H} + k_{ex,H}C_E - k_{ex,H}C_H \tag{15-30b}$$

with the following definitions (Table 15.6):

$k_{w,E} = \dfrac{Q}{V_E}$ — Flushing rate constant of epilimnion

$k_{g,E} = \dfrac{A_o}{V_E} v_{tot} = \dfrac{v_{tot}}{h_E}$ — Rate constant for air–water exchange for epilimnion

$k_{ex,E} = \dfrac{v_{th} \cdot A_{th}}{V_E}$ — Rate constant for water exchange of epilimnion by internal mixing, to hypolimnion

$k_{ex,H} = \dfrac{v_{th} \cdot A_{th}}{V_H}$ — Rate constant for water exchange of hypolimnion by internal mixing with epilimnion

Equation (15-30) represents a system of two first-order *coupled* linear inhomogeneous differential equations, a straightforward extension of the single FOLIDE to two dimensions. At first sight, the equation for the epilimnetic concentration, C_E (Eq. 15-30a), looks very similar to the one-box equation (15-6b). Yet, there are some important differences: (1) the time derivative of C_E is linked to the hypolimnetic concentration through the term $k_{ex,E}\ C_H$; (2) the homogeneous rate of change (the k of Eq. 15-7) includes an additional term ($k_{ex,E}$); and (3) all the rates are normalized by dividing the fluxes by the epilimnetic volume, not by the total volume. Since the former is smaller than the latter, the epilimnetic rates of the two-box model are larger than their one-box counterparts.

The general solution of the two-dimensional FOLIDE system is presented in Table 15.5. Each variable consists of an inhomogeneous term and two exponential functions of the form: constant $e^{-k_i t}$ (where i stands for 1 or 2). The new overall rate constants k_1 and k_2 (mathematicians call them "eigenvalues"), determine the time needed for the system to approach steady state. Since there are two k_i, we have two such times. The smaller k_i controls the overall response time of the system. By analogy

TABLE 15.5 Solution of System of Two First-Order Linear Inhomogeneous Differential Equations

General equations

(a)
$$\frac{dy_1}{dt} = J_1 - k_{11}y_1 + k_{12}y_2$$
$$\frac{dy_2}{dt} = J_2 + k_{21}y_1 - k_{22}y_2$$

Solution for initial values y_1^0, y_2^0:

(b)
$$y_1(t) = y_1^\infty + A_{11}e^{-k_1 t} + A_{12}e^{-k_2 t}$$
$$y_2(t) = y_2^\infty + A_{21}e^{-k_1 t} + A_{22}e^{-k_2 t}$$

The steady-state values:

(c)
$$y_1^\infty = \frac{k_{22}J_1 + k_{12}J_2}{k_{11}k_{22} - k_{12}k_{21}}; \qquad y_2^\infty = \frac{k_{21}J_1 + k_{11}J_2}{k_{11}k_{22} - k_{12}k_{21}}$$

Eigenvalues: $k_1 = \frac{1}{2}[(k_{11} + k_{22}) + q] \qquad k_2 = \frac{1}{2}[(k_{11} + k_{22}) - q]$

(d)
$$q = \{(k_{11} - k_{22})^2 + 4k_{12}k_{21}\}^{1/2}$$

The integration constants are defined by:

$$A_{11} = 1/q[(k_{11} - k_2)(y_1^\infty - y_1^0) - k_{12}(y_2^\infty - y_2^0)]$$
$$A_{12} = 1/q[(k_1 - k_{11})(y_1^\infty - y_1^0) + k_{12}(y_2^\infty - y_2^0)]$$
$$A_{21} = 1/q[(k_{21}(y_1^\infty - y_1^0) + (k_2 - k_{22})(y_2^\infty - y_2^0)]$$
$$A_{22} = 1/q[-k_{21}(y_1^\infty - y_1^0) + (k_{22} - k_1)(y_2^\infty - y_2^0)]$$

to Eq. 15-16 we have

$$t_{ss} = \frac{3}{\min\{k_i\}} \tag{15-31}$$

where $\min\{k_i\}$ is the smallest of all the rate constants k_i.

In a simplified manner we can visualize the eigenvalues to represent the two quasi-independent processes that govern the overall behavior of the two-box system. This becomes clearer when we look at a specific example such as the distribution of PER in Greifensee. From the rates listed in Table 15.6 we find that the sum of the removal rates ($k_{g,E} + k_{w,E}$) is significantly greater than the internal "adjustment" rates. In fact, in this case the eigenvalues can be approximated by the external and internal adjustments. The external adjustment describes the behavior of the total mass of PER in the system under the influence of external inputs and outputs. We expect this eigenvalue to be of the order of $k_{\text{external}} \sim k_{w,E} + k_{g,E} \sim 34 \cdot 10^{-3}\,\text{d}^{-1}$. The internal adjustment describes the redistribution of PER within the system, that is, between the two boxes. Since the hypolimnetic box is the larger and thus more inert system, we estimate the second eigenvalue to be of the order of $k_{\text{internal}} \sim k_{ex,H} \sim 4 \cdot 10^{-3}\,\text{d}^{-1}$.

In fact, the exact values are somewhat different, but we indeed find that they differ in size by about a factor of 10 (Table 15.6). It is the internal redistribution of PER which determines the response time of the system. The corresponding steady-state concentrations can be calculated either from Eq. 15-30a and b by replacing the differentials on the left-hand side by zero and then solving the set of two equations for the unknowns C_E and C_H, or from the general formula given in Table 15.5 (Eq. c). In both cases we have assumed the atmospheric PER concentrations C_a to be negligible.

Expressing the riverine input of PER in terms of a (hypothetical) mean input concentration, $C_{in} = I/Q$, leads to the following steady-state concentrations:

$$C_E^{\infty} = \frac{I/V_E}{k_{w,E} + k_{g,E}} = \frac{k_{w,E}}{k_{w,E} + k_{g,E}} C_{in} \tag{15-32a}$$

$$C_H^{\infty} = C_E^{\infty} + \frac{I_H/V_H}{k_{ex,H}} = C_E^{\infty} + \eta C_{in} \frac{k_{w,E}(V_E/V_H)}{k_{ex,H}} \tag{15-32b}$$

where $\eta = I_H/I$ is the relative fraction of PER introduced into the hypolimnion. First, the epilimnetic steady-state concentration C_E^{∞} is identical with the value calculated from the one-box model (Eq. 15-17). This result is reached because the only two output processes, outflow and air–water exchange, that are directly proportional to the epilimnetic concentration C_E operate at the lake surface and are thus independent of depth. Second, the hypolimnetic concentration C_H^{∞} is always greater than C_E^{∞}, except for $\eta = 0$ (no input into the hypolimnion) when both concentrations are equal. Two extreme values ($\eta = 0$ and 1, respectively) are used in Table 15.6.

TABLE 15.6 Dynamic Two-Box Model for PER in Greifensee

See values from Tables 15.1 and 15.2

Transfer velocities		
Air–water velocity	v_{tot}	$0.155\,m \cdot d^{-1}$
Turbulent exchange velocity across thermocline	v_{th}	$0.05\,m \cdot d^{-1}$
Rate constants		
Epilimnetic flushing rate	$k_{w,E} = Q/V_E$	$6.8 \cdot 10^{-3}\,d^{-1}$
Epilimnetic gas exchange rate	$k_{g,E} = v_{tot}/h_E$	$26.7 \cdot 10^{-3}\,d^{-1}$
Turbulent exchange rate across thermocline		
Epilimnion	$k_{ex,E} = v_{th} \cdot A_{th}/V_E$	$7.5 \cdot 10^{-3}\,d^{-1}$
Hypolimnion	$k_{ex,H} = v_{th} \cdot A_{th}/V_H$	$3.75 \cdot 10^{-3}\,d^{-1}$
Input functions (inhomogeneous terms)		
Total input of PER	I	$0.90\,mol \cdot d^{-1}$
Relative input to hypolimnion	$\eta = I_H/I$	$0 \leqslant \eta \leqslant 1$
Mean input concentration	$C_{in} = I/Q$	$2.6 \cdot 10^{-6}\,mol \cdot m^{-3}$

Input terms (see Table 15.5, Eq. a)

Epilimnion $J_E = I_E/V_E = (1-\eta)C_{in}k_{w,E} = (1-\eta) \cdot 0.018 \cdot 10^{-6}\,mol \cdot d^{-1}$

Hypolimnion $J_H = I_H/V_H = \eta C_{in}k_{w,E}(V_E/V_H) = \eta \cdot 0.009 \cdot 10^{-6}\,mol \cdot m^{-3} \cdot d^{-1}$

Steady-state concentrations and eigenvalues

Epilimnion $$C_E^{\infty} = \frac{k_{w,E}}{k_{w,E} + k_{g,E}} C_{in} = 0.20\,C_{in} \sim 0.54 \cdot 10^{-6}\,mol \cdot m^{-3}$$

$$C_H^{\infty} = C_E^{\infty} + \eta \frac{k_{w,E}(V_E/V_H)}{k_{ex,H}} C_{in} = (0.20 + 0.91\eta)C_{in}$$

$$= (0.54 + 2.4\eta) \cdot 10^{-6}\,mol \cdot m^{-3}$$

Eigenvalues	$k_1 \sim (k_{w,E} + k_{g,E})$	$41.7 \cdot 10^{-3}\,d^{-1}$
	$k_2 \sim (k_{ex,H})$	$3.0 \cdot 10^{-3}\,d^{-1}$
Overall time to steady state	$3/k_2$	1000 d

Specific values

(1) All input into epilimnion ($\eta = 0$)
$C_E^{\infty} = C_H^{\infty} = 0.54 \cdot 10^{-6}\,mol \cdot m^{-3}$

(2) All input into hypolimnion ($\eta = 1$)
$C_E^{\infty} = 0.54 \cdot 10^{-6}\,mol \cdot m^{-3}; C_H^{\infty} = 2.95 \cdot 10^{-6}\,mol \cdot m^{-3}$

Periodic Switching between One-Box and Two-Box Model

As a result of seasonal changes, many lakes in the midlatitudes undergo a typical annual cycle in their vertical mixing pattern. This cycle entails alternating between states of well-established vertical stratification and mixing to the bottom. (Note there are many, especially deep, lakes which do not completely mix to the deepest depths). For Greifensee we can approximate the mixing cycle by assuming four months of complete mixing (January to April) and eight months of stratification during which the lake is separated into an epilimnion and a hypolimnion. In terms of our models, the lake switches back and forward between a one-box state (winter) and a two-box state (summer). The onset of stratification poses no special problem. The initial values for the individual concentrations C_E^0 and C_H^0 are just equal to the end values of the preceding one-box phase. In contrast, switching from summer to winter requires us to calculate the weighted average using the end values C_E and C_H to get the initial

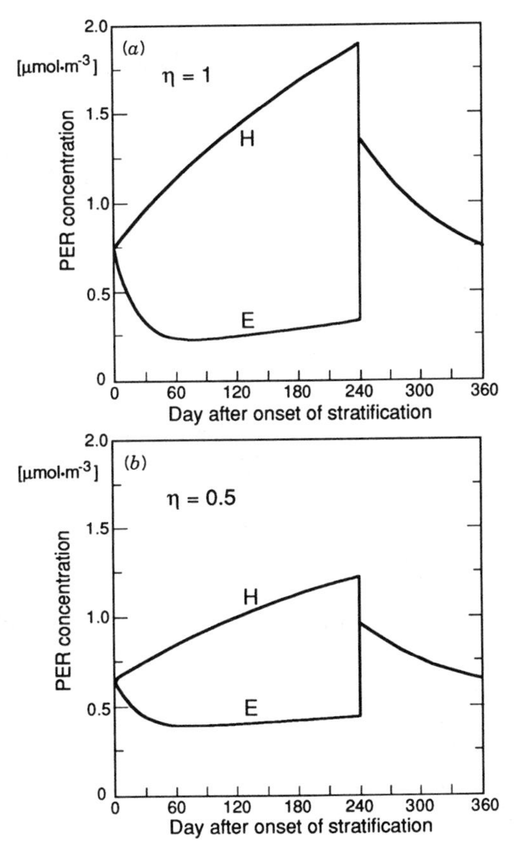

Figure 15.8 Periodic switching between one and two-box model serves to simulate the seasonal stratification cycle of lakes. Example PER in Greifensee, parameters as in Tables 15.4 and 15.6. (*a*) All input into hypolimnion (H) ($\eta = 1$). Steady-state concentrations during stagnation ($C_E^\infty = 0.54 \cdot 10^{-6}\ \text{mol} \cdot \text{m}^{-3}$, $C_H^\infty = 2.95 \cdot 10^{-6}\ \text{mol} \cdot \text{m}^{-3}$) are not fully reached before the onset of mixing after 8 months. (*b*) Half of input into each, H and E. Note that for $\eta = 0$ (all input into E) C_E and C_H would remain at $C_E^\infty = C_H^\infty = 0.54 \cdot 10^{-6}\ \text{mol} \cdot \text{m}^{-3}$ throughout the year.

value for the one-box model:

$$C^0_{1\,\text{box}} = \frac{C_E V_E + C_H V_H}{V_E + V_H} \tag{15-33}$$

In reality, of course, mixing between E and H occurs gradually, not at a single instant; but this simplification does not affect the central point of our discussion.

With the omnipresence of desk computers and standard simulation programs it is easy to program Eqs. 15-6 and 15-30 (or their analytical solutions), and to combine them in order to get the seasonal changes of PER in Greifensee. Examples of such simulations are shown in Figure 15.8.

Use of the Box Model Concept for Other Chemicals and Other Lakes

It may appear that we have devoted too much time to the discussion of the fate of PER in Greifensee and that only little profit can be gained for the remaining enormous variety of compounds and systems with which we actually are confronted. Although it is true that every situation asks for a new, "tailor-made" model, it is also true that the principal steps necessary to build and analyze these models can all be found in the preceding example. To demonstrate this potential we mention three paths along which one can extend these considerations:

1. How do different lakes behave?
2. How do different compounds behave?
3. How can we treat situations where either the system or the compound does not allow us to force the spatial variation into a simple one- or two-box scheme?

The last question is treated in the next section. Let us use the remaining part of this section to address questions 1 and 2.

First we want to examine how PER behaves in other lakes. From the preceding discussion, we can quickly identify the following internal parameters as being relevant for the level of PER in a lake: depth, water residence time, air–water exchange velocity, stratification cycle (length of circulation period, vertical mixing intensity, possible existence of some permanently stratified water layers at the bottom of the lake), and distribution of input with depth. Note that we have not listed the total input I; we consider it as the external driving force to which the concentrations in the lake are directly proportional provided that the models are linear, meaning that all the reactions are of zero or first order.

For a fixed flushing rate q, the relative importance of air–water exchange versus flushing as expressed by the ratio v_{tot}/q (Eq. 15-13), is independent of depth h but increases with wind exposure (i.e., with v_{tot}). Similarly, the steady-state concentrations (Eqs. 15-17 and 15-32) only depend on v_{tot} and q, but not on h. However, since the different rate constants k are inversely related to depth h, the response of the lake to changing loading conditions is slower in deep lakes.

Taking all the relevant expressions into account leads us to conclude that for a given loading per area, I/A_0, PER levels are expected to be largest in lakes with slow flushing, little wind, long stratification, and PER inputs directly to the hypolimnion. In contrast, shallow lakes with large q and v_{tot} should have generally small PER levels.

The second question referring to the behavior of other chemical compounds in the annually stratifying Greifensee can be approached by analyzing the different types of removal mechanisms found in lakes:

1. Surface removal (outflow, air–water exchange)
2. Sediment boundary reaction (sorption–desorption), molecular diffusion into or from pore water
3. Volumetric reaction either homogeneous in space (hydrolysis) or heterogeneous in space (biodegradation, photolysis)

Perchloroethylene served as an illustration of the surface removal case. In contrast, nitrilotriacetic acid (NTA)

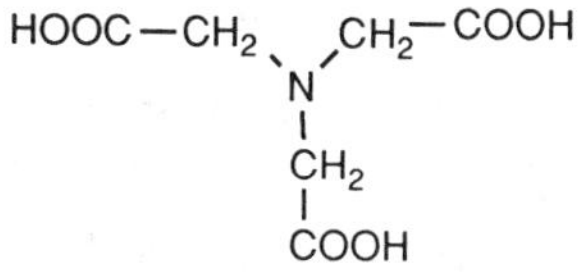

HOOC—CH₂–N(–CH₂—COOH)–CH₂–COOH

is used to demonstrate the behavior of a nonvolatile, reactive substance. As for PER, the fate of NTA in Greifensee, on one hand, is determined by the seasonal stratification cycle, that is, by the alternating one- and two-box situation. On the other hand, the removal mechanisms of the two compounds are very different. Therefore, we expect the onset of stratification to lead to different structures of the vertical concentration profiles. Indeed, field data show that the hypolimnetic NTA concentration drops below the surface values in summer (Table 15.7), whereas for PER we found the opposite effect (Fig. 15.6).

Measured NTA concentrations in winter are typically $C = 3.7 \cdot 10^{-6}\,\mathrm{mol \cdot m^{-3}}$, and in summer $C_E = 5.2 \cdot 10^{-6}\,\mathrm{mol \cdot m^{-3}}$ and $C_H = 2.1 \cdot 10^{-6}\,\mathrm{mol \cdot m^{-3}}$ (Table 15.7). They seem to reach a steady state, at least during the later part of each mixing regime. Based on a field investigation in the years 1982 and 1983 on the major rivers and sewage inlets of Greifensee, total NTA loading was estimated as $13\,\mathrm{mol \cdot d^{-1}}$. From laboratory experiments in which NTA was added to samples taken from natural waters (estuaries, rivers), first-order degradation rate constants k_r between 0.02 and $1\,\mathrm{d^{-1}}$ were determined for NTA concentrations between about $20 \cdot 10^{-6}$ and $5000 \cdot 10^{-6}\,\mathrm{mol \cdot m^{-3}}$ (Bartholomew and Pfaender, 1983; Larson and Davidson, 1982). Rate constants varied with temperature ($Q_{10} = 2$, i.e., k_r increases by a factor 2 for a temperature increase of 10°C) and were reduced for oxygen concentrations below 10^{-5} M (Larson et al., 1981). Data are not available regarding the degradation of NTA at concentrations below $20 \cdot 10^{-6}\,\mathrm{mol \cdot m^{-3}}$.

The following discussion shows how box models can be used to get initial estimates for the missing information. Once such coarse estimates are available, it is, of course,

TABLE 15.7 Dynamic Two-Box Model for Nitrilotriacetic Acid (NTA) in Greifensee

Lake-specific parameters as in Tables 15.1 and 15.6

Input		
Total input of NTA	I	$13\,\text{mol}\cdot\text{d}^{-1}$
Mean input concentration	$C_{\text{in}} = I/Q$	$38\cdot 10^{-6}\,\text{mol}\cdot\text{m}^{-3}$
Relative input to hypolimnion	$\eta = I_{\text{H}}/I$	$0 \leqslant \eta \leqslant 1$
NTA concentration in lake (measured)		
Winter (mixed lake)	C	$3.7\cdot 10^{-6}\,\text{mol}\cdot\text{m}^{-3}$
Summer: Epilimnion	C_{E}	$5.2\cdot 10^{-6}\,\text{mol}\cdot\text{m}^{-3}$
Hypolimnion	C_{H}	$2.1\cdot 10^{-6}\,\text{mol}\cdot\text{m}^{-3}$

Calculated *in situ* sink fluxes and reaction rates
(see Eqs. 15-35 to 15-37)

Winter	$R = 11.7\,\text{mol}\cdot\text{d}^{-1}$		$k_{\text{r}} = 0.021\,\text{d}^{-1}$
Summer			
(a) For	$\eta = 0$		(no input into hypolimnion)
	$R_{\text{E}} = 10.1\,\text{mol}\cdot\text{d}^{-1}$,		$k_{\text{r,E}} = 0.039\,\text{d}^{-1}$
	$R_{\text{H}} = 1.2\,\text{mol}\cdot\text{d}^{-1}$,		$k_{\text{r,H}} = 0.0057\,\text{d}^{-1}$
(b) For	$\eta = 0.3$		(30% input into hypolimnion)
	$R_{\text{E}} = 6.3\,\text{mol}\cdot\text{d}^{-1}$	}	$k_{\text{r,E}} = k_{\text{r,H}} = 0.024\,\text{d}^{-1}$
	$R_{\text{H}} = 5.0\,\text{mol}\cdot\text{d}^{-1}$		

Time to steady state (Summer case b)

Eigenvalues: $k_1 = 0.025\,\text{d}^{-1}$
$k_2 = 0.041\,\text{d}^{-1}$

Overall time to steady state: $t_{\text{ss}} = \dfrac{3}{k_1} = 120\,\text{d}$

always possible to use a dynamic model to investigate the influence of the seasonal stratification cycle on NTA in more detail.

Let us start with the winter situation and assume that the concentration $C = 3.7\cdot 10^{-6}\,\text{mol}\cdot\text{m}^{-3}$, measured at a time when no spatial gradients were observed, represents a steady-state situation. Three mass fluxes are relevant for the case of NTA: inflow I, outflow O, and *in situ* removal R. Thus, the mass balance equation (Eq. 15-2) reduces to

$$\frac{dM}{dt} = I - O - R = 0 \tag{15-34}$$

Using the outlet process model introduced earlier (Eq. 15-3) and solving for R yields

$$R = I - Q\cdot C \tag{15-35}$$

From the data shown in Table 15.7 it follows that R is $11.7\,\mathrm{mol \cdot d^{-1}}$. Put another way, 90% of the NTA entering the lake seems to disappear somewhere in the lake. Of course, the mass balance does not tell us whether the removal is by biodegradation, purely chemical decomposition, sedimentation, or even by loss to the atmosphere. Our knowledge of the physicochemical properties of NTA makes biodegradation the most likely candidate.

As mentioned before, to turn Eq. 15-35 into a dynamic model we still need information on how R depends on C and possibly on other parameters. From the laboratory experiments it is not obvious that degradation of NTA is first order. However, we can always interpret R in terms of a pseudo-first-order rate constant k_r, using the relation

$$k_r = \frac{R}{CV} \tag{15-36}$$

as long as we keep in mind that k_r may vary under different conditions. Thus, for the winter situation we get $k_r = 0.022\,\mathrm{d^{-1}}$, a value which is at the lower limit of the rates measured under laboratory conditions (albeit at larger concentrations).

A similar approach can be adopted for the stratification period. Using the corresponding modifications of Eqs. (15-28) and (15-29) and assuming steady-state yields:

$$\frac{dM_E}{dt} = (1-\eta)I - QC_E - v_{th}A_{th}(C_E - C_H) - R_E = 0 \tag{15-37a}$$

$$\frac{dM_H}{dt} = \eta I + v_{th}A_{th}(C_E - C_H) - R_H = 0 \tag{15-37b}$$

There are three unknowns (R_E, R_H, relative input into hypolimnion, η), but only two equations; thus the system has no unique solution. Let us explore the range of removal fluxes, R_E and R_H, by choosing extreme values for η. First, if all the NTA is entering the lake at the surface ($\eta = 0$), solution of Eq. 15-37 yields $R_E = 10.1\,\mathrm{mol \cdot d^{-1}}$ and $R_H = 1.2\,\mathrm{mol \cdot d^{-1}}$. This result corresponds to pseudo-first-order rates (Eq. 15-36) of $k_{r,E} = 0.039\,\mathrm{d^{-1}}$ and $k_{r,H} = 0.0057\,\mathrm{d^{-1}}$, values not implausible in view of the temperature and oxygen gradient between the epilimnion and hypolimnion.

For the other extreme ($\eta = 1$, all input into the hypolimnion), no solution with a positive R_E value exists since all the remaining terms of Eq. 15-37a are sinks for NTA, given the measured positive concentration difference $C_E - C_H$ of $3.1 \cdot 10^{-6}\,\mathrm{mol \cdot m^{-3}}$. Instead, we can solve the equations for the condition that the degradation rate is the same in both depth intervals. Again, the result looks plausible ($\eta = 0.3$, $k_{r,E} = k_{r,H} = 0.024\,\mathrm{d^{-1}}$).

With these results we can now check the validity of the steady-state assumption. For the winter situation, time to steady state is controlled by the total rate ($k_w + k_r$), where k_w is the one-box flushing rate (Table 15.4) and k_r the pseudo-first-order reaction rate (Table 15.7). By analogy to Eq. 15-16, we have $t_{ss} = 3/(k_w + k_r) \sim 120\,\mathrm{d}$. Since the winter situation holds for about 80 d, we see that steady state is almost reached.

For the summer situation, the overall rates of the composite two-box system follow from the eigenvalues. The values listed in Table 15.7 show that t_{ss} is the same as in the winter. The dynamic behavior of NTA is dominated by the *in situ* reaction rate k_r, which occurs in both boxes and thus provides both boxes with their own mechanisms to respond to changing input rates. Vertical transport between the two subsystems is less important for determining the size of t_{ss}. This contrasts with the case of PER, where air–water exchange at the water surface is the important elimination mechanism, and thus the hypolimnion only reaches a steady state via mixing with the epilimnion.

We add here an important note of caution. The examples given above were not put together with the intention of explaining the variation of PER or NTA concentration in Greifensee, once and for all. On the contrary—the main purpose of this kind of simple model calculation is to develop a "feeling" for the important mechanisms and to identify those areas where more knowledge is needed to fully understand the fate of the compound in the natural environment. For this task a simple model is more useful than a complicated computer model, which often contains such a large number of parameters that it becomes difficult to distinguish between important and marginal processes.

15.4 THE SOLID–WATER INTERFACE: MODELING THE ROLE OF PARTICLES

Partitioning between Dissolved and Particulate Phases

In the preceding sections no distinction was made between a compound being present as a dissolved species or sorbed to solid surfaces (e.g., suspended particles, sediment–water interface). We learned that several of the transport and transformation processes may selectively act on either the dissolved or the sorbed form of a constituent (Fig. 11.1). For instance, a molecule sitting on the surface of a sedimentary particle at the lake bottom does not feel the effect of turbulent flow in the lake water as does the dissolved species which is passively moved around by the currents. In contrast, a molecule sorbed to a suspended particle (e.g., an algal cell) can sink through the water column because of gravity, unlike its dissolved counterpart.

In this section the distinction between dissolved and sorbed species is introduced into the box model concept in the simplest possible manner, that is, by assuming an instantaneous and reversible linear equilibrium relationship between the dissolved concentration, $C_w(\text{mol}\cdot\text{m}_w^{-3})$, and the species sorbed on solids, $C_s(\text{mol}\cdot\text{kg}_s^{-1})$. (The units m_w^3 and kg_s refer to water volume and solid mass, respectively.) As discussed in Chapter 11, the (observed) solid–water distribution ratio K_d is defined by

$$K_d = \frac{C_s}{C_w} \qquad (\text{m}_w^3\cdot\text{kg}_s^{-1}) \qquad (11\text{-}2)$$

Note that in this chapter the volume is expressed in cubic meters and not, as in Chapter 11 and in most other chapters, in liters (L). Thus, K_d values are numerically smaller by 10^{-3}, densities and the solid-to-solution phase ratios r_{sw} are larger by 10^3.

By checking units during modeling calculations, such differences in expressing quantities like K_d should not cause a problem.

From Chapter 11 we remember that K_d may represent a sum of several kinds of sorption processes. For instance, for neutral, nonpolar chemicals only the weight fraction of the solid which is natural organic matter, f_{om}, may be relevant for sorption. In that case, we can use Eq. 11-16:

$$K_d = f_{om} \cdot K_{om} \qquad (m_w^3 \cdot kg_{om}^{-1}) \qquad (11\text{-}16)$$

where K_{om} is the sorption coefficient reflecting partitioning between particulate organic matter and solution phases.

In box models the concentration variables have to be related to the same reference volume. We use total (bulk) volume (indicated by m^3, without index). Thus, the dissolved and sorbed (particulate) concentrations per total volume, C_d and C_p, are

$$C_d = \frac{V_w}{V_t} C_w = \phi C_w \qquad (mol \cdot m^{-3}) \qquad (15\text{-}38)$$

$$C_p = C_s \cdot (M_s/V_t) \qquad (mol \cdot m^{-3}) \qquad (15\text{-}39)$$

where $\phi = V_w/V_t$ is the volumetric fraction occupied by water (in environments with many particles, such as sediments, ϕ is called porosity), and (M_s/V_t) is the mass of solids per total volume. We can express the "single phase concentrations" C_d and C_p in terms of the total concentration, $C_t = C_d + C_p$, and the fraction of the chemical in solution, f_w, introduced in Eq. 11-5:

$$f_w = \frac{1}{1 + r_{sw} K_d} \qquad (11\text{-}5)$$

or for neutral nonpolar chemicals:

$$f_w = \frac{1}{1 + f_{om} \cdot r_{sw} \cdot K_{om}} \qquad (15\text{-}40)$$

where r_{sw} is the solid-to-solution phase ratio (now in kilograms per cubic meter). Thus,

$$C_d = f_w \cdot C_t \quad \text{and} \quad C_p = (1 - f_w) C_t \qquad (15\text{-}41)$$

In open water bodies the volumetric fraction occupied by particles is so tiny that the "porosity" ϕ is equal to 1. Also, since $V_t \sim V_w$, the mass of solids per total volume, (M_s/V_t) is about equal to $r_{sw} = M_s/V_w$. Thus, we can often replace Eqs. 15-38 and 15-39 by

$$C_d \sim C_w \quad \text{and} \quad C_p \sim r_{sw} \cdot C_s \qquad (15\text{-}42)$$

The partition between the dissolved and particulate phases depends on both, system-specific properties (solid-to-solution phase ratio, composition of solids) and compound-specific properties (e.g., K_{om}). In the open water column, r_{sw} ranges typically between $10^{-5}\,kg_s \cdot m^{-3}$) (deep sea) and $10^{-3}\,kg_s \cdot m^{-3}$ (surface waters, lakes) with organic matter weight fractions f_{om} between 0.05 and 0.5. In Chapter 11 we saw that many organic compounds have K_{om} values between 0.1 and $10^3\,m^3 \cdot kg_{om}^{-1}$. Thus, using Eq. (15-40) we get f_w values between 1 and about 0.5. Assuming a maximum particulate organic matter to solution ratio $f_{om} \cdot r_{sw}$ of $10^{-3}\,kg_{om} \cdot m^{-3}$, compounds with k_{om} values below $1\,m^3 \cdot kg_{om}^{-1}$ have solid fractions $(1 - f_w)$ less than 10^{-3} and can thus, from the point of view of transport processes, be considered to be totally dissolved. However, for transformation processes the sorbed fraction, in spite of its small size, may be relevant for compounds with very fast sorbed-phase reactions.

In sediments and groundwaters $f_{om} \cdot r_{sw}$ can reach values up to $100\,kg_{om} \cdot m^{-3}$. This means that, even for "nonsorptive" compounds with K_{om} of $0.1\,m^3 \cdot kg_{om}^{-1}$, about 90% of the total concentration can be bound to particles ($f_w \sim 0.1$).

Particle Settling

To consider how the association of compounds with particles may influence their transport, we need to discuss how particles move in the water column. Besides the passive movement with the water in which they are suspended, particles feel the force of gravity. The settling velocity v_s of a particle with density ρ_s through a fluid with density $\rho_w < \rho_s$ can be described by Stokes' law (Lerman, 1979):

$$v_s = \alpha \cdot B \cdot r^2 \qquad (m \cdot s^{-1}) \tag{15-43}$$

where r is a linear dimension characterizing the particle, α is a nondimensional form factor, and B is a factor that depends on the nature of the fluid and the particle density:

$$B = \frac{2g(\rho_s - \rho_w)}{9\mu} \qquad (m^{-1} \cdot s^{-1}) \tag{15-44}$$

[$g = 9.81\,m \cdot s^{-2}$ is the acceleration due to gravity, μ is the dynamic viscosity of the fluid ($kg \cdot m^{-1} \cdot s^{-1}$)]. For a sphere the form factor α is 1 and r is the sphere radius. Using quartz spheres ($\rho_s = 2650\,kg \cdot m^{-3}$) and water at 20°C ($\mu = 10^{-3}\,kg \cdot m^{-1}\,s^{-1}$), we estimate the typical size of B to be $3.6 \cdot 10^6\,m^{-1}\,s^{-1}$. Thus for 1-$\mu$m radius spheres, v_s is $3.6 \cdot 10^{-6}\,m \cdot s^{-1}$ ($0.3\,m \cdot d^{-1}$) and v_s equals $3.6\,m \cdot s^{-1}$ for 1 mm spheres.

Strictly speaking, Stokes' law is only valid for laminar flow around the particle. The condition of the flow, that is, whether it is laminar or turbulent, can be expressed by the Reynolds number:

$$Re = \frac{2rv_s}{(\mu/\rho_w)} \tag{15-45a}$$

For laminar flow Re has to be smaller than the critical value 0.1. For turbulent conditions, that is, for $Re > 0.1$, the resisting force of the water on a settling particle is different from the Stokes' drag used to derive Eq. 15-43 (see e.g., Håkanson and Jansson, 1983). Inserting Eq. 15-43 into Eq. 15-45a yields the following condition for the validity of Stokes' law:

$$Re = \frac{2\alpha B r^3}{(\mu/\rho_w)} < 0.1 \qquad (15\text{-}45b)$$

In fact, for the quartz spheres the maximum particle radius compatible with Eq. 15-46b is only 24 μm. In contrast, biogenic particles have a much smaller excess density, $\rho_s - \rho_w$, and complicated shapes (α small). Typical settling velocities for these particles are 0.1 m·d^{-1} for 1-μm particles and 100 m·d^{-1} for 100-μm particles (Lerman, 1979).

Let us use the settling velocity v_s to calculate the removal of particles from a water body. First, consider a rectangular tank in which initially particles of equal size and density (i.e., of equal sinking velocity) are homogeneously suspended (Fig. 15.9a). The

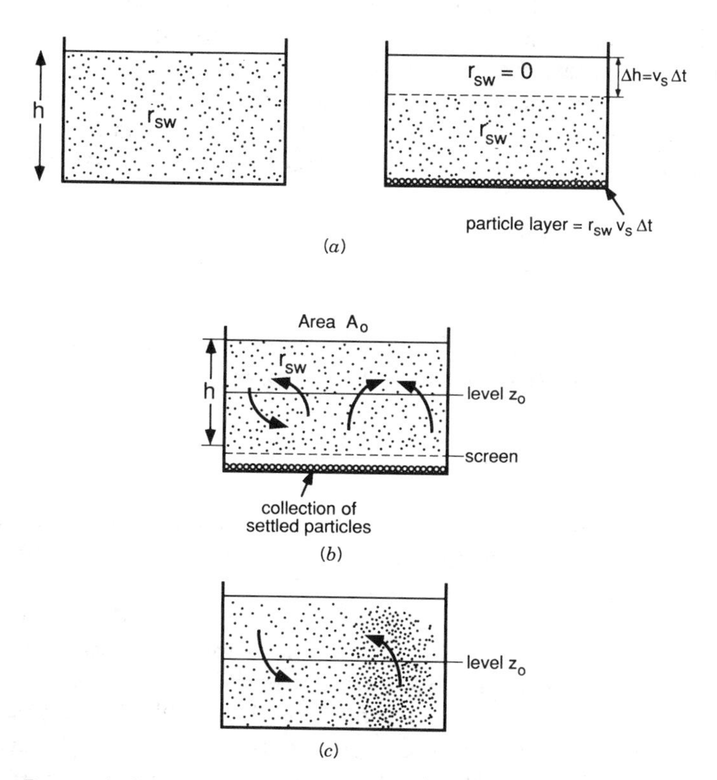

Figure 15.9 Removal of suspended particles (described by the solid-to-solution phase ratio r_{sw}) with uniform sinking velocity v_s. (*a*) No mixing: constant particle flux $F = r_{sw} \cdot v_s$ until upper "horizon" reaches the bottom after time $t = h/v_s$; (*b*) Homogeneously mixed system: exponential decrease of r_{sw}; (*c*) Change of mean particle flux across level z_0 for the case of heterogeneous distribution of particles and a spatially variable vertical velocity component.

initial suspended matter concentration is r_{sw}. In the absence of any currents in the tank, after some time Δt the top layer of depth $\Delta h = v_s \cdot \Delta t$ will be devoid of particles, and a particle layer of areal mass density $r_{sw}\Delta h = r_{sw}v_s \cdot \Delta t$ ($kg_s \cdot m^{-2}$) covers the bottom. Thus, the initial relative particle removal rate k_s is given by

$$k_s = \frac{\text{particles removed per area and time } \Delta t}{\text{particles in system per area}}$$

$$= \frac{r_{sw}v_s}{r_{sw}h} = \frac{v_s}{h} \qquad (s^{-1}) \tag{15-46}$$

The corresponding particle mass flux per unit area through any horizontal cross section is

$$F_s = r_{sw} \cdot v_s \qquad (kg_s \cdot m^{-2} \cdot s^{-1}) \tag{15.47}$$

Now, let us assume that the water body is kept well mixed by the production of turbulent fluid motion in the tank (Fig. 15–9*b*). The vertical motion of a particle is now given by the algebraic sum of the local vertical current velocity v_z and the setting velocity v_s pointing along the vertical axis; $v_{z,tot} = v_s + v_z$. The total particle flux, ΣF_s across an interface at an arbitrary depth z_0 is given by the integral of Eq. 15-47 over the area at z_0, A_0, but now we must replace v_s by the total vertical particle velocity $v_{z,tot}$:

$$\Sigma F_s = \int_{A_0} r_{sw}(x, y, z_0) \cdot [v_s + v_z(x, y, z_0)]\, dxdy \tag{15-48}$$

Note that the settling velocity v_s does not depend on the space coordinates (x, y, z) since we have assumed particles of equal size and mass. If the concentration of the suspended particles is spatially constant, $r_{sw}(x, y, z) = r_{sw}$, Eq. 15-48 becomes

$$\Sigma F_s = r_{sw}v_s A_0 + r_{sw} \int_{A_0} v_z(x, y, z_0)\, dxdy \tag{15-49}$$

In the absence of any subsurface source or sink of water, the law of mass conservation applied to the water itself requires the remaining integral in Eq. 15-49 to be zero. Thus, in spite of the water currents, the mean flux per unit area at z_0, $\bar{F}_s = \Sigma F_s / A_0$, is still given by Eq. 15-47, which was derived for a quiescent water body.

Let us assume that turbulence in the tank keeps the suspended particle concentration homogeneous, but that at the bottom of the tank, the particles can sink through some screen below which no water currents exist (Fig. 15.9*b*). In the absence of any external particle fluxes or *in situ* production/removal of particles, the mass balance equation for suspended particle mass is given by equating the rate of change of particle mass in the CV with time with the rate of loss due to settling:

$$V \frac{dr_{sw}}{dt} = -A_0 v_s r_{sw} \tag{15-50}$$

or after division by V:

$$\frac{dr_{sw}}{dt} = -k_s \cdot r_{sw} \quad \text{with } k_s = \frac{v_s}{h} \tag{15-51}$$

where $h = V/A_0$ is the mean depth and $k_s[\mathrm{T}^{-1}]$ is a first-order particle removal rate. Note the similarity between Eqs. 15-46 and 15-51. For the nonmixed system the sedimentation flux remains constant until all particles have disappeared from the system (this occurs after time h/v_s), where the removal for the mixed system is first order, leading to an exponential decrease of r_{sw}. There is a strong resemblance between the mixed-particle removal process and the process of air–water transfer. Therefore, the corresponding rates k_s and k_g (see Eq. 15-11) have the same form.

We have now developed the necessary equations to formulate box models in which a distinction is made between the dissolved and sorbed phases of a given compound. It seems that knowledge of the settling velocity v_s, together with the mean depth of the box, is all that is needed to calculate particle fluxes and thus fluxes of the compound sorbed to the particles. The settling velocities, in turn, can be determined by Stokes' law (Eq. 15-43). Unfortunately, nature is not so simple. Settling velocities, which have been determined indirectly by looking at the fate of some sorbed species in natural bodies (so-called apparent settling velocities), not only vary a great deal with time, but also are generally much smaller than the velocities calculated from Stokes' law (Bloesch and Sturm, 1986). There are several reasons for this discrepancy. One has to do with the omnipresence of mainly horizontal water currents. What we have discussed so far as a vertical displacement is, in fact, mainly a horizontal journey of particles with a tiny vertical net component. A typical sinking velocity of biogenic particles of $1\ \mathrm{m \cdot d^{-1}}$ or about $10^{-3}\ \mathrm{cm \cdot s^{-1}}$ coexists in a water body with typical horizontal velocities of the order of $1\ \mathrm{cm \cdot s^{-1}}$. Thus, the process of particle sinking looks like snow flakes during a wind storm! Snow does not accumulate homogeneously during a storm in which the wind velocities are 1000 times larger than the sinking velocity of the flakes.

To be more specific, let us assume that under the influence of some current pattern the particle concentration and the vertical current velocities v_z (see Eq. 15-48), are not quite homogeneous but show some covariance. For instance, above-average r_{sw} values may exist in those places where the currents are predominantly upward, while below-average r_{sw} values are located at downwelling areas (Fig. 15.9*c*). As a result, the mean particle flux is significantly reduced compared to the value calculated from Eq. 15-47. As first shown by Stommel (1949), such current structures supporting inhomogeneous particle distributions do indeed exist in lakes and oceans. Langmuir circulation is one example of a circular current system in which small particles can be trapped and exhibit apparent settling velocities of zero (Ledbetter, 1979).

We should also note that particles in natural water bodies do not come in a single size class; thus sinking velocities are not the same for all suspended solids. Colloidal particles do not sink at all, while fecal pellets can sink several hundred meters per day. Therefore, organisms have a strong influence on v_s since they constantly change the size and nature of the suspended solids. In addition, particle coagulation (O'Melia, 1985) and breakup constantly change the particle size spectrum.

In conclusion, the first-order particle removal model, Eq. 15-51, is a reasonable first approximation to describe the influence of suspended particles on sorbed chemical species. However, the sinking velocity is not necessarily identical with the Stokes' law velocity, but rather is an empirical parameter which may strongly depend on lake currents and biological processes and thus significantly vary over short time periods.

One-Box Model for Sorbing Species

Let us derive a one-box model for a chemical species for which dissolved and sorbed phases are important. To be more specific, the following processes are considered:

1. Acting on dissolved phase only:
   Air–water exchange, with rate k_g (Eq. 15-11)
2. Acting on sorbed phase only:
   Sedimentation, with rate k_s (Eq. 15-51)
3. Acting on both phases:
   Input, I_d and I_p, expressed as mass per unit time
   Loss through outlet, with rate k_w (Eq. 15-10)
   In situ reaction (sum of removal due to hydrolysis, photolysis, biodegradation), but with different rates for the two phases, $k_{r,d}$ and $k_{r,p}$

For both species concentrations, C_d and C_p, mass balance equations can be formulated by analogy to the example given in Table 15.4. Yet, since the compound can alter its speciation by sorption–desorption, the corresponding phase transfer rate has to be introduced into both equations. We arbitrarily define J_{trans} as the net transfer rate from the dissolved into the sorbed phase (per unit volume and time). Note that J_{trans} can also be negative. Thus, for the dissolved phase we have

$$\frac{dC_d}{dt} = \frac{I_d}{V} - k_g\left(C_w - \frac{C_a}{K'_H}\right) - k_w C_d - k_{r,d} C_d - J_{trans} \tag{15-52a}$$

where the right-hand terms describe input, gas exchange, outflow, in situ reaction, and dissolved-to-sorbed phase transfer. Note that for an open water body, but not necessarily for situations with large r_{sw} such as for water-saturated soil, C_w can be replaced by C_d (see Eq. 15-42). For the sorbed phase,

$$\frac{dC_p}{dt} = \frac{I_p}{V} - k_w C_p - k_s C_p - k_{r,p} C_p + J_{trans} \tag{15-52b}$$

where the right-hand terms are input, outflow, sedimentation, in situ reaction, and phase transfer.

To evaluate these two equations, a process model is needed for the sorption–desorption term J_{trans}. As discussed in Chapter 11, sorption can be slow, even seeming irreversible on certain time scales. In spite of the possible complexity, we adopt the simplest process model by assuming an instantaneous and linear equilibrium between C_d and C_p leading to Eqs. 15-40 and 15-41 with constant solute fraction f_w. This is equivalent to assuming (1) a fully reversible and infinitely fast sorption–desorption rate and (2) constant r_{sw} and K_d.

As a consequence, the two differential equations can be reduced to one. Let us express both C_d and C_p in terms of the total concentration C_t (Eq. 15-41) and add Eq. 15-52a and b:

$$\frac{dC_t}{dt} = \frac{dC_d}{dt} + \frac{dC_p}{dt} = \frac{I_t}{V} + k_g \frac{C_a}{K'_H} - k_w C_t - f_w(k_g + k_{r,d})C_t - (1 - f_w)(k_s + k_{r,p})C_t \quad (15\text{-}53)$$

where $I_t = I_d + I_p$ is the total input. This result is instructive from a general point of view since one can learn how to build "two-phase" models with instantaneous equilibrium. That is, we write a mass balance equation for the total concentration C_t in which all processes acting on both phases appear with their normal rate (e.g., flushing, k_w), all rates referring to processes acting on the dissolved phase only are multiplied by f_w (e.g., $f_w k_g$ and $f_w k_{r,d}$), and similarly for processes occurring in the particulate phase, we multiply by $(1 - f_w)$ [e.g., $(1 - f_w)\cdot(k_s + k_{r,p})$].

However, everything has its price! In the case discussed above, the price for combining two variables into one is hidden in the fractionation factor, f_w, which depends on the solid-to-solution phase ratio r_{sw}, which itself may be strongly time dependent and thus require its own mass balance model. Let us postpone the corresponding "payment" a little bit and first discuss an example for which a constant r_{sw} is assumed. As a case study we choose the fate of two different polychlorinated biphenyl congeners (PCBs) in Lake Superior (North America) (see Table 15.8 for characteristic data).

The PCBs usually come as mixtures of more than 35 important congeners, each having different physicochemical properties (Henry's Law constants, aqueous solubilities, etc.). One does not learn much by applying Eq. 15-53 to this mixture. Fortunately, in the Laurentian Great Lakes individual PCB congeners have been studied in the necessary detail to evaluate the mass balance equation for single congeners (e.g., Baker and Eisenreich, 1990). Two of them are chosen as examples here, 2′,3,4-trichlorobiphenyl and 2,2′,3,4,5,5′,6-heptachlorobiphenyl, the latter being much more hydrophobic than the former.

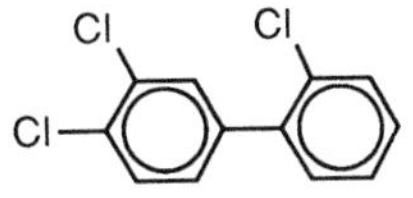

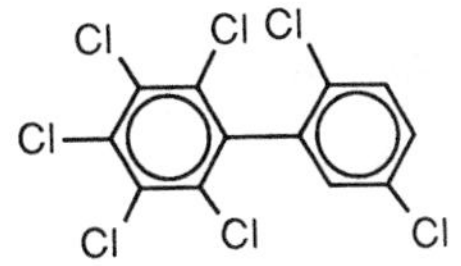

Their physicochemical properties and the individual factors appearing in Eq. 15-53 are listed in Table 15.9. Except for the measurements that are specific for Lake Superior (input rates, concentrations of PCBs, composition of the particles determin-

TABLE 15.8 Characteristic Data of Lake Superior (North America)[a]

Volume	V	12,230 km^3
Surface area	A_0	82,100 km^2
Mean depth	$h = V/A_0$	149 m
Maximum depth	h_{max}	406 m
Throughflow of water (including precipitation)	Q	71 km$^3 \cdot$yr^{-1}
Flushing rate	$k_w = Q/V$	$5.8 \cdot 10^{-3}$ yr^{-1}
Mean water residence time	$\tau_w = k_w^{-1}$	172 yr
Flushing velocity	$q = Q/A_0$	0.86 m$\cdot$yr^{-1}
Solid-to-solution phase ratio[b]	r_{sw}	$0.4 \cdot 10^{-3}$ kg$_s \cdot$m^{-3}
Organic matter content of suspended solids	f_{om}	0.4
Organic matter content of settled solids	f_{oms}	0.06
Porosity of surficial sediments	ϕ	0.9
Sediment accumulation		
On 50% of lake area		0.2 kg$_s \cdot$m$^{-2} \cdot$yr^{-1}
Mean for total lake	F_s	0.1 kg$_s \cdot$m$^{-2} \cdot$yr^{-1}
Mean particle settling velocity (Eq. 15-47)	$v_s = F_s/r_{sw}$	250 m$\cdot$yr^{-1}
Particle removal rate (Eq. 15-51)	$k_s = v_s/h$	1.7 yr^{-1}
Mean wind speed	u_{10}	5 m$\cdot$s^{-1}
Air–water transfer velocity for water vapor, Eq. 10-28	$v_a(H_2O)$	1100 m$\cdot$d^{-1}
Air–water transfer velocity for O_2, Eq. 10-32	$v_w(O_2)$	1.2 m$\cdot$d^{-1}

[a] From Eisenreich et al., 1989.
[b] For most surface waters, r_{sw} is equal to the suspended solid concentration.

ing K_d, etc.), all the data were derived from information given in this book either in tables (e.g., Henry's Law constants) or indirectly by approximative relationships (e.g., $K_d = f_{om} \cdot K_{om}$). More details are given in the footnotes to Table 15.9.

Since no effective *in situ* degradation of PCBs occurs ($k_{r,d} = k_{r,p} = 0$), three elimination pathways remain which, if described in terms of first-order reaction rates, can be directly compared with respect to their relative importance for the elimination of each PCB congener from the water column. As shown by the removal rates listed in Table 15.9, for 2′,3,4-trichlorobiphenyl the flux to the atmosphere is by far the most important process. In contrast, because of its larger K_d value, removal of the heptachlorobiphenyl to the sediments is predicted to be nearly as important as air–water exchange. By the way, from this simple model we would expect to find the heptachlorobiphenyl relatively enriched in the sediments compared to the trichlorobiphenyl. We shall see later whether this is true.

Given the PCBs' inputs and the atmospheric concentrations C_a, as listed in Table 15.9, we can now calculate the total steady-state concentration in the lake for the two congeners. Solving Eq. 15-53 for $C_t = C_t^\infty$ by taking $dC_t/dt = 0$ yields ($k_{r,d} = k_{r,p} = 0$):

$$C_t^\infty = \frac{I_t/V + k_g \cdot C_a/K'_H}{k_w + f_w k_g + (1 - f_w) k_s} \tag{15-54}$$

From C_t^∞ we can also calculate the dissolved concentration at steady state, C_d^∞, and compare it to concentrations measured in the surface waters of Lake Superior (Table 15.9). Actual values of the trichlorobiphenyl congener are about 30% larger than the calculated steady-state value, a remarkable consistency if we consider the many simplifying assumptions and estimates made to derive Eq. 15-54. A more severe discrepancy is found for the heptachlorobiphenyl congener; the measured values are 10 times larger than the calculated steady-state concentration. There are numerous reasons why the model calculation could be wrong, and it is certainly not possible to find the final answer to the PCB story of Lake Superior based on the presently available field data. However, the following discussion demonstrates how modelers should proceed from simple to more refined schemes by comparing their calculations to field data in order to decide which processes to include in their model.

Let us discuss some of the factors affecting the results yielded by use of Eq. 15-54:

1. The input I_t may be wrong. Since the input estimates are based on PCB mixtures and typical relative congener compositions, an error of 30% is not unlikely and could thus explain the discrepancy found for the trichlorobiphenyl. For the heavier congener the discrepancy seems to be too large to be solely explained by an input error.
2. The concentration measured in the surface waters may not represent the mean lake concentration. This hypothesis is supported by concentrations of sorbed PCBs, which are very different for solids collected from different depths (Baker and Eisenreich, 1989). This point is discussed in the next part of this section.
3. The air–water exchange rate k_g, the dominant removal rate for both congeners, may be overestimated by taking a mean wind velocity of $5\ m \cdot s^{-1}$. Although a decrease of k_g may indeed increase C_t^{∞}, it would affect the concentration of the trichlorobiphenyl more strongly since the heptachlorobiphenyl concentration is also controlled by the sediment removal rate. Thus, a change of k_g alone could not help to explain the larger deviation of the latter congener from the measured value.
4. The presence of organic colloids may give rise to a third fraction of the biphenyl molecules, the fraction sorbed to nonsettling microparticles and macromolecules. This fraction neither contributes to the air–water exchange equilibrium nor participates in the process of particle settling. By generalizing the notation introduced in Eq. 11-5 one can write

$$C_t = C_d + C_p + C_{COM} \tag{15-55}$$

and define

$$f_w = \frac{1}{1 + r_{sw}K_d + K_{om} \cdot [COM]} = C_d/C_t \tag{15.56a}$$

$$f_{COM} = \frac{K_{om}[COM]}{1 + r_{sw}K_d + K_{om} \cdot [COM]} = C_{COM}/C_t \tag{15.56b}$$

$$f_s = \frac{r_{sw}K_d}{1 + r_{sw}K_d + K_{om} \cdot [COM]} = C_p/C_t \tag{15-56c}$$

TABLE 15.9 Physicochemical Properties and Model Parameters for Two Selected PCB Congeners

			2′,3,4-Trichlorobiphenyl	2,2′,3,4,5,5′,6-Heptachlorobiphenyl
IUPAC No.			33	185
Molecular mass			257.5	395.3
Henry's Law constant at 15°C[a]	$K'_H = K_H/RT$	(1)	0.003	0.007
Air–water transfer velocity				
Air[b]	v_a	$(m \cdot d^{-1})$	450	400
Water[b]	v_w	$(m \cdot d^{-1})$	0.71	0.65
Total, Eq. 10-34	v_{tot}	$(m \cdot d^{-1})$	0.47	0.53
Air–water exchange rate	$k_g = v_{tot}/h$	(yr^{-1})	1.15	1.30
Atmospheric concentration[a]	C_a	$(mol \cdot m^{-3})$	$1.8 \cdot 10^{-13}$	$1.2 \cdot 10^{-14}$
Distribution coefficient				
Suspended solids[d]	K_d	$(m^3 \cdot kg_s^{-1})$	10	10^3
Settled solids[d]	K_{ds}	$(m^3 \cdot kg_s^{-1})$	1.5	150
Fraction dissolved in the water column				
Eq. 11-5	f_w	(1)	0.996	0.71
Total input rate of PCB congener[c]	I_t	$(mol \cdot yr^{-1})$	100	30
Terms of Eq. 15-53[e]				
Input per unit time and volume				
River, precipitation (wet + dry)	I_t/V	$(mol \cdot m^{-3} \cdot yr^{-1})$	$8.2 \cdot 10^{-12}$	$2.5 \cdot 10^{-12}$
From the atmosphere[f]	$k_g \cdot C_a/K'_H$	$(mol \cdot m^{-3} \cdot yr^{-1})$	$69 \cdot 10^{-12}$	$2.2 \cdot 10^{-12}$

Removal rates				
Flushing	k_w	(yr^{-1})	$5.8 \cdot 10^{-3}$	$5.8 \cdot 10^{-3}$
To the atmosphere	$f_w \cdot k_g$	(yr^{-1})	1.15	0.92
Sedimentation	$(1 - f_w)k_s$	(yr^{-1})	$6.8 \cdot 10^{-3}$	0.49
Steady-state concentration (calculated)				
Total, Eq. 15-54	C_t^∞	$(mol \cdot m^{-3})$	$6.7 \cdot 10^{-11}$	$3.3 \cdot 10^{-12}$
Dissolved	$C_d^\infty = f_w C_t^\infty$	$(mol \cdot m^{-3})$	$6.7 \cdot 10^{-11}$	$2.3 \cdot 10^{-12}$
On particles, Eq. 15-42	$C_s^\infty = \frac{1 - f_w}{r_{sw}} C_t^\infty$	$(mol \cdot kg_s^{-1})$	$67 \cdot 10^{-11}$	$240 \cdot 10^{-12}$
Measured concentrations				
Dissolved				
Surface waters[a]	C_w	$(mol \cdot m^{-3})$	$(1.0 \pm 0.9) \cdot 10^{-10}$	$(2.5 \pm 1.9) \cdot 10^{-11}$
Sorbed				
Epilimnetic particles[g]	C_s	$(mol \cdot kg_s^{-1})$	$(5.4 \pm 2.3) \cdot 10^{-9}$	$(1.6 \pm 0.8) \cdot 10^{-9}$
Sediment surface[g]	C_{ss}	$(mol \cdot kg_s^{-1})$	$(0.19 \pm 0.04) \cdot 10^{-9}$	$(0.48 \pm 0.13) \cdot 10^{-9}$
Apparent distribution coefficient	$K_d = C_s/C_w$	$(kg_s \cdot m^{-3})$	54 ± 54	64 ± 58

[a] From Baker and Eisenreich (1990).
[b] Calculated for wind speed $u_{10} = 5 \, m \cdot s^{-1}$ (see Table 15.8) using Eqs. 10-29 and 10-33 with $\alpha = 0.67$ and $\beta = 0.5$. Molecular diffusivities are approximated using molecular weights (see Section 9.3).
[c] From Eisenreich et al. (1989).
[d] K_d calculated from K_{ow} (Appendix) and f_{om} values (Table 15.8) using Eq. 11-16
[e] No *in situ* degradation ($k_{r,d,} = k_{r,p} = 0$).
[f] Note that for both congeners, given the measured lake concentration, the net input (input minus removal) is negative, that is, directed from the lake into the atmosphere.
[g] From Baker and Eisenreich (1989).

where C_{COM} is the amount of chemical sorbed to colloidal organic matter per unit bulk volume, [COM] is the concentration of organic colloids in the water, and K_{om} is the colloid–water distribution coefficient.

In Table 15.10 the characteristic parameters of the modified three-phase model for the two selected PCB congeners are listed. The sorption to colloids slightly reduces the dissolved fraction of the trichlorobiphenyl f_w, makes the air–water exchange $f_w \cdot k_g$ a little bit less effective, and thus increases the steady-state concentration C_t^{∞}. The changes for the heptachlorobiphenyl are more spectacular. Most of the congener is now sorbed to the colloids and "feels" neither the drive to participate in air–water exchange nor in sedimentation. Realizing that the "dissolved" fraction reported in the literature includes the colloidal fraction as well, we now have an excellent agreement between model calculation and measurements.

The three-phase model may help to explain another mystery, that is, the discrepancy between the theoretical K_d value and the apparent value calculated from comparing simultaneously measured concentrations C_s and C_w (see Table 15.9). For the three-

TABLE 15.10 Three-Phase Model for Two Selected PCB Congeners in Lake Superior[a]

	Unit	2′,3,4 Trichloro- IUPAC No. 33	2,2′,3,4,5,5′,6 Heptachloro- IUPAC No. 185
Concentration of colloidal organic matter [COM][b]	$(kg \cdot m^{-3})$	$3.2 \cdot 10^{-3}$	$3.2 \cdot 10^{-3}$
Distribution coefficient, K_{om}	$(m^3 \cdot kg^{-1})$	25	$2.5 \cdot 10^3$
$[COM] \cdot K_{om}$	(1)	0.08	8.0
$r_{sw} \cdot K_d$[c]	(1)	$4 \cdot 10^{-3}$	0.4
f_w	(1)	0.923	0.106
f_{COM}	(1)	0.074	0.851
$f_s = 1 - f_w - f_{COM}$	(1)	0.003	0.043
Modified terms of Eq. 15-53			
Removal rates			
Flushing k_w	(yr^{-1})	$5.8 \cdot 10^{-3}$	$5.8 \cdot 10^{-3}$
To the atmosphere $f_w k_g$	(yr^{-1})	1.06	0.14
Sedimentation $f_s k_s$	(yr^{-1})	$6.8 \cdot 10^{-3}$	0.073
Steady-state concentration (calculated)			
Total, Eq. 15-54, C_t^{∞}	$(mol \cdot m^{-3})$	$7.2 \cdot 10^{-11}$	$2.1 \cdot 10^{-11}$
"Dissolved," $(f_w + f_{COM}) C_t^{\infty}$	$(mol \cdot m^{-3})$	$7.2 \cdot 10^{-11}$	$2.0 \cdot 10^{-11}$
Apparent distribution K_d^{app}, calculated[d]	$(kg \cdot m^{-3})$	9.3	110

[a]If not stated otherwise, all parameter values are as in Tables 15.8 and 15.9.
[b]From Baker and Eisenreich (1989), assuming that [COM] is twice the measured dissolved organic carbon concentration, DOC.
[c]See Tables 15.8 and 15.9.
[d]Calculated from the theoretical K_d value of the "large" suspended solids (Table 15.9) according to Eq. 15-57, assuming that the fraction associated with colloids is measured as dissolved fraction.

phase model, the apparent K_d values can be written as

$$K_d^{app} = \frac{C_s}{C_w + C_{COM}} = \frac{C_s}{C_w} \cdot \frac{1}{1 + (f_{COM}/f_w)} \qquad (15\text{-}57)$$

The ratio C_s/C_w should be equal to the K_d value for the "large" suspended particles as listed in Table 15.9. The calculated apparent distribution coefficient for the heptachlorobiphenyl is only twice the measured value, whereas the "large particle" K_d value is nearly 20 times greater than observed. For the other congener the three-phase model has little effect on K_d and still leaves an unexplained discrepancy.

5. There is still another point to be discussed which may limit the calculations presented in Tables 15.9 and 15.10. In 1986, when the concentrations were measured, the lake may not have been at steady state. In fact, the PCB input, which mainly occurred through the atmosphere, dropped by about a factor of 5 between 1965 and 1980. However, the response time of Lake Superior (time to steady state, calculated according to Eq. 15-18 from the inverse sum of all removal rates listed in Table 15.9) for both congeners would be less than 3 years. This is quite short, expecially if we use Eq. 15-21 with $\alpha = -0.1\ \text{yr}^{-1}$ (the reduction rate which reduces the input by a factor of 5 within 15 years) to formally analyze the time delay between input and lake concentration.

However, there are indications that the lake as a whole may possess a much longer memory than the one that we calculated from taking account of the water column alone. The obvious candidate for the missing memory is the sediment bed. Remember that there is a significant flux to the sediment for at least the higher molecular mass congener. The lake water "feels" the sediment memory, which means that we have to consider a process that mediates the exchange of sedimentary constituents back into the free-water column. Let us thus discuss how to expand our box model to account for this additional process.

Exchange at the Sediment–Water Interface

As before, we are looking for the simplest model able to take into account the processes that turn out to be relevant to the problem. So we treat the sediment (or part of it) as a completely mixed box. (Obviously, there are a number of reasons why this may be not very realistic, but let us just disregard them at the moment.) More specifically, we define a "surface mixed sediment layer" (SMSL) by a mixing depth z_{mix}, or by the solid mass per unit area m (Fig. 15.10). Both are related:

$$m = z_{mix}(1 - \phi)\rho_s \qquad (\text{kg}_s \cdot \text{m}^{-2}) \qquad (15\text{-}58)$$

The mixing power can either originate from water currents or animals (bioturbation, see Robbins, 1986) or by a combination of both. In some "quiet" sediment environments z_{mix} may be so small that it just represents the sediment surface in contact with the overlying water. (The SMSL is indeed a somewhat hypothetical model parameter!) Below the SMSL lies the "permanent" sediment, that is, the zone from which no feedback into the water column is possible. Thus, the sedimentary memory of the chemical compound is assumed to be completely confined to the SMSL.

The chemical concentrations in the pore water and the solids of the sediment bed need to be linked. The sorbed compound concentration per solid mass in the sediment bed, $C_{ss}(\mathrm{mol \cdot kg_s^{-1}})$, and the dissolved concentration per unit pore water volume, C_{ws} $(\mathrm{mol \cdot m_w^{-3}})$, are related by an instantaneous sorption equilibrium (Eq. 11.2) whereby the distribution coefficient K_{ds} may be different from the one in the water column because of the different chemical composition of the two kinds of particles (e.g., organic carbon content):

$$K_{ds} = \frac{C_{ss}}{C_{ws}} \qquad (\mathrm{m_w^3 \cdot kg_s^{-1}}) \qquad (15\text{-}59)$$

We choose C_{ss} as the relevant variable to characterize the chemical compound in the sediments. However, to do a mass balance for the SMSL, we have to consider the total compound per unit bulk volume, C_{ts}. That is, we sum the fraction sorbed on the particles and the fraction dissolved in the pore water:

$$C_{ts} = \phi C_{ws} + (1 - \phi)\rho_s C_{ss} = C_{ss}\left[\frac{\phi}{K_{ds}} + (1 - \phi)\rho_s\right] \qquad (\mathrm{mol \cdot m^{-3}}) \qquad (15\text{-}60)$$

Obviously, the smaller K_{ds} is and the larger ϕ, the greater the relative contribution of the dissolved phase to C_{ts}. Let us take $\phi = 0.9$, $\rho_s = 2.5 \cdot 10^3\ \mathrm{kg \cdot m^{-3}}$, and a relatively small K_{ds} of $1.5\ \mathrm{m_w^3 \cdot kg_s^{-1}}$, like what we would estimate for trichlorobiphenyl. (Note that for even smaller K_{ds} values, not much of the compound is being removed to the sediments and thus the sediment model is not important anyway!) Then, ϕ/K_{ds} is about $0.6\ \mathrm{kg_s \cdot m_w^{-3}}$ and $(1 - \phi)\rho_s$ is equal to $250\ \mathrm{kg_s \cdot m^{-3}}$; thus the sorbed phase is still much larger than the contribution from the pore water. As a good approximation we therefore use

$$C_{ts} \sim (1 - \phi)\rho_s \cdot C_{ss} = \frac{m}{z_{mix}} \cdot C_{ss} \qquad (\mathrm{mol \cdot m^{-3}}) \qquad (15\text{-}61)$$

The following processes contribute to the mass balance of $m \cdot C_{ss}$, the sorbed chemical per unit sediment area (the same letters are used in Fig. 15.10):

A. Input as sorbed species on settling particles, $v_s \cdot r_{sw} \cdot C_s$.

B. Transfer into the permanent sediment layer, $\beta \cdot v_s \cdot r_{sw} \cdot C_{ss}$, where β is a preservation factor indicating the fraction of settled particulate matter eventually reaching the permanent sediment layer (i.e., $\beta \leqslant 1$). If total solids are used to describe the distribution ratio f_w, then β is about 1 (assuming no dissolution of carbonates or silica and return to the overlying water column). If particulate organic carbon (POC)

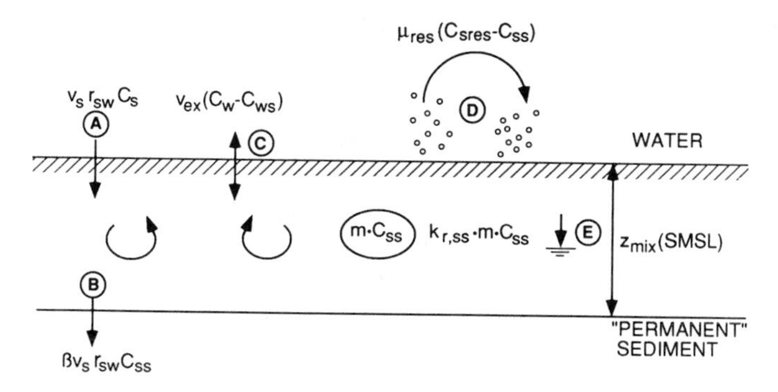

Figure 15.10 Exchange model between water and sediment column using a "surface mixed sediment layer" (SMSL) of thickness z_{mix} above a "permanent" sediment from which no feedback into the lake is possible. See text for explanation of processes designated by letters (A) to (E).

or suspended organic matter, $r_{sw} f_{om}$, are used together with K_{om} (Eq. 15-40), β is smaller than 1 because part of the organic material is degraded in the SMSL.

C. Diffusive exchange between the dissolved phase in the lake water and the pore water, $v_{ex}(C_w - C_{ws})$. The exchange velocity v_{ex} is similar is nature to the air–water exchange velocity; it can be interpreted as the (molecular) diffusivity divided by a boundary layer thickness, though the latter does not necessarily correspond to a "physical" laminar boundary layer.

D. Resuspension and fast resettling of sedimentary particles, $\mu_{res}(C_{sres} - C_{ss})$, where μ_{res} is the resuspended solid mass per unit area and time ($kg_s \cdot m^{-2} \cdot s^{-1}$). Note that $C_{sres} = K_{ds} \cdot C_w$ is the equilibrium concentration on the particles (characterized by the sediment distribution coefficient K_{ds}) exposed to the dissolved concentration in the lake, C_w. Thus, it is assumed that during the resuspension process the sorbed chemical immediately equilibrates with the lake water, and thus the particle has a different sorbed concentration when it falls back to the sediments.

E. Chemical or biochemical degradation in the SMSL, $k_{r,ss} \cdot m \cdot C_{ss}$. Since the dissolved compound is proportional to C_{ss}, the reaction rate may also account for degradation occurring in the dissolved fraction.

Putting all the processes into a mass balance equation yields

$$\frac{d}{dt}(m \cdot C_{ss}) = m \cdot \frac{dC_{ss}}{dt} \qquad (\text{mol} \cdot \text{m}^{-2}\,\text{s}^{-1})$$

$$= v_s \cdot r_{sw} \cdot C_s - \beta \cdot v_s \cdot r_{sw} \cdot C_{ss} + v_{ex}(C_w - C_{ws}) + \mu_{res}(C_{sres} - C_{ss}) - k_{r,ss} \cdot m \cdot C_{ss} \qquad (15\text{-}62)$$

Using the equilibrium expressions (Eqs. 15-56 and 15-59), the mass balance equation can be written in terms of just C_t and C_{ss}. Then Eq. 15-62 becomes, after division by m,

$$\frac{dC_{ss}}{dt} = \frac{v_s}{m} f_s C_t - \beta \frac{v_s \cdot r_{sw}}{m} C_{ss} + \frac{v_{ex}}{m}\left(f_w C_t - \frac{C_{ss}}{K_{ds}} \right) + \frac{\mu_{res}}{m}(K_{ds} f_w C_t - C_{ss}) - k_{r,ss} C_{ss}$$

The two exchange mechanisms, diffusion and resuspension (third and fourth terms on the right-hand side of the equation), turn out to depend on the same concentration difference, $f_w \cdot K_{ds} \cdot C_t - C_{ss}$. This allows them to be expressed by the combined exchange rate:

$$k_{ex} = \frac{v_{ex}}{K_{ds} \cdot m} + \frac{\mu_{res}}{m} \qquad [T^{-1}] \tag{15-63}$$

Thus,

$$\frac{dC_{ss}}{dt} = \frac{v_s}{m} f_s C_t - \beta \frac{r_{sw} \cdot v_s}{m} C_{ss} + k_{ex}(f_w \cdot K_{ds} \cdot C_t - C_{ss}) - k_{r,ss} \cdot C_{ss} \tag{15-64}$$

Since little is known about the size of the parameters in Eq. 15-63, it is often better to treat these two different processes as one, characterized by a single rate parameter k_{ex}. In most cases, k_{ex} would have to be calibrated by comparing the model to real data. This exchange term is of central importance to the model since it is responsible for the feedback from the sediment to the water. Recall that we need such a feedback to explain the apparently slow response of Lake Superior to a changing PCB input. In fact, we also need to add a corresponding exchange term to a modified version of the mass balance of the total concentration in the lake which takes organic colloids into account (see Eqs. 15-55 and 15-56):

$$\frac{dC_t}{dt} = \frac{I_t}{V} + k_g \frac{C_a}{K'_H} - k_w C_t - f_w \cdot (k_g + k_{r,d}) C_t - f_s C_t (k_s + k_{r,p}) - k^*_{ex}\left(f_w \cdot C_t - \frac{C_{ss}}{K_{ds}} \right) \tag{15-65}$$

where the rate

$$k^*_{ex} = k_{ex} \cdot \frac{m K_{ds}}{h} = \frac{v_{ex}}{h} + \frac{\mu_{res} K_{ds}}{h} \qquad [T^{-1}] \tag{15-66}$$

is not identical with k_{ex} since Eqs. 15-64 and 15-65 contain different kinds of concentration variables and refer to different "box sizes" (mean depth h and mixed layer mass m, respectively).

Remember the original motivation to look at the sediments was to explore their role in the response of the PCB concentration in the water column of Lake Superior to the decreasing PCB input. In Figure 15.11 the relevant processes are summarized in a schematic manner. To make the following discussion clearer, specific values for the new model parameters are introduced (Table 15.11), although these figures just represent order-of-magnitude estimates. Let us first look at the exchange between

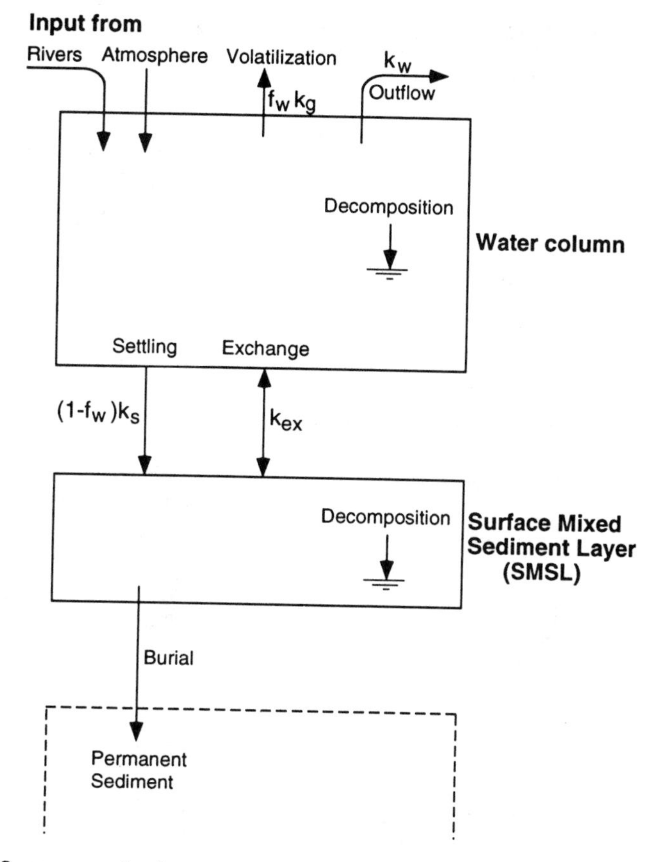

Figure 15.11 Summary of relevant processes for the combined lake–sediment model to describe the fate of PCBs in lakes.

the water column and the SMSL. Looking from the water side, a new rate, k^*_{ex}, is introduced into Eq. 15-65, which couples the chemical in the water to the SMSL. The corresponding timescale $3/k^*_{ex}$ decreases with increasing distribution coefficient K_{ds} from about 7 years for the trichlorobiphenyl to about 2 years for the heptachlorobiphenyl. In contrast, looking from the SMSL the relevant time scale $3/k_{ex}$ is inversely related to K_{ds}, that is, small for the trichlorobiphenyl (0.4 years) and much larger for the heptachlorobiphenyl (11 years). In other words, the SMSL is a slowly reacting compartment for the congeners with large K_{ds} values, but is close to a continuous exchange equilibrium with the open water column for the less sorptive congeners (K_{ds} small).

Given the strongly time-variable input history of the PCBs to the Great Lakes, it cannot reasonably be expected that both the open water and the surface sediments could be at steady state. Nevertheless, it is tempting to calculate such hypothetical stationary concentrations, C_t^∞ and C_{ss}^∞. This can easily be done with the tools that

TABLE 15.11 Sediment Exchange Model: Characteristic Parameters and Typical Values

Sediment mixed layer $m = z_{mix} \cdot \rho_s \cdot (1-\phi)$				
Mixing depth	z_{mix}	(m)	0.02	
Density of solids	ρ_s	$(kg_s \cdot m^{-3})$	$2.5 \cdot 10^3$	
Porosity	ϕ	(1)	0.9	
Mixed layer mass	m	$(kg_s \cdot m^{-2})$	5	
Resuspension	μ_{res}	$(kg_s \cdot m^{-2} \cdot yr^{-1})$	1	
Exchange velocity, approximated by $v_{ex} = D/\delta$				
Thickness of sediment boundary layer	δ	(m)	$5 \cdot 10^{-4}$	
Molecular diffusivity (Typical value)	D	$(m^2 \cdot s^{-1})$	10^{-9}	
Exchange velocity	v_{ex}	$(m \cdot yr^{-1})$	60	

Rates relevant for compound in sediment mixed layer (see also Table 15.9)			2′,3,4 Trichloro- IUPAC No. 33	2,2′,3,4,5,5′,6 Heptachloro- IUPAC No. 185
Sediment–water exchange				
For SMSL, Eq. 15-63	k_{ex}	(yr^{-1})	8.2	0.28
For water column, Eq. 15-66	k^*_{ex}	(yr^{-1})	0.41	1.4
Transfer to permanent sediment ($\beta = 1$)				
$\frac{r_{sw} \cdot v_s}{m}$		(yr^{-1})	0.02	0.02
Steady-state concentrations for combined system (Eqs. 15-64 and 15-65)				
Dissolved	C_t^∞	$(mol \cdot m^{-3})$	$72 \cdot 10^{-12}$	$28 \cdot 10^{-12}$
Sorbed on suspended particles	C_s^∞	$(mol \cdot kg_s^{-1})$	$0.72 \cdot 10^{-9}$	$3.2 \cdot 10^{-9}$
Sorbed on sediment particles	C_{ss}^∞	$(mol \cdot kg_s^{-1})$	$0.10 \cdot 10^{-9}$	$0.64 \cdot 10^{-9}$

we have developed so far. We need to calculate simultaneously the steady-state solutions of Eqs. 15-64 and 15-65 by means of the general formula listed in Table 15.5 (use Eq. c and substitute C_t and C_{ss} for the variables y_1 and y_2). The numerical results of such a calculation for the case of Lake Superior are listed at the end of Table 15.11.

Let us summarize our analysis of the two PCB congeners in Lake Superior by looking at Figure 15.12. We used three models of increasing complexity: (1) the one-box two-phase model, (2) the one-box three-phase model which includes organic colloids, and (3) the combined lake–sediment bed model. As shown in Figure 15.12, for the dissolved concentration the transition to the three-phase model was important for the heavier congener. The sorbed concentrations are more difficult to interpret. Yet the inclusion of the sediments helps to explain the large difference between the observed concentrations on epilimnetic and sedimentary particles, respectively, for the trichlorobiphenyl. Finally, we should not forget that probably the most important reason for adapting the combined lake–sediment model is related to the response time of the lake to changing external loadings. Obviously this effect cannot be observed if only steady-state values are discussed.

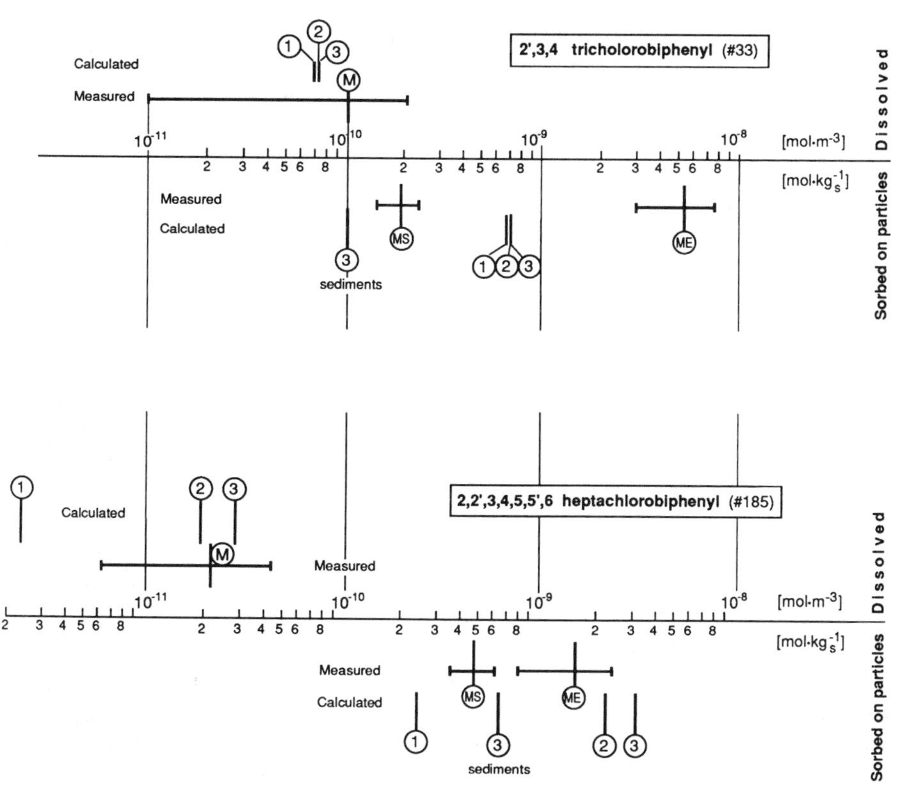

Figure 15.12 Comparison of different PCB model calculations to measured values (black bars show range) in Lake Superior (Baker and Eisenreich, 1989 and 1990) as listed in Table 15.9. The following letters refer to mean values of measured concentrations: M = dissolved, MS = sorbed to particles in surface sediments, ME = sorbed to epilimnetic particles. The following models are included: (1) one-box model with sorption on suspended particles (Table 15.9); (2) one-box model with sorption on particles and colloids (Table 15.10); (3) as model 2, including sediment bed (Table 15.11). Note that for model 3 the sorbed concentrations on sediment particles (designated "sediments") and on suspended particles are different.

That is about as far as we can reasonably go with a model as simple as this applied to compounds like PCBs in complicated environmental systems. Many questions remain. For instance, why is the observed concentration ratio of the two selected congeners so close to one (see Table 15.9) even though the relative importance of removal to the sediments is much greater for the heptachlorobiphenyl? This may have something to do with the history of the corresponding input ratio, or with the role of the SMSL, which is clearly different for the two congeners, or (most likely) with the dynamics of the compound in the water column.

Let us now leave the mystery of the PCBs and examine what a model should look like that allows us to describe systems that are not completely mixed and therefore exhibit a more complex spatial structure in the concentration of the relevant variables.

15.5 SPACE AND TIME: CONTINUOUS MODELS

Internal Transport versus Reaction and Boundary Fluxes

In natural systems there are basically two kinds of transport processes: (1) *direct* transport due to the motion of the surrounding medium (water, air) or external forces such as gravity (sinking of particles through the water column), and (2) *random* transport leading to molecular or turbulent diffusion. Both processes are described in Chapter 9, which gives simple expressions to calculate characteristic time scales needed to mix a system of linear dimension L by advection (t_{ad}) or diffusion (t_d), respectively (Eqs. 9-31 and 9-35).

In most aquatic systems, the sources and sinks of chemicals are not homogeneously distributed. For instance, PER enters a lake mainly through its inlets, that is, along the shore, whereas the main losses are due to air–water exchange at the lake surface. If we are interested in the PER concentration at midlake compared to the concentration close to the shore, we must therefore compare the time needed to transport PER from the shore to the center of the lake with the typical residence time of PER in the surface waters. If the horizontal transport time is short compared to the residence time, the concentration at the lake surface should be fairly constant from the shore to the center. From Table 15.2 we know the mean residence time of PER in the vertically mixed Greifensee is about 90 days. Typical horizontal length scales in this lake are $L = 1$ km. From Figure 9.10 we estimate the horizontal diffusivity E_h to be $5 \cdot 10^3\ cm^2 \cdot s^{-1}$. Although this diagram was derived from oceanic diffusion experiments, it appears that turbulent diffusion in lakes is of comparable magnitude. Thus, from Eq. 9-31 we get a characteristic transport time for diffusion, t_d, of 23 days. This value is probably too large for surface waters since the winds blowing across the lake produce a pattern of surface currents which make the transport from the shore to the center of the lake more "straightforward," that is more advective. If we take a horizontal advection velocity of $1\ cm \cdot s^{-1}$, the corresponding transport time, t_{ad}, turns out to be about 1 day. This is small compared to the residence time of 90 days. Thus, we expect to find no significant horizontal concentration gradients of PER in Greifensee.

However, if the lake is stratified, the situation may look different, since now the vertical transport may become the time-limiting step for complete mixing of PER. This was the reason for developing the two-box model (Table 15.6). Now we go one step further in the direction of an arbitrary spatial resolution of the concentration field. We consider a vertical water column of mean depth h with a constant vertical eddy diffusion coefficient E_z. The flux F_a of PER escaping to the atmosphere is given by Eq. 10-10 with $C_a = 0$:

$$F_a = v_{tot} \cdot C_w \tag{15-67}$$

Within the water column PER is transported upward to compensate for the loss at the surface. This flux is described by Fick's first law (Eq. 9-2):

$$F_d = -E_z \frac{dC}{dz} \tag{15-68}$$

Let us disregard the PER sources for the moment and calculate the vertical concentration gradient needed to compensate for the loss at the water surface. (Of course, at steady state we need a source to compensate for this loss, but this subtlety is not so important for what follows). Combining Eqs. (15-67) and (15-68) and solving for dC/dz yields

$$\frac{dC}{dz} = -\frac{v_{\text{tot}}}{E_z} \cdot C_w \tag{15-69}$$

where C_w is the concentration at the water surface. The vertical coordinate is chosen positive upward. Thus, the negative sign indicates that C is decreasing toward the surface.

Next, we can define homogeneity by requiring that the total concentration difference between lake bottom and lake surface, ΔC, may not be larger than 10% of C_w. Thus

$$\Delta C = h \cdot \left| \frac{dC}{dz} \right| \leqslant 0.1\, C_w \tag{15-70}$$

Inserting Eq. (15-69) into (15-70) and rearranging the terms yields

$$\frac{h \cdot v_{\text{tot}}}{E_z} \leqslant 0.1 \tag{15-71}$$

This inequality must be fulfilled to guarantee a (quasi-) homogeneous PER distribution within the vertical water column. Let us apply it to the case of Greifensee. From Tables 15.1 and 15.4 we have $h = 17.4\,\text{m}$, $v_{\text{tot}} = 0.155\,\text{m} \cdot \text{d}^{-1}$. From Figure 9.13 we get E_z in the hypolimnion to be about $0.20\,\text{cm}^2 \cdot \text{s}^{-1}$ ($1.7\,\text{m}^2 \cdot \text{d}^{-1}$) during the summer. Thus, the value of the nondimensional expression on the left-hand side of Eq. 15-71 is 1.6, indicating that (in accordance with the measurements, see Fig. 15.6) significant vertical gradients of the PER concentration are to be expected during the summer. In fact, since the supply of PER from the deeper parts of the lake to the surface is not sufficient, C_w drops, and so does F_a until a local balance between diffusion and air–water exchange is reached. In contrast, Eq. 15-71 is valid if E_z is larger than $3\,\text{cm}^2 \cdot \text{s}^{-1}$ ($27\,\text{m}^2 \cdot \text{d}^{-1}$), a value easily exceeded during the winter when the density stratification is weak.

This example serves to demonstrate a general principle regarding the spatial distribution of a chemical in an environmental system. The ratio of the time-scale of *internal transport* and the timescale of chemical *removal* (e.g., via transformation) determines the degree of spatial heterogeneity to be expected for a given chemical. Boundary fluxes (air–water, sediment–water) as well as chemical transformations are included in the group of "removals". The transport time scales may strongly vary from one direction (e.g., horizontal) to the other (e.g., vertical), but both are determined by the hydrodynamics of the aquatic system and usually not by the chemical. Note that this is not true if molecular diffusion is involved! In contrast, the time scales of removal are compound-specific. For a given pair of compounds the corresponding rate ratios can be written in terms of nondimensional expressions similar to the one appearing in Eq. 15-71. Since different compounds have very different removal rates, in a given system some compounds may be completely homogeneous, and others partially homogeneous (e.g., homogeneous only along a certain coordinate axis) or even completely heterogeneous. Thus, the question of whether a given lake can be considered to be fully mixed can only be addressed in relation to a specific compound.

In this section some tools are developed that are helpful for analyzing and describing the problem of spatial variation.

The One-Dimensional Transport–Reaction Model

In this section we combine the effect of the two typical transport mechanisms, diffusion and advection, with a simple in situ removal expression. To clearly indicate this restriction, the term "reaction" is used instead of the more general "removal," which includes boundary fluxes.

The combined action of advective and diffusive transport on the temporal variation of the concentration C follows from Eqs. 9-17 and 9-33:

$$\left(\frac{\partial C}{\partial t}\right)_{\text{Trans}} = -\left(v_x \frac{\partial C}{\partial x} + v_y \frac{\partial C}{\partial y} + v_z \frac{\partial C}{\partial z}\right) + \frac{\partial}{\partial x}\left(D_x \frac{\partial C}{\partial x}\right) + \frac{\partial}{\partial y}\left(D_y \frac{\partial C}{\partial y}\right) + \frac{\partial}{\partial z}\left(D_z \frac{\partial C}{\partial z}\right) \tag{15-72}$$

where v_x, v_y, v_z are the three Cartesian velocity components and D_x, D_y, D_z are the (molecular or turbulent) diffusion coefficients. Note that we have retained the D's inside the spatial derivatives to include the possibility that the diffusion coefficients vary in space. This is especially important for the case of turbulent diffusion for which we have introduced the notation E_x, E_y, E_z (see Eq. 9-37).

Though the mathematical structure of Eq. 15-72 written in three dimensions is not really different from the one-dimensional case, it is certainly easier to discuss the relative importance of transport versus reaction by retaining just one dimension, for example, the x axis. Furthermore, here we assume that D_x is spatially constant. The transformations of the species described by C are expressed as the combination of a zero-order production (J) and a first-order decay term (k_r):

$$\left(\frac{\partial C}{\partial t}\right)_{\text{React}} = J - k_r C \tag{15-73}$$

Thus, combining Eqs. 15-72 and 15-73 and using the simpler notation $D_x = D$, $v_x = v$ yields

$$\frac{\partial C}{\partial t} = \left(\frac{\partial C}{\partial t}\right)_{\text{Trans}} + \left(\frac{\partial C}{\partial t}\right)_{\text{React}} = D\frac{\partial^2 C}{\partial x^2} - v\frac{\partial C}{\partial x} - k_r C + J \tag{15-74}$$

This is a second order linear inhomogeneous partial differential equation. We will not deal with its time-dependent solutions; the interested reader is referred to the standard textbooks (e.g., Crank, 1975; Carslaw and Jaeger, 1959).

Let us concentrate on the following question. What does the steady-state concentration profile $C(x)$ between $x = 0$ and $x = L$ look like, provided that the concentrations at the two boundaries, $C(0)$ and $C(L)$, are kept constant? Such a situation is met, for instance, in the interior of the Pacific and Atlantic Oceans where the surface and bottom water concentrations are fixed by mixing processes that are fast compared to the slow mixing in the interior of the water column. Some groundwater systems may also be of this kind.

At steady state, the partial differential equation (Eq. 15-74) transforms into the ordinary differential equation:

$$D\frac{d^2C}{dx^2} - v\frac{dC}{dx} - k_r C + J = 0 \tag{15-75}$$

with the boundary conditions

$$C(0) = C_0, \qquad C(L) = C_L \tag{15-76}$$

Its solution, derived in Table 15.12, depends on two exponential terms of the form $A_i e^{\lambda_i(x/L)}$, which can be expressed in terms of the normalized length scale $\xi = x/L$ bound to the interval 0 to 1. Therefore the expressions $e^{\lambda_i \xi}$ vary between 1 (for $\xi = 0$) and e^{λ_i} (for $\xi = 1$). This means that the so-called eigenvalues λ_i determine the spatial structure of the concentration profile and the most important mechanisms (diffusion, advection, boundary fluxes, or transformations). For instance, if the absolute values of both λ_1 and λ_2 are much smaller than 1, the range of variation of the exponential terms of Eq. (h) (Table15.12) is small. Thus, $C(x)$ is nearly constant and depends only weakly on space. We may then use a box model with one or more boxes instead of the continuous model.

Consequently, the eigenvalues play a central role in the solution of Eqs. 15-74 and 15-75. Let us analyze Eq. (f) of Table 15.12 in more detail to find out which parameters determine λ_i. Two extreme cases characterized by the size of the nondimensional parameter $q = D \cdot k_r / v^2$ can be distinguished. First, if q is much larger than 1 (i.e., if $k_r \gg v^2/D$), the eigenvalues λ_i are approximated by

$$\lambda_i = \pm p\sqrt{q} = \pm L\left(\frac{k_r}{D}\right)^{1/2} \quad \text{for } q \gg 1 \tag{15-77}$$

an expression which no longer contains the advection velocity v. In this case the shape of the profile is solely influenced by diffusion and reaction, but not by advection. We call this the *diffusion–reaction regime* (Fig. 15.13); it can be further subdivided into the region of "slow removal" ($k_r \ll D/L^2$) and "fast removal" ($k_r \gg D/L^2$) corresponding to the case where the absolute values of λ_i are either much smaller or much greater than 1. In the former case the profile is linear and looks like the profile of a

TABLE 15.12 One-Dimensional Diffusion–Advection–Reaction Equation at Steady State

Basic equation (15-75):

(a) $$D\frac{d^2C}{dx^2} - v\frac{dC}{dx} - k_r C + J = 0$$

Boundary conditions (15-76):

(b) $C(0) = C_0 \qquad C(L) = C_L$

Replace x by the nondimensional coordinate $\xi = x/L$ $(0 \leqslant \xi \leqslant 1)$.

Note that $\dfrac{dC}{dx} = \dfrac{1}{L}\dfrac{dC}{d\xi'}$ and $\dfrac{d^2C}{dx^2} = \dfrac{1}{L^2}\dfrac{d^2C}{d\xi^2}$

Introduce the two nondimensional parameters:

(c) $p = \dfrac{Lv}{D}$ (Peclet) $\qquad q = \dfrac{Dk_r}{v^2} \geqslant 0$ (Damkohler)

and the "normalized" concentration $C^* = C - \dfrac{J}{k_r}$ which makes (a) homogeneous. Thus:

(d) $$\frac{d^2C^*}{d\xi^2} - p\frac{dC^*}{d\xi} - p^2 q C^* = 0$$

which has the solution:

(e) $C^*(\xi) = A_1 e^{\lambda_1 \xi} + A_2 e^{\lambda_2 \xi}$

where the eigenvalues λ_i are determined by the quadratic equation:

$\lambda^2 - p\lambda - p^2 q = 0$

with solutions:

(f) $$\lambda_{1,2} = \frac{p}{2}[1 \pm \sqrt{1 + 4q}]$$

The parameters A_i depend on the boundary conditions:

(g) $$A_1 = \frac{(C_L - J/k_r) - (C_0 - J/k_r)e^{\lambda_2}}{e^{\lambda_1} - e^{\lambda_2}} \qquad A_2 = \frac{(C_0 - J/k_r)e^{\lambda_1} - (C_L - J/k_r)}{e^{\lambda_1} - e^{\lambda_2}}$$

The solution for the original profile $C(x)$ is:

(h) $$C(x) = A_1 e^{\lambda_1 (x/L)} + A_2 e^{\lambda_2 (x/L)} + \frac{J}{k_r}$$

---

conservative substance; in the latter case strong concentration gradients appear close to the boundaries.

Now we consider a substance to be "conservative" when the time needed for diffusional transport across the system is much smaller than the time to "consume" a significant fraction of the compound by transformation. Since the former time is

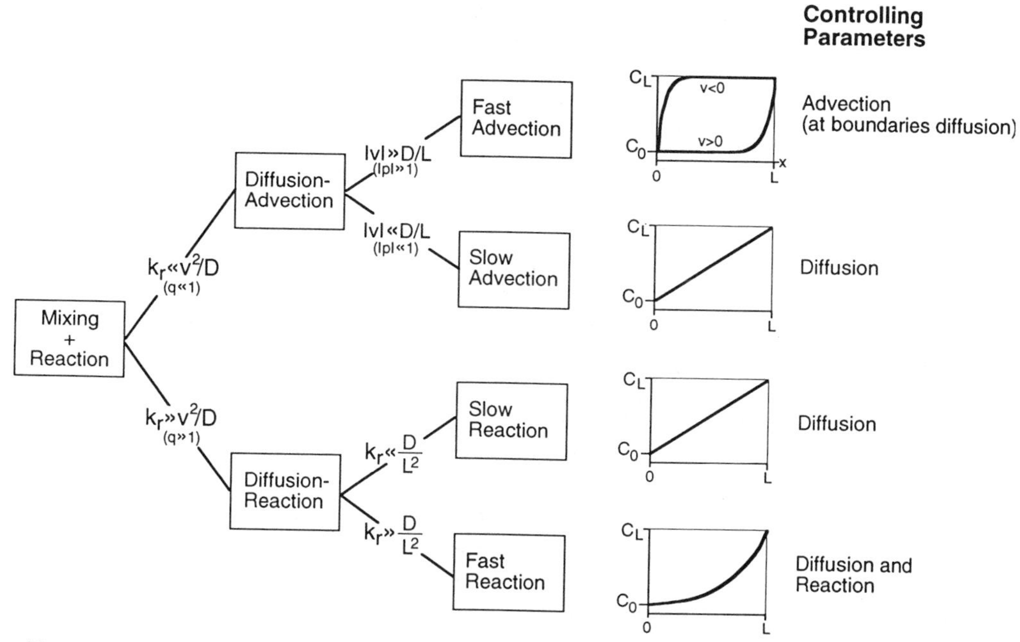

Figure 15.13 One-dimensional concentration profiles at steady state calculated from the diffusion–advection–reaction equation (Eq. 15-75) for different parameter values D (diffusivity), v (advection velocity), and k_r (first-order reaction rate constant). Boundary conditions at $x = 0$ and $x = L$ are C_0 and C_L, respectively. $p = Lv/D$ is the Peclet Number, $q = Dk_r/v_2$ is the Damkohler Number. See text for further explanations.

about L^2/D (see Eq. 9-31, where we have dropped the factor 2 for simplicity) and the latter k_r^{-1}, the condition for a parameter to be conservative is $k_r \ll D/L^2$. This situation corresponds to the "slow removal" case mentioned above.

Let us now look at the other extreme case, the so-called *diffusion–advection regime* characterized by the condition that q is much smaller than 1 ($k_r \ll v^2/D$). Then the eigenvalues are approximatively:

$$\lambda_1 \sim p = \frac{L \cdot v}{D} \qquad \lambda_2 \sim -pq = -\frac{Lk_r}{v} \tag{15-78}$$

A further subdivision is based on the size of p (note that p can have either sign!) If $|p|$ is much smaller than 1 (slow advection), the absolute value of the second eigenvalue is also much smaller than 1 (because $\lambda_2 = -\lambda_1 q$ with $q \ll 1$), and the profile again looks like the diffusive, "slow-removal" case. For $|p| \gg 1$, the spatial variation is completely shifted to the downstream end of the interval (fast advection, see Fig. 15.13.)

The lesson to be learned from this discussion is the following: it is not the absolute size of the parameters D, v, and k_r that determines the shape of a concentration distribution, but certain nondimensional ratios that can be formed from those parameters. Therefore, depending on reactivity, one species i can be in the diffusion–advection regime while in the same environmental system and at the same time another species j may be in the diffusion–reaction regime.

The nondimensional parameters p and q defined in Eq. (c) of Table 15.12 are of general significance. The first one, p, called the *Peclet Number*, measures the relative importance of advective versus diffusive transport within the interval of length L. The second one, q, sometimes called the *Damkohler Number* in the chemical engineering literature, expresses the relative importance of diffusion versus advection within the time scale k_r^{-1} of the species given by its "consumption" due to in situ transformation.

It is not easy to find simple and instructive examples for the one-dimensional stationary transport–reaction model which involve organic compounds. This is because in real systems diffusion is either at an extreme value, that is, either very small (molecular) or very large (full mixing) leading to extreme values of p and q; or the relevant mixing parameters and the boundary conditions do not remain constant long enough for the system to reach steady state.

Therefore, the chemically oriented reader may excuse our choice of an example from oceanography as a substitute. In fact, the theory of the vertical distribution of geochemical tracers in the deep waters of the Pacific by Munk (1966) is a classic and deserves some attention. In the deep ocean there exists a relatively calm zone, the deep water layer, bounded above by the thermocline and below by the so-called bottom water where mixing can be characterized by a vertical turbulent diffusion coefficient $D = 10^{-4}\,\mathrm{m^2\,s^{-1}} \sim 3\cdot10^3\,\mathrm{m^2\,yr^{-1}}$ and a slow upwelling velocity of about $v = +4\,\mathrm{m\,yr^{-1}}$ (Note: we use D instead of E for the turbulent diffusion coefficient in order to facilitate application of Table 15.12). The z axis is chosen positive upward, thus $v > 0$.

Two different radioisotopes are considered:

1. Radium-226, a product of the ^{238}U decay series, diffuses from the deep-sea sediments into the overlying water column. Its half-life is 1620 years, thus $k_{Ra} = 4.3\cdot10^{-4}\,\mathrm{yr^{-1}}$. The boundaries of the deep-water layer are chosen at 1 and 5 km depth, thus $L = 4\cdot10^3$ m. From Table 15.12 it follows that

$$p_{Ra} = 5 \qquad q_{Ra} = 0.08$$

Following the categories of Fig. 15.12 we find ^{226}Ra in the diffusion–advection regime with a tendency to fast advection. Thus, the vertical profile should be convex upward. Its shape is solely determined by D and v; the radioactive decay has no influence; ^{226}Ra looks like a conservative compound (Fig. 15.14).

2. Tritium (^3H), the radioactive hydrogen isotope, is produced in the atmosphere by natural nuclear reactions (cosmic rays). The nuclear weapon tests in the 1950s and 1960s led to a marked increase in the atmospheric tritium concentration. Tritium included in water molecules is removed from the air to the sea surface and diffuses into the ocean. Due to man-made perturbations, the tritium profile in the sea is

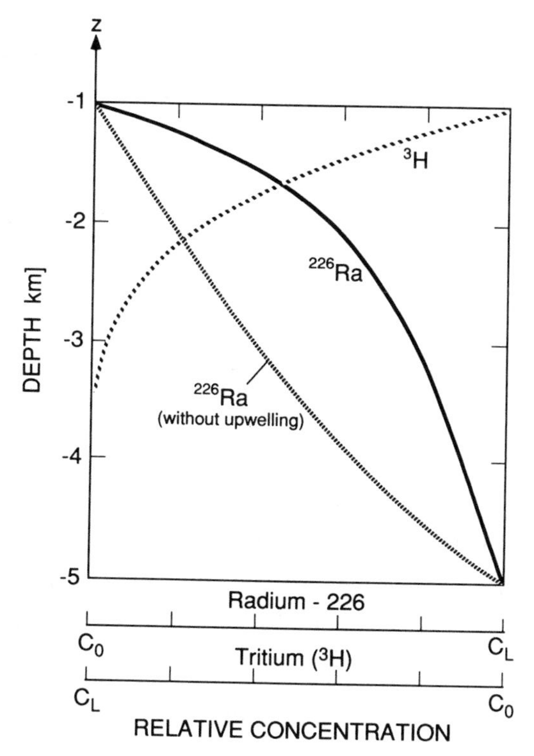

Figure 15.14 Vertical profiles of ^{226}Ra and ^{3}H (tritium) in the deep ocean calculated from the steady-state solution of Eq. 15-75 with $D = 3 \cdot 10^3\ m^2 \cdot yr^{-1}$ and $v = 4\ m \cdot yr^{-1}$. C_0 and C_L are the concentrations at the boundaries, defined by the thermocline and the bottom water, respectively. See text for further explanations.

certainly not at steady state; however, we can discuss its possible steady-state profile. The half-life of tritium is about 12 years, thus $k_T = 0.058\ yr^{-1}$. While the Peclet Number p is the same for both isotopes (Ra-226 and H-3), the Damkohler Number of tritium q_T is much larger because of the greater reactivity (smaller half-life) of this radioisotope:

$$p_T = 5 \qquad q_T = 11$$

According to Figure 15.12 tritium is in the diffusion–reaction regime. Since k_T is much greater than $d/L^2 = 1.9 \cdot 10^{-4}\ yr^{-1}$, the reaction is fast compared to diffusive transport, thus leading to a highly curved profile (Fig. 15.14). In fact, tritium in the ocean is confined to the surface and upper thermocline; only in regions of rapid deep water formation (e.g., North Atlantic) has tritium penetrated into the deep ocean.

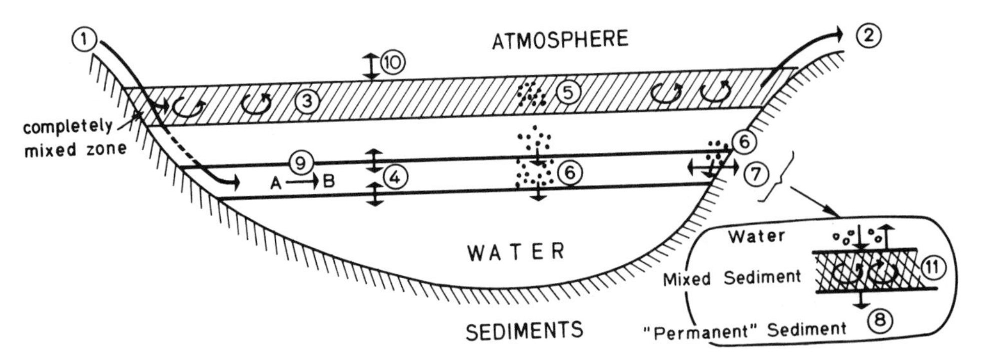

Figure 15.15 One-dimensional vertical transport–reaction model for lake with depth-dependent cross section (CHEMSEE). Processes are labeled by numbers and further explained in the text as well as in Table 15.13.

One-Dimensional Continuous Lake Model

In the preceding sections we discussed the elements needed to decompose the dynamic behavior of a chemical compound in the natural environment into the relevant transport and reaction processes. Let us now combine them and study the fate of a compound in some simple but typical aquatic environments. We can do this by exemplification only and again concentrate on lakes, the aquatic systems with which we are well acquainted.

We consider a lake of moderate size (surface area less than 100 km^2). As discussed in Section 9.4 and Table 9.6 because of the vertical density stratification of the water column, horizontal mixing by turbulent diffusion is usually much faster than vertical mixing. Therefore, as a reasonable approximation, a chemical compound whose mean life (or time of reaction) is not much shorter than the time needed to mix the lake horizontally (a few days), can be described by a one-dimensional vertical (1 DV) model as depicted in Figure 15.15 (Imboden and Schwarzenbach, 1985). A user-friendly program of this model was developed by Ulrich et al. (1991) for use on a personal computer. The basic assumption of the 1 DV-model is that horizontal mixing is fast compared to reaction and boundary exchange. However, to describe the interaction between the sediment and water column, the three-dimensional bottom topography of the lake cannot be completely ignored. Since the lake basin does not usually have the shape of a rectangular box, every horizontal water layer, not only the deepest one, is in direct contact with the sediments, albeit the contact areas are different for the different layers. There the compounds can be exchanged between the water and the sediment. The mechanism of sediment exchange at the sides turns out to be important, for instance, for the cycling of phosphorus from the sediments into a lake (Imboden and Gächter, 1978), for the fast removal of man-made redioisotopes (such as ^{137}Cs) to the sediments (Santschi et al., 1990) or for the long-term behavior of highly sorbing compounds such as PCBs, as discussed in the preceding part of this chapter. The ultimate aim of the 1 DV-model is to calculate the time-dependent continuous vertical concentration profile of a compound, $C(z,t)$, where the depth coordinate z is the height above the deepest point of the lake (thus the vertical coordinate z is chosen as positive upward). Let us consider a horizontal layer of thickness Δz confined by the cross sections at depth z and $z + \Delta z$, $A(z)$ and $A(z + \Delta z)$, respectively (Fig. 15.15). The volume of the layer ΔV can be approximated by $A(z) \cdot \Delta z$, and the sediment contact area ΔA by $[A(z + \Delta z) - A(z)]$. Note that bottom slopes of lakes are commonly so small that ΔA, the horizontal projection of the inclined sediment surface, is usually a good approximation; in drawings of lake basins (such as Fig. 15.15) the vertical dimensions are often strongly exaggerated so that the slopes look much steeper than they really are.

We define the specific flux rate F ($mol \cdot m^{-2} \cdot s^{-1}$) of a compound as positive if the flux is directed from the sediment into the water. Then the mean concentration in the layer of thickness Δz, $C_{\Delta z}$, changes at the rate of

$$\frac{dC_{\Delta z}}{dt} = \frac{\Delta A}{\Delta V} F = \frac{A(z + \Delta z) - A(z)}{A(z) \cdot \Delta z} F$$

If we decrease the thickness of the box Δz toward zero, we get

$$\frac{dC}{dt} = a(z) \cdot F \tag{15-79}$$

where the characteristic topographic function (remember the definition of the derivative.)

$$a(z) = \frac{1}{A} \left(\frac{A(z + \Delta z) - A(z)}{\Delta z} \right)_{\Delta z \to 0} = \frac{1}{A} \frac{dA}{dz} = \frac{dA}{dV} \tag{15-80}$$

is the change of lake cross section per lake volume. It is an important parameter to describe the effect of lake topography on the spatial distribution of chemical compounds. Frequently, the depth-dependent lake cross section $A(z)$ can be approximated by (Imboden, 1973):

$$A(z) = A_0 \left(\frac{z}{z_0} \right)^{\eta} \tag{15-81}$$

where z_0 is the maximum depth and the exponent η lies typically between 0.5 (flat littoral zone, "hole" at deepest part) and 2 (steep littoral zone, flat bottom). From Eqs. 15-80 and 15-81 follows

$$a(z) = \frac{\eta}{z} \tag{15-82}$$

Thus, the characteristic topographic function increases continuously from the surface

($z = z_0$) to the lake bottom ($z = 0$) where it becomes infinitely large. In fact, at the lake bottom a tiny lake volume stays in contact with a finite sediment area. This explains the great spatial and temporal gradients often found close to the bottom of lakes for compounds which are exchanged at the sediment-water interface (oxygen, phosphorus, methane, etc.).

The model combines the following lake-specific processes with properties of the chemical compound (numbers refer to Fig. 15.15):

Lake-Specific Processes

1. Input of chemicals and particles at the surface or at various depths which may vary with time: Inlets act as one potential source for chemicals and particles. It is assumed that inlets bring water only to the surface layer and that the input of chemicals to deeper layers occurs without (or without significant) water input.
2. Outflow from lake (usually from the surface): Outlets act as one sink for chemicals and particles.
3. Complete mixing of the surface water to the depth z_{mix} (epilimnion depth), which usually varies with time.
4. Vertical mixing of water by turbulence described by a (time and depth dependent) vertical eddy diffusion coefficient: Affects dissolved and particulate species.
5. In situ production of particulate matter, usually in the surface layer (e.g., growth of phytoplankton), but possibly also below the thermocline.
6. Settling of particles due to gravity either into the deeper water layer or onto the sediment surface at the side of each layer: This sink term acts only on the particulate or sorbed phase of the compound.
7. Exchange between sediment and water by diffusion or resuspension of particles.
8. Transfer of sediment material into the "permanent," noninteracting sediment column: Acts as a permanent sink for particulate or sorbed chemicals.

Compound-Specific Processes

9. *In situ* reaction of chemicals, for example, hydrolysis, photolysis, redox processes: Since such reactions often depend on key variables like water temperature, light intensity, or pH, the reaction rates are usually time and depth dependent.
10. Mass transfer at the lake surface.
11. Reaction of the compound in the sediments.

Most of the elements that we need to write down the equations of this model were introduced in the earlier part of this chapter. Unlike our earlier box models, the continuous model uses a new way to describe spatial transport. Remember that the first approach we used to quantify spatial transport involved introducing two adjacent boxes (Section 15.3) connected by a water exchange flux T_{EH} (Eq. 15-29). For the continuous model we use Fick's laws to describe the corresponding exchange. A second new element involves the way that the sediments influence the water column; the lake topography allows exchange fluxes between sediment and water at all depths.

Otherwise the description is pretty much the same, especially if we use the common technique in which the spatial gradients, $\partial C/\partial z$, are approximated by a finite difference of the form $(C_{i+1} - C_i)/\Delta z$ where C_i and C_{i+1} are concentrations at two adjacent "grid points" separated by the distance Δz. Application of the finite difference method is equivalent as subdividing the lake into horizontal layers of thickness Δz. The layers, like the well-mixed boxes introduced earlier, are characterized by mass balance equations for each state variable. For the lake model discussed here we need three state variables for each box (Table 15.13): (1) the total compound concentration, (2) the solid-to-water phase ratio, and (3) the sorbed compound concentration in the surface sediments in contact with the layer. Equations (a) and (c) of Table 15.13 correspond to Eqs. 15-65 and 15-64 introduced in Section 15.4. The mass balance equation (b) for the suspended solids is just a simplified version of Eq. (a).

There are a few points that need additional explanation. The first is the description of the Fickian flux. For instance, at the lower boundary of layer i (denoted by the index $i - 1/2$) the total diffusive flux of the compound into layer i is given by

$$- A_{i-1/2} \cdot E_{i-1/2} \cdot \left.\frac{\partial C}{\partial z}\right|_{i-1/2}$$

A similar expression can be written for the upper boundary $(i + 1/2)$. Thus, the concentration change in layer i due to diffusion alone is

$$\left(\frac{\partial C}{\partial t}\right)_{\text{diff}} = \left(- A_{i-1/2} E_{i-1/2} \left.\frac{\partial C}{\partial z}\right|_{i-1/2} + A_{i+1/2} E_{i+1/2} \left.\frac{\partial C}{\partial z}\right|_{i+1/2}\right) \frac{1}{V_i} \qquad (15\text{-}83)$$

Now we approximate the spatial gradients by finite differences:

$$\left.\frac{\partial C}{\partial z}\right|_{i-1/2} \sim \frac{C_i - C_{i-1}}{\Delta z} \quad \text{and} \quad \left.\frac{\partial C}{\partial z}\right|_{i+1/2} \sim \frac{C_{i+1} - C_i}{\Delta z} \qquad (15\text{-}84)$$

Furthermore, we use $A_{i-1/2} \sim A_{i+1/2} \sim A_i$ and $V_i \sim A_i \cdot \Delta z$. Combining all these simplifications yields

$$\left(\frac{\partial C}{\partial t}\right)_{\text{diff}} = - \frac{E_{i-1/2}}{\Delta z^2}(C_i - C_{i-1}) + \frac{E_{i+1/2}}{\Delta z^2}(C_{i+1} - C_i) \qquad (15\text{-}85)$$

These are the terms marked by (4) in Table 15.13. In a similar way we can treat the concentration change due to particle settling which results from the difference between

TABLE 15.13 Equations for the One-Dimensional Vertical Lake Model Describing a Chemical with Instantaneous Sorption Equilibrium with Suspended Particles and Exchange with the Sediment Column

Equations expressed for discretization into horizontal water layers of thickness Δz. Layers are numbered by index $i = 1,2\ldots.$ Surface mixed layer has index 0 and thickness Δz_0.
Each layer is characterized by three state variables:

$C_i(\mathrm{mol \cdot m^{-3}})$:	Total compound concentration in layer i (index t, used in Section 15.4, omitted for brevity)
$r_i(\mathrm{kg \cdot m^{-3}})$:	Solid-to-water phase ratio (concentration of suspended particles) in layer i (index sw used in Section 15.4 omitted)
$C_{ss,i}(\mathrm{mol \cdot kg^{-1}})$:	Sorbed compound concentration per solid mass in surface sediment in contact with layer i

The fraction of the chemical in solution is defined as (index w omitted for brevity)

$$f_i = \frac{1}{1 + r_i \cdot K_d}$$

Note: Indices $(i - 1/2)$ and $(i + 1/2)$ refer to value evaluated at boundary below and above layer i, respectively.

Dynamic equations for layers below surface mixed layer:

$$\frac{\partial C_i}{\partial t} = \underset{(1)}{\frac{I_i}{V_i}} - \underset{(4)}{\frac{E_{i-1/2}}{\Delta z^2}(C_i - C_{i-1})} + \underset{(4)}{\frac{E_{i+1/2}}{\Delta z^2}(C_{i+1} - C_i)} \qquad \text{(a)}$$

$$+ \underset{(6)}{(1 - f_i)\frac{v_s}{2 \cdot \Delta z}(C_{i+1} - C_{i-1})} + \underset{(7)}{a_i \cdot F_i} - \underset{(9)}{C_i\{f_i k_{r,d} + (1 - f_i)k_{r,p}\}}$$

$$\frac{\partial r_i}{\partial t} = \underset{(1)}{\frac{I_{r,i}}{V_i}} - \underset{(4)}{\frac{E_{i-1/2}}{\Delta z^2}(r_i - r_{i-1})} + \underset{(4)}{\frac{E_{i+1/2}}{\Delta z^2}(r_{i+1} - r_i)} - \underset{(6)}{\frac{v_s}{2 \cdot \Delta z}(r_{i+1} - r_{i-1})} \qquad \text{(b)}$$

$$\frac{\partial C_{ss,i}}{\partial t} = \underset{(6)}{\frac{v_s}{m}(1 - f_i)C_i} - \underset{(8)}{\beta \frac{r_i v_s}{m} C_{ss,i}} + \underset{(7)}{k_{ex}(f_i K_{ds} f_i C_i - C_{ss,i})} - \underset{(11)}{k_{r,ss} C_{ss,i}} \qquad \text{(c)}$$

For the surface mixed layer, Eqs. a and b are modified:

$$\frac{\partial C_0}{\partial t} = \underset{(1)}{\frac{I_0}{V_0}} - \underset{(4)}{\frac{E_{0-1/2}}{\Delta z \cdot \Delta z_0}(C_0 - C_{0-1})} - \underset{(6)}{(1 - f_0)\frac{v_s}{\Delta z_0} C_0} + \underset{(7)}{a_0 \cdot F_0}$$

TABLE 15.13 *(Continued)*

(a*)
$$-C_0\{f_0\cdot k_{r,d}+(1-f_0)k_{r,p}\}-\frac{Q\cdot C_0}{V_0}-\frac{A_0}{V_0}v_t\left(C_0-\frac{C_a}{K'_H}\right)$$
(9) (2) (10)

(b*)
$$\frac{\partial r_0}{\partial t}=\frac{I_{r,0}}{V_0}-\frac{E_{0-1/2}}{\Delta z\cdot\Delta z_0}(r_0-r_{0-1})-\frac{v_s}{\Delta z_0}r_0-\frac{Q\cdot r_0}{V_0}$$
(1) (4) (6) (2)

Definitions (see also Section 15.4):

$V_i=A_i\cdot\Delta z$	(m^3)	Volume of box i
I_i	$(mol\cdot s^{-1})$	Input of compound into layer i
$I_{r,i}$	$(kg\cdot s^{-1})$	Input of particles into layer i
$E_{i-1/2}$, $E_{i+1/2}$	$(m^2\cdot s^{-1})$	Turbulent eddy diffusivity at interface below or above layer i
F_i	$(mol\cdot m^{-2}\cdot s^{-1})$	Net flux of compound between water and sediment column defined as positive if directed from the sediment into the overlying water (see Eq. 15-88)

influx at the upper boundary and outflux at the lower boundary:

$$\left(\frac{\partial C_i}{\partial t}\right)_{\text{settling}}=(v_s\cdot A_{i+1/2}C_{i+1/2}-v_sA_{i+1/2}C_{i-1/2})\frac{1}{V_i}\qquad(15\text{-}86)$$

Since part of the particles leaving layer i sink onto the sediment surface of the sides, the larger area $A_{i+1/2}$ appears in both terms of Eq. 15-86. The concentrations at the boundary, $C_{i+1/2}$ and $C_{i-1/2}$, are replaced by the arithmetic mean value of the adjacent layer values, for example, $C_{i+1/2}=(C_i+C_{i+1})/2$. Thus,

$$\left(\frac{\partial C}{\partial t}\right)_{\text{settling}}=\frac{v_s}{2\cdot\Delta z}(C_{i+1}-C_{i-1})\qquad(15\text{-}87)$$

These are the terms marked by (6) in Table 15.13. Note the additional factor $(1-f_i)$ appearing in term (6) of the final equation; it is the result of sedimentation that acts only on the solid fraction $f_{si}=(1-f_i)$ of the total compound concentration C_i.

Finally, the sediment–water exchange terms need some explanation. Since Eq. (c) of Table 15.13 is analogous to Eq. 15-64, the exchange parameter k_{ex} keeps its meaning defined by Eq. 15-63. Similarly, Eq. (a) of Table 15.13 is analogous to Eq. 15-65 except that the sediment–water exchange flux is expressed differently, that is as a_iF_i in the former, and as $-k^*_{ex}\left(f_wC_t-\frac{C_{ss}}{K_{ds}}\right)$ in the latter. Thus,

$$F_i=-\frac{k^*_{ex}}{a_i}\left(f_{w,i}\cdot C_{t,i}-\frac{C_{ss,i}}{K_{ds}}\right)$$

The mean depth of the box, h, which appears in the box model approach, for example, in Eq. 15-66 relating k^*_{ex} to k_{ex}, has to be replaced by the corresponding parameter of the continuous model, that is $a^{-1} = dV/dA$ where a has been defined in Eq. 15-80.

Therefore,

$$k^*_{ex} = k_{ex} \cdot m \cdot K_{ds} \cdot a_i$$

Combining these expressions and adopting the simplified notation C_i and f_i (instead of $C_{t,i}$ and $f_{w,i}$) yields

$$F_i = k_{ex} \cdot m \cdot (C_{ss,i,} - f_i \cdot K_{ds} \cdot C_i) \tag{15-88}$$

The characteristics of the continuous lake model are discussed using two compounds with which we are already familiar. The first case continues the story of tetrachloroethylene (PER) in Greifensee. Remember that PER is a compound which in the water is quasi-conservative and is not sorbed significantly by particles. Besides flushing, exchange at the air–water interface is the only relevant process to be considered.

The results of a continuous model calculation for Greifensee during the years 1982–1984 have already been presented in Fig. 15.6. Here we discuss another chapter of this story, that is, the traces of another "input event" found in Greifensee in the subsequent year, 1985 (Fig. 15.16). During the winter of 1984–1985 and until May 1985 the PER content of Greifensee remained relatively low (between about 40 and 80 moles). On June 3, 1985 a PER content of more than 200 moles was found in the lake. The new PER was mainly detected in the surface mixed layer (about 4 m deep) and in the thermocline. On July 1, 1985 the PER content was still around 200 moles, but the concentration maximum had moved from the surface to the thermocline. This concentration peak remained visible throughout the summer and fall until the PER content had returned to its "normal" level. As shown in Figure 15.16, the continuous lake model nicely describes the concentration maximum, which slowly moved to greater depth due to the deepening of the surface mixed layer. From the model calculation we can conclude that the processes involved to produce this maximum are the combination of riverine PER input into the surface mixed layer and loss to the atmosphere by gas transfer. The extra input of PER into the lake between May 6 and July 1, 1985 had to be about 360 moles. Part of the compound was quickly and continuously lost to the atmosphere so that the PER content of the lake never increased beyond about 210 moles.

The second case deals with the behavior of sorbing and reacting species. The PCB congeners are good examples. Unfortunately, available data sets are usually too poor to match the kind of detailed information produced by the model. Thus, let us use the model in a different way, that is, as a tool to study the hypothetical situation of a lake which has undergone a period of heavy pollution and subsequent input reduction. From what we have learned, we expect the pollutant concentration in the surface sediments to control the decrease of the concentration in the lake water. This is indeed the case, as shown in Figure. 15.17 for the (hypothetical) pollution of Greifensee with 2,2′,3,4,5,5′,6-heptachlorobiphenyl (IUPAC #185). In the figure,

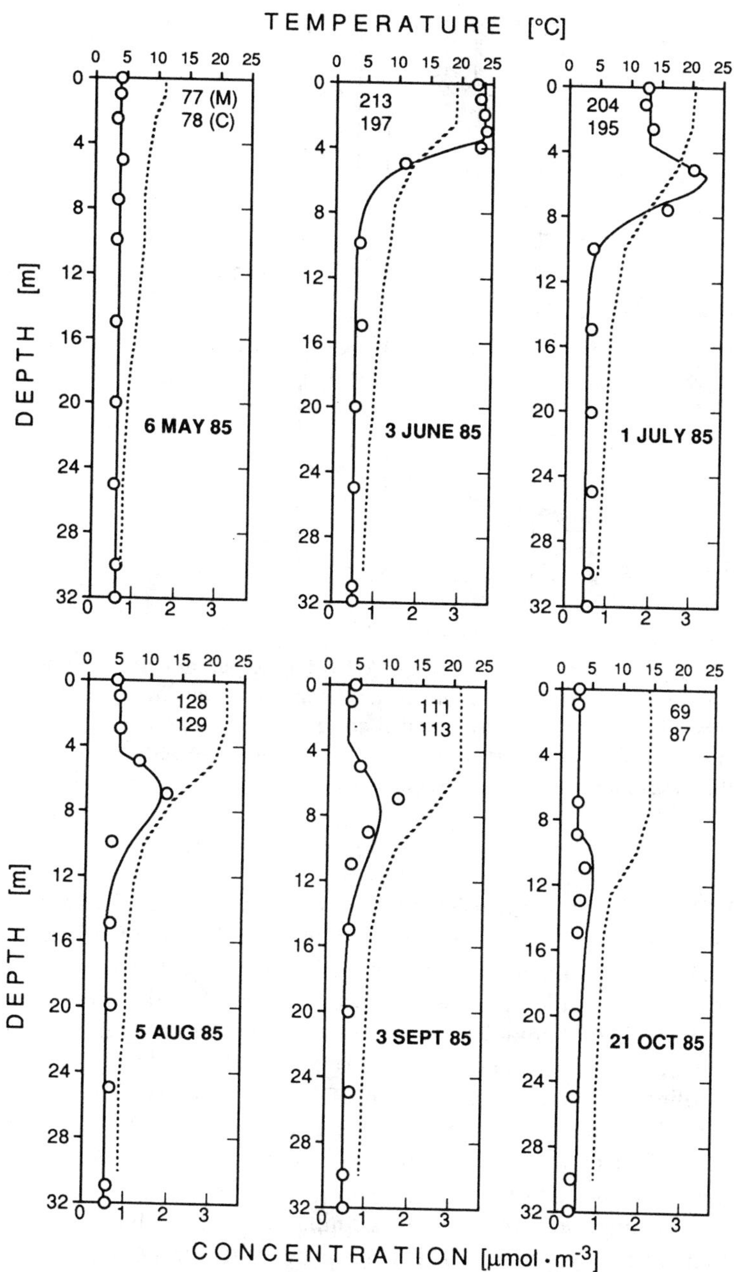

Figure 15.16 Vertical profiles of water temperature (dotted line) and of measured (circles) and calculated (solid line) PER concentration in Greifensee (Switzerland) for the period May to October 1985. Numbers give PER inventory in moles (M = measured, C = calculated). From the model calculation with CHEMSEE it can be concluded that between May 6 and July 1, 1985, about 360 mol of PER entered the lake, thus leading to a significant increase of the concentration in the lake during several months.

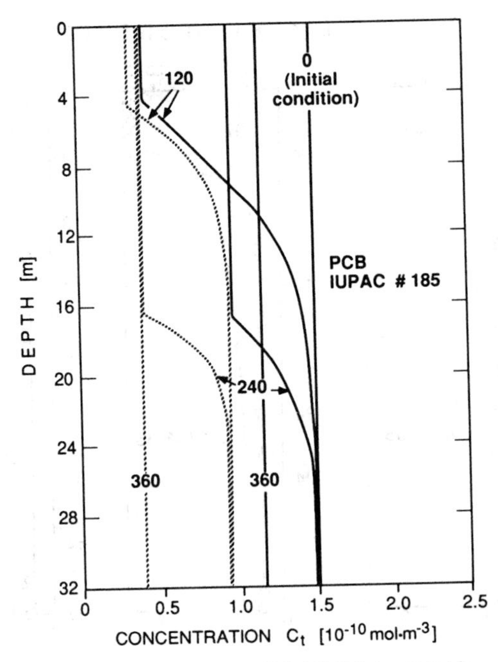

Figure 15.17 Hypothetical concentration of 2,2′,3,4,5,5′,6-heptachlorobiphenyl (IUPAC #185) in Greifensee resulting from a long-term PCB loading followed by a sudden input drop to zero, assumed to occur on January 1 (initial condition); 360 = profile at end of year. Broken line profiles show expected results assuming no removal to the sediments by particle settling and no diffusive exchange between sediment and water. Numbers = days of the year.

vertical PCB profiles are shown which were computed using CHEMSEE. The lake parameters (river input/output, vertical diffusivity, change of mixed layer depth) are the same as the values used in Figure 15.16. The air–water transfer velocity determined for PER (Table 15.6) was reduced to $v_{tot} = 0.1\ \mathrm{m \cdot d^{-1}}$ because of the larger molecular mass of the heptachlorobiphenyl. The suspended particle concentration (solid-to-water phase ratio) is assumed to have the constant value $r_{sw} = 1 \cdot 10^{-3}\ \mathrm{kg_s \cdot m^{-3}}$ and the settling velocity is $v_s = 0.68\ \mathrm{m \cdot d^{-1}}$ ($250\ \mathrm{m \cdot yr^{-1}}$). The physicochemical properties of the chlorinated biphenyl are taken from Table 15.9, except for the distribution coefficient which is assumed to be equal for both suspended and settled solids: $K_{ds} = K_d = 10^3\ \mathrm{m^3 \cdot kg_s^{-1}}$.

The model calculation starts at the beginning of the year when the water column is assumed to be vertically well mixed. The initial concentration C_t (sum of dissolved and sorbed PCB concentration) is $1.5 \cdot 10^{-10}\ \mathrm{mol \cdot m^{-3}}$; the initial concentration of the PCB congener sorbed to particles in the sediment mixed layer is $C_{ss} = 7.5 \cdot 10^{-8}\ \mathrm{mol \cdot kg_s^{-1}}$. These values correspond roughly to the hypothetical steady-state concentrations that one would find in Greifensee if the lake were exposed to a permanent loading of $0.2\ \mathrm{mol \cdot yr^{-1}}$ ($2.4 \cdot 10^{-8}\ \mathrm{mol \cdot m^{-2} \cdot yr^{-1}}$, i.e., about 60 times greater than the loading per unit area measured for Lake Superior, Table 15.9). It is assumed that the external PCB input instantaneously becomes zero when the model simulation begins. We are interested in the response of the PCB concentration in the lake to this sudden (and hypothetical!) change in the input.

As shown in Figure 15.17, the PCB concentration in the epilimnion is significantly reduced during the stratification period. The main removal mechanisms are air–water exchange, flushing, and absorption to settling particles. The last process is also effective in the deeper layers of the lake. However, removal on particles is counteracted by diffusion and resuspension from the still-polluted sediments, thus keeping the hypolimnetic concentrations large during the whole summer. At the end of the first year without external loading, the concentration in the lake drops to about $1.1 \cdot 10^{-10}\,\text{mol} \cdot \text{m}^{-3}$, that is, 70% of the initial value.

To demonstrate the role of particle settling and sediment–water exchange, we have added another set of PCB profiles calculated for the case in which the particle settling velocity v_s and the sediment–water exchange rate k_{ex} are both zero. Although settling to the sediments no longer acts as a removal mechanism, the PCB concentration drops to $0.4 \cdot 10^{-8}\,\text{mol} \cdot \text{m}^{-3}$ after 1 year, that is, about 25% of its initial value, since the sediment–water exchange is also stopped. The effect of the exchange becomes even more distinct if it is put to zero, and particle settling remains as in the original calculation. Then the PCB almost completely disappears from the lake after just 1 year (not shown in Fig. 15.17).

As mentioned before, the CHEMSEE model, although still relatively simple compared to the three-dimensional nature of real transport and reaction processes, produces predicted concentrations and inventories which in most cases are not matched by available field data in terms of chemical, spatial and temporal resolution. In fact, in a time when powerful computers are ubiquitously available, it is not unusual to find publications in which highly sophisticated model outputs are compared to poor data sets for which much simpler models would have been adequate. However, this is not an argument against the develeopment of good models, but a plea for their wise use. To compare measured and calculated data is not the only task for which models are good. Other reasons for using models are, for instance, to evaluate the relative importance of various processes for the dynamics of chemical compounds in aquatic systems. As an example, the PCB model described above could serve as a tool to assess the influence of changing particle concentrations on the PCB concentrations. The change of the trophic state of a water body, that is, the frequency and intensity of algal blooms in the water, greatly affect the concentration of sorbing chemical species. Thus, a model can help to find connections between different components of an aquatic ecosystem and to evaluate possible inadvertent effects of human interventions in complex systems. In addition, models that are more advanced than the present techniques of analytical chemistry can be used to find optimal strategies for the protection of aquatic systems against pollution by xenobiotic compounds. This can be done by identifying the mechanisms to which the system is most vulnerable, as well as by evaluating alternative restoration procedures in case the compound has already entered the system.

These considerations bring us to the end of a long and—as some readers may feel—complicated chapter. And yet, many stories have had to remain untold. We did not discuss models of other aquatic systems such as rivers and groundwaters. The scientific literature is full of examples from which the reader can learn once he or she has agreed on the view that models, although far from being able to explain all aspects of the real situation, may be as useful as other tools like spectrophotometers or chromatographs. Once one has reached a certain understanding of how models work for one specific environmental system such as lakes, it is much easier to master models for other environmental systems as well.

APPENDIX

Compound Name	Molecular Formula	Mol. Wt.	T_m (°C)	T_b (°C)	Vapor Pressure at 25°C (atm)		Aqueous Solubility at 25°C (mol L^{-1})		Henry's Law Constant at 25°C (L atm mol^{-1})	Octanol–Water Partition Coefficient at 25°C [(mol·L^{-1} octanol)·(mol·L^{-1} water)$^{-1}$]
					$-\log P^0$	$-\log P^0(L)$ (For Solids and Gases)	$-\log C_w^{sat}$	$-\log C_w^{sat}(L)$ (For Solids and Gases)	$\log K_H$ Calculated (Experimental)	$\log K_{ow}$
Saturated and Unsaturated Hydrocarbons										
Methane	CH_4	16.0	−182.5	−164.0		−2.44	2.82[a]	0.38[a]	2.82	1.09
Ethane	C_2H_6	30.1	−183.3	−88.6		−1.60	2.69[a]	1.09[a]	2.69	1.81
Propane	C_3H_8	44.1	−189.7	−42.1		−0.97	2.85[a]	1.88[a]	2.85	2.36
n-Butane	C_4H_{10}	58.1	−138.4	−0.5		−0.39	2.98[a]	2.59[a]	2.98	2.89
n-Pentane	C_5H_{12}	72.2	−129.7	36.1	0.16		3.25		3.09	3.62
n-Hexane	C_6H_{14}	86.2	−95.0	69.0	0.69		3.83		3.14	4.11
n-Heptane	C_7H_{16}	100.2	−90.6	98.4	1.21		4.51		3.30	4.66
n-Octane	C_8H_{18}	114.2	−56.8	125.7	1.73		5.20		3.47	5.18
n-Nonane	C_9H_{20}	128.3	−51.0	150.8	2.24		5.94		3.70	
n-Decane	$C_{10}H_{22}$	142.3	−29.7	174.1	2.76		6.57		3.81	
n-Dodecane	$C_{12}H_{26}$	170.3	−9.6	216.3	3.80		7.52		3.72	
n-Hexadecane	$C_{16}H_{34}$	226.4	18.2	287.0	5.73		7.80		2.07	
n-Octadecane	$C_{18}H_{38}$	254.4	28.2	316.1	6.67		8.08		1.41	
Cyclohexane	C_6H_{12}	84.2	6.6	80.7	0.90		3.15		2.25	3.44
1-Hexene	C_6H_{12}	84.2	−139.8	63.4	0.60		3.15		2.45	3.39
1-Octene	C_8H_{16}	112.2	−101.7	121.3	1.63		4.52		2.89	4.57
Miscellaneous Aliphatic Compounds[c]										
1-Butanol	$C_4H_{10}O$	74.1	−89.5	117.2	2.02		0.07		−1.95	0.79
1-Hexanol	$C_6H_{14}O$	102.2	−46.7	158.0	2.85		0.88			2.03
1-Octanol	$C_8H_{18}O$	130.2	−16.7	194.4			2.35			2.84
1-Nonanol	$C_9H_{20}O$	144.2	−5.5	213.5	4.00		3.13			3.77
2-Ethyl-1,3-hexanediol	$C_8H_{18}O_2$	146.2	−40.0	244.0			2.81			3.22
Cyclohexanol	$C_6H_{12}O$	100.2	25.1	161.1	2.70	2.70	0.42	0.42	−2.28	1.23
Acetone	C_3H_6O	58.1	−94.6	56.5	0.55		−1.13		−1.68	−0.24
									−1.44	

2-Octanone	$C_8H_{16}O$	128.2	−16.0	172.9	2.88		2.05		−0.83	2.76
2-Decanone	$C_{10}H_{20}O$	156.2	3.5	211.0	3.45		3.30		−0.15	3.81
n-Butylacetate	$C_6H_{12}O_2$	116.2	−77.9	126.5						1.78
1-Bromohexane	$C_6H_{13}Br$	165.1	−84.7	155.3						4.65
1-Bromoheptane	$C_7H_{15}Br$	179.1	−56.1	178.5						5.23
1-Bromooctane	$C_8H_{17}Br$	193.1	−55.0	200.8			5.06			5.82
				Substituted Benzenes[d]						
Benzene	C_6H_6	78.1	5.5	80.1	0.90		1.64		0.74 (0.75 exp.)	2.13
Toluene	C_7H_8	92.1	−95.0	110.6	1.42		2.25		0.83 (0.83 exp.)	2.69
Ethylbenzene	C_8H_{10}	106.2	−95.0	136.2	1.90		2.80		0.90 (0.93 exp.)	3.15
1,2-Dimethyl-benzene	C_8H_{10}	106.2	−25.2	144.4	2.05		2.76		0.71	3.12
1,4-Dimethyl-benzene	C_8H_{10}	106.2	13.2	138.0	1.93		2.77		0.84	3.18
n-Propylbenzene	C_9H_{12}	120.2	−101.6	159.2	2.35		3.34		0.99	3.63
1,2,4-Tri-methylbenzene	C_9H_{12}	120.2	−43.8	169.4	2.57		3.33		0.76	3.65
1,3,5-Tri-methylbenzene	C_9H_{12}	120.2	−44.7	164.7	2.48		3.40		0.92	3.42
n-Butylbenzene	$C_{10}H_{14}$	134.2	−88.0	183.0	2.86		3.97		1.13	4.28
n-Pentylbenzene	$C_{11}H_{16}$	148.3	−75.0	205.4	3.36		4.59		1.23	4.90
Chlorobenzene	C_6H_5Cl	112.6	−45.6	132.0	1.80		2.35		0.55 (0.54 exp.)	2.92
1,2-Dichloro-benzene	$C_6H_4Cl_2$	147.0	−17.0	180.0	2.71		3.20, 3.01		0.49, 0.29 (0.28 exp.)	3.38
1,4-Dichloro-benzene	$C_6H_4Cl_2$	147.0	53.1	174.0	3.04	2.76	3.39	3.11	0.35	3.38
1,2,4-Trichloro-benzene	$C_6H_3Cl_3$	181.5	16.9	213.5	3.21		3.65		0.44	4.00
1,3,5-Trichloro-benzene	$C_6H_3Cl_3$	181.5	63.0	208.0	3.49	3.11	4.53	4.15	1.04	4.02

Appendix (*Continued*)

Compound Name	Molecular Formula	Mol. Wt.	T_m (°C)	T_b (°C)	Vapor Pressure at 25°C (atm)		Aqueous Solubility at 25°C (mol L^{-1})		Henry's Law Constant at 25°C (L atm mol^{-1})	Octanol–Water Partition Coefficient at 25°C [(mol·L^{-1} octanol)·(mol·L^{-1} water)$^{-1}$]
					$-\log P^0$	$-\log P^0$(L) (For Solids and Gases)	$-\log C_w^{sat}$	$-\log C_w^{sat}$(L) (For Solids and Gases)	$\log K_H$ Calculated (Experimental)	$\log K_{ow}$
1,2,3,4-Tetrachlorobenzene	$C_6H_2Cl_4$	215.9	47.5	254.0	4.28	4.06	4.42	4.20	0.14	4.55
1,2,3,5-Tetrachlorobenzene	$C_6H_2Cl_4$	215.9	54.5	246.0	4.01	3.73	4.83	4.55	0.82	4.65
Pentachlorobenzene	C_6HCl_5	250.3	86.0	277.0	4.66	4.05	5.56	4.95	0.90	5.03
Hexachlorobenzene	C_6Cl_6	284.8	230.0	322.0	7.51	5.46	7.69	5.64	0.18	5.50
Bromobenzene	C_6H_5Br	157.0	−30.8	156.0	2.25		2.64		0.41 (0.39 exp.)	2.99
1,4-Dibromobenzene	$C_6H_4Br_2$	235.9	87.3	219.0	3.67	3.05	4.07	3.45	0.40	3.75
4-Bromochlorobenzene	C_6H_4BrCl	191.5	68.0	196.0	3.46	3.03	3.63	3.20	0.17	3.53
Fluorobenzene	C_6H_5F	96.1	−41.2	85.1	1.00		1.79		0.79	2.27
Phenol	C_6H_6O	94.1	43.0	181.7	3.59	3.41	0.20	0.02	−3.39	1.45
3-Methyl phenol (*m*-cresol)	C_7H_8O	108.6	11.5	202.2			1.59			1.96
2,4-Dimethyl phenol	$C_8H_{12}O$	122.2	27.5	210.0			1.19	1.17		2.35
2-Chlorophenol	C_6H_5ClO	128.6	9.0	174.9			1.05			2.16
Aniline	C_6H_7N	93.1	−6.3	184.0	2.89		0.41		−2.48	0.90
N-Methylaniline	C_7H_9N	107.2	−57.0	196.3			1.28			1.66

N-Dimethyl-aniline	$C_8H_{11}N$	121.2	2.5	194.0			2.04			2.31
Nitrobenzene	$C_6H_5NO_2$	123.1	5.7	210.8			1.77			1.83
Benzaldehyde	C_7H_6O	106.1	−26.0	178.0	2.90		1.51		−1.39	1.48
				Polycyclic Aromatic Hydrocarbons[e]						
Naphthalene	$C_{10}H_8$	128.2	80.6	217.9	3.98	3.43	3.61	3.06	−0.37 (−0.31 exp.)	3.36
Fluorene	$C_{13}H_{10}$	166.2	113.0	295.0	6.10	5.22	4.96	4.08	−1.14 (−1.00 exp.)	4.18
Phenanthrene	$C_{14}H_{10}$	178.2	99.5	340.2	6.79	6.05	5.20	4.46	−1.59 (−1.45 exp.)	4.57
Anthracene	$C_{14}H_{10}$	178.2	217.5	342.0	8.10	6.11	6.46	4.48	−1.64	4.54
Fluoranthrene	$C_{16}H_{10}$	202.3	110.8		7.91	7.06	5.93	5.08	−1.98	5.22
Pyrene	$C_{16}H_{10}$	202.3	156.0		8.22	7.40	6.17	5.35	−2.05 (−1.96 exp.)	5.13
Benz(a)anthra-cene	$C_{18}H_{12}$	228.3	159.8	435.0	9.55	8.20	7.31	5.96	−2.24	5.91
Benzo(a)pyrene	$C_{20}H_{12}$	252.3	176.5		11.14	9.63	8.22	6.71	−2.92	6.50
				Phthalates[f]						
Dimethyl phthalate	$C_{10}H_{10}O_4$	194.2	<25	283.7	4.62		1.66		−2.96	1.53
Diethyl phthalate	$C_{12}H_{14}O_4$	222.2	<25	298.0	5.08		2.38		−2.70	2.35
Di-*n*-propyl phthalate	$C_{14}H_{18}O_4$	250.3	<25	304.5			3.36			3.27
Diisobutyl phthalate	$C_{16}H_{22}O_4$	278.3	<25	296.5			4.13			4.11
Di-*n*-butyl phthalate	$C_{16}H_{22}O_4$	278.3	<25	340.0	7.02		4.47		−2.89	4.57

Appendix (*Continued*)

Compound Name	Molecular Formula	Mol. Wt.	T_m (°C)	T_b (°C)	Vapor Pressure at 25°C (atm)		Aqueous Solubility at 25°C (mol L^{-1})		Henry's Law Constant at 25°C (L atm mol^{-1})	Octanol–Water Partition Coefficient at 25°C [(mol·L^{-1} octanol)·(mol·L^{-1} water)$^{-1}$]
					$-\log P^0$	$-\log P^0(L)$ (For Solids and Gases)	$-\log C_w^{sat}$	$-\log C_w^{sat}(L)$ (For Solids and Gases)	$\log K_H$ Calculated (Experimental)	$\log K_{ow}$
					Halogenated C_1–C_4 Compounds[a]					
Chloromethane	CH_3Cl	50.5	−97.7	−24.2		−0.76	0.98[a]	0.22	0.98	0.91
Dichloromethane	CH_2Cl_2	84.9	−95.1	39.7	0.23		0.64		0.41 (0.43 exp.)	1.15
Trichloromethane	$CHCl_3$	119.4	−63.5	61.7	0.59		1.19		0.60 (0.59 exp.)	1.93
Tetrachloromethane	CCl_4	153.8	−22.9	76.5	0.82		2.20		1.38 (1.33 exp.)	2.73
1,1-Dichloroethane	$C_2H_4Cl_2$	99.0	−97.0	57.5	0.52		1.30		0.78	1.79
1,2-Dichloroethane	$C_2H_4Cl_2$	99.0	−35.4	83.5	1.04		1.07		0.03 (0.00 exp.)	1.47
1,1,1-Trichloroethane	$C_2H_3Cl_3$	133.4	−30.4	74.1	0.78		2.07		1.29	2.48
1,1,2,2-Tetrachloroethane	$C_2H_2Cl_4$	167.9	−36.0	146.2	2.06		1.74		−0.32	2.39
Chloroethene (vinyl chloride)	C_2H_3Cl	62.5	−153.8	−13.4		−0.59	1.35[a]	0.76	1.35	0.60
Trichloroethene	C_2HCl_3	131.4	−73.0	87.0	1.01		2.04		1.03	2.42
Tetrachloroethene	C_2Cl_4	165.8	−19.0	121.0	1.60		3.04		1.44	2.88

Hexachloro-butadiene	C_4Cl_6	260.8	−21.0	215.0	3.46 (calc)		4.90		1.44	4.90
Bromomethane	CH_3Br	94.9	−93.6	3.6		−0.26	0.79[a]	0.53	0.79	1.19
Tribromo-methane	$CHBr_3$	252.8	8.3	149.5	2.13		1.91		−0.22	
1,2-Dibromo-ethane	$C_2H_4Br_2$	187.9	−34.2	167.3	2.57		2.04		−0.53	
1,2-Dibromo-3-chloro-propane	$C_3H_5Br_2Cl$	236.4		178.0	2.90		2.44		−0.46	
Trichloro-fluoromethane	CCl_3F	137.4	−111.0	23.8		−0.02	2.10[a]	2.08	2.10	2.16
Dichlorodi-fluoromethane	CCl_2F_2	120.9	−158.0	−29.8		−0.80	2.60[a]	1.80	2.60	2.53
				Polychlorinated Biphenyls (PCBs)[h]						
Biphenyl	$C_{12}H_{10}$	154.2	71.0	255.9		5.00		3.88		4.09
2-CBP	$C_{12}H_9Cl$	188.6	32.1		4.66	4.60	4.57	4.51	−0.09	4.53
4-CBP	$C_{12}H_9Cl$	188.6	77.7		5.76	5.23	5.13	4.58	−0.63	4.40
2,5-CBP	$C_{12}H_8Cl_2$	223.1	23.0		5.53		5.06		−0.47	5.22
4,4′-CBP	$C_{12}H_8Cl_2$	223.1	149.0		7.32	6.08	6.53	5.29	−0.79	5.33
2,4,5-CBP	$C_{12}H_7Cl_3$	257.5	76.3		6.95	6.36	6.26	5.67	−0.69	5.74
2′,3,4-CBP	$C_{12}H_7Cl_3$	257.5	60.0		6.88	6.53	6.52	6.17	−0.36	5.78
2,2′,5,5′-CBP	$C_{12}H_6Cl_4$	292.0	87.0		7.60	6.64	7.06	6.10	−0.54	6.18
2,3′,4,4′-CBP	$C_{12}H_6Cl_4$	292.0	128.0		8.21	7.18	7.70	6.67	−0.51	6.31
2,2′,4,5,5′-CBP	$C_{12}H_5Cl_5$	326.4	77.0		8.02	7.40	7.40	6.78	−0.62	6.36
2,2′,4,4′,5,5′-CBP	$C_{12}H_4Cl_6$	360.9	103.0		8.97	8.15	7.82	7.00	−1.15	7.15
2,2′,3,3′,4,4′-CBP	$C_{12}H_4Cl_6$	360.9	151.9		9.65	8.38	8.72	7.45	−0.93	6.97
2,2′,3,3′,4,4′,6-CBP	$C_{12}H_3Cl_7$	395.3	122.3		9.40	8.53	8.26	7.48	−1.14	6.68
2,2′,3,3′,5,5′,6,6′-CBP	$C_{12}H_2Cl_8$	429.8	161.0		9.54	8.29	9.29	8.04	−1.25	7.12
Decachloro-BP	$C_{12}Cl_{10}$	498.7	305.8		12.28	9.85	10.55	8.12	−1.73	8.23

Appendix (*Continued*)

Compound Name	Molecular Formula	Mol. Wt.	T_m (°C)	T_b (°C)	Vapor Pressure at 25°C (atm): $-\log P^0$	Vapor Pressure at 25°C (atm): $-\log P^0(L)$ (For Solids and Gases)	Aqueous Solubility at 25°C ($mol\,L^{-1}$): $-\log C_w^{sat}$	Aqueous Solubility at 25°C ($mol\,L^{-1}$): $-\log C_w^{sat}(L)$ (For Solids and Gases)	Henry's Law Constant at 25°C ($L\,atm\,mol^{-1}$): $\log K_H$ Calculated (Experimental)	Octanol–Water Partition Coefficient at 25°C [($mol \cdot L^{-1}$ octanol)·($mol \cdot L^{-1}$ water)$^{-1}$]: $\log K_{ow}$
					Sulfur-Containing Compounds[i]					
Dimethyl sulfide	C_2H_6S	62.13	−98.0	37.3	0.20		0.45		(0.25)	
Diethyl sulfide	$C_4H_{10}S$	90.19	−103.8	92.1	1.12		1.34		(0.22)	1.95
Di-*n*-propyl sulfide	$C_6H_{14}S$	118.24	−102.5	142.4	2.10		2.58		(0.48)	
Diisopropyl sulfide	$C_6H_{14}S$	118.24	−78.1	120.0	1.72		2.24		(0.52)	
Dimethyl sulfide	$C_2H_6S_2$	94.20	−84.7	109.7	1.42		1.44		(0.02)	1.77
Diethyl sulfide	$C_4H_{10}S_2$	122.25	−101.5	154.0	2.20		2.42		(0.22)	
Thiophene	C_4H_4S	84.14	−38.2	84.2	0.98		1.33		(0.35)	1.81
2-Methyl thiophene	C_5H_6S	98.17	−63.4	112.6	1.46		1.85		(0.39)	
Methanethiol	CH_4S	48.11	−123.0	6.2		−0.300		0.091	(0.39)	
Ethanethiol	C_2H_6S	62.13	−144.4	35.0	0.16		0.71		(0.55)	
n-Propanethiol	C_3H_8S	76.17	−113.3	67.7	0.69		1.30		(0.61)	
					Miscellaneous Pesticides and Other Compounds[j]					
2,3,7,8-Tetrachlorodibenzo-*p*-dioxin	$C_{12}H_4O_2Cl_4$	322.0	305	421	11.6	8.8	10.3	7.5	−1.3	6.64
Lindane (γ-hexachlorocyclohexane)	$C_6H_6Cl_6$	290.8	112.9		7.08	6.20	4.59	3.71	−2.49	3.78
Dieldrin	$C_{12}H_8Cl_6O$	380.9	175.0		8.18	6.68	6.23	4.98	−1.95	5.48
p,p'-DDT	$C_{14}H_9Cl_5$	354.5	109.0		9.87	9.03	7.85	7.01	−2.02	6.36
Parathion	$C_{10}H_{14}NO_5PS$	291.3	6.1		7.65		4.23		−3.42	3.81

Malathion	$C_{10}H_{19}O_6PS_2$	330.4	2.9		3.36		2.89
Phosmit	$C_{11}H_{12}NO_4PS_2$	317.3	72.0	9.22 (20°C)	4.10	3.63	2.83
Dialifos	$C_{14}H_{17}ClNO_4PS_2$	393.9	68.0	10.08 (30°C)	6.34	5.91	4.69
Carbaryl	$C_{12}H_{11}NO_2$	201.2	142.0		3.70	2.53	2.36
Carbofuran	$C_{12}H_{15}NO_3$	221.3	151.0		2.73	1.47	1.60
Fluometuron	$C_{10}H_{11}F_3NO_2$	232.2	164.0		3.51	1.39	1.34
Atrazine	$C_8H_{14}ClN_5$	215.7	174.0		3.81	2.33	2.56
RDX (1,3,5-triaza-1,3,5-trinitrocyclohexane)	$C_3H_6N_6O_6$	222.1	205.0		4.57	2.77	0.87
Tributylphosphate	$C_{12}H_{27}PO_4$	266.3	<25		5.98		4.00
Tri-*o*-cresylphosphate	$C_{21}H_{21}PO_4$	368.4	77.0		6.01	5.49	5.11

[a] At 1 atm.

[b] Hansch and Leo (1979); Mackay and Shiu (1981); Stein (1981); Tewari et al. (1982).

[c] Hansch and Leo (1979); CRC 1985–1986; Tewari et al. (1982).

[d] Hansch and Leo (1979); Yalkowsky and Valvani (1979); Yalkowsky et al. (1979); Banerjee et al. (1980); CRC 1985–1986; Mackay and Shiu (1981); Chiou et al. (1982); Horvath (1982); Tewari et al. (1982); Miller et al. (1984).

[e] Wauchope and Getzen (1972); Mackay et al. (1980a); May et al. (1983); Sonnefeld et al. (1983); Whitehouse (1984).

[f] Wolfe et al. (1980); CRC 1985–1986; Leyder and Boulanger (1983).

[g] Hansch and Leo (1979); Banerjee et al. (1980); Mackay and Shiu (1981); Horvath (1982).

[h] Mackay et al. (1980b); Westcott et al. (1981); Bidleman (1984); Burkhard et al. (1984); Miller et al. (1984); Rapaport and Eisenreich (1984); Woodburn et al. (1984); Burkhard et al. (1985).

[i] Hansch and Leo (1979); CRC 1985–1986; Przyjazny et al. (1983).

[j] Banerjee et al. (1980); Mackay and Shiu (1981); Lyman (1982a,b); Monsanto Company (1985); Marple et al. (1986a,b); Eltzer and Hites (1988).

BIBLIOGRAPHY

Adamson, A. W., *Physical Chemistry of Surfaces*, Wiley, NY, 1982.

Alexander, D. M., "The solubility of benzene in water," *J. Am. Chem. Soc.*, **63**, 1021–1022 (1959).

Alexander, M., "Biodegradation of chemicals of environmental concern," *Science*, **211**, 132–138 (1981).

Amidon, G. L., S. H. Yalkowsky, and S. Leung, "Solubility of nonelectrolytes in polar solvents. II. Solubility of aliphatic alcohols in water," *J. Pharm. Sci.*, **63**, 1858–1866 (1974).

Appleton, H. T., S. Banerjee, and H. C. Sikka, "Fate of 3,3′-dichlorobenzidine in the aquatic environment." In *Dynamics, Exposure, and Hazard Assessment of Toxic Chemicals*, H. Rizwanul, Ed., Ann Arbor Science, Ann Arbor, MI, 1980, pp. 251–272.

Arbuckle, W. B., "Estimating activity coefficients for use in calculating environmental parameters," *Environ. Sci. Technol.*, **17**, 538–542 (1983).

Army, T. P., *Production and Transport of Biogenic Volatiles from a Freshwater, Floating-Mat Bog*, Ph.D. Thesis, Massachusetts Institute of Technology, Cambridge, MA, 1987, 325 pp.

Asher, W. E., and J. F. Pankow, "The interaction of mechanically generated turbulence and interfacial film with a liquid phase controlled gas/liquid transport process," *Tellus*, **38B**, 305–318 (1986).

Atkins, P. W., *Physical Chemistry*, W. H. Freeman, San Francisco CA, 1978.

Atlas, E., R. Foster, and C. S. Giam, "Air–sea exchange of high molecular weight organic pollutants: Laboratory studies," *Environ. Sci. Technol.*, **16**, 283–286 (1982).

Bachofer, R., F. Lingens, and W. Schafer, "Conversion of aniline into pyrocatechol by a *Nocardia sp.*: Incorporation of oxygen-18," *FEBS Lett.*, **50**, 288–290 (1975).

Backhus, D. A., *Colloids in Groundwater: Laboratory and Field Studies of their Influence on Hydrophobic Organic Contaminants*, Ph.D. Thesis, Massachusetts Institute of Technology, Cambridge, MA, 1990, 198 pp.

Backhus, D. A. and P. M. Gschwend, "Fluorescent polycyclic aromatic hydrocarbons as probes for studying the impact of colloids on pollutant transport in groundwater," *Environ. Sci. Technol.*, **24**, 1214–1223 (1990).

Bachmann, A., P. Walet, P. Wijnen, W. DeBruin, J. L. M. Huntjens, W. Roelossen, and A. J. B. Zehnder, "Biodegradation of α- and β-hexachlorocyclohexane in a soil slurry under different redox conditions," *Appl. Environ. Microbiol.*, **54**, 143–149 (1988).

Bailey, G. W., J. L. White, and T. Rothberg, "Adsorption of organic herbicides by montmorillonite: Role of pH and chemical character of adsorbate," *Soil Sci. Soc. Am. Proc.*, **32**, 222–234 (1968).

Baker, K. S. and R. C. Smith, "Bio-optical classification and model of natural waters," *Limnol. Oceanogr.*, **27**, 500–509 (1982).

Baker, J. E., P. D. Capel, and Eisenreich, S. J., "Influence of colloids on sediment–water partition coeffients of polychlorobiphenyl congeners in natural waters," *Environ. Sci. Technol.*, **20**, 1136–1143 (1986).

Baker, J. E. and S. J. Eisenreich, "PCBs and PAHs as tracers of particulate dynamics in large lakes," *J. Great Lakes Res.*, **15**, 84-103 (1989).

Baker, J. E. and S. J. Eisenreich, "Concentrations and fluxes of polycyclic aromatic hydrocarbons and polychlorinated biphenyls across the air–water interface of Lake Superior," *Environ. Sci. Technol.*, **24**, 342–352 (1990).

Balistrieri, L. S. and J. W. Murray, "The influence of the major ions of seawater on the adsorption of simple organic acids by goethite," *Geochim. Cosmochim. Acta*, **51**, 1151–1160 (1987).

Ball, W. P., Ch. Buehlar, T. C. Harmon, D. M. Mackay, and P. V. Roberts, "Characterization of a sandy aquifer material at the grain scale," *J. Contam. Hydrol.*, **5**, 253–295 (1990).

Ball, W. P. and P. V. Roberts, "Long-term sorption of halogenated organic chemicals by aquifer materials—Part 2. Intraparticle diffusion," *Environ. Sci. Technol.*, **25**, 1237–1249 (1991).

Banerjee, P., M. D. Piwoni, and K. Ebeid, "Sorption of organic contaminants to a low carbon subsurface core," *Chemosphere*, **8**, 1057–1067 (1985).

Banerjee, S., "Solubility of organic mixtures in water," *Environ. Sci. Technol.*, **18**, 587–591 (1984).

Banerjee, S., "Calculation of water solubility of organic compounds with UNIFAC-derived parameters," *Environ. Sci. Technol.*, **19**, 369–370 (1985).

Banerjee, S. and S. H. Yalkowsky, "Cosolvent-induced solubilization of hydrophobic compounds into water," *Anal. Chem.*, **60**, 2153–2155 (1988).

Banerjee, S., S. H. Yalkowsky, and S. C. Valvani, "Water solubility and octanol/water partition coefficents of organics. Limitations of the solubility–partition coefficient correlation," *Environ. Sci. Technol.*, **14**, 1227–1229 (1980).

Banks, R. B., "Some features of wind action on shallow lakes," *Proc. Am. Soc. Civil Eng.*, **101**, 813 (1975).

Bard, A. J. and H. Lund, Eds., *Encyclopedia of Electrochemistry of the Elements*, Volumes XI–XV, Marcel Dekker, New York, 1978–1984.

Barlin, G. B. and D. D. Perrin, "Prediction of the strengths of organic acids, " *Quart. Rev. Chem. Soc.*, **20**, 75–101 (1966).

Barnard, P. W. C., C. A. Burton, D. R. L'lewellyn, C. A. Vernon, and V. A. Welch, "The reactions of organic phosphates. Part V. The hydrolysis of triphenyl and trimethyl phosphate," *J. Chem. Soc.*, 2670–2676 (1961).

Bartha, R., "Fate of herbicide-derived chloronilines in soil," *J. Agric. Food Chem.*, **19**, 385–387 (1971).

Bartholomew, G. W. and F. K. Pfaender, "Influence of spatial and temporal variations on organic pollutant biodegradation rates in an estuarine environment," *Appl. Environ. Microbiol.*, **45**, 102–109 (1983).

Bartnicki, E. W., N. O. Belser, and C. E. Castro, "Oxidation of heme proteins by alkyl halides: A probe for axial inner sphere redox capacity in solution and in whole cells," *Biochemistry*, **17**, 5582–5586 (1978).

Bell, R. P. and P. G. Evans, "Kinetics of the dehydration of methylene glycol in aqueous solution," *Proc. Roy. Soc., London, Ser. A*, **241**, 297–323 (1966).

Bell, R. P., and A. O. McDougall, "Hydration equilibrium of some aldehydes and ketones," *Trans. Faraday Soc.*, **56**, 1281–1285 (1960).

Bender, M. L. and R. B. Homer, "The mechanism of the alkaline hydrolysis of *p*-nitrophenol *N*-methylcarbamate," *J. Org. Chem.*, **30**, 3975–3978 (1965).

Benz, R. and F. McLaughlin, "The molecular mechanism of action of the proton ionophore FCCP (Carbonyl cyanide *p*-trifluoromethoxyphenylhydrazone)," *Biophys. J.*, **41**, 381–398 (1983).

Berezin, I. V., A. V. Levashov, and K. Martinek, "On the modes of interaction between competitive inhibitors and the alpha-chymotrypsin active centre," *FEBS Lett.*, **7**, 20–22 (1970).

Betterton, E. A. and M. R. Hoffman, "Henry's Law constants of some environmentally important aldehydes," *Environ. Sci. Technol.*, **22**, 1415–1418 (1988).

Bidleman, T. F., "Estimation of vapor pressures for nonpolar organic compounds by capillary gas chromatography", *Anal. Chem.*, **56**, 2490–2496 (1984).

Bidleman, T. F., "Atmospheric processes," *Environ. Sci. Technol.*, **22**, 361–367 (1988).

Birkett, J. D., "Heat capacity." In W. J. Lyman, W. F. Reehl, and D. H. Rosenblatt (Eds.), *Handbook of Chemical Property Estimation Methods: Environmental Behavior of Organic Compounds*, McGraw-Hill, New York, 1982, Chapter 23.

Blanchard, D. C. and A. H. Woodcock, "Bubble formation and modification in the sea and its meteorological significance," *Tellus*, **9**, 145–158 (1957).

Bloesch, J. and M. Sturm, "Settling flux and sinking velocities of particulate phosphorus (PP) and particulate organic carbon (POC) in Lake Zug, Switzerland." In P. G. Sly, Ed., *Sediments and Water Interactions*, Springer, New York, 1986, p. 481–490.

Blumer, M. "Organic compounds in nature: Limits of our knowledge," *Angewandte Chemie*, **14**, 507–514 (1975).

Bohon, R. L. and W. F. Claussen, "The solubility of aromatic hydrocarbons in water," *J. Am. Chem. Soc.*, **73**, 1571–1578 (1951).

Bosshardt, H., "Entwicklungstendenzen in der Bekämpfung von Schadorganismen in der Landwirtschaft," *Vierteljahresschr. Naturforsch. Ges. Zurich*, **133**, 225–240 (1988).

Boucher, F. R. and G. F. Lee, "Adsorption of lindane and dieldrin pesticides on unconsolidated aquifer sands," *Environ. Sci. Technol.*, **6**, 538–543 (1972).

Bouwer, E. J. and P. L. McCarty, "Transformations of 1- and 2-carbon halogenated aliphatic organic compounds under methanogenic conditions," *Applied Environ. Microbiol.*, **45**, 1286–1294 (1983a).

Bouwer, E. J. and P. L. McCarty, "Transformations of halogenated organic compounds under denitrification conditions," *Appl. Environ. Microbil,.*, **45**, 1295–1299 (1983b).

Boyd, S. A., Mortland, M. M., and Chiou, C. T.., "Sorption characteristics of organic compounds on hexadecyltrimethyl ammonium–smectite," *Soil Sci. Soc. Am. J.*, **52**, 652–657 (1988).

Boyle, M., "The environmental microbiology of chlorinated aromatic decomposition," *J. Environ. Qual.*, **18**, 395–402 (1989).

Broecker, H. C., J. Petermann, and W. Seims, "The influence of wind on CO_2-exchange in a wind-wave tunnel, including the effects of monolayers," *J. Mar. Res.*, **36**, 595–610 (1978).

Broecker, W. S., "An application of natural radon to problems in ocean circulation." In T. Ichiye Ed., *Diffusion in Oceans and Fresh Waters*, Lamont Doherty Geol. Obs., Palisades, NY, 1965, pp. 116–145.

Broecker, W. S. and T.-H. Peng, "The vertical distribution of radon in the Bomex area," *Earth Planet. Sci. Lett.*, **11**, 99–108 (1971).

Broecker, W. S. and T.-H. Peng, "Gas exchange rates between air and sea," *Tellus*, **XXVI**, 21–35 (1974).

Brønsted, J. N. and K. Pederen, "Die katalytische Zersetzung des Nitramids und ihre physikalisch-chemische Bedeutung," *Z. Phys. Chem.*, **108**, 185–235 (1924).

Brownawell, B. J., *The Role of Colloid Organic Matter in the Marine Geochemistry of PCBs*, Ph.D. Dissertation, Woods Hole–MIT Joint Program, Cambridge, MA, 1986, 318 pp.

Brownawell, B. J., H. Chen, J. M. Collier, and J. C. Westall, "Adsorption of organic cations to natural materials," *Environ. Sci. Technol.*, **24**, 1234–1241 (1990).

Bruice, T. C., T. H. Fife, J. J. Bruno, and N. E. Brandon, "Hydroxyl group catalysis. II. The reactivity of the hydroxyl group to serine. The nucleophilicity of alcohols and the ease of hydrolysis of their acetyl esters nucleophilicity as related to their pK's," *Biochemistry*, **1**, 7–12 (1962).

Brunner, W., D. Staub, and T. Leisinger, "Bacterial degradation of dichloromethane," *Appl. Environ. Microbiol.*, **40**, 950–958 (1980).

Brusseau, M. L., R. E. Jessup, and P. S. C. Rao, "Nonequilibrium sorption of organic chemicals: Elucidation of rate-limiting processes," *Environ. Sci. Technol.*, **25**, 134–142 (1991).

Brusseau, M. L. and P. S. C. Rao, "The influence of sorbate–organic mater interactions on sorptive nonequilibrium," *Chemosphere*, **18**, 1691–1706 (1989).

Brusseau, M. L. and P. S. C. Rao, "Influence of sorbate structure on nonequilibrium sorption of organic compounds," *Environ. Sci. Technol.*, **25**, 1501–1506 (1991).

Buffle, A. J. and R. S. Altmann, "Interpretation of metal complexation of heterogeneous complexants," In W. Stumm Ed., *Aquatic Surface Chemistry*, Wiley, New York, 1987.

Bumpus, J. A., M. Tier, D. Wright, and S. D. Aust, "Oxidation of persistent environmental pollutants by a white rot fungus," *Science*, **228**, 1434–1436 (1985).

Burkhard, L. P., A. W. Andren, and D. E. Armstrong, "Estimation of vapor pressures for polychlorinated biphenyls: A comparison of eleven predictive methods," *Environ. Sci. Technol.*, **19**, 500–507 (1985).

Burkhard, L. P., D. E. Armstrong, and A. W. Andren, "Vapor pressures for biphenyl, 4-chlorobiphenyl, 2,2′,3,3′,5,5′,6,6′-octachlorobiphenyl, and decachlorobiphenyl," *J. Chem. Eng. Data*, **29**, 248–250 (1984).

Burlinson, N. E., L. A. Lee, and D. H. Rosenblatt, "Kinetics and products of hydrolysis of 1,2-dibromo-3-chloropropane," *Environ. Sci. Technol.*, **16**, 627–632 (1982).

Butler, J. A. C. and C. N. Ramchandani, "The solubility of non-electrolytes. Part II: The influence of the polar group on the free energy of hydration of aliphatic compounds," *J. Chem. Soc.*, 952–955 (1935).

Butler, J. A. V., D. W. Thomson, and W. H. Maclennan, "The free energy of the normal aliphatic alcohols in aqueous solution. Part I. The partial vapour pressures of aqueous solutions of methyl, *n*-propyl, and *n*-butyl alcohols. Part II. The solubilities of some normal aliphatic alcohols in water. Part III. The theory of binary solutions, and its application to aqueous-alcoholic solutions," *J. Chem. Soc.*, **1933**, 674–686 (1933).

Butte, W., C. Fooken, R. Klussman, and D. Schuller, "Evaluation of lipophilic properties for a series of phenols using reversed-phase high-performance liquid chromatography and high performance thin-layer chromatography," *J. Chromatogr.*, **214**, 59–67 (1981).

Calvert, J. G. and J. N. Pitts, Jr., *Photochemistry*, Wiley, New York, 1966.

Campbell, R. R., R. G. Luthy, and M. J. T. Carrondo, "Measurement and prediction of distribution coefficients for wastewater aromatic solutes," *Environ. Sci. Technol.*, **17**, 582–590 (1983).

Carslaw, H. S. and J. C. Jaeger, *Conduction of Heat in Solids*, 2nd ed., Oxford University Press, Oxford, UK, 1959.

Carter, C. W. and Suffet, "Binding of DDT to dissolved humic materials," *Environ. Sci. Technol.*, **16**, 735–740 (1982).

Castro, C. E. and E. W. Bartnicki, "Conformational isomerism and effective redox geometry in the oxidation of heme proteins by alkyl halides, cytochrome C, and cytochrome oxidase," *Biochem.*, **14**, 498–503 (1975).

Castro, C. E., R. S. Wade, and N. O. Belser, "Biodehalogenation. Reductive dehalogenation of the biocides ethylene dibomide, 1,2-dibromo-3-chloropropane, and 2,3-dibromobutane in soil," *Environ. Sci. Technol.*, **2**, 779–783 (1968).

Castro, C. E., R. S. Wade, and N. O. Belser, "Biodehalogenation: Reactions of cytochrome P-450 with polyhalomethanes," *Biochem.*, **24**, 204–210 (1985).

Castro, C. E., W. H. Yokoyama, and N. O. Belser, "Biodehalogenation. Reductive reactivities of microbial and mammalian cytochromes P-450 compared with heme and whole-cell models," *J. Argric. Food Chem.*, **36**, 915–919 (1988).

Chandar, S., D. W. Fuerstenau, and D. Stigler, "On hemimicelle formation at oxide/water interfaces." In R. H. Ottewill, Ed., *Adsorption from Solution, Academic*, 1983, pp. 197–210.

Chandar, P., P. Somasundaran, and N. J. Turro, "Fluorescence probe studies on the structure of the absorbed layer of dodecyl sulfate at the alumina-water interface, *J. Colloid Int. Sci.*, **117**, 31–46 (1987).

Chaudhry, G. R. and A. N. Ali, "Bacterial metabolism of carbofuran," *Appl. Environ. Microbiol.*, **54**, 1414–1419 (1988).

Chaudhry, G. R. and S. Chapalamadugu, "Biodegradation of halogenated organic compounds," *Microbiol. Rev.*, **55**, 59–79 (1991).

Chin, Y.-P. and W. J. Weber, Jr., "Estimating the effects of dispersed organic polymers on the sorption of contaminants by natural solids. 1. A predictive thermodynamic humic substance–organic solute interaction model," *Environ. Sci. Technol.*, **23**, 978–984 (1989).

Chiou, C. T., V. H. Freed, D. W. Schmedding, and R. L. Kohnert, "Partition coefficients and bioaccumulation of selected organic chemicals," *Environ. Sci. Technol.* **11**, 475–478 (1977).

Chiou, C. T., D. E. Kile, T. I. Brinton, R. L. Malcolm, J. A. Leenheer, and P. MacCarthy, "A comparison of water solubility enhancements of organic solutes by aquatic humic materials and commercial humic acids," *Environ. Sci. Technol.*, **21**, 1231–1234 (1987).

Chiou, C. T., J.-F. Lee, and S. A. Boyd, "The surface area of soil organic matter," *Environ. Sci. Technol.*, **24**, 1164–1166 (1990).

Chiou, C. T., R. L., Malcolm, T. I., Brinton, and D. E. Kile, "Water solubility enhancement of some organic pollutants and pesticides by dissolved humic and fulvic acids," *Environ. Sci. Technol.*, **20**, 502–508 (1986).

Chiou, C. T., L. J. Peters, and V. H. Freed, "A physical concept of soil–water equilibria for nonionic compounds," Science, **206**, 831–832 (1979).

Chiou, C. T., P. E. Porter, and D. W. Schmedding, "Partition equilibria of nonionic organic compounds between soil organic matter and water," *Environ. Sci. Technol.*, **17**, 227–231 (1983).

Chiou, C. T., D. W. Schmedding, and M. Manes, "Partitioning of organic compounds in octanol–water systems," *Environ. Sci. Technol.*, **16**, 4–10 (1982).

Chisaka, H. and P. C. Kearney, "Metabolism of propanil in soils," *J. Agric. Food. Chem.*, **18**, 854–858 (1970).

Chou, J. T. and P. C. Jurs, "Computer assisted computation of partition coefficients from molecular structures using fragment constants," *J. Chem. Inf. Computer Sci.*, **19**, 172–178 (1979).

Clark, J. and D. D. Perrin, "Prediction of the strengths of organic bases," *Quart. Revs. Chem. Soc.*, **18**, 295–320 (1964).

Clark, W. M., *Oxidation–Reduction Potentials of Organic Systems*, Williams and Wilkins, Baltimore, MD 1960.

Coates, J. T. and A. W. Elzerman, "Desorption kinetics for selected PCB congeners from river sediments," *J. Contam. Hydrology*, **1**, 191–210 (1986).

Collander, R. M., "The permeability of plant protoplasts to small molecules," *Physiol. Plant.*, **2**, 300–311 (1949).

Connolly, J. P., "The Effect of Sediment Suspension on Adsorption and Fate of Kepone," Ph.D. Dissertation, Univ. of Texas, Austin, 209 pp. (1980).

Conrad, H. E., K. Lieb, and I. C. Gunsalus, "Mixed function oxidation. III. An electron transport complex in camphor ketolactonization," *J. Biol. Chem.*, **240**, 4029–4037 (1965).

Cooper, W. J., R. G. Zika, R. G. Petasne, and A. M. Fischer, "Sunlight induced photochemistry in natural waters: Major reactive species." In *Aquatic Humic Substances: Influence on Fate and Treatment of Pollutant*, I. H. Suffet and P. MacCarthy, Eds., Advances in Chemistry Series 219, American Chemical Society, Washington, DC, 1989, pp. 333–362.

Cowan, C. T. and D. White, "The mechanism of exchange reactions occurring between sodium montmorillonite and various *n*-primary aliphatic amine salts," *Trans. Faraday Soc.*, **54**, 691–697 (1958).

Craig, H. and T. Hayward, "Oxygen supersaturation in the ocean: Biological versus physical contributions," *Science*, **235**, 199–202 (1987).

Crank, J., *The Mathematics of Diffusion*, 2nd Ed., Clarendon Press, Oxford, UK, 1975.

CRC Handbook of Chemistry and Physics, 66th Ed., CRC Press, Boca Raton, FL, 1985–1986.

Criddle, C. S., P. L. McCarty, M. C. Elliot, and J. F. Barker, "Reduction of hexachloroethane to tetrachloroethylene in groundwater," *J. Contam. Hydrol.*, **1**, 133–142 (1986).

Csanady, G. T., *Turbulent Diffusion in the Environment*, Reidel, Dordrecht, Holland, 1973.

Dalton, J., "Experimental essays on the constitution of mixed gases; on the force of steam or vapor from waters and other liquids, both in a Torricellian vacuum and in air; on evaporation; and on the expansion of gases by heart," *Mem. Proc. Manchester Lit. Phil. Soc.*, **5**, 535–602 (1802).

Danckwerts, P. V., "Significance of liquid–film coefficients in gas absorption," *Ind. Eng. Chem.*, **43**, 1460–1467 (1951).

Davis, J. A., "Adsorption of natural dissolved organic matter at the oxide/water interface," *Geochim. Cosmochim. Acta*, **46**, 2381–2393 (1982).

Dean, J. A., Ed., *Lange's Handbook of Chemistry*, 12th Ed., McGraw-Hill, New York, 1979.

Dexter, R. N. and S. P. Pavlou, "Mass solubility and aqueous activity coefficients of stable organic chemicals in the marine environment: Polychlorinated biphenyls," *Mar. Chem.*, **6**, 41–53 (1978).

Dickerson, R. E., *Molecular Thermodynamics*, Benjamin, Menlo Park, CA 1969, 452 pp.

Dilling, W. L., "Interphase Transfer Processes. II. Evaporation rates of chloromethanes, ethanes, ethylenes, propanes, and propylenes from dilute aqueous solutions. Comparisons with theoretical predictions," *Environ. Sci. Technol.*, **11**, 405–409 (1977).

Dilling, W. L., N. B. Tefertiller, and G. J. Kallos, "Evaporation rates and reactivities of methylene chloride, chloroform, 1,1,1-trichloroethane, trichloroethylene, tetrachloroethylene, and other chlorinated compounds in dilute aqueous solutions", *Environ. Sci. Technol.*, **9**, 833–838 (1975).

Dittert, L. W. and T. Higuchi, "Rates of hydrolysis of carbonate and carbonate esters in alkaline solution" *J. Pharm. Sci.*, **52**, 852–857 (1963).

Dolfing, J. and J. M. Tiedje, "Hydrogen cycling in a three-tiered food web growing on the methanogenic conversion of 3-chlorobenzoate," *FEMS Microbiol. Ecol.*, **38**, 293–298 (1986).

Downing, A. L. and G. A. Truesdale, "Some factors affecting the rate of solution of oxygen in water," *J. Appl. Chem.*, **5**, 570–581 (1955).

Drever, J. I., *The Geochemistry of Natural Waters*, 2nd. ed., Prentice Hall, Englewood Cliffs, NJ, 1988.

Drost-Hansen, W., "Structure of water near solid interfaces," *Ind. Eng. Chem.*, **61**, 10–47 (1969).

Dulin, D. and T. Mill, "Development and evaluation of sunlight actinometers," *Environ. Sci. Technol.*, **16**, 815–820 (1982).

Duran, A. P. and H. F. Hemond, "Dichlorodifluoromethane (Freon-12) as a tracer for nitrous oxide release from a nitrogen-enriched river." In *Gas Transfer at Water Surfaces*, W. Brutsaert and G. H. Jirka, Eds., Reidel, Holland, 1984, pp. 421–429.

Dzombak, D. A. and F. M. M. Morel, *Surface Complexation Modeling*, Wiley-Interscience, New York, 1990.

Eadsforth, C. V. and P. Moser, "Assessment of reverse-phase chromatographic methods for determining partition coefficients," *Chemosphere*, **12**, 1459–1475 (1983).

Eberson, L., *Electron Transfer Reactions in Organic Chemistry*, Springer, Berlin, 1987.

Eganhouse, R. P. and J. A. Calder, "The solubility of medium molecular weight aromatic hydrocarbons and the effects of hydrocarbon co-solutes and salinity," *Geochim. Cosmochim. Acta*, **40**, 555–561 (1976).

Egli, C. R. Scholtz, A. M. Cook, and T. Leisinger, "Anaerobic dechlorination of tetrachloromethane and 1,2-dichloroethane to degradable products by pure cultures of *Desulfobacterium* sp. and *Methanobacterium* sp.," *FEMS Microbiol. Lett.*, **43**, 257–261 (1987).

Eisenreich, S. J. and B. B. Looney, "Evidence for the atmospheric flux of polychlorinated biphenyls to Lake Superior." In *Physical Behavior of PCBs in the Great Lakes*, D. Mackay, S. Patternson, S. J. Eisenreich, and M. S. Simmons, Eds., Ann Arbor Science, Ann Arbor, MI, 1983.

Eisenreich, S. J., W. A. Willford, and W. M. J. Strachan, "The role of atmospheric deposition in organic contaminant cycling in the Great Lakes." In D. Allen, Ed., *Intermedia Pollutant Transport: Modelling and Field Measurements*, Plenum, New York, 1989.

El-Amamy, M. M. and T. Hill, "Hydrolysis kinetics of organic chemicals on montmorillonite and kaolinite surfaces as related to moisture content," *Clays and Minerals*, **32**, 67–73 (1984).

Eltzer, B. D., and R. A. Hites, "Vapor pressures of chlorinated dioins and dibenzofurans," *Environ. Sci. Technol.*, **22**, 1362–1364 (1988).

Emerson, S. "Gas exchange rates in small Canadian Shield lakes." *Limnol. Oceanogr.*, **20**, 754–761 (1975).

Emerson, S., W. S. Broecker, and D. W. Schindler, "Gas exchange rate in a small lake as determined by the radon method," *J. Fish. Res. Bd. Can.*, **30**, 1475–1484 (1973).

Enfield, C. G., G. Bengtsson, and R. Lindqvist, "Influence of macromolecules on chemical transport," *Environ. Sci. Technol.*, **23**, 1278–1286 (1989).

Englehardt, G., P. R. Wallnofer, and R. Plapp, "Purification and properties of an aryl acylamidase of *Bacillus sphaericus*, catalyzing the hydrolysis of various phenylamide herbicides and fungicides," *Appl. Microbiol.*, **26**, 709–718 (1973).

Epand, R. M. and I. B. Wilson, "Evidence for the formation of hippuryl chymotrypsin during the hyrolysis of hippuric acid esters," *J. Biol. Chem.*, **238**, 1718–1723. (1963).

Estes, T. J., R. V. Shah, and V. L. Vilker, "Adsorption of low molecular weight halocarbons by montmorillonite," *Environ. Sci. Technol.* **22**, 377–381 (1988).

Etzler, F. M. and P. J. White, "The heat capacity of water in silica pores," *J. Colloid Int. Sci.*, **120**, 94–99 (1987).

Farhataziz, and A. B. Ross. In *Selected Specific Rates of Reactions of Transients from Water in Aqueous Solution. III. Hydroxyl Radical and Perhydroxyl Radicals and Their Radical Ions.* National Bureau of Standards, Report No. NSRDS-NBS 59, Washington, DC, (1977).

Faust, B. C. and J. Hoigne, "Sensitized photo-oxidation of alkylphenols by fulvic acid and natural water," *Environ. Sci. Technol.*, **21**, 957–964 (1987).

Faust, S. D. and H. M. Gomaa, "Chemical hydrolysis of some organic phosphorus and carbamate pesticides in aquatic environments," *Environ. Lett.*, **3**, 171–201 (1972).

Feller, W., *An Introduction to Probability Theory and Its Applications*, Vol. 1, Wiley, New York, 1957.

Fersht, A., *Enzyme Structure and Mechanism*, Freeman, N.Y., 1985, 475 pp.

Finlayson-Pitts, B. J. and J. N. Pitts, Jr., *Atmospheric Chemistry: Fundamental and Experimental Techniques*," Wiley-Interscience, New York (1986).

Fishtine, S. H., "Reliable latent heats of vaporization," *Ind. Eng. Chem.*, **55**, 47–56 (1963).

Fitzgerald, D., "Evaporation," *Trans. ASCE*, **15**, 581–646 (1886).

Fleck, G. M., *Equilibria in Solution*, Holt, Rinehart, and Winston, *Inc*, New York, 1966.

Fogel, M. N., A. R. Taddeo, and S. Fogel, "Biodegradation of chlorinated ethenes by a methane-utilizing mixed culture," *Appl. Environ. Microbiol.*, **54**, 720–724 (1986).

Fowkes, F. M., "Attractive forces at interfaces," *Ind. Eng. Chem.*, **56**, 40–52 (1964).

Frank, H. S. and M. W. Evans, "Free volume and entropy in condensed systems. III. Entropy in binary liquid mixtures; partial molal entropy in dilute solutions; structure and thermodynamics in aqueous electrolytes," *J. Chem. Phys.*, **13**, 507–532 (1945).

Frank, R., and H. Rau, "Photochemical transformation in aqueous solution and possible environmental fate of ethylenediaminetetraacetic acid (EDTA)," *Ecotoxicology Envrion. Safety*, **19**, 55–63 (1990).

Fredenslund, A., R. L., Jones, and J. M. Prausnitz, "Group-contribution estimation of activity coefficients in nonideal liquid mixtures", *AIChE J.*, **21**, 1086–1099 (1975).

Frost, A. A. and R. G. Pearson, *Kinetics and Mechanism*, Wiley-Interscience, New York, 1961.

Fuerstenau, D. W., "Streaming potential studies on quartz in solutions of ammonium acetates in relation to the formation of hemimicelles at the quartz-solution interface", *J. Phys. Chem.*, **60**, 981–985 (1956).

Fuerstenau, D. W. and T. Wakamatsu, "Effect of pH on the adsorption of sodium dodecane-sulphonate at the alumina/water interface," *Farady Disc. Chem. Soc.*, **59**, 157–168 (1975).

Fujita, T., J. Iwasa, and C. Hansch, "A new substituent constant, π, derived from partition coefficients," *J. Am. Chem. Soc.*, **86**, 5175–5180 (1964).

Fuller, E. N., Schettler, P. D., and Giddings, J. C., "A new method for prediction of binary gas-phase diffusion coefficient," *Ind. Eng. Chem.*, **58**, 19–27 (1966).

Garbarini, D. R. and L. W. Lion, "Evaluation of sorptive partitioning of nonionic pollutants in closed systems by headspace analysis," *Environ, Sci. Technol.*, **19**, 1122–1128 (1985).

Garbarini, D. R. and L. W. Lion, "The influence of the nature of soil organics on the sorption of toluene and trichloroethylene," *Environ. Sci. Technol.*, **20**, 1263–1269 (1986).

Gargett, A. E. and G. Holloway, "Dissipation and diffusion by internal wave breaking," *J. Mar. Res.*, **42**, 15–27 (1984).

Garrett, W. D., "Damping of capillary waves at the air–sea interface by organic surface active material," *J. Mar. Res.*, **25**, 279–291 (1967).

Gauthier, T. D., E. C. Shane, W. F. Guerin, W. R. Seltz, and C. L. Grant, "Fluorescence quenching method for determining equilibrium constants for polycyclic aromatic hydrocarbons binding to dissolved humic materials," *Environ. Sci. Technol.*, **20**, 1162–1166 (1986).

Gauthier, T. D., W. R. Seitz, and C. L. Grant, "Effect of structural and compositional variations of dissolved humic materials on pyrene K_{oc} values," *Environ. Sci. Technol.*, **21**, 243–248 (1987).

Genereux, D. P., *Field Studies of Stream flow Generation Using Natural and Injected Traces on Bickford and Walker Branch Watersheds*, Ph.D. Thesis, Massachusett, Institute of Technology, Cambridge, MA, 1991, 288 pp.

Gerrard, W., *Gas Solubilities, Widespread Applications*, Pergamon, New York 1980.

Giam, C. S., E. Atlas, M. A. Powers, Jr., and J. E. Leonard, "Phthalic acid esters." In: O. Hutzinger (Ed.), *The Handbook of Environmental Chemistry*, Vol. 3, Part C, Springer, Berlin, pp. 67–142 (1984).

Gibbs, J. W., XI. Graphical methods in the thermodynamics of fluids, *Trans. Conn. Acad.*, **2**, 309–342 (1873).

Gibbs, J. W., V. On the equilibrium of heterogeneous substances, *Trans. Conn. Acad.*, **3**, 108–248 (1876).

Gibson, D. T., J. R. Koch, and R. E. Kallio, "Oxidative degradation of aromatic hydrocarbons by microorganisms. I. Enzymatic formation of catechol from benzene. *Biochemistry*, **7**, 2653–2662 (1968).

Gibson, D. T., M. Hensley, H. Yoshioka, and T. J. Mabrys, "Formation of (+)-*cis*-2,3-dihydroxy-1-methyl cyclohexa-4,6-diene from toluene by *Pseudomonas putida. Biochem.*, **9**, 1626–1630 (1970).

Giger, W., P. H. Brunner, and C. Schaffner, "4-Nonylphenol in sewage sludge: Accumulation of toxic metabolites from nonionic surfactants," Science, **225**, 623–625 (1984).

Gilliland, E. R., "Diffusion coefficients in gaseous systems," *Ind. Eng. Chem.*, **26**, 681–685 (1934).

Gmehling, J., P. Rasmussen, and A. Fredenslund, "Vapor–liquid equilibria by UNIFAC group contribution. Revision and extension 2," *Ind. Eng. Chem. Proc. Des. Dev.*, **21**, 118–127 (1982).

Goldfinger, P. and G. Martens, "Elementary rate constants in atomic chlorination reactions. Part 3: Bond dissociation energies and entropies of the activated state," *Faraday Soc. Trans.*, **57**, 2220–2225 (1961).

Goldman, P., "The enzymatic cleavage of the carbon-fluorine bond in fluoroacetate," *J. Biol. Chem.*, **240**, 3434–3438 (1965).

Gordon, J. E. and R. L. Thorne, "Salt effects on the activity coefficient of naphthalene in mixed aqueous electrolyte solutions. I. Mixtures of two salts," *J. Phys. Chem.*, **71**, 4390–4399 (1967a).

Gordon, J. E. and R. L. Thorne, "Salt effects on non-electrolyte activity coefficients in mixed aqueous electrolyte solutions. II. Artificial and natural sea waters," *Geochim. Cosmochin. Acta,*, **31**, 2433–2443 (1967b).

Gottschalk, G., *Bacterial Metabolism*, Springer-Verlag, New York, 1986, 359 pp.

Grain, C. F., "Vapor Pressure," In: W. J. Lyman, W. F. Reehl, and D. H. Rosenblatt (Eds.), *Handbook of Chemical Property Estimation Methods; Environmental Behavior of organic Compounds*, Chapter 14, McGraw-Hill, New York, NY (1982a).

Grain, C. F., "Activity Coefficient." In W. J. Lyman, W. F. Reehl, and D. H. Rosenblatt, Eds., *Handbook of Chemical Property Estimation Methods*; *Environmental Behavior of Organic Compounds*, Chapter 11, McGraw-Hill, New York, NY (1982b).

Gray, P. H. H. and H. G. Thornton, "Soil bacteria that decompose certain aromatic compounds," *Zentralblatt Bakteriologie Parasitenkunde Infektionskrankheiten*, **3**, 74–96 (1928).

Gregg, M. C., "Diapycnal mixing in the thermocline: A review," *J. Geophys. Rev.*, **92**, 5249–5286 (1987).

Grim, R. E., *Clay Mineralogy*, McGraw-Hill, New York, 1968.

Gschwend, P. M., and S.-C. Wu, "On the constancy of sediment–water partition coefficients of hydrophobic organic pollutants," *Environ. Sci. Technol.*, **19**, 90–96 (1985).

Guengerich, F. P. and T. L. MacDonald, "Chemical mechanisms of catalysis by cytochromes P-450: A unified view," *Acc. Chem. Res.*, **17**, 9–16 (1984).

Haag, W. R. and J. Hoigné, "Photo-sensitized oxidation in natural water via HO· radicals," *Chemosphere*, **14**, 1659–1671 (1985).

Haag, W. R. and J. Hoigné, "Singlet oxygen in surface waters. 3. Photochemical formation and steady-state concentrations in various types of waters," *Environ. Sci. Technol.*, **20**, 341–348 (1986).

Haag, W. R., J. Hoigné, E. Gassmann, and A. M. Braun, "Singlet oxygen in surface waters—Part I: Furfuryl alcohol as a trapping agent," *Chemosphere*, **13**, 631–640 (1984a).

Haag, W. R., J. Hoigné, E. Gassmann, and A. M. Braun, "Singlet oxygen in surface waters—Part II: Quantum yields of its production by some natural humic materials as a function of wavelength," *Chemosphere*, **13**, 641–650 (1984b).

Haag, W. R. and T. Mill, "Direct and indirect photolysis of water-soluble azodyes: Kinetic measurements and structure–activity relationship," *Environ. Toxicol. Chem.*, **6**, 359–369 (1987).

Haag, W. R. and T. Mill, "Effect of subsurface sediments on hydrolysis of haloalkanes and epoxides," *Environ. Sci. Technol.*, **22**, 658–663 (1988).

Haigler, B. E., S. F. Nishino, and J. C. Spain, "Degradation of 1,2-dichlorobenzene by a *Pseudomonas sp.*" *Appl. Environ. Microbiol.* **54**, 294–301 (1988).

Håkanson, L. and M. Jansson, *Principles of Lake Sedimentology*, Springer Verlag, Heidelberg, 1983.

Hamilton, D. J., "Gas chromatographic measurement of volatility of herbicide esters," *J. Chromatogr.*, **195**, 75–83 (1980).

Hammel, K. E., B. Kalyanaraman, T. K. Kirk, "Oxidation of polycyclic aromatic hydrocarbons and dibenzo(p) dioxins by *Phanerochaete chrysosporium ligninase.*" *J. Biol. Chem.*, **261**, 16984–16952 (1986).

Hammett, L. P., *Physical Organic Chemistry*, McGraw-Hill, New York, NY (1940).

Hansch, C., J. E. Quinlan, and G. L. Lawrence, "The linear free-energy relationship between partition coefficients and the aqueous solubility of organic liquids," *J. Org. Chem.*, **33**, 347–350 (1968).

Hansch, C. and A. Leo, *Substituent Constants for Correlation Analysis in Chemistry and Biology*, Wiley, New York, 1979.

Hanselmann, K., "Microbially-mediated processes in environmental chemistry," *Chimia*, **40**, 146–159 (1986).

Hanzlik, R. P., "Reactivity and toxicity among halogenated methanes and related compounds. A physicochemical correlate with predictive value," *Biochem. Pharmacol.*, **30**, 3027–3030 (1981).

Haque, R., F. T. Lindstrom, V. H. Freed, and R. Sexton, "Kinetic study of the sorption of 2,4-D on some clays," *Environ. Sci. Technol.*, **2**, 207–211 (1968).

Harris, J. C. and M. J. Hayes, "Acid dissociation constant," In W. J. Lyman W. F. Reehl, and D. H. Rosenblatt, Eds., *Handbook of Chemical Property Estimation Methods; Environmental Behavior of Organic Compounds*, Chapter 6, McGraw-Hill, New York, NY (1982).

Harris, J. M. and C. C. Wamser, "Fundamentals of Organic Reaction Mechanisms," Wiley, New York, 1976.

Harvey, G. R., D. A. Boran, L. A. Chesal, and J. M. Tokar, "The structure of marine fulvic and humic acids," *Marine Chem.*, **12**, 119–132 (1983).

Harvey, G. R. and W. G. Steinhauer, "Atmospheric transport of polychlorinated biphenyls to the North Atlantic," *Atmos. Environ.*, **9**, 777–782 (1973).

Hashimoto, Y., K. Tokura, H. Kishi, and W. M. J. Strachan, "Prediction of seawater solubility of aromatic compounds," *Chemosphere*, **13**, 881–888 (1984).

Hassett, J. J., J. C. Means, W. L. Banwart, S. G. Wood, S. Ali, and A. Khan, ''Sorption of dibenzothiophene by soils and sediments,'' *J. Environ. Qual.*, **9**, 184–186, (1980).

Hayduk, W. and H. Laudie, "Prediction of diffusion coefficients for non-electrolytes in dilute aqueous solutions," *AIChE, J.*, **20**, 611–615 (1974).

Hedegard, J. and I. C. Gunsalus, "Mixed function oxidation. IV. An induced methylene hydroxylase in camphor oxidation," *J. Biol. Chem.*, **240**, 4038–4043 (1965).

Hedges, J. I., "The formation and clay mineral reactions of melanoidins," *Geochim. Cosmochim. Acta*, **42**, 69–76 (1977).

Hedges, J. I. and P. L. Parker, "Land-derived organic matter in surface sediments from the Gulf of Mexico," *Geochim. Cosmochim. Acta*, **40**, 1019–1029 (1976).

Hedges, J. I., H. J. Turin, and J. R. Ertel, "Sources and distributions of sedimentary organic matter in the Columbia River drainage basin, Washington and Oregon," *Limnol. Oceanogr.*, **29**, 35–46 (1984).

Helfferich, F. *Ion Exchange*. McGraw-Hill, New York 1962.

Hendrickson, J. B., D. J. Chram, and G. S. Hammond, *Organic Chemistry*, 3rd Ed., McGraw-Hill, New York, NY (1970).

Henglein A., "Pulse radiolysis and polarogrphy: Electrode reactions of short-lived free radicals." In *Electroanalytical Chemistry*, Vol. 9, A. J. Bard, Ed., Marcel Dekker, New York, 1976, pp. 163–244.

Hermann, R. B., "Theory of hydrophobic bonding. I. The correlation of hydrocarbon solubility in water with solvent cavity surface area," *J. Phys. Chem.*; **76**, 2754–2759 (1972).

Higbie, R., "The rate of adsorption of a pure gas into a still liquid during short periods of exposure," *Trans. Am. Inst. Chem. Eng.*, **35**, 365–389 (1935).

Hill, D. W. and P. L. McCarty, "Anaerobic degradation of selected chlorinated hydrocarbon pesticides," *J. Water Pollution Control Fed.*, **39**, 1259–1277 (1967).

Hine, J., *Physical Organic Chemistry*, McGraw-Hill, New York (1962).

Hine, J. and P. K. Mookerjee, "The intrinsic hydrophilic character of organic compounds. Correlations in terms of structural contributions," *J. Org. Chem.*, **40**, 292–298 (1975).

Hoffman, M., "Catalysis in aquatic environments." In W. Stumm, Ed., *Aquatic Chemical Kinetics*, Wiley-Interscience, New York, pp. 71–111, 1990.

Hoigné, J. and H. Bader, ''Ozonation of water: Kinetics of oxidation of ammonia by ozone and hydroxyl radical,'' *Environ. Sci. Technol.*, **12**, 79–84 (1978).

Hoigné, J., H. Bader, W. Haag, and J. Stahelin, ''Rate constants of reactions of ozone with organic and inorganic compounds in water-III,'' *Water Res.*, **19**, 993–1004 (1985).

Hoigné, J., B. C. Faust, W. R. Haag, F. E. Scully, Jr., and R. G. Zepp, ''Aquatic humic substances as sources and sinks of photochemically produced transient reactants,'' In *Aquatic Humic Substances: Influence on Fate and Treatment of Pollutants*, I. H. Suffet and P. MacCarthy Eds., Advances in Chemistry Series *219*, American Chemical Society, Washington, D.C. 1989, pp. 363-381.

Holmén, K. and P. Liss, "Models for air–water gas transfer: An experimental investigation," *Tellus*, **36B**, 92–100 (1984).

Holmstead, R. L., "Studies of the degradation of mirex with an iron (II) prophyrin model system," *J. Agric. Food Chem.*, **24**, 620–624 (1976).

Horvath, A. L., *Halogenated Hydrocarbons. Solubility–Miscibility with Water*, Marcel Dekker, New York, 1982.

Hsu, T. S. and R. Bartha, "Interaction of pesticide-derived chloroaniline residues with soil organic matter," *Soil Sci.*, **116**, 444–452 (1974).

Hsu, T.-S. and R. Bartha, "Hydrolyzable and nonhydrolyzable 3,4-dichloroaniline-humus complexes and their respective rates of biodegradation," *J. Agric. Food Chem.*, **24**, 118–122 (1976).

Hughes, E. A. M., *The Chemical Statics and Kinetics of Solutions*, Academic, London, 1971.

Hunter-Smith, R. J., P. W. Balls, and P. S. Liss, "Henry's Law constants and the air–sea exchange of various low molecular weight halocarbons gases," *Tellus*, **35B**, 170–176 (1983).

Ilan, Y. A., G. Czapski, and D. Meisel, "The one-electron tansfer redox potential of free radicals: 1. The oxygen/superoxide system," *Biochim. Biophys. Acta*, **430**, 209–244 (1987).

Imboden, D. M., "Limnologische Transport—und Nährstoffmodelle," *Schweiz. Z. Hydrol.*, **35**, 29–68 (1973).

Imboden, D. M., B. S. F. Eid, T. Joller, M. Schuster, and J. Wetzel, "MELIMEX, an experimental heavy-mental pollution study. 2. Vertical mixing in a large limnocorral. *Schweiz. Z. Hydrol.*, **47**, 177–189 (1979).

Imboden, D. M., and S., Emerson, "Natural radon and phosphorus as limnological tracers: Horizontal and vertical eddy diffusion in Greifensee," *Limnol. Oceanogr.*, **23**, 77–90 (1978).

Imboden, D. M. and R. Gächter, "A dynamic lake model for trophic state prediction,"*Ecol. Modelling*, **4**, 77–98 (1978).

Imboden, D. M. and R. P. Schwarzenbach, "Spatial and temporal distribution of chemical substances in lakes: Modeling concepts." In W. Stumm Ed., *Chemical Processes in Lakes*, Wiley-Interscience, New York, 1985, pp. 1–30.

Jafvert, C. T., "Sorption of organic acid compounds to sediments: Initial model development," *Environ. Toxicol. Chem.*, **9**, 1259–1268 (1990).

Jafvert, C. T., J. C. Westall, E. Grieder, and R. P. Schwarzenbach, "Distribution of hydrophobic ionogenic organic compounds between octanol and water: Organic acids," *Environ. Sci. Technol.*, **24**, 1795–1803 (1990).

Jafvert, C. T. and N. L. Wolfe, "Degradation of selected halogenated ethanes in anoxic sediment-water systems," *Environ. Toxicol. and Chem.*, **6**, 827–837 (1987).

Jähne, B., K. H. Fisher, J. Ilmberger, P. Libner, W. Weiss, D. Imboden, U. Lemin, and J. M. Jaquet, "Parameterization of air/lake exchange." In *Gas Transfer at Water Surfaces*, W. Brutsaert and G. H. Jirka, Eds., D. Reidel, Boston, 1984, pp. 459–466.

Jannasch, H. W., "Enrichments of aquatic bacteria in continuous culture," *Arch. Mikrobiol.*, **59**, 165–173 (1967).

Jannasch, H. W., "Competitive elimination of *Enterobacteriaceae* from seawater," *Appl. Microbiol.*, **16**, 1616–1618 (1968).

Jarvis, N. L., W. D. Garrett, M. A. Scheiman, and C. O. Timmons, "Surface chemical characterization of surface-active material in seawater," *Limnol. Oceanogr.*, **12**, 88–96 (1967).

Jassby, A. and T. Powell, "Vertical patterns of eddy diffusion during stratification in Costle Lake, California," *Limnol. Oceanogr.*, **20**, 530–543 (1975).

Jirka, G. H. and W. Brutsaert, " Measurements of wind effects on water–side controlled gas exchange in riverine systems." In *Gas Transfer at Water Surfaces*, W. Brutsaert and G. H. Jirka Eds., D. Reidel, Boston, 1984, pp. 437–446.

Johnson, B. D. and R. C. Cooke, "Bubble populations and spectra in coastal waters: A photographic approach," *J. Geophys. Res.*, **84**, 3761–3766 (1979).

Johnson, C. A. and J. C. Westall, "Effect of pH and KCl concentration on the octanol–water distribution of methyl anilines," *Environ. Sci. Technol.*, **24**, 1869–1875 (1990).

Jones, A. S., "Metabolism of aldicarb by five soil fungi," *J. Agric. Food Chem.*, *24*, 115–117 (1976).

Jung, R. F., R. O. James, T. W. Healy, "Adsorption, precipitation, and electrokinetic processes in the iron oxide (goethite)–oleic acid-oleate system," *J. Colloid Int. Sci.*, **118**, 463–472 (1987).

Junge, C. E., "Basic considerations about trace constituents in the atmosphere as related to the fate of global pollutants." In: *Fate of Pollutants in the Air and Water Environments, Section I, Mechanisms of Interaction Between Environments*, I. H. Suffet (Ed.), Wiley-Interscience, New York, 1977, pp. 7–26.

Kanwisher, J., "Effect of wind on CO_2 exchange across the sea surface," *J. Geophys. Res.*, **68**, 3921–3927 (1963).

Kanwisher, J., "On the exchange of gases between the atmosphere and the sea," *Deep-Sea Res.*, **10**, 195–207 (1963).

Karickhoff, S. W., "Sorption kinetics of hydrophobic pollutants in natural sediments." In *Contaminants and Sediments*, Vol. 2, Analysis, Chemistry, and Biology, R. A. Baker, Ed., Ann Arbor Science, Ann Arbor, MI, 1980, pp. 193–205.

Karickhoff, S. W., "Semi-empirical estimation of sorption of hydrophobic pollutants on natural sediments and soils," *Chemosphere*, **10**, 833–846 (1981).

Karickhoff, S. W., "Organic pollutant sorption in aquatic system," *J. Hydraulic Eng.*, **110**, 707–735 (1984).

Karickhoff, S. W., D. S. Brown, and T. A. Scott, "Sorption of hydrophobic pollutants on natural sediments," *Water Res.*, **13**, 241–248 (1979).

Kearny, P. C., and D. D. Kaufman, Eds., *Herbicides: Chemistry, Degradation, and Mode of Action*, Vol. 2, 2nd ed., Marcel Dekker, New York, 1976.

Kemula, W. and T. M. Krygowski, "Nitro compounds." In: *Encyclopedia of Electrochemistry of the Elements*, Vol. XIII, Bard, A. J., and H. Lund, Eds., Marcel Dekker, New York, 1979.

Keuning, S., D. B. Janssen, and B. Witholt, "Purification and characterization of hydrolytic haloalkane dehalogenase from *Xanthobacter autophicus* GJ10," *J. Bacteriol.*, **163**, 635–639 (1985).

Khan, S. U., *Pesticides in the Soil Environment*, Elsevier, Amsterdam, 1980, 240 pp.

Kirby, A. J., "Hydrolysis and formation of esters of organic acids." In C. H. Brandford and C. F. H. Tipper Eds., *Comprehensive Chemical Kinetics*. Vol. 10, Elsevier, Amsterdam, 1972, pp. 57–207.

Kirby, A. J. and S. G. Warren, *The Organic Chemistry of Phosphorus*, Elsevier, New York, 1967.

Kistiakowsky, W., "Über Verdampfungswärme und einige Gleichungen, welche die Eigenschaften der unassoziierten Flüssigkeiten bestimmen," *Z. Phys. Chem.*, **107**, 65–73 (1923).

Klecka, G. M. and S. J. Gonsior, "Reductive dechlorination of chlorinated methanes and ethanes by reduced iron (II) porphyrins," *Chemosphere*, **13**, 391–402 (1984).

Knackmuss, H.-J., "Degradation of halogenated and sulfonated hydrocarbons." In *Microbial Degradation of Xenobiotics and Recalcitrant Compounds*, T. Leisinger, R. Hütter, A. M. Cook, and J. Nüesch, Eds., Academic New York, 1981, pp. 189–212.

Kohler-Staub, D. and T. Leisinger, "Dichloromethane dehalogenase of *Hyphomicrobium* sp. strain DM2," *J. Bacteriol.*, **162**, 676–681 (1985).

Kolovayev, D. A., "Investigation of the concentration and statistical size distribution of wind-produced bubbles in the near-surface ocean," *Oceanology*, **15**, 659–661 (1976).

Konings, W. N., K. J., Hellingwerf, and G. T. Robillard, "Transport across bacterial membranes." In *Membrane Transport*, S. L. Bonting and J.J.H.H.M. dePont, Eds., Elsevier,, New York, 1981, pp. 257–283.

Kortüm, G., W. Vogel, and K. Andrussow, *Dissociation Constants of Organic Acids in Aqueous Solution*, Butterworths, London, 1961.

Krone, K. E., R. K. Thauer, and H. P. C. Hogenkamp, "Reductive dehalogenation of chlorinated C_1-hydrocarbons mediated by corrinoids," *Biochemistry*, **28**, 4908–4914 (1988a).

Krone, U. E., K. Laufer, and R. K. Thauer, "Coenzyme F_{430} as a possible catalyst for the reductive dehalogenation of chlorinated C_1-hydrocarbons in methanogenic bacteria," *Biochemistry*, **28**, 10061–10065 (1989b).

Kummert, R. and W. Stumm, "The surface complexation of organic acids on hydrous γ-Al_2O_3," *J. Colloid Interface Sci.*, **75**, 373–385 (1980).

Kung, K.-H. and M. B. McBride, "Adsorption of para-substituted benzoates on iron oxides," *Soil Sci. Soc. Am., J.*, **53**, 1673–1678 (1989).

Laidler, K. J., *Chemical Kinetics*, McGraw-Hill, New York, 1965.

Larson, R. and D. Davidson, "Acclimation to and biodegradation of NTA at trace concentrations in natural waters," *Water Res.*, **16**, 1597–1604 (1982).

Larson, R. A. and R. G. Zepp, "Reactivity of the carbonate radical with aniline derivatives," *Environ. Toxicol. Chem.*, **7**, 265–274 (1988).

Larson, R. J., G. G. Clinckemaillie, and L. VanBelle, "Effect of temperature and dissolved oxygen on biodegradation and nitrilotriacetate," *Water Res.*, **15**, 615–620 (1981).

Ledbetter, M., "Langmuir circulations and plankton patchiness," *Ecol. Modeling*, **7**, 289–310 (1979).

Ledwell, J. J., "The variation of the gas transfer coefficient with molecular diffusivity." In *Gas Transfer at Water Surfaces*, W. Brutsaert and G. H. Jirka, Eds., Reidel, Boston, 1984, pp. 293–302.

Leenheer, J. A. and J. L. Ahlrichs, "A kinetic and equilibrium study of the adsorption of carbaryl and parathion upon soil organic matter surfaces," *Soil Sci. Soc. Am., Proc.*, **35**, 700–704 (1971).

Lehninger, A. L., *Biochemistry*, Worth, Publishers Inc., New York, 1970.

Leifer, A., *The Kinetics of Environmental Aquatic Photochemistry*, ACS Professional Reference Book, American Chemical Society, Washington, DC, 1988.

Leinonen, P. J., and D. Mackay, "The multicomponent solubility of hydrocarbons in water," *Can. J. Chem. Eng.*, **51**, 230–233 (1973).

Lemaire, J., J. A. Guth, O. Klais, J. Leahey, W. Merz, J. Philp, R. Wilmes, and C. J. M. Wolff, "Ring test of a method for assessing the phototransformation of chemicals in water," *Chemosphere*, **14**, 53–77 (1985).

Leo, A., C. Hansch, and D. Elkins, "Partition coefficients and their uses," *Chem. Rev.*, **71**, 525–553 (1971).

Lerman, A., *Geochemical Processes. Water and Sediment Environments*, Wiley, New York, 1979.

Leuenberger, C., M. P. Ligocki, and J. F. Pankow, "Trace organic compounds in rain. 4. Identities, concentrations, and scavenging mechanisms for phenols in urban air and rain," *Environ. Sci. Technol.*, **19**, 1053–1058 (1985).

Leversee, G. J, P. F. Landrum, J. P. Giesy, and T. Fannin, "Humic acids reduce bioaccumulation of some polycyclic aromatic hydrocarbons," *Can. J. Fish. Aquat. Sci.*, **40**, (Suppl. 2), 63–69 (1983).

Lewis, G. N., "The law of physico-chemical change" *Proc. Am. Acad.*, **37**, 49–69 (1901).

Lewis, G. N., "Das Gesetz physiko-chemischer Vorgänge" Z. Phys. Chem., **38**, 205–226 (1901).

Lewis, G. N. and M. R. Randall, *Thermodynamics*, 2nd ed., McGraw-Hill, New York, 1961, 723 pp.

Leyder, F. and P. Boulanger, Ultraviolet absorption, aqueous solubility, and octanol–water partition coefficients for several phthalates," *Bull. Environ. Contam. Toxicol.*, **30**, 152–157 (1983).

Liao, W., R. F. Christman, J. D. Johnson, D. S. Millington, and J. R. Hass, "Structural characterization of aquatic humic material," *Environ. Sci. Technol.*, **16**, 403–410 (1982).

Ligocki, M. P. and J. F. Pankow, "Measurements of the gas/particle distribution of atmospheric organic compounds," *Environ Sci. Technol.*, **23**, 75–83 (1989).

Lincoff, A. H. and J. M. Gossett, "The determination of Henry's Law constant for volatile organics by equilibrium partitioning in closed systems. In W. Brutsaert, and G. H. Jirka Eds., *Gas Transfer at Water Surfaces*, Reidel, Publishing Co., Germany, 1984, pp. 17–25.

Lion, L. W., T. B. Stauffer, and W. G. MacIntyre, "Sorption of hydrophobic compounds on aquifer materials: Analysis methods and the effect of organic carbon," *J. Contam. Hydrol.*, **5**, 215–234 (1990).

Liss, P. S., "Processes of gas exchange across an air–water interface," *Deep Sea Res.*, **20**, 221–238 (1973).

Liss, P. S. and P. G. Slater, "Flux of gases across the air-sea interface," *Nature*, **247**, 181–184 (1974).

Little, C. D., A. V. Palumbo, S. E. Herbes, M. E. Lidstrom, R. L. Tyndall, and P. J. Gilmer, "Trichloroethylene biodegradation by a methane-oxidizing bacterium," *Appl. Environ. Microbiol.*, **54**, 951–956 (1988).

Lopez-Avila, V. and R. A. Hites, "Oxidation of phenolic antioxidants in a river system," *Environ. Sci. Technol.*, **15**, 1386–1388 (1981).

Lövgren, L., "Complexation reactions of phthalic acid and aluminium (III) with the surface of goethite," *Geochim. Cosmochim. Acta*, **55**, 3639–3645 (1991).

Lowry, T. H. and K. Schueller-Richardson, *Mechanisms and Theory in Organic Chemistry*, 2nd ed., Harper and Row, New York, 1981.

Lyman, W. J., "Octanol/water partition coefficient." In W. J. Lyman, W. F. Reehl, and D. H. Rosenblatt, Eds., *Handbook of Chemical Property Estimation Methods*; *Environmental Behavior of Organic Compounds*, Chapter 1, McGraw-Hill, New York, NY (1982a).

Lyman, W. J., "Solubility in water." In W. J. Lyman, W. F. Reehl, and D. H. Rosenblatt, Eds., *Handbook of Chemical Property Estimation Methods*; *Environmental Behavior of Organic Compounds*, Chapter 2, McGraw-Hill, New York, NY (1982b).

Mabey, W. and T. Mill, "Critical review of hydrolysis of organic compounds in water under environmental conditions," *J. Phys. Ref. Data*, **7**(2), 383–415 (1978).

Mabey, W. R., D. Tse, A. Baraze, and T. Mill, "Photolysis of nitroaromatics in aquatic systems. I. 2,4,6,-Trinitrotoluene," *Chemosphere*, **12**, 3–16 (1983).

Macalady, D. L., P. G. Tratnyek, and T. J. Grundl, "Abiotic reduction reactions of anthropogenic organic chemicals in anaerobic systems: A critical review," J. Contam. Hydrol., **1**, 1–28 (1986).

Macalady, D. L., P. G. Tratnyek, and N. L. Wolfe, "Influence of natural organic matter on the abiotic hydrolysis of organic contaminants in aqueous systems." In I. M. Suffet, and P. MacCarthy, Eds., *Aquatic Humic Substances. Influence of Fate and Treatment of Pollutants*, Advances in Chemistry Series 219, American Chemical Society, .Washington, DC, 1989.

Mackay, D., A, Bobra, D. W. Chan, and W. Y. Shiu, "Vapor pressure correlations for low-volatility environmental chemicals," *Environ. Sci. Technol.*, **16**, 645–649 (1982).

Mackay, D., A. Bobra, W. Y. Shiu, and S. H. Yalkowsky, "Relationships between aqueous solubility and octanol–water partition coefficients," *Chemosphere*, **9**, 701–711 (1980a).

Mackay, D., R. Mascarenhas, W. Y. Shiu, "Aqueous solubility of polychlorinated biphenyls," *Chemosphere*, **9**, 257–264 (1980b).

Mackay, D. and W. Y. Shiu, "Aqueous solubility of polynuclear aromatic hydrocarbons," *J. Chem. Eng. Data*, **22**, 399–402 (1977).

Mackay, D. and W. Y. Shiu, "A critical review of Henry's Law constants for chemicals of environmental interest," *J. Phys. Chem. Ref. Data*, **10**, 1175–1199 (1981).

Mackay, D., W. Y. Shiu, and R. P. Sutherland, "Determination of air–water Henry's Law constants for hydrophobic pollutants," *Environ. Sci. Technol.*, **13**, 333–337 (1979).

Mackay, D. and A. T. K. Yeun, "Mass transfer coefficients correlations for volatilization of organic solutes from water," *Environ Sci. Technol.*, **17**, 211–233 (1983).

Macknick, A. B., and J. M. Prausnitz, "Vapor pressures of high-molecular-weight hydrocarbons," *J. Chem. Eng. Data*, **24**, 175–178 (1979).

Mahler, H. R. and E. H. Cordes, *Biological Chemistry*, 2nd ed., Harper and Row, New York, 1971, 1099 pp.

Maloney, S. E., A. Maule, and A. R. W. Smith, "Microbial transformation of pyrethroid insecticides: Permethrin, deltamethrin, fastac, fenvalerate, and fluvalinate," *Appl. Environ. Microbiol.*, **54**, 2874–2876 (1988).

Mannervik, B., "The isoenzymes of glutathione transfers," *Adv. Enzymol.*, **57**, 357–417 (1985).

March, J., *Advanced Organic Chemistry*, 3rd ed., McGraw-Hill New York, 1985.

Marks, T. S., J. D. Allpress, and A. Maule, "Dehalogenation of lindane by a variety of porphyrins and corrins," *Appl. Environ. Microbiol.*, **55**, 1258–1261 (1989).

Marple, L., R. Brunck, and L. Throop, "Water solubility of 2,3,7,8-tetrachlorodibenzo-*p*-dioxin," *Environ. Sci. Technol.*, **20**, 180–182 (1986a).

Marple, L., B. Berridge, and L. Throop, "Measurement of the water-octanol partition coefficient of 2,3,7,8-tetrachlorodibenzo-*p*-dioxin," *Environ. Sci. Technol.*, **20**, 397–399 (1986b).

Martell, A. E. and R. M. Smith, *Critical Stability Constants*, Vol. 3, Plenum, New York, 1977.

May, W. E., "The solubility behavior of polycyclic aromatic hydrocarbons in aqueous systems." In L. Petrakis and F. T. Weiss, Eds., *Petroleum in the Marine Environment*, Chapter 7, Advances in Chemistry Series No. 185, American Chemical Society, Washington, DC, 1980, pp. 143–192.

May, W. E., S. P. Wasik, and D. H. Freeman, "Determination of the aqueous solubility of polynuclear aromatic hydrocarbons by a coupled column liquid chromatographic techniques," *Anal. Chem.*, **50**, 175–179 (1978a).

May, W. E., S. P. Wasik, and D. H. Freeman, "Determination of the solubility behavior of some polycyclic aromatic hydrocarbons in water," *Anal. Chem.*, **50**, 997–1000 (1978b).

May, W. E., S. P. Wasik, M. M. Miller, Y. B. Tewari, J. M. Brown-Thomas, and R. N. Goldberg, "Solution thermodynamics of some slightly soluble hydrocarbons in water," *J. Chem. Eng. Data*, **28**, 197–200 (1983).

McAuliffe, C. D., "Solubility in water of paraffin, cycloparaffin, olefin, acetylene, cycloolefin, and aromatic hydrocarbons," *J. Phys. Chem.*, **70**, 1267–1275 (1966).

McAuliffe, C. D., "GC determination of solutes by multiple phase equilibration," *Chem. Tech.*, **1**, 46–51 (1971).

McDevit, W. F., and F. A. Long, "The activity coefficient of benzene in aqueous salt solutions," *J. Am. Chem. Soc.*, **74**, 1773–1777 (1952).

McGovern, E. W., "Chlorohydrocarbons solvents," *Ind. Eng. Chem.*, **35**, 1230–1239 (1943).

Means, J. C., S. G. Wood, J. J. Hassett, and W. L. Banwart, "Sorption of polynuclear aromatic hydrocarbons by sediments and soils," *Environ. Sci. Technol.*, **14**, 1524–1528 (1980).

Meisel, D. and G. Czapski, "One-electron transfer equilibria and redox potentials of radicals studied by pulse radiolysis," *J. Phys. Chem.*, **79**, 1503–1509 (1975).

Meisel, D. and P. Neta, "One-electron redox potentials of nitro compounds and radiosensitizers. Correlation with spin densities of their radical anions," *J. Am. Chem. Soc.*, **97**, 5198–5203 (1975).

Meyer, H., "Zur Theorie der Alkohol-narkose. I. Welche Eigenschaft der Anesthetica bedingt ihre narkotische Wirkung," *Arch. Exp. Pathol. Pharmakol.*, **42**, 109–118 (1899).

Mihelcic, J. R. and R. G. Luthy, "Degradation of polycyclic aromatic hydrocarbon compounds under various redox conditions in soil–water systems," *Appl. Environ. Microbiol.*, **54**, 1182–1187 (1988).

Mikhail, R. S., S. Brunauer, and E. E. Broder, "Investigations of a complete pore structure analysis. I. Analysis of macropores," *J. Colloid Int. Sci.*, **26**, 45–53 (1968a).

Mikhail, R. S., S. Brunauer, and E. E. Broder, "Investigations of a complete pore structure analysis. II. Analysis of four silica gels," *J. Colloid Int. Sci.*, **26**, 54–61 (1968b).

Mill, T., "Structure–activity relationships for photooxidation processes in the environment," *Environ. Toxicol. Chem.*, **8**, 31–43 (1989).

Mill, T. and W. Mabey, "Photochemical transformations." In W. B. Neely and G. E. Blau, Eds., *Environmental Exposure from Chemicals*, Vol. 1, CRC Press, Boca Raton, FL 1985.

Mill, T., W. M. Mabey, B. K. Lan, and A. Baraze, "Photolysis of polycyclic aromatic hydrocarbons in water," *Chemosphere*, **10**, 1281–12 (1981).

Miller, D. H., *Water at the Surface of the Earth. An Introduction to Ecosystem Hydrodynamics*, *Academic*, New York, 1977.

Miller, G. C. and R. G. Zepp, "Photoreactivity of aquatic pollutants sorbed on suspended sediments," *Environ. Sci. Technol.*, **13**, 860–853 (1979a).

Miller, G. C. and R. G. Zepp, "Effects of suspended sediments on photolysis rates of dissolved pollutants," *Water Res.*, **13**, 543–459 (1979b).

Miller, M. M., S. Ghodbane, S. P. Wasik, Y. B. Tewari, and D. E. Martire, "Aqueous solubilities, octanol/water partition coefficients, and entropies of melting of chlorinated benzenes and biphenyls," *J. Chem. Eng. Data*, **29**, 184–190 (1984).

Mills, A. C. and J. W. Biggar, "Solubility–temperature effect on the adsorption of gamma- and beta-BHC from aqueous and hexane solutions by soil materials," *Soil Sci. Soc. Am. Proc.*, **33**, 210–216 (1969a).

Mills, A. C. and J. W. Biggar, "Adsorption of 1,2,3,4,5,6-hexachlorocyclohexane from solution: The differential heat of adsorption applied to adsorption from dilute solutions on organic and inorganic surfaces," *J. Coll. Int. Sci.*, **29**, 720–731 (1969b).

Molina, M. J., and F. S. Rowland, "Stratospheric sink for chlorofluoromethanes: Chlorine atom-catalyzed destruction of ozone," *Nature* (*London*), **249**, 810–812 (1974).

Monahan, E. C., "Oceanic whitecaps," *J. Phys. Oceanogr.* **1**, 139–144 (1971).

Monod, J., "The growth of bacterial cultures," *Ann. Rev. Microbiol.*, **3**, 371–394 (1949).

Monsanto Company, Physical Property Research, S. C. Cheng, F. E. Hileman, and J. M. Schroy, Dayton, OH and St. Louis, MO, 1983 to 1984.

Morel, F. M. M., *Principles of Aquatic Chemistry*, Wiley-Interscience, New York, 1983, 446 pp.

Morris, K. R., R. Abramowitz, R. Pinal, P. Davis, and S. H. Yalkowsky, "Solubility of aromatic pollutants in mixed solvents," *Chemosphere*, **17**, 285–298 (1988).

Muhlmann, R. and A. Schrader, "Hydrolyse der insektiziden Phosphorsäurester," *Z. Naturforsch*, **12b**, 196–208 (1957).

Munk, W., "Abyssal recipes," *Deep-Sea Res.*, **13**, 707–730 (1966).

Munnecke, D. M. "Enzymatic hydrolysis of organophosphate insecticides, a possible pesticide disposal method," *Appl. Environ. Microbiol.*, **32**, 7–13 (1976).

Münnich, K. O., W. B. Clarke, K. H. Fischer, D. Flothmann, B. Kromer, W. Roether, U. Siegenthaler, Z. Top, and W. Weiss, "Gas exchange and evaporation studies in a circular wind tunnel, continuous radon-222 measurements at sea, and tritium/helium-3 measurements in a lake." In H. Favre and K. Hasselmann, Eds., *Turbulent Fluxes through the Sea Surface, Wave Dynamics and Predictions*, Plenum, New York, 1978, pp. 151–165.

Munz, C. H. and P. V. Roberts, "The effects of solute concentration and cosolvents on the aqueous activity coefficients of low molecular weight halogenated hydrocarbons," *Environ. Sci. Technol.* 20, 830–836 (1986).

Munz, C. H. and P. V. Roberts, "Air–water phase equilibria of volatile organic solutes," *J. AWWA*, **79**, 62–69 (1987).

Murphy, T. J., J. C. Pokojowczyk, and M. D. Mullin, "Vapor exchange of PCBs with Lake

Michigan: The atmosphere as a sink for PCBs." In *Physical Behavior of PCBs in the Great Lakes*, D. Mackay, S. Patterson, S. J. Eisenreich, and M. S. Simmons, Eds., Ann Arbor Science Ann Arbor, MI, 1983, pp. 49–58.

Neely, W. B., G. E. Blau, T. Alfrey, Jr., "Mathematical models predict concentration–time profiles resulting from chemical spill in a river," *Environ. Sci. Technol.*, **10**, 72–76 (1976).

Neta, P. "Redox properties of free radicals," *J. Chem. Ed.*, **58**, 110–113 (1981).

Neta, P. and D. Meisel, "Substituent effects of nitroaromatic radical anions in aqueous solution," *J. Phys. Chem.*, **80**, 519–524 (1976).

Nicholson, B., B. P. Maguire, and D. B. Bursill, "Henry's Law constants for the trihalomethanes: Effects of water composition and temperature," *Environ. Sci. Technol.*, **18**, 518–521 (1984).

Nikaido, H., "Nonspecific transport through the outer membrane." In *Bacterial Outer Membranes, Biogenesis and Functions*, M. Inouye, Ed., Wiley, New York 1979, pp. 361–407.

Nkedi-Kizza, P., P. S. C. Rao, and A. G. Hornsby, "Influence of organic cosolvents on sorption of hydrophobic organic chemicals by soils," *Environ. Sci. Technol.*, **19**, 975–979 (1985).

Nys, G. G. and R. F. Rekker, "Statistical analysis of a series of partition coefficients with special reference to the predictability of folding drug molecules. The introduction of hydrophobic fragment constants (f values)," *Chim. Therap.*, **8**, 521–535 (1973).

O'Connor, D. J. and W. E. Dobbins, "Mechanisms of reaeration in natural streams," *Trans. Am. Soc. Civ. Eng.*, **123**, 641–684 (1958).

O'Melia, C. R., "The influence of coagulation and sedimentation on the fate of particles, associated pollutants, and nutrients in lakes." In W. Stumm, Ed., *Chemical Processes in Lakes*, Wiley, New York, 1985, p. 207–224.

OECD, *OECD Guidelines for Testing of Chemicals*, Section 104–105, Organization of Economic Cooperation and Development, Paris, 1981a.

OECD, *Guidelines for Testing of Chemicals*, No. 107, Organization of Economic Cooperation and Development 1981b.

Ogram , A. V., R. E. Jessup, L. T. Ou, and P. S. C. Rao, "Effects of sorption on biological degradation rates of (2,4-dichlorophenoxy) acetic acid in soils," *Appl. Environ. Microbiol.*, **49**, 582–587 (1985).

Okubo, A., "Oceanic diffusion diagrams," *Deep-Sea Res.*, *18*, 789–802 (1971).

Ong, J. H. and C. E. Castro, "Oxidation of iron (II) porphyrins and hemoproteins by nitro aromatics," *J. Am. Chem. Soc.*, **99**, 6740–6745 (1977).

Osborn T. R., and C. S. Cox, "Oceanic fine Structure," *Geophys. Fluid Dyn.*, **3**, 321–345 (1972).

Oser, B. L., Ed., *Hawk's Physiological Chemistry*, 14th ed., McGraw-Hill, New York, 1965.

Othmer, D. F. and M. S. Thakar, "Correlating diffusion coefficients in liquids," *Ind. Eng. Chem.*, **45**, 589–593 (1953).

Overton, E., "Üeber die allgemeinen osmotischen Eigenschaften der Zelle, ihre vermutlichen Ursachen und ihre Bedeutung für die Physiologie," *Vierteljahresschr. Naturforsch. Ges. Zurich.*, **44**, 88–135 (1899).

Pankow, J. F., "Magnitude of artifacts caused by bubbles and headspace in the determination of volatile compounds in water," *Anal. Chem.*, **58**, 1822–1826 (1986).

Paris, D. F., D. L. Lewis, and N. L. Wolfe, "Rates of degradation of malathion by bacteria isolated from aquatic systems," *Environ. Sci. Technol.*, **9**, 135–138 (1975).

Paris, D. F., W. C. Steen, G. L. Baughman, and J. T. Barnett, Jr., "Second-order model to predict microbial degradation of organic compounds in natural waters," *Appl. Environ. Microbiol.*, **41**, 603–609 (1981).

Parks, G. A., "The isoelectric points of solid oxides, solid hydroxides, and aqueous hydroxo complex systems," *Chem. Rev.*, **65**, 177–198 (1965).

Parlar, H., "Photochemistry at surfaces and interfaces." In O. Hutzinger, Ed., *The Handbook of Environmental Chemistry*, Vol. 2, Springer, Berlin, 1980.

Parr, J. F. and S. Smith, "Degradation of toxaphene in selected anaerobic soil environments," *Soil Sci.*, **121**, 52–57 (1976).

Pauling, L. *The Nature of the Chemical Bond.* Cornell University Press, Ithaca, New York, 1960.

Pavelich, W. H. and R. W. Taft, Jr., "The evaluation of inductive and steric effects on reactivity. The methoxide ion-catalyzed rates of methanolysis of *l*-menthyl esters in methanol," *J. Am. Chem. Soc.*, **79**, 4935–4940 (1957).

Pavlostathis, S. G. and K. Jaglal, "Desorptive behavior of trichloroethylene in contaminated soil," *Environ. Sci. Technol.*, **25**, 274–279 (1991).

Pearson, C. R., "C_1- and C_2-Halocarbons." In O. Hutzinger (Ed.), *The Handbook of Environmental Chemistry*, Vol. 3, Part B, Springer, Berlin, pp. 69–88 (1982a).

Pearson, C. R., "Halogenated aromatics," In O. Hutzinger (Ed.), *The Handbook of Environmental Chemistry*, Vol. 3, Part B, Springer, Berlin, pp. 89–116 (1982b).

Peng, T.-H., W. S. Broecker, G. G. Mathieu, Y.-H. Li, and A. E. Bainbridge, "Radon evasion rates in the Atlantic and Pacific Oceans as determined during the GEOSECS program," *J. Geophys. Res.*, **84**, 2471–2586 (1979).

Penman, H. L., "Natural evaporation from open water, bare soil, and grass," *Proc. Roy. Soc., London. Ser. A.*, **193**, 120–146 (1948).

Perrin, D. D., *Dissociation Constants of Organic Bases in Aqueous Solution: Supplement 1972*, Butterworths, London, 1972.

Perrin, D. D., "Prediction of pK_a values." In S. H. Yalkowsky, A. A. Sinkula, and S. C. Valvani, Eds., *Physical Chemical Properties of Drugs*, Dekker, New York, 1980.

Picel, K. C., V. C. Stamoudis, and M. S. Simmons, "Distribution coefficients for chemical components of a coal-oil/water system," *Water Res.*, **22**, 1189–1199 (1988).

Pierotti, G. J., C. H. Deal, and E. L. Derr, "Activity coefficients and molecular structure," *Ind. Eng. Chem.*, **51**, 95–102 (1959).

Piorr, R., "Structure and Application of Surfactants," In: J. Falbe (Ed.), *Surfactants in Consumer Products*, Springer, Heidelberg, pp. 5–22 (1987).

Piwoni, M. D. and P. Banerjee, "Sorption of vòlatile organic solvents from aqueous solution onto subsurface solids," *J. Contam. Hydrol.*, **4**, 163–179 (1989).

Prausnitz, J. M., *Molecular Thermodynamics of Fluid-Phase Equilibria*, Prentice-Hall, Englewood Cliffs, NJ, 1969.

Pretsch, E., J. T. Clerc, J. Seibl and W. Simon, *Spectral Data for Structure Determination of Organic Compounds*, Springer, Berlin, 1983.

Przyjazny, A., W. Janicki, W. Chrzanowski, and R. Staszewki, "Headspace gas chromatographic determination of distribution coefficients of selected organosulphur compounds and their dependence of some parameters," *J. Chromatogr.*, **280**, 249–260 (1983).

Quirke, J. M. E., A. S. M. Marci, and G. Eglinton, "The degradation of DDT and its degradative products by reduced iron(II) porphyrins and ammonia," *Chemosphere*, **3**, 151–155 (1979).

Ramanand, K., M. Sharmila, and N. Sethunathan, "Mineralization of carbofuran by a soil bacterium," *Appl. Environ. Microbiol.*, **54**, 2129–2133 (1988).

Rao, P. S. C., R. E. Jessup, and T. M. Addiscott, "Experimental and theoretical aspects of solute diffusion in spherical and nonspherical aggregates," *Soil Sci.*, **133**, 342–349 (1982).

Rao, P. S. C., R. E. Jessup, D. E. Rolston, J. M. Davidson, and D. P. Kilcrease, "Experimental and mathematical description of nonadsorbed solute transfer by diffusion in spherical aggregates," *Soil Sci. Soc. Am. J.*, **44**, 684–688 (1980).

Rapaport, R. A. and S. J. Eisenreich, "Chromatographic determination of octanol–water partition coefficients for 58 polychlorinated biphenyl congeners," *Environ. Sci. Technol.*, **18**, 163–170 (1984).

Rathbun, R. E. and D. Y. Tai, "Gas–film coefficients for streams," *J. Environ. Eng. ASCE*, **109**, 1111–1127 (1983).

Rathbun, R. E. and D. J. Tai, "Volatilization of organic compounds from streams," *J. Environ. Eng. Div., ASCE*, **108**, 973–989 (1982).

Ratledge, C., "Microbial conversions of alkanes and fatty acids," *A. Am. Oil Chem. Soc.*, **61**, 447–453 (1984).

Rechsteiner, C. E., Jr., "Boiling point." In W. J. Lyman, W. F. Reehl, and D. H. Rosenblatt, Eds., *Handbook of Chemical Property Estimation Methods; Environmental Behavior of Organic Compounds*, Chapter 12, McGraw-Hill, New York, NY (1982).

Reichardt, P. B., B. L. Chadwick, M. A. Cole, B. R. Robertson, and D. K. Button, "Kinetic study of the biodegradation of biphenyl and its monochlorinated analogues by a mixed marine microbial community," *Environ. Sci. Technol.*, **15**, 75–79 (1981).

Reid, E. E., *Organic Chemistry of Bivalent Sulfur*, Vol. I, Chemical Publishing, New York, 1958.

Reid, R. C., J. M. Prausnitz, and T. K. Sherwood, *The Properties of Gases and Liquids*, 3rd ed., McGraw-Hill, New York, 1977, 688 pp.

Reineke, W. and H.-J. Knackmuss, "Microbial metabolism of haloaromatics: Isolation and properties of chlorobenzene-degrading bacterium," *Appl. Environ. Microbiol.*, **47**, 395–402 (1984).

Reineke, W. and H.-J. Knackmuss, "Microbial degradation of haloaromatics" *Ann. Rev. Microbiol.*, **42**, 263–287 (1988).

Reiner, A. M., "Metabolism of benzoic acid by bacteria: 3,5-cyclohexadiene-1,2-diol-1-carboxylic acid is an intermediate in the formation of catechol," *J. Bacteriol.*, **108**, 89–94 (1971).

Reiner, A. M., and G. D. Hegeman, "Metabolism of benzoic acid by bacteria: Accumulation of (-)-3,5-cyclohexadiene-1,2-diol-1,2-carboxylic acid by a mutant strain of Alcaligenes eutrophus," *Biochemistry*, **10**, 2530–2536 (1971).

Rekker, R. F., *The Hydrophobic Fragment Constant*, Elsevier, New York, 1977.

Reynolds, O., "On the dynamical theory of incompressible viscous fluids and the determination of the criterion," *Phil. Trans. Roy. Soc.* **186**, 123 (1894).

Riddick, J. A. and W. B. Bunger, *Organic Solvents*, Wiley, New York, 1970.

Rijnaarts, H. H. M., A. Bachmann, J. C. Jumelet, and A. J. B. Zehnder, "Effect of desorption and intraparticle mass transfer on the aerobic biomineralization of α-hexachlorocyclohexane in a contaminated calcareous soil," *Environ. Sci. Technol.*, **24**, 1349–1354 (1990).

Rittmann, B. E., P. L., McCarty, and P. V. Roberts, "Trace-organics biodegradation in aquifer recharge," *Ground Water*, **18**, 326–343 (1980).

Robbins, J. A., "A model for particle-selective transport of tracers in sediments with conveyor belt deposit feeders," *J. Geophys. Res.*, **91**, 8542–8558 (1986).

Roberts, P. V. and P. G. Dändliker, "Mass transfer of volatile organic contaminants from aqueous solution to the atmospheric during surface aeration. *Environ. Sci. Technol.*, **17**, 484–489 (1983).

Roberts, P. V., M. N. Goltz, and D. M. Mackay, "A natural gradient experiment on solute transport in a sand aquifer. 3. Retardation estimates and mass balances for organic solutes," *Water Resources Res.*, **22**, 2047–2058 (1986).

Rohwer, E., "Evaporation from free water surfaces," US Department of Agriculture, Technical Bulletin No. 271, 1931.

Roof, A. A. M. "Basic principles of environmental photochemistry," In O. Hutzinger, Ed., *The Handbook of Environmental Chemistry*, Vol. 2, Part B, Springer, Berlin, 1982.

Rosenberg, A. and M. Alexander, "Microbial cleanage of various organophosphorus insecticides," *Appl. Environ. Microbiol.*, **37**, 886–891 (1979).

Rossi, S. S. and W. H. Thomas, "Solubility behavior of three aromatic hydrocarbons in distilled water and natural seawater," *Environ. Sci. Technol.*, **15**, 715–716 (1981).

Rounds, S. A. and J. F. Pankow, "Application of a radial diffusion model to describe gas/particle sorption kinetics," *Environ. Sci. Technol.*, **24**, 1378–1386 (1990).

Russow, J. "Fluorocarbons," in: O. Hutzinger, Ed., *The Handbook of Environmental Chemistry*, Vol. 3, Part A, Springer, Berlin, pp. 133–148 (1980).

Sada, E., S. Kito, and Y. Ito, "Solubility of toluene in aqueous salt solutions," *J. Chem. Eng. Data*, **20**, 373–375 (1975).

Santschi, P. H., S. Bollhalder, S. Zingg, A. Lück, and K. Farrenkothen, "The self-cleaning capacity of surface waters after radioactive fallout. Evidence from European waters after Chernobyl, 1986–1988," *Environ. Sci. Technol.*, **24**, 519–527 (1990).

Satterfield, C. N., C. K. Colton, and W. H. Pitcher, Jr., "Restricted diffusion in liquids within fine pores," *AIChE J.*, **19**, 628–635 (1973).

Schecker, H. and G. Schulz, "Untersuchungen zur Hydratationskinetik von Formaldehyd in wässriger Lösung," *Z. Phys., Chem. Neue Folge*, **65**, 221–224 (1969).

Schellenberg, K., C. Leuenberger, and R. P. Schwarzenbach, "Sorption of chlorinated phenols by natural sediments and aquifer materials," *Environ. Sci. Technol.*, **18**, 652–657 (1984).

Schindler, P. W. and W. Stumm, "The surface chemistry of oxides, hydroxides, and oxide minerals." In *Aquatic Surface Chemistry*, W. Stumm, Ed., Wiley-Interscience, New York, 1987, pp. 83–110.

Schlegel, H. G. *General Microbiology*, Cambridge University Press, Cambridge, UK, 1988, pp. 587.

Schmidt, W., "Wirkungen der ungeordneten Bewegungen im Wasser der Meere und Seen." *Ann. Hydrogr. Marit. Meteorol.*, **45**, 367–381 (1917).

Schnitzer, M. and S. U. Khan, *Humic Substances in the Environment*, Dekker, New York, 1972.

Schoder, A., *Metal Exchange of Porphyrin Complexes* (in German), Ph.D. Thesis No. 5491, Swiss Federal Institute of Technology (ETH), Zurich, 1975.

Scholtz, R., A. Schmuckle, A. M. Cook, and T. Leisinger, "Degradation of eighteen 1-monohaloalkanes by *Arthobacter* sp. strain HA1," *J. Gen Microbiol.*, **133**, 267–274 (1987a).

Scholtz, R., T. Leisinger, F. Suter, and A. M. Cook, "Characterization of 1-chlorohexane halidohydrolase, a dehalogenase of wide substrate range from an Arthrobacter sp.," *J. Bacteriol.*, **169**, 5016–5021 (1987b).

Schwarzenbach, R. P., *Assessing the Behavior and Fate of Hydrophobic Organic Compounds in the Aquatic Environment—General Concepts and Case Studies Emphasizing Volatile Halogenated Hydrocarbons*, Habilitation Thesis, Swiss Federal Institute of Technology, Zurich, 1983, 132 pp.

Schwarzenbach, R. P., W. Giger, C. Schaffner, and O. Wanner, "Groundwater contamination by volatile halogenated alkanes: Abiotic formation of volatile sulfur compounds under anaerobic conditions," *Environ. Sci. Technol.*, **19**, 322–327 (1985).

Schwarzenbach, R. P., R. Stierli, K. Lanz, and J. Zeyer, "Quinone and iron porphyrin mediated reduction of nitroaromatic components in homogeneous aqueous solution," *Environ. Sci. Technol.*, **24**, 1566–1574 (1990).

Schwarzenbach, R. P. and J. Westall, "Transport of nonpolar organic compounds from surface water to groundwater: Laboratory sorption studies," *Environ. Sci. Technol.*, **15**, 1360–1367 (1981).

Scully, F. E., Jr. and J. Hoigné, "Rate constants for reactions of singlet oxygen and phenols and other compounds in water," *Chemosphere*, **16**, 681–694 (1987).

Segel, I. H., *Enzyme Kinetics. Behavior and Analysis of Rapid Equilibrium and Steady-State Enzyme Systems.* Wiley, New York, 1975, 957 pp.

Senior, E., A. T. Bull, and J. H. Slater, "Enzyme evolution in a microbial community growing on the herbicide Dalapon," *Nature*, **263**, 476–479 (1976).

Serjeant, E. P. and B. Dempsey, *Ionization Constants of Organic Acids in Aqueous Solution*, Pergamon, New York, 1979.

Setschenow, J., "Über die Konstitution der Salzlösungen auf Grund ihres Verhaltens zu Kohlensäure," *Z. Phys. Chem., Vierter Band*, **1**, 117–125 (1889).

Shinoda, K., "Iceberg formation and solubility," *J. Phys. Chem.*, **81**, 1300–1302 (1977).

Shorter, J., *Correlation Analysis in Organic Chemistry: An Introduction to Linear Free-Energy Relationships*, Clarendon Press, Oxford, 1973.

Sigleo, A. C., T. C. Hoering, and G. R. Helz, "Composition of estuarine colloidal organic matter: Organic components," *Geochim. Cosmochim. Acta*, **46**, 1619–1626 (1982).

Simmons, M. S. and R. G. Zepp, "The influence of humic substances on the photolysis of nitroaromatic compounds in aqueous systems," pers. comm.

Slinn, W. G. N., L. Hasse, B. B. Hicks, A. W. Hogan, D. Lal, P. S. Liss, K. O. Munnich, G. A. Sehmel, and O. Vittori, "Some aspects of the transfer of atmospheric trace constituents past the air–sea interface," *Atmos. Environ.*, **12**, 2055–2087 (1978).

Smith, J. H., D. C. Bomberger, Jr., and D. L. Haynes, "Prediction of the volatilization rates of high-volatility chemicals from natural water bodies," *Environ. Sci. Technol.*, **14**, 1332–1337 (1980).

Smith, J. H., W. R. Mabey, N. Bohonos, B. R. Hoh, S. S. Lee, T.-W. Chou, D. C. Bomberger, and T. Mill, "Environmental pathways of selected chemicals in freshwater systems. Part II. Laboratory Studies," EPA-600/7-78-074, Washington, DC, 1978, 348 pp.

Smith, R. C. and J. E. Tyler, "Transmission of solar radiation into natural waters," *Photochem. Photobiol. Rev.*, **1**, 117–155 (1976).

Smith, S. D. and E. P. Jones, "Evidence for wind pumping of air–sea gas exchange based on direct measurements of CO_2-fluxes," *J. Geophys. Res.*, **90**, 869–875 (1985).

Somasundaran, P. and G. E. Agar, "The zero point of charge of calcite," *J. Colloid Interface Sci.*, **24**, 433–440 (1967).

Somasundaran, P., T. W. Healy, and D. W. Fuerstenau, "Surfactant adsorption at the solid–liquid interface—Dependence of mechanism on chain length," *J. Phys. Chem.*, **68**, 3562–3566 (1964).

Somasundaran, P., R. Middleton, and K. V. Viswanathan., "Relationship between surfactant structure and adsorption," In M. J. Rosen, Ed., *Structure/Performance Relationship in Surfactants*, ACS Symposium Series 253, American Chemical Society, Washington, DC, 1984, pp 269–290.

Sonnefeld, W. J., W. H. Zoller, and W. E. May, "Dynamic coupled-column liquid chromatographic determination of ambient temperature vapor pressures of polynuclear aromatic hydrocarbons," *Anal. Chem.*, **55**, 275–280 (1983).

Spencer, W. F. and M. M. Cliath, "Measurement of pesticide vapor pressures," *Residue Rev.*, **85**, 57–71 (1983).

Stackelberg, M. and W. Stracke, "Das polarographische Verhalten ungesättigter und halogenierter Kohlenwasserstoffe," *Z. Electrochem.*, **53**, 118–125 (1949).

Staehelin, J. and J. Hoigné, "Decomposition of ozone in water in the presence of organic solutes acting as promotors and inhibitors of radical chain reactions," *Environ. Sci. Technol.*, **19**, 1206–1213 (1985).

Stanier, R. Y., E. A. Adelberg, and J. L. Ingraham, "The Effect of Environment on Microbial Growth," *The Microbial World*, Chapter 10, Prentice-Hall, Englewood Cliffs, NY, 1976.

Stanier, R. Y. and L. N. Ornston, "The β-ketoadipate pathway," *Adv. Microbiol. Physiol.*, **9**, 89–151 (1973).

Stein, B. E., "Application of additive estimation methods to vaporization properties of liquids, *n*-Alkanes," *J. Chem. Soc. Faraday Trans. I*, **77**, 1457–1467 (1981).

Steinberg, S. M., J. J. Pignatello, and B. L. Sawhney, "Persistence of 1,2-dibromoethane in soils: Entrapment in intraparticle micropores," *Environ. Sci. Technol.*, **21**, 1201–1208 (1987).

Stephenson, R., J. Stuart, and M. Tabak, "Mutual solubility of water and aliphatic alcohols," *J. Chem. Eng. Data*, **29**, 287–290 (1984).

Stevenson, F. J., "Organic matter reactions involving pesticides in soil." In D. D. Kaufman, G. G. Still, G. D. Paulson, and S. K. Bandal, Eds., *Bound and Conjugated Pesticide Residues*, ACS Symposium Series 29, American Chemical Society, Washington, DC, 1976, pp. 180–207.

Stommel, H., "Trajectories of small bodies sinking slowly through convection cells," *J. Mar. Res.*, **8**, 24–29 (1949).

Stone, A. T., "Reductive dissolution of manganese (III/IV) oxides by substituted phenols," *Environ. Sci. Technol.*, **21**, 979–988 (1987).

Stone, A. T., "Enhanced rates of monophenyl terephthalate hydrolysis in aluminium oxide suspensions," *J. Colloid. Interface Sci.*, **127**, 429–441 (1989).

Stradins, J. P. and V. T. Glezer, "Azo, azoxy and diazo compounds," In *Encyclopedia of Electrochemistry of the Elements*, Vol. XIII, A. J., Bard, and H. Lund, Eds., Dekker, New York, 1979.

Stucki, G., R. Gälli, H.-R. Ebersold, and T. Leisinger, "Dehalogenation of dichloromethane by cell extracts of *Hyphomicorbium* DM2," *Arch. Microbiol.*, **130**, 366–376 (1981).

Stuermer, D. H. and J. R. Payne, "Investigations of seawater and terrestrial humic substances with C-13 and proton nuclear magnetic resonance," *Geochim. Cosmochim. Acta*, **40**, 1109–1114 (1976).

Stumm, W., R. P. Schwarzenbach, and L. Sigg, "From environmental analytical chemistry to ecotoxicology—A plea for more concepts and less monitoring and testing," *Angewandte Chemie*, **22**, 380–389 (1983).

Stumm, W., R. Kummert, and L. Sigg, "A ligand exchange model for adsorption of inorganic and organic ligands at hydrous oxide interfaces," *Croat. Chem. Acta*, **53**, 291–312 (1980).

Stumm, W. and J. J. Morgan, *Aquatic Chemistry*, Wiley-Interscience, New York, 1981.

Suatoni, J. C., R. E. Snyder, and R. O. Clark, "Voltametric studies of phenol and aniline using ring substitution." *Anal. Chem.*, **33**, 1894–1897 (1961).

Suflita, J. M., J. A. Robinson, and J. M. Tiedje, "Kinetics of microbial dehalogenation of haloaromatic substrates in methanogenic environments," *Appl. Environ. Microbiol.*, **45**, 1466–1473 (1983).

Sutton, C. and J. A. Calder, "Solubility of alkylbenzenes in distilled water and seawater at 25°C," *J. Chem. Eng. Data*, **20**, 320–322 (1975).

Sutton, C. and J. A. Calder, "Solubility of higher-molecular weight *n*-paraffins in distilled water and seawater," *Environ. Sci. Technol.*, **8**, 654–657 (1974).

Sverdrup, H. U., M. W. Johnson, and R. H. Fleming, *The Oceans: Their Physics, Chemistry, and General Biology*, Prentice-Hall, Englewood Cliffs, NJ, 1942, pp. 115–124.

Swain, C. G. and C. B. Scott, "Quantitative correlation of relative rates. Comparison of hydroxide ion with other nucleophilic reagents toward alkyl halides, esters, epoxides, and acyl halides," *J. Am. Chem. Soc.*,. **75**, 141–147 (1953).

Swisher, R. D., Surfactant Biogradation, 2nd ed., Dekker, New York, 1987, 1085 pp.

Szecsody, J. E. and R. C. Bales, "Sorption kinetics of low-molecular weight hydrophobic organic compounds on surgace-modified silica," *J. Contam. Hydrol.*, **4**, 181–203 (1989).

Taft, R. W., Jr., *Steric Effects in Organic Chemistry*, Wiley, New York, 1956.

Talbot, R. J. E., "The hydrolysis of carboxylic acid derivatives," In C. H. Branford and C. F. H. Tipper, Eds., *Comprehensive Chemical Kinetics*, Vol. 10, Elsevier, Amsterdam, 1972, pp. 209–293.

Tanford, C., *The Hydrophobic Effect: Formation of Micelles and Biological Membranes*, Wiley-Interscience, New York, 1984.

Tewari, Y. B., M. M. Miller, S. P. Wasik, and D. E. Martine, "Aqueous solubility and octanol–water partition coefficient of organic compounds at 25°C," *J. Chem. Eng. Data*, **27**, 451–454 (1982).

Thauer, R. K., K. Jungermann, and K. Decker, "Energy conservation in chemotrophic anaerobic bacteria," *Bacteriol. Rev.*, **41**, 100–180 (1977).

Thomas, R. G., "Volatilization from soil." In W. J. Lyman, W. F. Reehl, and D. H. Rosenblatt, Eds., *Handbook of Chemical Properties Estimation Methods*, Chapter 16, McGraw-Hill, New York, NY (1982).

Thomson, G. W., "Determination of vapor pressure," In *Physical Methods of Organic Chemistry*, Part I, A. Weissberger, Ed., Wiley Interscience, New York, 1959.

Thorpe, S. A., "The role of bubbles produced by breaking waves in super-saturating the near-surface ocean mixing layer with oxygen," *Ann. Geophys.*, **2**, 53–56 (1984).

Thurman, E. M., *Organic Geochemistry of Natural Waters*, Martinus Nijhoff, Boston, 1985.

Tinsley, I. J., *Chemical Concepts in Pollutant Behavior*, Wiley-Interscience, New York, 1979.

Tipping, E., "The adsorption of aquatic humic substances by iron oxides," *Geochim. Cosmochim. Acta*, **45**, 191–199 (1981).

Tomlinson, E., "Chromatographic hydrophobic parameters in correlation analysis of structure–activity relationship," *J. Chromatogr.*, **113**, 1–45 (1975).

Tratnyek, P. G. and D. L. Macalady, "Abiotic reduction of nitroaromatic pesticides in anaerobic laboratory systems," *J. Agric. Food Chem.*, **37**, 248–254 (1989).

Trouton, F., "IV. On molecular latent heat," *Phil. Mag.*, **18**, 54–57 (1884).

Tucker, E. E. and S. D. Christian, "A prototype hydrophobic interaction. The dimerization of benzene in water," *J. Phys. Chem.*, **83**, 426–427 (1979).

Turro, N. J., *Modern Molecular Photochemistry*, Benjamin/Cummings, Menlo Park, CA, 1978.

Ullrich, V., "Enzymatic hydroxylations with molecular oxygen," *Angew. Chem. Int. Ed.*, **11**, 701–712 (1972).

Ulrich, H.-J. and A. T. Stone, "Oxidation of chlorophenols adsorbed to manganese oxide surfaces," *Environ. Sci. Technol.*, **23**, 421–428 (1989).

Ulrich, M., R. P. Schwarzenbach, and D. M. Imboden, "MASAS—Modelling of anthropogenic substances in aquatic systems on personal computers," *Environ. Software*, **6**, 34–38 (1991).

Valvani, S. C., S. H. Yalkowsky, and G. L. Amidon, "Solubility of nonelectrolytes in polar solvents. VI. Refinements in molecular surface area computations," *J. Phys. Chem.*, **80**, 829–835 (1976).

van Bladel, R. and A. Moreale, "Adsorption of fenuron and monuron (substituted ureas) by two montmorillonite clays," *Soil Sci. Soc. Am. Proc.*, **38**, 244–249 (1974).

Veith, G. D., N. M. Austin, and R. T. Morris, "A rapid method for estimating log P for organic chemicals," *Water Res.*, **13**, 43–47 (1979).

Vogel, T. M., C. S. Criddle, and P. L. McCarty, "Transformation of halogenated aliphatic compounds," *Environ. Sci. Technol.*, **21**, 722–736 (1987).

Vontor, T., J. Socha, and M. Vecera, "Kinetics and mechanisms of hydrolysis of 1-naphthyl-*N*-methyl- and *N,N*-dimethylcarbamates," *Coll. Czechoslov. Chem. Commun.*, **37**, 2183–2196 (1972).

Wade, R. S. and C. E. Castro, "Oxidation of iron (II) prophyrins by alkyl halides," *J. Am. Chem. Soc.*, **95**, 226–230 (1973a).

Wade, R. S. and C. E. Castro, "Oxidation of heme proteins by alkyl halides," *J. Am. Chem. Soc.*, **95**, 231–234 (1973b).

Wagman, D. D., W. H. Evans, V. B. Parker, R. H. Schumm, I. Halow, S. M. Bailey, K. L. Churney, and R. L. Nuttall, "The NBS tables of chemical thermodynamic properties," *J. Phys. Chem. Ref. Data*, **11**, Suppl. No. 2 (1982).

Wahid, P. A., C. Ramakrishna, and N. Sethunathan, "Instantaneous degradation of parathion in anaerobic soils," *J. Environ. Qual.*, **9**, 127–130 (1980).

Walsh, C., *Enzyme Reaction Mechanisms*, Freeman, San Fransisco, CA, 1979, 978 pp.

Wanner, O., T. Egli, T. Fleischmann, K. Lanz, P. Reichert, and R. P. Schwarzenbach, "The behavior of the insecticides disulfoton and thiometon in the Rhine River—A chemodynamic study," *Environ. Sci. Technol.*, **23**, 1232–1242 (1989).

Wanner, O. and W. Gujer, "A multispecies biofilm model," *Biotechnol. Bioeng.* **XXVIII**, 314–328 (1986).

Wanninkhof, R., J. P. Ledwell, and W. S. Broecker, "Gas exchange on Mono Lake and Crowely Lake, California," *J. Geophys. Res.* **92**, 14567–14580 (1987).

Wardman, P., "The use of nitroaromatic compounds as hypoxic cell radiosenitizers," *Curr. Top. Radiat. Res. Q.* **11**, 347–398 (1977).

Wardman, P., "Reduction potentials of one-electron couples involving free radicals in aqueous solution," *J. Phys. Chem. Ref. Data*, **18**, 1637–1755 (1989).

Wauchope, R. D. and F. W. Getzen, "Temperature dependence of solubilities in water and heats of fusion of solid aromatic hydrocarbons," *J. Chem. Eng. Data*, **17**, 38–41 (1972).

Weber, A. of BUWAL (Swiss EPA), personal communication based on data from the United Nations Environmental Program (1987).

Weber, K., "Degradation of parathion in seawater," *Water Res.*, **10**, 237–241 (1976).

Weber, W. J., Jr., P. M. McGinley, and L. E. Katz, "Sorption phenomena in subsurface systems: Concepts, models and effects on contaminant fate and transport," *Water Res.*, **25**, 499–528 (1991).

Weber, W. J., Jr. and R. R. Rumer, Jr., "Intraparticle transport of sulfonated alkylbenzenes in a porous solid: Diffusion with nonlinear adsorption," *Water Resources Res.*, **1**, 361–373 (1965).

Weber, E. J. and N. L. Wolfe, "Kinetics studies of the reduction of aromatic azo compounds in anaerobic sediment/water systems," *Environ. Toxicol. Chem.*, **6**, 911–919 (1987).

Weiner, J. H., D. P. MacIsaac, R. E. Bishop, and P. T. Bilons, "Purification and properties of *Escherichia coli.* dimethyl sulfoxide reductase, an iron–sulfur molybdoenzyme with broad substrate specificity," *J. Bacteriol.*, **170**, 1505–1510 (1988).

Welander, P., "Theoretical forms for the vertical exchange coefficients in a startified fluid with application to lakes and seas," *Geophys. Gothoburg*, **1**, 1–27 supplement (1968).

Wells, P. R., "Linear free-energy relationships," *Chem. Rev.*, **63**, 171–219 (1963).

Westall, J. C., "Properties of organic compounds in relation to chemical binding." In *Biofilm Processes in Groundwater Research.*, Symposium, Proceedings Stockholm, Sweden, 1983, pp. 65–90.

Westall, J. C., "Adsorption mechanisms in aquatic surface chemistry." In *Aquatic Surface Chemistry*, W. Stumm, Ed., Wiley-Interscience, New York, 1987, pp. 3–32.

Westall, J. C., C. Leuenberger, and R. P. Schwarzenbach, "Influence of pH and ionic strength on the aqueous–nonaqueous distribution of chlorinated phenols," *Environ. Sci. Technol.*, **19**, 193–198 (1985).

Westcott, J. W., C. G. Simon, and T. F. Bidleman, "Determination of polychlorinated biphenyl vapor pressures by a semimicro gas saturation method," *Environ. Sci. Technol.*, **15**, 1375–1378 (1981).

White, R. E. and M. J. Coon, "Oxygen activation by cytochrome P-450," *Ann. Rev. Biochem.*, **49**, 315–356 (1980).

Whitehouse, B. G., "The effects of temperature and salinity on the aqueous solubility of polynuclear aromatic hydrocarbons," *Mar. Chem.*, **14**, 319–332 (1984).

Whitman, W. G., "The two-film theory of gas absorption," *Chem. Metal. Eng.*, **29**, 146–148 (1923).

Wilcock, R. J., "Reaction studies on some New Zealand rivers using methyl chloride as a gas tracer." In *Gas Transfer at Water Surfaces*, W. Brutsaert and G. H. Jirka, Eds., Reidel, Boston, 1984, pp. 413–420.

Williams, A., "Alkaline hydrolysis of substituted phenylcarbamates—Structure–reactivity relationships consistent with an E1B mechanism," *J. Chem. Soc. Perkins II*, 808–812 (1972).

Williams, A., "Participation of an elimination mechanism in alkaline hydrolysis of alkyl *N*-phenylcarbamates," *J. Chem. Soc. Perkins II*, 1244–1247 (1983).

Williams, A., "Free-energy correlations and reaction mechanisms." In *The Chemistry of Enzyme Action*, M. I. Page, Ed., Elsevier, Amsterdam, 1984.

Williams, D. H. and I. Fleming, *Spectroscopic Methods in Organic Chemistry*, 3rd ed., McGraw-Hill, New York, 1980.

Williams, D. J. A. and K. P. Williams, "Electrophoresis and zeta potential of kaolinite," *J. Colloid Int. Sci.*, **65**, 79–87 (1978).

Wilson, J. T. and B. H. Wilson, "Biotransformation of trichloroethylene in soil," *Appl. Environ. Microbiol.*, **49**, 242–243 (1985).

Winterle, J. S., D. Tse, and W. R. Mabey, "Measurement of attenuation coefficients in natural water columns," *Environ. Toxicol. Chem.*, **6**, 663–672 (1987).

Wisegarver, D. P. and J. D. Cline, "Solubility of trichlorofluoromethane (F-11) and dichlorodifluoromethane (F-12) in seawater and its relationship to surface concentrations in the North Pacific," *Deep-Sea Res.*, **32**, 97–106 (1985).

Wolfe, N. L., "Organophosphate and organophosphothionate esters: Application of LFER's to estimate hydrolysis rate constants for use in environmental fate assessement," Chemosphere, **9**, 571–579 (1980).

Wolfe, N. L., L. A. Burns, and W. C. Steen, "Use of linear free-energy relationships and an evaluative model to assess the fate and transport of phthalate esters in the aquatic environment," *Chemosphere*, **9**, 393–402 (1980a).

Wolfe, N. L., B. E. Kitchens, D. L. Macalady, and T. J. Grundl, "Physical and chemical factors that influence the anaerobic degradation of methyl parathion in sediment systems," *Environ. Toxicol. Chem.*, **5**, 1019–1026 (1986).

Wolfe, N. L., D. F. Paris, W. C. Steen, and G. L. Baughman, "Correlation of microbial degradation rates with chemical structure," *Environ. Sci. Technol.*, **14**, 1143–1144 (1980c).

Wolfe, N. L., W. C. Steen, and L. A. Burns, "Phthalate ester hydrolysis: Linear free-energy relationships," *Chemosphere*, **9**, 403–408 (1980b).

Wolfe, N. L., R. G. Zepp, and D. F. Pairs, "Use of structure–reactivity relationships to estimate hydrolytic persistence of carbamate pesticides," *Water Res.*, **12**, 561–563 (1978).

Wood, J. M., F. S. Kennedy, and R. S. Wolfe, "The reaction of multihalogenated hydrocarbons with free and bound reduced vitamin B_{12}," *Biochemistry*, **7**, 1707–1713 (1968).

Wood, W. W., T. F. Kreemer, and P. P. Hearn, Jr., "Intragranual diffusion: An important mechanism influencing solute transport in clastic aquifers?" *Science*, **247**, 1569–1572 (1990).

Woodburn, K. B., W. J. Doucette, and A. W. Andren, "Generator column determination of octanol/water partition coefficients for selected polychlorinated biphenyl congeners," *Environ. Sci. Technol.*, **18**, 457–459 (1984).

Wu, J., "Bubble populations and spectra in near-surface ocean: Summary and review of field measurements," *J. Geophys. Res.*, **86**, 457–463 (1981).

Wu, S.-C. and P. M. Gschwend, "Numerical modeling of sorption kinetics of organic compounds to soil and sediment particles," *Water Resources Res.*, **24**, 1373–1383 (1988).

Wu, S.-C. and P. M. Gschwend, "Sorption kinetics of hydrophobic organic compounds to natural sediments and soils," *Environ. Sci. Technol.*, **20**, 717–725 (1986).

Wüest, A., *Ursprung und Grösse von Mischungsprozessen im Hypolimnion von Seen*, Dissertation ETH, Zurich, 1987.

Yalkowsky, S. H., "Estimation of entropies of fusion of organic compounds," *Ind. Eng. Chem. Fundam.*, **18**, 108–111 (1979).

Yalkowsky, S. H., R. J. Orr, and S. C. Valvani, "Solubility and partitioning. 3. The solubility of halobenzenes in water," *Ind. Eng. Chem. Fundam.*, **18**, 351–353 (1979).

Yalkowsky, S. H., S. C. Valvani, and G. L. Amidon, ''Solubility of nonelectrolytes in polar solvents. IV. Nonpolar drugs in mixed solvents,'' *J. Pharm. Sci.*, **65**, 1488–1494 (1976).

Yalkowsky, S. H. and S. C. Valvani, ''Solubilities and partitioning. 2. Relationships between aqueous solubilities, partition coefficients, and molecular surface areas of rigid aromatic hydrocarbons,'' *J. Chem. Eng. Data*, **24**, 127–129 (1979).

Yaron, B., A. R. Swoboda, and G. W. Thomas, "Aldrin adsorption by soils and clays," *J. Agric. Food Chem.*, **15**, 671–675 (1967).

Yeh, W. K., D. T. Gibson, and E. Liug, "Toluene dioxygenase: A multicomponent enzyme system," *Biochem. Biophys. Res. Comm.*, **78**, 401–410 (1977).

Yost, E. C., M. I. Tejedor-Tejedor, and M. A. Anderson, "*In situ* CIR–FTIR characterization of salicylate complexes at the goethite/aqueous solution interface," *Environ. Sci. Technol.*, **24**, 822–828 (1990).

Zachara, J. M., C. C. Ainsworth, L. J. Felice, and C. T. Resch, "Quinoline sorption to subsurface materials: Role of pH and retention of the organic cation," *Environ. Sci. Technol.*, **20**, 620–627 (1986).

Zafiriou, O. C., "Sources and reactions of OH and daughter radicals in seawater," *J. Geophys. Res.*, **79**, 4491–4497 (1974).

Zafiriou, O. C., "Reactions of methyl halides with seawater and marine aerosols," *J. Marine Res.*, **33**, 75–81 (1975).

Zafiriou, O. C., "Natural water photochemistry," In *Chemical Oceanography*, Vol. 8, Academic Press, London, pp. 339–379 (1983).

Zafiriou, O. C., J. Joussot-Dubien, R. G. Zepp, and R. G. Zika, "Photochemistry of natural waters," *Environ. Sci. Technol.*, **18**, 358A–371A (1984).

Zepp, R. G., "Quantum yields for reaction of pollutants in dilute aqueous solution," *Environ. Sci. Technol.*, **12**, 327–329 (1978).

Zepp, R. G., "Assessing the photochemistry of organic pollutants in aquatic environments." In *Dynamics, Exposure and Hazard Assessment of Toxic Chemicals*, R. Haque, Ed., Ann Arbor Science, Ann Arbor, MI, 1980, pp. 69–109.

Zepp, R. G., "Experimental approaches to environmental photochemistry." In *The Handbook of Environmental Chemistry*, Vol. 2, Part B, O. Hutziner, Ed., Springer, Berlin, 1982.

Zepp, R. G., "Environmental photoprocesses involving natural organic matter," In *Humic Substances and Their Role in the Environment*, F. H. Frimmel and R. F. Christman, Eds., Wiley, New York, 1988, pp. 193–214.

Zepp, R. G., G. L. Baughman, and P. F. Schlotzhauer, "Comparison of photochemical behavior of various humic substances in water: I. Sunlight induced reactions of aquatic pollutants photosensitized by humic substances," Chemosphere, **10**, 109–117 (1981).

Zepp, R. G. and G. L. Baughman, "Prediction of photochemical transformation of pollutants in the aquatic environment." In O. Hutzinger, I. M. van Lelyveld, and B. C. J. Zoeteman, Eds., *Aquatic Pollutants, Transformations and Biological Effects*. Pergamon, New York, 1978, pp. 237–263.

Zepp, R. G., A. M. Braun, J. Hoigné, and J. A. Leenheer, "Photoproduction of hydrated electrons from natural organic solutes in aquatic environments," *Environ. Sci. Technol.*, **21**, 485–490 (1987a).

Zepp, R. G. and D. M. Cline, "Rates of direct photolysis in aquatic environment," *Environ. Sci. Technol.*, **11**, 359–366 (1977).

Zepp, R. G., J. Hoigné, and H. Bader, "Nitrate-induced photooxidation of trace organic chemicals in water," *Environ. Sci. Technol.*, **21**, 443–450 (1987b).

Zepp, R. G. and P. F. Schlotzhauer, "Photoreactivity of selected aromatic hydrocarbons in water," In *Polynuclear Hydrocarbons*, T. W. Jones and T. Leber, Eds., Ann Arbor Science, Ann Arbor, MI, 1979.

Zepp, R. G., P. F. Schlotzhauer, and R. M. Sink, "Photosensitized transformations involving electronic energy transfer in natural waters: Role of humic substances," *Environ. Sci. Technol.*, **19**, 74–81 (1985).

Zepp, R. G. and N. L. Wolfe, "Abiotic transformation of organic chemicals at the particle–water interface." In *Aquatic Surface Chemistry–Chemical Processes at the Particle–Water Interface*, W. Stumm, Ed., Wiley-Interscience, New York, 1987.

Zepp, R. G., N. L. Wolfe, G. L. Baughman, and R. C. Hollis, "Singlet oxygen in natural waters," *Nature*, **267**, 421–423 (1977).

Zerner, B., R. P. M. Band, and M. L. Bender, "Kinetic evidence for the formation of acylenzyme intermediates in the α-chymotrypsin-catalyzed hydrolyses of specific substrates," *J. Am. Chem. Soc.*, **86**, 3674–3679 (1964).

Zeyer, J., and H. P. Kocher, "Purification and characterization of bacterial nitrophenol oxygenase which converts *ortho*-nitrophenol to catechol and nitrite," *J. Bacterial.* **170**, 1789–1794 (1988).

Zeyer, J., A. Wasserfallen, and K. N. Timmis, "Microbiol mineralization of ring-substituted anilines through an ortho-cleavage pathway," *Appl. Environ. Microbiol*, **50**, 447–453 (1985).

Zhang, Z. Z., P. E. Low, J. H. Cushman, and C. B. Roth, "Adsorption and heat of adsorption of organic compounds on montmorillonite from aqueous solutions," *Soil Sci. Soc. Am.*, **54**, 59–66 (1990).

Zinder, S. H. and T. D. Brock, "Dimethyl sulphoxide reduction by microorganisms," *J. Gen. Microbial.*, **105**, 335–342 (1978).

Zoro, J. A., J. M. Hunter, G. Englinton, and G. C. Ware, "Degradation of *p,p*'-DDT in reducing environments," *Nature*, **247**, 235–237 (1974).

Zullig, J. J. and J. W. Morse, "Interaction of organic acids with carbonate mineral surfaces in seawater and related solutions. I. Fatty acid adsorption," *Geochim Cosmochim. Acta*, **52**, 1667–1678 (1988).

Zuman, P., *Substituent Effects in Organic Polarography*, Plenum, New York, 1967.

INDEX

SOME USEFUL PROPERTIES OF WATER (mw = 18.0153 $g \cdot mol^{-1}$)

T(°C)	ρ, Density ($kg \cdot m^{-3}$)	μ Viscosity ($kg \cdot m^{-1} \cdot s^{-1}$)	σ, Surface Tension against Air ($J \cdot m^{-2}$)	ε Dielectric Constant ($C \cdot V^{-1} \cdot m^{-1}$)	pK_w, Ionization Constant ($mol^2 \cdot L^{-2}$)
0	999.868	0.001787	0.0756	88.28	14.9435
5	999.992	0.001519	0.0749	86.3	14.7338
10	999.726	0.001307	0.07422	84.4	14.5346
15	999.125	0.001139	0.07349	82.5	14.3463
20	998.228	0.001002	0.07275	80.7	14.1669
25	997.069	0.0008904	0.07197	78.85	13.9965
30	995.671	0.0007975	0.07118	77.1	13.8330

SOME USEFUL PROPERTIES OF THE EARTH

Atmospheric Composition (exclusive of water, by volume)

78.08% N_2
20.95% O_2
0.934% Ar
0.035% CO_2
~ 2 ppm CH_4

	Typical River Water Composition	Typical Seawater Composition
pH	6–8	8
Na^+	4×10^{-4} M	5×10^{-1} M
K^+	6×10^{-5} M	1×10^{-2} M
Ca^{2+}	4×10^{-5} M	1×10^{-2} M
Mg^{2+}	2×10^{-4} M	5×10^{-2} M
$Cl^{\ominus}$	2×10^{-4} M	6×10^{-1} M
SO_4^{2-}	1×10^{-4} M	3×10^{-2} M
$HCO_3^{\ominus}$	1×10^{-4} M	2×10^{-3} M

Atmospheric mass	52×10^{17} kg
Water in atmosphere	0.105×10^{17} kg
Ocean mass	$13{,}700 \times 10^{17}$ kg
Water in lakes, rivers	0.34×10^{17} kg
Water in ice	165×10^{17} kg
Water in pores of sediments, soil, and rocks	3200×10^{17} kg
Annual runoff to seas	0.32×10^{17} $kg \cdot yr^{-1}$
Annual precipitation	4.5×10^{17} $kg \cdot yr^{-1}$

SOME USEFUL ESTIMATION EQUATIONS

$\ln K' = -\Delta G/RT$ (see Eq. 3-20)

$\Delta G = \Delta H - T\Delta S$ (see Eq. 3-21)

$\Delta S_{vap}(\text{in J}\cdot\text{mol}^{-1}\cdot\text{K}^{-1}) \simeq (36.6 + 8.31 \ln T_b)$ (see Eq. 4-11)

$\ln P^0(\text{l, L}) \simeq -K_F(4.4 + \ln T_b)\cdot\left[1.8\left(\frac{T_b}{T} - 1\right) - 0.8\ln\left(\frac{T_b}{T}\right)\right]$ (see Eq. 4-18)

$\Delta S_{melt}(\text{in J}\cdot\text{mol}^{-1}\cdot\text{K}^{-1}) \simeq [56.6 + 10.5(n-5)]$ (see Eq. 4-20)

$P^0(s) \simeq P^0(L)\cdot\exp\left\{-[6.8 + 1.26(n-5)]\left(\frac{T_m}{T} - 1\right)\right\}$ (see Eq. 4-21)

$\Delta G_s^e = RT\ln\gamma_w$ (see Eq. 5-16)

Setschenow: $\log C_{w,salt}^{sat} = \log C_w^{sat} - K^s[\text{salt}]_t$ (see Eq. 5-22)

Henry: $K_H = \frac{P_i}{C_w}$ (see Eq. 6-1)

$K_H \simeq \gamma_w^{sat} V_w P^0(\text{l, L})$ (for low-solubility compounds) (see Eq. 6-9)

$\log K_{ow} \simeq -a\log C_w^{sat}(\text{l, L}) + b$ (see Table 7.2)

$\log K_{ow} = \sum_i f_i + \sum_j F_j$ (see Eq. 7-17)

Hammett: $\log K_a = \log K_{aH} + \rho\cdot\sum_i \sigma_i$ (see Eq. 8-23)

Fick: $F_x = -D\cdot\frac{dC}{dx}$ (see Eq. 9-2)